Distributed Parameter Control Systems

International Series on

SYSTEMS AND CONTROL, Volume 6

Other Titles of Interest

AKASHI
Control Science and Technology for the Progress of Society (8th Triennial IFAC World Congress)

BROADBENT & MASUBUCHI
Multilingual Glossary of Automatic Control Technology

ELLOY & PIASCO
Classical and Modern Control with Worked Examples

EYKHOFF
Trends and Progress in System Identification

HASEGAWA
Real-Time Programming 1981

MAHALANABIS
Theory and Application of Digital Control

MAHMOUD & SINGH
Large Scale Systems Modelling

MILLER
Distributed Computer Control Systems

MORRIS
Communication for Command and Control Systems

PATEL & MUNRO
Multivariable System Theory and Design

SINGH *et al*
Applied Industrial Control: An Introduction

SINGH & TITLI
Systems: Decomposition, Optimization and Control

Pergamon Related Journals (Free specimen copies gladly sent on request)

AUTOMATICA
COMPUTERS & INDUSTRIAL ENGINEERING
COMPUTERS & OPERATIONS RESEARCH
JOURNAL OF THE OPERATIONAL RESEARCH SOCIETY

Distributed Parameter Control Systems

Theory and Application

Edited by

SPYROS G. TZAFESTAS

Control Systems Laboratory, University of Patras, Greece

PERGAMON PRESS

OXFORD · NEW YORK · TORONTO · SYDNEY · PARIS · FRANKFURT

U.K.	Pergamon Press Ltd., Headington Hill Hall, Oxford OX3 0BW, England
U.S.A.	Pergamon Press Inc., Maxwell House, Fairview Park, Elmsford, New York 10523, U.S.A.
CANADA	Pergamon Press Canada Ltd., Suite 104, 150 Consumers Rd., Willowdale, Ontario M2J 1P9, Canada
AUSTRALIA	Pergamon Press (Aust.) Pty. Ltd., P.O. Box 544, Potts Point, N.S.W. 2011, Australia
FRANCE	Pergamon Press SARL, 24 rue des Ecoles, 75240 Paris, Cedex 05, France
FEDERAL REPUBLIC OF GERMANY	Pergamon Press GmbH, 6242 Kronberg-Taunus, Hammerweg 6, Federal Republic of Germany

First edition 1982

Library of Congress Cataloging in Publication Data

Main entry under title:
Distributed parameter control systems.
(International series on systems and
control ; v. 6)
Includes bibliographies and index.
1. Automatic control. 2. Control theory.
I. Tzafestas, S. G., 1939- . II. Series.

TJ213.D545 1982 629-8'312 82-13208
ISBN 0-08-027624-5

In order to make this volume available as economically and as rapidly as possible the authors' typescripts have been reproduced in their original forms. This method unfortunately has its typographical limitations but it is hoped that they in no way distract the reader.

Printed and bound in Great Britain by
Redwood Burn Ltd., Trowbridge, Wiltshire.

FOREWORD

Distributed parameter control systems (i.e. systems the state equation of which is a partial differential equation) are encountered in an everincreasing variety of areas. More and more situations in Physics, in Chemistry, in Biomedicine, in Econometrics, etc.. are recognized to be adequately modelled by a distributed system.

These models are often non linear - and even in the classical "linear-quadratic" case, one meets very interesting theoretical questions, the solution of which is of practical importance. Presently the theory of these systems has reached a well advanced state and many important results are available.

It is really with great pleasure that I write the prologue of this volume which is offering a distinguished service towards the end of documenting some of the most recent results in the field. The book covers in a well balanced way such areas as simulation, identification, state estimation, stability, control (optimal, stochastic, coordinated), numerical approximation methods, optimal sensor and actuator positioning, and five applications works which include chemical reactors, heat exchangers, petroleum reservoirs/aquifers, and nuclear reactors. Of course, it is not possible to include in one volume all available results or to cover all fields of application, but the results and applications involved in this book are representative and cover some of the most important aspects of the distributed parameter systems area.

To give an idea of the vitality of this field, let us mention *some* important topics not covered in this book in a systematic manner: control of free surfaces, optimum design, control of systems with instabilities, multi-states and bifurcations.

Several books are now in preparation on these topics; this volume - which will be complemented by the publication of the Encyclopedia of Systems and Control, edited by Professor M.G. Singh - provides a very significant addition to the current literature on distributed parameter systems and will no doubt stimulate further research and further applications. One must heartily thank the Editor, Professor S.G. Tzafestas, for the work he achieved with the co-authors of this book.

Academician J. L. LIONS
College de France and I.N.R.I.A.

PREFACE

During the last two decades a great deal of research work was carried out by control theorists and practitioners in the direction of advancing the general field of distributed-parameter systems. The results derived involve either genuine extensions of available results from finite-dimensional multivariable control theory or rigorous mathematical formulations based on the infinite-dimensional structure of distributed-parameter systems.

This book involves ten theoretical and five application oriented chapters contributed by well-known workers in the distributed-parameter systems area. Special care was taken to cover most aspects of the theory, namely modelling, simulation, stability, identification filtering, optimal control, coordinated control, sensor/actuator location, and a set of important applications. To this end, each contributor was asked to write a chapter reflecting his own work, or reviewing recent results in his field of expertise, or both. The outcome is that the book provides a wide coverage of the entire field by comprehensive contributions. Hence it is my hope that this volume will be a useful addition to the existing literature on distributed-parameter systems, and that it may serve as one of the references for both graduate and professional research work in the field.

I wish to thank my colleagues for their excellent contributions, and for their help in the process of cross-refereeing the chapters of the book.

Spyros G. Tzafestas

CONTENTS

Chapter 5
STABILITY AND OPTIMAL CONTROL OF STOCHASTIC EVOLUTION EQUATIONS
A. Ichikawa

Chapter 6
ADAPTIVE AND ROBUSTNESS PROPERTIES OF CERTAIN BILINEAR DISTRIBUTED PARAMETER CONTROL SYSTEMS
D. Franke

PART II APPLICATIONS

Chapter 11
EXPERIMENTAL APPLICATION OF DISTRIBUTED PARAMETER CONTROL THEORY
G. K. Lausterer and Ed. Eitelberg

Chapter 12
DISTRIBUTED PARAMETER IDENTIFICATION IN GEOPHYSICS-PETROLEUM RESERVOIRS AND AQUIFERS
J. H. Seinfeld and C. Kravaris

Chapter 13
IDENTIFICATION FOR DISTRIBUTED PARAMETER CHEMICAL ENGINEERING SYSTEMS-FIXED-BED TUBULAR REACTORS
W. Fred Ramirez

Chapter 14
DISTRIBUTED PARAMETER HEAT EXCHANGERS-MODELING, DYNAMICS AND CONTROL
H. Kanoh

CONTRIBUTORS

Prof. M. AMOUROUX	Laboratoire de Physique Appliquée, University of Perpignan, Perpignan, France.
Dr. J. P. BABARY	Laboratoire d'Automatique et d'Analyse des Systèmes, Toulouse, France.
Prof. L. CAROTENUTO	Department of Systems, University of Calabria, Arcavacata, Italy.
Dr. E. EITELBERG	Institut für Regelungs-und Steuerungssysteme, University of Karlsruhe, Karlsruhe, W.Germany.
Prof. D. FRANKE	Hochschule der Bundeswehr Hamburg, Faculty of Electrotechnics, Hamburg, W. Germany.
Prof. A. ICHIKAWA	Faculty of Engineering, Shizuoka University, Hamamatsu, Japan.
Prof. G. JUMARIE	Department of Mathematics, University of "Québec à Montréal", Montreal, Canada.
Prof. H. KANOH	Faculty of Engineering, Osaka University, Yamada-Oka, Suita, Osaka, Japan.
Dr. C. KRAVARIS	Department of Chemical Engineering, California Institute of Technology, Pasadena, California, 91125, USA.
Dr. G. K. LAUSTERER	Siemens AG, Systemtechnische Entwicklung, Karlsruhe,W. Germany.
Prof. K.Y. LEE	Department of Electrical Engineering, University of Houston, Houston, Texas, 77004, USA.
Prof. S. OMATU	Department of Information Science and Systems Engineering, University of Tokushima, Tokushima, Japan.

Prof. G. RAICONI	Department of Systems, University of Calabria, Arcavacata, Italy.
Prof. W. Fred RAMIREZ	Department of Chemical Engineering, University of Colorado, Boulder, Colorado, 80309,USA.
Prof. T. SEIDMAN	Department of Mathematics, University of Maryland Baltimore County, Catonsville, Maryland, 21228, USA.
Prof. J. H. SEINFELD	Department of Chemical Engineering, California Institute of Technology, Pasadena, California 91125, USA.
Prof. Y. SUNAHARA	Department of Mechanical Engineering, Kyoto Institute of Technology, Matsugasaki, Kyoto 606, Japan.
Prof. S. G. TZAFESTAS	Control Systems Laboratory, Department of Electrical Engineering, University of Patras, Patras, Greece.
Prof. D. M. WIBERG	Department of System Science, University of California, Los Angeles, USA.

EDITORIAL INTRODUCTION

Distributed-parameter control systems occupy single-and multi-dimensional domains and are modelled by partial differential equations or integral equations. Some of the earliest investigators of distributed-parameter systems are Brogan (1967), Butkovskiy and Lerner (1960), Lions (1966), and Wang (1964). Presently there exist numerous theoretical and technical publications dealing with such aspects as modelling, stability, simulation, identification, estimation, and control, and describing important applications in chemical systems, mechanical systems, environmental systems, plasma control, space antennas, air- and space-crafts, etc. For convenience, the books and survey papers on distributed systems are provided at the end of this introduction.

The intent of this book is to collect and present in a single source some of the recent results in the field, covering as far as possible the whole range of problems as well as some specific applications. Thus naturally the book is divided in two parts. The first part presents the theoretical results, and the second provides the applications.

Part I involves the first ten chapters. In Chapter 1, by Tzafestas, a survey of the hybrid simulation techniques for distributed-parameter models is provided, along with an exposition of the principal implementation aspects. Some numerical results are also included. The main goal of the effort for hybrid simulation in the distributed-parameter field was to reduce the cost of simulation while achieving acceptable levels of accuracy. In nowdays the entire field must be reconsidered to take into account the availability of cheap and versatile microprocessors.

Chapter 2, by Sunahara, begins with the exhibition of a number of stochastic distributed-parameter models subject to various boundary conditions. Then it presents two examples. The first example shows a method for identifying parameters representing locations of air pollution sources based on the Bayesian approach. The second example is concerned with the identification of a class of distributed-parameter systems involving unknown stochastic coefficients in the differential operator. The solution of the identification problem is found using the maximum likelihood technique. Some computational experimental results for the two examples are also presented.

Chapter 3, by Omatu, gives a survey of the rigorous approach to distributed-parameter state estimation theory. A formal derivation of the various estimators, based on Wiener-Hopf theory, is first given, and then the various rigorous estimation methods are presented.

In Chapter 4, Jumarie, provides a guide to the stability and hyperstability analysis of distributed systems. The systems are successively represented by 2-dimensional Laplace tranforms, Laplace-Green kernels and state-space models. Both continuous and discrete-time systems are considered, and the space sampling effects are studied. The case where the system involves random disturbances is also investigated.

Chapter 5, by Ichikawa, is concerned with the stability and optimal control of stochastic evolution equations through a utilisation of the theory of semigroups. The deterministic quadratic control problem is first considered, and is then extended to the stochastic case by using the concepts of Wiener processes and stochastic integrals in Hilbert spaces together with the Ito's differential rule. The chapter closes with a study of the mean square stability of stochastic evolution equations and quadratic cost control, along with some results on nonlinear stability.

Chapter 6, by Franke, is concerned with variable structure control of distributed parameter systems, described by bilinear partial differential state equations. Due to the multiplicative mode of control, such a system offers a number of advantages compared to strictly linear control. For example, the transient response can be improved. Moreover, variable structure control exhibits adaptive and robustness properties in case of plant uncertainties. The design and analysis is essentially based on Lyapunov's direct method. Some important features of the results of the chapter are adaptivity without the need of parameter identification, asymptotic stability forthe closed-loop control system for a specified class of parametric disturbances, and a solution to the model following problem for distributed parameter systems.

Chapter 7, by Lee, introduces a unified approach to coordinated control of distributed-parameter systems. The distributed systems are regarded as large-scale systems which are decomposed by three techniques, namely finite difference, finite element, and integration decomposition technique. The application of decentralised and coordinated control methods is straightforward. The chapter discusses both the theoretical and computational aspects. It is noted that complexity and high dimensionality of distributed-parameter systems require the application of efficient decomposition and coordination schemes as necessity.

Chapter 8, by Carotenuto and Raiconi, presents the finite element methodology with emphasis on its capability to solve a large class of distributed-parameter estimation and control problems. First, the construction of finite element approximating spaces is described, and the Ritz-Galerkin finite-element method for elliptic problems is studied. Then the case of parabolic initial value problems is considered. The chapter includes an investigation of the properties of the approximate eigenvalues and eigenfunctions of elliptic operators computed by the finite-element method, and closes with the presentation of a set of distributed-parameter control theory problems treated by the finite-element method.

Chapter 9, by Seidman, discusses some principal concepts and topics in the general field of approximation theory for distributed-parameter systems, and presents a few approximation approaches. The material is given in the form of a number of examples related to parabolic systems. These examples concern the identification, signal recovering, Dirichlet null control, and approximate Neuman control problems.

Chapter 10, by Amouroux and Babary, is concerned with the problem of optimal positioning of sensors and control actuators in distributed-parameter systems. Several cases are considered, namely, open loop-control, closed-loop control and estimation problems. In each case the development of the solution methodology is presented, and some simple illustrative examples are given, along with a brief survey of the existing results and literature.

Part II starts with Chapter 11 on experimental applications by Lausterer and Eitelberg. In this chapter some experimental results, obtained with practical implementation of algorithms derived from distributed-parameter control theory, are presented. Linear one-and two-dimensional examples as well as a moving boundary example are described in detail. Even though only very few practical experiments applying distributed-parameter systems theory have been reported, it can be concluded that real-time implementation is indeed feasible. However, it must be emphasised that the processes studied are very good-natured, except for the moving boundary problem. Thus for applications to other types of processes, e.g. chemical reactors, problems relating to stability and sensitivity cannot be excluded.

Chapter 12, by Seinfeld and Kravaris, studies the classical identification problem arising in the modelling of petroleum reservoirs and subsurface aquifers, namely the problem of estimating the parameter $\alpha(x,y)$ in the corresponding parabolic partial differential equation model. The problem is ill-posed, and the chapter is devoted to a study of the problem with special attention to the alleviation of the ill-posedness. The ill-posedness of the identification problem can be attacked by incorporating additional a priori information into the problem beyond that available in the model and the observational data. Particularly applicable to the present problem are the regularisation approaches for solving general ill-posed problems. These regularisation methods are reviewed in the chapter, and an approach for the present problem based on the regularisation concept is proposed.

In Chapter 13, by Ramirez, a review of the scope of distributed-parameter identification and filtering applications to chemical engineering, systems is first given. These applications include problems concerning tubular reactors, heat transfer, oil recovery, water pollution, and air pollution. Then the class of fixed-bed tubular reactors is considered in detail, in order to illustrate the practical implementation of dynamic steady-state distributed-parameter filters.

Chapter 14, by Kanoh, is a guide to mathematical modelling and dynamics of heat exchangers. Firstly a historical review of modelling and dynamics studies is given, and the basic framework for modelling is constructed. Secondly, various models are presented, including the transfer function models of typical heat exchangers. Thirdly the dynamic characteristics of heat exchangers are discussed and the lumped-parameter simulation of them is studied. The chapter closes with a survey of several control problems, and an extensive collection of the related bibliography.

The final chapter of the book is by Wiberg, and deals with the academic application of distributed-parameter techniques to nuclear reactor systems. Starting with a review of the general distributed-parameter nuclear reactor state equations, various aspects of the methodology are studied such as linearisation, incorporation of the power control system, noise modelling, and nuclear reactor controllability. Then, nointeracting modal expansions are investigated and used for both filtering and control problems, with an application to LMFB reactors. Finally, it is shown how premature discretisation of the delay equation leads to erroneous results.

From the applications included in this book and the related bibliography, one can observe that most of the efforts in the application area were devoted to the identification and state estimation problems. However it is anticipated that it is only a matter of time to see important industrial applications of distributed-parameter control as well.

Books and Survey Papers on Distributed Parameter Systems

Ahmed, N.U. and K.L. Teo (1981). Optimal control of distributed-parameter systems. Elsevier/North Holland, Amsterdam.

Aziz, A.K., J.W. Wingate and M.J. Balas (1977). Control theory of systems governed by partial differential equations. Academic Press.

Banks, S. and A. Pritchard, Editors (1978). Control of distributed parameter systems (Proc. IFAC Symp.), Pergamon Press, Oxford.

Bensoussan, A. (1971). Filtrage optimal des systèmes lineaires. Dunod.

Brogan, W.L. (1967). Theory and application of optimal control for distributed parameter, I-Theory, Automatica, Vol. 4, pp. 107-120; II-Computational Results, ibid, pp. 121-137.

Butkovskiy, A.G. (1965). Distributed control systems. American Elsevier, New York (Translation from Russian).

Butkovskiy, A.G., A.I. Egorov, and K.A. Lurie (1968). Optimal control of distributed systems. SIAM J. Control, Vol. 6, p. 437 (Survey of Russian works).

Butkovskiy, A.G. and A.Y. Lerner (1960). The optimal control of systems with distributed parameters. Autom. Remote Control, Vol. 21, pp. 472-477.

Franke, D. (1976). Regelung von systemen mitverteilten parametern. Regelungstechnik, Vol. 24, pp. 109-152.

Gilles, E.D. (1973). Systeme mitverteilten parametern. Oldenburg.

Gustavsson, I. (1975). Survey of applications of identification in chemical and physical processes. Automatica, Vol. 11, p.3.

Lions, J.L. (1966). Sur le contrôle optimal de systèmes decrits par des equations aux derivées partielles lineaires. C.R. Acad. Sc. Paris, Vol. 263, I:pp. 661-663, II:pp. 713-715.

Lions, J.L. (1971). Optimal control of systems governed by partial differential equations. Springer Verlag, Berlin (Translation from French).

Polis, M.P. and R.E. Goodson (1976). Parameter identification in distributed systems:A synthesising overview. Proc. IEEE, Vol. 64, p. 47.

Phillipson, G.A. (1971). Identification of distributed systems. American Elsevier, New York.

Ray, W.H. (1978). Some recent applications of distributed-parameter systems theory- A survey. Automatica. Vol. 14, pp. 281-287.

Ray, W.H. and D.G. Lainiotis (1978). Distributed-parameter systems: Identification, estimation and control. Marcel Dekker, New York.

Robinson, A.G. (1971). A survey of optimal control of distributed-parameter systems. Automatica, Vol. 7, p. 371.

Ruberti, A. (1978). Distributed-parameter systems:Modelling and identification (Proc. IFIP Working Conf., Rome). Springer Verlag (Lecture Notes in Control and Information Sciences).

Stavroulakis, P. (1982). Distributed parameter systems theory. Vol. I: Control; Vol. II:Estimation. Benchmark Papers in Electrical Engineering and Computer Science, Doweden, Hutchinson and Ross. Inc.

Tzafestas, S.G. (1972). Parameter estimation in distributed parameter dynamic models. In Decision, Design and the Computer, I. Chem.E. Symp. Series No. 35, pp. 5:43-5:50.

Tzafestas, S.G., Ed. (1980). Simulation of distributed parameter and large-scale systems (Proc. IMACS Meeting, Patras, Greece). North Holland, Amsterdam.

Tzafestas, S.G. Ed. (1982). Special issue on Distributed-parameter systems, J. Franklin Institute.

Wang, P.K.C. (1964). Control of distributed parameter systems. In Advances in Control Systems, Vol. 1, (C. Leondes, Ed.).

Wang, P.K.C. (1968). Theory of stability and control for distributed parameter systems (A Bibliography). Int. J. Control, Vol. 7, p. 101.

Williams, J.C. (1969). A survey of distributed parameter systems. In Control of Distributed Parameter Systems, Proc. AACC-JACC Symp.

Part I

Theory

CHAPTER 1

HYBRID SIMULATION OF DISTRIBUTED PARAMETER SYSTEMS

S. G. Tzafestas

Control Systems Laboratory, Patras University, Patras, Greece

INTRODUCTION

Hybrid computers have been developed during the last two decades and offer high accuracy at high speed (Bekey and Karplus, 1968). They were initially designed to meet the demands of aerospace applications, (Connely and Fedoroff, 1965) but currently they are used in a variety of applications such as parameter optimisation (Gran, Sabroff and Wong, 1966), optimal control (Andrews and Korn, 1975; Kahne, 1968), system identification (Piceni and Eykhoff, 1975), simulation of control algorithms (Humeau and Grolleau, 1975), (Tzafestas, 1970a), etc. Usually simulation is employed as part of other more general goals, e.g. for investigating the validity of models or the effectiveness of control procedures, but in many situations simulation may be the goal itself. Actually, hybrid simulation is based on the general idea to decompose the simulation problem at hand into parts appropriate for the digital and analogue components of a hybrid (analogue/digital) computer system. In this way one obtains very much shorter simulation times than those obtained by a digital computer alone, and also considerable analogue hardware savings compared to a pure analogue simulation.

The traditional methods for simulating distributed-parameter systems (DPS), i.e. for integrating the corresponding partial differential equations (PDEs), employ some kind of finite-difference approximation for the time and/or space derivatives involved (Fox, 1962). Hybrid DPS simulation methods are essentially based on approximations of the same type (Vichnevetsky, 1971). The only difference is that instead of discretising all of the n+1 space and time variables, only n of them are discretised, and a smaller number of difference equations are obtained. The high-speed analogue computer integrates the equation for the nondiscretised variable very fast, while the digital computer, combined with the analogue memory, allows the utilisation of the time sharing principle by which the number of the required analogue elements can be reduced to an acceptable level.

The purpose of the present chapter is to provide a survey of the principal hybrid DPS simulation methods, including some implementation aspects. Due to space limitation, the numerical stability/error propagation problem is very briefly discussed. A number of selected illustrative examples are also given.

DISTRIBUTED-PARAMETER MODELS

Distributed-parameter models are classified as parabolic, hyperbolic and elliptic. The first two types are dynamic, whereas the third one is static and, usually, describes the steady-state of dynamic models. The majority of the works dealing with hybrid DPS simulation concern parabolic and hyperbolic models. Since a general and unique model covering all DPS does not exist, here the main models, studied in the context of hybrid simulation, are written down.

Parabolic Models

Most of the works on DPS hybrid simulation were devoted to the simulation of the simplest parabolic model, namely the linear heat (diffusion) equation model with space-time independent parameter and various boundary conditions. This model is

$$\partial X(x,t)/\partial t = c\partial^2 X(x,t)/\partial x^2, \quad 0<x<1, \quad t\geq 0 \tag{1}$$

$$X(x,0)=X^o(x), \quad 0\leq x\leq 1 \tag{2}$$

with boundary conditions of the form

$$X(0,t)=X_o(t), \quad X(1,t)=X_1(t), \quad t\geq 0, \tag{3a}$$

or of the form

$$X(0,t)=X_o(t), \quad [\partial X(x,t)/\partial x]_{x=1}=X_1(t), \quad t\geq 0 \tag{3b}$$

etc. The parameter $c>0$ may constant, or function of x and/or t, or function of $X(x,t)$ in which case the model is nonlinear.

In the multi-dimensional spatial domain case, the heat equation model is $(m\geq 2)$:

$$\frac{\partial X(x,t)}{\partial t} = c\nabla^2 X(x,t), \quad \nabla^2 = \sum_{i=1}^{m} \frac{\partial^2}{\partial x_i^2}, \quad x\varepsilon D \tag{4}$$

where x_i, $i=1,2,\ldots,m$ are the coordinates of the spatial domain D. The general parabolic-like model is

$$\partial X(x,t)/\partial t = N\{X(x,t)\}+U(x,t), \quad x\varepsilon D, \quad t\geq 0 \tag{5a}$$

$$\beta\{X(x,t)\}=0, \quad x\varepsilon\partial D; X(x,0)=X^o(x), \quad x\varepsilon D x \partial D \tag{5b}$$

where $N\{.\}$ and $\beta\{.\}$ are well posed linear or nonlinear spatial differential operators over D and its boundary surface ∂D respectively, and $U(x,t)$ is the forcing function which may also be a function of the system response $X(x,t)$, i.e. a feedback function.

By letting $X(x,t)$ and $U(x,t)$ to be vector valued variables, and $N\{.\}$,

$\beta\{.\}$ to be vector-valued operators, the model (5a,b) can describe a wide class of DPS including both parabolic and hyperbolic systems. For example, the scalar convection-diffusion model

$$\partial X(x,t)/\partial t=-A\partial X(x,t)/\partial x+B\partial^2 X(x,t)/\partial x^2 \tag{6}$$

is of the form (5a), and the wave equation model {see (8a)} can be written in this form as

$$\frac{\partial}{\partial t}\begin{bmatrix}X_1\\X_2\end{bmatrix}=\begin{bmatrix}0&0\\c&0\end{bmatrix}\frac{\partial^2}{\partial x^2}\begin{bmatrix}X_1\\X_2\end{bmatrix}+\begin{bmatrix}0&1\\0&0\end{bmatrix}\begin{bmatrix}X_1\\X_2\end{bmatrix} \tag{7}$$

Hyperbolic Models

The first typical hyperbolic model is the wave equation

$$\partial^2 X(x,t)/\partial t^2=c\partial^2 X(x,t)/\partial x^2,\quad x\varepsilon D,\quad t\geqslant 0 \tag{8a}$$

$$X(x,0)=X^o(x),\ [\partial X(x,t)/\partial x]_{t=0}=X^1(x),\quad x\varepsilon D \tag{8b}$$

where $D=(-\infty,\infty)$ or $D=(0,\infty)$ or $D=(0,1)$. The boundary conditions depend on the form of D. For example, in the finite domain case $D=(0,1)$, one may have the conditions

$$X(0,t)=X_o,\quad [\partial X(x,t)/\partial x]_{x=1}=0,\quad t\geqslant 0 \tag{8c}$$

An other typical hyperbolic model is

$$\partial X(x,t)/\partial t=A(X,x,t)\partial X(x,t)/\partial x+U(X,x,t),\quad x\varepsilon D,\quad t\geqslant 0 \tag{9}$$

where X,U are vectors, and A is a matrix-valued function.

This model involves the wave equation, the fluid transport equation, the heat exchanger systems, etc. For example, the wave equation (8a) can be written in the form (9) as

$$\partial X_1/\partial t=-\sqrt{c}\partial X_1/\partial x+X_2,\quad \partial X_2/\partial t=\sqrt{c}\partial X_2/\partial x, \tag{10}$$

the simple fluid transport equation is

$$\partial X(x,t)/\partial t=-\upsilon\partial X(x,t)/\partial x+U(X,x,t) \tag{11}$$

and the counter-current heat exchanger (with velocity-independent heat transfer coefficient, and negligible diffusion effects) is modelled by:

$$\partial X_1/\partial t+\upsilon_1\partial X_1/\partial x=\mu(X_2-X_1),\partial X_2/\partial t-\upsilon_1\partial X_2/\partial x=\mu(X_1-X_2) \tag{12}$$

Elliptic Models

Setting $\partial X(x,t)/\partial t=0$ in (4) gives the steady-state model $\nabla^2 X(x)=0$,

$x\varepsilon D$, which is the standard elliptic model (Laplace equation). Associated with this model we have the Dirichlet problem in which $X(x)$ is specified over the boundary ∂D, and the Neumann problem in which the normal derivative dX/dn along the direction of the outward normal of ∂D is specified. In addition we have the so-called Helmhotz equation $\nabla^2 X(x)+\lambda^2 X(x)=0$.

The general elliptic model is

$$\mathcal{L}\,X(x)=0,\quad x\varepsilon D;\quad \beta X(x)=0,\quad x\varepsilon\partial D \tag{13}$$

$$\mathcal{L} = \sum_{i=1}^{m} \left(a_i\frac{\partial}{\partial x_i} + b_i\frac{\partial^2}{\partial x_i^2}\right),\, x\varepsilon D \tag{14}$$

with a_i, b_i being or not being dependent on $x\varepsilon D$, and β being a well-posed linear spatial boundary operator including the Dirichlet and Neumann conditions. The model (13) is homogeneous with homogeneous boundary conditions. The corresponding nonhomogeneous model is

$$\mathcal{L}\,X(x)=U_d(x),\quad x\varepsilon D,\quad \beta X(x)=U_b(x),\quad x\varepsilon\partial D \tag{15}$$

where $U_d(x)$, $U_b(x)$ are known functions with appropriate smoothness properties.

A General Scalar Model

This model (Ruppel and Zeitz, 1975) involves the fluid transport, the heat conduction, the convection-diffusion and the wave equation models, and in 1-dimensional spatial domain $D=(0,1)$, is described by

$$\mathcal{L}_x X(x,t)+\mathcal{L}_t X(x,t)=U_d(X,x,t),\quad x\varepsilon D,\quad t>t_o \tag{16}$$

$$\mathcal{L}_x=\sum_{i=1}^{m} b_i(x)\frac{\partial^i}{\partial x^i},\, m=1,2,4;\;\mathcal{L}_t=\sum_{i=0}^{n} a_i(x,t)\frac{\partial^i}{\partial t^i},\quad n=1,2 \tag{17}$$

$$\left[\partial^i X/\partial t^i\right]_{t=0}=X_o^i(x),\quad x\varepsilon\,[0,1],\quad i=0,\ldots,n-1 \tag{18}$$

$$\beta_x^i X(x,t)=U_{b_i}(X,x,t),\, x\varepsilon\{0,1\},\, i=1,2,\ldots,m \tag{19}$$

where the spatial differential boundary operators β_x^i are of maximum order m. The inputs $U_d(X,x,t)$, $U_{b_i}(X,x,t)$ are known external or natural feedback functions.

HYBRID DISTRIBUTED-PARAMETER SIMULATION METHODS

The simulation methods of DPS are classified in
- (i) discrete-time discrete-space (DTDS) methods.
- (ii) discrete-time continuous-space (DTCS) methods.
- (iii) continuous- time discrete-space (CTDS) methods.
- (iv) function expansion (modal) methods.
- (v) other methods (method of characteristics, implicit method, power series method, Monte Carlo method).

The DTDS methods in which both the time and space variables are discretised are the subject of numerical analysis and cannot be implemented on a hybrid computer but on a digital one. These methods will not be considered here (Evans, 1965;Fox, 1962;Vichnevetsky, 1969b). The CTDS methods, also known as parallel methods, are more appropriate for the analogue computer, but they can also be implemented on a hybrid computer. The purpose of the spatial discretisation is to convert the DPS into a model described by a set of simultaneous ODEs along the time axis. The DTCS methods are actually the ones which are appropriate for direct hybrid implementation. The time variable is discretised and the space variable is kept in continuous form. This leads to a sequence of ODEs integrated on the analogue part of the computer, while the transition form one time instant to the next is performed by the digital computing part. The method of characteristics, in which the integration path is parallel to some characteristic axis (not parallel to the time or space axis) leads to ODEs and can be implemented with a hybrid computer. Finally, the function or modal expansion methods (Fourier, Ritz-Galerkin, weighted residuals) reduce the PDE models to ODE ones and can be implemented with a hybrid computer. The Monte Carlo simulation technique, originally developed for elliptic models, is appropriate for computing the system response at a small number of spatial points. This is because in order to compute the system response at a point $x\varepsilon D$, the corresponding Markovian (random walk) process has to be simulated a large number of times (say 1000) on the analogue computer and the result to be averaged on the digital part.

Discrete-Time Continuous-Space Methods

The simplest case of the DTCS simulation concerns the fluid-transport model (1) which involves first-order partial derivatives only. Discretising the time in (11) yields

$$dX^{k+1}(x)/dx+(1/\upsilon\Delta t)X^{k+1}(x)-(1/\upsilon)U^{k+1}(X^{k+1},x)=(1/\upsilon\Delta t)X^{k}(x) \tag{20}$$

for $k=0,1,2,3,\ldots$, where $X^k(x)=X(x,k\Delta t)$ and $U^k(X^k,x)=U(X^k(x),x,k\Delta t)$. At $t=0$ (i.e. for $k=0$) we know $X^o(x)$, $x\varepsilon D$. Using this as forcing function, Eq. (20) is integrated on the analogue computer to give $X^1(x)$, $x\varepsilon D$. Then using $X^1(x)$ as forcing function one can determine $X^2(x)$, $x\varepsilon D$. etc. As initial conditions for integrating the sequence of ODEs (20) we use the given boundary values, i.e. $X^{k+1}(0)=X(0,(k+1)\Delta t)$. The most important property of (20) is that, to integrate the whole sequence of ODEs, only one analogue cell is required.

The stability of (20) depends on the form of $U^{k+1}(X^{k+1},x)$, but most practical problems lead to stable situations. This is not the case for DPS involving second (or higher) order spatial derivatives.

Discretising the time in the heat-equation model (1), and using the implicit approximation

$$\partial^2X/\partial x^2\simeq\Theta d^2X^{k+1}(x)/dx^2+(1-\Theta)d^2X^k(x)/dx^2,\quad 0\leqslant\Theta\leqslant 1 \tag{21}$$

yields

$$d^2X^{k+1}(x)/dx^2-(1/c\Theta\Delta t)X^{k+1}(x)=-(1/c\Theta\Delta t)U^k(x) \tag{22a}$$

where the forcing term $U^k(x)$ is given by

$$U^k(x)=X^k(x)+(1-\Theta)c\Delta t\frac{d^2X^k(x)}{dx^2}=U^{k-1}+\frac{1}{\Theta}(X^k-U^{k-1})\simeq X^{k+1-\Theta}(x) \tag{22b}$$

with $X^{k+1-\Theta}(x)=X(x,,(k+1-\Theta)\Delta t)$. The boundary conditions (3a) yield

$$X^{k+1}(0)=X_o((k+1)\Delta t)\stackrel{\Delta}{=}X_o^{k+1},X^{k+1}(1)=X_1((k+1)\Delta t)\stackrel{\Delta}{=}X_1^{k+1} \tag{22c}$$

The ODE (22a) with initial/final conditions (22c) constitute a standard two-point boundary-value problem (TPBVP). The difficulty in this TPBVP is that one knows one condition at x=0 and the other at x=1, and so (22a) cannnct be integrated as an initial-value problem. Also one can observe that (22a-c) is unstable in both the forward and backward x-directions.

The basic methods for circumventing the difficulty of having two-point boundary conditions and/or numerical instability conditions are:

a) Decomposition method
b) Green function method
c) Integral equation method
d) optimal control method
e) invariant imbedding method

Decomposition method. Using this method (Vichnevetsky, 1968;1969a) the second-order model (22a) is decomposed into two-first-order models, each one bein stable in one direction of integration. Consider for simplicity $\Theta=1$, and decompose the operator $L=d^2/dx^2-(1/c\Delta t)$ into the product

$$L=L_FL_B=(d/dx-\lambda_F)(d/dx-\lambda_B) \tag{23}$$

where λ_F and λ_B are the eigenvalues of L, i.e. $\lambda_F=-1/\sqrt{c\Delta t}$ and $\lambda_B=1/\sqrt{c\Delta t}$. Clearly, $L_F=d/dx-\lambda_F$ is stable in the forward direction of x, and $L_B=d/dx-\lambda_B$ is stable in the backward direction of x.

The decomposition simulation procedure is the following

i) Integrate

$$L_FZ_F^{k+1}(x)=-(1/c\Delta t)U^k(x),\quad U^k(x)=X^k(x) \tag{24a}$$

in forward direction with $Z_F^{k+1}(0)=0$. Any other initial value could be used.

ii) Integrate

$$L_BZ_B^{k+1}(x)=Z_F^{k+1}(x),\quad Z_B^{k+1}(1)=0 \tag{24b}$$

in backward time.

One can observe that $LZ_B^{k+1}(x)=L_FL_BZ_B^{k+1}(x)=L_FZ_F^{k+1}(x)=-(1/c\Delta t)U^k(x)$, i.e. that $Z_B^{k+1}(x)$ satisfies (22a).

iii) Integrate the input-free equations

$$L_F Y_F^{k+1}(x)=0 \quad \text{in forward direction} \tag{25}$$

$$L_B Y_B^{k+1}(x)=0 \quad \text{in backward direction} \tag{26}$$

Since $L_F L_B = L_B L_F$ (i.e. L_F, L_B commute) we have $LY_F^{k+1}(x)=LY_B^{k+1}(x)=0$. This implies that $Y_F^{k+1}(x)$ and $Y_B^{k+1}(x)$ are solutions of the input-free part of Eq. (22a).

iv) Determine the parameters α_F^{k+1} and α_B^{k+1} such that $X^{k+1}(x)$ given by

$$X^{k+1}(x)=Z_B^{k+1}(x)+\alpha_F Y_F^{k+1}(x)+\alpha_B Y_B^{k+1}(x) \tag{27}$$

satisfies the boundary conditions (22c), in which case $X^{k+1}(x)$ is the required solution of (22a-c) at the time step k+1. The resulting parameter values are given by $\alpha^{k+1}=Y_{k+1}^{-1}\zeta^{k+1}$, where

$$\alpha^{k+1}=\begin{bmatrix}\alpha_F^{k+1}\\ \hline \alpha_B^{k+1}\end{bmatrix},\quad Y_{k+1}=\begin{bmatrix}Y_F^{k+1}(0) & Y_B^{k+1}(0)\\ \hline Y_F^{k+1}(1) & Y_B^{k+1}(1)\end{bmatrix},\quad \zeta^{k+1}=\begin{bmatrix}X_o^{k+1}-Z_B^{k+1}(0)\\ \hline X_1^{k+1}-Z_B^{k+1}(1)\end{bmatrix}$$

If we consider $\Theta\neq1$, the starting value of $U^k(x)$ must be set equal to $U^o(x)=X^{1-\Theta}(x)$, and a separate starting procedure is needed for computing $X^{1-\Theta}(x)$. The hybrid set-up for implementing this method is shown in Fig.1. We observe that the same analogue cell is used twice, for integrating (24a) in forward time (F), and (24b) in backward time (B). The same set-up can be used for integrating (25) and (26) by setting the forcing terms to zero.

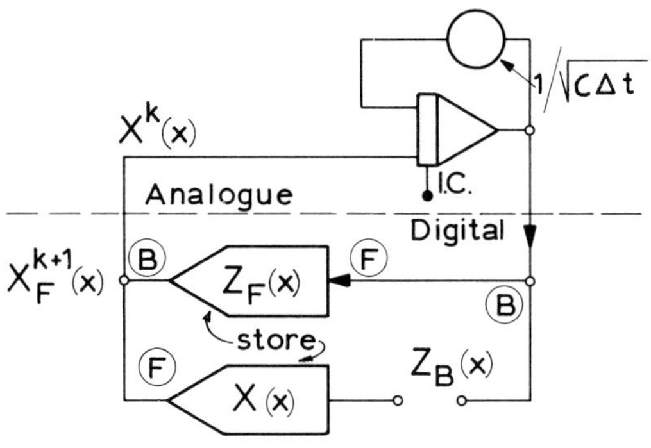

Fig. 1. Hybrid computer set-up for the decomposition method.

If $c=c(x)$, $x\varepsilon D$ the commuting property $L_F L_B = L_B L_F$ does not generally hold. In this case we have (Vichnevetsky, 1969a):

$$L=L_B L_F=(d/dx-\lambda_B(x))(d/dx-\lambda_F(x))$$

$$=d^2/dx^2-\{\lambda_B(x)+\lambda_F(x)\}d/dx+\{-d\lambda_F(x)/dx+\lambda_B(x)\lambda_F(x)\} \quad (28)$$

from which it follows that $\lambda_B(x)+\lambda_F(x)=0$, and $-d\lambda_F/dx+\lambda_B(x)\lambda_F(x)=-\sigma(x)$, or equivalently

$$d\lambda_F/dx+\lambda_F^2=\sigma(x),\ \lambda_B(x)=-\sigma_F(x) \quad (29)$$

where $\sigma(x)=1/\Delta tc(x)$. Therefore, prior to solving (24a-b), the Riccati equation (29) has to be solved for $\lambda_F(x)$. This is stable in the backward x-direction if we choose $\lambda_F(1)\leqslant 0$, in which case $\lambda_F(x)\leqslant 0$ for all $x\varepsilon D=(0,1)$, and the operator $L_F=d/dx-\lambda_F(x)$ is stable in the forward direction, as required. Similarly, $\lambda_B(x)\geqslant 0$, $x\varepsilon D$, which ensures that $L_B=d/dx-\lambda_B(x)$ is stable in the backward direction.

The decomposition method is also applicable to DPS models of the type (16). Applying the DTCS approximation leads to a sequence of ODEs of the form $L\{X^{k+1}\}=H^k(x)$, $x\varepsilon(0,1)$, $k=0,1,...$ for which some of the boundary conditions are given at $x=0$ and the rest at $x=1$. The key point is again the decomposition $L\{X^{k+1}\}=L_B L_F\{X^{k+1}\}$ where L_F is stable when integrated in the forward direction and L_B is stable in the backward direction. The procedure is simplified if L_B and L_F are permutable.

Finally, the decomposition method was used in cases where $c=c(X,x)$ (Vichnevetsky, 1969a) and also for simulating hyberbolic models of the type (9). To treat the $c=c(X,x)$ case a "predicted" approximation of X^{k+1} is formed as $\hat{X}^{k+1}=\sum_{i=0}^{M}\mu_i X^{k-i}$ where μ_i are constants, and M depends on the past values of X^k used (e.g. when $M=1$, $\hat{X}^{k+1}=2X^k-X^{k-1}$). Then, a corrected approximation of X^{k+1} is obtained from

$$X^{k+1}(x)-X^k(x)=\Delta tc(\hat{X}^{k+1},x)\{\Theta d^2X^{k+1}/dx^2+(1-\Theta)d^2X^k/dx^2\}$$

using the decomposition method as described for the space-dependent case. Introducing the corrected estimate $c(\hat{X}^{k+1},x)$ one can obtain further improved estimates in a repetitive way. If $c(X,x)$ is a mild function of X one can choose the initial approximation of $\hat{X}^{k+1}$ as $\hat{X}^{k+1}=X^k$.

Green Function Method. This method is equivalent to the well-known method of parameter variation of differential equations theory. Applying the transformation $Y(x,t)=X(x,t)-\{X_0(t)+(X_1(t)-X_0(t))x\}$, the DTCS model (22a,c) reduces to,

$$d^2Y^{k+1}(x)/dx^2-(1/c\Delta t)Y^{k+1}(x)=-(1/c\Delta t)Y^k(x),$$
$$Y^{k+1}(0)=Y^{k+1}(1)=0 \tag{30}$$

Let $G(x,\xi)$ be the Green's function of the operator $L=d^2/dx^2-1/c\Delta t$ satisfying the homogeneous boundary conditions. Then

$$Y^{k+1}(x)=\int_0^1 G(x,\xi)\left[-\frac{1}{c\Delta t}Y^k(\xi)\right]d\xi \tag{31a}$$

$$G(x,\xi)=\begin{cases}\sinh(\beta\xi)\sin\beta(1-x)/\beta\sinh\beta, & 0\leqslant\xi\leqslant x\leqslant 1\\ \sinh(\beta x)\sin\beta(1-\xi)/\beta\sinh\beta, & 0\leqslant x\leqslant\xi\leqslant 1\end{cases} \tag{31b}$$

where $\beta^2=1/c\Delta t$. The implementation of (31a-b) on the hybrid computer is straightforward (Chan, 1967;1969;Witsenhausen,1965). However the speed of simulation is slower than that of the decomposition. This is because, for obtaining the solution at k+1 for each one spatial point x, a complete integration in ξ is required. Hence, if N point samples are used, N integrations are needed per time step. However writing $G(x,\xi)$ as

$$G(x,\xi)=\begin{cases}AG_1(x)G_2(\xi), & 0\leqslant\xi\leqslant x\leqslant 1\\ AG_2(x)G_1(\xi), & 0\leqslant x\leqslant\xi\leqslant 1\end{cases}$$

where the two directions of causality are separated, we obtain

$$Y^{k+1}(x)=AG_1(x)\int_0^x G_2(\xi)U^k(\xi)d\xi+AG_2(x)\int_x^1 G_1(\xi)U^k(\xi)d\xi \tag{32}$$

where $U^k(\xi)=-(1/c\Delta t)Y^k(\xi)$. Using (32) provides a truly high-speed simulation which is actually faster than any pure digital method. Of course (32) requires $G_1(x)$ and $G_2(x)$ to be known. An experiment showed that this method based on (31a) with 25 sample points in $G(x,\xi)$ is five times slower than a simple target shooting method worked with 5 iterations (Chan, 1969). The hybrid computer set-up is shown in Fig. 2.

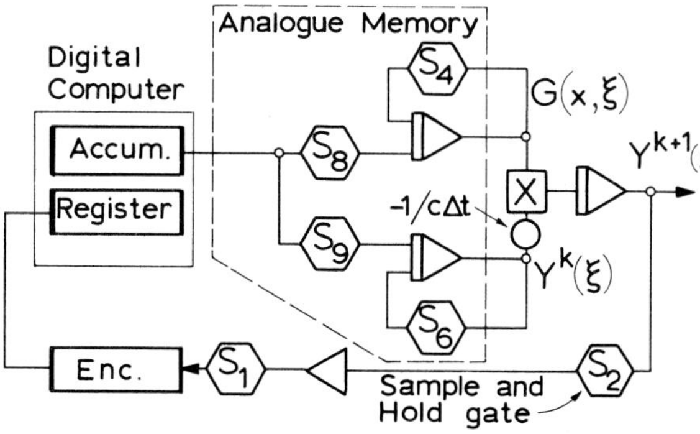

Fig. 2. Hybrid set up for Green's function method.

Integral Equation Method. In this method, instead of using the Green's function for the operator $L=d^2/dx^2-1/c\Delta t$ {cf. (30)}, we use the Green's function $K(x,\xi)$ of d^2/dx^2, which is found to be

$$K(x,\xi)=\begin{cases} x(1-\xi), & 0\leq x\leq \xi<1 \\ \xi(1-x), & 0\leq \xi\leq x\leq 1 \end{cases}$$

Then, the solution of (30) is given by (Chan 1967;1969):

$$Y^{k+1}(x)=\frac{1}{c\Delta t}\int_o^1 K(x,\xi)Y^k(\xi)d\xi - \frac{1}{c\Delta t}\int_o^1 K(x,\xi)Y^{k+1}(\xi)d\xi \qquad (33)$$

The difference between (31a) and (33), is that (33) involves the unknown response function $Y^{k+1}(x)$ both on the left hand side and under the integral sign. This means that (33) is an integral equation with fixed limits (Fredholm integral equation of the second kind). To solve it use is made of the modified Neumann series. Defining the operator P as $PX=\int_o^1 K(x,\xi)X(\xi)d\xi$, the standard Neumann series solution of (33) is

$$Y^{k+1}(x)=\Gamma\Xi(x) \quad , \quad \Gamma=I+\lambda P+\lambda^2P^2+\ldots.. \qquad (34)$$

where $\lambda=-1/\beta^2$, $\beta^2=c\Delta t$, and

$$\Xi(x)= \frac{1}{\beta^2}\int_o^1 K(x,\xi)Y^k(\xi)d\xi, P^{m+1}\Xi(x)=\int_o^1 K(x,z)\{P^m\Xi(z)\}dz, \quad P^o=I \qquad (35)$$

The nearest, to the origin, pole of Γ is $\lambda=\lambda_1=\pi^2$. The Neumann series solution converges if Δt is such that $|\lambda|=1/\beta^2=1/|c\Delta t|<|\lambda_1|=\pi^2$.

Otherwise it does not converge. For example when c=1 and t=0.04 we have $1/|c\Delta t|=25$ i.e. $|\lambda|>|\lambda_1|=\pi^2$. Hence in this case the method does not converge. The minimum value of Δt for which convergence is assured, when c=1, is $\Delta t_{min}=1/\pi^2=0.102$. To improve the convergence limits, we form the modified Neumann series

$$(\lambda-\lambda_1)\Gamma=(\lambda-\lambda_1)(I+\lambda P+\ldots)=-\lambda_1 I+\lambda(I-\lambda_1 P)+\lambda^2(P-\lambda_1 P^2)+\ldots. \qquad (36)$$

which is convergent when $|\lambda|<|\lambda_2|=4\pi^2$. The solution takes now the form

$$Y^{k+1}(x)=\{1/(\lambda-\lambda_1)\}\{-\lambda_1 I+\lambda(I-\lambda_1 P)+\lambda^2(P-\lambda_1 P^2)+\ldots\}\Xi(x)$$

$$=(1/(\lambda-\lambda_1)\}\{-\lambda_1\Xi+\lambda[(\Xi-\lambda_1P\Xi)+\lambda P(\Xi-\lambda_1P\Xi)+\lambda^2P^2(\Xi-\lambda_1P\Xi)+\ldots]\}$$

$$=-\{\lambda/(\lambda-\lambda_1)\}\{-Y^k+[(Y^k-\lambda_1PY^k)+\lambda P(Y^k-\lambda_1PY^k)+\ldots.]\} \quad (37)$$

and is convergent when $1/|c\Delta t|<4\pi^2$, i.e. when $\Delta t_{min}=1/4\pi^2=0.0255$.

Hence, the value $\Delta t=0.04$, which makes the standard Neumann solution nonconvergent, leads now to a convergent situation. The hybrid computer set-up is shown in Fig. 3. For further widening the convergence range, one can use more factors $(\lambda-\lambda_2)$, $(\lambda-\lambda_3)$ etc, as it was done in (36).

It is noted that both the Green's function and the integral equation methods can readily be used when the boundary conditions are time dependent, or when a forcing term is present. Also, the integral equation method is applicable when the heat equation concerns the case of an inhomogeneous material, and has the form, $\partial X/\partial t=\partial\{c(x)\partial X/\partial x\}/\partial x+U(x,t)$.

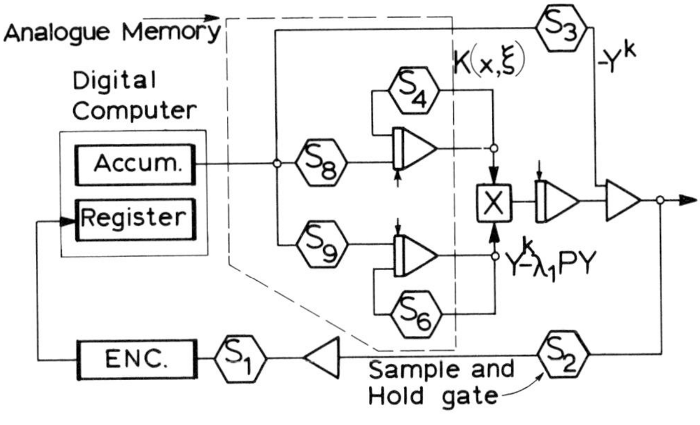

Fig. 3. Hybrid simulation set-up for the integral equation method.

The basic requirement for safely using the integral equation method with a small time step is to know at least the first eigenvalue λ_1.

Optimal Control Method.

By this method (Hara and Karplus, 1968) the solution of the DTCS model (22a) is converted to the solution of a distributed-parameter optimal control problem. Consider the case $\Theta=1$, and write (22a) as

$$d^2X^{k+1}(x)/dx^2=(1/c\Delta t)\{2\hat{X}^{k+1}(x)-X^{k+1}(x)-x^k(x)\} \quad (38a)$$

which is a stable equation. Here $\hat{X}^{k+1}(x)$ is an approximation to $X^{k+1}(x)$, and is regarded as a control variable, which must be selected so as to minimise the error (cost) functional

$$J=\int_0^1\{X^{k+1}(x)-\hat{X}^{k+1}(x)\}^2dx \tag{38b}$$

This is a standard DP optimal control problem and can be solved by various techniques. For example, we may reduce it into a final-value control problem, by using the state variable Z_o^{k+1} defined as

$$dZ_o^{k+1}/dx=\{X^{k+1}(x)-\hat{X}^{k+1}(x)\}^2,\quad Z_o^{k+1}(0)=0$$

in which case J takes the form $J=Z_o^{k+1}(1)$. Then defining the state variables $Z_1^{k+1}=X^{k+1}$, $Z_2^{k+1}=dX^{k+1}/dx$, the model (38a) takes the form

$$d\underline{Z}^{k+1}/dx=F(\underline{Z}^{k+1},\hat{X}^{k+1}),\quad \underline{Z}=[Z_o,Z_1,Z_2]^T$$

Of course the additional constraint $\Psi=\{Z_1^{k+1}(x)\}_{x=1}-X_1^{k+1}=0$ must be imposed to satisfy the boundary condition of the original problem at x=1. The resulting final-value control problem can be solved by the Pontryagin principle enriched with the penalty method (Tzafestas, 1970b). A similar approach can be applied to the DTCS version of the Crank-Nicholson method $(X^{k+1}-X^k)/\Delta t=(c/2)(d^2X^{k+1}/dx^2+d^2X^k/dx^2)$. Introducing the new variable $W^{k+1}=(X^{k+1}+X^k)/2$, i.e. $X^k=2W^{k+1}-X^{k+1}$, the above equation becomes $d^2W^{k+1}/dx^2=(2/c\Delta t)(X^{k+1}-W^{k+1})$, and the cost functional to be minimised is $J=\int_0^1\{(2W^{k+1}-X^{k+1})-X^k\}^2dx$ with X^{k+1} as the control input.

<u>Invariant Imbedding Method</u>. This method was developed as an alternative to the decomposition method for converting the TPBVP of the DTCS equations (22a-c) to an initial-value problem (Alton and Nelson, 1970; Nelson, 1970; Nelson and Altom, 1971).

To start with, consider the homogeneous initial-value problem (set $\Theta=1$ in (22a)):

$$d^2Y(x)/dx^2-(1/c\Delta t)Y(x)=0,\quad Y(0)=0,\quad dY(0)/dx=1, \tag{39a}$$

and the sequence of nonhomogeneous initial-value problems

$$d^2Y_*^k/dx^2-(1/c\Delta t)Y_*^k=-U_*^k,\ Y_*^k(0)=X_o^{k+1},\ dY_*^k(0)/dx=0,\quad k=0,1,2,... \tag{39b}$$

where $U_*^k(x)\triangleq U^k(x)/c\Delta t$. Then, the solution of (22a) is given by

$$X^{k+1}(x)=\{X_1^k-Y_*^k(1)\}Y(x)/Y(1)+Y_*^k(x) \tag{39c}$$

Now, defining the functions

$$R(x)=Y(x)/\{dY(x)/dx\},\quad P_*^k(x)=-\{dY_*^k(x)/dx\}/\{dY(x)/dx\}$$
$$S(x)=1/\{dY(x)/dx\},\quad P^k(x)=Y_*^k(x)-Y(x)\{dY_*^k(x)/dx\}/\{dY(x)/dx\} \tag{40}$$

one can see from (22a-c) and (39a-c), that they satisfy the following ODEs

$$dR/dx=1-(1/c\Delta t)R^2(x),\ R(0)=0;\ dS/dx=-(1/c\Delta t)R(x)S(x),S(0)=1$$
$$dP^k/dx=[-(1/c\Delta t)P^k(x)+U_*^k(x)]R(x),\ P^k(0)=X_o^k \qquad (41)$$
$$dP_*^k/dx=[-(1/c\Delta t)P^k(x)+U_*^k(x)]S(x),\ P_*^k(0)=0$$

In terms of these functions, (39c) is written as

$$X^{k+1}(x)=[P_*^k(1)-P_*^k(x)]R(x)/S(x)+[S(1)/R(1)][R(x)/S(x)]X_1^k$$
$$+P^k(x)-[S(1)/R(1)][R(x)/S(x)]P^k(1),\ k=0,1,2,... \qquad (42)$$

This implies that the TPBVP (22a-c) was reduced to the initial-value problem (41). The only drawback is that the equation for S(x) in (41) cannot be integrated accurately near x=1. This is overcome by using the variable $\hat{S}(x)=S(1)/S(1-x)$ and expressing $X^{k+1}(x)$ in (42) in terms of $\hat{S}(x)$. Note that the ODEs in (41) are stable in the forward x direction. Concerning the analogue scaling aspect of these equations see Nelson and Altom (1971).

Continuous-Time Discrete-Space Methods.

By the CTDS approximation the PDE model of our DPS is converted to a set of coupled ODEs, which have to be solved simultaneously (in parallel) by the analogue part of the hybrid computer. Normally, due to the large number of ODEs, this would require a prohibitive amount of analogue equipment. Hence, the effort was mainly concentrated in finding ways of reducing this equipment, without slowing very much the simulation speed. The result of this effort is a variety of methods based on some kind of multiplexing and iteration.

Here the following methods will be described which are not to be considered as exhaustive.

i) relaxation (cell multiplexing) method
ii) regional multiplexing (pyramid) method
iii) differential-difference method
iv) block time-sharing method
v) partial multiplexing method
vi) other multiplexing methods

Cell Multiplexing Method. Consider the heat-equation model (1). Subdividing the spatial domain D=[0,1] into N+1/2 equal subintervals Δx, yields the following basic CTDS model (Hsu and Howe, 1968):

$$dX_i(t)/dt=(c/\Delta x^2)\{X_{i-1}(t)-2X_i(t)+X_{i+1}(t)\},\ i=1,2,...,N-1 \qquad (43a)$$
$$dX_N(t)/dt=(c/\Delta x^2)\{X_{N-1}(t)-X_N(t)\},\ X_o(t)=X(0,t)=0 \qquad (43b)$$

The straight-forward approach is to integrate this set of N coupled ODEs, simultaneously, by using an analogue cell for each one of them. By the cell multiplexing method only one analogue computing cell (including integrators and other required elements) is used in a time-sharing way for integrating these equations iteratively, as follows:

(1) Find an initial guess $X_i^o(t)$ of $X_i(t)$, $i=1,2,\ldots,N$.

(2) Integrate (43a) for i=1 using the digitally stored value of $X_o(t)$ and $X_2^o(t)$, via D/A converters, as inputs.

(3) Convert the resulting $X_1(t)$ into digital form, via the A/D converter, and call it $X_1^1(t)$. Replace $X_1^o(t)$ by $X_1^1(t)$.

(4) Do the same for i=2, and replace $X_2^o(t)$ by $X_2^1(t)$, and so on up to $X_N^1(t)$.

(5) Repeat the steps 2 to 4 and find the next approximations of the solution at the N stations, i.e. $X_i^2(t), X_i^3(t),\ldots,X_i^k(t)$.

(6) Terminate the procedure when $|X_N^{k+1}(t)-X_N^k(t)|<\eta, t\varepsilon[0,t_f]$, where η is the prescribed accuracy (say $\eta=5\text{x}10^{-4}$).

This method can also be used for simulating hyperbolic and nonlinear systems.

For example, the CTDS model of the wave equation (8a) gives the following multiplexing (iterative) form.

$$d^2X_i^k/dt^2=[c/(\Delta x)^2]\,[X_{i-1}^k-2X_i^k\;X_{i+1}^{k-1}],$$

$$d^2X_N^k/dt^2=[c/(\Delta x)^2]\,[X_{N-1}^k-X_N^k]$$

where the upper index k indicates the approximation step. Correspondingly, the heat equation in two space dimensions:
$\vartheta X/\vartheta t=c(\vartheta^2X/\vartheta x_1^2+\vartheta^2X/\vartheta x_2^2)$, $x_1,x_2\varepsilon(0,1)$; $X(x_1,x_2,0)=1$, $x_1,x_2\varepsilon(0,1)$;
$X(x_1,x_2,0)=1$, $x_1,x_2\varepsilon(0,1)$; $X(x_1,x_2,t)=0$, x_1,x_2 over the boundary, gives the single-cell-multiplexing model $(i=1,2,\ldots,N)$:

$$dX_{i,j}^k/dt=\{c/(\Delta x)^2\}(X_{i-1,j}^k+X_{i+1,j-1}^k+X_{i+1,j}^k+X_{i,j+1}^{k-1}-4X_{i,j}^k) \quad (44)$$

which is solved for $j=1,2,\ldots,N$, i.e. from line to line in the x_2-direction.

As a nonlinear example we give the DPS $\vartheta X/\vartheta t=\vartheta(X\vartheta X/\vartheta x)/\vartheta x$, $X(x,0)=1$, $X(0,t)=0$, $(\vartheta X/\vartheta x)_{x=1}=0$, which for N=6 yields $(i=1,2,3,4,5)$:

$$dX_i^k/dt=21.125\{(X_{i-1}^k)^2-2(X_i^k)^2+(X_{i+1}^{k-1})^2\},$$
$$dX_6^k/dt=21.125\{(X_5^k)^2-(X_6^k)^2\}$$

Regional Multiplexing Method. This is a compromise between complete parallel simulation and single cell multiplexing. The spatial domain is divided into N regions and I intervals. In each region there are i=I/N intervals. The DPS considered by Saphier (1972) is the parabolic neutron-transport model of an infinite-slab nuclear reactor.

The method proceeds as follows. All equations of each region are solved in parallel. The computation marches from region to region from "left" to "right" using the same group of analogue cells and the same digital software. The boundary conditions for each region are at first chosen to be the initial conditions at the boundary meshpoints.

The computations at each region provide the new boundary conditions for the regions to its right and to its left. The iteration is performed on the regional boundary values which are the only values stored, and then played back to the analogue units. The iteration is terminated when the desired accuracy is achieved.

The number of iterations required for a certain accuracy can be reduced by using better approximations of the starting regional boundary values. This has a beneficial effect on the cummulative errors caused by the analogue components (offset errors, etc.). A method for obtaining such improved regional boundary values is as follows. Consider a coarse subdivision of the spatial domain D in N intervals (N is the number of regions). The set of equations obtained is solved for a single region. No iterations are needed, since the boundary values are well defined.

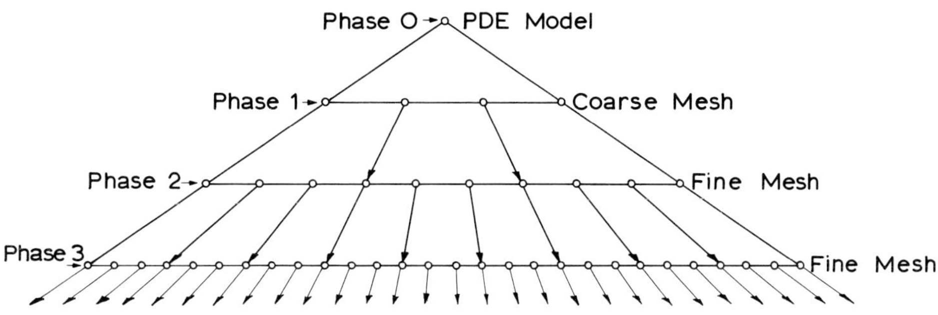

Fig. 4. Schematic representation of the regional simulation (pyramidal) method.

Next, each of these N intervals is subdivided into I/N subintervals, and use is made of the previous result as time dependent regional boundary condition, in place of the fixed, in time, values used previously. It is possible to further subdivide each of these intervals, and proceed in a "pyramidal" way until the required fine mesh division is obtained (Fig. 4). This is called the improved pyramidal method, and obviously can be extended to DPS in two and three space dimensions,

In examples computed, it was found that the price per problem solved, using a hybrid set-up, was cheaper by an order of magnitude in comparison with the pure digital solution.

Differential-Difference Method. In this method the system is simulated in a sequential manner over consecutive small time intervals, and the same analogue cell is employed for the integration over these successive intervals (Petrov, 1975).

Consider the heat equation model (1) with an external input term $U(X,x,t)$. Its CTDS representation is

$$dX_i(t)/dt=(c/\Delta x^2)\{X_{i-1}(t)-2X_i(t)+X_{i+1}(t)\}+U(X_i(t),x_i,t) \quad (45)$$

Now, divide the time interval $T=[0,t_f]$ of interest, in small subintervals $[t_k,t_k+\Delta t]$, $k=0,1,2,\ldots$, and assume that all values $X_i(t)$, $i=0,1,2,\ldots,N$ at $t=t_k$ are known. If Δt is sufficiently small the expression $X_{i-1}(t)-2X_i(t)+X_{i+1}(t)$ can be replaced by the constant $X^k_{i-1}-2X^k_i+X^k_{i+1}$ where $X^k_i \overset{\Delta}{=} X_i(t_k)$. Thus (45) can be written as

$$dX_i(t)/dt=(c/\Delta x^2)(X^k_{i-1}-2X^k_i+X^k_{i+1})+U(X_i(t),x_i,t)$$

$$k\Delta t\leqslant t\leqslant (k+1)\Delta t \quad (46a)$$

with given data

$$[X^k_i]_{k=0}=X_i(0),\ [X^k_{i-1}]_{i=1}=X^k_o,\ [X^k_{i+1}]_{i=N-1}=X^k_N \quad (46b)$$

We see that (46a) can be integrated separately for each node i, with initial value $X_i(t_k)=X^k_i$ to give an approximate value $X^{k+1}_i=X_i(t_{k+1})$ of the response at the next time interval, which in turn provides the initial conditions for the integration over $[t_{k+1},t_{k+1}+\Delta t]$, and so on. The name of the scheme (46a) as "differential-difference" is due to the above successive continuous integration procedure over small intervals of length Δt. Since only the initial conditions change from interval to interval, the same analogue cell can be used for all these intervals. At each time step the solution is transfered via A/D converters on the digital computer, to be stored. The digital computer. calculates the term $X^k_{i-1}-2X^k_i+X^k_{i+1}$ at the beginning of each integration interval, and controls the data flow from one interval to the next. The simulation speed depends on the repetitive properties of the analogue part, and also on the A/D and D/A interface.

Block Time-Sharing Method. In this method (Neundorf and Schoenefeld, 1980) the iterative multiplexing (time sharing) principle is used, such that one block with m lines is integrated on space stations from left to right. The digital computer is again used for function storage and playback.

Consider the heat--equation model (1) with $c=c(x,t)\geqslant a>0$ for $x\in[0,1]$,

$t \varepsilon [0,t_f]$, and a forcing function $U(X,x,t)$. Then its CTDS representation is given by (45) with $c=c_i(t)$ and given initial-boundary data $X_i^o=X_i(0)$, $[X_{i-1}(t)]_{i=1}=X_o(t)$, and $[X_{i+1}(t)]_{i=N-1}=X_N(t)$. Now, (45) can be written in the following block-iterative form (k=iteration index):

$$d\hat{X}_p(t)^{k+1}/dt=(c_p(t)/\Delta x^2)\{P-2\hat{X}_p(t)^{k+1}+\hat{X}_{p+1}(t)^{k+1}\}+U_p(t)^{k+1} \quad (47a)$$

$$d\hat{X}_i(t)^{k+1}/dt=(c_i(t)/\Delta x^2)\{\hat{X}_{i-1}(t)^{k+1}-2\hat{X}_i(t)^{k+1}+\hat{X}_{i+1}(t)^{k+1}\}+U_i(t)^{k+1},\ i=p+1,\dots,q-1=p+m-2 \quad (47b)$$

$$d\hat{X}_q(t)^{k+1}/dt=(c_q(t)/\Delta x^2)\{\hat{X}_{q-1}(t)^{k+1}-2\hat{X}_q(t)^{k+1}+Q\}+U_q(t)^{k+1} \quad (47c)$$

Equations (47a-c) are integrated over $[t_r,t_{r+1}]$ with initial conditions $\hat{X}_i(t_r)^{k+1}=\hat{X}_i(t_r)=\hat{X}_{i,r}$, $i=1,2,\dots,N-1$, for which the first (starting) guesses $X_i(t^o)$ are given on certain time points, and with the exact exterior boundary conditions

$$\hat{X}_o(t)^k=X_o(t),\quad \hat{X}_N(t)^k=X_1(t)$$

Here, $\Delta t_r=t_{r+1}-t_r=t_f/R$ is the time increment, $\tau=t_{r,j+1}-t_{r,j}=t_f/R_L$ is the sample interval (playback period), and $L=\Delta t_r/\tau$ is the number of storages of solutions $\hat{X}_p(t)^{k+1}$, $\hat{X}_{p+1}(t)^{k+1},\dots,\hat{X}_q(t)^{k+1}$ and playbacks of $\hat{X}_{p-1}(t)^{k+1}$, $\hat{X}_{q+1}(t)^{k+1}$ on the points $t_{r,j}=t_r+j\tau$. The playback of interior boundary values is made through interpolation/extrapolation formulas of higher order. The final values $\hat{X}_i(t_{r+1})^{k+1}$, $i=1,2,\dots,N-1$ of the interval $[t_k,t_{k+1}]$ serve as initial values (inputs) for the next interval. The iteration may be carried out by Jacobi, Gauss-Seidel, or relaxation methods. The iteration is terminated when

$$\max_{1\leqslant i\leqslant N-1}\left|X_i(t_{r+1})^{k+1}-X_i(t_{r+1})^k\right|<\eta,$$

where $\eta=0(\Delta x^2)$ is determined from the accuracy of the analogue computer. When $m=L=1$ we have $\tau=\Delta t_r$, and the method leads to the following single time-sharing scheme $(i=1,2,\dots,N-1)$:

$$d\hat{X}_i(t)^{k+1}/dt=\{c_i(t)/\Delta x^2\}\{P-2\hat{X}_i(t)^{k+1}+Q\}+U_i(t)^{k+1},\quad \hat{X}_i(t_r)^{k+1}=X_{i,k}.$$

The integration for each line (station) i is carried out on the analogue part, and the result is transfered via the A/D converters into the

digital memory, replacing the previous values. If we have available initial guesses $\hat{X}^o_{i,k+1}$ of suitable starting points, then the number of iterations needed for a given accuracy may be reduced.
Using first-order interpolation we obtain

$$P=\hat{X}_{i-1,r}+(1/\Delta t_r)\{\hat{X}^o_{i-1,r+1}-\hat{X}^o_{i-1,r}\}(t-t_r),$$

$$Q=\hat{X}_{i+1,r}+(1/\Delta t_r)\{\hat{X}^o_{i+1,r+1}-\hat{X}^o_{i+1,r}\}(t-t_r)$$

which is not far from the true solution (weak consistency of order $O(\tau+\tau^2/\Delta x^2)$). Note that if we use in P and Q extrapolation the consistency order is $O(\tau+\tau/\Delta x^2)$.

Under the assumption of continuous $c(x,t)$, and $\partial U/\partial X$ with $\partial U(X,x,t)/\partial X \geq 0$ for $x\varepsilon[0,1]$, $t\varepsilon[0,t_f]$ this method is always stable. Note that the differential-difference scheme (46a), which is conditionally stable and noniterative, can be used for determining the initial guess $\hat{X}_i(t)^o$. The next iterations can be made using (47a-c) with the appropriate values for P and Q.

<u>Partial Multiplexing Method</u>. In this case only part of the analogue circuitry is multiplexed (usually the most expensive one) for example the part which computes the nonlinear functions. However, depending on the availability and speed of the analogue and digital hardware, the multiplexed operations can be performed either on the analogue or on the digital part of the hybrid computer.

In many cases the multiplexing equipment is used for calculating correction terms and thus improving the analogue solution. These correction terms may be computed in digital form and transfered to the analogue part via the D/A converters (Nomura and Deiters, 1968). To illustrate how to compute correction terms, consider the heat equation model (1) with $1/c=(\pi/2)^2$, and nondimensional time variable $\tau=(\pi/2)^2(\Delta x)^2 t$. Also assume that X=1 for $\tau<0$ and X=0 at x=1 for $\tau\geq 0$. This actually describes the cooling of an homogeneous metal sheet when the temperature of both surfaces is suddenly dropped.

The CTDS model, with N+1/2 spatial data points, is a usual

$$dX_i/dt=(c/\Delta x^2)(X_{i-1}-2X_i+X_{i+1})+E_i \quad (48a)$$

$$dX_o/dt=(c/\Delta x)^2(X_1-X_o)+E_o,\ i=1,2,\ldots,N-1 \quad (48b)$$

where E_i are correcting inputs to be discussed later.

Assuming $E_i=0$, $i=0,1,2,\ldots,N-1$ and N=6, (48a,b) gives the analogue simulation results of Fig.5.

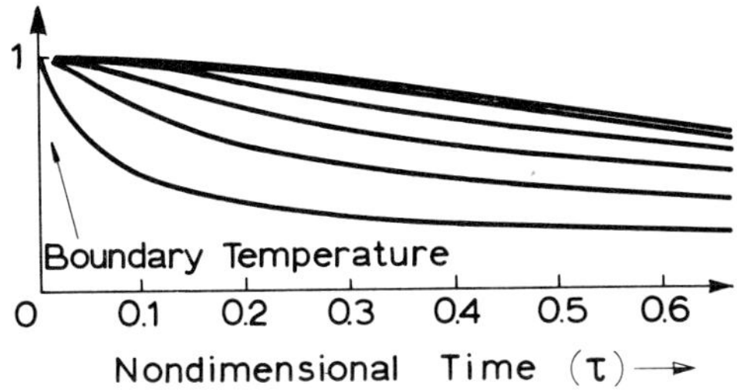

Fig. 5. Analogue simulation results (CTDS) of the cooling process for N=6.

At the beginning of the transient, where the variation of X is high, the error in the finite-difference approximation of the spatial derivatives is large. On the other hand whenever X is nearly flat the spatial sampling is denser than required. Thus, one has to look for correction terms and adaptive space sampling procedures.

Some ways for digitally computing correction terms E_i, $i=0,1,2,...,N$, added as inputs, via the D/A interface, on the right hand sides of (48a,b) are the following.

(i) Using higher-order approximation for the spatial derivatives. This method has many drawbacks (difficult analogue implementation, etc).

(ii) Using a least squares polynomial fit of $X(x,t)$ over D. The derivatives, which are computed using the resulting fitting curve, may have large errors between the sample points.

(iii) Using an infinite Fourier series to fit the spatial data points via the least squares criterion. This method does not have the drawbacks of the polynomial fit, i.e. the resulting approximation, together with its derivatives, is good not only at the sample points, but also between them. Of course one has to design a truncated series having the properties of the infinite Fourier series.

In our example the interval $0\leqslant x\leqslant 1$ can be considered as a quarter cycle of a curve with halfwave symmetry, in which case the least-squares fit to 2N-1 equidistant spatial points has the cosine form

$$X(x_i) = \sum_{n=1,3,...}^{2N-1} A_n \cos(\frac{n\pi}{2}x_i), \quad i=0,1,2,...,N-1 \tag{49a}$$

$$A_n = \frac{2}{N}\sum_{i=0}^{N-1} X(x_i)\cos(\frac{n\pi}{2}x_i), \quad n=1,3,...,2N-1 \tag{49b}$$

with $X(x_o)=X(0)/2$, and N being the number of sections in $[0,1]$.

If the components with frequencies higher than $(2N-1)\pi/2$ have considerable magnitudes, the truncated series (49a) possesses an error appearing as a high frequency ripple on $X(x_i)$. This error becomes larger in the derivatives of $X(x)$. The ripple in $dX(x)/dx$ can be smoothed out if A_n in (49a) is replaced by $A'_n=A_n\{\sin(n\pi/m)\}/(n\pi/m)$, $n=1,3,5,\ldots,p$ where m is the lowest omitted frequency, i.e. $m=2N+1$, and $p=2N-1$. Similarly, the ripple in $d^2X(x)/dx^2$ is reduced by using the coefficients $A'_n=2A_n[1-\cos(n\pi/m)]/(n\pi/m)^2$, $n=1,3,\ldots,p$. The adaptive sampling must be such that to bunch the samples in the region where the response $X(x)$ has a rapidly varying second spatial derivative, and spread them outside this region. This can be done by transforming the real-world x-domain into a domain y such that y is equidistant along a single-valued continuous curve $y=f(x)$, $x\varepsilon[0,1]$. In this case the distance y_i corresponding to x_i is $y_i=\int_0^{x_i}\{1+(df/dx)^2\}^{1/2}dx$, whence $dy/dx=\{1+(df/dx)^2\}^{1/2}$ and $d^2y/dx^2=\{1+(df/dx)^2\}^{-1/2}(df/dx)d^2f/dx^2$. The y-axis is normalised as $y'=y/y_T$, where y_T is the total y-distance corresponding to $x=1$. In the transformed domain y' the spatial derivatives are approximated by the standard methods for equidistant sampling (e.g. first-order differences, truncated series approximation, etc). The real-world second derivative of X, involved in the heat-equation model, is given by

$$\vartheta^2X/\vartheta x^2=\{(dy/dx)^2/y_T^2\}\vartheta^2X/\vartheta y'^2+\{(d^2y/dx^2)/y_T\}\vartheta X/\vartheta y'$$

where the factors in brackets depend on the choice of $y=f(x)$ and can be stored in the digital memory prior to the hybrid simulation. The pot settings can be calculated such that the implementation in the analogue part corresponds to the real-world x-domain. Note that it was found experimentally, that the errors due to A/D conversion (even with an A/D of only 10-bit accuracy) are negligible and have not any significant effect on the computation of the E'_is. A suitable transformation for the heat-equation model is $y=ax^2-bx$ with $a=6$ and $b=2$.

Other Multiplexing Methods. In case of a multidimensional spatial problem, a worthwile reduction in analogue equipment is obtained by using analogue integration only along one spatial line in a time-multiplexed way (Vichnevetsky, 1971).

For example, the line multiplexed model of the heat-equation, in two space dimensions, is

$$dX_{ij}/dt=c\{(\delta_x^2X/\Delta x^2)_{ij,t}+(\delta_y^2X/\Delta y^2)_{ij,k}\},k\Delta t\leqslant t\leqslant(k+1)t \qquad (50)$$

where k denoted discrete time $k\Delta t$, $X_{ij}\simeq X(x_i,y_j,t)$, $x_i=i\Delta x$, $y_j=j\Delta y$, $\delta_x^2X=X_{i+1,j}-2X_{ij}+X_{i-1,j}$, etx. The stability condition is found to be $\Delta t/(\Delta y)^2<$ const. The solution procedure is to solve (50) sequentially for $j=1,2,\ldots,m$ keeping i fixed, and then repeat the same for the next time intervals $(k+1)\Delta t\leqslant t\leqslant(k+2)\Delta t$, etc.

To make the procedure unconditionally stable for $\Delta x=\Delta y$, one must alternate the spatial line direction, i.e. from y=constant to x=constant. The resulting equations are (Vichnevetsky, 1970b):

$$dX_{ij}/dt=c\{(\delta_x^2X/\Delta x^2)_{ij,t}+(\delta_y^2X/\Delta y^2)_{ij,k}\},\quad k\Delta t\leqslant t\leqslant(k+\tfrac{1}{2})\Delta t$$

$$dX_{ij}/dt=c\{(\delta_x^2X/\Delta x^2)_{ij,k+\frac{1}{2}}+(\delta_y^2X/\Delta y^2)_{ij,t}\},(k+\tfrac{1}{2})\Delta t\leqslant t\leqslant(k+1)\Delta t$$

This is equivalent to the alternating direction implicit finite-difference method, and possesses better stability characteristics even when $\Delta x\neq\Delta y$. Now suppose that the right hand sides of the above equations are replaced by $2c(\delta_x^2X/\Delta x^2)_{ij,t}$ and $2c(\delta_y^2X/\Delta y^2)_{ij,t}$ respectively. Clearly, over one Δt, the computed $X_{ij}(t)$ is, on the average, the same, but now the coupling terms dissapear, and the required D/A interfacing equipment is reduced significantly.

We close by mentioning the digital oriented hybrid multiplexing method of Karplus (1964) in which the time integration of the simultaneous ODEs (say (43a,b)) is performed digitally, and the analogue computer is used for the parallel solution of the algebraic equations resulting from the application of usual inplicit finite-difference approximations. The time saving in this case is due to the fast parallel solution of the implicit algebraic equations by the special purpose analogue equipment.

Modal Expansion Methods

Currently, a variety of modal or functional expansion methods are available including the classical separation, the assumed mode, and the Walsh function methods (Ashley and Bullock, 1968; Neuman and Strauss, 1968; 1969; Vichnevetsky, 1969c). In all cases the PDEs are reduced to ODEs and, particularly, in the Walsh expansion case to algebraic equations (Tzafestas 1978a,b).

Ritz-Galerkin:Homogeneous Boundary Conditions. Consider the general parabolic-like model (5a,b), and an approximate (trial) solution of the form

$$\hat{X}(x,t)=\sum_{i=1}^{N}H_i(x)\Phi_i(t) \tag{51}$$

where $H_i(x)$ are prescribed orthogonal functions (called "assumed modes"). The orthogonality condition is expressed as

$$\int_D H_i(x)H_j(x)dx=h_i\delta_{ij},\quad (\delta_{ij}\ \text{Kronecker's delta}) \tag{52}$$

Associated with $H_i(x)$ one has the residuals {see (5a,b)}:

$$R(x,t)=\partial\hat{X}/\partial t-N[\hat{X}(x,t)]-U(x,t),\quad x\varepsilon D,\quad t\geqslant0 \tag{53a}$$

$$B(x,t)=\beta[\hat{X}(x,t)]-\beta[X(x,t)],\quad x\varepsilon\partial D,\quad t\geqslant0 \tag{53b}$$

$$I(x)=[\hat{X}(x,t)]_{t=0}-X^o(x),\quad x\varepsilon D \tag{53c}$$

called the *interior*, *boundary*, and *initial residual*, respectively.

If the assumed modes satisfy the boundary conditions, in which case $B(x,t)=0$, $x\varepsilon\vartheta D$, $t\geqslant 0$, then one has to choose $\Phi_i(t)$, $i=1,2,\ldots,N$ so as to minimise $R(x,t)$, $x\varepsilon D$, $t\geqslant 0$ (interior method). If the $H_i(x)$s are such that $R(x,t)=0$, $x\varepsilon D$, $t\geqslant 0$, then the $\Phi_i(t)$s must be selected so as to minimise $B(x,t)$, $x\varepsilon\vartheta D$, $t\geqslant 0$ (boundary method). If both $R(x,t)$ and $B(x,t)$ are to be minimised in some sense, we have the mixed method.

The Ritz approach for a mixed method consists in choosing $\Phi_i(t)$ so as to minimise

$$J=J_D+J_{\vartheta D},\quad J_D=\int_D \langle R(x,t),R(x,t)\rangle dx,\; J_{\vartheta D}=\int_{\vartheta D}\langle B(\xi,t),B(\xi,t)\rangle d\xi \tag{54}$$

Now, consider for simplicity that $B(x,t)=0$, $x\varepsilon\vartheta D$, $t\geqslant 0$. Galerkin suggested to choose $\Phi_i(t)$ so as to orthogonalise the residual with respect to the assumed modes, i.e.

$$\int_D \langle R(x,t),H_i(x)\rangle dx=0,\quad i=1,2,\ldots,N \tag{55}$$

This is equivalent to Ritz's approach if the minimisation of $J=J_D$ is made with respect to $d\Phi_i(t)/dt$.

Introducing $\hat{X}(x,t)$ from (51) into the system (5a) gives

$$\sum_{i=1}^{N} H_i(x)\frac{d\Phi_i(t)}{dt}=N\left[\sum_{i=1}^{N} H_i(x)\Phi_i(t)\right]+U(x,t)+R(x,t) \tag{56a}$$

which, in view of (52) and (55), gives

$$h_i d\Phi_i(t)/dt=\int_D \langle N\left[\sum_{i=1}^{N} H_i\Phi_i\right]+U,H_i\rangle dx,\quad i=1,2,..,N \tag{56b}$$

We see that (56b) involves N ODEs in the $\Phi_i(t)$s, which have to be solved in parallel. The initial conditions $\Phi_i(0)$ are determined by minimising the initial cost $J_o(x)=\int_D \langle I(x),I(x)\rangle dx$, and are given by $\Phi_i(0)=(1/h_i)\int_D \langle X^o(x),H_i(x)\rangle dx$.

If the $H_i(x)$s are not orthogonal, (52) is not satisfied, and one obtains

$$\frac{d}{dt}\begin{bmatrix}\Phi_1\\ \vdots\\ \Phi_N\end{bmatrix}=F^{-1}\int_D \langle N\left[\sum_{j=1}^{N} H_j\Phi_j\right]+U,\begin{bmatrix}H_1\\ \vdots\\ H_N\end{bmatrix}\rangle dx,\quad \begin{bmatrix}\Phi_1(0)\\ \vdots\\ \Phi_N(0)\end{bmatrix}=F^{-1}\begin{bmatrix}a_1\\ \vdots\\ a_N\end{bmatrix}$$

where the (i,j)th element f_{ij} of the NxN matrix F, and a_i are given by

$$f_{ij}=\int_D \langle H_i(x),H_j(x)\rangle dx,\quad a_i=\int_D \langle X^o(x),H_i(x)\rangle dx$$

Of course, the matrix F is assumed to be invertible. The only difference from the orthogonal case is the requirement to premultiply by F^{-1}, which can be computed by the digital computer prior to the actual simulation.

When the operator $N[.]$ is linear, the integral term in (56b) takes the simplified form

$$\int_D \langle N\left[\sum_{j=1}^{N} H_j\Phi_j\right]+U,.\rangle dx=\sum_{j=1}^{N}\left(\int_D \langle N[H_j],.\rangle dx\right)\Phi_j+\int_D \langle U,.\rangle dx$$

Differential and nonhomogeneous boundary conditions. For simplicity suppose that $D=(0,1)$, and consider the boundary conditions $L_bX(x,t)=0$, $x=0$; $X(1,t)=0$, where L_b is a linear differential operator. To determine the assumed modes $H_i(x)$ which satisfy these boundary conditions one first selects the modes $H_i(x)$ which satisfy the homogeneous conditions ($L_bH_i(x)=0$, $x=0$ and $H_i(1)=0$), and then solves the ODE $L_b\{G_i(x)\}=H_i(x)$, $G_i(1)=0$. For many types of $H_i(x)$, $G_i(x)$ can be found in closed form. Then it is easily verified that

$$X^*(x,t)=\sum_{i=1}^{N} G_i(x)\Phi_i(t)$$

satisfies the differential boundary conditions considered. The functions $G_i(x)$ must then be orthogonalised. Now, consider the case of nonhomogeneous boundary conditions $L_b[X]_{x=0}=U_{bo}(t)$, $X(1,t)=U_{b1}(t)$. To treat this case one determines a function $X_b(x,t)$ such that $L_b[X_b]_{x=0}$ $U_{bo}(t)$, $X_b(1,t)=U_{b1}(t)$, and defines $X'(x,t)$ by $X(x,t)=X'(x,t)+X_b(x,t)$. Then the model (5a,b) reduces to

$$\partial X'(x,t)/\partial t=N[X'+X_b]+U(x,t)-\partial X_b(x,t)/\partial t,$$

$$L_b[X']_{x=0}=0,\ X'(1,t)=0,\ X'(x,0)=X^o(x)-X_b(x,0)$$

which has homogeneous boundary conditions and can be dealt with as described above.

Finally, consider the case of nonlinear boundary conditions $L_b[X]_{x=0}=N_b[X]_{x=0}$, $X(1,t)=U_{b1}(t)$, where L_b is linear and N_b is nonlinear of order less than or equal to that of L_b. Defining $\xi(t)=N_b[X]_{x=0}$ and $X'(x,t)=X(x,t)-X_b(x,t)$, we find $L_b[X']_{x=0}=\xi(t)$, $X'(1,t)=0$ which can be treated as above.

Some Remarks. The Galerkin approach is a special case of the *"weighted residuals"* method (Finlayson and Scriven, 1966) in which the conditions $\int_D \langle R(x,t),\ W_i(x)\rangle dx=0$, $i=1,2,\ldots,N$ must be satisfied, and the weighting functions $W_i(x)$ can be selected arbitrarily. The most natural selection is $W_i(x)=H_i(x)$, $x\varepsilon D$ where the $H_i(x)$s are the assumed modes themselves, as it was done by Galerkin. Other special cases of the method of weighted residuals are the *method of moments*, the *collocation method*, and the *method of subdomains*.

In the method of moments, $W_i(x)$, $i=1,2,\ldots,N$ are members of a set of linearly independent polynomials, e.g. $W_i(x)=(x-\bar{x})^i$, or preferably members of a set of orthogonal polynomials. In the collocation method $W_i(x)$, $i=1,2,\ldots,N$ are unit impulse functions $W_i(x)=\delta(x-x_i)$ located at a set of spatial points x_i, $i=1,2,\ldots,N$ called the *collocation points*. In this case we obtain $f_{ij}=\int_D<\delta(x-x_i),\ H_j(x)>dx=H_j(x_i)$, and the collocation points must be selected so that $F=[f_{ij}]$ is invertible.

The method of subdomains consists in dividing D in N, not necessarily disjoint subdomains D_j, $j=1,2,\ldots,N$ selecting $W_i(x)$ as $W_i(x)=1$ when $x\varepsilon D_i$ and $W_i(x)=0$ when $x\not\in D_i$, and setting $\int_{D_i} RW_i dx=\int_{D_i} Rdx=0$. As D_i is reduced in size, this method approaches the collocation method.

If the system is linear self-adjoint, i.e. $N(.)=\mathcal{L}(.)$ and $\beta(.)$ are linear, we can select $H_i(x)=\Psi_i(x)$, $x\varepsilon D$, $i=1,2,3,\ldots$, where $\Psi_i(x)$ are the orthonormalised eigenfunctions of the *Sturm-Liouville* problem $\mathcal{L}\Psi_i(x)=\lambda_i\Psi_i(x)$, $x\varepsilon D$; $\beta\Psi_i(x)=0$, $x\varepsilon\partial D$, where λ_i are the associated eigenvalues. In this case we have $h_i=1$, and (56b) reduces to:

$$d\Phi_i/dt=\lambda_i\Phi_i+u_i,\quad u_i=\int_D U(x,t)\Psi_i(x)dx \tag{57}$$

for $i=1,2,\ldots,N$. Equations (56b) or (57) can be solved by the analogue or by the digital or by the hybrid computer. Also use can be made of Walsh series expansions in the time domain to reduce these equations into algebraic form (Tzafestas, 1978b). Alternatively, one can use double Walsh series expansions, i.e. choose the $H_j(x)$s to be Walsh functions over $D=(0,1)$ and expand the $\Phi_i(t)$s in time Walsh series (Shih and Han, 1978).

Extended Space Technique. The aim of this technique (Neuman and Strauss, 1969) is to absorb part of the residual $R(x,t)$ by appropriately using two more functions $H_{N+1}(x)$ and $H_{N+2}(x)$. To understand the method, recall that (55) simply ensures that $R(x,t)$ does not involve any modes from the first N ones, but since $H_i(x)$, $i=1,2,\ldots,N$ satisfy the homogeneous boundary conditions, they do not form a good basis for R. A useful expression for investigating the residual is found by combining (56a) with (56b), namely

$$R(x,t)=\sum_{i=1}^{N}\frac{H_i(x)}{h_i}\int_D<N\left[\sum_{i=1}^{N}H_i\Phi_i\right]+U,H_i(x)>dx-N\left[\sum_{i=1}^{N}H_i\Phi_i\right]-U(x,t)$$

Expanding $U(x,t)$ in terms of the modes $H_i(x)$, i.e. as $U(x,t)=\sum_{i=1}^{N}H_i(x)u_i(t)$, one possibly omits interesting modes existing in $U(x,t)$. Similarly, approximating $N[X(x,t)]$ by $N[\hat{X}(x,t)]$ one may have errors, not due to the omission of modes that are in $X(x,t)$, but due to the errors in the $\Phi_i(t)$s. These errors are not eliminated rapidly by adding more terms in (51).

To eliminate much of the errors due to $N[\hat{X}(x,t)]$, we determine two functions $H_{N+1}(x)=g_1^o(x)$ and $H_{N+2}(x)=g_2^o(x)$, which are orthogonal to $H_i(x)$, $i=1,2,\ldots,N$, and, together with $H_i(x)$, $i=1,2,\ldots,N$, form a basis for $N[X(x,t)]$. Then, in addition to the orthogonality conditions (55) we also have

$$\int_D <R(x,t),g_i^o(x)>dx=0, \quad i=1,2, \tag{58}$$

This extended function space $H_i(x)$, $i=1,2,\ldots,N+2$ is more appropriate for describing R than the original space $H_i(x)$, $i=1,2,\ldots,N$. In fact using the expression

$$\sum_{i=1}^{N+2} H_i(x)\Phi_i(t)=\hat{X}(x,t)+H_{N+1}(x)\Phi_{N+1}(t)+H_{N+2}(x)\Phi_{N+2}(t) \tag{59}$$

one obtains better practical results, than those obtained by the conventional Galerkin method. For parabolic PDEs the functions $g_1^o(x)$ and $g_2^o(x)$ are constructed from the functions 1 and x. Actually, $N[\hat{X}(x,t)]$ involves a first part $N[\hat{X}(x,t)]$, a second part having the same basis as $\hat{X}(x,t)$, but generated by functions not involved in $\hat{X}(x,t)$, and a third part which has a basis different than $N[\hat{X}(x,t)]$. The third part has no effect on (58), and so it does not cause any error in the $\Phi_i(t)$s. But the second part causes such errors. The extended space technique, defined by (58) and (59), minimises the erros due to the absence of the second part, and can be generalised to cases of multiple boundary conditions and nonpolynomial modes.

Using the extended space approximation (59) along with the orthogonality conditions (55) and (58) one obtains an improvement over the standard Galerkin method at the expense of some more algebra. Clearly, more than two functions $H_i(x)$, $i=N+1,\ldots,N+m$ may be obtained for even more accuracy.

Further Hybrid Methods

Here four methods will be described, namely the method of characteristics, the implicit hybrid method, the power series expansion method, and the Monte Carlo method, which do not fall in the preceding categories.

Method of Characteristics. This is applicable to hyperbolic systems (Moroney, 1963; Paul and Ahmad, 1970; Vichnevetsky, 1962). Hybrid implementations of this method were proposed by Chan (1967) and McAvoy (1972). Consider the ith component of $X(x,t)$ in the model (9). This, after the diagonalisation, is described by

$$\partial X_i/\partial t=A_{ii}(X,x,t)\partial X_i/\partial x+U_i(X,x,t) \tag{60a}$$

The characteristic direction is described by

$$dx/dt=-A_{ii}(X,x,t) \tag{60b}$$

Along the characteristic direction (line 1→2 of Fig.6), (60a) becomes $\partial X_i/\partial t+(\partial X_i/\partial x)(dx/dt)=U_i(x,t)$, i.e. $dX_i/dt=U_i(x,t)$, which by Euler's finite-difference approximation gives (see Fig. 6):

$$X_i(x_2,t+\Delta t)=X_i(x_1,t)+\Delta t U_i(X(x_1,t),x_1,t) \tag{61a}$$

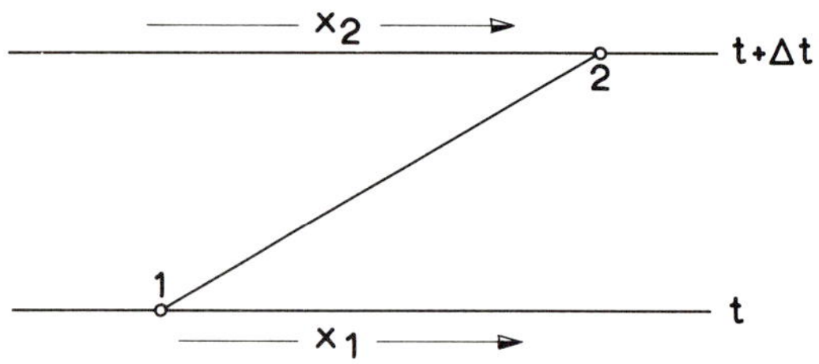

Fig. 6. Illustration of a characteristic line between t and t+Δt.

Equation (61a) can be solved by playing back $X_i(x_1,t)$ as a function of x_1 and solving for $X_i(x_2,t+\Delta t)$ as a function of x_2. To determine the storage point x_2, one solves (60b) again by the Eulers method:

$$x_2=x_1-\Delta t A_{ii}(X(x_1,t),x_1,t) \tag{61b}$$

The computation from the time level t to the one at t+Δt is made at the same spatial points. To this end, a comparator is used to decide whether x_2 is equal to one of these stored points, and if so, the solution (61a) is sampled and the comparator updated. Clearly, for each X_i one has to solve the two equations (61a,b). Note that playback is synchronous, but storage is not, since the storage points are determined by the comparator (see Fig. 11).

The method requires special attention at the boundaries for determining the region of applicability of (61a,b) (see Fig. 7).

In general the method of characteristic needs more equipment and is slower than DTCS methods due to the additional boundary computation required. However it leads to algebraic equations which can be solved in the fastest available analogue scale. The solution via the method of characteristics has perfect frequency response relative to the exact solution (i.e. the propagation velocity of a step is perfect). This is not true if a DTCS method is used, where the step response is in error and may cause troubles when studying transient responses and system dynamics (McAvoy, 1972).

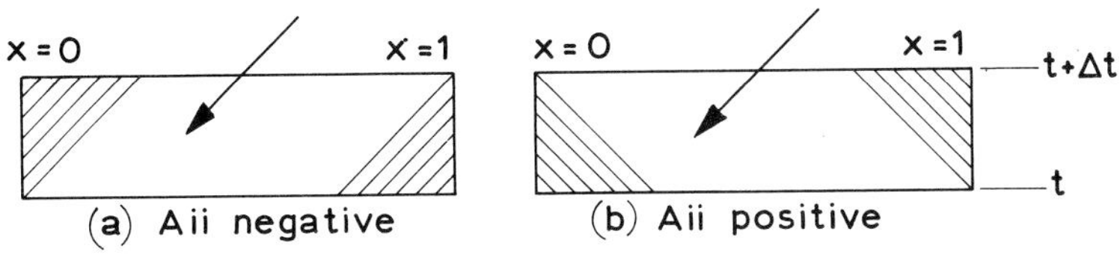

Fig. 7. Equations (61a,b) are applicable in the non-dashed regions.

Implicit simulation method. This method (Ruppel and Zeitz, 1975) was developed for the generalised model (18), and is based on the standardised boundary-value problem of order m obtained by keeping on the left-hand side the operator $\tilde{\mathcal{L}}_x = \partial^m/\partial x^m + \gamma^m$ (γ arbitrary), and using the new input

$$\tilde{U}(X,x,t) = \frac{1}{b_m(x)}\{U_d(X,x,t) - \sum_{i=0}^{m-1} \frac{\partial^i X}{\partial x^i}\} + \gamma^m X, b_m(x) \neq 0$$

Then the PDE (16), with nonhomogeneous boundary conditions (18)-19), can be written as an implicit integral equation

$$X(x,t) = \int_0^1 G(x,\xi)\{\tilde{U}(X,\xi,t) - \mathcal{L}_t X(\xi,t)\}d\xi + X_b(U_{bi})$$

where the term $X_b(U_{bi})$ is due to the boundary forcing functions $U_{bi}(X,x,t)$ for $x\varepsilon\{0,1\}$. This equation is written as

$$\int_0^1 G(x,\xi)\mathcal{L}_t X(\xi,t)\,d\xi = -X(x,t) + \int_0^1 G(x,\xi)\tilde{U}(X,\xi,t)\,d\xi + X_b(U_{bi}) \tag{62}$$

and can be solved using the following steps.

a) Compute the right-hand side $Y^*(x,t) = -X(x,t) + L\tilde{U}(X,x,t) + X_b(U_{bi})$ of (62) where $L = \int_0^1 G(x,\xi,t)\,[\cdot]\,d\xi$.

b) Compute $X(x,t)$ using $Y^*(x,t)$, i.e. from $L\mathcal{L}_t X(x,t) = Y^*(x,t)$ find

$$X(x,t) = \mathcal{L}_t^{-1} Y(x,t),\quad Y(x,t) = L^{-1} Y^*(x,t) \tag{63}$$

subject to the given initial conditions.

The hybrid implementation of these steps, is made as follows (see Fig. 8).

i) Discretise the time as $t_k = k\Delta t$ $(k=0,1,2,\ldots)$, and compute the variable $W(x,t_k) = L\tilde{U}(X,x,t_k) + X_b(U_{bi})$ by solving the boundary-value problem $(i=1,2,\ldots,m)$:

$$\tilde{\mathcal{L}}_x W(x,t_k) = \tilde{U}(X,x,t_k),\quad x\varepsilon(0,1);\ \beta_{xi} W(x,t_k) = U_{bi}(X,x,t_k),\ x\varepsilon\{0,1\} \tag{64}$$

This can be done on the analogue computer by using the integration results $\beta_{xi} X(1,t_{k-1})$ for converting the two-point boundary values to initial ones.

ii) To determine $Y(x,t_k)$ apply the relations {see (62),(63)}:

$$\tilde{\mathcal{L}}_x Y^*(x,t_k)=Y(x,t_k),\ x\varepsilon(0,1);\beta_{xi}Y^*(x,t_k)=0,\ x\varepsilon\{0,1\}$$

at $x=x_j=(j-1)\Delta x$, $j=1,2,\ldots,N$ where $\Delta x=1/(N-1)$, and use a suitable approximation of (64). This leads to the algebraic equation

$$\underline{Y}(t_k)=H\underline{Y}^*(t_k),\underline{Y}^*(t_k)=[Y^*(x_o,t_k),\ldots,Y^*(x_N,t_k)]^T,$$

$\underline{Y}(t_k)=[Y(x_o,t_k),\ldots,Y(x_N,t_k)]^T$ where H is a NxN matrix.

iii) Apply the first equation (63) and obtain $\underline{X}(t_{k+1})=\mathcal{L}_k^{-1}H\underline{Y}^*(t_k)$, subject to the given initial conditions, where $\mathcal{L}_k$ is an explicit difference approximation of $\mathcal{L}_t$, and $\underline{X}(t_k)=[X(x_o,t_k),\ldots,X(x_N,t_k)]^T$.

iv) Generate the profiles $X(x,t_{k+1})$ using some interpolating circuit. These profiles are used in computing $\tilde{U}(X,x,t_k)$, $U_b(X,x,t)$ and $W(x,t_k)$ on the analogue computer.

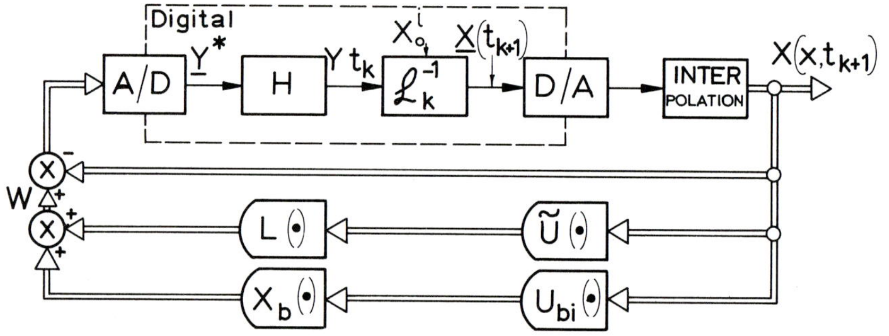

Fig. 8. Hybrid computer set-up for the implicit method.

The initial profiles $X_o^i(x)$ and transformed functions $Y^*(x,t_k)$ are fed to the digital computer, where $\underline{X}(t_{k+1})$ is computed using the algebraic operations H and $\mathcal{L}_k^{-1}$ as indicated in step (iii). The analogue integrations of $\mathcal{L}_k$ are made in high-speed repetitive mode. The A/D inputs and D/A outputs are initiated by I/O pulses corresponding to the spatial discretisation points $x_j=(j-1)\Delta x$. The vectors $\underline{Y}(t_k)$ and $\underline{X}(t_{k+1})$ are computed digitally in the pause between the I/O operations. The total accuracy of the method depends on the spatial discretisation. Note that N should not exceed a maximum number so as to satisfy the corresponding stability condition $\Delta t\leqslant f(\Delta x)$.

Power series expansion method. This is applicable to elliptic systems (Ashley and Bullock, 1968). Consider the two-dimensional Laplace equation in cylindrical coordinates:

$$\nabla^2 X(r,z)=0;\ X(b(z),z)=U_b(z),\ 0<z<a \quad (65a)$$

$$X(r,0) \text{ given for } 0<r<b(0);X(r,a) \text{ given for } 0<r<b(a) \quad (65b)$$

where b(z) is a boundary surface.

Expanding $X(r,z)$ in power series of r, for small r around the axis $(0,z)$, gives

$$X(r,z)=a_o(z)+a_1(z)r+\frac{1}{2}a_2(z)r^2+.... \tag{66}$$

Introducing it in $\nabla^2X(r,z)=0$, and equating the coefficients of like powers r^k yields

$$d^2a_k(z)/dz^2+\{1+k!/(k+1)!\}a_{k+2}(z)=0,\ k=0,1,2,.... \tag{67}$$

with $a_1(z)=0$. From (67) and the condition $a_1(z)=0$ it follows that all odd coefficients are zero. To solve the remaining equations we proceed as follows. From (66) we have $a_o(z)=X(0,z)$. Then for $k=0,2,....$ equation (67) gives

$$a_2(z)=-\frac{1}{2}\partial^2X(0,z)/\partial z^2,\ a_4(z)=\frac{3}{8}\partial^4X(0,z)/\partial z^4,....$$

Hence

$$X(r,z)=X(0,z)-(r^2/4)\partial^2X(0,z)/\partial z^2+(r^4/64)\partial^4X(0,z)/\partial z^4....$$

To satisfy the boundary conditions (65a) the following equation must hold

$$U_b(z)=X(0,z)-(b^2(z)/4)\partial^2X(0,z)/\partial z^2+(b^4(z)/64)\partial^4X(0,z)/\partial z^4+... \tag{68}$$

Truncating this at the qth derivative, we obtain an ODE of order q which must be solved for $X(0,z)$. To this end, we need q boundary conditions. These conditions are determined by matching (65b) at a set of arbitrary points $(r_1,0),(r_2,0),...,(r_{q/2},0);(r_1,a),(r_2,a),...,(r_{q/2},a)$. For $q=6$, the resulting equations are

$$X(r_i,z)=X(0,z)-\frac{r_i^2}{4}\frac{\partial^2X(0,z)}{\partial z^2}+\frac{r_i^4}{64}\frac{\partial^4X(0,z)}{\partial z^4},\ z=\{0,a\},i=1,2,3 \tag{69}$$

which, when $r_1=0$, reduce to a system of four equations in the four unknowns $\partial^2X(0,z)/\partial z^2$, $\partial^4X(0,z)/\partial z^4$, $z=\{0,a\}$.
The ODE (68), together with the boundary conditions resulting from (69), constitute a TPBVP which can be solved by iteration, or analytically, since (68) is linear. This problem, is highly involved and suggests to use the analogue part of a hybrid computer for its solution. The usual Dirichlet conditions specify $X(0,0)$ and $X(0,a)$, but not $\partial X(0,z)/\partial z$ at $z=0$. The method is also applicable to other elliptic systems such as $\nabla^2X(r,z)+\mu X(r,z)=U_d(r,z)$. In this case we use the expansions for small r and z:

$$U_d(r,z)=\sum_k u_k(z)r^k/k!,\ X(r,z)=\sum_k a_k(z)r^k/k!,$$

and (67) is replaced by $(1+\mu)d^2a_k/dz^2+2a_{k+2}(z)=u_k(z)$, $(i=2,3,...)$ with $a_1(z)=0$, which leads to the solution

$$X(r,z)=X(0,z)+\frac{1}{4}\left[u_1(z)-(1+\mu)\partial^2X(0,z)/\partial z^2\right]r^2+\frac{1}{12}u_2(z)r^3$$
$$+\frac{1}{48}\left[u_2(z)-(1+\mu)d^2u_2(z)/dz^2+(1+\mu)^2\partial^4X(0,z)/\partial z^4\right]r^4+\ldots$$

The results of this method are comparable in accuracy with those obtained by using finite-difference digital methods, provided that the region studied is several times longer than its maximum radius. The boundary values are easily inserted using diode analogue generators, and the solution obtained is continuous in the radial direction and can be incremented along the axial line at the cost of computing time. The gradient of $X(r,z)$ is also available from the computer.

Monte Carlo method. The Monte Carlo simulation method is based on the relation between the PDE model and an appropriate Markovian process over the domain D of the system (Chuang and others, 1960; Cimorelli and others, 1968; Fahrmeir, 1973; Handler, 1967; Jerman and others, 1969, 1971; King, 1951; Little, 1966; Soudack and Little, 1965; Swillens and Paiva, 1975). The method was applied to elliptic and parabolic systems. Consider the Markovian stochastic process $dx(t)/dt= A(x,t)=B(x,t)w(t)$ where $x(t)=[x_1,x_2,\ldots,x_n]^T\varepsilon D$, and $w(t)=[w_1,w_2,\ldots,w_n]^T$ is a Gaussian white noise vector with $E\{w(t)w^T(t+\tau)\}=\text{diag.}[2\Sigma_1,\ldots,2\Sigma_n]\}\delta(\tau)$ where $\delta(.)$ is the delta function. The matrices $A(x,t)=\text{diag.}[A_1,\ldots,A_n]$ and $B(x,t)=\text{diag.}[B_1,\ldots,B_n]$ are assumed slowly varying continuous functions of $x(t)$ and t.

Associated with the above Markovian process we have the following Fokker-Planck-Kolmogorov equations

$$-\partial p(x_2,t_2|x_o,t_o)/\partial t=\mathcal{L}_{(x_o,t_o)}p(x_2,t_2|x_o,t_o) \qquad (70a)$$

$$-\partial q(x_b,t_b|x_o,t_o)/\partial t=\mathcal{L}_{(x_o,t_o)}q(x_b,t_b|x_o,t_o) \qquad (70b)$$

where $\mathcal{L}_{(x_o,t_o)}$ is the backward diffusion operator

$$\mathcal{L}_{(x_o,t_o)}=\sum_{i=1}^{n}\{a_i\frac{\partial}{\partial x_i}+b_i\frac{\partial^2}{\partial x_i^2}\}_{x=x_o}$$

and

$p(x_2,t_2|x_o,t_o)dx_2$=transition propability that x is in $dx_2\varepsilon D$ at $t=t_2$ given that $x=x_o$ at $t=t_o$.

$q(x_b,t_b|x_o,t_o)dx_b dt_b$=propability that x is in ∂D for the first time in $dx_b\varepsilon\partial D$ within dt_b given that $x=x_o$ at $t=t_o$.

The initial-boundary conditions for (70a,b) are

$$p(x_2,t_2|x_o,t_o)\to\delta(x_2-x_o) \text{ as } t_o\to t_2; p(x_2,t_2|x_o,t_o)\to 0 \text{ as } x_o\to x_2$$

$$q(x_b,t_b|x_o,t_o)\to 0 \text{ as } t_o\to t_b; q(x_b,t_b|x_o,t_o)\to\delta(x_o-x_b)\delta(t_b-t_o) \text{ as } x_o\to x_b$$

The coefficients a_i and b_i are related to the coefficients A_i and B_i of the original Markovian process equation by

$$a_i(x_o,t)=A_i(x_o,t),\ b_i(x_o,t)=\Sigma_i B_i^2(x_o,t_o).$$

Now, consider the basic problem of simulating the DPS

$$-\vartheta X(x_o,t_o)/\vartheta t_o=\mathcal{L}_{(x_o,t_o)}X(x_o,t_o),\ x\varepsilon D \tag{71a}$$

$$X(x_o,0)=X_o(x_o),\ x_o\varepsilon D;X(x_b,t_o)=X_b(x_b,t_o),x_b\varepsilon\vartheta D \tag{71b}$$

defined over the bounded spatial domain D with boundary surface ϑD, where the initial and boundary functions $X_o(x_o)$ and $X_b(x_b,t_o)$ are piece-wise continuous. The time t_o is assumed negative, and the final time is assumed at t=0. The value $X(x_o,t_o)$ is obtained by forming the average

$$X_N(x_o,t_o)=\frac{1}{N}\sum_{i=1}^{N}X_i \tag{72}$$

of the terminating values X_i, i=1,2,...,N of N random walks resulting by simulating the Markovian process x(t) starting at (x_o,t_o) and ending when the space-time boundary is reached, i.e. when $(x,t)=(x_b,0)$, $x_b\varepsilon\vartheta D$.

The average $X_N(x_o,t_o)$ tends to the exact solution when $N\to\infty$. This can be established by applying the operator $\vartheta/\vartheta t_o+\mathcal{L}_{(x_o,t_o)}$ upon the exact average value

$$X(x_o,t_o)=\int_D X_o(x_2)p(x_2,0|x_o,t_o)dx_2$$
$$+\int_{t_o}^{0}\int_{\vartheta D}X_b(x_b,t_b)q(x_b,t_b|x_o,t_o)dx_b dt_b \tag{73}$$

and using (70a,b). Then, obviously, if only N values X_i are available for $X_o(x_2)$ and $X_b(x_b,t_b)$, the exact average (73) can be approximated by the sum (72).

By letting $t_o\to-\infty$, the sum (72) provides the solution of the associated elliptic PDE, i.e. of

$$\mathcal{L}_{x_o}X(x_o)=0,\ x_o\varepsilon D;\ X(x_b)=X_b(x_b),\ x_b\varepsilon\vartheta D \tag{74}$$

In practice, for solving (74) the random walk simulations are started at t=0 and terminated when a boundary point is reached. Of course it is always possible to reach the boundary, provided that we let the time increase sufficiently. Now consider the nonhomogeneous elliptic PDE

$$\mathcal{L}_{x_o} X(x_o)+U(x_o)=0,\ x_o \varepsilon D;\ X(x_b)=X_b(x_b),\ x_b \varepsilon \partial D$$

where $U(x_o)$ and $X_b(x_b)$ are piece-wise continuous functions. Its solution is given by

$$X(x_o)=X^*(x_o)+\int_0^{\infty} X^{**}(x_o,t)\,dt,\ x_o \varepsilon D$$

where $X^*(x_o)$ is the solution of (74), and $X^{**}(x_o,t)$ is the solution of (71a,b) with $X(x_o,0)=U(x_o)$, $x_o \varepsilon D$ and $X_b(x_b,t_o)=0, x_b \varepsilon \partial D$.

Note that one can also treat the case where a term $f(x_o)X(x_o,t_o)$ is included on the right-hand side of (71a) (Handler, 1967; Little, 1966). To find the solution of (71a,b) with a constant forcing term-K the Monte Carlo solution of the homogeneous PDE is added to a term proportional to the average time of the random walks. Fahmeir (1973) extends the above results to the elliptic model

$$\mathcal{L}_x X(x)=U(x),\ \mathcal{L}_x= \sum_{i=1}^{n} a_i(x) \frac{\partial}{\partial x_i} + \frac{1}{2} \sum_{i,j=1}^{n} b_{ij}(x)\frac{\partial^2}{\partial x_i \partial x_j}+c(x),\ x \varepsilon D$$

where the coefficients $b_{ij}(x)$, $i,j=1,2,\ldots,n$ are not zero for $i \neq j$, or slowly varying, as assumed before. Here $b_{ij}(x)$ are assumed to have continuous first derivatives, which, together with the functions $a_i(x)$, $c(x)$, $U(x)$ satisfy on $D x \partial D$ a Holder condition with exponent λ, i.e. $|a_i(x)-a_i(y)| \leq L||x-y||^{\lambda}$ etc, where $||x-y||$ is the distance of x and y and L is constant. Let $\bar{B}(x)$ denote the symmetric matrix defined by $\bar{B}^2(x)=B(x)$ where the matrix $B(x)=[b_{ij}(x)]$ is assumed symmetric. If $B(x)$ is diagonal, as assumed in our previous development, then $\bar{b}_{ii}(x) =\sqrt{b_{ii}(x)}$ and $\bar{b}_{ij}(x)=0$, $i \neq j$. Now if $w_i(t)$, $i=1,2,\ldots,n$, are white noise processes with power spectral densities σ_i^2, and Σ is the diagonal matrix with elements $\Sigma_i=\sigma_i^2/2$, then the associated Markovian process is

$$\frac{dx_i}{dt}=a_i(x)- \frac{1}{2} \sum_{j,k=1}^{n} \bar{b}_{kj}(x) \frac{\partial \bar{b}_{ij}(x)}{\partial x_k}+ \sum_{j=1}^{n} b^*_{ij}(x)w_i(t),\ i=1,2,\ldots,n$$

where b^*_{ij} are the elements of the matrix $B^*(x)=\bar{B}(x)\Sigma^{-1}$ and $\bar{b}_{ij}(x)$ the elements of $\bar{B}(x)$. Clearly, if $B(x)$ is diagonal, i.e. $b^{*2}_{ii}(x)=b_{ii}(x)/\sigma_i^2$, and slow varying (ideally constant), the derivatives $\partial \bar{b}_{ij}(x)/\partial x_k$ are negligible (ideally zero), and the above Markovian process reduces to the one associated with (70a,b).

To implement the Monte Carlo method (Korn, 1965) one must have means for

i) simulating the appropriate Markovian process.
ii) determining and storing the terminating values X_i, $i=1,2,\ldots,N$.
iii) computing $X_N(x_o,t_o)$ using (72).
iv) resetting (x,t) to the initial pair (x_o,t_o) for $i<N$.

In a hybrid system step i) is performed by the analogue computer. The storage of the X_i's and the computation of $X_N(x_o,t_o)$ is done on the digital computer. The time-space boundary detection for the determination of the X_i's is made by using appropriate comparators.

GENERAL IMPLEMENTATION ASPECTS

For implementing a DTCS hybrid simulation method one needs some form of analogue memory (Biorge, 1966; Conelly and Fedoroff, 1965) which has the capability to store and playback functions. Obviously, the high speed of analogue integration is limited by the time required to read information in and out of the digital computer. Thus, any DTCS simulation method becomes of interest mainly for relatively complex DPS (e. g. once through boilers of nuclear power plants, etc).

One form of sample stored-continuous playback analogue memory uses an A/D converter to encode samples of the analogue signal at a desired rate, which are then passed sequentially to the core memory of the digital computer. To play back the signal, this sample sequence is fed into the analogue domain, with the aid of an analogue integrator which performs a linear interpolation between points. The gain of this integrator determines the rate at which we want to put-out the sampled data.

The rate of playing back the signal determines the time increment Δt between any two successive samples. This increment Δt is kept constant throughout, but is can be changed from signal to signal. The problem is how to exchange speed for accuracy by reducing Δt. The accuracy which can be obtained is bounded by limitations caused by round-off and error accumulation effects. This phenomenon is called *"stalling effect"*, and resembles the round-off errors of fixed-point numerical methods (Frank and Lapidus, 1965). To reduce the stalling effect one transfers from analogue to digital only increments of order Δt, such as $\Delta X^{k+1}=X^{k+1}-X^k$ or X^k-U^{k-1} {see (22a) and (22c)}. For example, subtracting $dX^k/dx+(1/\upsilon\Delta t)X^k$ from both sides of (20) yields

$$d\Delta X^{k+1}/dx+(1/\upsilon\Delta t)\Delta X^{k+1}-(1/\dot{\upsilon})U^{k+1}(X^{k+1},x)=-dX^k/dx$$

where the dependent variable is the increment ΔX^k. In this case dX^k/dx may be computed either by tabular differences on the digital part, or as shown in the implementation circuit of Fig. 9 on the analogue part.

Another way for overcoming the stalling effect is described in (Chareton and others 1970), and we remark that the error correcting subroutines of Karplus and Russel (1971) are based on a similar idea. The utilisation of incremental methods, i.e. the transfer of increments rather than the system variables themselves, leads to more accurate results in both directions of data transfer, namely the analogue-to-digital and the digital-to-analogue directions.

Continuous playback is obtained through analogue interpolation. Suppose that $\Delta X^{k+1}=X^{k+1}-X^k$ is the increment computed on the digital part. This is then converted by the D/A converter and stored at the input

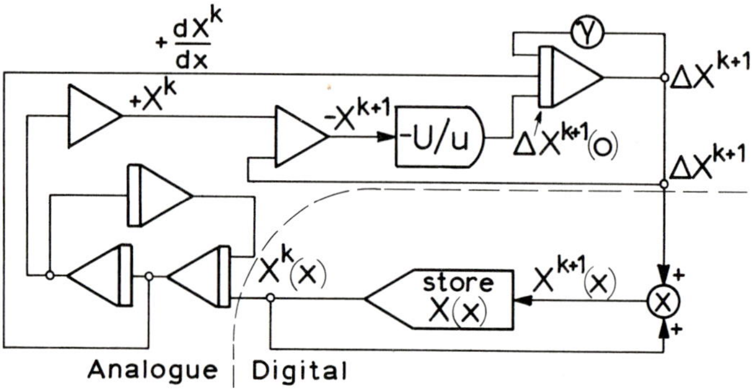

Fig. 9. Incremental implementation of (20).

of the analogue integrator having a gain $1/\Delta t$. After the elapse of time Δt at the output of the integrator, which has the initial value X^k,

$$X^k+(1/\Delta t)\int_{k\Delta t}^{(k+1)\Delta t}\Delta X^{k+1}dt=X^k+\Delta X^{k+1}=X^{k+1}$$

Thus a linear interpolation between X^k and X^{k+1} is obtained. This is actually an open-loop scheme and suffers from cumulative build-up of errors due to component imperfections (drift, offset, etc). This is avoided using a closed loop arrangement as shown in Fig. 10.

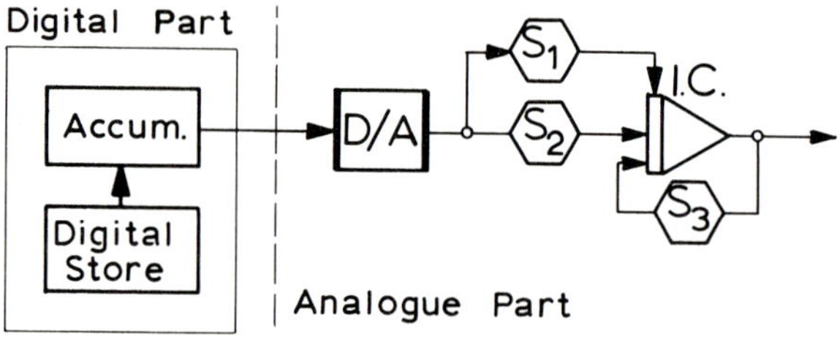

Fig. 10. Closed-loop arrangement of sample stored-continuous playback memory.

The initial condition X^o is stored via the sample-hold gate S_1. The accumulator is loaded with the successive values X^k, which are converted by the D/A and stored on the sample-hold gate S_2. The net input of the integrator is $X^{k+1}-X^{k}=\Delta X^{k+1}$. The errors are self-correcting.

One of the beneficial features of the DTCS approach is that the number of equations in the system model can be increased easily; one only needs more amplifiers integrators for each mesh point (or group of mesh points) represented in the analogue part. However, no increase in simulation time occurs since all equations are solved in parallel. The experiment showed that using a hybrid simulation there may be a cost saving factor between 5 and 20, when compared with the cost of a similar pure digital simulation using a traditional computer. This is expected to change drastically by the utilisation of microprocessors for the simulation.

For implementing the CTDS methods we use many stations (mesh points or mesh groups) which require many A/D and D/A transfers and large memory capacity. In general, the required arithmetic operations on the digital computer are limited. Hence, the utilisation of a general purpose digital computer in a hybrid set-up is not the best solution. It seems better to use a small digital computer together with a special parallel-in parallel-out memory coupled with an A/D and D/A conversion facility of sufficient size. The details of the implementation depend on the particular method used (cell multiplexing, regional multiplexing, etc). Of course, by carefully implementing a given method we may have profitable effects on the actual convergence speed, on the effective use of the analogue units, etc.

In the method of characteristics, the digital computer is used for function storage and playback, and the analogue computer for parallel (simultaneous) solution of algebraic equations and decisions on whether a storage point has been reached (Fig. 11).

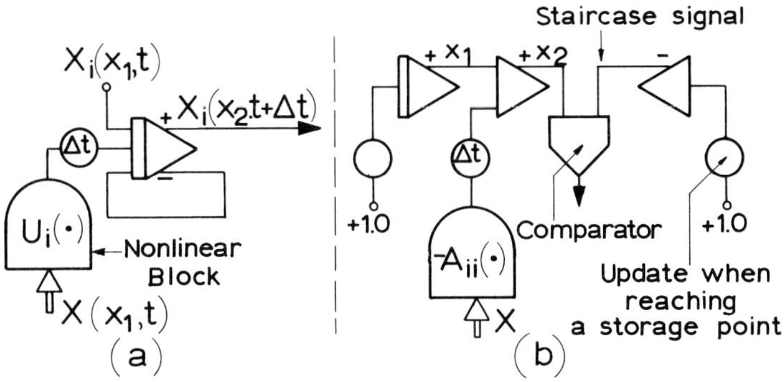

Fig. 11. Implementation set-ups for the method of characteristics.
(a) Playback of $X_i(x_1,t)$ and solution of (61a) for $X_i(x_2,t+\Delta t)$.
(b) Determination of storage point x_2 using a comparator {see (61b)}.

Similar combinations of analogue and digital computation have been used in the hybrid implementation of the modal expansion methods, where the analogue computer is used for solving in parallel (or through multiplexing techniques) the ODEs, and the digital computer for storage purposes and for computing the modal expansion approximations of the solution {see (51)}.

STABILITY ANALYSIS OF HYBRID SIMULATION METHODS

Most of the methods presented here are supported by convergence, stability, and accuracy investigations. Some works devoted entirely to the stability/error propagation aspect are those of Bosgra and Buis (1973, 1974), Frank and Lapidus (1965), and Vichnevetsky (1974). The theory of Bosgra and Buis concerns the error/stability characteristics of DTCS methods based on first-order playback, and is successfully applied to convection-diffusion and wave-equation models.

The traditional approaches to error/stability analysis of numerical methods for solving PDEs are based on polynomial approximations and Taylor series expansions. But these methods give error estimates in the small, not in the large, as it is usually required in practice. By errors in the large we mean errors occuring when the spatial and time step sizes are increased with a view to reduce to computational effort. The parallel CTDS methods can be analysed via standard results of analogue computation or finite-difference methods for ODEs (Fox, 1962; Friedman, 1956). The truncation errors introduced in the parallel approximations affect the transient state and reduce to zero in the steady state.

Galerkin expansions are very effective when using the first few modes, but the performance may be worse for many-mode solutions. Actually, the Galerkin method with small number of modes gives higher accuracy than other methods for the same number of ODEs. In general the convergence rate of Galerkin expansions is slow. An exception to this is the expansion in Walsh function (Tzafestas, 1978a, 1978b). The following discussion is based on the results of Bosgra and Buis (1974) and Vichnevetsky (1974).

Consider the convection-diffusion model {see (6)}:

$$\partial X/\partial t+A\partial X/\partial x=-B\partial^2 X/\partial x^2+KX \tag{75}$$

which modelises estuaries, water quality in rivers, etc.

The largest numerical errors are due to approximating the left hand side by finite differences. Thus a good tool for error analysis of (75) may be obtained by studying the pure advection model $\partial X/\partial t+A\partial X/\partial x=0$. The general six-point implicit difference scheme for this model is

$$(X_i^{k+1}-X_i^k)/\Delta t+\{\Theta(X_{i+1}^{k+1}-X_{i-1}^{k+1})+(1-\Theta)(X_{i+1}^k-X_{i-1}^k)\}(A/2\Delta x)=0 \tag{76}$$

where $X_i^k=X(i\Delta x,k\Delta t)$. The error analysis of this equation is difficult in the general case. When $A\Delta t/\Delta x<<1$ one can ignore the errors due to the time discretisation and set $dX_i(t)/dt=-(A/2\Delta x)(X_{i+1}-X_{i-1})$ where $X_i(t)=X(i\Delta x\ ,t)$. When $A\Delta t/\Delta x>>1$ one can ignore the errors due to space discretisation, and set

$$(X^{k+1}-X^k)/\Delta t=-A\{\Theta dX^{k+1}/dx+(1-\Theta)dX^k/dx\}$$

where $X^k(x)=X(x,k\Delta t)$. The numerical experiments showed that in these cases the theoretical error predictions are correct. Finally, when $A\Delta t/\Delta x=1$ the errors are minimal. In practice, the case $A\Delta t/\Delta x>>1$ must be preferred, since the approximation is implicit, and the method may remain stable for large values of Δt. On the contrary, all explicit schemes become unstable in this case since the fundamental stability condition (Courant-Friedrichs-Lewis) is violated. In (Vichnevetsky, 1974) the above cases are studied in the frequency domain, and hyperbolic systems of higher order are also studied. It is shown that by a proper choice of Δx, Δt and Θ one may go to the point of relying entirely upon spurious diffusion, to use (76) with the intent of obtaining solutions of (75) with $B\neq 0$.

The results of Bosgra and Buis (1974) are based on the concept of *"equivalent hybrid difference scheme"*. Consider again the operator $\partial/\partial t+A\partial/\partial x$. The basic operations leading to a hybrid DTCS method are the discretisation of $\partial/\partial t$, and the sampling and storage of the integration result along the x-direction to be used for playback which is usually interpolation and may be regarded as a non-ideal filter. This filter allows all frequencies up to half the sample frequency to pass into the analogue circuitry and blocks the passage of the periodic spectra extensions. The per-time-step passage through the DTCS hybrid cycle from one line of storage to the next, represents a difference scheme in (x,t) grid, and is the "equivalent hybrid difference scheme".

Any DTCS scheme can be written as

$$dY(x)/dx=CY(x)+GX^k(x),\quad X^{k+1}(x)=FY(x)+EX^k(x) \tag{77}$$

augmented with the appropriate boundary conditions. Here $X^k(x)$ is the solution (system response) at time step k, and Y(x) is an intermediate (state) variable.

In an elementary grid area $(\Delta x,\Delta t)$, (77) yields the following equivalent hybrid difference scheme,

$$Y_i=e^{\Delta xC}Y_{i-1}+C^{-1}\{(\Delta xC)^{-1}(e^{\Delta xC}-1)-I\}GX_i^{k-1}$$

$$+C^{-1}[\{I-(\Delta xC)^{-1}\}(e^{\Delta xC}-1)+I]GX_{i-1}^{k-1} \tag{78a}$$

$$X_i^k=FY_i+EX_i^{k-1} \tag{78b}$$

If F is invertivle (78b) gives $Y_i=F^{-1}(X_i^k-EX_i^{k-1})$ which can be inserted in (78a). From (78a,b) one can easily determine the frequency transfer function $H(j\omega)$ which describes the per-time-step transfer of the spatial frequency content of the approximated system response vector X.

The main result of Bosgra and Buis is: "The DTCS scheme (77) is a second-order approximation of $\partial X/\partial t+A\partial X/\partial x+B\partial^2 X/\partial x^2=0$, if and only if

$$E=I+FC^{-1}G,\ FC^{-2}G=A\Delta t,\ B-(A^2/2)\Delta t=(\Delta x/2\Delta t)F\ \coth(\Delta xC/2)C^{-2}G" \tag{79}$$

The first condition (79) ensures the elimination of the zero-order error, the second ensures the elimination of the first-order error, and the third guarrantees the elimination of the second-order error. This can be shown by Taylor series expansion of X aournd X_i^k.

In case of $\partial/\partial t+A\partial/\partial x$, the matrices (79) are equal to

$$C=-A^{-1}/(1-\Theta)\Delta t,\ G=A^{-1}/(1-\Theta)\Delta t,\ F=I/(1-\Theta),\ E=\Theta I/(1-\Theta) \tag{80}$$

where Θ is the parameter indicating the degree of implicitness of the finite-difference scheme {see (21)}. Using (80) one can see that (77) is an approximation having first-order accuracy, with the corresponding second-order error being given by

$$B=(A^2/2)\Delta t+(\Delta x/2\Delta t)F\coth(\Delta xC/2)C^{-2}G \tag{81}$$

Actually, this second order error may be regarded as a spurious diffusion term. The influence of the sampling interval Δt can be determined from the condition of numerical stability in the time direction. In terms of the frequency transfer function $H(j\omega)$, this condition is $|H(j\omega)|<1$, and for the case of $\partial/\partial t+A\partial/\partial x$ is found to be

$$\Theta<1-z/2\text{arctgh}(z),\ z=\Delta x/A\Delta t \tag{82}$$

where z is the dimensionless mesh size parameter. The stability condition (82) is illustrated in Fig. 12a. For $\Theta=1$ a completely explicit method is obtained, and the condition for stability is $z=\Delta x/A\Delta t>1$. When $z\text{arctgh}\ (z)\simeq z$, i.e. for infinitesimal z, the condition of stability is $\Theta<1/2$. In general, the stability limit for Θ increases and becomes $\Theta=1$ for $z\to 1$. It is noted that actually (82) gives the stability limit of $\partial X/\partial t+A\partial X/\partial x+B\partial^2 X/\partial x^2=0$ for the case $B=0$, and corresponds to the second-order condition (79) {or (81)} with $B=0$.

The second-order condition for the present case when $B\neq 0$, is

$$\hat{B}=\frac{1}{2}z\coth\{z/2(1-\Theta)\}-\frac{1}{2},\ \hat{B}=-B/A^2\Delta t \tag{82a}$$

which reduces to

$$\Theta=1-z/2\text{arctgh}\{z/2(1+2\hat{B})\} \tag{82b}$$

The stable situation corresponds to $B<0$, whereas positive values of B represent negative spurious diffusion coefficient {see (75)} and are in the unstable region. The plot of the computational diffusion coefficient $\hat{B}$ versus Θ for various values of $z=\Delta x/A\Delta t$ is shown in Fig. 12b. We close this section with a few comments on the stability of iterative implementations of the CTDS method. Actually, each iterative scheme has its own properties, and a comparative study should be quite useful.

Petrov (1975) shows that an error $\xi^k(t)$ of $X_i(t)$ occuring at the kth time step of the difference-differential scheme (46a) will not increase at the (k+1)th step provided that $||\xi^k||<\delta$, where δ is an existing appropriate small number, and that the upper limit τ_0 of the time step Δt, which ensures numerical stability, is greater than the corresponding upper limit of the classical explicit method. This is due to the negative feedback term $U(X_i, x_i, t)$ which gives always a stable analogue implementation circuit, if $u_i(t)=\partial U(X_i, x_i, t)/\partial X_i>0$. The error of the differential-difference scheme is of the same order as that of the classical finite-difference schemes, i.e. of order $O(\Delta t+(\Delta x)^2)$, but the permissible time step Δt is larger. Analogous results hold for the block time-sharing method (Neundorf and Schoenefeld, 1980).

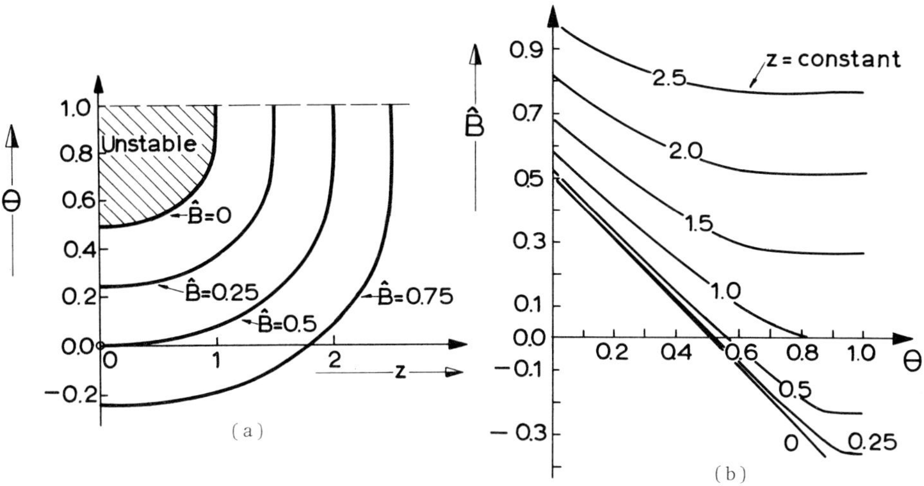

Fig.12. (a) Illustration of the stability condition (82).
(b) Plot of $\hat{B}=f(\Theta, z)$ versus Θ for various values.

EXPERIMENTAL RESULTS

Here a set of DPS examples are provided which were simulated by the hybrid methods discussed. These examples do not exhaust the experimental work carried out in the field, but they are sufficient for illustrative purposes. More results may be found in (Bosgra and Buis, 1973, 1974; Moroney, 1963; Sanathanan and others, 1967; Saphier, 1972; Vichnevetsky, 1970a; Wozny and others, 1970) where a variety of applications are studied such as nuclear reactors, heat flow in rods, forced steam flow, hydraulic systems, chemical reactors, evaporating systems, river pollution problems, etc.

Application of DTCS Methods

Example 1, Decomposition method. This method, when applied to the scalar convection-diffusion system $\partial X/\partial t+A\partial X/\partial x=B_0\partial^2 X/\partial x^2$, yields the DTCS scheme

$$\frac{d}{dt}\begin{bmatrix} z_1 \\ z_2 \end{bmatrix} = \begin{bmatrix} \lambda_B & 0 \\ 0 & \lambda_F \end{bmatrix} \begin{bmatrix} z_1 \\ z_2 \end{bmatrix} + \begin{bmatrix} -\mu^2 \\ 0 \end{bmatrix} X^k, \quad X^{k+1} = \begin{bmatrix} 0, & \frac{1}{1-\Theta} \end{bmatrix} \begin{bmatrix} z_1 \\ z_2 \end{bmatrix} - \frac{\Theta}{(1-\Theta)} X^k \qquad (83)$$

where $\lambda_B>0$, $\lambda_F<0$, $\mu=1/\sqrt{(1-\Theta)B_o\Delta t}$, and $\lambda_{B,F}=A/2B_o\pm\sqrt{(A/2B_o)^2+\mu^2}$. Using the linear transformation $Y=QZ$, where $Q=[q_{ij}]$, $q_{11}=q_{21}=1/(\lambda_B-\lambda_F)$, $q_{12}=0$ and $q_{22}=1$, the model (83) can be transformed to an equivalent DTCS of the form (77) with

$$C=\begin{bmatrix}\lambda_B & 0\\ 0 & \lambda_F\end{bmatrix}, G=\frac{\mu^2}{\lambda_B-\lambda_F}\begin{bmatrix}-1\\ 1\end{bmatrix}, \quad F=\frac{1}{1-\Theta}[1,1], \quad E=-\frac{\Theta}{1-\Theta} \quad Y=\begin{bmatrix}Y_1\\ Y_2\end{bmatrix} \tag{84}$$

We observe that the first two stability conditions in (79) are satisfied. The third condition gives, for $A=0$, $-B=B_o(\mu\Delta x/2)\coth(\mu\Delta x/2)$, where $\mu=\lambda_B=-\lambda_F$, and the condition $|H(j\omega)|<1$ gives

$$\Theta<1-\{\sin(\mu\Delta x)/\mu\Delta x\}/\{1+\cosh(\mu\Delta x)\}$$

which is the condition for numerical stability of the DTCS scheme for the heat equation (Fig. 13).

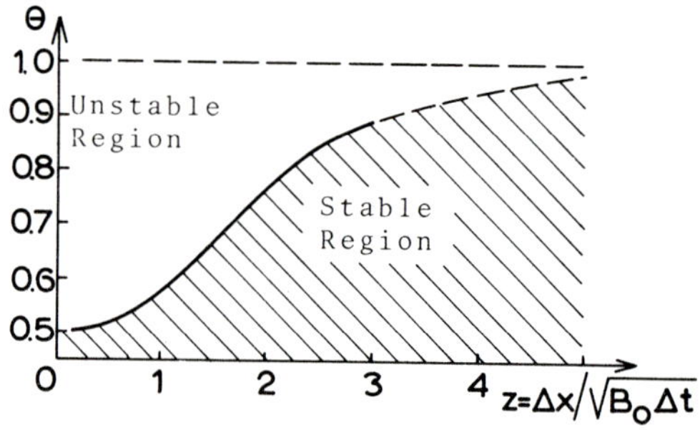

Fig. 13. Stability region for the heat equation DTCS model.

The actual simulation of the heat equation (i.e. $A=0$) was made by Bosgra and Buis using a Gauss-shaped initial condition over $x\varepsilon[0,1]$ for many parameter sets, e.g. (a) $\Delta t=\Delta x=0.01$, $\Theta=0$, $z=0.32$, $B_o=0.1$, $\hat{B}=1.01$, and (b) $\Delta t=0.001, \Delta x=0.04$, $\Theta=0$, $z=4$, $B_o=0.1$, and $\hat{B}=1.98$.

In case (a) the error between the actual and true diffusion coeefficient is 1%, but in case (b) this error is about 100%. This affects the simulated time response and suggests that the implementation DTCS parameter must be selected carefully, if a confident simulation model is desired.

Example 2: Green's function and integral equation methods. Again the heat-equation model was simulated for testing these methods (Chan, 1967, 1969), using the initial-boundary conditions $X^o(x)=\sin\pi x$, $X_o(t)=0$, $X_1(t)=1$. Applying the transformation $Y(x,t)=X(x,t)-\{X_o(t)+(X_1(t)-X_o(t))x\}$ a model with homogeneous boundary conditions and initial condition $\bar{X}^{o}(x)=\sin\pi x-x$ is obtained. The simulated responses at t=0.04 and t=0.28 are shown in Fig. 14, along with the exact response.

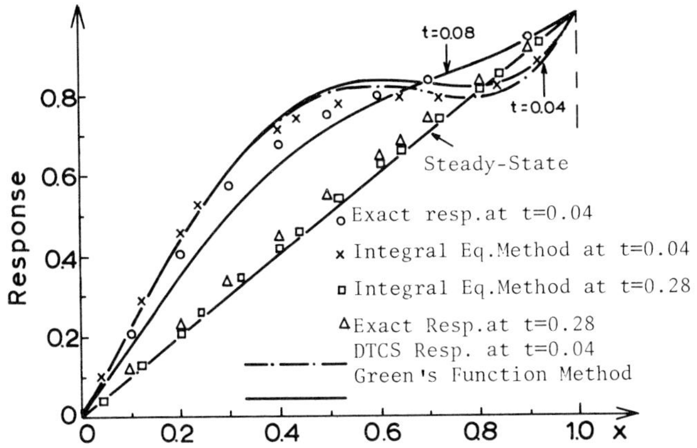

Fig. 14. Response of the heat-equation model.

One observes that the DTCS response agrees quite well with the exact one for large t (say t=0.28). The difference appearing for small t is due to the fact that time quantisation affects the value of the exact (analytic) solution. Actually the DTCS is also accurate for t=0. For $\Delta t=0.04$ the DTCS is not too accurate, and for $\Delta t=0.1$ is very incaccurate. However the integral equation method based on the modified Neumann series is reasonably accurate for both t=0.04 and t=0.1. The Green's function and integral equation methods have comparable accuracies (Fig. 14).

Example 3: Invariant imbedding method. This method was implemented on a hybrid system involving a PDP-10 digital computer and an AD-4 analogue computer (Nelson, 1970). Zero-order hold was used to reconstruct the functions from the D/A converter, with the reconstructed value over each of the N equal time intervals representing the mean value of the samples on either end of the interval. Table 1 shows the simulation results for the heat equation model. One observes that in this problem the invariant imbedding method is a little more accurate than the standard DTCS method.

Application of CTDS Methods

Example 4: Cell Multiplexing method. Several examples were simulated using this method (Hsu and Howe, 1968). The integration of the simultaneous ODEs was made using a 4th-order Runge-Kutta technique on an IBM 7090 computer. The D/A and A/D converters were included in simulated form. The CTDS model (43a,b) was implemented with N=6, accuracy parameter $\eta=5\times10^{-4}$ and converter sampling period $\tau=0.01$ units of t. Using zero-order extrapolation in the playback, convergence is ve-

TABLE 1 Simulated Results Compared

t	x	Imbedding	DTCS	Exact	t	x	Imbedding	DTCS	Exact
	.2	.5326	.5325	.5325		.2	.2192	.2189	.2191
.01	.4	.8616	.8616	.8617	.10	.4	.3544	.3542	.3545
	.6	.8615	.8616	.8617		.6	.3542	.3542	.3545
	.2	.4826	.4824	.4825		.2	.0498	.0497	.0498
.02	.4	.7807	.7806	.7807	.25	.4	.0806	.0805	.0807
	.6	.7805	.7806	.7807		.6	.0806	.0805	.0807
	.2	.3590	.3587	.3588		.2	.0042	.0042	.0042
.05	.4	.5806	.5804	.5806	.50	.4	.0068	.0068	.0068
	.6	.5803	.5804	.5806		.6	.0068	.0068	.0068

ry nice but not to the exact CTDS solution (Fig. 15a). However advancing the playback function by half the sampling period gives better results (1% maximum error) as shown in Fig. 15b. Even better accuracy is expected with smaller playback sampling period. In Fig. 15(a,b) the-

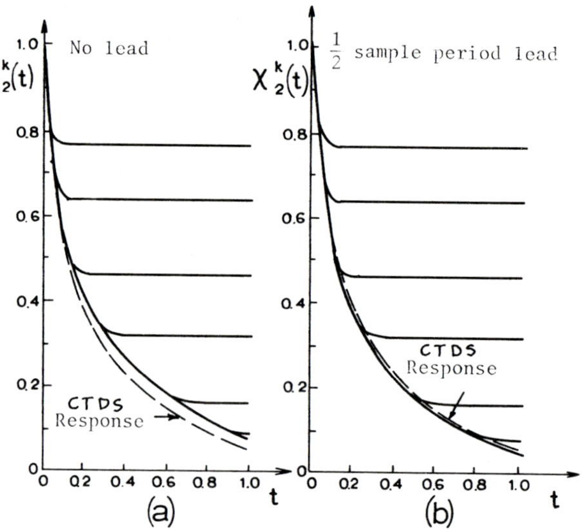

Fig. 15. Iterative simulation of the heat equation
(a) No lead in playback,
(b) With lead in playback.

re are also shown the number of iterations at each interval for convergence withing a maximum error $\eta=0.005$.

To reduce the total computing time, the interval $[0,t_f]$ is divided into equal subintervals (say 10 or 20), and the iterative scheme is applied succesively at these subintervals. The number of iterations at each subinterval is reduced by using linear extrapolation, i.e.

$$X_i^o(t)=\bar{X}_i(T)+\{d\bar{X}_i/dt\}_{t=T}(t-T), \quad T\leqslant t<T+\Delta T,$$

where $\bar{X}_i(t)$ is the approximation of the CTDS response obtained in the previous subinterval $T-\Delta T\leqslant t<T$.

Example 5: Regional multiplexing method. This method was applied to an infinite-slab nuclear reactor fueled with uranium oxide enriched to 16% with plutonium, and to a 3-region reactor model (Saphier,1972). Some of the results, obtained in the 3-region model case by using a 6% perturbation $\Delta\Sigma_a$ in the fission cross-section near mesh points 15 and 16, are given in Fig. 16.

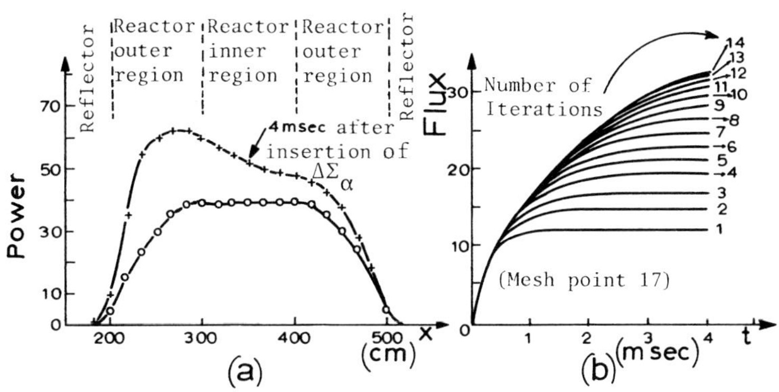

Fig. 16. Simulated response of a 3-region reactor
(a) Power profile at t=4 msec,
(b) Flux at mesh point 17.

Example 6: Differential-difference method. This method was tested on a model of the cardial arrythmia mechanism under myocardial infraction (Petrov, 1975). The aim of the simulation was to determine the mechanisms of the genesis and maintenance of self-subtained chaotic activity (fibrillation) preventing the heart from working as a pump. A square domain D of dimension 1.75 cm was considered with $\Delta x=\Delta y=0.5$mm i.e. 35x35=1225 nodes.

The ODE at each node was integrated by an analogue cell (of the HCS-100 USSR-Yugoslavian hybrid system) with repetition frequency 1kHz. The time scale was chosen five times faster than the real-time scale.

The time increment used is $\Delta t=2.5$ msec for ensuring computational stability. The simulation time of the differential-difference method in an experiment was found 11 seconds, while the corresponding time of the explicit finite-difference method is 68 seconds, and that of the implicit method is 115 seconds, for the same accuracy.

Example 7: Effect of simulation correction terms. The truncated Fourier series (49a,b) was employed for correcting the first-order difference approximation (Nomura and Deiters, 1968). The smoothing coefficients A_n' were also used.

Fig. 17a compares the exact second derivative response, with its values computed by a difference approximation, a truncated Fourier series, and a smoothed Fourier series. It is seen that the smoothed series is superior.

Fig. 17b presents the time response d^2X/dx^2 at $x=0.8368$ using adaptive sampling determined by the axis transformation $y=6x^2-2x$. Here all approximations reach the exact value much earlier, i.e. for

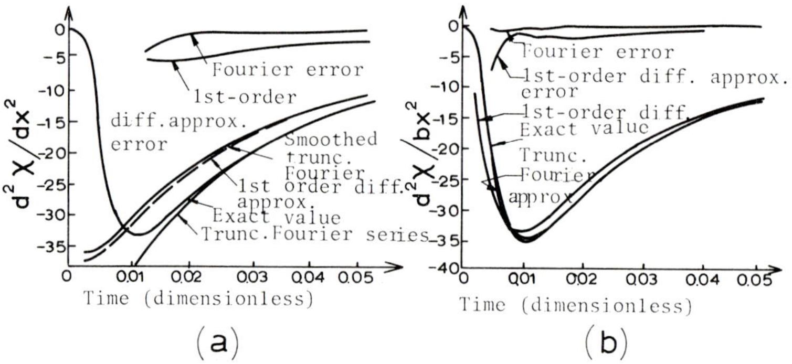

Fig. 17. Time response of d^2X/dx^2 at $x=x_n$ the errors are also shown.
(a) Equidistant sampling ($x_n=x_5=0.8333$, $N=6$),
(b) Adaptive sampling ($x_n=x_4=0.8368$, $N=6$).

$t \geqslant 0.005$. Actually, the difference and the Fourier approximations are indistinguishable in the scales of Fig. 17b. The error reduction obtained by the truncated Fourier series was found to be about 10-15% in the early part of the transient response.

Application of Modal and Other Hybrid Methods

Example 8: Galerkin and extended space methods. The cooling process of a metal fin was studied using these two methods (Neumann and Strauss, 1969). The model $k\partial^2X/\partial x^2=\mu\partial X/\partial t$ with $k=0.1$ and $\mu=1$, where k represents the conductivity and μ the heat capacity of the metal. The boundary condition is of the Newton cooling type, i.e. $\partial X/\partial x= a(X(x,t))-s(t)$ at fin surface, where $s(t)$ is the cooling fluid temperature. The initial temperature is $X^0(x)=100$ degrees throughout the cross-section, and s(t)=100 degrees for t=0, s(t)=0 degrees for t>0. Cooling water 0° begins to circulate on both sides x=0 and x=2 at t=0. Since the model has symmetry about x=1, only half of the fin needs to be considered in the simulation study. The modes $H_i(x)$ are found by orthogonalising the functions $H_i^*(x)=(i+1)(x+1)-x^{i+1}$, i=1,2, ...,N. The coefficients of the linear ODEs obtained using (56b) are as shown in Table 2. The ODEs for the extended space technique are obtained using (58) and (59). Again, these equations are linear, and their coefficients for two and three modes are given in Table 3.

TABLE 2

0.074074	$4.2867_{10}{-4}$	$1.8292_{10}{-5}$	$1.4610_{10}{-6}$	$-1.5616_{10}{-7}$
0.14341	-1.1769	$7.7397_{10}{-3}$	$-3.8192_{10}{-3}$	$7.5717_{10}{-5}$
-0.30674	0.38796	-19.084	0.020270	-0.034980
0.71994	-5.6258	0.59565	-9.4932	0.033540
-1.8134	2.6283	-24.223	0.79036	-17.439

TABLE 3

Two-Mode Case		Three-Mode Case		
-0.0740726	0.000426177	-0.0740726	0.000427785	$-1.799_{10}{-5}$
0.142572	-1.174004	-0.1431100	-1.174666	0.0741377
		-0.301688	0.371627	-4.15921

Fig. 18a shows the simulated response at t=5,20,100 for both the Galerking and extended space methods. Fig. 18b gives the Galerkin error on an expanded scale (full scale of the figure is about 0.20% of the solution). Fig. 18c shows the errors obtained using the extended space method. The 3-mode response is almost identical with the exact one. In this case the error is comparable to that resulting by the numerical integration of ODEs.

Example 9: Method of characteristics. A counter-current heat exchanger with negligible wall effects was simulated by this method (McAvoy, 1970) with $\upsilon_2=2\upsilon_1$. For t=0 the exchanger is assumed at steady-state, and $X_1(0,t)$, $X_2(1,t)$ are subject to step changes from their steady state values $X_1(0)_{ss}=100^o$ and $X_2(1)_{ss}=0^o$ to the values $X_1(0,t)=120^o$, t>0 and $X_2(1,t)=-20^o$, t>0, respectively. Fig. 19 shows the temperature profiles obtained at $\tau=0$, 0.25, 0.50 and 0.75. First-order playback and 100 storage points in x were used. The same profiles were also determined by the standard DTCS method, and it was found that the DTCS simulation results are very poor since the step nature of the response is destroyed, whereas it is preserved by the method of characteristics. Also the case of nonconstant velocities was tested, and good agreement between the exact and approximate responses was obtained. In Fig. 19, τ is normalised time $\tau=t\upsilon_1/L$, where L is the length of the

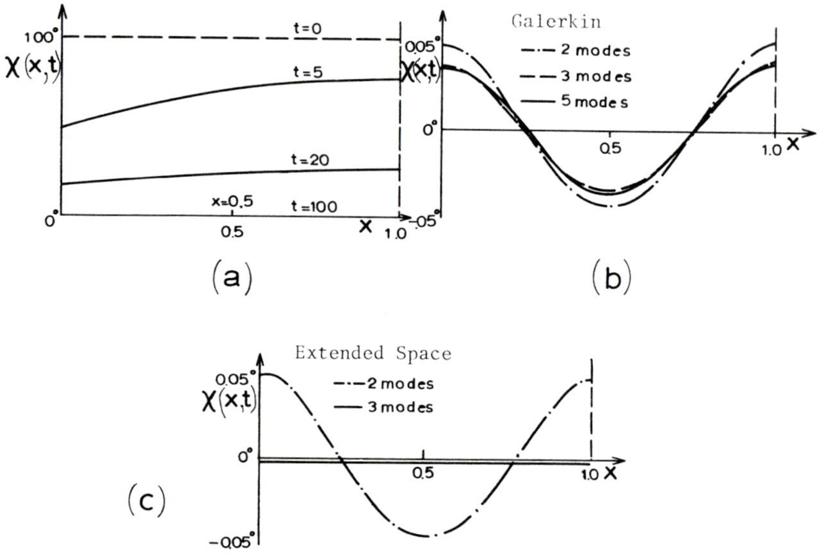

Fig. 18. Simulated results with Galerkin and extended-space methods.
(a) Response at t=5,20,100, (b) Galerkin error at t=5, (c) Extended-space method error at t=5.

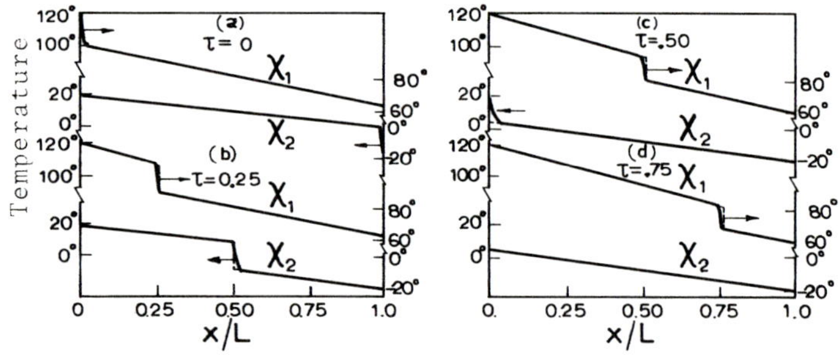

Fig. 19. Heat-exchanger simulation by the method of characteristics.

exchanger (here L=1). The parameter μ is given by $\mu L/\upsilon_1=0.5$ and the time step was chosen as $\Delta\tau=0.05$.

Example 10: Hybrid implicit method. This method was tested on a catalytic-bed reactor converting a substance A to another substance B with the aid of a catalyst K. The reactor is modelled by the fluid phase F and the catalyst phase K, both of which are connected by an intensive exchange of mass and heat. Let $T_F, T_K; C_F$ and C_K denote the temperatures and concentrations of the phases F and K respectively. Since the PDEs of C_F, T_F and C_K can be considered as quasi-stationary, it suffices to study the PDE for $T_K^*(x,t)=X(x,t)$, namely

$$\alpha_1 \partial X/\partial t+\partial^2 X/\partial x^2+\beta^2 X=b_1 U(C_F^*,X)+b_2(X-T_F^*)+\beta^2 X$$

$$\partial X(0,t)/\partial x=\partial X(1,t)/\partial x=0, \quad X(x,0)=T_w^*(x,0)$$

where $U(C_F^*,X)$ is the reaction rate $r(C_K,T_K)$ in which C_K is eliminated, and T_F^*, T_w^*, C_F^* represent normalised variable. Here, (64) gives

$$d^2W_1/dx^2+\beta^2 W_1=\tilde{U}(C_F^*,T_F^*,X), \quad W_1(0,t_k)=dW_1(0,t_k)/dx=0$$

$$d^2W_2/dx^2+\beta^2 W_2=0, \quad W_2(0,t_k)=W_1(1,t_{k-1})/\beta\sin\beta, dW_2(0,t_k)/dx=0$$

and we set $W(x,t_k)=W_1(x,t_k)+W_2(x,t_k)$. The equations for $Y^*(x,t_k)$ are found to be

$$d^2Y^*(x,t_k)/dx^2+\beta^2 Y^*(x,t_k)=Y(x,t_k),$$

$$dY^*(0,t_k)/dx=dY^*(1,t_k)/dx=0$$

which, by a simple difference approximation of L^{-1} yield the following matrix H

$$H=\frac{2}{(\Delta x)^2}\begin{bmatrix} H_1 & 0 \\ 0 & H_2 \end{bmatrix}, H_1=\begin{bmatrix} h_1 & h_2 & 0 \\ h_3 & h_1 & h_3 \end{bmatrix}, H_2=\begin{bmatrix} h_3 & h_1 & h_3 \\ 0 & h_2 & h_1 \end{bmatrix}, \quad \begin{matrix} h_1=\beta^2(\Delta x)^2/2-1 \\ h_2=2h_3=1 \end{matrix}$$

The inverse of $\mathcal{L}_t=a_1\partial/\partial t$ is realised by the explicit first-order difference formula $\underline{X}(t_{k+1})=(\Delta t/a_1)\underline{Y}(t_k)+\underline{X}(t_k)$ which, due to the small time increments is relatively exact. The parameter values used are $\beta=0.75\pi$, N=28 (number of space increments), $T_R=7\times10^{-3}$ sec (simulation cycle period), and $\alpha=\Delta t/T_R=100$ (time scale factor), where Δt is the real time increment (Ruppel and Zeitz, 1975). Some typical simulation results are shown in Fig. 20.

Example 11: Power series hybrid method. This method was applied for solving the Laplace equation model of an electron gun (Ashley and Bullock, 1968). Fig. 21a shows the sectional view of the gun, and Fig. 21b shows the approximate boundary functions $U_b(z)$ and $b(z)$, generated using the 10-segment diode function generators available on the iterative analogue computer AD-80. The boundary was chosen as indicated by the dashed line of Fig. 21a for making both $b(z)$ and $U_b(z)$ single valued.

The ODE (68), truncated to two terms, was solved on the analogue part of the hybrid computer. The starting value $X(0,z)\overset{\Delta}{=}V_z(z)$ was set ma-

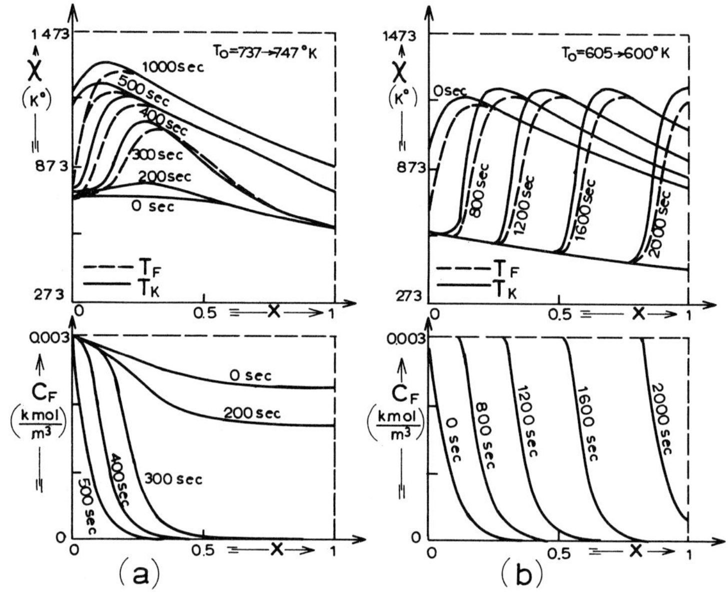

Fig. 20. Hybrid implicit simulation of the fixed-bed reactor.
(a) Ignition of reaction,
(b) Blow-out of reaction.

nually by adjusting the appropriate potentiometer. The resulting solution is shown in Fig. 22a. An agreement with maximum discrepancy 1% of the axial potential determined by using the hybrid and the pure digital technique is found. The sharp corner of $V_z''=d^2V_z/dz^2$ near z=43 is due to the discontinuity of $dU_b(z)/dz$ at this point. Around z=43 the solution far from the axis is inaccurate.

The variables V_z and V_z'' are fed to the digital memory via A/D converters. The simulated results (Fig. 22b) are formed by solving

$$\vartheta^2 X(r,z)/\vartheta r^2 = -(1/2)d^2V_z(z)/dz^2, \quad X(0,z)=V_z(z), \quad X(0,z)/\vartheta r=0$$

on the analogue computer, keeping z fixed, for r in the interval $0<r<b(z)$. The quantities z, $V_z(z)$ and $V_z''(z)$ are supplied by the digi-

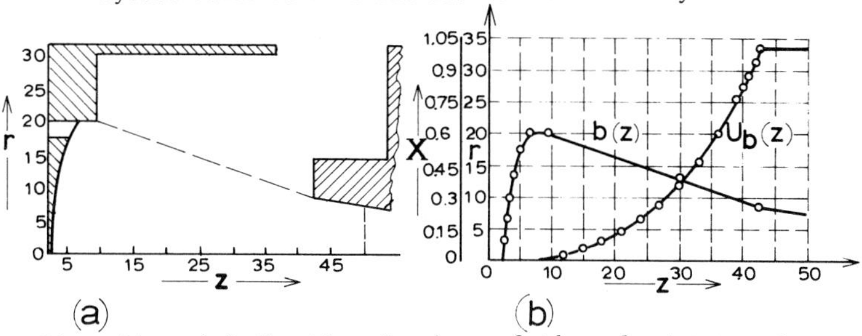

Fig. 21. (a) Sectional view of the electron gun,
(b) Analogue approximation of boundary conditions.

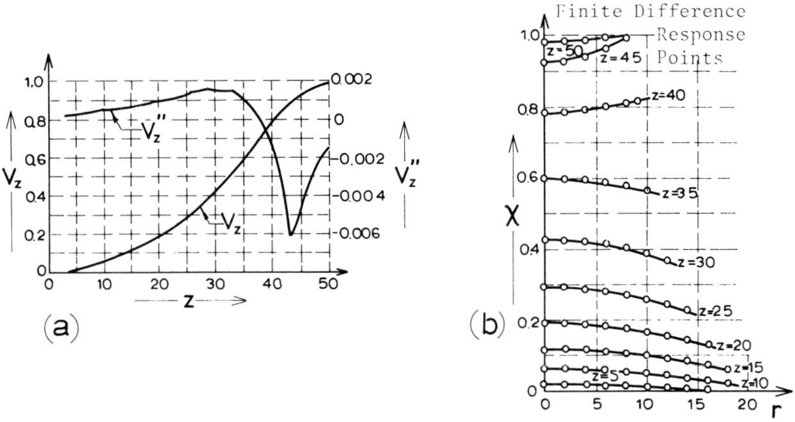

Fig. 22. (a) Axial potential,
(b) Radial potential profiles at various z.

tal computer. Note that a pure analogue method for storing V_z, V_z'' in diode function generators could also be used.

Example 12: Monte Carlo Method. Both elliptic and parabolic models were simulated by this method (Handler, 1967; Jerman, 1971). As a first problem we present the case of $Ad^2X/dx^2+BdX/dx=-E$, $X(-1)=-1$, $X(1)=1$, and $E/A=1$. The solution is given by

$$X(x_o)=X^*(x_o)+G^2PET(x_o)$$

where $X^*(x_o)$ is the solution of the homogeneous PDE (set E=0) with the same boundary conditions, $T(x_o)$ is the average random walk time for the point x_o, 2P is the power spectral density of the noise source, and G is the gain of the integrator. The solution $X^*(x_o)$ based on 1000 random walks is shown in Fig. 23a for B/A=0, 0.5, 2. The random walk generation speed was about 10 random walks per second on the "ALWAK III digital/EAI 231 2V analogue system", and 50-100 random walks per second on the "PDP 8/I digital/EAI TR-480 analogue system". The solution $X(x_o)$ determined by the above formula is shown in Fig. 23b.

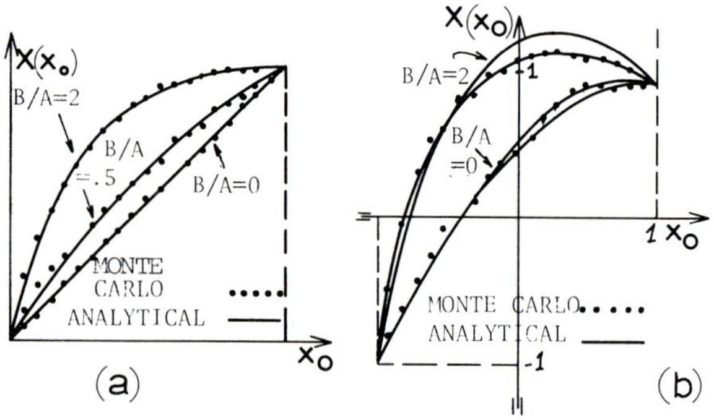

Fig. 23. Monte Carlo simulation of elliptic systems (N=1000)
(a) Homogeneous system
(b) Forced system.

As a second case consider the parabolic system $\partial X/\partial t=\partial^2 X/\partial x^2$, $X(x,0)=0$, $-10<x<10$, $X(-10,t)=-10$, $X(10,t)=10$. The Monte Carlo solution was obtained using the ASTRAC 11 DDA and is shown in Fig. 24. The averaging of the terminal values was made using an analogue filter. The maximum systematic error is 6.4% and the maximum statistical error is 2.2%.

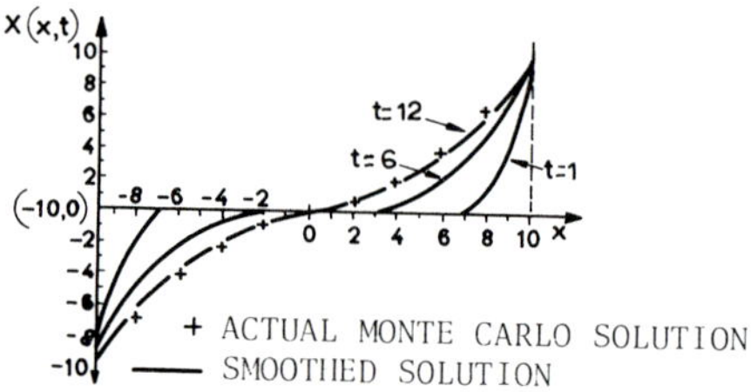

Fig. 24. Monte Carlo solution of the parabolic system.

Finally, the following elliptic DPS was solved on the hybrid system of the Technical University of Munich, with noise having $\sigma=\sigma_1=\sigma_2=0.096$ (Fahrmeir, 1973):

$$(\sigma^2/2)\{(x^2+y^2)(\partial^2X/\partial x^2+\partial^2X/\partial y^2)+4xy\partial^2X/\partial x\partial y-4x\partial X/\partial x\}=0$$

$$X(x,y)_{\partial D}=\begin{cases} x^3-(x-0.1)^3/3-0.1x^2+1, & (x,y)\in\partial D_1 \\ (x-0.1)^3/3-x^3+0.1x^2+1, & (x,y)\in\partial D_2 \\ y-y^3/3+1 & , (x,y)\in\partial D_3 \end{cases}$$

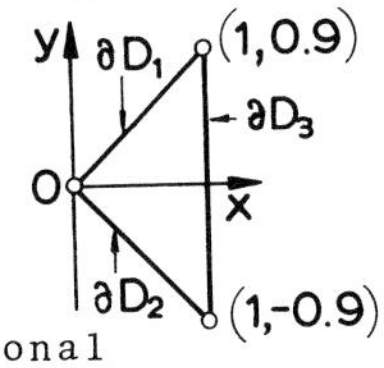

TABLE 4 Simulated Results for the 2-Dimensional Elliptic System

y \ x	.4	.5	.6	.7	
.4			1.128	1.140	Monte Carlo
			1.123	1.175	Exact
.2	1.045	1.049	1.065	1.070	Monte Carlo
	1.031	1.047	1.069	1.096	Exact
.1	1.018	1.032	1.044	1.044	Monte Carlo
	1.016	1.025	1.036	1.049	Exact
.0	1.009	0.999	1.011	1.005	Monte Carlo
	1.000	1.000	1.000	1.000	Exact
-.2		0.957	0.952	0.931	Monte Carlo
		0.953	0.930	0.905	Exact
-.4			0.887	0.857	Monte Carlo
			0.887	0.804	Exact

The exact solution is $X(x,y)=x^2y-y^3/3+1$. Using

$$B=\sigma^2\begin{bmatrix} x^2+y^2 & 2xy \\ 2xy & x^2+y^2 \end{bmatrix} \text{ and } \Sigma=\begin{bmatrix} \sigma & 0 \\ 0 & \sigma \end{bmatrix} \text{ we get } \bar{B}=\sigma\begin{bmatrix} y & x \\ x & y \end{bmatrix} \text{ and } B^*=\begin{bmatrix} y & x \\ x & y \end{bmatrix}$$

Then it follows that the random walks are governed by

$$dx/dt=-3\sigma^2x+N_1(t)y+N_2(t)x,\quad dy/dt=-\sigma^2y+N_1x+N_2y$$

The results are shown in Table 4.

CONCLUSIONS

The hybrid simulation field of DPS has been expanded during the last twenty years, and a variety of methods have been developed with the principal goal to reduce the cost of simulation. On the hardware side the search for more powerful DPS simulators has led to the development of efficient general purpose hybrid computers by many industries, as well as to the design of a number of specialised hybrid computation facilities in many research laboratories. Our purpose in this chapter was to give a comprehensive presentation of the principal methods highlighting their key points and features. Of course, due to space limitation, it was impossible to include all available results and references. These methods are useful in all cases where some hybrid computer or digital/analogue facility is available. As it is evidenced by the examples, one can cheaply simulate any practical DPS by combining a medium-size digital computer with a number of high speed analogue computing elements and D/A-A/D converters. Today, with the advent of microprocessors, it is possible to have cheap parallel digital computation, and so new types of hybrid DPS simulation set-ups are expected to appear. In closing the chapter we remark that there are also available many digital super computers possessing various combinations of parallel and pipe-line capabilities not encountered in standard serial computers. With these computers one can have fast and cheap digital simulation of DPS. To facilitate the actual simulation from the user's point of view several software packages and simulation languages have been developed for all general types of DPS (parabolic, hyperbolic, elliptic). A brief survey of these software methods and packages was provided by Karplus (1980). The present chapter is actually an expansion of the material presented at the IASTED Simulation Course (Tzafestas, 1980).

REFERENCES

Altom, D., and P. Nelson, Jr. (1970). On the use of invariant imbedding within the continuous-space discrete-time method:Digital simulation. Proc. 6th AICA Congress, pp. 88-92.

Andrews, M. and G. Korn (1975). Hybrid computer methods for direct functional optimisation. IEEE Trans. Computers, Vol. C-24, pp. 958-965.

Ashley, J., and T. Bullock (1968). Hybrid computer integration of PDEs by use of an assumed sum separation of variables. Proc. 1968 FJCC, Vol. 33, pp. 585-591.

Bekey, G., and W. Karplus (1968). Hybrid Computation. John Wiley, New York.

Bjørge, P. (1966). Experimental analog memory using sampled data digital technique, MIT 1st Memo. DSR 5161, Electron. Syst. Lab.

Bosgra, O., and J. Buis (1973). Hybrid simulation of multi-component chemical reactions in multifunctional catalyst systems Simulation, Vol. 19, pp. 37-48.

Bosgra, O., and J. Buis (1974). Fundamental approach to the selection of implementation parameters in the hybrid serial solution of partial differential equations. AICA Annales, pp. 24-34.

Chan, S. (1967). Analog-digital methods of solving PDEs. MIT Tech. Report AD-679-741, Electron. Syst. Lab.

Chan, S. (1969). The serial solution of the diffusion equation using nonstandard hybrid techniques, IEEE Trans. Computers, Vol. C-18, pp. 786-799.

Chareton, H., J. Corriol, and P. Poulard (1970). Etude d'un enchageur chaleur par calcul hybride. Proc. AICA-IFIP Conf. on Hybrid Computation, Munich, Aug. 31-Sept. 4.

Chuang, K., L. Kazda, and T. Windekneght (1960). A stochastic method of solving PDEs using an electronic analog computer. Project Michigan Report 290-91-T., Univ. of Michigan.

Cimorelli, L., R. Caparaso, and A. Federico (1968). Hybrid stochastic Monte Carlo methods for solving heat transfer problems in nuclear reactors. AICA Annals, pp. 195-205.

Connelly, M., and O. Fedoroff (1965). A demonstration hybrid computer for real-time flight simulation, AF. Tech. Report AMLR-TR-65-97, Springfield, Virginia.

Evans, D. (1965). A stable explicit method for the finite-difference solution of fourth-order parabolic PDEs. Computer J., Vol. 8, pp. 280-287.

Fahrmeir, L. (1973). Extension of the hybrid Monte Carlo method for boundary-value problems. Simulation, Vol. 19, pp. 177-178.

Finlayson, B., and L. Scriven (1966). The method of weighted residuals: A review. Appl. Mech. Reviews, Vol. 19, p. 735.

Fox, L. (1962). Numerical solution of ordinary and partial differential equations. Pergamon Press, Oxford.

Frank, A., and L. Lapidus (1965). The effect of error propagation on hybrid computation to chemical engineering problems. Proc. 1965 IFIP Congress, Vol. 2, Spartan Books. Also Trans. Am. Nucl. Soc., Vol. 13, p. 623.

Friedmann, N. (1956). The truncation error in semi-discrete analog of the heat equation. J. Math. Physics, Vol. 35, p. 299.

Gran, M., A. Sabroff and A. Wong (1966). Parameter optimization by random search using hybrid computer techniques. Proc. FJCC, Vol. 31, pp. 191-199.

Handler, H. (1967). Monte Carlo solution of PDEs using a hybrid computer. IEEE Trans. Computers, Vol. EC-16, pp. 603-610.

Hara, H., and W. Karplus, (1968). Application of functional optimization techniques for the serial hybrid computer solution of PDEs. Proc. AFIPS, Vol. 39, p. 565.

Hsu, S., and R. Howe (1968). Preliminary investigation of a hybrid method for solving PDEs. Proc. 1968 EJCC, Vol. 33, pp. 601-609.

Humeau, J., and M. Grolleau (1975). Hybrid simulation of a control problem for a DPS. AICA Annales, No. 2, pp. 110-115.

Jerman, W. (1969). Hybrid Monte Carlo techniques for the solution of PDEs. Res. Report, 1969 ASEE-NASA Summer Faculty Fellowship Program, Marshall Space Flight Center, pp. 198-216.

Jerman, W., M. Calhoun, and R. Thomas (1971). Hybrid Monte Carlo techniques with a minimal interface. Simulation, Vol. 17, pp. 225-233.

Kahne, S. (1968). Computational questions in optimal control, quasilinearization, and hybrid computers, Simulation, Vol. 14, pp. 81-86.

Karplus, W. (1964). A hybrid computer technique for treating nonlinear PDEs. IEEE Trans. Computers, Vol. C-13, pp. 597-605.

Karplus, W. and Russel, R.A. (1971). Increasing digital computer efficiency with the aid of error-correcting subroutines, IEEE Trans. Computers, Vol. C-20, pp. 831-837.

Karplus, W. (1980). Software for distributed system simulation, Proc. IASTED Simulation Course, Interlaken, Switzerland, June 23-24, ACTA Press, Calgary.

King, G. (1951). Monte Carlo method for solving diffusion problems. Industr. Engrg. Chem., pp. 2475-2478.

Korn, G. (1965). Hybrid Monte Carlo techniques, Simulation, Vol. 5, pp. 9-11.

Little, W. (1966). Hybrid computer solutions of PDEs by Monte Carlo methods. Proc. 1966 FJCC, Vol. 31, pp. 181-190.

McAvoy, T. (1972). Solution of hyperbolic PDEs via a hybrid implementation of the method of characteristics. Simulation, Vol. 18, pp. 91-98.

Moroney, R. (1963). The method of characteristics. Memos DSR 9128WP-6,7, Electron. Syst. Lab., MIT.
Nelson, P. (Jr.) (1970). Application of invariant imbedding to the solution of PDEs by the CSDT method, Proc. 1970 SJCC, Vol. 36, pp. 397-402.
Nelson, P., (Jr.) and D. Alton (1971). Hybrid solution of PDEs by application of invariant imbedding to the serial method. Simulation, Vol. 17, pp. 147-153.
Neundorf, W. and Re. Schoenefeld (1980). Stable iterative hybrid method for the solution of weak nonlinear PDEs. Maths. Computers in Simulation (IMACS Trans.), Vol. XXII, pp. 1-6.
Newman, D., and J. Strauss (1968). Hybrid assumed mode solution of nonlinear PDEs., Proc. 1968 FJCC, Vol. 33, pp. 575.
Newman, D., and J. Strauss (1969). The extended space technique for hybrid computer solution of PDEs. Proc. 1969 FJCC, Vol. 34, pp. 751-760.
Nomura, T., and R. Deiters (1968). Improving the analog simulation of PDEs by hybrid computation. Simulation, Vol. 14, pp. 73-80.
Paul, A. and A. Ahmad (1970). Analogue solution of 2nd-order hyperbolic PDEs by the method of continuous characteristics., Proc. IEE, Vol. 117, pp. 1175-1184.
Petrov, A. (1975). A differential-difference technique for the hybrid computer solution of parabolic PDEs, AICA Annales, pp. 104-109.
Piceni, H., and P. Eykhoff (1975). The use of hybrid computers for systems parameter estimation. AICA Annales, pp. 9-22.
Ruppel, W., and M. Zeitz (1975). Implicit solution method for DPS. AICA Annales, pp. 171-177.
Sanathanan, C., J. Carter, L. Bryant, and L. Amiot (1967). The application of a hybrid computer to the analysis of transient phenomena in a fast reactor core., Nucl. Sci. Eng., Vol. 28, pp. 82-92.
Saphier, D. (1972). A hybrid method for solving the reactor space-time-dependent equations, Nucl. Sci. Eng., Vol. 47, pp. 275-289.
Shih, Y., and J. Han (1978). Double Walsh series solution of first-order partial differential equations, Int. J. Systems Sci., Vol. 9, pp. 569-578.
Soudack, A., and W. Little (1965). An economical hybridizing scheme for applying Monte Carlo methods to the solution of PDEs. Simulation, Vol. 5, pp. 9-11.
Swillens, S., and M. Paiva (1975). A stochastic method for the solution of the diffusion equation, Simulation, Vol. 21, pp. 33-41.
Tzafestas, S. (1970a). Hybrid computation in optimal DPS control. Proc. 6th AICA-IFIP Conf. Hybrid Computation, pp. 171-179, Munich.
Tzafestas, S. (1970b). Final-value control of nonlinear composite DP and LP systems. J. Franklin Inst., Vol. 290, pp. 439-451.
Tzafestas, S. (1978a). Design of DP optimal controllers and filters via Walsh-Galerkin expansions. In S. Banks, and A. Pritchards (Eds.), Control of DPS, pp. 201-217. Pergamon Press, Oxford.
Tzafestas, S. (1978b). Walsh series approach to lumped and distributed system identification. J. Franklin Inst., Vol. 305, pp. 199-220.
Tzafestas, S. (1980). Hybrid computation in DPS: The state of art. Proc. IASTED Simulation Course, Interlaken Switzerland, June 23-24, ACTA Press, Calgary.
Vichnevetsky, R. (1962). Method of characteristics in the hybrid solution of 1st-order PDEs. Electronic Assoc. Rept. No. 60, Princeton, N.J.
Vichnevetsky, R. (1968). A new stable computing method for the serial

hybrid computer integration of PDEs. Proc. 1968 SJCC, Vol. 32, pp. 143-150.

Vichnevetsky, R. (1969a) Serial solution of parabolic PDEs: The decomposition method for non-linear and space-dependent problems., Simulation, Vol. 13, pp. 47-48.

Vichnevetsky, R. (1969b). Generalized finite difference approximations for the parallel solution of initial-value problems. Simulation, Vol. 13, pp. 233-237.

Vitchnevetsky, R. (1969c). Use of functional approximation methods in the computer solution of initial value PDE problems, IEEE Trans. Comput., Vol. C-18, pp. 499-512.

Vichnevetsky, R. (1970a). A hybrid computer method for the analysis of time-dependent river pollution problems. Proc. 1970 SJCC, Vol. 36, pp. 43-49, AFIPS Press, N.J.

Vichnevetsky, R. (1970b). Line multiplexing, alternating directions and splitting hybrid procedures for parabolic equations in two spatial dimensions. Electronic Assoc. Rept. TR-70-14, Princeton, N.J.

Vichnevetsky, R. (1971). Hybrid methods for PDEs. Simulation, Vol. 17, pp. 168-180.

Vichnevetsky, R. (1974). Physical criteria in computer methods for PDEs., AICA Annales, pp. 3-16.

Witsenhausen, H. (1965). Hybrid techniques for solution of PDEs. Proc. 1965 IFIP Congress, Vol. 2, pp. 425-431, Spartan Books.

Wozny, M., P. Bartells, and R. Bailey (1970). Hybrid computer simulation of xenon spatial oscillations, Simulation, Vol. 14, No.5.

CHAPTER 2

IDENTIFICATION OF DISTRIBUTED-PARAMETER SYSTEMS

Y. Sunahara

Kyoto Institute of Technology, Matsugasaki, Kyoto, Japan

INTRODUCTION

The first concern with determining the dynamical structure for a practical system is the problem of system parameter identification. Up to the present time, the principal line of research activities is directed to treat lumped parameter systems described by ordinary differential equations. A lumped parameter representation of a physical system implies that its local spatial variations are disregarded. However, dynamic behaviors of all real physical systems are, in fact, distributed. A distributed parameter system is characterized by a class of partial differential equations, integral equations or integro-differential equations. In this chapter, we restrict our attention to physical systems described by partial differential equations, because the great majority of research including efforts of practical applications is concerned only with this kind of mathematical models. Although a class of partial differential equations may usually be derived by considering basic conservation principles, system parameters remain to be determined in many examples. From practical viewpoints, determination of heat transfer coefficients, specific heats, electro-magnetic properties, gas properties, chemical reaction rate constants, diffusion constants, etc. is a requisite effort throughout state estimates and/or optimal controls of distributed parameter systems.

Retrospective Review

It has become customary to open the identification chapter in the field of distributed parameter systems by considering the problem to determine a set of unknown parameters involved in the mathematical model of a real system from acquired data. The fact that a surprisingly high rate of publication of papers in this field is continuing may be recognized in several excellent survey papers.

For lumped parameter systems, Åström and Eykhoff (1971) surveyed such aspects as general properties of identification problems, classification of identification methods and choice of model structure. Robinson (1971) included in his survey the identification aspect of optimal control of distributed parameter systems. Polis and Goodson (1976) showed a step by step approach to solve the distributed system parameter identification problem in his synthesizing overview. In a survey by Kubrusly (1977), the various identification methods are grouped into the following three classes: (1) direct method, (2) reduction to a lumped parameter system, and (3) reduction to an algebraic equation. His survey brings us a useful information with many references included. Ray (1978) made a brief survey of some recent applications of the theory of system identification and parameter estimation. Two full-length books edited by Ruberti (1978) and Ray and Lainiotis (1978) and currently more are in preparation. Since the major body of references concerning parameter identification problems for distributed parameter systems is beautifully summarized by Goodson and Polis (1978), the reappearance of historical aspects is abandoned in this chapter.

Scope of Chapter

In spite of a wide range of recent applications of the theory of distributed parameter systems, it has been only in the last $10 \sim 15$ years that readily applicable techniques have been developed which allow us to determine unknown parameters using noisy observation data. In modelling procedures, identification methods using noisy observations are, in fact, powerful tools for systems operating in a random environment. Examples of applications include structural analysis and design, acoustical problems, design and control in industrial processes, geophysical analysis, earthquake study, seismic profile identification in oil exploration survey. Thus, a special emphasis is, in this chapter, placed on parameter identification problems of stochastic distributed parameter systems under noisy observations. While this chapter tries to cover some variety of problems, there are other related ones which are excluded.

Regarding the arrangement of this chapter, there are a number of ways in which a piece of work in this field could be characterized. First of all, for convenience of descriptions, several abbreviations are introduced in the sequel. They are

DPS : Distributed parameter system
SDPS : Stochastic distributed parameter system
PDE('s) : Partial differential equation(s)
BC('s) : Boundary condition(s)
IC('s) : Initial condition(s)
LPS : Lumped parameter system

The chapter is organized as follows: In order to show the existing status of unknown parameters, an exhibition of mathematical models of DPSs is shown in Section II. With this exhibition as the first step, successive approaches are developed to identify unknown parameters involved in DPSs. Overviews related to each of the steps are briefly described in the section. Particular results developed by the author and his colleagues are summarized in Section III. The list of references closes this chapter.

EXHIBITION OF MATHEMATICAL MODELS

Physical systems of interest are distributed over some spatial domain $\bar{D}$ which is contained in the n-dimensional Euclidean space $E^{(n)}$. In most works, this domain is fixed, closed and bounded. Occasionally, it is infinite or semi-infinite in extent and some of recent studies consider a domain with movable or free boundaries. There is also a time domain T of interest, usually a finite interval or a semi-infinite one. Denoting an open n-dimensional spatial domain by D and its boundary by ∂D, the system considered is usually characterized by one of partial differential equations of parabolic or hyperbolic type which must be satisfied on $T \times D$ with boundary conditions on $T \times \partial D$. The answer for the question whether a DPS representation is adequate or not is not simple. An intuitive answer is that, if the system behavior exhibits a significant dependence on a spatial direction, then a DPS model is preferable. However, a general and unique model which is capable of covering various kinds of DPSs does not exist.

DPS Subjected to Additive Noise

Let $u(t,x)$ be the scalar system state. In case of linear businesses, many authors adopt the form (Tzafestas, 1978)

$$\frac{\partial u(t,x)}{\partial t} = \mathcal{L}_x u(t,x) + f(t,x) + G(t,x)\gamma(t,x), \quad t \in T =]0, t_f[, \; x \in D, \tag{1}$$

with initial and boundary conditions

$$\text{I.C.} \qquad u(0,x) = u_0(x), \quad x \in D, \tag{2}$$

$$\text{B.C.} \qquad \mathcal{B}_x u(t,x) = 0, \qquad t \in T =]0, t_f[, \; x \in \partial D, \tag{3}$$

where both $\mathcal{L}_x$ and $\mathcal{B}_x$ are well-posed linear spatial differential operators, $\gamma(t,x)$ is the zero-mean Gaussian white (with respect to t) distributed noise process, f and G are known functions and $u_0(x)$ is the initial state function which is usually assumed to be the Gaussian stochastic variable for each x independent of $\gamma(t,x)$. There are many cases where one or more parameters of $\mathcal{L}_x$ are unknown.

A mathematical model of a general class of nonlinear DPSs is written by using a conventional function F in the form

$$\frac{\partial u(t,x)}{\partial t} = F(t, x, u, u_x, u_{xx}, \cdots) + G(t,x)\gamma(t,x), \tag{4}$$

where, for the mathematical security, the function F is usually assumed to be sufficiently smooth with respect to its arguments, and where u_x is the partial derivative with respect to x. In order to handle practical problems, Eq.(4) is written in a more concrete form, e.g.,

$$\frac{\partial u(t,x)}{\partial t} = \mathcal{N}_s(t,x,u) + G(t,x)\gamma(t,x) \tag{5}$$

with the boundary condition which is considered to be also nonlinear, e.g.,

$$\text{B.C.} \qquad \mathcal{N}_b(t,x,u) = 0, \qquad t \in T =]0, t_f[, \; x \in \partial D \tag{6}$$

and with the same initial condition as given by (2). It should be noted that, if there exists an additive noise at the system boundary, then the noise term $\xi_b(t,x)$ appears on the right hand side of both Eqs.(3) and (6).

DPS with Stochastic Coefficients

In many practical problems, one or more coefficients in differential operators are random because measurements of physical properties of the system considered inherently exhibit greater uncertainty. It may, therefore, be more realistic to consider system parameters as stochastic variables with a probability distribution. Many different forms have been proposed for DPSs and each is useful for different tasks. A general form using the parabolic type, which arises directly from fundamental physical axioms, is as follows:

$$\frac{\partial u(t,x)}{\partial t} = \mathcal{A}(t,x,\omega;\, D_x)u(t,x) + f(t,x), \quad (t,x) \in T \times D, \tag{7}$$

together with the boundary condition

$$\mathcal{B}(t,x;\, D_x)u(t,x) = 0, \qquad (t,x) \in T \times \partial D \tag{8}$$

and with the same initial condition as given by (2), where $\mathcal{A}$ is a partial differential operator containing stochastic coefficients, $\mathcal{B}$ is a boundary operator with deterministic nonvanishing coefficients and ω is the generic point of the sample space. Since unknown stochastic coefficients are involved in $\mathcal{A}$, the problem at hand belongs to the context of stochastic inverse problems and is motivated by the oil exploration survey, the experimental physiology, cardiology, etc.

DPS with Free Boundary

Recent advances in control technique have stirred a great deal of enthusiasm in the development of parameter identification and state estimate of DPSs with free boundaries. A number of systems of practical interest involve boundaries moving by phase change such as melting or solidification, chemical reaction, heat transfer and so on. Considering a spatial region consisting of water and ice, and choosing the system state as the temperature distribution of the water region, the water-ice interface becomes a moving boundary denoted by $y(t)$. For example, consider a mathematical model of the system

$$\frac{\partial u(t,x)}{\partial t} = a\frac{\partial^2 u(t,x)}{\partial x^2}, \quad 0 < x < y(t), \quad 0 < t < t_f \tag{9}$$

with

$$\text{I.C.} \qquad u(0,x) = u_0(x) \geq 0, \quad \text{for } 0 < x < y(0)(=y_0). \tag{10}$$

On the boundary, i.e., the ice-water interface, by considering the latent heat of fusion and the rate at which ice is converted into water, it follows that, for $0 < t < t_f$,

$$au(t,0) = g(t), \; u(t,y(t)) = 0, \; a\frac{\partial u(t,y(t))}{\partial x} = -\frac{dy(t)}{dt}, \tag{11}$$

where a is a positive constant and where the ice region is assumed to be bounded. Figure 1 illustrates the construction of a DPS with free boundary. A fundamental difference between regular boundary problems and free boundary ones is that the domain of solutions to the basic state equation is not known but should be determined by additional information through underlying properties of physical systems under consideration.

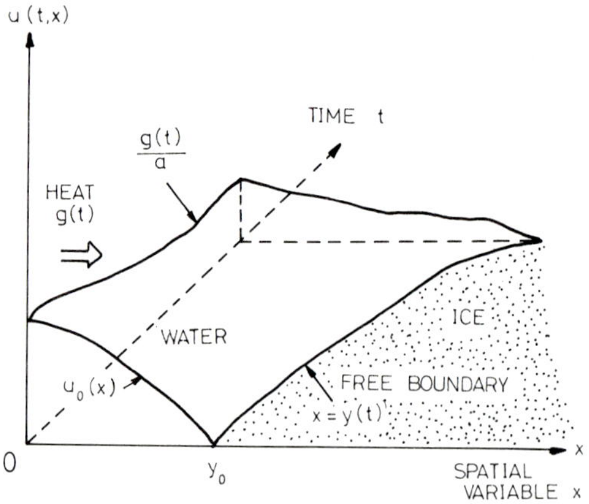

Fig. 1. An illustration of the DPS with free boundary.

DPS with Inequality Boundary Conditions

There has recently been much practical interest shown in problems of the DPS with a permeable wall as its system boundary. Identification problems of system parameters in which the osmotic pressure dues to the nature of membrane are of practical importance. More basically, in the field of synovial joint biomechanics, the permeability plays an important role of analyzing locomotion of the musculoskeletal system. From theoretical viewpoints, a remarkable feature of the problem considered here relates to a mathematical modelling of the system boundary.

Suppose that the region D is filled up by the liquid as shown in Fig. 2. Then, the state variable u(t,x) is reasonably selected to be the pressure of the fluid. The boundaries of D are respectively denoted by Γ_0 at x = 0 and by Γ_1 at x = 1. The boundary conditions peculiar to the system shown in Fig. 2 are considered as follows. At the boundary Γ_0, the well-known Neumann condition is set in a form of

$$- a\frac{\partial u(t,x)}{\partial x} = f(t), \qquad \text{on } \Gamma_0, \tag{12}$$

where a is a positive constant expressing the diffusion coefficient and f(t) is the known input function applied to the boundary Γ_0.

On the other hand, the boundary Γ_1 consists of a semi-permeable membrane whose thickness is supposed to be negligible. This implies that, as shown in Fig. 2, although the fluid flow leaving D freely passes Γ_1, the counter flow is prevented by the existence of the membrane. Let h(t) be the given fluid pressure in the outside of D.

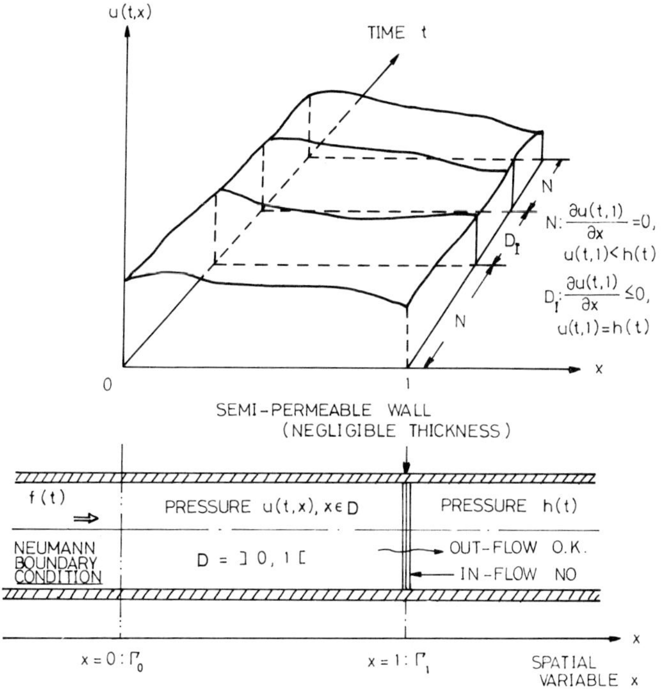

Fig. 2. An illustration of the system with a semi-permeable wall.

First suppose that

Case-1: $u(t,l) < h(t)$; There exists a fight where the semi-peameable wall prevents the fluid flow against the fluid trying to enter into D. A success of the fight allows us to write

$$\frac{\partial u(t,x)}{\partial x} = 0, \qquad \text{on } \Gamma_1. \tag{13a}$$

Secondly consider the inverse case, i.e.,

Case-2: $u(t,l) \geq h(t)$; This case implies that the fluid tries to leave D. Hence

$$\frac{\partial u(t,x)}{\partial x} \leq 0, \qquad \text{on } \Gamma_1. \tag{13b}$$

However, since the thickness of the semi-permeable wall is negligibly thin, $u(t,l)$ is not greater than $h(t)$, i.e., $u(t,l) = h(t)$. Summarizing the results mentioned above, with the same state equation as given by Eq. (9) and the same initial conditions as given by (2), the boundary conditions are modelled by (12), (13a) and (13b).

The problems actually encountered in trying to identify parameters in DPS or more practically SDPS as follows:

(1) *Completion of the SDPS as a mathematical model of the system with unknown parameters*: One of SPDEs might be adopted as the system state equation where, in most cases, unknown parameters are involved in the linear operator or nonlinear one. A special care must be exercised so that the boundary condition represents the boundary state of the physical system under consideration.

(2) *Decision on measuring methods and spatial allocation of sensors*: In the past scheme, identification of SDPS is performed by making measurements distributed spatially or by making discrete measurements at many locations in the spatial domain. However, the requirement of a large number of sensors makes the data processing for identification impractical. From viewpoints of capital investment limits and of maintenance businesses, the simultaneous decision of number of sensors and their optimal allocation brings up a central question in parameter identification, state estimation and further the optimal control for SDPS.

(3) *Identification of noise statistics*: A very conventional way is to assume *a priori* that both system noise and observation one are Gaussian white noise with zero mean. Naturally, in practice, there are many cases where the assumptions mentioned above are violated. Since the theoretical aspect is developed along the line of Gaussian white noise statistics, in order to make the theoretical tool available, an additional procedure must be introduced, for example, a shaping filter technique generating the colored noise from the Gaussian white noise. However, even if we introduce a shaping filter, another problem of deciding its dynamics appears.

(4) *Examination of mathematical aspects of the system model and the observation dynamics*: Conditions for existence of the unique continuous solution to PDE or SPDE must be examined for the mathematical security. During recent years, the greater portion of theoretical studies in this area has centered on function space methods. Highly mathematical discussions are developed by abstract formulations in which the solution to PDE is represented in a suitable Hilbert, Banach or Sobolev space. Through these function space methods, it is possible to obtain very compact and elegant statements of conditions for existence of solutions to PDE, which can be applied not only to identification problems but to state estimation and control problems. The function space point of view was initiated in the field of deterministic PDEs. A recent survey of this aspect of the subject by Mitter (1969) includes considerably detail. Stochastic versions involving PDE with stochastic

coefficient have not had a great deal of attention, but more than a start has been made (Sunahara and others, 1981; Sunahara, 1981).

(5) *Choice of a criterion for identification*: Although we are free to choose a performance criterion, effects of our choice on the whole aspect of the problem must be evaluated as a category of the sensitivity analysis. In fact, the solution to the problem, "What type of criterion should be adopted?" is a kind of uncertainties. The designer's strategy must have a very high assurance of success regardless of this uncertainty. At present, although no unique choice of performance criterion can be said *a priori* to be the best choice for a given identification problem, a type of the least squares performance criterion is commonly used.

(6) *Choice of an identification technique*: The most pertinent techniques used for identification of DPS and SDPS are listed below;

(i) Gradient methods (conjugate gradient method, steepest descent method)
(ii) Least squares (weighted least squares)
(iii) Kalman filter in the version of SDPS
(iv) Maximum likelihood
(v) Stochastic approximation
(vi) Nonlinear filtering (by Bayesian approach)
(vii) Quasi-linearization
(viii) Nonlinear programming
(ix) Pattern search (Hooke-Jeeves pattern search)

Detailed aspects of optimization methods are beautifully tabulated by Goodson and Polis (1978) and for concise versions, see a survey paper by Robinson (1971) and also a paper by Polis, Goodson and Wozny (1973).

(7) *Error evaluation and sensitivity analysis*: For evaluating effects on system performance due to changes in parameter values very little has appeared in the field of DPS. An interesting try by Polis, Goodson and Wozny (1973) has been reported in their survey paper, including a couple of references.

Detailed aspects of literature survey are omitted to write here because a couple of excellent survey papers cited previously have already completed this task.

SPECIFIC RESULTS FOR IDENTIFICATION IN STOCHASTIC DISTRIBUTED PARAMETER SYSTEMS

In this section, specific results of the author and his colleagues are given. There are a wide variety of important engineering problems which arise in physical systems described by a class of SPDE. However, we shall be able to describe here a couple of research works motivated by practical engineering requirements.

Example 1: Parameter Identification Involved in the Spatially Distributed Input

The motivation of the first example was in the research project of the air pollution prevention in environmental systems.

(1) *Mathematical model*: Consider a physical system modelled by the following one-dimensional linear partial differential equation:

$$\frac{\partial u(t,x)}{\partial t} = L_x u(t,x) + f(t,x,\theta) + G(t,x)\gamma(t,x), \quad (t,x) \in T \times D \tag{14a}$$

with

$$\text{I.C.} \qquad u(0,x) = u_0(x), \qquad x \in D \tag{14b}$$

$$\text{B.C.} \qquad u(t,x) = 0, \qquad t \in T =]0, t_f[, \; x \in \partial D \tag{14c}$$

where θ is an unknown time-invariant parameter considered to be a stochastic variable and the statistics of the function $u_0(x)$ is assumed to be known. Equation (14) is a simplified model for the diffusion of air pollution levels in which the system state is, for example, considered to be the SO_2 concentration level and the function $f(t,x,\theta)$ represents the external influence due to the sources at n discrete locations in the spatial domain D. Another example can be found out in nuclear engineering where the system state is considered as the field of temperature (e.g., Wiberg, 1967). In this case, the first term of the right-hand side is due to diffusion, the second term is the generated heat in a nuclear reactor where θ characterizes the location of the heat source. In these cases, the third term of the right-hand side in Eq. (14a) represents an additive system noise.

(2) *Measuring methods*: Measurements are made on the system state $u(t,x)$. We shall define the observation process $y(t)$ by

$$\frac{dy(t)}{dt} = \int_D H(t,z)u(t,z)dz + R(t)\gamma_0(t), \quad y(0) = 0, \tag{15}$$

where H is a scalar function and $R(t)$ is a continuous positive coefficient on $]0, t_f[$ and $\gamma_0(t)$ is the normalized white Gaussian process independent of both $u(t,x)$ and $\gamma(t,x)$. The choice of the mathematical model (15) implies that observations are continuously made on the system state with respect to time and spatial points. This is only for mathematical convenience to develop the theoretical aspect in the continuous parameter process.

(3) *Mathematical aspects including noise statistics*: A mathematical difficulty is caused by the existence of the white Gaussian noise processes $\gamma(t,x)$ so that Eq. (14a) is meaningless. A more rigorous version of Eq. (14a) may be expressed by Itô differential form,

$$du(t,x) = L_x u(t,x)dt + f(t,x,\theta)dt + G(t,x)dw(t,x), \quad (t,x) \in T \times D \tag{16}$$

where w is the Brownian motion process with covariance

$$E\{dw(t,x)dw(t,z)\} = Q(x,z)dt. \tag{17}$$

In (17), the symbol $E\{\cdot\}$ denotes a mathematical expectation and Q is a symmetric nonnegative function for all $x,z \in D$. Although many arguments are the subject of current researches (Falb, 1967; Kushner, 1968), we shall assume that all conditions hold on f and G so as to assure the existence of the solution to Eq. (16).

Along a similar procedure to Eq. (16), Eq. (15) is converted into

$$dy(t) = \left[\int_D H(t,z)u(t,z)dz\right]dt + R(t)dv(t), \; y(0) = 0. \tag{18}$$

(4) *An assumption associated with unknown parameters*: In a practical problem, we have an additional task of determining the forcing term f. We assume that there are N input sources so that the function f is identified with

$$f(t,x,\theta) = \sum_{i=1}^{N} f_i(t,x,\theta^{(i)}), \tag{19}$$

where $f_i(t,x,\theta^{(i)})$ is a function expressing the i-th input. In this situation, it is of great interest to derive the identification method for the unknown parameters $\theta^{(i)}$. To fix the ideas, we shall first consider the case where $N = 1$, because an extension to the case of N inputs is straightforward with few changes.

Although, in most practical cases, the unknown parameter θ possibly takes any value of the admissible set, as might be expected, computational requirements are, in general, excessive. Consequently, the *a priori* probability $P(\theta)$ is assumed to be

$$P(\theta) = \sum_{i=1}^{K} P(\theta_i)\tilde{\delta}(\theta - \theta_i), \tag{20}$$

that is, the parameter θ takes its value of $\theta_1, \theta_2, \cdots, \theta_K$, where, in (20), $\tilde{\delta}$ is the function such that $\tilde{\delta}(\theta - \theta_i) = 1$ $(\theta = \theta_i)$ and $= 0$ $(\theta \neq \theta_i)$.

Let Y_t be the observation data up to the present time t. The problem is to find the best estimate of the unknown parameter θ based on the observed data $Y_t = \{y(s), 0 \le s \le t\}$.

(5) *Basic assumptions for the mathematical security*: Besides the assumptions associated with Eqs. (16) and (18), the essential assumptions are made:

(H-1) The operator L_x is second-order elliptic with positive coefficient of the second-order partial derivatives.

(H-2) The coefficients of L_x and their first and second derivatives are continuous in $T \times \bar{D}$.

(H-3) The initial value $u_0(x)$ is Gaussian in x and this has at least up to the continuous second derivative.

(H-4) $$\int_0^T\int_D |G(t,x)|^2 dx\, dt < \infty \quad \text{and} \quad \sup_\theta \int_0^T\int_D |f(t,x,\theta)|^2 dx\, dt < \infty.$$

(6) *Some basic preliminaries*: Our first task is to express the system state equation corresponding to the i-th location of the unknown parameter θ. To do this, let $A^{(i)}$ be the event such that

$$A^{(i)} = \{\omega: \theta(\omega) = \theta_i\}, \quad (i = 1, 2, \cdots, K). \tag{21}$$

With the event $A^{(i)}$, Eq. (16) may be expressed by

$$A^{(i)} : du_i(t,x) = L_x u_i(t,x)dt + f(t,x,\theta_i)dt + G(t,x)dw(t,x) \tag{22a}$$

The associated initial and boundary conditions are

$$\text{I.C.} \qquad u_i(0,x) = u_0(x), \quad x \in D \tag{22b}$$

$$\text{B.C.} \qquad u_i(t,x) = 0, \qquad t \in T,\ x \in \partial D \tag{22c}$$

respectively.

With the event $A^{(i)}$, Eq. (18) can be expressed by

$$A^{(i)} : dy(t) = [\int_D H(t,z)u(t,z)dz]dt + R(t)dv(t). \tag{23}$$

An elegant way to derive the dynamics for parameter estimate is the introduction of Radon-Nikodym derivative. For this purpose, let $P_i (i = 1,2,\cdots,K)$ and P_0 denote respectively the measures induced in the space of continuous functions by the observation Y_t under $A^{(i)}$ and by the observation

$$A_0 : d\tilde{y}(t) = R(t)dv(t), \ \tilde{y}(0) = 0. \tag{24}$$

The Radon-Nikodym derivative is given by the following lemma (Kailath, 1969; Sunahara and others, 1973).

[Lemma 1] Let P_i and P_0 be the two measures induced by Eqs. (18) and (24) respectively. Then it follows that

(i) $P_i \ll P_0$, i.e., P_i is absolutely continuous with respect to P_0 (Skorokhod, 1965)

(ii) the Radon-Nikodym derivative of P_i with respect to P_0 is given by

$$\frac{dP_i}{dP_0} = \exp\left[\int_0^T \hat{h}_i(t,u_t)R^{-2}(t)dy(t) - \frac{1}{2}\int_0^T \hat{h}_i^{\,2}(t,u_t)R^{-2}(t)dt\right], \tag{25}$$

where u_t is included in the following definition;

$$\hat{h}_i(t,u_t) = E\{\int_D H(t,x)u(t,x)dx \mid Y_t, A^{(i)}\} = \int_D H(t,x)\hat{u}_i(t,x)dx \tag{26}$$

and $\hat{u}_i(t,x)$ is defined by

$$\hat{u}_i(t,x) = E\{u(t,x) \mid Y_t, A^{(i)}\}, \ (i = 1, 2, \cdots, K). \tag{27}$$

Thus, we need the state estimator equation to compute $\hat{u}_i(t,x)$. The equations comprising the state estimate under $A^{(i)}$ take the form (Kushner, 1970; Sunahara and others, 1973)

$$d\hat{u}_i(t,x) = L_x\hat{u}_i(t,x)dt + f(t,x,\theta_i)dt$$

$$+ \left[\int_D H(t,\xi)S_i(t,\xi,x)d\xi\right]R^{-2}(t)\left[dy(t) - \{\int_D H(t,\xi)\hat{u}_i(t,\xi)d\xi\}dt\right], \tag{28a}$$

$$\hat{u}_i(0,x) = \hat{u}_0(x), \tag{28b}$$

for all i, where $S_i(t,\xi,x)$ is the associated covariance defined by

$$S_i(t,\xi,x) \triangleq E\{[u(t,\xi) - \hat{u}_i(t,\xi)][u(t,x) - \hat{u}_i(t,x)] \mid Y_t, A^{(i)}\} \tag{29}$$

and this may be found from the solution to the partial differential equation,

$$\frac{\partial}{\partial t}S_i(t,\xi,x) = (L_\xi + L_x)S_i(t,\xi,x) + G(t,\xi)Q(\xi,x)G(t,x)$$

$$- \left[\int_D H(t,z)S_i(t,z,\xi)dz\right]R^{-2}(t)\left[\int_D H(t,z)S_i(t,z,x)dz\right], \tag{30}$$

with the condition $S_i(0,\xi,x) = S_0(\xi,x)$ for all i, so that $S_i(t,\xi,x) = S_j(t,\xi,x)$ for all i and j.

It is well-known that the likelihood-ratio function $\Lambda_i(T)$ is equal to

$$\Lambda_i(T) = \frac{p(Y_T \mid A^{(i)})}{p(\tilde{Y}_T \mid A^{(0)})} = \frac{dP_i}{dP_0}, \tag{31}$$

where $\tilde{Y}_T$ is the observation data obtained by Eq. (24), i.e., $\tilde{Y}_T = \{\tilde{y}(s),\ 0 \le s \le T\}$.

(7) *Version of the stochastic differential of the likelihood-ratio function*: Letting

$$\zeta_i(t) \triangleq \int_0^t \hat{h}_i(s,u_s)R^{-2}(s)dy(s) - \frac{1}{2}\int_0^t \hat{h}_i^{\,2}(s,u_s)R^{-2}(s)ds, \tag{32}$$

then, it is straightforward to write the stochastic differential of the $\zeta_i(t)$-process in the form,

$$d\zeta_i(t) = -\frac{1}{2}\hat{h}_i^2(t,u_t)R^{-2}(t)dt + \hat{h}_i(t,u_t)R^{-2}(t)dy(t). \tag{33}$$

From (32), since the likelihood-ratio function $\Lambda_i(T)$ given by (25) is expressed as

$$\Lambda_i(t) = \exp[\zeta_i(t)], \tag{34}$$

an application of Itô's chain rule to the function $\Lambda_i(t)$ brings us

$$d\Lambda_i = \frac{\partial\Lambda_i}{\partial t}dt + \frac{\partial\Lambda_i}{\partial\zeta_i}d\zeta_i + \frac{1}{2}\frac{\partial^2\Lambda_i}{\partial\zeta_i^2}(d\zeta_i)^2. \tag{35}$$

Simply, we have

$$\frac{\partial\Lambda_i}{\partial t} = 0, \qquad \frac{\partial\Lambda_i}{\partial\zeta_i} = \Lambda_i, \qquad \frac{\partial^2\Lambda_i}{\partial\zeta_i^2} = \Lambda_i \tag{36}$$

and, from Eq. (33),

$$(d\zeta_i)^2 = \hat{h}_i^2(t,u_t)R^{-2}(t)dt. \tag{37}$$

Combining (36) and (37) with (35), it follows that

$$d\Lambda_i(t) = \Lambda_i(t)\hat{h}_i(t,u_t)R^{-2}(t)dy(t), \quad \Lambda_i(0) = 1. \tag{38}$$

(8) *Choice of a criterion*: As the author already pointed out in Lemma-1, a simultaneous information of the state estimate is required. From this point of view, we define a vector β by $\beta \triangleq [u,\theta']'$, where "'" denotes its transpose and adopt the minimal conditional mean square performance criterion,

$$J(\hat{\beta}) \triangleq E\{[\beta(t,x) - \hat{\beta}(t,x)]'[\beta(t,x) - \hat{\beta}(t,x)]|Y_t\}. \tag{39}$$

It is easily seen that $J(\hat{\beta})$ defined by (39) is reduced to

$$J(\hat{\theta},\hat{u}) = E\{\|\theta - \hat{\theta}(t)\|^2|Y_t\} + E\{[u(t,x) - \hat{u}(t,x)]^2|Y_t\}, \tag{40}$$

where $\|\cdot\|$ is the Euclidean norm. The criterion (40) implies that the conditional mean square error with respect to both parameter identification and state estimation becomes minimal separately.

(9) *Identification technique*:

(A) Parameter identification: The parameter identification technique is concerned with the first term of the RHS in the criterion (40). We write the conditional probability and probability density associated with the event $A^{(i)}$ conditioned by Y_t by $P(A^{(i)}|Y_t)$ and $p(A^{(i)}|Y_t)$, respectively. From (20), it is apparent that

$$\hat{\theta}(t) \triangleq E\{\theta|Y_t\} = \sum_{i=1}^{K}\theta_i P(A^{(i)}|Y_t). \tag{41}$$

The *a posteriori* probability $P(A^{(i)}|Y_t)$ in (41) can be obtained by Bayes law in the form,

$$P(A^{(i)}|Y_t) = \frac{p(Y_t|A^{(i)})P(A^{(i)})}{\sum_{\nu=1}^{K} p(Y_t|A^{(\nu)})P(A^{(\nu)})}, \tag{42}$$

where $P(A^{(i)})$ is the *a priori* probability appeared in (7) and $P(A^{(i)}) = P(\theta_i)$ from (21). Introducing the modified likelihood-ratio function $\Lambda_{\nu i}(t)$ defined by

$$\Lambda_{\nu i}(t) = \frac{p(Y_t|A^{(\nu)})}{p(Y_t|A^{(i)})}, \quad (i, \nu = 1, 2, \cdots, K) \tag{43}$$

and setting as

$$\alpha_{\nu i} = \frac{P(A^{(\nu)})}{P(A^{(i)})}, \tag{44}$$

the *a posteriori* probability given by (42) yields

$$P(A^{(i)}|Y_t) = [\sum_{\nu=1}^{K} \alpha_{\nu i}\Lambda_{\nu i}(t)]^{-1} \triangleq M_i(t). \tag{45}$$

Combining this with (41), we have

$$\hat{\theta}(t) = \sum_{i=1}^{K} \theta_i M_i(t). \tag{46}$$

For computer implementation, the differential version of $\hat{\theta}(t)$ is useful and this is obtained to be

$$d\hat{\theta}(t) = \sum_{i=1}^{K} \theta_i dM_i(t). \tag{47}$$

Hence, we make a short excursion to compute the $dM_i(t)$-function. This starts with an application of the Itô's chain rule to (45). To make the description more compact, the result is given by the following theorem.

[Theorem 1] The sample process of the $M_i(t)$-process is determined by

$$\begin{aligned} dM_i(t) = &- \sum_{\nu=1}^{K} \alpha_{\nu i}\Lambda_{\nu i}(t)M_i^2(t)\{\hat{h}_\nu(t,u_t) - \hat{h}_i(t,u_t)\}R^{-2}(t)\{dy(t) - \hat{h}_i(t,u_t)dt\} \\ &+ \sum_{\nu=1}^{K} \sum_{k=1}^{K} \alpha_{\nu i}\alpha_{ki}\Lambda_{\nu i}(t)\Lambda_{ki}(t)M_i^3(t) \\ &\times \{\hat{h}_\nu(t,u_t) - \hat{h}_i(t,u_t)\}\{\hat{h}_k(t,u_t) - \hat{h}_i(t,u_t)\}R^{-2}(t)dt \end{aligned} \tag{48}$$

where $i = 1, 2, \cdots, K$.

Before proving Theorem-1, we need the following lemma which determines the sample process of the $\Lambda_{\nu i}(t)$-process in (48).

[Lemma 2] The modified likelihood-ratio function $\Lambda_{\nu i}(t)$ defined by (43) is determined by

$$d\Lambda_{\nu i}(t) = \Lambda_{\nu i}(t)\{\hat{h}_\nu(t,u_t) - \hat{h}_i(t,u_t)\}R^{-2}(t)\{dy(t) - \hat{h}_i(t,u_t)dt\} \tag{49a}$$

$$\Lambda_{\nu i}(0) = 1 \qquad \text{for } i, \nu = 1, 2, \cdots, K. \tag{49b}$$

Proof of Lemma 2: Noting that from (31) and (43),

$$\Lambda_{\nu i}(t) = \frac{\Lambda_\nu(t)}{\Lambda_i(t)} \tag{50}$$

and using (25), we have

$$\Lambda_{\nu i}(t) = \exp[\int_0^t [\hat{h}_\nu(s,u_s) - \hat{h}_i(s,u_s)]R^{-2}(s)dy(s)$$

$$- \frac{1}{2}\int_0^t [\hat{h}_\nu{}^2(s,u_s) - \hat{h}_i{}^2(s,u_s)]R^{-2}(s)ds]. \tag{51}$$

It is a direct consequence that $\Lambda_{\nu i}(t)$ given by (51) has the stochastic differential,

$$d\Lambda_{\nu i}(t) = \Lambda_{\nu i}(t)[\exp\{(\hat{h}_\nu(t,u_t) - \hat{h}_i(t,u_t)]R^{-2}(t)dy(t)$$

$$- \frac{1}{2}(\hat{h}_\nu{}^2(t,u_t) - \hat{h}_i{}^2(t,u_t))R^{-2}(t)dt\} - 1]. \tag{52}$$

Expanding the exponential function in (52) and deleting the terms of higher order than $(dt)^{3/2}$, the final result can be obtained.

Proof of Theorem 1: By applying the Itô's chain rule to (45), we have

$$dM_i(t) = \frac{\partial M_i}{\partial t}dt + \sum_{\nu=1}^{K} \frac{\partial M_i}{\partial \Lambda_{\nu i}} \frac{\partial \Lambda_{\nu i}}{\partial \zeta_{\nu i}} d\zeta_{\nu i} + \frac{1}{2}\sum_{\nu=1}^{K} \sum_{k=1}^{K} \frac{\partial^2 M_i}{\partial \zeta_{\nu i}\partial \zeta_{ki}} d\zeta_{\nu i} d\zeta_{ki} \tag{53}$$

where, from (51), the $\zeta_{\nu i}(t)$-process is defined by

$$\zeta_{\nu i}(t) = \int_0^t [\hat{h}_\nu(s,u_s) - \hat{h}_i(s,u_s)]R^{-2}(s)dy(s)$$

$$- \frac{1}{2}\int_0^t [\hat{h}_\nu{}^2(s,u_s) - \hat{h}_i{}^2(s,u_s)]R^{-2}(s)ds \tag{54}$$

and this has the stochastic differential

$$d\zeta_{\nu i}(t) = - \frac{1}{2}(\hat{h}_\nu{}^2 - \hat{h}_i{}^2)R^{-2}dt + (\hat{h}_\nu - \hat{h}_i)R^{-2}dy. \tag{55}$$

In (53), it can be shown from (45) that

$$\frac{\partial M_i}{\partial t} = 0, \qquad \frac{\partial M_i}{\partial \Lambda_{\nu i}} = - \alpha_{\nu i}M_i{}^2, \qquad \frac{\partial \Lambda_{\nu i}}{\partial \zeta_{\nu i}} = \Lambda_{\nu i} \tag{56a}$$

and

$$\left.\begin{aligned} \frac{\partial^2 M_i}{\partial \zeta_{\nu i}\partial \zeta_{ki}} &= - \alpha_{\nu i}\Lambda_{\nu i}M_i{}^2 + 2\alpha_{\nu i}{}^2\Lambda_{\nu i}{}^2M_i{}^3 && \text{for } \nu = k \\ &= 2\alpha_{\nu i}\alpha_{ki}\Lambda_{\nu i}\Lambda_{ki}M_i{}^3 && \text{for } \nu \neq k. \end{aligned}\right\} \tag{56b}$$

Substituting (55) and (56) into (53), we obtain

$$dM_i = - \sum_{\nu=1}^{K} \alpha_{\nu i}\Lambda_{\nu i}M_i{}^2 d\zeta_{\nu i} - \frac{1}{2}\sum_{\nu=1}^{K} \alpha_{\nu i}\Lambda_{\nu i}M_i{}^2(d\zeta_{\nu i})^2$$

$$+ \sum_{\nu=1}^{K} \sum_{k=1}^{K} \alpha_{\nu i}\alpha_{ki}\Lambda_{\nu i}\Lambda_{ki}M_i{}^3(d\zeta_{\nu i})(d\zeta_{ki}). \tag{57}$$

It is easy to obtain from (55) that

$$(d\zeta_{\nu i})^2 = (\hat{h}_\nu - \hat{h}_i)^2 R^{-2} dt \tag{58a}$$

and

$$(d\zeta_{\nu i})(d\zeta_{ki}) = (\hat{h}_\nu - \hat{h}_i)(\hat{h}_k - \hat{h}_i)R^{-2}dt. \tag{58b}$$

Hence, the substitution of (55) and (58) into (57) yields (48).

(B) State estimate: The optimal estimate $\hat{u}(t,x)$ is assumed to be defined formally by

$$\hat{u}(t,x) = \int_{-\infty}^{\infty} up(t,x,u|Y_t)du. \tag{59}$$

Bearing (21) in mind, the conditional probability density in (59) becomes

$$p(t,x,u|Y_t) = \sum_{i=1}^{K} p(t,x,u|Y_t,A^{(i)})P(A^{(i)}|Y_t). \tag{60}$$

Hence, the optimal state estimate $\hat{u}$ by (59) is computed to be

$$\hat{u}(t,x) = \sum_{i=1}^{K} P(A^{(i)}|Y_t)\int_{-\infty}^{\infty} up(t,x,u|Y_t,A^{(i)})du. \tag{61}$$

Furthermore, introducing the definitions (21) and (45), the optimal state estimate is simply given by

$$\hat{u}(t,x) = \sum_{i=1}^{K} M_i(t)\hat{u}_i(t,x), \tag{62}$$

where the i-th optimal state estimate $\hat{u}_i$ may be computed by Eq. (28).

The error covariance associated with the optimal state estimate $\hat{u}$ is defined by

$$\begin{aligned} S(t,x,z) &\triangleq E\{[u(t,x) - \hat{u}(t,x)][u(t,z) - \hat{u}(t,z)]|Y_t\} \\ &= \sum_{i=1}^{K} E\{u(t,x)u(t,z)|Y_t,A^{(i)}\}P(A^{(i)}|Y_t) - \hat{u}(t,x)\hat{u}(t,z). \end{aligned} \tag{63}$$

In (63), recalling (27), it follows that

$$E\{u(t,x)u(t,z)|Y_t,A^{(i)}\} = S_i(t,x,z) + \hat{u}_i(t,x)\hat{u}_i(t,z). \tag{64}$$

Hence, from (45) and (64), the error covariance is finally given by

$$S(t,x,z) = \sum_{i=1}^{K} [S_i(t,x,z) + \hat{u}_i(t,x)\hat{u}_i(t,z)]M_i(t) - \hat{u}(t,x)\hat{u}(t,z) \tag{65a}$$

with

$$\text{I.C. } S(0,x,z) = \sum_{i=1}^{K} [S_i(0,x,z) + \hat{u}_i(0,x)\hat{u}_i(0,z)]M_i(0) - \hat{u}(0,x)\hat{u}(0,z). \tag{65b}$$

An entire aspect of computing the identification is performed by use of Eqs. (28), (30), (49a) and (63). Their preassigned initial conditions are $E\{u_i(0,x)\} = \hat{u}_i(0,x)$ for (28), $S_i(0,\xi,x)$ for (30), (49a) and (65b) for (65a).

(10) *A digital simulation experiment*:

(A) System model: As an example, consider the one-dimensional distributed parameter system which has an input generated at the spatial point x = θ so that

$$f(t,x,\theta) = c\tilde{\delta}(x - \theta), \tag{66}$$

where c is a known constant. For the reduced spatial domain $x \in]0, 1[$ and $t \in]0, t_f[$, the mathematical model is given by

$$du(t,x) = [B\frac{\partial^2 u(t,x)}{\partial x^2}]dt + c\tilde{\delta}(x - \theta)dt + Gdw(t,x) \tag{67a}$$

with the initial and boundary conditions,

$$\text{I.C.} \qquad u(0,x) = A\sin^2 \pi x, \quad x \in]0, 1[\tag{67b}$$

$$\text{B.C.} \qquad u(t,x) = 0, \qquad x = 0, 1 \tag{67c}$$

where A, B, c and G are all constants. Naturally, the true system state u(t,x) is, in practice, inaccessible. However, for the purpose of comparative discussions with sample runs of the unknown parameter and state estimates, Eqs. (67) were simulated on a digital computer with the numerical values of the known parameters set as A = 1.0, B = 1.0, c = 500 and G = 0.45, where partitions of the spatial region and the time intervals were Δx = 1/24 and Δt = 0.0005, respectively. In simulation studies, investigators are free to choose the number of numerical classes of unknown parameter

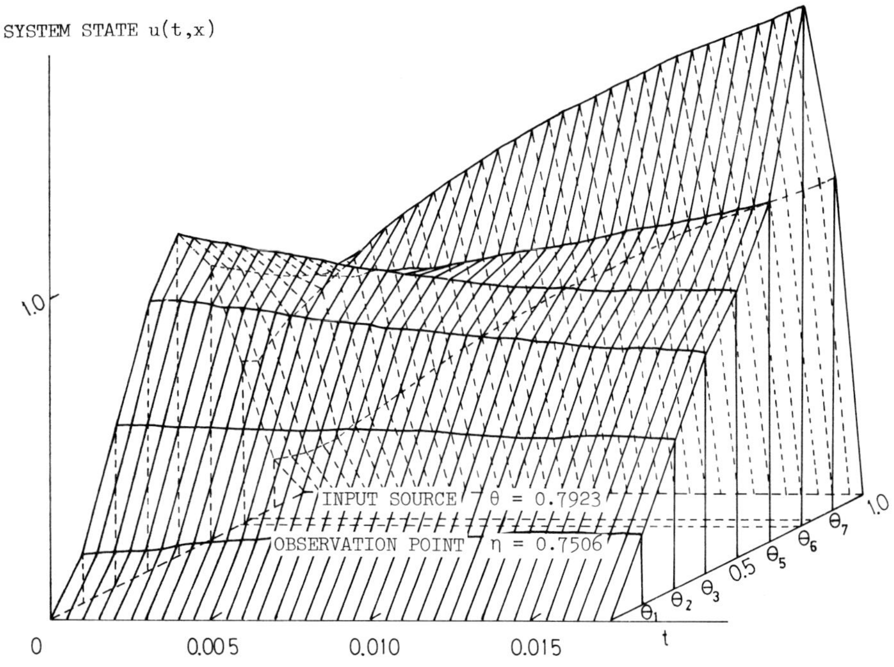

Fig. 3. A sample run of the true system state u(t,x) determined by Eq. (67) with the value of θ = 0.7923.

θ. A particular choice of θ_i's depends on the situation of the problems which are being considered. In the simulation experiment presented here, the value of the parameter θ was set as $\theta = 0.7923$. A choice of $K = 7$ in (21) was given in the simulation experiments. The value of θ_i in (19) was taken as $\theta_i = 1/8$, where $i = 1, 2, \cdots, 7$. Figure 3 shows a sample run of the system state $u(t,x)$ determined by Eq. (67) with the initial and boundary conditions $u(0,x) = \sin^2\pi x$ and $u(t,0) = u(t,1) = 0$.

(B) <u>Observation mechanism</u>: The observation mechanism is given by

$$dy(t) = [\int_0^1 H\delta(z - \eta)u(t,z)dz]dt + Rdv, \quad y(0) = 0, \tag{68}$$

where the symbol $\delta(\cdot)$ is the Dirac delta function, both H and R are respectively constants and η is the preassigned location of the sensor. Observations were currently made with respect to time, where three trials were made on the choice of sensor's location, i.e., $\eta = 0.6255$, 0.7506 and 0.8757. The values of H and R were set as $H = 4.0$ and $R = 0.2$ respectively. Furthermore, we assume that $Q(x,z) = \tilde{\delta}(x - z)$.

(C) <u>State estimator</u>: By using Eqs. (28) and (29), the dynamics of the state estimator and the error covariance can be obtained as

$$d\hat{u}_i(t,x) = [B\frac{\partial^2}{\partial x^2}\hat{u}_i(t,x)]dt + c\tilde{\delta}(x - \theta_i)dt + HS_i(t,\eta,x)R^{-2}\{dy(t) - H\hat{u}_i(t,\eta)dt\} \tag{69}$$

and

$$\frac{\partial}{\partial t}S_i(t,x,z) = B(\frac{\partial^2}{\partial x^2} + \frac{\partial^2}{\partial z^2})S_i(t,x,z) + G^2\tilde{\delta}(x - z) - H^2R^{-2}S_i(t,\eta,x)S_i(t,\eta,z). \tag{70}$$

In this example, the initial values of (69) and (70) were respectively set as

$$\hat{u}_i(0,x) = 0 \text{ and } S_i(0,x,z) = \sin^2\pi x \sin^2\pi z. \tag{71}$$

(D) <u>Parameter estimator</u>: The modified likelihood-ratio function (52) can be obtained as

$$d\Lambda_{\nu i}(t) = \Lambda_{\nu i}(t)H\{\hat{u}_\nu(t,\eta) - \hat{u}_i(t,\eta)\}R^{-2}\{dy(t) - H\hat{u}_i(t,\eta)dt\} \tag{72}$$

with $\Lambda_{\nu i}(0) = 1$. For the convenience of simulation experiments, we assume that, in (44), the *a priori* probability of the event $A^{(i)}$ is uniformly distributed, i.e., $P(A^{(i)}) = P(A^{(\nu)})$. According to this assumption, we have $\alpha_{\nu i} = 1$ for all i, ν. Furthermore, the definition (45) yields

$$M_i(t) = [\sum_{\nu=1}^{K} \Lambda_{\nu i}(t)]^{-1}. \tag{73}$$

The optimal estimate $\hat{\theta}(t)$ can thus be computed by combining (46) with (73) or by (47) and

$$\begin{aligned} dM_i(t) = &- \sum_{\nu=1}^{K} \Lambda_{\nu i}(t)M_i^2(t)H\{\hat{u}_\nu(t,\eta) - \hat{u}_i(t,\eta)\}R^{-2}\{dy(t) - H\hat{u}_i(t,\eta)dt\} \\ &+ \sum_{\nu=1}^{K}\sum_{k=1}^{K} \Lambda_{\nu i}(t)\Lambda_{ki}(t)M_i^3(t) \\ &\times H^2\{\hat{u}_\nu(t,\eta) - \hat{u}_i(t,\eta)\}\{\hat{u}_k(t,\eta) - \hat{u}_i(t,\eta)\}R^{-2}dt. \end{aligned} \tag{74}$$

Equations (69), (70), (47) and (74) are simultaneously simulated on a digital com-

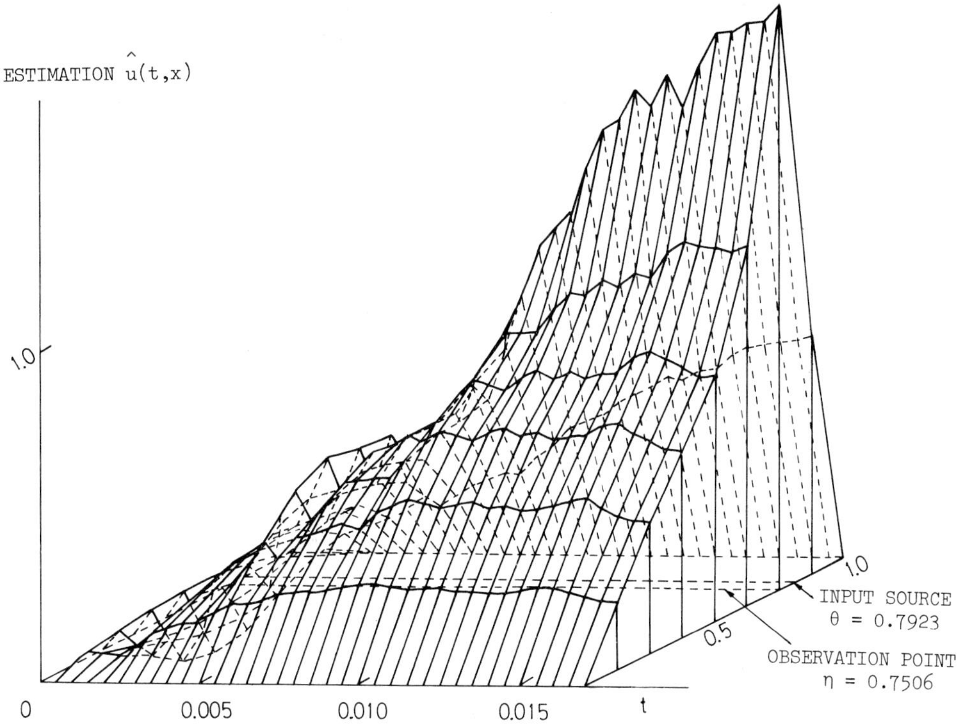

Fig. 4. A sample run of the state estimate $\hat{u}(t,x)$ determined by Eq. (69) with Eq. (70) as its error covariance.

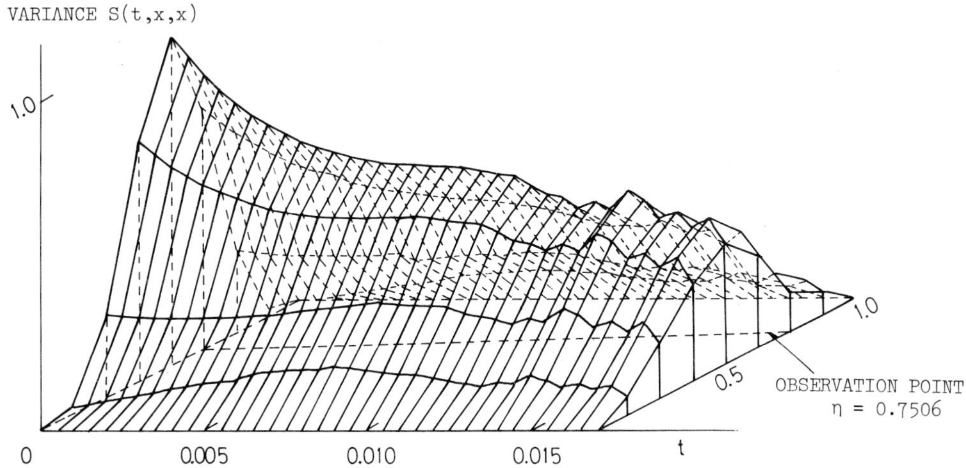

Fig. 5. The S(t,x,x)-run.

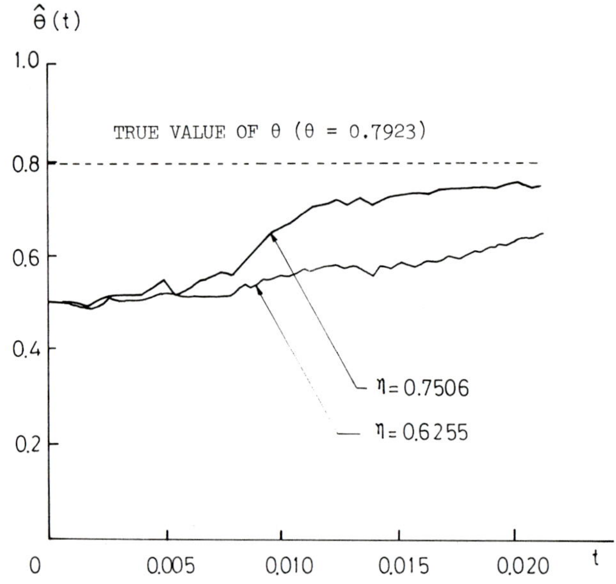

Fig. 6. Sample runs of the parameter estimate $\hat{\theta}(t)$.

puter. Figure 4 depicts a sample run of the system state estimate $\hat{u}(t,x)$ in the case where $\eta = 0.7506$ with the initial condition given by (71), while Fig. 5 shows that of the associated error covariance $S(t,x,x)$. Two sample runs of the θ-process are demonstrated in Fig. 6 corresponding to the two kinds of sensor's location, i.e., $\eta = 0.7506$ and $\eta = 0.6255$.

Example 2: Parameter Identification Involved in the System Dynamics of Parabolic Type - Stochastic Inverse Problem

The second example is motivated by the problem of estimating the temperature distribution of a heating system and is one of typical examples which attribute to the stochastic inverse problems.

(1) *Mathematical model*: Consider a physical system modelled by the following one-dimensional heat equation:

$$\frac{\partial u(t,x)}{\partial t} - \frac{\partial}{\partial x}\left[a(x)\frac{\partial u(t,x)}{\partial x}\right] = 0, \qquad (t,x) \in T =]0,\, t_f[\times D =]0,\, n[\qquad (75a)$$

with

$$\text{I.C.} \qquad u(0,x) = u_0(x) \qquad x \in D \qquad (75b)$$

$$\text{B.C.} \qquad -a(0)\frac{\partial u(t,0)}{\partial x} = g(t) \quad \text{and} \quad u(t,n) = 0, \qquad t \in T \qquad (75c)$$

where n is a positive integer, $g(t)$ is a non-zero time function which is assumed to be known and $a(x)$ is an unknown parameter which depends on the spatial variable and this is to be determined.

(2) *Observation methods*: Observations of the system are assumed to be made at n

discrete locations in the spatial domain [0, n]. From practical viewpoints, we consider the following two types of observation mechanisms:

(Type-1) Spatially distributed observations: We shall consider a finite number of partitioned domain in $D =]0, n[$ by defining D_i and ∂D_i respectively as

$$D_i \triangleq \{x | i - 1 < x < i;\ i = 1, 2, \cdots, n\} \tag{76a}$$

and

$$\partial D_i \triangleq \{x | x = i;\ i = 0, 1, \cdots, n\}. \tag{76b}$$

Furthermore, as shown in Fig. 7(a), set the sub-region D_i^0 in the region D_i, i.e., $D_i^0 \subset D_i$. In each sub-region D_i^0, the system output is taken as

$$[h_1\int_{D_1^0} u(t,x)dx,\ h_2\int_{D_2^0} u(t,x)dx,\ \cdots,\ h_n\int_{D_n^0} u(t,x)dx]', \tag{77}$$

where h_i $(i = 1, 2, \cdots, n)$ are known constants. There is another possibility in which two subregions D_{2i-1}^0 and D_{2i}^0 are considered in the alternative region D_{2i-1}, i.e., D_{2i-1}^0, $D_{2i}^0 \subset D_{2i-1}$. Besides these two methods for acquiring the output data, various kinds of setting subregions are considered. However, our attention is placed on the method expressed by (77) in the sequel.

(Type-2) Pointwise observations: Set a spatial point x_i^p as shown in Fig. 7(b), i.e., $x_i^p \in D_i$. Then, the system output is taken as

$$[h_1\int_D \delta(x - x_1^p)u(t,x)dx,\ h_2\int_D \delta(x - x_2^p)u(t,x)dx,\ \cdots,\ h_n\int_D \delta(x - x_n^p)u(t,x)dx]'. \tag{78}$$

Naturally, it is possible to set two spatial points x_{2i-1}^p and x_{2i}^p in the alternative region D_{2i-1}, i.e., x_{2i-1}^p, $x_{2i}^p \in D_{2i-1}$.

A joint expression of (77) with (78) might be convenient for further theoretical development, i.e., denoting the observation process by $y(t)$, we express

$$y(t) = \int_0^t Hu(s)ds + v(t), \tag{79}$$

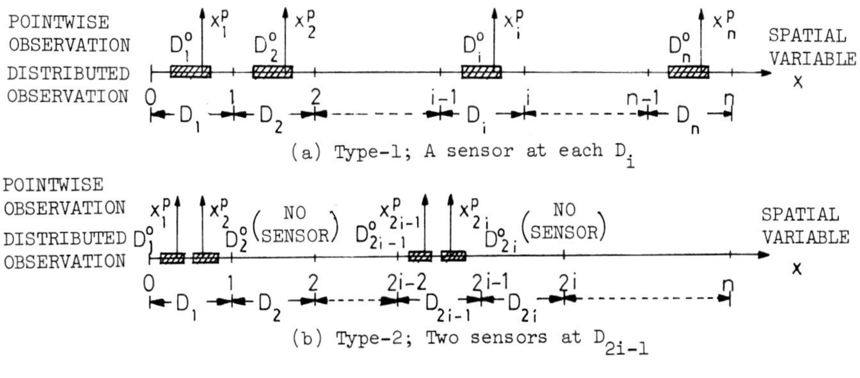

(a) Type-1; A sensor at each D_i

(b) Type-2; Two sensors at D_{2i-1}

Fig. 7. An illustration of sensors' allocation.

where

$$H(\cdot) = [H_1(\cdot), H_2(\cdot), \cdots, H_n(\cdot)]' \tag{80a}$$

$$H_i(\cdot) = \begin{cases} h_i \int_{D_i^0} (\cdot)dx & (i = 1, 2, \cdots, n) \\ h_i \int_D \delta(x - x_i^p)(\cdot)dx & (i = 1, 2, \cdots, n) \end{cases} \tag{80b}$$

and $v(t)$ is the n-dimensional standard Brownian motion process.

(3) *Assumptions associated with an unknown parameter and other known functions*: In Eqs. (75), the following assumptions are made:

(H-5) $a \in L^\infty(D)$ and $a(x) \geq \alpha > 0$ for $\alpha > 0$, $\forall x \in D$

(H-6) $u_0 \in L^2(D)$

(H-7) $g \in L^2(T)$

(4) *Mathematical security by calculus in function spaces*: A precise formulation of Eqs. (75) is performed by considering its solution at time t as an element of a function space $H^1(D)$. Before developing mathematical aspects, a couple of symbolic conventions are introduced below:

H :	Space of square integrable functions on D, i.e., $L^2(D)$
H^1 :	Sobolev space of order 1
$(\cdot,\cdot)_*$, $\|\cdot\|_*$:	An inner product and a norm in the space "*", in particular, we set $(\cdot,\cdot)_H = (\cdot,\cdot)$ and $\|\cdot\|_H = \|\cdot\|$, respectively
$L^2(X;Y)$:	Space of the function f defined on the space X with the value in the space Y and $\int_X \|f(\cdot)\|_Y^2 dx < \infty$
$C(\bar{T};H)$:	Space of continuous functions on $[0,t_f]$ with the value in H.

We introduce the following Hilbert space

$$V = \{\phi | \phi \in H^1(D),\ \phi(n) = 0\}. \tag{81}$$

If we identify $H = L^2(D)$ with its dual space H', then we have the inclusion, $V \subset H = H' \subset V'$, where V' denotes the dual space of V.

The operator $A(a)$ is defined by

$$\langle A(a)\phi,\ \psi\rangle \triangleq \int_0^n a(x) \frac{\partial\phi(x)}{\partial x} \frac{\partial\psi(x)}{\partial x} dx \tag{82}$$

and this is a linear continuous mapping from V into V'. With the definition (82), Eqs. (75) can be represented as the following integral equation in the variational form:

$$(u(t), \phi) + \int_0^t \langle A(a)u(s), \phi\rangle ds - \int_0^t (g(s), \phi)_{\Gamma_0} ds = (u_0, \phi) \text{ for } {}^\forall\phi \in V \tag{83}$$

where $(\phi, \psi)_{\Gamma_0} = \phi(0)\psi(0)$.

Existence of a unique continuous solution to Eq. (83) is guaranteed by the following theorem.

[Theorem 3] (Lions, 1971) With (H-5) to (H-7), there exists a unique continuous solution to Eq. (83) such that

$$u \in L^2(T;V) \cap C(\bar{T};H).$$

The following theorem states the existence of a unique stationary solution to Eq. (83).

[Theorem 4] Suppose that

$$\text{(H-8)} \qquad \lim_{t\to\infty} g(t) = \bar{g}.$$

With (H-5) to (H-8), there exists a unique stationary solution, $\lim_{t\to\infty} u(t) = \bar{u}$, to Eq. (83), satisfying

$$\langle A(a)\bar{u}, \phi\rangle - (\bar{g}, \phi)_{\Gamma_0} = 0 \qquad \text{for } ^\forall\phi \in V. \tag{84}$$

Proof: Let $\bar{u}$ be the solution to Eq. (84). Introduce the new process $\xi(t)$ defined by $\xi(t) \triangleq u(t) - \bar{u}$. Hence, from Eqs. (83) and (84), the function $\xi(t)$ satisfies the following variational form:

$$(\xi(t), \phi) + \int_0^t \langle A(a)\xi(s), \phi\rangle ds = (\xi_0, \phi) + \int_0^t (\tilde{g}(s), \phi)_{\Gamma_0} ds, \tag{85}$$

where

$$\xi_0 \triangleq u_0 - \bar{u} \qquad \text{and} \qquad \tilde{g}(t) = g(t) - \bar{g}. \tag{86}$$

Using the differential formula with respect to $\|\xi(t)\|^2$, we obtain

$$\|\xi(t)\|^2 + 2\int_0^t \langle A(a)\xi(s), \xi(s)\rangle ds = \|\xi_0\|^2 + 2\int_0^t (\tilde{g}(s), \xi(s))_{\Gamma_0} ds. \tag{87}$$

On the other hand, from (H-5) and the definition (82), the following coercivity condition holds:

$$\langle A(a)\psi, \psi\rangle \geq \alpha\int_0^n \left|\frac{\partial\psi(x)}{\partial x}\right|^2 dx = \alpha\|\psi\|_V^2 \geq 0. \tag{88}$$

Combining this with (87), and using the Poincare's inequality, it follows that, for any $\varepsilon > 0$,

$$\begin{aligned} \|\xi(t)\|^2 + 2\alpha\int_0^t \|\xi(s)\|_V^2 ds &\leq \|\xi_0\|^2 + \varepsilon\int_0^t \|\xi(s)\|_{\Gamma_0}^2 ds + \frac{1}{\varepsilon}\int_0^t |\tilde{g}(s)|^2 ds \\ &\leq \|\xi_0\|^2 + \varepsilon n\int_0^t \|\xi(s)\|_V^2 ds + \frac{1}{\varepsilon}\int_0^t |\tilde{g}(s)|^2 ds. \end{aligned} \tag{89}$$

Choosing some $\varepsilon > 0$ as $\tilde{\alpha} = 2\alpha - \varepsilon n > 0$, from the inequality (89), we have

$$\|\xi(t)\|^2 + \tilde{\alpha}\int_0^t \|\xi(s)\|_V^2 ds \leq \|\xi_0\|^2 + C_1\int_0^t |\tilde{g}(s)|^2 ds, \tag{90}$$

where $C_1 \equiv 1/\varepsilon$. Applying again the Poincare's inequality, $\sqrt{2}\|\xi(s)\|/n \leq \|\xi(s)\|_V$, to the second term of the LHS in (90), we obtain

$$\|\xi(t)\|^2 \leq \|\xi_0\|^2 + C_1\int_0^t |\tilde{g}(s)|^2 ds - C_2\int_0^t \|\xi(s)\|^2 ds, \tag{91}$$

where $C_2 \equiv 2\tilde{\alpha}/n^2$. Using the Gronwall's inequality, the inequality (91) yields

$$\| \xi(t) \|^2 \leq \| \xi_0 \|^2 \exp(-C_2 t) + C_1 \int_0^t |\tilde{g}(s)|^2 \exp\{-C_2(t-s)\} ds. \tag{92}$$

Since

$$\lim_{t\to\infty} \frac{C_1 \int_0^t |\tilde{g}(s)|^2 \exp(C_2 s) ds}{\exp(C_2 t)} = \frac{C_1}{C_2} \lim_{t\to\infty} |\tilde{g}(t)|^2 = 0, \tag{93}$$

it can easily be shown that

$$\lim_{t\to\infty} \| \xi(t) \|^2 = 0 \tag{94}$$

and this implies that

$$\lim_{t\to\infty} u(t) = \bar{u} \quad \text{in } H. \tag{95}$$

(5) *Some basic preliminaries and a class of admissible parameters*: Along the partitioned state D_i, the unknown parameter $a(x)$ is assumed to be

(H-9) $$a(x) = a_{i+1} \text{ for } i \leq x \leq i+1, \tag{96}$$

where $i = 0, 1, 2, \cdots, n-1$. We further assume that

(H-10) $$0 < \alpha \leq a_i \leq \beta < \infty \quad \text{for } \forall i. \tag{97}$$

Hence, we define the admissible class of the unknown parameters by

$$\mathcal{U}_{ad} = \{a | a(x) = a_{i+1} \text{ for } i \leq x < i+1\}. \tag{98}$$

In what follows, our attention is placed on the aspect of the stationary state of $\partial u(t)/\partial a_i \triangleq u_{ai}(t)$ $(i = 1, 2, \cdots, n)$. First, the $u_{ai}(t)$-process is determined by the equation shown in the following proposition.

[Proposition 1] The $u_{ai}(t)$-process is determined by

$$(u_{ai}, \phi) + \int_0^t \{\langle A(a)u_{ai}(s), \phi\rangle + (\chi_i \frac{\partial u(s)}{\partial x}, \frac{\partial \phi}{\partial x})\} ds = 0 \tag{99}$$
$$(i = 1, 2, \cdots, n)$$

for $\forall \phi \in V$, where χ_i is a characteristic function in $]i-1, i[$ and

$$u_{ai} \in L^2(T;V) \cap C(\bar{T};H). \tag{100}$$

[Proposition 2] There exists a stationary solution $\bar{u}_{ai}$ to Eq. (99) satisfying

$$\langle A(a)\bar{u}_{ai}, \phi\rangle + (\chi_i \frac{\partial \bar{u}}{\partial x}, \frac{\partial \phi}{\partial x}) = 0. \tag{101}$$

for $\bar{u}_{ai} \in V$.

Proofs of Propositions 1 and 2 are easily performed from Theorem 5.

Let the stationary state $\bar{u}_i$ $(i = 1, 2, \cdots, n)$ be the restriction of $\bar{u}(x)$ in the domain D_i defined by (76a). Then, the stationary version of Eq. (84) becomes

$$\frac{d^2\bar{u}_i(x)}{dx^2} = 0 \qquad \text{in } D_i \tag{102}$$

with the conditions on ∂D_i such that

$$\bar{u}_i(x)\Big|_{\partial D_i} = \bar{u}_{i+1}(x)\Big|_{\partial D_i} \qquad \text{for } i = 1, 2, \cdots, n-1 \tag{103a}$$

and

$$a_i \frac{d\bar{u}_i(x)}{dx}\Bigg|_{\partial D_i} = a_{i+1}\frac{d\bar{u}_{i+1}(x)}{dx}\Bigg|_{\partial D_i} \qquad \text{for } i = 1, 2, \cdots, n-1. \tag{103b}$$

The states $\bar{u}_i$ have the boundary conditions

$$-a_1\frac{d\bar{u}_1(x)}{dx}\Bigg|_{\partial D_0} = \bar{g} \qquad \text{in } D_0 \tag{104a}$$

and

$$-\bar{u}_n(x)\big|_{\partial D_n} = 0 \qquad \text{in } D_n. \tag{104b}$$

Solving Eq. (102) with the conditions given by Eqs. (103) and (104), we obtain

$$\bar{u}_i(x) = \{(i - x)\frac{1}{a_i} - \sum_{k=i+1}^{n}\frac{1}{a_k}\}\bar{g} \tag{105}$$

from which we have

$$\bar{u}(x) = \sum_{i=1}^{n}\chi_i\bar{u}_i(x). \tag{106}$$

(6) *Identification technique*:

(A) Maximum likelihood estimator: The problem is to estimate an unknown parameter vector $a^* = [a_1, a_2, \cdots, a_n]'$ in (98). For the purpose of deriving the estimator, we shall first compute the Radon-Nikodym derivative.

Let μ_1 and μ_2 be the measures on $C(\bar{T}\,;R^n)$ by the y(t)-process determined by (79) and by the v(t)-process such that $\hat{y}(t) = v(t)$ respectively. Since the derivation is quite lengthy and the method is identical to that used in the reference (Sunahara and others, 1981), we shall proceed to show directly the result. From Girsanov's theorem (1960), the Radon-Nikodym derivative is obtained to be

$$\frac{d\mu_1}{d\mu_2}(t) = \exp[\int_0^t [Hu(s)]'dy(s) - \frac{1}{2}\int_0^t [Hu(s)]'Hu(s)ds] \tag{107}$$

and this equals to the likelihood-ratio function $\Lambda(a^*;t,y)$. The identification algorithm is to seek a root of the gradient of the logarithmic likelihood-ratio function with respect to a^*. Thus the maximum likelihood estimator satisfies

$$\nabla_{a^*}\log\Lambda(a^*;t,y)\Big|_{a^* = \hat{a}^*(t)} = 0, \tag{108}$$

where $\hat{a}^*(t) = [\hat{a}_1(t), \hat{a}_2(t), \cdots, \hat{a}_n(t)]'$ and

$$\nabla_{a^*} = [\partial/\partial a_1, \partial/\partial a_2, \cdots, \partial/\partial a_n]'. \tag{109}$$

From Eq. (107), Eq. (108) can be represented by

$$\left[\int_0^t \nabla_{a*}[Hu(s)]'dy(s) - \int_0^t [\nabla_{a*}[Hu(s)]']Hu(s)ds\right]_{a* = \hat{a}*(t)} = 0. \quad (110)$$

The following theorem states that the maximum likelihood estimate derived here is the consistent estimate (in strong sense) of $a(x) \in \mathcal{U}_{ad}$ under certain conditions associated with the sensors' allocation.

[Theorem 5] If the allocation of sensors satisfies either Type-1 or Type-2, then

$$\lim_{t\to\infty} \hat{a}*(t) = a* \quad \text{w.p.1.} \quad (111)$$

Proof: Following Balakrishnan (1973), it suffices to show that

$$\lim_{t\to\infty} \frac{1}{t^2}\int_0^t [H(\nabla_{a*}u(s))']^2 ds\Big|_{a* = \hat{a}*(t)} = 0 \quad (112)$$

and that the Fisher's information matrix $\bar{Q}(a*)$ is positive definite, that is,

$$\bar{Q}(a*) = \lim_{t\to\infty} \frac{1}{t}\int_0^t \{\nabla_{a*}(Hu(s))'(\nabla_{a*}(Hu(s))')'\}ds \geq \lambda I > 0, \quad (113)$$

where $[\cdot]^2$ denotes the matrix norm and where I is an $n \times n$ identity matrix.

From Theorem 4, we have

$$\lim_{t\to\infty} \nabla_{a*}(Hu(t))' = \nabla_{a*}(H\bar{u})'. \quad (114)$$

Using (114), the LHSs of Eqs. (112) and (113) become

$$\lim_{t\to\infty} \frac{1}{t^2}\int_0^t [H(\nabla_{a*}u(s))']^2 ds\Big|_{a* = \hat{a}*(t)} = \lim_{t\to\infty} \frac{\bar{Q}(a*)}{2t} \quad (115)$$

and

$$\bar{Q}(a*) = [\nabla_{a*}(H\bar{u})'][\nabla_{a*}(H\bar{u})']'. \quad (116)$$

Hence, in order to prove the equalities (112) and (113), it is sufficient to show that the matrix $\bar{Q}(a*)$ given by (116) is positive definite in the sensors' allocation of both Types 1 and 2.

For example, let us adopt the allocation of Type-2. From (78) and (105), since

$$H_i\bar{u}(x) = h_i\bar{g}\{(i - x_i^p)\frac{1}{a_i} - \sum_{k=i+1}^{n}\frac{1}{a_k}\}, \quad (i = 1, 2, \cdots, n) \quad (117)$$

we have

$$\frac{\partial}{\partial a_j}[H_i\bar{u}(x)] = 0 \qquad \text{for } j < i \quad (118a)$$

$$= h_i a_i^{-2}\bar{g}(x_i^p - i) \qquad \text{for } j = i \quad (118b)$$

$$= h_i a_j^{-2}\bar{g} \qquad \text{for } j > i. \quad (118c)$$

Using (118), it can be shown that the matrix $\nabla_{a*}(H\bar{u})'$ becomes a lower triangular one with non-zero elements and that

$$\det[\nabla_{a*}(H\bar{u})'] = \prod_{i=1}^{n} h_i a_i^{-2}\bar{g}(x_i^{\,p} - i) \neq 0. \tag{119}$$

Consequently, from the fact that $\nabla_{a*}(H\bar{u})'$ is regular, $\bar{Q}(a*)$ is positive definite and from (115), the equality (112) holds. By a similar procedure to the case of Type-2, this fact is easily affirmed in the case of Type-1.

(B) <u>Parameter estimation algorithm</u>: In order to circumvent massive computations, we replace the system state $u(t,x)$ in Eq. (110) by the stationary state $\bar{u}(x)$, i.e., for Eq. (110), we have

$$\left[\int_0^t \nabla_{a*}(H\bar{u})'dy(s) - \int_0^t \nabla_{a*}(H\bar{u})'H\bar{u}ds\right]_{a* = \bar{a}*(t)} = 0, \tag{120}$$

where $\bar{a}*(t) = [\bar{a}_1(t), \bar{a}_2(t), \cdots, \bar{a}_n(t)]'$. Using the fact that, from Theorem 5, the matrix $\nabla_{a*}(H\bar{u})'$ is invertible, from (120), we obtain

$$H\bar{u}t\big|_{a* = \bar{a}*(t)} = y(t). \tag{121}$$

The equality (117) is rewritten by

$$h_i\bar{u}_i(x_i^{\,p})t\big|_{a* = \bar{a}*(t)} = y_i(t), \quad \text{for } i = 1, 2, \cdots, n. \tag{122}$$

Hence, applying (117) to (121), $\bar{a}_i(t)$ can be successively be obtained as

$$\left.\begin{aligned}
\bar{a}_n(t) &= h_n\bar{g}(n - x_n^{\,p})ty_n^{-1}(t)\\
\bar{a}_{n-1}(t) &= h_{n-1}\bar{g}(n - 1 - x_{n-1}^{\,p})\{t^{-1}y_{n-1}(t) - h_{n-1}\bar{g}\bar{a}_n^{-1}(t)\}^{-1}\\
&\ \ \vdots \\
\bar{a}_i(t) &= h_i\bar{g}(i - x_i^{\,p})\{t^{-1}y_{i-1}(t) - h_i\bar{g}\sum_{k=i+1}^{n}\bar{a}_k^{-1}(t)\}^{-1}\\
&\ \ \vdots \\
\bar{a}_1(t) &= h_1\bar{g}(1 - x_1^{\,p})\{t^{-1}y_1(t) - h_1\bar{g}\sum_{k=2}^{n}\bar{a}_k^{-1}(t)\}^{-1}.
\end{aligned}\right\} \tag{123}$$

The following theorem states the consistency property of the estimator $\bar{a}_i(t)$ where $i = 1, 2, \cdots, n$.

[Theorem 6] With (H-5) to (H-10), if the allocation of sensors is of Type-2, then

$$\lim_{t\to\infty} \bar{a}*(t) = a*. \tag{124}$$

Proof: Let $y(t)$ be defined by $y(t) \triangleq [y_1(t), y_2(t), \cdots, y_n(t)]'$. From the observation mechanism (79), we have

$$\begin{aligned}
\lim_{t\to\infty} t^{-1}y_i(t) &= \lim_{t\to\infty} t^{-1}h_i\int_0^t u(s,x_i^{\,p})ds + \lim_{t\to\infty} t^{-1}v_i(t)\\
&= h_i\bar{u}_i(x_i^{\,p}) \quad \text{w.p.1} \quad \text{for } i = 1, 2, \cdots, n.
\end{aligned} \tag{125}$$

Hence, from (123) and (125), we obtain

$$\lim_{t\to\infty} \bar{a}_n(t) = h_n\bar{g}(n - x_n^{\,p})\{h_n\bar{u}_n(x_n^{\,p})\}^{-1} = a_n, \tag{126}$$

where the relation (117) has been used. Similarly, let us compute $\lim_{t\to\infty}\bar{a}_{n-1}(t)$. The result is

$$\lim_{t\to\infty} \bar{a}_{n-1}(t) = h_{n-1}\bar{g}(n - 1 - x_{n-1}^{p})\{h_{n-1}\bar{u}_{n-1}(x_{n-1}^{p}) - h_{n-1}a_n^{-1}\bar{g}\}^{-1} = a_{n-1}. \tag{127}$$

Furthermore, it is a simple exercise to show that

$$\lim_{t\to\infty} \bar{a}_i(t) = h_i\bar{g}(i - x_i^{p})\{h_i\bar{u}_i(x_i^{p}) - h_i\bar{g}\sum_{k=i+1}^{n} a_k^{-1}\}^{-1} = a_i. \tag{128}$$

Thus, the proof has been completed, because the consistency properties of $\bar{a}_n(t)$, $\bar{a}_{n-1}(t)$, $\cdots$, $\bar{a}_1(t)$ have successively been shown.

(7) *A digital simulation experiment*: For computer implementation, we set the time and space partitions as $\Delta t = t_f/m$, $t_i = i\Delta t$ $(i = 0, 1, 2, \cdots, m)$ and $\Delta x = n/\ell$ and $x_j = (j\ell/n)\Delta x$ $(j = 0, 1, 2, \cdots, n)$. Based on these partitions, instead of the definitions (76), the following new symbolic conventions are introduced in the sequel:

$$D_k = \{j | (k - 1)\ell/n < j < (k/n)\ell\} \qquad (k = 1, 2, \cdots, n) \tag{129a}$$

$$\partial D_k = \{j | j = (k/n)\ell\} \qquad (k = 1, 2, \cdots, n-1) . \tag{129b}$$

Denoting the discrete version of $u(t,x)$ by $u(i,j)$, our approximations to $\partial u(t,x)/\partial t$, $\partial u(t,x)/\partial x$ and $\partial^2 u(t,x)/\partial x^2$ are respectively as follows:

$$\frac{\partial u(t,x)}{\partial t} \cong \frac{u(i+1, j) - u(i,j)}{\Delta t} \tag{130a}$$

$$\frac{\partial u(t,x)}{\partial x} \cong \frac{u(i,j) - u(i, j-1)}{\Delta x}, \text{ for } i \in T,\ j \in \partial D_k,\ (k = 1, 2, \cdots, n) \tag{130b}$$

$$\frac{\partial^2 u(t,x)}{\partial x^2} \cong \frac{1}{2(\Delta x)^2}[\{u(i, j+1) - 2u(i,j) + u(i, j-1)\}$$
$$+ \{u(i+1, j+1) - 2u(i+1, j) + u(i+1, j-1)\}]$$
$$\text{for } i \in T = \{i\},\ (i = 1, 2, \cdots, m-1),\ j \in D_k,\ (k = 1, 2, \cdots, n) \tag{130c}$$

Using (98), we define the new variable r_k by $r_k = a_k\Delta t/2(\Delta x)^2$, Eq. (75a) is expressed in the form,

$$r_k u(i+1, j-1) - (1+2r_k)u(i+1, j) + r_k u(i+1, j+1)$$
$$= - r_k u(i, j-1) - (1-2r_k)u(i,j) - r_k u(i, j+1). \tag{131}$$

The condition on ∂D_k is given by

$$a_k\frac{u(i,j) - u(i, j-1)}{\Delta x} = a_{k+1}\frac{u(i, j+1) - u(i,j)}{\Delta x}$$
$$\text{for } i \in T,\ j \in \partial D_k,\ (k = 1, 2, \cdots, n-1). \tag{132}$$

The initial and boundary conditions given by (75b) and (75c) are respectively expressed by

$$u(0,j) = \sqrt{30-j} \qquad (j = 0, 1, 2, \cdots, \ell) \tag{133}$$

$$- a_1\frac{u(i, 1) - u(i,0)}{\Delta x} = \exp(-2i\Delta t) + \bar{g} \qquad \text{for } i \in T \tag{134a}$$

$$u(i,\ell) = 0 \qquad \text{for } i \in T, \tag{134b}$$

where the value of n has been set as n = 3 and the RHS of Eq. (134a) came out from the preassigned form of the g(t)-function as $g(t) = \exp(-2t) + \bar{g}$. Consequently, the generating routine of the state u(i,j) is given recursively by solving the following vector equations:

$$A_k U(i+1) = -B_k U_k, \tag{135}$$

where both A_k and B_k are respectively the $\{(n/\ell)+1\}\{(n/\ell)+1\}$ matrices defined as follows:

$$A_k = \begin{bmatrix} 1 & 0 & & & & \\ r_k & -(1+2r_k) & r_k & & 0 & \\ & \ddots & \ddots & \ddots & & \\ & 0 & & r_k & -(1+2r_k) & r_k \\ & & & & 0 & 1 \end{bmatrix}, \quad B_k = \begin{bmatrix} 1 & 0 & & & & \\ r_k & 1-2r_k & r_k & & 0 & \\ & \ddots & \ddots & \ddots & & \\ & 0 & & r_k & 1-2r_k & r_k \\ & & & & 0 & 1 \end{bmatrix} \tag{136a}$$

and where

$$\begin{aligned} U_1(i) &= [u(i,0),\ u(i,1),\ \cdots,\ u(i,\ell/n)]', \\ U_2(i) &= [u(i,\ell/n),\ u(i,(\ell/n)+1),\ \cdots,\ u(i,2\ell/n)]', \\ &\cdots\cdots \\ U_n(i) &= [u(i,(n-1)\ell/n),\ u(i,\{(n-1)\ell/n\}+1),\ \cdots,\ u(i,\ell)]'. \end{aligned} \tag{136b}$$

In simulating the observation mechanism given by (78), the white noise process dv(t)/dt in (79) is approximated by $n(j)/\sqrt{\Delta t}$ as usual, where n(j) is the Gaussian random sequence as N(0,1). From (78) and (79), a difference form of the observation mechanism is constructed to be

$$y_k(i+1) = y_k(i) + h_k u(i+1,\ j_k{}^p)\Delta t + n(i)\sqrt{\Delta t} \quad (k = 1,\ 2,\ \cdots,\ n), \tag{137}$$

where $(i,\ j_k{}^p)$ is a symbolic convention to be understood as $(t_i,\ x_k{}^p)$.

The spatial region [0, 3] is, in the experiment, divided into three sub-regions, i.e., $D_1 =]0,\ 1[$, $D_2 =]1,\ 2[$ and $D_3 =]2,\ 3[$. The unknown parameter was considered to have three stepwise values. This situation implies that the assumption (96) was set as

$$a(x) = a_1 \ \text{for } D_1, \quad a(x) = a_2 \ \text{for } D_2, \quad a(x) = a_3 \ \text{for } D_3. \tag{138}$$

The estimates of these parameters, $\bar{a}_1(t)$, $\bar{a}_2(t)$ and $\bar{a}_3(t)$ can be obtained by solving Eq. (135) in the case where k = 3. The solutions are as follows:

$$\bar{a}_3(i) = \bar{g}(\ell - j_3{}^p)(\Delta t)(\Delta x)y_3{}^{-1}(i), \tag{139a}$$

$$\bar{a}_2(i) = (\tfrac{2}{3}\ell - j_2{}^p)(\Delta x)\{\frac{y_2(i)}{\bar{g}i(\Delta t)} - \frac{1}{\bar{a}_3(i)}\}^{-1}, \tag{139b}$$

$$\bar{a}_1(i) = (\frac{\ell}{3} - j_1{}^p)(\Delta x)\{\frac{y_1(i)}{\bar{g}i(\Delta t)} - (\frac{1}{\bar{a}_2(i)} + \frac{1}{\bar{a}_3(i)})\}^{-1}. \tag{139c}$$

Throughout digital simulation experiments, we preassigned the true value of parameters as $a_1 = 0.3$, $a_2 = 0.2$ and $a_3 = 0.1$. The partitioned time interval and spatial variable were taken as $\Delta t = 1.0$ and $\Delta x = 0.1$ respectively. The values of h_k and $\bar{g}$ were

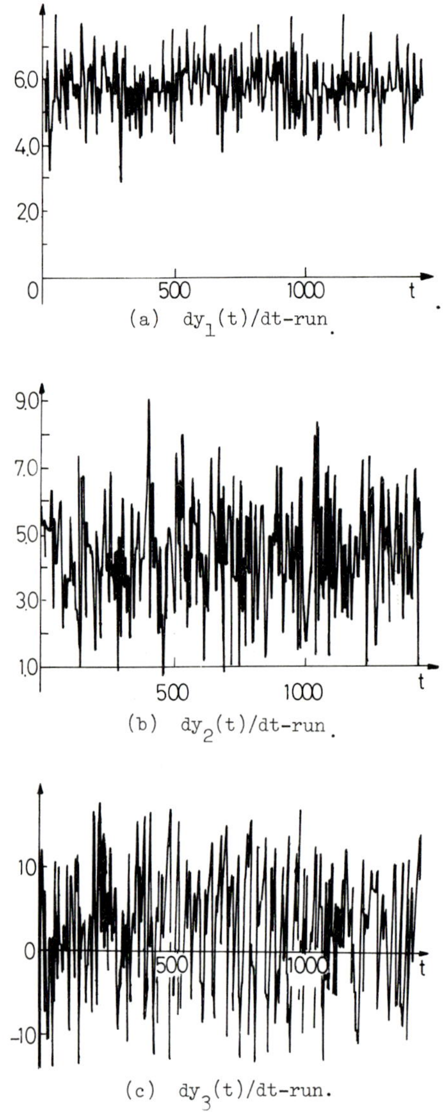

(a) $dy_1(t)/dt$-run.

(b) $dy_2(t)/dt$-run.

(c) $dy_3(t)/dt$-run.

Fig. 8. The three-dimensional observation data $dy(t)/dt$.

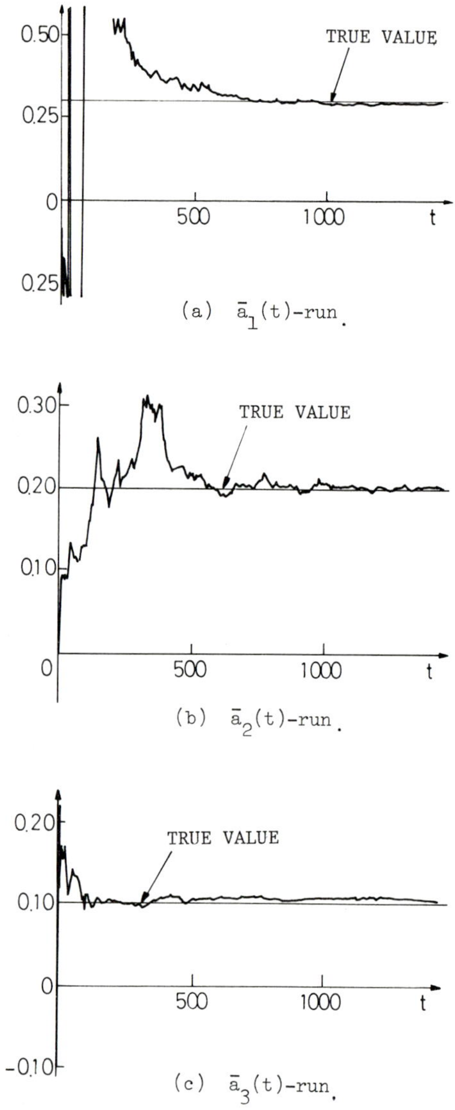

(a) $\bar{a}_1(t)$-run.

(b) $\bar{a}_2(t)$-run.

(c) $\bar{a}_3(t)$-run.

Fig. 9. A sample run of the parameter estimate $\bar{a}(t)$.

set as $h_k = 1$ and $\bar{g} = 15$, respectively. Furthermore, three sensors were located at $x_1{}^p = 0.5$, $x_2{}^p = 1.5$, and $x_3{}^p = 2.5$. The observation data $dy_1(t)/dt$, $dy_2(t)/dt$ and $dy_3(t)/dt$ are shown in Fig. 8. The estimated runs obtained by Eq. (139) are demonstrated in Fig. 9.

CONCLUSIONS

In this chapter, we have presented a brief survey of methods for the parameter identification of DPSs, in particular, SDPSs. In order to show the motivation of studies on the parameter identification an exhibition of various kinds of mathematical models available in the current literature has been developed. Mathematical models have been given by stochastic partial differential equations in which unknown parameters appear in the forcing terms as disturbances to the system or in the differential operator as physical parameters of the system.

It is inevitable for us to set an assumption on mathematical descriptions on unknown parameters. Whether or not an unknown parameter may be assumed to be a random constant? If it is not, then it should be mathematically described by a time function or by a function of spatial variables from more realistic viewpoints. Furthermore, there are a number of practical examples, in which unknown parameters depend on the system state. Problems of identifying unknown parameters involved in the differential operator belong to the context of the inverse problem. Although the second example in this chapter relates apparently to the stochastic inverse problem, this is one of very simple examples using a familiar partial differential equation of parabolic type with a straightforward boundary condition.

In spite of practical problems which remain to be solved, the ideas developed in this chapter have obvious applications in a wide variety of areas concerning the parameter identification. An extensive study on the stochastic inverse problem associated with the oil exploration survey was already reported (Sunahara and others, 1981), where a class of partial differential equation of hyperbolic type was used for generating the wavelet by an artificial earthquake.

The elegant body of mathematical theory pertaining to examine the existence and uniqueness properties of the basic stochastic differential equation is now being constituted by a highly mathematical tool with the function space concept as the central figure. However, turning our eyes to the version of identification techniques, we may conclude that we have two principal line of attacks, i.e., one is the least squares estimate and another the Bayesian estimate or the maximum likelihood estimate. In any case, the problem of circumventing massive computations is still opened.

ACKNOWLEDGEMENTS

The author wishes to express his many thanks for assistance by Prof. A. Ohsumi, Dr. Sh. Aihara, Dr. Y. Morita, Mr. F. Kojima and Mr. T. Yasuda. Special thanks should be extended to Mr. Y. Kumagawa and Mrs. M. Date. The work presented here was supported by the Educational Ministry Grant of Japan, No. 56550289.

REFERENCES

Åstrom, K. J., and P. Eykhoff (1971). System identification: a survey. *Automatica*, 7, 123-162.

Balakrishnan, A. V. (1973). Stochastic differential systems. *Lecture Notes in economics and Mathematical Systems*, Vol. 84. Springer, New York. pp. 192-222.

Falb, P. L. (1967). Infinite-dimensional filtering: the Kalman-Bucy filtering in Hilbert space. *Inf. & Control*, 11, 102-137.

Girsanov, I. V. (1960). On transforming a certain class of stochastic processes by absolutely continuous substitution of measures. *Theory of Prob. Appl.* 5, 285-301.

Goodson, R. E., and M. P. Polis (1978). Identification of parameters in distributed systems. In W. H. Ray and D. G. Lainiotis (Ed.), *Distributed Parameter Systems*. Marcel Dekker, Inc. New York and Basel. pp. 47-122.
Kailath, T. (1969). A general likelihood-ratio formula for random signals in Gaussian noise. *IEEE Trans. Inf. Theory*, 15, 350-361.
Kubrusly, C. S. (1977). Distributed parameter system identification: a survey. *Int. J. Control*, 26, 509-535.
Kushner, H. J. (1968). On the optimal control of a system governed by a linear parabolic equation with noise inputs. *SIAM J. Control*, 6, 596-614.
Kushner, H. J. (1970). Filtering for linear distributed parameter systems. *SIAM J. Control*, 8, 346-359.
Lions, J. L. (1971). *Optimal Control of Systems Governed by Partial Differential Equations*. Springer-Verlag, Berlin.
Mitter, S. K. (1969). Optimal control of distributed parameter systems. *Proceedings of 1969 JACC*. Colorado. 620.
Polis, M. P., R. E. Goodson, and J. L. Wozny (1973). On parameter identification for distributed systems using Galerkin's criterion. *Automatica*, 9, 53-64.
Polis, M. P., and R. E. Goodson (1976). Parameter identification in distributed systems: a synthesizing overview. *Proceedings of the IEEE*, 64, 45-61.
Ray, W. H. (1978). Some recent applications of distributed parameter control theory: a survey. *Automatica*, 14, 281-287.
Ray, W. H., and D. G. Lainiotis (Ed.) (1978). *Distributed Parameter Systems*, 6. Marcel Dekker, Inc. New York and Basel.
Robinson, A. C. (1971). A survey of optimal control of distributed-parameter systems. *Automatica*, 7, 371-388.
Ruberti, A. (Ed.) (1978). Distributed parameter systems: modelling and identification. *Lecture Notes in Control and Information Sciences*. Springer-Verlag, Berlin.
Skorokhod, A. V. (1965). *Studies in the Theory of Random Processes*. Addison-Wesley, Reading, Mass.
Sunahara, Y., A. Ohsumi, and M. Imamura (1973). A filtering theory for nonlinear distributed parameter systems. *Trans. Soc. Instrm. Control. Engng.*, 9, 446-455. (in Japanese).
Sunahara, Y., Sh. Aihara, and F. Kojima (1981). A method of parameter estimation for stochastic distributed parameter systems and its applications to seismic profiles identification. *Proceedings of the 8th IFAC Triennial World Congress*, Kyoto.
Sunahara, Y. (1981). Recent trends in optimal control of stochastic distributed parameter systems. *Proceedings of 10th IFIP Conference on System Modelling and Optimization*. New York.
Tzafestas, S. G. (1978). Distributed parameter state estimation. In W. H. Ray and D. G. Lainiotis (Ed.), *Distributed Parameter Systems*. Marcel Dekker, Inc. New York and Basel. pp. 135-208.
Wiberg, D. M. (1967). Optimal control of nuclear reactor systems. In C. T. Leondes (Ed.), *Advances in Control Systems*, Vol 5. Academic Press, New York.

CHAPTER 3

FILTERING, SMOOTHING, AND PREDICTION OF DISTRIBUTED PARAMETER SYSTEMS: SURVEY OF RIGOROUS APPROACH

S. Omatu

Department of Information Science and Systems Engineering, Faculty of Engineering, University of Tokushima, Tokushima, Japan

1. INTRODUCTION

The problem of estimating the state of a stochastic dynamical system from noisy observation is one of central importance in engineering. There are major avenues which have been followed in estimation theory depending on whether the system under study is assumed to be concentrated at a single spatial point(lumped parameter system, LPS) or it is assumed to occupy a certain spatial domain(distributed parameter system,DPS). The former LPS is modeled by an ordinary differential equation(ODE), whereas the latter DPS is modeled by a partial differential equation(PDE). The physical examples of DPS are nuclear reactors, heat exchangers, diffusion processes, chemical reactors, fluid systems, vibration systems, atmospheric phenomena, drying systems, petroleum, steel, papermaking, and glass processes, magnetrohydrodynamic systems, etc.

Recently, there have been many theoretical papers on DP estimation theory, which cover a wide class of systems, including linear and nonlinear systems with partial-integro-differential operators, with delays and then coupled with LPS. In the largest part of the literature on DP estimation problems, the choice of an appropriate mathematical model for the random disturbance or noise is an essential one. These authors have assumed that the "undisturbed" dynamical system is described by a well-posed partial-integro-differential equation and have arbitrarily tacked on some white noise term, without ever analyzing the fundamental problem of modeling infinite-dimensional white noise. This approach is called the "formal approach" in contrast to the "rigorous approach" in which the DP stochastic processes are defined in appropriate rigorous ways. Although the word "formal approach" seems to be thoughtless, it is more valuable in engineering fields than the rigorous approach since the estimators by using both approaches are identical. Furthermore, an important problem for engineers is how to construct and use the DP estimation theory. But it is also preferable to survey the rigorous approach to the DP estimation theory from the mathematical and engineering viewpoints.

The purpose in this chapter is to give a survey of the estimation theory for DPS from rigorous mathematical viewpoint. It is also important to get an intuitive meaning of the rigorous derivation. Therefore, we show a formal derivation based on the Wiener-Hopf theory in Section 2. Then we show various rigorous approaches to the DP estimation problems in Section 3.

2. FORMAL APPROACH BASED ON THE WIENER-HOPF THEORY

The purpose in this section is to give a comprehensive survey of the DP state estimation theory developed thus far by using the formal approach. Up to now the largest part of the literature on estimation theory of DPS is a formal analysis of the filtering problem for the linear DPS. These approaches are captured in several survey paper(Tzafestas, 1978, Bencala and Seinfeld, 1979, and Curtain,1975c). In this section we consider the optimal state estimation problems, that is, prediction, filtering, and smoothing estimation problems for the linear DPS based on the Wiener-Hopf theory. Then we will discuss various approaches to the optimal DP estimation theory.

2.1. Description of the dynamical systems

Let D be a connected bounded open domain of an r-dimensional Euclidean space R^r and let ∂D be a sufficiently smooth boundary of D. The spatial coordinate vector is denoted by $x=(x_1,\cdots,x_r) \in D$. Consider a linear stochastic DPS which is described by the PDE

$$\frac{\partial u(t,x)}{\partial t} = A_x u(t,x) + b(t,x)w(t,x) \tag{2.1}$$

defined for all $t \geq t_0$ on the spatial domain D where $u(t,x)=\mathrm{col}[u_1(t,x),\cdots,u_n(t,x)]$ is an n-dimensional state vector function of the system, $w(t,x)$ is a vector-valued white Gaussian process, A_x is a matrix linear spatial differential operator whose parameter may depend on x, and $b(t,x)$ is a known matrix function. The operator A_x is assumed to be elliptic and self-adjoint and of the form

$$A_x u = \sum_{i=1}^{r} \frac{\partial}{\partial x^i}\Big(\sum_{j=1}^{r} a_{ij}(x)\frac{\partial u}{\partial x^j} \Big) + a_0(x)u$$

where a_{ij} and a_0 are $n \times n$ symmetric matrix functions satisfying $a_{ij} = a_{ji}$. The initial and boundary conditions are given by

$$u(t_0,x) = u_0(x) \tag{2.2}$$

$$\Gamma_\xi u(t,\xi) = 0, \quad \xi \in \partial D \tag{2.3}$$

$$\Gamma_\xi u = f_1(\xi)u + f_2(\xi)\frac{\partial u}{\partial n}, \quad \frac{\partial u}{\partial n} = \sum_{j=1}^{r}\Big(\sum_{i=1}^{r} a_{ij}(\xi)\cos(n,x^i)\Big)\frac{\partial u}{\partial x^j}$$

where n is an exterior normal to the surface ∂D at a point $\xi \in \partial D$ and $\cos(n,x^i)$ is the i-th direction cosine of n. The mean and covariance functions of $u_0(x)$ are

$$E[u_0(x)] = 0 \tag{2.4}$$

$$E[u_0(x)u_0'(y)] = p_0(x,y).$$

Let us assume that measurement data are taken at fixed m points $x^1,\cdots,x^m$ of $\bar{D} = D \cup \partial D$. Furthermore, let us define an mn-dimensional column vector $u_m(t)$ by

$$u_m(t) = \mathrm{col}[u(t,x^1),\cdots,u(t,x^m)]. \tag{2.5}$$

Then the observation equation is given by

$$z(t) = c(t)u_m(t) + v(t) \tag{2.6}$$

where $z(t)$ is a p-dimensional measurement vector at the points $x^1,\cdots,x^m$, $c(t)$ is a known $p \times mn$ matrix function, and $v(t)$ is a p-dimensional Gaussian process. The white Gaussian processes $w(t,x)$ and $v(t)$ are assumed to be statistically independent of each other and also independent of the stochastic initial condition $u_0(x)$. Their mean and covariance functions are given by

$$E[w(t,x)] = 0, \quad E[V(t)] = 0,$$

$$E[w(t,x)w'(s,y)] = Q(t,x,y)\delta(t-s), \quad E[v(t)v'(s)] = R(t)\delta(t-s) \tag{2.7}$$

where $R(t)$ is a symmetric positive-definite matrix and $Q(t,x,y)$ is an nonnegative matrix function such that

$$Q(t,x,y) = Q'(t,y,x).$$

Let us assume the following:

(A1) The problem given by (2.1)-(2.3) is well-posed in the sense of Hadamard, that is, the solution exists uniquely and depends continuously on the initial and boundary data.

(A2) There exists an $n \times n$ matrix fundamental solution $U(t,s,x,y)$ of (2.1)-(2.3) such that

$$U(t,s,x,y) = U(t,s,y,x),$$

$$\frac{\partial U(t,s,x,y)}{\partial t} = A_x U(t,s,x,y), \tag{2.8}$$

$$U(s,s,x,y) = I\delta(x-y),$$

$$\Gamma_\xi U(t,s,\xi,y) = 0, \quad \xi \in \partial D, \quad y \in D$$

and that the solution of (2.1)-(2.3) is expressed as

$$u(t,x) = \int_D U(t,t_0,x,y)u_0(y)dy + \int_{t_0}^{t} ds \int_D U(t,s,x,y)b(s,y)w(s,y)dy. \tag{2.9}$$

In the case where $u(t,x)$ is a scalar function and

$$A_x = \frac{\partial^2}{\partial x_1^2} + \cdots + \frac{\partial^2}{\partial x_r^2} - q(x), \quad q(x) = \text{Hölder continuous function}$$

it is well-known that the above assumptions are satisfied(Ito,1966).

2.2. Estimation problems and Wiener-Hopf theory

Let us denote the estimate of $u(\tau,x)$ based on the observation data $z(s)$, $t_0 \leq s \leq t$, by $\hat{u}(\tau,x/t)$ which is given by the following linear transformation of $z(s)$, $t_0 \leq s \leq t$,

$$\hat{u}(\tau,x/t) = \int_{t_0}^{t} \tilde{F}(\tau,x,s)z(s)ds \tag{2.10}$$

where $\tilde{F}(\tau,x,s)$ is an unknown $n \times p$ matrix function called the estimation kernel. Furthermore, we denote the estimation error and error covariance functions by $\tilde{u}(\tau,x/t)$ and $p(\tau,x,y/t)$, respectively, where $\tilde{u}(\tau,x/t) = u(\tau,x) - \hat{u}(\tau,x/t)$ and $p(\tau,x,y/t) = E[\tilde{u}(\tau,x/t)\tilde{u}'(\tau,y/t)]$. The estimate $\hat{u}(\tau,x/t)$ that minimizes $J(\hat{u}) = E[\tilde{u}'(\tau,x/t)\tilde{u}(\tau,x/t)]$ is said to be optimal. Note that by using $p(\tau,x,y/t)$, $J(\hat{u})$ can be rewritten as $J(\hat{u}) = tr[p(\tau,x,x/t)]$

In order to clarify the differences between the prediction, filtering, and smoothing problems, we express $\tilde{F}(\tau,x,s)$ differently for each problems:

(i) Prediction problem($\tau > t$).

$$\hat{u}(\tau,x/t) = \int_{t_0}^{t} A(\tau,x,s)z(s)ds. \tag{2.11}$$

(ii) Filtering problem($\tau = t$).

$$\hat{u}(t,x/t) = \int_{t_0}^{t} F(t,x,s)z(s)ds. \tag{2.12}$$

(iii) Smoothing problem($\tau < t$).

$$\hat{u}(\tau,x/t) = \int_{t_0}^{t} B(\tau,t,x,s)z(s)ds. \tag{2.13}$$

Then the following theorem can be proved by the similar way to finite dimensional estimation problems(Sakawa,1972, Sawaragi, Soeda, and Omatu,1978).

[Theorem 2-1] (Wiener-Hopf theorem)

A necessary and sufficient condition for the estimate $\hat{u}(\tau,x/t)$ to be optimal is that the following Wiener-Hopf equation holds for $t_0 \leqq \zeta \leqq t$ and $x \in \bar{D}$.

$$\int_{t_0}^{t} \tilde{F}(\tau,x,s)E[z(s)z'(\zeta)]ds = E[u(\tau,x)z'(\zeta)], \tag{2.14}$$

or equivalently

$$E[\tilde{u}(\tau,x/t)z'(\zeta)] = 0. \tag{2.15}$$

[Corollary 2-1] (Orthogonal projection lemma)

The orthogonality condition between the estimate and estimation error vectors holds, that is, for any time ζ and x, $y \in \bar{D}$, $E[\tilde{u}(\tau,x/t)\hat{u}'(\zeta,y/t)] = 0$.

[Lemma 2-1] (Uniqueness of the optimal kernel)

Let $\tilde{F}(\tau,x,s)$ be the optimal kernel satisfying the Wiener-Hopf equation (2.14) and let $\tilde{F}(\tau,x,s) + F_\Delta(\tau,x,s)$ be also the optimal kernel satisfying the Wiener-Hopf equation (2.14). Then it follows that $F_\Delta(\tau,x,s) = 0$, $t_0 \leqq s \leqq t$ and $x \in \bar{D}$, that is, the optimal kernel is unique.

In order to consider the prediction, filtering, and smoothing problems, separately, we rewrite (2.14) using the notation of (2.11)-(2.13).

[Corollary 2-2] The Wiener-Hopf equation (2.14) is rewritten for the prediction, filtering, and smoothing problems as follows:

(i) Prediction problem($\tau > t$).

$$\int_{t_0}^{t} A(\tau,x,s)E[z(s)z'(\zeta)]ds = E[u(\tau,x)z'(\zeta)],\ t_0 \leqq \zeta \leqq t \text{ and } x \in \bar{D}. \tag{2.16}$$

(ii) Filtering problem($\tau = t$).

$$\int_{t_0}^{t} F(t,x,s)E[z(s)z'(\zeta)]ds = E[u(t,x)z'(\zeta)],\ t_0 \leqq \zeta \leqq t \text{ and } x \in \bar{D}. \tag{2.17}$$

(iii) Smoothing problem($\tau < t$).

$$\int_{t_0}^{t} B(\tau,t,x,s)E[z(s)z'(\zeta)]ds = E[u(\tau,x)z'(\zeta)],\ t_0 \leqq \zeta \leqq t \text{ and } x \in \bar{D}. \tag{2.18}$$

2.3. Derivation of the optimal prediction estimator

In this section we derive the optimal prediction estimator by using the Wiener-Hopf theory in the previous section.

[Theorem 2-2] The optimal prediction estimator is given by

$$\frac{\partial \hat{u}(\tau,x/t)}{\partial \tau} = A_x \hat{u}(\tau,x/t), \quad \tau > t \tag{2.19}$$

$$\Gamma_\xi \hat{u}(\tau,\xi/t) = 0, \quad \xi \in \partial D \tag{2.20}$$

where the initial value is $\hat{u}(t,x/t)$ of the optimal filtering estimate.

[Proof] Differentiating (2.16) with respect to τ and substituting (2.1) yields

$$\int_{t_0}^{t} \frac{\partial A(\tau,x,s)}{\partial \tau} E[z(s)z'(\zeta)]ds = A_x E[u(\tau,x)z'(\zeta)]$$

where the independence of $w(\tau,x)$ and $z(\zeta)$, $\zeta < \tau$ is used.
Substituting (2.16) into the above equation yields

$$\int_{t_0}^{t} F_\Delta(t,x,s)E[z(s)z'(\zeta)]ds = 0$$

where

$$F_\Delta(t,x,s) = \frac{\partial A(\tau,x,s)}{\partial \tau} - A_x A(\tau,x,s).$$

From Lemma 2-1 we get

$$\frac{\partial A(\tau,x,s)}{\partial \tau} = A_x A(\tau,x,s). \tag{2.21}$$

Differentiating (2.11) with respect to τ and substituting (2.21) yields (2.19). Multiplying each side of (2.16) by $f_1(\xi)$ and $f_2(\xi)\partial/\partial n$ and using (2.3) yields

$$\int_{t_0}^{t} \Gamma_\xi A(\tau,\xi,s)E[z(s)z'(\zeta)]ds = E[\Gamma_\xi u(\tau,\xi)z'(\zeta)] = 0.$$

Thus, we get

$$\Gamma_\xi A(\tau,\xi,s) = 0, \quad \xi \in \partial D.$$

Multiplying each side by $z(s)$ and integrating from $s = t_0$ to t yields (2.20). Q.E.D.

[Theorem 2-3] The optimal prediction error covariance function $p(\tau,x,y/t)$ is governed by

$$\frac{\partial p(\tau,x,y/t)}{\partial \tau} = A_x p(\tau,x,y/t) + (A_y p(\tau,x,y/t))' + \tilde{Q}(\tau,x,y) \tag{2.22}$$

$$\Gamma_\xi p(\tau,\xi,y/t) = 0, \quad \xi \in \partial D, \quad \tilde{Q}(\tau,x,y) = b(\tau,x)Q(\tau,x,y)b'(\tau,y) \tag{2.23}$$

where the initial value is $p(t,x,y/t)$ of the optimal filtering error covariance matrix function.

[Proof] From (2.1) and (2.19) we get

$$\frac{\partial\tilde{u}(\tau,x/t)}{\partial\tau} = A_x\tilde{u}(\tau,x/t) + b(\tau,x)w(\tau,x) \tag{2.24}$$

and from (2.3) and (2.20)

$$\Gamma_\xi\tilde{u}(\tau,\xi/t) = 0, \quad \xi \in \partial D. \tag{2.25}$$

Differentiating the definition of $p(\tau,x,y/t)$ with respect to τ and using (2.24) yields

$$\frac{\partial p(\tau,x,y/t)}{\partial\tau} = A_x p(\tau,x,y/t) + (A_y p(\tau,x,y/t))' + \Sigma(\tau,x,y)$$

where

$$\Sigma(\tau,x,y) = b(\tau,x)E[w(\tau,x)\tilde{u}'(\tau,y/t)] + E[\tilde{u}(\tau,x/t)w'(\tau,y)]b'(\tau,y).$$

The solution of (2.24) can be represented in terms of the fundamental solution $U(\tau,\sigma,x,y)$ as follows:

$$\tilde{u}(\tau,x/t) = \int_D U(\tau,t,x,\alpha)\tilde{u}(t,\alpha/t)d\alpha + \int_t^\tau \int_D U(\tau,\sigma,x,\alpha)b(\sigma,\alpha)w(\sigma,\alpha)d\alpha d\sigma. \tag{2.26}$$

Substituting (2.26) into $\Sigma(\tau,x,y)$ and using (2.7) yields $\Sigma(\tau,x,y) = \tilde{Q}(\tau,x,y)$. Multiplying each side of (2.25) by $\tilde{u}'(\tau,y/t)$ and taking the expectation yields (2.23).
Q.E.D.

[Corollary 2-3] The optimal prediction estimate $\hat{u}(\tau,x/t)$ and prediction error covariance function $p(\tau,x,y/t)$ can be represented as

$$\hat{u}(\tau,x/t) = \int_D U(\tau,t,x,\alpha)\hat{u}(t,\alpha/t)d\alpha \tag{2.27}$$

and

$$p(\tau,x,y/t) = \int_D \int_D U(\tau,t,x,\alpha)p(t,\alpha,\beta/t)U'(\tau,t,y,\beta)d\alpha d\beta$$
$$+ \int_t^\tau \int_D \int_D U(\tau,\sigma,x,\alpha)\tilde{Q}(\sigma,\alpha,\beta)U'(\tau,\sigma,y,\beta)d\alpha d\beta d\sigma. \tag{2.28}$$

2.4. Derivation of the optimal filter

Let us derive the optimal filter by using the Wiener-Hopf theory. Differentiating (2.17) with respect to t and substituting (2.1) yields

$$\int_{t_0}^t \frac{\partial F(t,x,\sigma)}{\partial t} E[z(\sigma)z'(\zeta)]d\sigma + F(t,x,t)E[z(t)z'(\zeta)]$$

$$= E[\frac{\partial u(t,x)}{\partial t} z'(\zeta)] = A_x E[u(t,x)z'(\zeta)] = \int_{t_0}^t A_x F(t,x,\sigma)E[z(\sigma)z'(\zeta)]d\sigma$$

where (2.17) and the relation $E[w(t,x)z'(\zeta)] = 0$, $t > \zeta$ are used.
Since $E[v(t)z'(\zeta)] = 0$, $t > \zeta$, it follows that

$$F(t,x,t)E[z(t)z'(\zeta)] = F(t,x,t)c(t)E[u_m(t)z'(\zeta)]$$

$$= F(t,x,t)c(t) \int_{t_0}^t F_m(t,\sigma)E[z(\sigma)z'(\zeta)]d\sigma$$

where

$$F_m(t,\sigma) = \begin{bmatrix} F(t,x^1,\sigma) \\ \vdots \\ F(t,x^m,\sigma) \end{bmatrix}$$

Thus, we get

$$\int_{t_0}^{t} F_\Delta(t,x,\sigma)E[z(\sigma)z'(\zeta)]d\sigma = 0$$

where

$$F_\Delta(t,x,\sigma) = \frac{\partial F(t,x,\sigma)}{\partial t} - A_x F(t,x,\sigma) + \Gamma(t,x,t)c(t)F_m(t,\sigma).$$

Since it is clear that $F(t,x,\sigma) + F_\Delta(t,x,\sigma)$ is also an optimal kernel satisfying the Wiener-Hopf equation (2.17), from Lemma 2-1 $F_\Delta(t,x,\sigma) = 0$. Thus, the following lemma holds.

[Lemma 2-2] The optimal kernel $F(t,x,\sigma)$ of the filter is given by

$$\frac{\partial F(t,x,\sigma)}{\partial t} = A_x F(t,x,\sigma) - F(t,x,t)c(t)F_m(t,\sigma). \tag{2.29}$$

[Theorem 2-4] The optimal filtering estimate $\hat{u}(t,x/t)$ is given by

$$\frac{\partial \hat{u}(t,x/t)}{\partial t} = A_x \hat{u}(t,x/t) + F(t,x,t)\nu(t) \tag{2.30}$$

$$\hat{u}(t_0,x/t_0) = 0, \quad \Gamma_\xi \hat{u}(t,\xi/t) = 0, \; \xi \in \partial D \tag{2.31}$$

where

$$\nu(t) = z(t) - c(t)\hat{u}_m(t/t) \tag{2.32}$$

$$\hat{u}_m(t/t) = \text{col}[\hat{u}(t,x^1/t),\cdots,\hat{u}(t,x^m/t)]. \tag{2.33}$$

[Proof] Using (2.12) and (2.29) we get

$$\frac{\partial \hat{u}(t,x/t)}{\partial t} = A_x \hat{u}(t,x/t) + F(t,x,t)(z(t) - c(t)\hat{u}_m(t/t)).$$

From the Wiener-Hopf equation we get

$$\int_{t_0}^{t} \Gamma_\xi F(t,\xi,\sigma)E[z(\sigma)z'(\zeta)]d\sigma = E[\Gamma_\xi u(t,\xi)z'(\zeta)] = 0.$$

Thus, from Lemma 2-1 we get

$$\Gamma_\xi F(t,\xi,\sigma) = 0, \quad \xi \in \partial D.$$

From (2.12) and the above equation we get (2.31). Q.E.D.

[Theorem 2-5] The optimal filtering gain $F(t,x,t)$ is given by

$$F(t,x,t) = p_m(t,x/t)c'(t)R^{-1}(t) \tag{2.34}$$

where

$$p_m(t,x/t) = [p(t,x,x^1/t),\cdots,p(t,x,x^m/t)]. \tag{2.35}$$

[Proof] From (2.6) we get

$$E[z(\sigma)z'(\zeta)] = E[z(\sigma)u_m'(\zeta)]c'(\zeta) + R(\sigma)\delta(\sigma-\zeta)$$

$$E[u(t,x)z'(\zeta)] = E[u(t,x)u_m'(\zeta)]c'(\zeta).$$

Substituting the above equations into (2.17) yields

$$\int_{t_0}^{t} F(t,x,\sigma)E[z(\sigma)u_m'(\zeta)]c'(\zeta)d\sigma + F(t,x,\zeta)R(\zeta) = E[u(t,x)u_m'(\zeta)]c'(\zeta),\ t_0\leq\zeta<t.$$

By virtue of (2.12), the above equation can be rewritten as

$$F(t,x,\zeta)R(\zeta) = E[u(t,x)u_m'(\zeta)]c'(\zeta) - E[\hat{u}(t,x/t)u_m'(\zeta)]c'(\zeta)$$

$$= E[\tilde{u}(t,x/t)u_m'(\zeta)]c'(\zeta) = E[\tilde{u}(t,x/t)\tilde{u}_m'(\zeta/\zeta)]c'(\zeta) = p_m(t,x/\zeta)c'(\zeta)$$

where $\tilde{u}_m(\zeta/\zeta) = u_m(\zeta) - \hat{u}_m(\zeta/\zeta)$ and $E[\tilde{u}(t,x/t)\hat{u}_m'(\zeta/\zeta)] = 0$ are employed. Letting $\zeta \to t$ in the above equation yields

$$F(t,x,t) = p_m(t,x/t)c'(t)R^{-1}(t).$$

Q.E.D.

[Theorem 2-6] The optimal filtering error covariance function p(t,x,y/t) is given by

$$\frac{\partial p(t,x,y/t)}{\partial t} = A_x p(t,x,y/t) + (A_y p(t,x,y/t))' + \tilde{Q}(t,x,y)$$

$$- p_m(t,x/t)\tilde{R}(t)p_m'(t,y/t), \qquad \tilde{R}(t) = c'(t)R^{-1}(t)c(t) \tag{2.36}$$

$$p(t_0,x,y/t_0) = p_0(x,y), \tag{2.37}$$

$$\Gamma_\xi p(t,\xi,y/t) = 0, \quad \xi\in\partial D. \tag{2.38}$$

[Proof] From (2.1) and (2.30) we get

$$\frac{\partial\tilde{u}(t,x/t)}{\partial t} = A_x\tilde{u}(t,x/t) - p_m(t,x/t)c'(t)R^{-1}(t)\nu(t) + b(t,x)w(t,x) \tag{2.39}$$

and from (2.3) and (2.31)

$$\Gamma_\xi\tilde{u}(t,\xi/t) = 0, \quad \xi\in\partial D. \tag{2.40}$$

Differentiating p(t,x,y/t) with respect to t yields

$$\partial p(t,x,y/t)/\partial t = E[\partial\tilde{u}(t,x/t)/\partial t\ \tilde{u}'(t,y/t)] + E[\tilde{u}(t,x/t)\partial\tilde{u}'(t,y/t)/\partial t].$$

From (2.39) we get

$$E[\partial\tilde{u}(t,x/t)/\partial t\ \tilde{u}'(t,y/t)] = A_x p(t,x,y/t) - p_m(t,x/t)c'(t)R^{-1}(t)E[\nu(t)\tilde{u}'(t,y/t)]$$

$$+ b(t,x)E[w(t,x)\tilde{u}'(t,y/t)].$$

But we can easily derive

$$E[\nu(t)\tilde{u}'(t,y/t)] = \frac{1}{2} c(t)p_m'(t,y/t), \quad E[w(t,x)\tilde{u}'(t,y/t)] = \frac{1}{2} Q(t,x,y)b(t,y).$$

Thus, we get

$$\frac{\partial p(t,x,y/t)}{\partial t} = A_x p(t,x,y/t) + (A_y p(t,x,y/t))' + \tilde{Q}(t,x,y) - p_m(t,x/t)\tilde{R}(t)p_m'(t,y/t).$$

Multiplying each side of (2.40) by $\tilde{u}'(t,y/t)$ and taking the expectation yields (2.38). Q.E.D.

2.5. Derivation of the optimal smoothing estimator

Let us now derive the optimal smoothing estimator based on the Wiener-Hopf theory.

[Lemma 2-3] The optimal kernel $B(\tau,t,x,\sigma)$ of the smoothing estimator is given by

$$\frac{\partial B(\tau,t,x,\sigma)}{\partial t} = - B(\tau,t,x,t)c(t)F_m(t,\sigma), \quad t_0 \leqq \sigma < t. \tag{2.41}$$

[Proof] Differentiating (2.18) with respect to t yields

$$B(\tau,t,x,t)E[z(t)z'(\zeta)] + \int_{t_0}^{t} \frac{\partial B(\tau,t,x,\sigma)}{\partial t} E[z(\sigma)z'(\zeta)]d\sigma = 0.$$

From the Wiener-Hopf equation (2.17) of the filtering problem we get

$$E[u_m(t)z'(\zeta)] = \int_{t_0}^{t} F_m(t,\sigma)E[z(\sigma)z'(\zeta)]\, d\sigma.$$

From (2.6) and the above equation we get

$$E[z(t)z'(\zeta)] = c(t)E[u_m(t)z'(\zeta)] = \int_{t_0}^{t} c(t)F_m(t,\sigma)E[z(\sigma)z'(\zeta)]d\sigma, \quad t_0 \leqq \zeta < t.$$

Hence, we get

$$\int_{t_0}^{t} F_\Delta(\tau,t,x,\sigma)E[z(\sigma)z'(\zeta)]d \;= 0$$

where

$$F_\Delta(\tau,t,x,\sigma) = \frac{\partial B(\tau,t,x,\sigma)}{\partial t} + B(\tau,t,x,t)c(t)F_m(t,\sigma).$$

Using Lemma 2-1 yields $F_\Delta(\tau,t,x,\sigma) = 0$, $t_0 \leqq \sigma < t$, that is, (2.41). Q.E.D.

[Theorem 2-7] The optimal smoothing estimate $\hat{u}(\tau,x/t)$ is given by

$$\frac{\partial \hat{u}(\tau,x/t)}{\partial t} = B(\tau,t,x,t)\nu(t) \tag{2.42}$$

$$B(\tau,t,x,t) = D_m(\tau,x/t)c'(t)R^{-1}(t) \tag{2.43}$$

$$\Gamma_\xi \hat{u}(\tau,\xi/t) = 0, \qquad \xi \in \partial D \tag{2.44}$$

where the initial condition of $\hat{u}(\tau,x/t)$ is $\hat{u}(\tau,x/\tau)$ and

$$D(\tau,x,y/t) = E[\tilde{u}(\tau,x/t)u'(t,y)], \quad D_m(\tau,x/t) = (D(\tau,x,x^1/t),\cdots,D(\tau,x,x^m/t)). \tag{2.45}$$

[Proof] Differentiating (2.18) with respect to t and using (2.41) yields

$$\frac{\partial\hat{u}(\tau,x/t)}{\partial t} = B(\tau,t,x,t)z(t) + \int_{t_0}^{t} \frac{\partial B(\tau,t,x,\sigma)}{\partial t} z(\sigma)d\sigma$$
$$= B(\tau,t,x,t)(z(t) - c(t)\int_{t_0}^{t} F_m(t,\sigma)z(\sigma)d\sigma) = B(\tau,t,x,t)\nu(t).$$

It follows from the Wiener-Hopf equation (2.18) and (2.3) that

$$\int_{t_0}^{t} B(\tau,t,x,\sigma)E[z(\sigma)z'(\zeta)]d\sigma = E[u(\tau,x)z'(\zeta)] = E[u(\tau,x)u_m'(\zeta)]c'(\zeta).$$

But from (2.3) we egt

$$E[z(\sigma)z'(\zeta)] = E[z(\sigma)u_m'(\zeta)]c'(\zeta) = R(\zeta)\delta(\zeta-\sigma).$$

Thus, we get

$$E[\tilde{u}(\tau,x/t)u_m'(\zeta)]c'(\zeta) + B(\tau,t,x,\zeta)R(\zeta).$$

Letting $\zeta \to t$ yields (2.43). (2.44) can be obtained by the same procedure as (2.31). Q.E.D.

If the equation satisfied by $D(\tau,x,y/t)$ is derived, then the optimal smoothing estimator can be determined. Using the orthogonality condition of Corollary 2.1 yields

$$D(\tau,x,y/t) = E[\tilde{u}(\tau,x/t)\tilde{u}'(t,y/t)] = E[\tilde{u}(\tau,x/\tau)\tilde{u}'(t,y/t)]. \tag{2.46}$$

Let us define $V(t,\sigma,y,z)$ as follows:

$$\frac{\partial V(t,\sigma,y,z)}{\partial t} = A_y V(t,\sigma,y,z) - p_m(t,y/t)\tilde{R}(t)V_m'(t,\sigma,z), \tag{2.47}$$

$$V(\sigma,\sigma,y,z) = I\delta(y-z),$$

$$\Gamma_\xi V(t,\sigma,\xi,z) = 0, \quad \xi \in \partial D$$

where

$$V_m(t,\sigma,z) = (V(t,\sigma,x^1,t),\cdots,V(t,\sigma,x^m,z)). \tag{2.48}$$

Then the solution $\tilde{u}(t,y/t)$ of (2.39) can be represented as

$$\tilde{u}(t,y/t) = \int_D V(t,t_0,y,\alpha)u_0(\alpha)d\alpha + \int_{t_0}^{t}\int_D V(t,\sigma,y,\alpha)[b(\sigma,\alpha)w(\sigma,\alpha) - p_m(\sigma,\alpha/\sigma)c'(\sigma)R^{-1}(\sigma)v(\sigma)d\alpha d\sigma. \tag{2.49}$$

[Theorem 2-8] The optimal smoothing gain $D(\tau,x,y/t)$ is given by

$$\partial D(\tau,x,y/t)/\partial t = A_y D(\tau,x,y/t) - D_m(\tau,x/t)\tilde{R}(t)p_m'(t,y/t) \tag{2.50}$$

or

$$\frac{\partial D(\tau,x,y/t)}{\partial \tau} = A_x D(\tau,x,y/t) + \int_D \tilde{Q}(\tau,\alpha,x)V'(t,\tau,y,\alpha)d\alpha \tag{2.51}$$

$$D(\tau,x,y/\tau) = p(\tau,x,y/\tau) \tag{2.52}$$

$$\Gamma_\xi D(\tau,\xi,y/t) = 0, \quad \xi \varepsilon \partial D. \tag{2.53}$$

[Proof] Using the orthogonality condition of Corollary 2-1 yields

$D(\tau,x,y/t) = E[u(\tau,x)\tilde{u}'(t,y/t)]$.

Differentiating the above equation with respect to τ and using (2.1) yields

$$\frac{\partial D(\tau,x,y/t)}{\partial \tau} = A_x D(\tau,x,y/t) + b(\tau,x)\Delta_\tau, \Delta_\tau = E[w(\tau,x)\tilde{u}'(t,y/t)].$$

Using (2.49) yields

$$\Delta_\tau = \int_{t_0}^{t} \int_D E[w(\tau,x)w'(\sigma,\alpha)]b'(\sigma,\alpha)V'(t,\sigma,y,\alpha)d\alpha d\sigma = \int_D Q(\tau,x,\sigma)b'(\tau,\alpha) V'(t,\tau,y,\alpha)d\alpha.$$

Thus, we get (2.51). Differentiating (2.46) with respect to t yields

$$\frac{\partial D(\tau,x,y/t)}{\partial t} = E[\tilde{u}(\tau,x/\tau)\frac{\partial \tilde{u}'(t,y/t)}{\partial t}].$$

Substituting (2.39) into the above equation and using the same procedure as (2.51) yields (2.50). From the definition (2.46) of $D(\tau,x,y/t)$ it is clear that $D(\tau,x,y/\tau) = p(\tau,x,y/\tau)$ and $\Gamma_\xi D(\tau,\xi,y/t) = 0$. Q.E.D.

[Corollary 2-3] $D(\tau,x,y/t)$ is given by

$$D(\tau,x,y/t) = \int_D V(t,\tau,y,\alpha)p(\tau,\alpha,x/\tau)d\alpha. \tag{2.54}$$

[Proof] Since (2.50) or (2.51) is a linear PDE, it is clear that there exists a unique solution of (2.50) or (2.51) with (2.52) and (2.53). By direct differentiation of (2.54) with respect to t, it is clear that (2.54) satisfies (2.50). Since $D(\tau,x,y/t)$ defined by (2.54) satisfies (2.52) and (2.53), the proof of the corollary is complete. Q.E.D.

[Theorem 2-9] The optimal smoothing error covariance function $p(\tau,x,y/t)$ is given by

$$\frac{\partial p(\tau,x,y/t)}{\partial t} = - D_m(\tau,x/t)\tilde{R}(t)D_m'(\tau,y/t), \quad \tau < t \tag{2.55}$$

$$\Gamma_\xi p(\tau,\xi,y/t) = 0, \quad \xi \varepsilon \partial D \tag{2.56}$$

where the initial condition is $p(\tau,x,y/\tau)$.

[Proof] Using the orthogonality condition of Corollary 2-1 yields

$$p(\tau,x,y/t) = E[\tilde{u}(\tau,x/t)u'(\tau,y)].$$

Differentiating the above equation with respect to t and using the relation that

$$\partial\tilde{u}(\tau,x/t)/\partial t = - \partial\hat{u}(\tau,x/t)/\partial t,$$

we get from (2.42)

$$\frac{\partial p(\tau,x,y/t)}{\partial t} = - B(\tau,t,x,t)c(t)E[\tilde{u}_m(t/t)u'(\tau,y)].$$

But from the orthogonality condition of Corollary 2-1 we egt

$$E[\tilde{u}_m(t/t)u'(\tau,y)] = E[u_m(t)\tilde{u}'(\tau,y/t)] = D_m'(\tau,y/t).$$

Thus, we get (2.55). From (2.44) we get

$$\Gamma_\xi \tilde{u}(\tau,\xi/t) = 0, \quad \xi \in \partial D.$$

Multiplying each side of the above equation by $\tilde{u}'(t,y/t)$ and taking the expectation yields (2.56). Q.E.D.

Thus, we get the following theorem.

[Theorem 2-10] The optimal smoothing estimate $\hat{u}(\tau,x/t)$ and the smoothing error covariance function $p(\tau,x,y/t)$ are given by

$$\hat{u}(\tau,x/t) = \hat{u}(\tau,x/\tau) + \int_\tau^t D_m(\tau,x/\sigma)\nu_m(\sigma)d\sigma \tag{2.57}$$

where

$$\nu_m(\sigma) = c'(\sigma)R^{-1}(\sigma)\nu(\sigma)$$

and

$$p(\tau,x,y/t) = p(\tau,x,y/\tau) - \int_\tau^t D_m(\tau,x/\sigma)\tilde{R}(\sigma)D_m'(\tau,y/\sigma)d\sigma. \tag{2.58}$$

From this theorem we can derive three types optimal smoothing estimators, that is, fixed-point, fixed-interval, and fixed-lag smoothing estimators(Sawaragi, Soeda, and Omatu,1978).

2.6. Historical notes and related works

The state estimation theory of DPS dates back to 1967(Balakrishnan and Lions, 1967, Falb,1967, and Kwakernaak,1967). In 1968 and 1969 many works on DP estimation problems were published(Tzafestas and Nightingale,1968a,b, Thau,1969, Phillipson and Mitter,1969, and Tzafestas,1969) and in later years one finds many papers(Tzafestas, 1972a-c, 1973, Meditch,1970, 1971, Sakawa,1972, Atre and Lamba,1972a-d, Bensoussan, 1969, 1971a,b, Kushner,1970, Seinfeld,1969, 1970, Omatu, Soeda, and Tomita,1977a,b, Omatu and Soeda, 1976, Omatu, Ohe, and Soeda,1979, 1980, Sawaragi, Soeda, and Omatu, 1978, Omatu and Seinfeld, 1981, Curtain,1975b, 1976, Curtain and Ichikawa, 1977 etc.). The estimation principally used are (1)orthogonal projection method(Tzafestas and Nightingale,1968a, Thau,1969, Atre and Lamba,1972a); (2)least squares method (Tzafestas,1973, Meditch,1970,1971, Balakrishnan and Lions,1967, Phillipson and Mitter,1969, Seinfeld,1969, Seinfeld and Gavalas,1971, Hwang, Seinfeld, and Gavalas, 1972); (3)conditional characteristic function method(Tzafestas and Nightingale,1968b, Tzafestas,1972b); (4)Wiener-Hopf method(Falb,1967, Sakawa,1972, Curtain,1975a,Omatu, Tomita, and Soeda,1977a, Omatu, Ohe, and Soeda, 1979, 1980, Sawaragi, Soeda, and Omatu, 1978, Omatu and Seinfeld,1981); (5)Bayesian method(Kushner,1970, Tzafestas, 1972a); (6)maximum likelihood method(Tzafestas and Nightingale,1969); (7)innovation method(Tzafestas,1973, Atre and Lamba,1972a, Curtain,1976); and (8)maximum principle approach(Atre and Lamba,1972a).

3. RIGOROUS APPROACHES TO OPTIMAL ESTIMATION PROBLEMS

An early paper which posed an infinite-dimensional filtering problem was that by Falb(1967) and further or independently explored rigorous state estimation theories were reported in many papers(Bensoussan, 1969, 1971a,b, Kushner, 1970,Curtain and Falb, 1971, Curtain 1975a,b, Curtain and Ichikawa, 1977, Omatu, Soeda, and Tomita,1977a, Omatu, Tomita, and Soeda,1977b, Sawaragi, Soeda, and Omatu, 1978). In this section we show a unified approach to the infinite-dimensional estimation problems based on the Wiener-Hopf theory in Hilbert spaces.

3.1. Description of the Dynamical Systems

Let us suppose that H and K are real Hilbert spaces, (Ω,S,μ) is an underlying probability space, and L(H,K) is a family of bounded linear operators from H into K. Let the state $u(t,\omega)$ be generated by the stochastic linear differential equation in the Hilbert space H

$$du(t,\omega) = a(t)u(t,\omega)dt + b(t)dw(t,\omega) \tag{3.1}$$

$$u(t_0,\omega) = u_0(\omega) \tag{3.2}$$

and let the observation value $z(t,\omega)$ be generated by the stochastic linear differential equation in the Hilbert space K

$$dz(t,\omega) = c(t)u(t,\omega)dt + dv(t,\omega) \tag{3.3}$$

where $b(\cdot)\varepsilon\ L^{\infty}(t_0,t_f;L(H,H))$, $c(\cdot)\varepsilon\ L^{\infty}(t_0,t_f;L(H,K))$, and $u_0(\omega)$ is an H-valued random variable independent of $w(t,\omega)$ and $v(t,\omega)$ and has zero mean and covariance operator p_0. $w(t,\omega)$ and $v(t,\omega)$ are independent of Wiener processes on H and K with covariance operators Q(t) and R(t), respectively, that is,

$$\text{cov}[w(t,\omega)-w(s,\omega),\ w(t,\omega)-w(s,\omega)] = \int_s^t Q(\tau)d\tau, \tag{3.4}$$

$$\text{cov}[v(t,\omega)-v(s,\omega),\ v(t,\omega)-v(s,\omega)] = \int_s^t R(\tau)d\tau, \tag{3.5}$$

where Q(t) and R(t) are nuclear, positive, and symmetric operators for almost all $t\ \varepsilon(t_0,t_f)$ and

$$\int_{t_0}^{t_f} \text{tr}Q(t)dt < \infty.$$

Note that the covariance operator $\text{cov}[\cdot,\cdot]$ is defined by

$$\text{cov}[x,y] = E[x\circ y] - E[x]\circ E[y]$$

where $E[\cdot]$ and $\text{tr}[\cdot]$ denote the expectation and trace of the operator, respectively, and $\circ$ is given by the following relation(Falb, 1967):

$$(x\circ y)z = x\ \langle y,z\rangle \quad \text{for fixed } x\ \varepsilon\ K,\ y\ \varepsilon\ H, \text{ and } \forall z\ \varepsilon\ H. \tag{3.6}$$

Here $\langle\cdot,\cdot\rangle$ denotes the inner product on H and a(t) is a linear operator in H, satisfying the following:

(i) The domain D(a) of a(t) is dense in H and independent of t, and a(t) is a closed operator;

(ii) $(\lambda I - a(t))^{-1}$ exists for $\text{Re}\lambda \geqq 0$ and $\|(\lambda I - a(t))^{-1}\| \leqq c/(1+|\lambda|)$, where c is constant and I denotes the identity operator;

(iii) $\|(a(t)-a(\tau))a^{-1}(\tau)\| \leq c_1|t-\tau|^{\alpha}$ for all $t,\tau > 0$ where c_1 and α are positive constants.

Then note that there exists the evolution operator $U(t,s)$ with the following properties (Kato and Tanabe, 1962):

(1) $U(t,s)$: $T\times T \to L(H,H)$ and is strongly continuous in s and t for $0 \leq s \leq t \leq t_f$ where $T = [t_0,t_f]$;

(2) $U(t,s) = U(t,\tau)U(\tau,s)$ if $0 \leq s \leq \tau \leq t \leq t_f$ and $U(s,s) = I$;

(3) $U(t,s)$ is strongly continuously differentiable in t for $t > s$ and

$$\frac{\partial U(t,s)}{\partial t} = A(t)U(t,s) \tag{3.7}$$

where

$$\|A(t)U(t,s)\| \leq c_2/(t-s) \quad \text{for } 0 \leq s < t \leq t_f \text{ and some constant } c_2.$$

Furthermore, the unique solution of (3.1) can be given by

$$u(t,\omega) = U(t,t_0)u_0(\omega) + \int_{t_0}^{t} U(t,s)b(s)dw(s,\omega). \tag{3.8}$$

We assume that $R(t)$ is invertible. Thus, K is a finite dimensional space since $R(t)$ denotes the invertible nuclear operator (Yosida, 1968). From now on we assume that the random variables considered here have expected values and we often delete the explicit ω-dependence of these variables.

3.2. Preliminary Results

Let us recapitulate the properties of stochastic integral of Ito type required in this paper which have beeen proved by Falb (1967).

[Lemma 3-1] Let $e(s)$ and $f(s)$ be elements of $L^2(T;L(H,H))$ and $L^2(T;L(K,K))$, respectively. Then

$$\operatorname{cov}\left[\int_{t_0}^{t} e(s)dw(s), \int_{t_0}^{t} f(s)dv(s)\right] = 0 \tag{3.9}$$

and

$$\operatorname{cov}\left[\int_{t_0}^{t} e(s)dw(s), \int_{t_0}^{t} f(s)dw(s)\right] = \int_{t_0}^{t} e(s)Q(s)f^*(s)ds. \tag{3.10}$$

[Proof] Assume that $e(s)$ and $f(s)$ are step functions such that

$$e(s) = \sum_{j=0}^{n} e_j\chi_j(s), \quad f(s) = \sum_{j=0}^{n} f_j\chi_j(s)$$

where $t_0 < t_1 < \cdots < t_n = t_f$ and

$$\chi_j(s) = \begin{cases} 1, & t_j \leq s < t_{j+1} \\ 0, & \text{otherwise.} \end{cases}$$

Then stochastic integrals of Ito type are

$$\int_{t_0}^{t} e(s)dw(s) = \sum_{j=0}^{n-1} e_j(w(t_{j+1}) - w(t_j))$$

and

$$\int_{t_0}^{t} f(s)dv(s) = \sum_{j=0}^{n-1} f_j(v(t_{j+1}) - v(t_j)).$$

Since $E[w(t)]=0$ and $E[v(t)]=0$, we get

$$E[\int_{t_0}^{t} e(s)dw(s)] = 0, \quad E[\int_{t_0}^{t} f(s)dv(s)] = 0. \tag{3.11}$$

Thus, we get

$$\text{cov}[\int_{t_0}^{t} e(s)dw(s), \int_{t_0}^{t} f(s)dw(s)] = \sum_{j,k=0}^{n-1} e_j E[(w(t_{j+1}) - w(t_j))\circ(w(t_{k+1})$$

$$- w(t_k))]f_k^* = \sum_{j=0}^{n-1} e_j Q(t_{j+1}) f_j^*(t_{j+1} - t_j) = \int_{t_0}^{t} e(s)Q(s)f^*(s)ds.$$

For $e(s)\varepsilon L^2(T;L(H,H))$ and $f(s)\varepsilon L^2(T;L(K,K))$ we get the same results by using the approximation argument(Falb, 1967). Q.E.D.

[Lemma 3-2] Suppose that $\text{cov}[w(t),u_0] = 0$ for all t and $k(t,s)$ is in $L^2(T\times T;L(H,H))$. Then

$$\text{cov}[\int_{t_0}^{t} k(t,s)dw(s), \Delta z_\sigma] = \int_{t_0}^{\sigma} k(t,s)Q(s)b^*(s)\phi_\sigma^*(s)ds \tag{3.12}$$

where $t_0 \leqq \sigma \leqq t$ and

$$\Delta z_\sigma = z(\sigma) - z(t_0) \tag{3.13}$$

$$\phi_\sigma(s) = \int_s^\sigma c(\tau)U(\tau,s)d\tau. \tag{3.14}$$

[Proof] From (3.8) and (3.13) we get

$$\Delta z_\sigma = \int_{t_0}^{\sigma} c(s)u(s)ds + \int_{t_0}^{\sigma} dv(s) = \int_{t_0}^{\sigma} c(s)U(s,t_0)u_0\, ds + \int_{t_0}^{\sigma} dV(s)$$

$$+ \int_{t_0}^{\sigma}[\int_{\tau}^{\sigma} c(s)U(s,\tau)ds]b(\tau)dw(\tau) = \int_{t_0}^{\sigma} c(s)U(s,t_0)u_0 ds + \int_{t_0}^{\sigma} dv(s)$$

$$+ \int_{t_0}^{\sigma} \phi_\sigma(\tau)b(\tau)dw(\tau). \tag{3.15}$$

Setting $\Theta(\tau) = \phi_\sigma(\tau)b(\tau)$ for $\tau \leqq \sigma$ and $\Theta(\tau) = 0$ for $a > \sigma$ and noting that

$$\text{cov}[\int_{t_0}^{t} k(t,s)dw(s),u_0] = 0,$$

we get from Lemma3-1

$$\text{cov}[\int_{t_0}^{t} k(t,s)dw(s),\Delta z_\sigma] = \text{cov}[\int_{t_0}^{t} k(t,s)dw(s), \int_{t_0}^{t}\Theta(s)dw(s)]$$

$$= \int_{t_0}^{t} k(t,s)Q(s)\Theta^*(s)ds = \int_{t_0}^{\sigma} k(t,s)Q(s)b^*(\tau)\,\phi_\sigma^*(\tau)d\tau.$$ Q.E.D.

Then we get the following lemma.

[Lemma 3-3] If $\text{cov}[w(t),u_0]=0$, $k(t,s)$ is in $L^2(T\times T;L(H,H))$, and there exists $\partial k(t,s)/\partial t$ which is regulated in t and is L^2 in s, then for $\sigma<t$

$$\frac{d}{dt}\,\mathrm{cov}[\int_{t_0}^{t} k(t,s)dw(s),\Delta z_\sigma] = \mathrm{cov}[\int_{t_0}^{t} \frac{\partial k(t,s)}{\partial t}\,dw(s),\Delta z_\sigma]. \tag{3.16}$$

[Proof] Differentiating each side of (3.12) with respect to t and using again (3.12) yields

$$\frac{d}{dt}\,\mathrm{cov}[\int_{t_0}^{t} k(t,s)dw(s),\Delta z_\sigma] = \int_{t_0}^{\sigma} \frac{\partial k(t,s)}{\partial t}\,Q(s)b^*(s)\phi_\sigma^*(s)ds$$

$$= \mathrm{cov}[\int_{t_0}^{t} \frac{\partial k(t,s)}{\partial t}\,dw(s),\Delta z_\sigma].$$

Q.E.D.

[Lemma 3-4] If $\mathrm{cov}[v(t),u_0]=0$, then for $\sigma < t$

$$\frac{d}{dt}\,\mathrm{cov}[u(t),\Delta z_\sigma] = \mathrm{cov}[a(t)u(t),\Delta z_\sigma]. \tag{3.17}$$

[Proof] Letting $k(t,s)=U(t,s)b(s)$ and using (3.7), (3.8), and (3.16) yields (3.17).

Q.E.D.

[Lemma3-5] Suppose that $k(t,s)$ is regulated. Then

$$\mathrm{cov}[\int_{t_0}^{t} k(t,s)dz(s),\Delta z_\sigma] = \int_{t_0}^{t} \psi(t,s)b(s)Q(s)b^*(s)\phi_\sigma^*(s)ds$$

$$+ \int_{t_0}^{t} k(t,s)R(s)ds + \int_{t_0}^{t} k(t,s)c(s)U(s,t_0)ds\,p_0\,\phi_\sigma^*(t_0) \tag{3.18}$$

where

$$\psi(t,s) = \int_{s}^{t} k(t,\alpha)c(\alpha)U(\alpha,s)d\alpha \tag{3.19}$$

and

$$p_0 = \mathrm{cov}[u_0,u_0]. \tag{3.20}$$

[Proof] First of all, we note that

$$\int_{t_0}^{t} k(t,s)dz(s) = \int_{t_0}^{t} k(t,s)c(s)(U(s,t_0)u_0 + \int_{t_0}^{s} U(s,\alpha)b(\alpha)dw(\alpha))ds$$

$$+ \int_{t_0}^{t} k(t,s)dv(s).$$

However, in view of (3.15), Lemma 3-1, the independence of u_0 and both $w(t)$ and $v(t)$, and Lemma 3-5, we readily see that

$$\mathrm{cov}[\int_{t_0}^{t} k(t,s)dz(s),\Delta z_\sigma] = \mathrm{cov}[\int_{t_0}^{t} \psi(t,s)b(s)dw(s),\Delta z_\sigma] + \int_{t_0}^{t} k(t,s)R(s)ds$$

$$+ \int_{t_0}^{t} k(t,s)c(s)U(s,t_0)ds\,p_0\,\phi_\sigma^*(t_0).$$

Then the lemma follows immediately from Lemma 3-2. Q.E.D.

[Lemma 3-6] If $\partial k(t,s)/\partial t$ exists, is regulated in t, and is L^2 in s, then for $\sigma < t$

$$\frac{d}{dt}\,\mathrm{cov}[\int_{t_0}^{t} k(t,s)dz(s),\Delta z_\sigma] = \mathrm{cov}[\int_{t_0}^{t} \frac{\partial k(t,s)}{\partial t}\,dz(s) + k(t,t)c(t)u(t),\Delta z_\sigma]. \tag{3.21}$$

[Proof] Differentiating each side of (3.18) with respect to t and using (3.18) again yields (3.21). Q.E.D.

3.3. Problem Formulation

Let us denote the estimate of $u(\tau)$ based on the observation data $\{z(s),\ t_0 \leq s \leq t\}$, by $\hat{u}(\tau/t)$ which is given by the following linear transformation of $\{z(s),\ t_0 \leq s \leq t\}$,

$$\hat{u}(\tau/t) = \int_{t_0}^{t} \tilde{F}(\tau,s)dz(s) \tag{3.22}$$

where $\tilde{F}(\tau,s)$ is a bounded linear transformation and is integrable in both arguments. Furthermeore, we denote the estimation error and error covariance functions by $\tilde{u}(\tau/t)$ and $p(\tau/t)$, respectively, that is,

$$\tilde{u}(\tau/t) = u(\tau) - \hat{u}(\tau/t) \tag{3.23}$$

and

$$p(\tau/t) = E[\tilde{u}(\tau/t)\circ\tilde{u}(\tau/t)] = \mathrm{cov}[\tilde{u}(\tau/t),\tilde{u}(\tau/t)]. \tag{3.24}$$

The estimate $\hat{u}(\tau/t)$ that minimizes

$$J(\hat{u}) = E[\langle x^*,\tilde{u}(\tau/t)\rangle^2] = x^*p(\tau/t)x, \quad x^* \in H^*=H \tag{3.25}$$

is said to be optimal where x is x* viewed as an element of H.

To clarify the difference between the prediction, filtering, and smoothing problems, we express $\tilde{F}(\tau,s)$ differently for each problem:

(i) Prediction problem($\tau > t$)

$$\hat{u}(\tau/t) = \int_{t_0}^{t} A(\tau,s)dz(s). \tag{3.26}$$

(ii) Filtering problem($\tau=t$)

$$\hat{u}(t/t) = \int_{t_0}^{t} F(t,s)dz(s). \tag{3.27}$$

(iii) Smoothing problem($\tau < t$)

$$\hat{u}(\tau/t) = \int_{t_0}^{t} B(\tau,t,s)dz(s). \tag{3.28}$$

Thus, the problems are how to determine the optimal kernels $A^o(\tau,s)$, $F^o(t,s)$, and $B^o(\tau,t,s)$ of (3.26)-(3.28) which minimize $J(\hat{u})$.

3.4. Wiener-Hopf Theory

Let us derive the Wiener-Hopf theorem which gives the basic necessary and sufficient condition for the estimate $\hat{u}(\tau/t)$ to be an optimal solution of the above estimation problems.

[Theorem 3-1] (Wiener-Hopf theorem)
A necessary and sufficient condition for the estimate $\hat{u}(\tau/t)$ to be optimal is that the following Wiener-Hopf equation holds for all α and β, $t_0 \leq \beta < \alpha \leq t$

$$\text{cov}[\int_{t_0}^{t} F(t,s)dz(s),\ z(\alpha) - z(\beta)] = \text{cov}[u(\tau),\ z(\alpha) - z(\beta)], \tag{2.39}$$

or equivalently, if and only if

$$E[\tilde{u}(\tau/t),\ z(\alpha) - z(\beta)] = 0 \tag{3.30}$$

for all α and β, $t_0 \leqq \beta < \alpha \leqq t$.

[Proof] Let x^* be an element of $H^* = H$ and let $X(x^*)$ be the space of all real random variables of the form $<x^*,x(\cdot)>$ where $x(\cdot) \in L^2(\Omega,H)$. The inner product on $X(x^*)$ is given by $E[<x^*,x(\cdot)><x^*,y(\cdot)>]$. Let $U(x^*)$ denote the subspace of $X(x^*)$ generated by elements of the form

$$<x^*,\ y(a)> = <x^*,\ \int_{t_0}^{a} D(\tau,s)dz(s)>,\quad a \leqq t,$$

where $D(\tau,s)$ is integrable. By the well-known orthogonal projection lemma(Yosida, 1968), $\hat{u}(\tau/t)$ will be a solution of the estimation problems if and only if $\tilde{u}(\tau/t)$ is orthogonal to $U(x^*)$ in $X(x^*)$ for every x^*. In other words, $\hat{u}(\tau/t)$ is a solution of the estimation problems if and only if

$$E[<x^*,\ \tilde{u}(\tau/t)><x^*,y(a)>] = 0$$

for all x^*, where

$$y(a) = \int_{t_0}^{a} D(\tau,s)dz(s),\quad a \leqq t. \tag{3.31}$$

Let us suppose that (3.30) holds. We observe that, in view of (3.6)

$$E[<x^*,\tilde{u}(\tau/t)><x^*,y(a)>] = x^*E[\tilde{u}(\tau/t)\circ y(a)]x.$$

However, $y(a)$ is given by (3.31); thus, if $D(\tau,s)$ is a step function, then

$$E[\tilde{u}(\tau/t)\circ y(a)] = \textstyle\sum_j E[\tilde{u}(\tau/t)\circ B_j(z(\alpha)-z(\beta))] = \sum_j E[\tilde{u}(\tau/t)\circ(z(\alpha)-z(\beta))]B_j^* = 0.$$

It follows from Schwarz's inequality that $E[\tilde{u}(\tau/t)\circ y(a)] = 0$ for any $y(a)$ given by (3.31), and hence that

$$x^*E[\tilde{u}(\tau/t)\circ y(a)]x = 0$$

for all x^*. Therefore, $\tilde{u}(\tau/t)$ is a solution of the estimation problems.

On the other hand, let us assume that $\hat{u}(\tau/t)$ is a solution of the estimation problems. Now assume that (3.30) does not hold, so that

$$E[\tilde{u}(\tau/t)\circ(z(\alpha) - z(\beta))] = \text{cov}[\tilde{u}(\tau/t),\ z(\alpha) - z(\beta)] \neq 0$$

for some α and β, $t_0 \leqq \beta < \alpha \leqq t$. If we define $D(\tau,s)$ by setting

$$D(\tau,s) = \begin{cases} \text{cov}[\tilde{u}(\tau/t),\ z(\alpha) - z(\beta)], & \beta \leqq s \leqq \alpha \\ 0, & \text{otherwise} \end{cases}$$

then $D(\tau,s)$ is integrable and

$$y(t) = \int_{t_0}^{t} D(\tau,s)dz(s) = \text{cov}[\tilde{u}(\tau/t),\ z(\alpha) - z(\beta)](z(\alpha) - z(\beta)).$$

Thus,

$$E[\langle x^*,\tilde{u}(\tau/t)\rangle\langle x^*,y(t)\rangle] = x^*E[\tilde{u}(\tau/t)\circ y(t)]x$$

$$= x^*E[\tilde{u}(\tau/t)\circ \mathrm{cov}[\tilde{u}(\tau/t),z(\alpha)-z(\beta)](z(\alpha)-z(\beta))x$$

$$= x^*\mathrm{cov}[\tilde{u}(\tau/t),z(\alpha)-z(\beta)](\mathrm{cov}[\tilde{u}(\tau/t),z(\alpha)-z(\beta)])^*x.$$

But (3.30) implies that the right-hand side of the above equation is not zero for some x; this is contradiction. The quivalence between (3.29) and (3.30)is clear and the proof of the theorem is complete. Q.E.D.

[Cororally 3-1] (Orthogonal projection lemma)
The orthogonality condition

$$E[\tilde{u}(\tau/t)\circ\hat{u}(\zeta/t)] = 0 \tag{3.32}$$

holds for any time ζ.
[Proof] Setting $z(\alpha) - z(\beta) = dz(\beta)$ in (3.30) and operating $\tilde{F}(\zeta,\beta)^*$ to (3.30) yields

$$E[\tilde{u}(\tau/t)\circ dz(\beta)]F(\zeta,\beta)^* = 0.$$

Integrating each side from t_0 to t yields

$$E[\tilde{u}(\tau/t)\circ \int_{t_0}^{t} \tilde{F}(\zeta,\beta)dz(\beta)] = E[\tilde{u}(\tau/t)\circ \hat{u}(\zeta/t)] = 0. \qquad \text{Q.E.D.}$$

[Lemma 3-7] (Uniqueness of the optimal kernel)
Let $\tilde{F}(\tau,s)$ be the optimal kernel satisfying the Wiener-Hopf equation (3.29) and let $\tilde{F}(\tau,s)+F_\Delta(\tau,s)$ be also the optimal kernel function satisfying the Wiener-Hopf equation (3.29). Then it follows that $F_\Delta(\tau,s)=0$, for $t_0 \leqq s \leq t$, that is, the optimal kernel function is unique.
[Proof] Let us define $\hat{u}_\Delta(\tau/t)$ as

$$\hat{u}_\Delta(\tau/t) = \int_{t_0}^{t} (\tilde{F}(\tau,s) + F_\Delta(\tau,s))dz(s) = \hat{u}(\tau/t) + \int_{t_0}^{t} F_\Delta(\tau,s)dz(s).$$

From Corollary 3-1 we get

$$E[(\hat{u}_\Delta(\tau/t) - \hat{u}(\tau/t))\circ(\hat{u}_\Delta(\tau/t) - \hat{u}(\tau/t))] = 0.$$

In other words,

$$\Delta_t = \mathrm{cov}[\int_{t_0}^{t} F_\Delta(\tau,s)dz(s), \int_{t_0}^{t} F_\Delta(\tau,s)dz(s)] = 0.$$

But from Lemma 3-1 we get

$$\Delta_t = \mathrm{cov}[\int_{t_0}^{t} F_\Delta(\tau,s)c(s)x(s)ds, \int_{t_0}^{t} F_\Delta(\tau,s)c(s)x(s)ds]$$

$$+ \int_{t_0}^{t} F_\Delta(\tau,s)R(s)F_\Delta^*(\tau,s)ds.$$

Since R(s) is positive definite for all s, we immediately conclude that $F_\Delta(\tau,s) = 0$. Q.E.D.

[Cororally 3-2] The Wiener-Hopf equation (3.29) is rewritten for the prediction, filtering, and smoothing problems as follows:

(i) Prediction problem($\tau > t$)

$$\text{cov}[\int_{t_0}^{t} A(\tau,s)dz(s), z(\alpha)-z(\beta)] = \text{cov}[u(\tau), z(\alpha)-z(\beta)] \tag{3.33}$$

for $t_0 \leqq \beta < \alpha < t$.

(ii) Filtering problem($\tau = t$)

$$\text{cov}[\int_{t_0}^{t} F(t,s)dz(s), z(\alpha)-z(\beta)] = \text{cov}[u(t), z(\alpha)-z(\beta)] \tag{3.34}$$

for $t_0 \leqq \beta < \alpha < t$.

(iii) Smoothing problem($\tau < t$)

$$\text{cov}[\int_{t_0}^{t} B(\tau,t,s)dz(s), z(\alpha)-z(\beta)] = \text{cov}[u(\tau), z(\alpha)-z(\beta)] \tag{3.35}$$

for $t_0 \leqq \beta < \alpha < t$.

3.5. Derivation of the optimal prediction estimator

In this section we derive the optimal prediction estimator by using the Wiener-Hopf equation in the preceding section.

[Theorem 3-2] The optimal prediction estimator is given by

$$d\hat{u}(\tau/t) = a(\tau)\hat{u}(\tau/t)d\tau, \quad \tau > t \tag{3.36}$$

where the initial value is $\hat{u}(t/t)$ of the optimal filtering estimate.

[Proof] From (3.33) we get

$$\frac{d}{d\tau}\text{cov}[u(\tau), \Delta z_\alpha] = \frac{d}{d\tau}\text{cov}[\int_{t_0}^{t} A(\tau,s)dz(s), \Delta z_\alpha]$$

$$\Delta z_\alpha = z(\alpha) - z(\beta).$$

It follows from Lemma 3-3 and independence of $w(\tau)$ and Δz_α that

$$\text{cov}[a(\tau)u(\tau), \Delta z_\alpha] = \text{cov}[\int_{t_0}^{t} \frac{\partial A(\tau,s)}{\partial\tau} dz(s), \Delta z_\alpha].$$

Applying the Wiener-Hopf equation (3.33) to the above equation we get

$$\text{cov}[\int_{t_0}^{t} (a(\tau)A(\tau,s) - \frac{\partial A(\tau,s)}{\partial\tau})dz(s), \Delta z_\alpha] = 0.$$

Hence, from Lemma 3-7 we get

$$\frac{\partial A(\tau,s)}{\partial\tau} = a(\tau)A(\tau,s). \tag{3.37}$$

Thus, from (3.26) it follows that

$$\frac{d\hat{u}(\tau/t)}{d\tau} = \int_{t_0}^{t} (\frac{\partial A(\tau,s)}{\partial\tau})dz(s) = a(\tau)\hat{u}(\tau/t). \qquad \text{Q.E.D.}$$

[Corollary 3-3] The optimal prediction error $\tilde{u}(\tau/t)$ is governed by

$$d\tilde{u}(\tau/t) = a(\tau)\tilde{u}(\tau/t)dt + b(\tau)dw(\tau) \tag{3.38}$$

with the initial condition $\tilde{u}(t/t)$ of the optimal filtering error.

[Theorem 3-3] The optimal prediction error covariance operator $p(\tau/t)$ is given by

$$\frac{dp(\tau/t)}{d\tau} = a(\tau)p(\tau/t) + p(\tau/t)a^*(\tau) + \tilde{Q}(\tau), \quad \tau > t \tag{3.39}$$

$$\tilde{Q}(\tau) = b(\tau)Q(\tau)b^*(\tau)$$

where the initial condition is the optimal filtering error covariance operator $p(t/t)$, or equivalently

$$p(\tau/t) = U(\tau,t)p(t/t)U^*(\tau,t) + \int_t^\tau U(\tau,s)\tilde{Q}(s)U^*(\tau,s)ds. \tag{3.40}$$

[Proof] The solution of (3.38) can be represented by

$$\tilde{u}(\tau/t) = U(\tau,t)\tilde{u}(t/t) + \int_\tau^t U(\tau,s)b(s)dw(s).$$

Thus, from Lemma 3-1 we get

$$p(\tau/t) = \text{cov}[\tilde{u}(\tau/t),\tilde{u}(\tau/t)] = U(\tau,t)p(t/t)U^*(\tau,t) + \int_\tau^t U(\tau,s)\tilde{Q}(s)U^*(\tau,s)ds.$$

Q.E.D.

3.6 Derivation of the optimal filtering estimator

Let us derive the optimal filtering estimator by using the Wiener-Hopf equation (3.34).

[Lemma 3-8] The optimal filtering kernel $F(t,s)$ is given by

$$\frac{\partial F(t,s)}{\partial t} = a(t)F(t,s) - F(t,t)c(t)F(t,s) \tag{3.41}$$

for $t_0 \leq s \leq t$.

[Proof] Since $\hat{u}(t/t)$ is a solution of the filtering problem, by virtue of (3.34) we get

$$\frac{d}{dt}\text{cov}[u(t),z(\alpha)-z(\beta)] = \frac{d}{dt}\text{cov}\left[\int_{t_0}^t F(t,s)dz(s),z(\alpha)-z(\beta)\right].$$

It follows from (3.17), (3.21), and (3.34) that

$$\text{cov}\left[a(t)\int_{t_0}^t F(t,s)dz(s),z(\alpha)-z(\beta)\right] = \text{cov}\left[\int_{t_0}^t \frac{\partial F(t,s)}{\partial t}dz(s)\right.$$

$$\left. + F(t,t)c(t)\int_{t_0}^t F(t,s)dz(s),z(\alpha)-z(\beta)\right]$$

or equivalently, that

$$\text{cov}\left[\int_{t_0}^t \left[a(t)F(t,s) - \frac{\partial F(t,s)}{\partial t} - F(t,t)c(t)F(t,s)\right]dz(s),z(\alpha)-z(\beta)\right] = 0.$$

Setting

$$F_\Delta(t,s) = a(t)F(t,s) - \partial F(t,s)/\partial t - F(t,t)c(t)F(t,s),$$

we observe that $F(t,s)+F_\Delta(t,s)$ also satisfies the Wiener-Hopf equation (3.34). Thus, from Lemma 3-7 we get (3.41). Q.E.D.

[Theorem 3-4] The optimal filtering estimate $\hat{u}(t/t)$ is given by

$$d\hat{u}(t/t) = a(t)\hat{u}(t/t)dt + F(t,t)d\nu(t),\ \hat{u}(t_0/t_0) = 0 \tag{3.42}$$

$$d\nu(t) = dz(t) - c(t)\hat{u}(t/t)dt. \tag{3.43}$$

[Proof] We know that

$$\hat{u}(t/t) = \int_{t_0}^{t} F(t,s)dz(s) = \int_{t_0}^{t} F(t,s)c(s)u(s)ds + \int_{t_0}^{t} F(t,s)dv(s).$$

Now, let us observe that

$$\int_{t_0}^{t} (a(s) - F(s,s)c(s))\hat{u}(s/s)\, ds = \int_{t_0}^{t} (a(s) - F(s,s)c(s)) \int_{t_0}^{s} F(s,\alpha)c(\alpha)$$

$$u(\alpha)d\alpha ds + \int_{t_0}^{t} (a(s) - F(s,s)c(s)) \int_{t_0}^{s} F(s,\alpha)dv(\alpha)ds.$$

By the rigorous versions of the Fubini theorem(Falb, 1967), we get

$$\int_{t_0}^{t} (a(s) - F(s,s)c(s)) \int_{t_0}^{s} F(s,\alpha)c(\alpha)u(\alpha)d\alpha ds$$

$$= \int_{t_0}^{t} (\int_{\alpha}^{t} (a(s) - F(s,s)c(s))F(s,\alpha)ds)c(\alpha)u(\alpha)d\alpha,$$

$$\int_{t_0}^{t} (a(s) - F(s,s)c(s)) \int_{t_0}^{s} F(s,\alpha)dv(\alpha)ds$$

$$= \int_{t_0}^{t} (\int_{\alpha}^{t} (a(s) - F(s,s)c(s))F(s,\alpha)ds)dv(\alpha),$$

and hence, in view of (3.14),

$$\int_{t_0}^{t} (a(s) - F(s,s)c(s))\hat{u}(s/s)ds = \int_{t_0}^{t} (\int_{\alpha}^{t} \frac{\partial F(s,\alpha)}{\partial s} ds)dz(\alpha)$$

$$= \int_{t_0}^{t} (F(t,\alpha) - F(\alpha,\alpha))dz(\alpha) = \hat{u}(t/t) - \int_{t_0}^{t} F(\alpha,\alpha)dz(\alpha).$$

Differentiating each side with respect to t yields

$$d\hat{u}(t/t) = a(t)\hat{u}(t/t)dt + F(t,t)(d z(t) - c(t)\hat{u}(t/t)dt). \qquad \text{Q.E.D.}$$

[Corollary 3-4] The optimal filtering error $\tilde{u}(t/t)$ is governed by

$$d\tilde{u}(t/t) = (a(t) - F(t,t)c(t))\tilde{u}(t/t)dt + b(t)dw(t) - F(t,t)dv(t)$$

with the initial condition $\tilde{u}(t_0/t_0) = u_0$.

[Theorem 3-5] The optimal filtering gain $F(t,t)$ is given by

$$F(t,t) = p(t/t)c^*(t)R^{-1}(t) \tag{3.45}$$

where $p(t/t)$ is a solution of the Riccati equation

$$\frac{dp(t/t)}{dt} = a(t)p(t/t) + P(t/t)a^*(t) + \tilde{Q}(t) - p(t/t)\tilde{R}(t)p(t/t), \tag{3.46}$$

$$\tilde{R}(t) = c^*(t)R^{-1}(t)c(t)$$

with $p(t_0/t_0) = p_0 (= \text{cov}[u_0,u_0])$.

[Proof] Let us set

$$y(\sigma) = \int_{t_0}^{\sigma} c(s)u(s)ds = z(\sigma) - \int_{t_0}^{\sigma} dv(s).$$

Then by direct computation we get

$$\frac{d\text{cov}[\tilde{u}(t/t),y(\sigma)]}{d} = U(t,t_0)P_0U^*(\sigma,t_0)c^*(\sigma)$$

$$+ \int_{t_0}^{t} U(t,s)\tilde{Q}(s)U^*(t_0,s)dsU^*(\sigma,t_0)c^*(\sigma)$$

$$- \psi(t,t_0)P_0U^*(\sigma,t_0)c^*(\sigma) - \int_{t_0}^{t} \psi(t,s)\tilde{Q}(s)U^*(\sigma,t_0)c^*(\sigma)ds$$

and from (3.34)

$$\text{cov}[\hat{u}(t/t),y(\sigma)] = \int_{t_0}^{\sigma} F(t,s)R(s)ds$$

where

$$\psi(t,s) = \int_{s}^{t} F(t,\alpha)c(\alpha)U(\alpha,s)d\alpha. \tag{3.47}$$

Thus, it follows that for $\sigma < t$

$$F(t,\sigma)R(\sigma) = (\Delta(t,t_0)p_0 + \int_{t_0}^{t} \Delta(t,s)\tilde{Q}(s)U^*(t_0,s)ds)U^*(\sigma,t_0)c^*(\sigma)$$

where

$$\Delta(t,s) = U(t,s) - \psi(t,s).$$

Since any regulated function is equivalent to a function continuous on the left in an almost everywhere sense, we can take limits as σ goes to t from below in the above equation and thus, deduce that

$$F(t,t)R(t) = p(t/t)c^*(t)$$

where

$$p(t/t) = [\Delta(t,t_0)p_0 + \int_{t_0}^{t} \Delta(t,s)Q(s)U^*(t_0,s)ds]U^*(t,t_0). \tag{3.48}$$

Clearly, $p(t_0/t_0) = p_0$.

Now, we note that from (3.7)

$$\frac{\partial U(t,s)}{\partial t} = a(t)U(t,s)$$

and from (3.41) and (3.47)

$$\partial\psi(t,s)/\partial t = F(t,t)c(t)\Delta(t,s) + a(t)\psi(t,s). \tag{3.49}$$

Thus, we get

$$\frac{\partial\Delta(t,s)}{\partial t} = a(t)\Delta(t,s) - F(t,t)c(t)\Delta(t,s). \tag{3.50}$$

Differentiating (3.48) with respect to t and using (3.48) and (3.50) yields

$$\frac{dp(t/t)}{dt} = a(t)p(t/t) + p(t/t)a(t) + \tilde{Q}(t) - p(t/t)\tilde{R}(t)p(t/t).$$

It is clear from (3.48) that the initial condition of p(t/t) is given by p_0. Q.E.D.
[Corollary 3-5] p(t/t) denotes the optimal filtering error covariance operator, that is,

$$p(t/t) = \text{cov}[\tilde{u}(t/t),\tilde{u}(t/t)]. \tag{3.51}$$

[Proof] Using the orthogonality condition $\text{cov}[\tilde{u}(t/t),\hat{u}(t/t)] = 0$ yields

$$\text{cov}[\tilde{u}(t/t),\tilde{u}(t/t)] = \text{cov}[\tilde{u}(t/t),u(t)] = \text{cov}[u(t),u(t)] - \text{cov}[\hat{u}(t/t),u(t)].$$

But

$$\text{cov}[u(t),u(t)] = (U(t,t_0)p_0 + \int_{t_0}^{t} U(t,s)\tilde{Q}(s)U^*(t_0,s)ds)U^*(t,t_0)$$

and

$$\text{cov}[\hat{u}(t/t),u(t)] = (\psi(t,t_0)p_0 + \int_{t_0}^{t} \psi(t,s)\tilde{Q}(s)U^*(t_0,s)ds)U^*(t,t_0).$$

Thus, we get

$$\text{cov}[\tilde{u}(t/t),\tilde{u}(t/t)] = [\Delta(t,t_0)p_0 + \int_{t_0}^{t} \Delta(t,s)\tilde{Q}(s)U^*(t,s)ds]U^*(t,t_0) = p(t/t).$$

Q.E.D.

Note that p(t/t) is symmetric.

3.7. Derivation of the optimal smoothing estimator

In this section we derive the optimal smoothing estimator by using the Wiener-Hopf equation in the preceding section.
[Lemma 3-9] The optimal smoothing kernel $B(\tau,t,s)$ is given by

$$\frac{\partial B(\tau,t,s)}{\partial t} = - B(\tau,t,t)c(t)F(t,s) \tag{3.52}$$

where F(t,s) is given by (3.41).
[Proof] Since $\hat{u}(\tau/t)$ is an optimal solution of the smoothing problem, by virtue of the Wiener-Hopf equation (3.35), we get

$$\frac{d}{dt}\text{cov}[u(\tau),z(\alpha) - z(\beta)] = \frac{d}{dt}\text{cov}[\int_{t_0}^{t} B(\tau,t,s)dz(s),z(\alpha) - z(\beta)].$$

It follows from Lemma 3-6 and (3.34) that

$$\text{cov}[\int_{t_0}^{t} \frac{\partial B(\tau,t,s)}{\partial t} dz(s) + B(\tau,t,t)c(t)u(t), z(\alpha) - z(\beta)]$$

$$= \text{cov}[\int_{t_0}^{t} (\frac{\partial B(\tau,t,s)}{\partial t} + B(\tau,t,t)c(t)F(t,s))dz(s), z(\alpha) - z(\beta)] = 0.$$

Hence, from Lemma 3-7 we get

$$\frac{\partial B(\tau,t,s)}{\partial t} = - B(\tau,t,t)c(t)F(t,s). \qquad \text{Q.E.D.}$$

[Theorem 3-6] The optimal smoothing estimate $\hat{u}(\tau/t)$ satisfies the following stochastic differential equation;

$$d\hat{u}(\tau/t) = B(\tau,t,t)d\nu(t) \tag{3.53}$$

$$d\nu(t) = dz(t) - c(t)\hat{u}(t/t)dt.$$

[Proof] Applying the standard Fubini theorem(Falb, 1967) and Lemma 3-9 yields

$$- \int_{t_0}^{t} B(\tau,s,s)c(s)\hat{u}(s/s)ds = - \int_{t_0}^{t} B(\tau,s,s)c(s)[\int_{t_0}^{s} F(s,\alpha)dz(\alpha)]ds$$

$$= \int_{t_0}^{t} [\int_{\alpha}^{t} \frac{\partial B(\tau,s,\alpha)}{\partial s} ds] dz(\alpha) = \int_{t_0}^{t} [B(\tau,t,\alpha) - B(\tau,\alpha,\alpha)]dz(\alpha)$$

$$= \hat{u}(\tau/t) - \int_{t_0}^{t} B(\tau,\alpha,\alpha)dz(\alpha).$$

Hence, the theorem follows from this relation by direct differentiation. Q.E.D.

[Corollary 3-6] The optimal smoothing error $\tilde{u}(\tau/t)$ satisfies

$$d\tilde{u}(\tau/t) = - B(\tau,t,t)d\nu(t), \tau < t \tag{3.54}$$

with the initial condition $\tilde{u}(\tau/\tau)$.
Then we get the following theorem.

[Theorem 3-7] The optimal smoothing gain $B(\tau,t,t)$ is given by

$$B(\tau,t,t) = D(\tau/t)c^*(t)R^{-1}(t), \tag{3.55}$$

$$D(\tau/t) = \text{cov}[\tilde{u}(\tau/t),\tilde{u}(t/t)] \tag{3.56}$$

where $D(\tau/t)$ satisfies the following equation

$$\frac{dD(\tau/t)}{dt} = D(\tau/t)(a(t) - p(t/t)\tilde{R}(t))^* \tag{3.57}$$

$$D(\tau/\tau) = p(\tau/\tau).$$

[Proof] Let us set

$$y(\sigma) = \int_{t_0}^{\sigma} c(s)u(s)ds = z(\sigma) - \int_{t_0}^{\sigma} dv(s).$$

By the direct computation we get

$$\frac{d}{d\sigma} \text{cov}[\tilde{u}(\tau/t),y(\sigma)] = \text{cov}[\tilde{u}(\tau/t),c(\sigma)u(\sigma)] = \text{cov}[\tilde{u}(\tau/t),u(\sigma)]c^*(\sigma)$$

and by Theorem 3-1

$$\text{cov}[\tilde{u}(\tau/t),y(\sigma)] = \text{cov}[\tilde{u}(\tau/t),z(\sigma) - \int_{t_0}^{\sigma} dv(s)] = \text{cov}[\hat{u}(\tau/t), \int_{t_0}^{\sigma} dv(s)]$$

$$= \mathrm{cov}[\int_{t_0}^{t} B(\tau,t,s)dz(s), \int_{t_0}^{\sigma} dv(s)] = \int_{t_0}^{\sigma} B(\tau,t,s)R(s)ds.$$

Hence, it follows that for $\sigma < t$

$$B(\tau,t,\sigma)R(\sigma) = \mathrm{cov}[\tilde{u}(\tau/t),u(\sigma)]c^*(\sigma).$$

Since any regulated function is equivalent in an almost everywhere sense to a function continuous on the left, we can take limits as σ goes to t from below in the above relation and thus deduce that

$$B(\tau,t,t)R(t) = \mathrm{cov}[\tilde{u}(\tau/t),u(t)]c^*(t). \tag{3.58}$$

But from Corollary 3-1 we get

$$\mathrm{cov}[\tilde{u}(\tau/t),u(t)] = \mathrm{cov}[\tilde{u}(\tau/t),\tilde{u}(t/t)] = D(\tau/t). \tag{3.59}$$

Thus, we get

$$B(\tau,t,t) = D(\tau/t)c^*(t)R^{-1}(t).$$

From Corollary 3-1 and (3.59) we get

$$D(\tau/t) = \mathrm{cov}[u(\tau),\tilde{u}(t/t)]. \tag{3.60}$$

But from Corollary 3-4 we get

$$\tilde{u}(t/t) = V(t,\tau)\tilde{u}(\tau/\tau) + \int_{\tau}^{t} V(t,s)b(s)dw(s) - \int_{\tau}^{t} V(t,s)F(s,s)dv(s) \tag{3.61}$$

where

$$\frac{\partial V(t,\tau)}{\partial t} = (a(t) - F(t,t)c(t))V(t,\tau)$$

$$V(\tau,\tau) = I.$$

Then we get

$$D(\tau/t) = \mathrm{cov}[u(\tau),\tilde{u}(\tau/\tau)]V^*(t,\tau) = p(\tau/\tau)V^*(t,\tau). \tag{3.62}$$

By direct differentiation of each side in the above equation we get

$$\frac{dD(\tau/t)}{dt} = p(\tau/\tau)V^*(t,\tau)(a(t) - F(t,t)c(t))^*.$$

Using (3.62) and (3.54) we get

$$\frac{dD(\tau/t)}{dt} = p(\tau/t)(a(t) - p(t/t)\tilde{R}(t))^*$$

$$D(\tau/\tau) = p(\tau/\tau).$$

Q.E.D.

[Theorem 3-8] The optimal smoothing error covariance operator $p(\tau/t) = \mathrm{cov}[\tilde{u}(\tau/t), \tilde{u}(\tau/t)]$ is given by

$$\frac{dp(\tau/t)}{dt} = - D(\tau/t)\tilde{R}(t)D^*(\tau/t) \tag{3.63}$$

with the initial condition $p(\tau/\tau)$.
[Proof] From Corollary 3-6 we get

$$\tilde{u}(\tau/t) = \tilde{u}(\tau/\tau) - \int_{\tau}^{t} B(\tau,s,s)d\nu(s)$$

and from Corollary 3-1

$$p(\tau/t) = \mathrm{cov}[u(\tau),\tilde{u}(\tau/t)].$$

Thus, we get

$$p(\tau/t) = \mathrm{cov}[u(\tau),\tilde{u}(\tau/\tau)] - \mathrm{cov}[u(\tau), \int_{\tau}^{t} B(\tau,s,s)d\nu(s)].$$

But

$$d\nu(s) = dz(s) - c(s)\hat{u}(s/s)ds = c(s)\tilde{u}(s/s)ds + dv(s).$$

Then we get

$$p(\tau/t) = p(\tau/\tau) - \int_{\tau}^{t} \mathrm{cov}[u(\tau),\tilde{u}(s/s)]c^*(s)B^*(\tau,s,s)ds$$

$$= p(\tau/\tau) - \int_{\tau}^{t} D(\tau/s)c^*(s)B^*(\tau,s,s)ds$$

and from (3.55)

$$p(\tau/t) = p(\tau/\tau) - \int_{\tau}^{t} D(\tau/s)c^*(s)R^{-1}(s)c(s)D^*(\tau/s)ds$$

$$= p(\tau/\tau) - \int_{\tau}^{t} D(\tau/s)\tilde{R}(s)D^*(\tau/s)ds.$$

By direct differentiation of the above equation with respect to t we get

$$\frac{dp(\tau/t)}{dt} = - D(\tau/t)\tilde{R}(t)D^*(\tau/t).$$ Q.E.D.

Therefore, we get the following theorem.
[Theorem 3-9] The optimal smoothing estimate $\hat{u}(\tau/t)$ is given by

$$\hat{u}(\tau/t) = \hat{u}(\tau/\tau) - \int_{\tau}^{t} D(\tau/s)c^*(s)R^{-1}(s)d\nu(s) \tag{3.64}$$

$$D(\tau/t) = p(\tau/\tau)V^*(t,\tau) \tag{3.65}$$

$$\frac{\partial V(t,\tau)}{\partial t} = (a(t) - F(t,t)c(t))V(t,\tau) \tag{3.66}$$

$$V(\tau,\tau) = I \tag{3.67}$$

and the optimal smoothing error covariance operator $p(\tau/t)$ is given by

$$p(\tau/t) = p(\tau/\tau) - \int_{\tau}^{t} D(\tau/s)\tilde{R}(s)D^*(\tau/s)ds. \tag{3.68}$$

For the optimal smoothing estimation problem there are three types of the smoothing problem, that is, (i) fixed-point, (ii) fixed-interval, and (iii) fixed-lag smoothing estimation problems. These types can be easily derived by using Theorem 3-9.
(i) Fixed-point smoothing problem

$$d\hat{u}(t_1/t_1) = D(t_1/t)c^*(t)R^{-1}(t)d\nu(t) \tag{3.69}$$

$$dD(t_1/t)/dt = D(t_1/t)(a(t) - p(t/t)\tilde{R}(t))^* \tag{3.70}$$

$$D(t_1/t_1) = p(t_1/t_1) \tag{3.71}$$

$$dp(t_1/t)/dt = -D(t_1/t)\tilde{R}(t)D^*(t_1/t). \tag{3.72}$$

(ii) Fixed-interval smoothing problem

$$d\hat{u}(t/T) = a(t)\hat{u}(t/T)dt + \tilde{Q}(t)p^{-1}(t/t)(\hat{u}(t/T) - \hat{u}(t/t))dt \tag{3.73}$$

$$\frac{dp(t/T)}{dt} = (a(t) + \tilde{Q}(t)p^{-1}(t/t))p(t/T) + p(t/T)(a(t) + \tilde{Q}(t)p^{-1}(t/t))^* - \tilde{Q}(t). \tag{3.74}$$

(iii) Fixed-lag smoothing problem

$$d\hat{u}(t/t+\nabla) = a(t)\hat{u}(t/t+\nabla)dt + G(t+\nabla)p(t+\nabla/t+\nabla)c^*(t+\nabla)R^{-1}(t+\nabla)d\nu(t+\nabla) + \tilde{Q}(t)p^{-1}(t/t)(\hat{u}(t/t+\nabla) - \hat{u}(t/t))dt \tag{3.75}$$

$$\frac{dG(t+\nabla)}{dt} = \tilde{Q}(t)p^{-1}(t/t)G(t+\nabla) - G(t+\nabla)\tilde{Q}(t+\nabla)p^{-1}(t+\nabla/t+\nabla) + a(t)G(t+\nabla) - G(t+\nabla)a^*(t+\nabla) \tag{3.76}$$

$$G(t_0+\nabla) = D(t_0/t_0+\nabla) \tag{3.77}$$

$$\frac{dp(t/t+\nabla)}{dt} = (a(t) + \tilde{Q}(t)p^{-1}(t/t))p(t/t+\nabla) + p(t/t+\nabla)(a(t) + \tilde{Q}(t)p^{-1}(t/t))^* - D(t/t+\nabla)\tilde{R}(t)D^*(t/t+\nabla) - \tilde{Q}(t) \tag{3.78}$$

with the initial condition $p(t_0/t_0+\nabla)$.

3.8. Related works

The other rigorous approaches to the optimal estimation problems for distributed parameter systems will be reviewed in this section.

(i) Kushner's approach: Some special cases of distributed parameter systems

Kushner(1970) considers the filtering problem for two classes of distributed parameter systems. The first is a second order parabolic partial differential equation with an integral operator in the state equation and Dirichlet-type boundary conditions. The observation is a weighted average of the state vector over space and time, and the disturbance processes are Ito differentials of Wiener processes in time only, i.e., Kushner considers the degenerate case where the noise is uniformly distributed in space. Taking his best estimate to be the conditional expectation of the state, given the observation process, he derives filtering equations for this estimate and for the covariance of the error, rigorously establishing existence and uniqueness of the solutions. The second problem is a second order parabolic partial differential equation with mixed boundary conditions. Again the disturbance processes are only random in time, but now Kushner allows noise on the boundary as well as in the state equation. Taking his observations averaged over the boundary this time, he does a similar analysis to the first problem.

(ii) Bensoussan: A generalized random variable approach to linear filtering problem

Bensoussan(1971a) has solved the filtering problem for a very wide class of linear distributed parameter systems by an ingenious application of Lions theory of the linear quadratic control problem for distributed parameter systems (Lions, 1968), Schwaz theory of random measures (Schwarz, 1973), and Gel'fand and Vilenkin's concept of generalized random variables(Gel'fand and Vilenkin, 1964). In order to put the theory of generalized random variables on a firm mathematical basis, he introduces a random linear functional and develops a linear estimation theory for the random linear functional along with similar lines to the real-valued theory.

(iii) Bensoussan: A stochastic differential equation theory

Although Bensoussan's approach in (ii) provides some mathematical justification for a linear estimation problem for distributed parameter systems, it does have the weakness that the state u(t) is not a proper stochastic process. In order to remedy this weakness, Bensoussan(1971a)introduces an infinite-dimensional Wiener process and defines the generalized Ito integral in an infinite-dimensional system. Then he derives the optimal filter by using the conditional expectation operator.

(iv) Curtain: A generalized infinite-dimensional theory

A more recent generalization of the Kalman-Bucy filter to infinite dimensions is by Curtain(1975a,b, 1976). The idea is similar to that of Falb(1976)'s and Bensoussan(1971a)'s approaches. Again she uses Hilbert space probability theory and defines a Wiener process and the Ito integral in infinite-dimensional spaces. By using the evolution operator theory by Kato(1961) Curtain derives the optimal filter which is an extension of Falb(1967)'s result.

(v) Curtain; Vinter and Mitter: Filtering of stochastic evolution equations and applications to delay equations

A very recent extension of the ideas in the work by Curtain(1975a) has been made simultaneously and independently by Curtain(1974) and by Vinter and Mitter (1974) to study filtering for differential delay systems. Although differential delay systems are not intrinsically infinite-dimensional, they may be formulated as abstract equations on a Hilbert space. In this setting, Curtain(1974) and Vinter and Mitter(1974) derive an abstract Kalman-Bucy filter similar to the distributed parameter case. Furthermore, Vinter and Mitter consider stability properties of the filter.

(vi) Omatu: A generalized parameter optimization method

Omatu derives the optimal smoothing estimators by using functional analysis and generalized Ito stochastic differential rules (Omatu and Soeda, 1976). In this case, the optimal estimate is determined based on the unbiased and minimum variance estimation error criterion. Furthermore, this method is extended to derive the optimal prediction, filtering, and smoothing estimators by a unified approach(Omatu and Soeda, 1978).

(vii) Balakrishnan: Finitely additive white noise appraoch

This approach is essentially a combination of the generalized random variable approach and the abstract evolution approach. Balakrishnan(1974) considers stochastic differential equations on a Hilbert space where he introduces a white noise process in Hilbert spaces. Then he uses the estimation theory of second order weak random variables to obtain a best mean squares filter and obtains the actual Kalman-Bucy equations for the filter using techniques from integral equation theory. With these very weak assumptions on the white noise process, Balakrishnan can extend these filtering results to very general systems, including those with boundary noise. He also considers the problems of stochastic control and identification for infinite dimensional systems.

REFERENCES

Atre, S. R., and S. S. Lamba (1972a). A note on Kalman-Bucy filtering for linear distributed-parameter systems, IEEE Trans. Auto. Contr., 17, 712-713.
Atre, S. R., and S. S. Lamba (1972b). Filtering for linear distributed-parameter systems via boundary measurements, Proc. IEE, 119, 757-761.
Atre, S. R., and S. S. Lamba (1972c). Optimal estimation in distributed processes using the innovation approach, IEEE Trans. Auto. Contr., 17, 710-712.
Atre, A. R., and S. S. Lamba (1972d). Derivation of an optimal estimator for distributed-parameter systems via maximum principle, IEEE Trans. Auto. Contr., 17, 388-390.
Balakrishnan, A.V. (1974). Stochastic optimization theory in Hilbert spaces 1, Appl. Math. and Optimization, 1, 97-120.
Balakrishnan, A. V.,and J. L. Lions (1967). State estimation for infinite dimensional systems, J. Comput. Systems Science, 1, 391-403.
Bencala, K. E., and J. H. Seinfeld (1979). Distributed parameter filtering: Boundary noise and discrete observations, Int. J. Systems Science, 10, 493-512.
Bensoussan, A. (1969). L'identification et le filtrage, Thesis IRIA.
Bensoussan, A. (1971a). Filtrage Optimal des Systemes Lineaires, Dunod, Paris.
Bensoussan, A. (1971b). Filtering theory: A comparison between lumped and distributed systems, Mem. Contr. Int. Sobre Sistemas Redes y Computadores, Mexico, 160-163.
Curtain, R. F. (1974). A Kalman-Bucy filtering theory for affine hereditary differential equations, Lecture Notes in Economics and Math. Systems, 107, 22-43.
Curtain, R. F. (1975a). Infinite-dimensional filtering, SIAM J. Control, 13, 89-104.
Curtain, R. F. (1975b). Estimation theory for abstract evolution equations excited by general white noise processes, Report No.40, Control Theory Centre, Univ. of Warwick, England, 1-35.
Curtain, R. F. (1975c). A survey of infinite-dimensional filtering, SIAM Review, 17, 395-411.
Curtain, R. F. (1976). Estimation theory for abstract evolution equations excited by general white noise processes, SIAM J. Control and Optimiz., 14, 1124-1150.
Curtain, R. F., and P. L.Falb(1971). Stochastic differential equations in Hilbert space, J. Differential Equations, 10, 412-430.
Curtain, R. F., and A. Ichikawa (1977). The separation principle for stochastic evolution equations, SIAM J. Control and Optimiz., 15, 367-383.
Duncan, T. L. (1972). Some Banach valued processes with applications, Lecture Notes in Mathematics, 294, Springer, New York, 85-96.
Falb,P.L.(1967) Infinite dimensional filtering: The Kalman-Bucy filter in Hilbert space, Information and Control, 11, 102-137.
Gel'fand, I. M., and N. Ya. Vilenkin (1964) Generalized Functions, vol.4: Applications to Harmonic Analysis, Academic Press, New York.
Hwang, M., J. H. Seinfeld, and G. R. Gavalas (1972) Optimal least square filtering and interpolation in distributed-parameter systems, J. Math. Anal. Appl., 39, 49-74.
Ito, S. (1966) Partial Differential Equations (in Japanese), Baifukan, Tokyo.
Kato,T. (1961) Abstract evolution equation of parabolic type in Banach and Hilbert spaces, Nagoya Math. J., 19, 93-125.
Kato, T., and H. Tanabe (1962) On the abstract evolution equation, Osaka J. Math., 14, 107-133.
Kushner, H. J. (1970) Filtering for linear distributed parameter systems, SIAM J. Control, 8, 346-359.
Kwakernaak, H. (1967) Optimal filtering in linear system with time delays, IEEE Trans. Auto. Contr., 12, 169-173.
Lions, J. L. (1968) Optimal Control of Systems Governed by Partial Differential Equations, Springer, New York.
Meditch, J. S. (1970) On state estimation for distributed-parameter systems, J. Franklin Inst., 290, 49-59.

Omatu, S., and T. Soeda (1976). Estimation problems for a linear stochastic differential equation in Hilbert space, Proc. of Int. Symp. on Stochastic Differential Equations, K. Ito ed. , Kinokuniya Book-Store, Tokyo, 341-365.

Omatu, S., and T. Soeda (1978). Optimal estimation problems for a linear distributed parameter system, Int. J. Systems Science, 9, 753-771.

Omatu, S., S. Ohe, and T. Soeda (1979). Fixed-interval smoothing for a linear distributed parameter system, Int. J. Systems Science, 10, 1219-1234.

Omatu, S., S. Ohe, and T. Soeda (1980). Fixed-lag smoothing estimator for a linear distributed parameter system, Information and Control, 44, 1-15.

Omatu, S., and J. H. Seinfeld (1981). A unified approach to discrete-time distributed parameter estimation by the least-squares method, to appear in Int. J. Systems Science, 12.

Omatu, S., T. Soeda, and Y. Tomita (1977a). Fixed-point smoothing for a stochastic distributed parameter system by using functional analysis, IEEE Trans. Auto. Contr., 22, 9-18.

Omatu, S., Y. Tomita, and T. Soeda (1977b). Fixed-point smoothing in Hilbert spaces, Information and Control, 34, 324-338.

Phillipson, G. A., and Mitter S. K. (1969). State identification of a class of linear distributed systems, Proc. 4th IFAC Congress, Warsaw.

Sakawa, Y. (1972). Optimal filtering in linear distributed-parameter systems, Int. J. Control, 16, 115-127.

Sawaragi, Y., T. Soeda, and S. Omatu (1978). Modeling, Estimation, and Their Applitions for Distributed Parameter Systems, Lecture Notes in Control and Information Sciences, 11, Springer, New York.

Schwarz, L. (1973). Random Measures on Arbitrary Topological Spaces and Cylindrical Measures, Oxford Univ. Press, London.

Seinfeld, J. H. (1968). Nonlinear estimation for partial differential equations, Chem. Eng. Science, 24, 75-83.

Seinfeld, J. H. (1971). Nonlinear filtering in distributed parameter systems, Trans. ASME, Series G 93, 157-168.

Seinfeld, J. H., and G. R. Gavalas (1969). Control of nonlinear stochastic systems, Ind. Eng. Chem. Fundam., 8, 257-261.

Thau, F. E. (1969). On optimal filtering for a class of linear distributed parameter systems, Trans. ASME, Series D 91, 173-178.

Tzafestas, S. G. (1969). Boundary and volume filtering of linear distributed-parameter systems, Electronics Letters, 5, 199-200.

Tzafestas, S. G. (1972a). Bayesian approach to distributed parameter filtering and smoothing, Int. J. Control, 15, 273-295.

Tzafestas, S. G. (1972b). On optimum distributed-parameter filtering and fixed-interval smoothing for a colored noise, IEEE Trans. Auto. Contr., 17, 448-458.

Tzafestas, S. G. (1972c). Concerning controllability and observability of linear sequential machines, Int. J. Systems Science, 3, 197-208.

Tzafestas, S. G. (1973). On the distributed parameter least squares state estimation theory, Int. J. Systems Science, 4, 273-295.

Tzafestas, S. G., (1978). Distributed Parameter Systems, ed. W. H. Ray, and D. G. Lainiotis, Dekker, New York.

Tzafestas, S. G. , and J. M. Nightingale (1968a). Optimal filtering, smoothing and prediction in linear distributed parameter systems, Proc. IEE, 115, 1207-1212.

Tzafestas, S. G., and J. M. Nightingale (1968b). Concerning optimal filtering theory of linear distributed-parameter systems, Proc. IEE, 115, 1737-1742.

Tzafestas, S. G., and J. M. Nightingale (1969). Maximum likelihood approach to the optimum filtering of distributed-parameter systems, Proc. IEE, 116, 1085-1093.

Vinter, R., and S. K. Mitter (1974). Filtering for linear stochastic hereditary differential systems, Lecture Notes in Economics and Math. Systems, 107, 1-21.

Yosida, K. (1968). Functional Analysis, Springer, New York.

CHAPTER 4

SOME METHODS FOR ANALYZING THE STABILITY AND THE HYPERSTABILITY OF DISTRIBUTED PARAMETER SYSTEMS

G. Jumarie

Department of Mathematics, University of "Québec à Montréal", Montreal, Canada

ABSTRACT

While one may think of using standard methods to analyze the stability and the hyperstability of lumped parameter systems, one cannot expect such results for distributed systems, and the approaches which we may develop must necessarily be specific in the way that they depend upon the boundary conditions, the considered domain, the class of the differential equation, the type of representation, etc... . In the present report, distributed systems will be successively defined by means of Laplace transforms in space and time, Laplace-Green kernels, and state vector representation; in each case are stated stability and hyperstability conditions the main feature of which is that they enlighten the averaging behaviour of the distributed parameter. The effects on the stability with respect to the boundary conditions are exhibited. Both continuous and sampled data systems are considered, and the consequences of space sampling are analyzed. Problems of stability and hyperstability in the presence of random disturbances are only outlined and a new practical model for distributed white noises is suggested.

KEYWORDS

Stability; hyperstability; nonlinear distributed systems; distributed sampled data; space sampling; transfer function, Laplace-Green kernel; distributed white noises.

FOREWORD

Limitations of space lead to an unavoidable trade-off between the details of the results we can state and the number of questions we may consider, in this important and broad question that is the stability of distributed systems. Thus we had to make a choice regarding the topics to be dealt with, and we have chosen to fully described the problem and the specific features of stability and hyperstability in distributed systems, together with various ways by means of which they can be tackled. The results are given without proof but with the references where the reader can find more details. The present article is an applied mathematical one, in the engineering sense of this term, and all the theoretical machinery we shall need is engineering mathematics, only, without abstract mathematical spaces.

INTRODUCTION

While the population of lumped parameter systems on the whole may be fully described with only a few models, this is not so for distributed parameter systems (DPS in the following), mainly because of their diversity. Thus most of the papers which deal with the stability of DPS are generally related to very particular systems and in effect state specific results. It is very difficult to characterize the general trends of these works. One may classify DPS's by whether they have distributed control or boundary control or both ones. Or else, one can consider DPS's classified as in parabolic, hyperbolic and elliptic systems, see for instance Bellman (1948), Prodi (1951), Narasimhan (1954), Friedman (1959), Gelig (1965). Also, one can think of DPS as distributed lumped parameter systems or, in contrast, as infinite dimensional systems as, for instance in Belman (1947) and Persidskii (1958). Kastenberg (1969, 1970, 1973) made the distinction between those DPS's which are purely dissipative, (that is to say without energy production) and those which are active (in the way that they produce energy) and directed his approach to comparison functions.

In an attempt to build up a unified approach to the question, one can generalize the well known circle criterion so useful in the theory of lumped parameter. In this trend, Freedman, Falb and Zames (1969) gave an extension which applies to a general locally compact group. Freedman, Falb and Anton (1971) combined this generalized circle criterion with the coerciveness theory for studying elliptic problems involving nonlinearities which are bounded by constant gains. The frequency domain stability criterion, originally developed by Popov (1962) for investigating the stability of lumped nonlinear systems, has also been extended to study nonlinear distributed systems (Gelig (1966); Blodgett and King (1967); Khatri (1970); Nagarajan and Khatri (1971)).

A general common feature of these works is that they deal with DPS's in which the nonlinearities f(x,z,t) satisfy such conditions as

$$a < f(x,z,t)/x < b$$

where a and b are given constants. In the present article, we have chosen to concentrate our attention on a general approach to the stability and hyperstability of the DPS in which the above limits a and b, are not constant, but rather are explicitly dependent upon the distributed parameter z.

As a first step, a brief background on stability and hyperstability describes the framework of the paper. We then consider various definitions for the class of DPS we shall investigate, and pass to the statements of the main results. We shall restrict ourselves to deterministic systems, with only a few comments regarding stochastic systems.

STABILITY AND HYPERSTABILITY

Broadly speaking *stability* is related to the behaviour of a system the input of which is subject to perturbations, and *hyperstability* deals with the behaviour of the system when its structure is disturbed as well. *Asymptotic stability* indicates that the deviations which are caused by the perturbations converge to zero; *uniform stability* is related to the dependence of the stability properties upon the initials conditions, and *stability on the whole* corresponds to the invariance with regard to the state space on the whole. In the following we shall consider distributed parameter systems (DPS) defined by the vector equation

$$\dot{x}(t,z) = f\left(x, \frac{\partial x}{\partial z_1}, \ldots, \frac{\partial x}{\partial z_m}, \ldots, \frac{\partial^K x}{\partial z_1^{k_1} \partial z_2^{k_2} \ldots \partial z_m^{k_m}}, u, z, t\right) \tag{1}$$

with given initial and boundary conditions for $x, dx/dz_1, \partial x/\partial z_m, \partial^K x/\partial z_1^{k_1} \partial z^{k_2} \ldots \partial z_m^{k_m}$. $x \in E^n$ is the state vector, u, say $u(t,z) \in E^r$, is the control vector and $z^T = (z_1, z_2, \ldots z_m) \in \Gamma_1 \subset R^m$ is the distributed parameter vector. When u=0, the system is said to be *autonomous*. An *equilibrium* position x_e for the autonomous system is such that $f(\cdot) = 0$ when $x \equiv x_e$. It is evident that we need some mathematical assumptions (continuity, differentiability,...) in order that there be a solution to Equ. (1); we shall implicitly assume that these are fulfilled without explicit specification, as these matters are not directly related to the questions of stability and hyperstability.

Autonomous Systems

We assume that E^n is endowed with a scalar product v(x,y) and we denote by N(x) the norm

$$N(x) := [\int_{\Gamma_1} v(x,x)dz]^{1/2} \tag{2}$$

(the symbol:= means that the left-hand side is defined by the right hand side; ALGOL-PASCAL symbolic code)

Definition 1. The equilibrium position $x_e = 0$ is stable in the Liapunov sense, if for every $\varepsilon > 0$, there exists $\delta(\varepsilon,t_o)>0$ such that $N[x(t_o,z)] < \delta(\varepsilon,t_o)$ implies $N[x(t,z)] < \varepsilon$ for every $t \geq t_o$. □

Definition 2. The DPS defined by Equ. (1) is asymptotically stable at $x_e = 0$ whenever it is stable in the Liapunov sense and

$$\lim N(x) = 0 \ , \ t \to \infty \tag{3}$$

Definition 3. The DPS defined by Equ. (1) is *quasi-asymptotically* stable at $x_e = 0$ provided that the following conditions are satisfied.

(i) For every $\varepsilon > 0$, there exists $\delta(\varepsilon,t_o)$ and T, with $t_o < T < \infty$, such that $N[x(t_o,z)] < \delta(\varepsilon,t_o)$ implies $N[x(t,z)] < \varepsilon$, $\forall t \geq T$;

(ii) $\lim N(x) = 0, \ t \to \infty$. □

Definition 4. The *domain of asymptotic stability* of the DPS (1) is defined as the set of the initial conditions for which the DPS is asymptotically stable. □

Definition 5. The DPS (1) has the property of *global stability* when its domain of stability is the whole space. □

Definition 6. The DPS (1) is uniformly stable for a given type of stability if its stability properties are independent of the initial instant considered. □

Definition 7. The DPS (1) is said to be absolutely stable if it is globaly stable for every functional f(.) which pertains to the admissible given class F.

Non-Autonomous Systems

Definition 8. Let H denote a given Banach space with its norm $||(\cdot)||_H$. The

DPS (1) is input-output stable provided there exists M, $0<M<\lambda<\infty$ such that

$$||x(t,z)||_H \leq M\ ||u(t,z)||_H \tag{4}$$

In some instances, when the system is subject to unknown disturbing input, it may be interesting to modify the above definition as follows:

Definition 9. The DPS (1) is modified input-output stable whenever there exist two constants M_1 and M_2, $0 < M_1, M_2 < \lambda < \infty$ such that

$$||x(t,z)||_H \leq M_1 + M_2\ ||u(t,z)||_H . \quad \square \tag{5}$$

These concepts of stability can be extended by noticing that the basic properties of conditions (4) and (5) are

$$\text{bounded } ||u||_H \rightarrow \text{bounded } ||x||_H . \tag{6}$$

It follows that every functional which implies (6) provides a possible extent of the input-output stability, therefore the following first statement.

"The DPS (1) is input-output stable provided there exists a functional Q(x,u) with

$$Q(x,u) \leq 0 \quad \forall \text{ admissible } u \tag{7}$$

such that (7) implies (6)".

With such a definition, the study of the stability of the DPS is reduced to the determination of Q(x,u) which is not easy. In contrast, if Q(x,u) along with condition (7) are given as defining the system, we need only to check that (7) is satisfied for every admissible u, to conclude that the system is stable. This notion of stability supposes that the structure of the DPS is defined by the condition

$$Q(x,u) \leq 0 . \tag{8}$$

therefore the name of *hyperstability* which has been given it by Popov. (1970).

Definition 10. The DPS defined by the equation (1) and the structural inequality (8) is hyperstable provided that condition (8) is satisfied for every admissible u. □

We refer the reader to Šiljack (1969), for instance, where he will find a complete list of various stability concepts applied to lumped parameter systems. □

DEFINITION OF DISTRIBUTED PARAMETER SYSTEMS

A Few Notations

The notations are standard; but nevertheless we bear in mind the main ones.

Let $z \in \Gamma_1 \subset \mathbb{R}^m$; $z^t = (z_1, z_2, .., z_m)$. $\Gamma_2 := \mathcal{B}(\Gamma_1)$, boundary of Γ_1. Also $\bar{\Gamma}_1 := \Gamma_1 \cup \Gamma_2$. In the same way $x(t,z) \in \mathbb{C}^n$ is a vector whose the components are $x_i(t,z)$, $(i=1,2,...,n)$. Most of the functions which we shall consider are mappings like $\{x: R_+ \times \bar{\Gamma}_1 \rightarrow \mathbb{C}^n; (t,z) \rightarrow x(t,z)\}$, and they will be precewise continuous. The scalar or inner product of $x \in \mathbb{C}^n$ with $y \in \mathbb{C}^n$ is $x^T y$. The *small norm* $|x(t,\cdot)|$ of $x(t,z)$ is

$$|x(t,\cdot)| := (\int_{\Gamma_1} x^T(t,z)x(t,z)dz)^{1/2} \qquad (9)$$

Sometimes we shall use also the *great norm* defined as

$$||x|| := (\int_0^\infty |x(t,\cdot)|^2 dt)^{1/2}. \qquad (10)$$

The two integrals above may be taken in the Riemann sense which will be largely sufficient for our purpose. □

<u>Differential Equations</u>

We introduce the following sequence of notations.

(i) $x(t,z) \in \mathbb{R}$ denotes the state of the DPS, with

$$\partial^K x/\partial z^k := \partial^{k_1+\cdot+k_m}/\partial z_1^{k_1}\, \partial z_2^{k_2} \ldots \partial z_m^{k_m}.$$

(ii) $\{a_i(z), z\in\bar{\Gamma}_1,\ i = 1,2,\ldots,n_1\}$, $\{b_i(z), z\in\Gamma_1,\ i= 1,2,\ldots,n_2\}$

and $\{c_{ij}(z'),\ z' \in \Gamma_2,\ i = 1,2,\ldots,n_3,\ j = 1,2,\ldots,n_4\}$ denote three spaces of real valued, continuous, bounded functions. (iii) L denotes a vector differential operator defined by means of the coefficients $c_{ij}(z)$, that is, $L = \sum_{ij} c_y(z)d^{i+j}/dt^i dz^j$. (iv) $f_1(x,z,t)$, $\mathbb{R}\times\Gamma_1\times\mathbb{R}_+ \to \mathbb{R}$; and $f_2(x,z',t)$, $\mathbb{R}\times\Gamma_2\times\mathbb{R}_+\to\mathbb{R}$, denote two nonlinear functions of the argument x. We shall assume that they satisfy suitable mathematical conditions in order that the partial differential equation of the DPS has a solution. (v) $e_1(t,z)$, $\mathbb{R}_+\times\Gamma_1 \to \mathbb{R}$; and $e_2(t,z)$, $\mathbb{R}_+\times\Gamma_2\to\mathbb{R}$, denote two real valued mappings which are infinitely differentiable in z and piecewise infinitely differentiable in t.

With these notations, the DPS is defined by the partial differential equation

$$\sum_{i=1}^{i_1} a_i(z)\, \partial^{k_i+K_i} x(t,z)/\partial t^{k_i} \partial z^{K_i} = v_1(t,z) \qquad (11)$$

$$v_1(t,z) = \sum_{i=1}^{i_2} b_i(z)\, \partial^{n_i+N_i} f_1(e_1-x,z)/\partial t^{n_i} \partial z^{N_i} \qquad (12)$$

$$L.x(t,z') = 0 \qquad (13)$$

with zero initial conditions.

As so defined, the DPS is a distributed control system via $v_1(t,z)$; it does not involve direct boundary control, and there is only an indirect boundary control which is defined by equation (13). Thus if we set

$$u_1(t,z) := f_1[e_1(t,z) - x(t,z),z], \qquad (14)$$

we shall have

$$x(t,z) = H_1\cdot u_1(t,z) + H_2\cdot u_2(t,z) \qquad (15)$$

where H_1 denotes the linear operator associated with $u_1(t,z)$ and H_2 is that one associated with $u_2(t,z)$. The presence of a boundary control will then appear in Condition (13) which is then resolved into two parts, that is

$$\sum_{i=1}^{i_3} d_i(z')[\partial^{l_i+L_i} x(t,z)/\partial t^{l_i} dz^{L_i}]_{\Gamma_2} = v_2(t,z') \qquad (16)$$

with

$$v_2(t,z') := \sum_{i=1}^{i_4} g_i(z')[\partial^{r_i+R_i} f_2(e-x,z)/\partial t^{r_i} \partial z^{R_i}]_{\Gamma_2}; \tag{17}$$

and

$$L_1 \cdot x(t,z') = 0, \tag{18}$$

where $d_i(z')$, $g_i(z')$, $f_2(\cdot)$ and L_1 have definitions similar to those of $c_{ij}(z')$, $f_1(\cdot)$ and L respectively. Here we shall set

$$u_2(t,z') := f_2[e_2(t,z') - x(t,z'),z'] \tag{19}$$

so that we shall also have

$$x(t,z) = H_1 \cdot u_1(t,z) + H_2 \cdot u_2(t,z'), \tag{20}$$

where u_1 is the distributed control in Γ_1 and u_2 is the distributed boundary control at Γ_2.

Integral Equation

Another way to describe DPS is by a mixed representation using a so-called Green function w.r.t. z and a convolution product in t. Thus, let $h_1(t,z,y)$, $\mathbb{R}_+ \times \bar{\Gamma}_1 \times \bar{\Gamma}_1 \to \mathbb{R}$ and $h_2(t,z,y)$, $\mathbb{R}_+ \times \bar{\Gamma}_1 \times \Gamma_2 \to \mathbb{R}$, denote two continuous functions; the DPS is then defined by the integral equation

$$x(t,z) = \int_{\Gamma_1} \int_o^t h_1(t-\tau,z,y) f_1[\varepsilon_1(\tau,y),y,\tau] d\tau\, dy + \\ + \int_{\Gamma_2} \int_o^t h_2(t-\tau,z,y') f_2[\varepsilon_2(\tau,y'),y',\tau] d\tau\, dy \tag{14}$$

$$\varepsilon_i(t,z) := e_i(t,z) - x(t,z) \quad , \ i=1,2. \tag{15}$$

Here h_1 and h_2 may be thought of as the *impulse responses* of the linear parts of the DPS, with a physical meaning similar to that of lumped parameter systems.

Laplace-Green Function

We also could use a mixed representation by means of Green function in z and Laplace transformation in t. We define

$$x(s,z) := L\{x(t,z)\} \\ := \int_o^\infty e^{-st}\, x(t,z) dt$$

with $s \in \mathbb{C}$. Also we define

$$G_i(s,z,y) := L\{h_i(t,z,y)\} \ , \ i=1,2. \tag{16}$$

Taking the Laplace transformation of Equation (14), we get the desired model, that is

$$x(s,z) = \int_{\Gamma_1} G_1(s,z,y)\ L\{f_1(\varepsilon_1,y,t)\}\, dy +$$

$$+\int_{\Gamma_2} G_2(s,z,y)\ L\{f_2(\varepsilon_2,y',t)\}dy'. \tag{17}$$

The functions G_1 and G_2 are the *distributed transfer functions* of the DPS. From a practical standpoint, numerical approximation can be obtained by discretizing w.r.t. z, to yield (we assume here that $G_2 \equiv 0$)

$$\tilde{x}(s,z_n) = h \sum_i G_1(s,z_n,z_i)\ L\{f(\varepsilon_1,z_i,t)\}$$

where h denotes the discretization span (mesh size).

Transfer Function in Space and Time

This is another standard extension of the useful transfer function for lumped parameter systems. The two-variable Laplace transformation $L^2\{x(t,z)\}$ of $x(t,z)$ is

$$x(s,r) = L^2\{x(t,z)\}$$

$$= \int_{\mathbb{R}_+^{m+1}} e^{-st-r^T z}\ x(t,z)dt\ dz \tag{17}$$

where $r^T \in \mathbb{C}^m$ is the vector $r^T := (r_1,r_2,..,r_m)$. When $\Gamma_1 \equiv \mathbb{R}_+^m$, (17) is a straightforward extension of $L\{x(t)\}$ so that the same calculation rules apply. If Γ_1 is any domain in $\mathbb{R}^m$ it is quite impossible to arrive at an operational calculus with some practical interest. However, if Γ_1 is a hyper-rectangle, then the two-variables Laplace transformation can be valuable.

As an example, consider the equation

$$\frac{\partial^2 x}{\partial j^2} - \omega_o^2 \frac{\partial^2 x}{\partial z^2} = u_1(t,z) \quad , \quad 0 \le z \le L \tag{18}$$

$$x(t,0) = x(t,L) = 0$$

with initial zero conditions. Define

$$W(s,r) := 1/(s^2-\omega_o^2 r^2)$$

Taking the two-variable Laplace transformation of (18) yields immediately

$$x(s,r) = W(s,r)\ u(s,r)$$

$$u(s,r) := u_1(s,r) + \omega_o^2[e^{-rL}\ L_r\{\tfrac{\partial x}{\partial z}|_L\} - L_r\{\tfrac{\partial x}{\partial z}|_o\}$$
$$+ \omega_o^2 r\ [e^{-rL} L_r\{x|_L\} - L_r\{x|_o\}]\ .$$

We shall see later how to take account of the inputs which are induced by the boundary conditions when one analyses the stability of the system.

State Vector

We introduce the following notations. (i) d/dz denotes the derivative operator w.r.t. z; so that

$$\partial^K x(t,z)/\partial z^K = P(\partial/\partial z) \cdot x(t,z)$$

where $P(\cdot)$ is a polynomial in d/dz_i. (ii) $A(z,d/dz)$ denotes a $n\times n$ matrix whose elements $a_{ij}(z,d/dz)$ are

$$a_{ij}(z,\partial/\partial z) := \sum_{k=1}^{n_{ij}} a_{ij,k}(z)(\partial^{K_k}/\partial z^{K_k})$$

with the $a_{ij,k}(z)$' s continuous and bounded on $\bar{\Gamma}_1$. (iii) $x^T:=(x_1,x_2,..x_n)\in \mathbb{C}^n$ denotes the n-state-vector of the system. (iv) $b_1^T(z) := [b_{11}(z),b_{12}(z),...,b_{1n}(z)] \in \mathbb{R}^n$ is continuous and bounded on Γ_1. (v) $b_2^T(z'):=[b_{21}(z'),b_{22}(z'),...,b_{2n}(z')] \in \mathbb{R}^n$ is continuous and bounded on Γ_2.

The DPS is then defined by the partial differential vector equation

$$\dot{x}(t,z) = A(z,\frac{\partial}{\partial z})\cdot x(t,z) + b_1(z)u_1(t,z)+b_2(z)u_2(t,z) \tag{19}$$

where u_1 and u_2 are the scalar valued controls.

Important Remark. All these representations are equivalent subject to the condition that the problem is "well posed" in the sense of Hadamard, that is to say (i) there exists a solution, (ii) this solution is unique and (iii) it is continuous w.r.t. the initial and boundary conditions. □

A GENERAL RESULT ON THE STABILITY OF DPS

For brevity we consider only DPS with zero boundary conditions. However, in the present section we note the general model which show how to deal with boundary control or with non zero boundary conditions.

Systems Without Boundary Feedback

The equation of the DPS is

$$x(t,z) = H_1\cdot f_1(e_1-x,z,t) + H_2\cdot e_2(t,z'); \tag{20}$$

and this latter is I-0 stable provided there exist M_1 and M_2, $0<M_1,M_2<\infty$, such that

$$||x|| < M_1||e_1|| + M_2||e_2||. \tag{21}$$

x is resolved into two components x_1 and x_2, $x = x_1+x_2$, such that

$$x_1 = H_1\cdot f[(e_1-x_2)-x_1,z,t] \tag{22}$$

$$x_2 = H_2\cdot e_2 ; \tag{23}$$

and we resolve (21) into the two inequalities

$$||x_1|| \le M_1||e_1|| + M_1'||e_2|| \tag{24}$$

$$||x_2|| \le M_2||e_2|| \tag{25}$$

Condition (25) is the I-0 stability condition for the open-loop DPS defined by Equ. (23); and Condition (24) is the I-0 stability condition for the closed-loop DPS defined by Equ (22) *subject to the condition that the* DPS (23) *is stable.*□

Systems with Boundary Control

Analogously, the DPS defined by the equation

$$x(t,z) = H_1 \cdot e_1(t,z) + H_2 \cdot f_2(e_2 - x, z', t) \tag{26}$$

is I-0 stable provided that the two DPS

$$x_1(t,z) = H_1 \cdot e_1(t,z) \tag{27}$$

$$x_2(t,z) = H_2 \cdot f_2(c_2', z', t), \tag{28}$$

one both I-0 stable. □

Systems with Distributed Control and Boundary Control

The DPS defined by the equation

$$x(t,z) = H_1 \cdot f_1(e_1 - x, z, t) + H_2 \cdot f_2(e_2 - x, z, t) \tag{29}$$

is I-0 stable provided the following conditions are satisfied.
i) The two DPS' whose the equations are

$$x_1(t,z) = H_1 \cdot f_1(e_1' - x_1, z, t) \tag{30}$$

$$x_2(t,z) = H_2 \cdot f_2(e_2' - x_2, z', t) \tag{31}$$

both ones are I-0 stable; and

ii) In addition, the coefficients M_i, $(i=1,2)$ of the conditions $||x_i|| \le M_i ||e_i||$ satisfy the inequality $0 < M_i < 1$, $i = 1,2$. □

Indeed, here we have

$$||x_1|| \le M_1 ||e_1|| + M_1 ||x_2||$$

$$||x_2|| \le M_2 ||e_2|| + M_2 ||x_1||$$

so

$$(1-M_1)||x_1|| + (1-M_2)||x_2|| \le M_1||e_1|| + M_2||e_2||.$$

SPACE MEAN SQUARE STABILITY CRITERIA FOR A CLASS OF DISTRIBUTED SYSTEMS WITH CONSTANT COEFFICIENTS

Definitions and Notations

The DPS is defined by the equation

$$P(D_t, D_z) \cdot x(t,z) - Q(D_t, D_z) \cdot f[\varepsilon(t,z), z, t] = 0 \tag{32}$$

$$\varepsilon(t,z) := e(t,z) - x(t,z) \ , \ z \in [0,L] \in \mathbb{R}_+ \tag{33}$$

with boundary conditions independent of $x(t,z)$. We make the following assumptions. (H5.1), the function $f(x,z,t)$ is continuous with regard to x,z and t. (H5.2), $f(0,z,t) = 0$ and satisfies the sector condition

$$a(z) \le f(x,z,t)/x \le b(z) \tag{34}$$

where $a(z)$ and $b(z)$ are continuous in $[0,L]$. $P(\cdot)$ and Q denote polynomials in the operators $D_t := d/dt$ and $D_z := d/dz$. We define the $L_{\tilde{p}}$ -norm ove $\mathbb{R}_+ \times \Gamma_1$

$$||x(t,z)||_{\tilde{p}} := [\frac{1}{L} \int_o^L \int_o^\infty |x(t,z)|^p dtdz]^{1/p} \tag{35}$$

and the various expressions

$$<a(z)> := \frac{1}{L} \int_o^L a(z)dz \tag{36}$$

$$<<a(z)>> := [\frac{1}{L} \int_o^L a^2(z)dz]^{1/2} \tag{37}$$

$$g(a,b) := \frac{<<a>>^2 - <<b>>^2}{2(<a>-<b>)} \tag{38}$$

$$G(a,b) := <<a-g(a,b)>> \tag{39}$$

$$= <<b-g(a,b)>>$$

The transfer function of the system is $W(s,r) = Q(s,r)/P(s,r)$; we also define

$$\tilde{W}(iw,i\Omega;a,b) := \frac{W(i\omega,i\Omega)}{1+g(a,b)W(i\omega,i\Omega)} \tag{40}$$

and the set $\{\Omega\} := \{\Omega_k | \Omega_k = k\pi/L;\ k\in N\}$. We state :

Proposition 1. With the definitions and assumptions in subsection 5.1, the nonlinear DPS defined by Equations (32)-(34) is $L_{\tilde{\infty}}$ and $L_{\tilde{2}}$ input-output stable provided that the following conditions hold:

(i) $\tilde{W}(i\omega,i\Omega;a,b)$ is open-loop stable;

(ii) $$\sup_{\omega\in\mathbb{R},\Omega\in\{\Omega\}} |\tilde{W}(i\omega,i\Omega;a,b)| \cdot G(a,b) < 1. \quad \square \tag{41}$$

Proposition 2. Space mean square Popov criterion.

Define X_o as $-g(a,b) + G(a,b)$ and assume that the conditions of Proposition 5.1 are satisfied. Then the system (32)-(34) is $L_{\tilde{\infty}}$ and $L_{\tilde{2}}$ input-output stable provided that the following conditions hold:

(i) $\tilde{W}(i\omega,i\Omega;\ a,b)$ is open-loop stable;

(iii) $$R_e\{[W^{-1}(i\omega,i\Omega)-X_o]^{-1}\}-X_o + \frac{1}{2G(a,b)} > 0. \quad \square \tag{42}$$

For the proofs, see Jumarie (1973a, 1973b)

A Few Comments

The following comment clarifies the spatial averaging effect of the criteria above. Indeed, at first glance, it is straightforward to extend the so-called circle criterion by using the following new inequality derived from (34), that is

$$\underline{a} < f(x,z,t)/x < \bar{b}$$

where $\underline{a}$ and $\bar{b}$ denote respectively the lower bound of $a(z)$ and the upper bound of $b(z)$. But

$$|<<a>>^2 - <<b>>^2| \le |\underline{a}^2 - \bar{b}^2|,$$

$$|\langle a\rangle - \langle b\rangle| \leq |\underline{a}-\bar{b}|,$$

$$G(a,b) \leq G(\underline{a},\bar{b});$$

in other words, the criterion that we would get by this is more pessimistic than the one obtained earlier in that it may not conclude to the statibility of the system while Proposition (5.1) does it. □

STABILITY OF DISTRIBUTED SYSTEMS WITH VARIABLE COEFFICIENTS BY REDUCING TO DISTRIBUTED SYSTEMS WITH CONSTANT COEFFICIENTS

We consider the DPS defined by the partial differential equation

$$\sum_{ij} a_{ij}(z) D_f^{n_i} D_s^{n_j} . x(t,z) + \sum_{k\ell} b_{k\ell}(z) D_t^{n_k} D_z^{n_\ell} . f[u(t,z)-x(t,z),z] = 0 \quad (43)$$

in which the coefficients $a_{ij}(z)$ and $b_{k\ell}(z)$ may now depend explicitly upon z; and this is perhaps the major source of difficulties to analyze DPS. In a first approach, one might try to derive from Equ (43) a new DPS with constant coefficient in such a way that if the latter is stable, then so will be the former. We herein suggest two very simple methods which may be of practical use in engineering although they are somewhat pessimistic.

Derivation of an Associated System

We define the bounds of a_{ij} and $b_{k\ell}$ by the conditions $\underline{a}_{ij} \leq a_{ij}(z) \leq \bar{a}_{ij}$ and $\underline{b}_{ij} \leq b_{ij}(z) \leq \bar{b}_{ij}$, and we define a family of admissible coefficients α_{ij} and β_{ij} by the inequalities

$$\underline{a}_{ij} \leq \alpha_{ij} \leq \bar{a}_{ij}$$

$$\underline{b}_{ij} \leq \beta_{ij} \leq \bar{b}_{ij}$$

We consider the constant-coefficients DPS defined by the following equation:

$$\sum_{ij} \alpha_{ij} D_t^{n_i} D_z^{n_j} \cdot y(t,z) + \sum_{k\ell} \beta_{k\ell} D_t^{n_k} D_z^{n_\ell} \cdot f[u(t,z)-x(t,z),z] = 0 \quad (44)$$

Proposition 3. The DPS defined by Equ. (43) is L_2-input-output stable provided the associated DPS defined by Equ. (44) is itself L_2-imput-output stable for every admissible set of coefficients $\{\alpha_{ij}, \beta_{k\ell}\}$. □

The proof of this intuitive result is very simple and is left to the reader for brevity.

Successive Integrations

As an illustrative example, consider the equation

$$\frac{d^2x}{dt^2} + a(z)\frac{dx}{dz} + b(z)\frac{d^2x}{dz^2} = u(t,z) \; ; \; 0<z<L \in \mathbb{R} \quad (45)$$

On integrating by parts, one gets the new equation

$$\frac{d^2}{dt^2}\langle x(t,z)\rangle = \frac{1}{4}\int_0^L c(z)x(t,z)dz + \langle u(t,z)\rangle + u(t,z') \quad (46)$$

with

$$Lu(t,z') := -[a(z)x(t,z)]_o^L -[b(z)\frac{dx(t,z)}{dz}]_o^L + [x(t,z)\frac{db(z)}{dz}]_o^L$$

$$e(z) := (da(z)/dz) -(d^2b(z)/dz^2).$$

Define

$$\bar{c} := \{\sup c(z)/L,\ z\in[0,L]\};$$

equation (46) then yields the differential inequality

$$(d^2\langle x\rangle/dt^2) \le \bar{c}\langle x\rangle + \langle u\rangle + u(t,z') \tag{47}$$

which can be compared to

$$(d^2y/dt^2) = \bar{c}\ y(t) + \langle u\rangle + u(t,z'). \tag{48}$$

It follows that if the lumped parameter system defined by Equation (48) is I-0 stable, then the system described by (46) will be also I-0 stable. □

The reader will easily generalize the method to distributed vector parameter by using the divergence theorem (Green-Ostrograski formula).

The statement of Proposition 7.1 is given is the reference (Jumarie, 1975). The proof in this paper contains a mistake which can be easily rectified by the reader. □

STABILITY CRITERIA FOR NONLINEAR DISTRIBUTED SYSTEMS DEFINED BY LAPLACE-GREEN FUNCTIONS.

One of the sources of difficulties in DPS is the presence of coefficients which depend explicitly upon the space parameter. A way to circumvent this problem consists of describing the system by means of its Laplace-Green function. Given a partial differential equation with space dependent coefficients, it is always possible to numerically determine its Laplace-Green representation numerically with as good an accuracy as desired, so that the approach can thus be of practical interest. In this section we shall restrict ourselves to continuous DPS in space and time, refering the reader to other references for the results related to sampled data systems.

The Model

The DPS is defined by the functional equation

$$x(s,z) = \int_{\Gamma_1} G(s,z,y)\ \mathcal{L}\{f(\varepsilon,y,t)\}\ dy \tag{49}$$

$$\varepsilon(t,z) := e(t,z)-x(t,z)\ . \tag{50}$$

$\mathcal{L}\{f\}$ is the Laplace transform of f w.r.t. time t, x(s,z) is the Laplace transformation of x(t,z) w.r.t time t, and f(u,z,y) again satisfies the sector condition (34). G(s,z,y) is called the Laplace-Green function of the DPS. The imput is e(t) so that the I-0 stability is defined with e(t,z) and x(t,z).

Another system of interest is described by the equation

$$x(s,z)\ \int_{\Gamma_1} G(s,z,y)\ \mathcal{L}\{k(t,y)\varepsilon(t,y)\}dy \tag{51}$$

where $k(t,z)$ denotes a function which is periodic in t and continuous in z. In order to study this DPS, we shall introduce the semi-norm

$$||x(t,z)||_M := \lim_{T\to\infty} [\frac{1}{T} \int_o^T \int_{\Gamma_1} |x(t,z)|^2 dtdz]^{1/2}. \tag{52}$$

This semi-norm will be referred to hence forth as the M_2-semi norm in the following.

Main Result

We make the following assumptions.

(H8.1), the integral in Equ (49) is a Riemann integral. This assumption is required to use the technique of Fredholm in the proof. (H8.2), for every $\omega\in\mathbb{R}$, $|G(i\omega,z,y)|$ is bounded in $\bar{\Gamma}_1\cdot\bar{\Gamma}_1$. (H8.3), the impulse response $h(t,z,y)$ of $G(s,z,y)$, considered as a transfer function in s, is such that

$$|h(t,z,y)| \le K_1(z,y)\exp[-tK_2(z,y)] \tag{53}$$

where K_1 and K_2 denote two bounded functions over $\bar{\Gamma}_1\cdot\bar{\Gamma}_1\cdot$(H8.4), the quantity $[a(y)+b(y)]/2$ pertains to the Routh-Hurwitz sector, or else similarly, to the closed-loop stability sector of $G(s,z,y)$ for every $z\in\Gamma_1$. (H8.5), $f(x,z,t)$ is continuous w.r.t. $(x,z)\in\mathbb{R}\cdot\bar{\Gamma}_1$. (H8.6), $f(0,z,t) = 0$. (H6.7), $f(x,z,t)$ is that

$$a(z) \le f(x,z,t)/x \le b(z) \ ; \ (x,z,t)\in\mathbb{R}\cdot\bar{\Gamma}_1\cdot\mathbb{R}_+ \tag{54}$$

where $a(z)$ and $b(z)$ are both continuous on $\bar{\Gamma}_1$.

Definition 11 The solving kermel of the Fredholm's integral equation

$$\Phi(s,z)=g(s,z) + \lambda\int_{\Gamma_1} G(s,z,y)\,\frac{a(y)+b(y)}{2}\,\Phi(s,y)dy \tag{55}$$

is denoted by $R(s,z,y;\lambda)$; in other words, one has

$$\Phi(s,z)=g(s,z) + \lambda\int_{\Gamma_1} R(s,z,y;\lambda)g(s,y)dy. \tag{56}$$

We then have the following result.

Proposition 4 Assume that the hypotheses (H8.1)-(H8.7) are satisfied. Then the DPS defined by Equ. (49) is L_∞- and L_2 input-output stable provided that the following condition hold.

$$\sup_{\omega\in\mathbb{R}} ||R(i\omega,z,y;-1)\cdot G(i\omega,z,y)\cdot|\frac{a(y)+b(y)}{2}|\,||<1. \ \square \tag{57}$$

Proposition 5. Assume that the t-periodic function $k(t,z)$ is continuous w.r.t. z for every $z\in\Gamma_1$. Then the DSP defined by Equ. (51) is M_2 input-output stable, provided the following condition holds:

$$\sup_{\omega_o\in\mathbb{R}} ||L^{-1}\{\int_{\Gamma_1} G(s,z,y)L\{k(t,z)e^{i\omega_o t}\}||_M < 1. \ \square \tag{58}$$

A Few Corollaries

The following corollaries are of practical interest in applications and to state them simply, we shall refer to the above DPS as the systems (49) and (51) according to the number of their equation. In addition, the above assumptions are conserved.

Corollary 1. The DPS (49) is L_∞- and L_2 I-0 stable provided that

$$\sup_{\omega\in\mathbb{R}} ||R(i\omega,z,y;-1)G(i\omega,z,y)||\cdot||\frac{a(z)+b(z)}{2}|| < 1; \tag{59}$$

and the DPS (51) is M_2 I-0 stable provided that

$$\sup_{\omega\in\mathbb{R}} ||G(i\omega,z,y)||\cdot||k(t,z)||_M < 1. \quad \square \tag{60}$$

Corollary 2. Assume that $b(z) + a(z) = 0$. The DPS (49) is L_∞ and L_2 I-0 stable provided that

$$\sup_{\omega\in\mathbb{R}} ||G(i\omega,z,y)||\cdot||b(z)|| < 1. \quad \square \tag{61}$$

For the proofs of Propositions 8.1 and 8.2, as well as for those of the corollaries, see Jumarie, 1974. In this paper, the results are first obtained for sampled data systems, and then, by making the sampling period tend to zero, results for continuous systems are

Extensions

We now indicate how these results can be directly extended to sampled data systems and/or systems with boundary control.

(i) The case of systems with sampling errors is straightforward. We first define the step-function

$${}_1u(t,z) := u(nT,z) \;;\; \forall t,\; nT\le t<(n+1)T \tag{62}$$

where T denotes the sampling period *which may depend upon the distributed parameter* z, and we define

$$G^*(s,z,y):=((1-e^{-sT})/s) \sum_{n=0}^{\infty} h(nT,z,y)e^{-nsT} . \tag{63}$$

With these notations, the equations of the sampled data system are

$$x^*(s,z) = \int_{\Gamma_1} G^*(s,y,t)\mathcal{L}\{[f(\varepsilon,y,t)\}dy \tag{64}$$

$$\varepsilon^*(s,z) = e^*(s,z) - x^*(s,z), \tag{65}$$

and Propositions 8.1 and 8.2 apply, but stated with $R^*(i\omega,z,y;-1)$, $G^*(i\omega,z,y)$, $x^*(t,z)$ and $e^*(t,z)$, Ref[23]. $\square$

(ii) We now assume that there is a non-zero boundary condition, which is nevertheless independent of the state $x(t,z)$; say

$$x(s,z) = \int_{\Gamma_1} G_1(s,z,y)u_1(s,y)dy + \int_{\Gamma_2} G_2(s,z,z')u_2(s,z')dz' \tag{66}$$

with

$$u_1(t,z) := f[e_1(t,z)-x(t,z)]. \tag{67}$$

According to Section 4, System (66) is I-0 stable provided that the following two systems

$$x_1(s,z) = \int_{\Gamma_1} G_1(s,z,y)u_1(s,y)dy \quad (68)$$

$$u_1(t,z) = L\{f_1(e_1-x_1,z,t)\} \quad (69)$$

and

$$x_2(s,z) = \int_{\Gamma_2} G_2(s,z,z')u_2(s,z')dz' \quad (70)$$

are each I-0 stable. But, Equation (70) provides

$$||x_2(t,z)|| < ||G_2(i\omega,z,z')||\cdot||u_2(t,z')|| \quad (71)$$

so that the corresponding system will be I-0 stable if one has

$$\sup_{\omega\in\mathbb{R}} ||G_2(i\omega,z,z')|| \le M < \infty. \quad (72)$$

(iii) Next, assume that

$$u_2(s,z') = L\{f_2[e_2(t,z')-x(t,z'),z',t]\} \quad (73)$$

where $e_2(t,z')$ denotes the boundary control. Again, following Section 4, the DPS defined by Equs (66) and (73) is I-0 stable provided that the DPS

$$x(s,z') = \int_{\Gamma_2} G_2(s,z',z')u_2(s,z')dz' \quad (74)$$

is I-0 stable, and provided that

$$\sup_{\omega\in\mathbb{R}} ||G_1(i\omega,z,y)|| \le M < \infty . \;\square \quad (75)$$

Example 1.

Consider the continuous system of hyperbolic type defined by the damped wave equation

$$\frac{\partial^2 y}{\partial t^2} + \frac{\partial y}{\partial t} = \frac{\partial^2 y}{\partial z^2} + f(e,z,t)$$

together with the boundary conditions

$$y(t,0) = y(t,L) = 0, \; L > 0 .$$

The Laplace transform Y(s,z) is the solution for the equation

$$\frac{\partial^2 Y(s,z)}{\partial z^2} - (s+s^2)Y(s,z) = -L\{f(\cdot)\}$$

with the additional condition

$$Y(s,0) = Y(s,L) = 0 .$$

Calculation of the Green's function of the above equation is straightforward, though slightly lengthy. The result is

$$Y(s,z) = \frac{\sinh[(L-2)\phi(s)]}{\phi(s)\sinh[L\phi(s)]} \int_o^z \sinh[\xi\phi(s)L\{f(\cdot)\}d\xi \;+$$

$$+ \sinh[z\phi(s)]\int_z^1 \sinh[(L-\xi)\phi(s)]L\{f(\cdot)\}d\xi$$

with

$$\phi^2(s) := s+s^2 .$$

Since the system is a low-pass one, the supremum of $G(i\omega,z,\xi)$ will be reached for a small value of ω. In such a case, we may write

$$Y(s,z) \cong \frac{L-z}{L} \int_0^z \xi L\{f(\cdot)\}d\xi + \frac{z}{L} \int_z^L (L-\xi)L\{f(\cdot)\}d\xi$$

and by calculating the norm of the approximated kernel so obtained, one has

$$\sup_{\omega\in\mathbb{R}} ||G(i\omega,z,\xi)|| \cong L^3/3\sqrt{10} .$$

As a result, a condition for L_∞ and L_2-stability is

$$||b(z)|| < 3\sqrt{10}/L^2. \quad \square$$

HYPERSTABILITY OF NONLINEAR DISTRIBUTED SYSTEMS

Popov (1970) initiated a powerful theory to investigate the stability of lumped parameter systems, the "hyperstability theory", and we now give an extension of this approach to DPS.

Definition of the Considered System

The DPS is defined by the Equ.

$$\dot{x}(t,z) = A(z,d/dz)\cdot x(t,z) + b(z)u(t,z) \tag{76}$$

(see Subsection 3.6 for more details), with zero boundary condition. In addition to (76), the DPS is defined by the *structural condition*

$$g(t_o,T_o) \le 0 \ , \ 0 < t_o < T_o \tag{77}$$

$$g(t_o,T_o) := [\int_{\Gamma_1} x^*J(z)xdz]_{t_o}^{T_o} + \int_{t_o}^{T_o} \int_{\Gamma_1} [\chi(z)u^*u+ +u^*l^*(z) + x^*l(z)u + x^*M(z)x] \, dzdt. \tag{78}$$

We shall assume that $b(z)$ is a scalar control and that the matrices $\chi(z)$, $l(z)$, $J(z)$ and $M(z)$ are suitably related w.r.t. their dimensions.

Polarized System

Following the terminology of Popov, the *polarized system* is defined by (i) Equation (76); (ii) the polarized equation

$$\dot{\tilde{x}}(t,z) = \tilde{x}(t,z)\cdot A^*(z,\partial/\partial z) + \tilde{u}(t,z)b^* \tag{79}$$

and (iii) the polarized criterion

$$\tilde{g}(t_o,T_o) := [\int_0^L \tilde{x}Jxdz]_{t_o}^{T_o} + \int_{t_o}^{T_o} \int_{\Gamma_1} [\chi(z)\tilde{u}u+\tilde{u}l(z)x + + \tilde{x}l(z)u + \tilde{x}\, M(z)x] \, dz\, dt. \tag{80}$$

Characteristic Function

We consider elementary scalar controls in the form

$$u(t,z) = u_o \exp(st), \tag{81}$$

$$\tilde{u}(t,z) = u_o^* \exp(rt), \tag{82}$$

where u_o denotes a constant. Given such controls, Equs. (76) and (78) have the particular solutions

$$x(t,z) = u_o w(s,z) \exp(st), \tag{83}$$

$$\tilde{x}(t,z) = u_o^* \tilde{w}(r,z) \exp(rt), \tag{84}$$

where the vectors $w(s,z)$ and $\tilde{w}(r,z)$ are solutions of the differential equations

$$sw(s,z) = A(z,\partial/\partial z)\circ w(s,z) + b(z), \tag{85}$$

$$r\tilde{w}(r,z) = \tilde{w}(r,z)\circ A^*(z,\partial/\partial z) + b^*(z), \tag{86}$$

with boundary conditions derived from the boundary conditions of the initial equation. The expression for $\tilde{g}(t_o,T_o)$ then becomes

$$\tilde{g}(t_o,T_o) = u_o^* u_o \cdot h_z(r,s) \int_{t_o}^{T_o} \exp(r+s)t.\ dt \tag{87}$$

with

$$h_z(r,s) := \int_{\Gamma_1} h_1(z,r,s)dz, \tag{88}$$

and

$$h_1(z,r,z) := \chi(z) + l^*(z)w(s,z)+\tilde{w}(r,s)l(z)+$$

$$+\tilde{w}(r,s)[M(z)+(r+s)J(z)]w(s,z).\quad \square \tag{89}$$

Main Result

Proposition 6. Assume that the following conditions are satisfied: (i) the coefficient $b(z)$ is not identically zero; (ii) the criterion $g(t_o,T_o)$ is not identically zero, and (iii) the DPS has the property of minimal stability, that is, for every initial condition $x(t_o,z)$, there exist a pair $[x(t,z),u(t,z)]$ which satisfies equations (76)-(78) with the additional condition (see Definition 9)

$$\lim|x(t,z)| = 0\ ,\ t \to \infty.$$

Then the DPS is hyperstable provided that at least one of the following conditions are satisfied.

(i) The inequality $h_z(-i\omega,+i\omega) \geq 0$ holds for every $\omega\in R$ which is not an eigenvalue of the operator $A(z,d/dz)$ taken with the boundary conditions of (76); or (ii) the inequality $h_z(\bar{s},s) \geq 0$ is satisfied for every $s\in\mathbb{C}$, $\mathrm{Re}s>0$ which is not an eigenvalue for the operator $A(z,d/dz)$ taken with the boundary conditions of (76). □

Extensions

(i) *Distributed systems with vector control.*

Shortly the DPS is described by the equation

$$\dot{x}(t,z) = A(z,\partial/\partial z)\cdot x(t,z) + B(z)u(t,z) \tag{90}$$

with the criterion

$$g_m(t_o,T_o) := [\int_{\Gamma_1} x^*J(z)x\, dz]_{t_o}^{T_o} + \int_o^L[u^*K(z)u+u^*L^*(z)x +$$

$$+ x^*L(z)x + x^*M(z)x]\, dzdt \tag{91}$$

$$(J^*=J,\ K^*=K,\ M^*=M)$$

The matrices J,K,M are hermitian, and B(z) now is an n×m matrix. Again, via the polarized system concept, we can define the *characteristic function of the system* as

$$H_z(u_o^*,r,s,u_c) := \int_{\Gamma_1} \{u_o^*Ku_o+\widetilde{W}(u_o^*,z,r)Lu_o+u_o^*L^*W(z,s,u_o)$$

$$+\widetilde{W}(u_o^*,z,r)[M(z)+(r+s)J(z)]W(z,s,u_o)\}dz \tag{92}$$

in which now $W(s,z,u_o)$ is given by the counterpart of Equation (83), that is to say,

$$x(t,z) = W(s,z,u_o)\cdot\exp(st). \tag{93}$$

With these notations, Proposition 9.1 holds again for the DPS described by Equs (90-91), but stated with $H_z(u_o^*,r,s,u_o)$ instead of $h_z(r,s)$.

For further details, see Jumarie (1974,1975).

(ii) *Distributed systems with boundary conditions.*

There remains the task of generalizing the hyperstability theory of Popov to DPS involving non-zero boundary conditions, say

$$F(z',\partial/\partial z')\cdot x(t,z') = 0\ ,\ z' \in \Gamma_2. \tag{94}$$

It is clear that such an extension could be performed as follows. a) The criterion should be modified to the form

$$g(t_o,T_o) = g_{\Gamma_1}(t_o,T_o) + g_{\Gamma_2}(t_o,T_o) \tag{95}$$

where $g_{\Gamma_1}(t_o,T_o)$ is exactly the right-hand side term of (78) and $g_{\Gamma_2}(t_o,T_o)$ has a similar expression but with Γ_2. b) The elementary controls again would be defined by (81) and (82). c) In this way it would be possible to build up a characteristic function which would be defined on both Γ_1 and Γ_2. □

To conclude this section related to hyperstability, we shall point out that this latter presents an elegant and poweful approach to stability on the one hand, and that it may be thought of as an intermediate step between the theory of stability and the modern theory of singularities. □

Example 2.

As an illustrative example, we consider the equation

$$\frac{\partial^2 y}{\partial t^2} + \frac{\partial y}{\partial t} + \frac{\partial^2 y}{\partial z^2} = u \ , \quad z \in [0,L]$$

$$y(0,t) = y(L,t) = 0$$

$$u = -f(c_1(z)y + c_2(z)\,\frac{\partial y}{\partial t}\ ,\ z)$$

where f satisfies Condition 34. Here we have

$$x' := (y, \partial y/\partial t) := (x_1, x_2)\ ,$$

$$b' := (0,1),$$

$$A(\partial/\partial z) := \begin{bmatrix} 0 & 1 \\ -\partial^2/\partial z^2 & -1 \end{bmatrix},$$

and

$$g(t_o, T_o) := \int_{t_o}^{T_o} \int_o^L [u^2 + (a+b)(c'x)u + ab(c'x)^2]dzdt.$$

The vector $w := (w_1, w_2)$ is the solution of the equations

$$sw_1(s,z) = w_2(z,s)$$

$$sw_2(z,s) = -\partial^2 w_1(z,s)/\partial z^2 - w_2(z,s) + 1$$

$$w_1(0,s) = w_1(L,s) = 0\ ;$$

therefore

$$w_1(z,s) = \frac{\sin[(z-L)\sqrt{\Theta(s)}] - [\sin z\sqrt{\Theta(s)} - \sin L\sqrt{(s)}]}{\Theta(s)\ \sin L\sqrt{\Theta(z)}}$$

with

$$\Theta(s) := s^2 + s.$$

The characteristic function is

$$h_z(\bar{s},s) = \int_o^L \{1 + (a+b)\mathcal{R}[c'w(z,s)] + \alpha\beta|c'w(z,s)|^2\}dz.$$

If we define the average value of a function a(z) as

$$\langle a\rangle_z := \frac{1}{L}\int_o^L a(z)dz;$$

then, a hyperstability condition derived for Proposition 9.1 is

$$\min_{i\omega\notin\Lambda} \langle 1 + (a+b)\mathcal{R}[c_1'w_1(z,i\omega)] + ab|c_1'w_1(z,i\omega)|^2\rangle_z \geq 0.$$

where Λ denotes the set

$$\Lambda := \{\lambda_n : \sin(L\sqrt{\Theta(\lambda_n)}) = 0\}.$$

SPACE SAMPLING EFFECT

In number of cases, it is difficult or even impossible to continuously apply controls in space, or else, the sensors can be located only on a finite net in the region where the DPS is defined. Of course, this spacing discretization affects the stability of the system, but we ask:in which way, and how can one estimate this effect?

In this section we give three models, directly derived from the preceding ones, which provide a first approach to the question.

Transfer Function in Space and Time

(i) Similarly to the "z-transform" technique used for sampled-data systems, we introduce the complex variable $R := \exp(Zr)$ and define the "s-R transform" of a space sampled function $x^*(t,z)$ by the expression

$$x(s,R) := \sum_{n=0}^{\infty} L\{x(t,nZ)\}R^{-n} \tag{96}$$

where again the same x is used for the function and its transformation, since there is no possible confusion. Also we may define $x(s,R) =: x^*(s,r)$.

(ii) Let $h(t,z)$ denote the "impulse response" of $W(s,r)$, and define

$$W(s,R) := \sum_{n=0}^{\infty} L\{h(t,nZ)\}R^{-n} \tag{97}$$

(iii) Define the transfer function $ZOH(r)$ of the zero-order hold in space as

$$ZOH(r) := (1-\exp(-Zr))/r. \tag{98}$$

The space-pulse transfer function of $W(s,r)$ together with the zero-order hold is denoted by $[W(s,r)ZOH(r)]^*$.

(iv) With these notations the equation of the DPS* is

$$x_o^*(s,r) = [W(s,r)\cdot ZOH(r)]^* v^*(s,r) \tag{99}$$

$$x_1(s,r) = W(s,r)v_1(s,r) \tag{100}$$

where $v^*(s,r)$ is the control of the linear component $W(s,r)$, and $v_1(s,r)$ is the control induced by the boundary conditions.

The I-O stability conditions in Section 4 apply to Equations (99)-(100).

Systems Defined by Laplace-Green Functions

We proceed in a like manner. Assume that the feedback control is

$$v(t,z) = f[\,e(t,z), nZ, t\,] \tag{101}$$

where $e(t,z)$ is the step function defined as

$$e(t,z) = e(t,nZ),\ nZ \le z < (n+1)Z, \tag{102}$$

while the boundary condition $v_1(t,z)$ is continuous in z'. For definiteness, assume that Γ_1 is the range [0,L] (in the more general case, of course, z is a vector). Then the equation of the DPS* is

$$x^*(s,z) = \int_{\Gamma_2} G_1(s,z,z')v_1(s,z')dz' + \qquad\qquad + \sum_n G(s,z,nZ)v(s,nZ). \tag{103}$$

This being so, let $\Delta(\xi)$ denote the summation

$$\Delta(\xi) := \sum_n \delta(\xi - nZ\cdot); \tag{104}$$

and re-write Equ. (103) in the form

$$x^*(s,z) = \int_{\Gamma_2} G_1(s,z,z')v_1(s,z')dz' + \qquad + \int_{\Gamma_1} G(s,z,\xi)\Delta(\xi)v(s,\xi)d\xi \;. \tag{105}$$

The I-0 stability criteria in Section 6 apply to Equ. (105).

Systems Defined by State-Vector

Here again, the result of Section 9 can be adapted easily, and this extension is based on the following two remarks.

(i) Since the DPS* is sampled in z and is not sampled in t, the basic control $\exp(i\omega t)$ can be used again.

(ii) As a consequence, the main features of the sampling are introduced in the criterion g(t,T), therefore the definition of the new *sampled characteristic function*.

A Few Remarks

The first approach involves simple calculations and is easier to use, but in contrast it deals with partial differential equations with constant coefficient only. In addition it is pessimistic due to the fact that it involves a kind of stability in space which is not truly meaningful from a practical standpoint. Sharper results are obtained by the second and third methods which, however, require numerical computations to determine the Laplace-Green's functions and the distributed transfer functions. These may conclude to the stability of the sampled-data system when the first approach is inapplicable.

For more details, see Jumarie (1980,1981)

Example 3.

Let Ω_o denote the space sampling frequency, then one has

$$[W(s,r)ZOH(r)]^* = \frac{1}{Z}\sum_{-\infty}^{+\infty} Ze^{-iZ\frac{\Omega+n\Omega_o}{2}} \frac{\sin\frac{\Omega+n\Omega_o}{2}Z}{\frac{\Omega+n\Omega_o}{2}Z} \cdot W(s,i(\Omega+n\Omega_o))$$

If Z is small enough and $W(i\omega, i\Omega)$ is low-pass in Ω, all the harmonics in Ω can be neglected and one has the approximate equation

$$[W(s,r)\cdot ZOH(r)]^* \cong W(i\omega, i\Omega)\exp(-i\frac{\Omega Z}{2})$$

since

$$\sin(\Omega z/2)/(\Omega z/2) \cong 1.$$

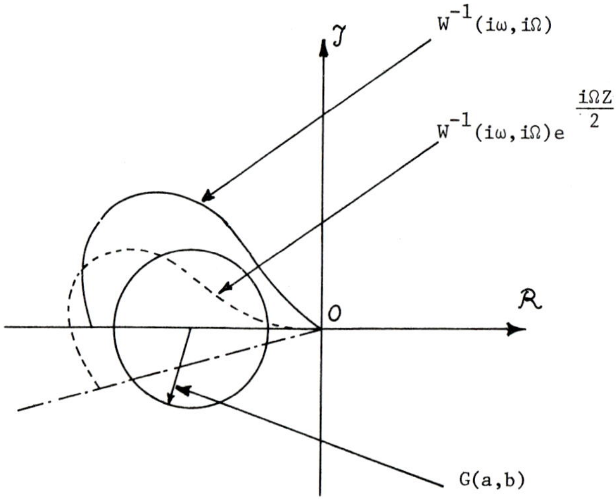

Fig 1. Destabilizing effect of space sampling

The result is pictured on Fig. 1.

The continuous system, i.e. with no space sampling, is input-output stable when $W_1(s,r)$ is open-loop stable, and when the Nyquist's locus of $W^{-1}(i\omega, i\Omega)$ is plotted outside the circle with radius G(a,b) (see Equ. 39) and centre

$$X = -g(a,b) \ ; \ Y = 0$$

for every fixed $\Omega \in \mathbb{R}$.

Now $W^{-1}(i\omega, i\Omega)\exp(i\Omega Z/2)$ is derived from $W^{-1}(i\omega, i\Omega)$ by using a positive counter clockwise rotation with the centre 0 and the angle $\Omega Z/2$, and its Nyquist's locus may in intersect the limiting circle when $W^{-1}(i\omega, i\Omega)$ does not. □

STABILITY OF DPS IN THE PRESENCE OF RANDOM DISTURBANCES

As the knowledgeable reader might guess, we cannot deal properly with such a question in a mere section of an article. So our purpose herein is only to comment on some aspects of the problem, referring to appropriate articles for further details.

Mathematics or Physics?

If one believes mathematicians one cannot suitably deal with distributed random noises without using the modern mathematical machinery, say a suitable abstract infinite dimensional space, a good measure, Gaussian mesure, Radon measure, Bochner integral, spaces L^{α}, cylindrical measure etc... . For instance, if one believes Curtain (1975b), "it is important to ask the mathematically difficult question of what is actually meant by "distributed noise" and to develop a rigorous mathematical infinite-dimensional" theory for a large class of problems.

The problem perhaps is of importance, but it is a concern for mathematicians only, and a long time ago, physicians (Einstein) and engineers had already understood what they mean by distributed noises. We did not wait for abstract mathematicians to design the bodies of airplanes and these fly nevertheless despite the presence of acrodynamic noises!

Practical models for distributed white noises can be found, for instance, in Tzafestas and Nightingale (1968,1969), Tzafestas (1972,1973) and Sakawa (1972). The validity of these models has been checked by using numerical simulation techniques, and as a result they appear to be quite satisfactory in engineering problems. Of course, one can argue that these models have not been confronted with actual concrete problems; as this has been contended by abstract mathematicians; but we can reply that they are good at least for this kind of reality which is represented by the simulations in question.

A *distributed system* is conceptually different from an *infinite-dimensional system*, although they may be considered as modelling the same physical pattern. Rather than considering stochastic DPS as infinite dimensional systems, we shall define a mixed model which uses the Blanc-Lapierre and Fortet approach (1965) w.r.t. the distributed parameter itself, and the Doob approach (1942) in time. Such a describing is directly suggested by the analysis of practical examples (strings subject to aerodynamical turbulence, vibration of elastic roads subject to random initial conditions, shot effect in electronics, rainfall on a country, etc.) so that we shall refer to it as a *phenomenological approach*

A Model For Distributed White Noises

We summarize here a model for distributed white noise that we have defined elsewhere (1977a, 1977b) in our phenomenological approach.

We define two Wiener-Levy processes $\alpha_z(t)$ and $\beta_t(z)$ where the subscript variably are stochastic while the variables in parentheses are deterministic. Immediate extension of the result of Parzen (1962) yields

$$\mathrm{Var}\{\alpha_z(t)-\alpha_\xi(t)\} := \eta^2(t)(z-\xi);\ z > \xi \tag{106}$$

$$\mathrm{Var}\{\beta_t(z)-\beta_\tau(z)\} := \sigma^2(z)(t-\tau);\ t > \tau \tag{107}$$

Moreover, their correlation functions are

$$\gamma_\alpha(t,z;t,\xi) = \eta^2(t)\min(z,\xi)\ ;\ z,\xi > 0 \tag{108}$$

$$\gamma_\beta(t,z;\tau,z) = \sigma^2(z)\min(t,\tau)\ ;\ t,\tau > 0\ . \tag{109}$$

Formally, we then have

$$\partial\alpha_z(t)/dz = v_2(t) \tag{110}$$

$$\partial\beta_t(z)/dt = w_t(z)\ . \tag{111}$$

$\{w_t(z),\ z>0\}$ is a family of Gaussian noises indexed by z, and likewise $\{v_2(t),\ t\geq 0\}$ is a family of Gaussian noises in z but indexed by t.

Definition 12 The distributed Gaussian noise n_{tz} generated by the Wiener-Levy processes $\alpha_z(t)$ and $\beta_t(z)$ is defined by the Jacobian

$$n_{tz} := \partial[\alpha_z(t), \beta_t(z)]/\partial(t,z). \quad \square \tag{112}$$

An important special case is obtained as follows. Assume that there are no stochastic errors in the determination of the position z at which x_{tz} changes its value, for a given t. Assume further that the deterministic errors if there are any, are reduced to the variation of z itself; we then have the identity $d\alpha_z(t) = dz$ therefore

$$d\beta_t(z)dz = |w_t(z)|dtdz \tag{113}$$

As a result, if we consider the stochastic integral

$$\int_Z dz \int_{t_o}^{t} x_t(z)d\beta_t(z) := \int_Z \int_{t_o}^{t} x_t d\beta_t(z)dz \tag{114}$$

in the mean-square calculus; it is meaningful as a direct extension of the I to integral. □

Stochastic Distributed Systems

We consider a class of distributed systems defined by the equation

$$x(t,z) = x(t_o,z) + \int_{t_o}^{t} \int_{\Gamma_1} h_1(t-\tau,z,\xi)u(\tau,\xi)d\xi d\tau +$$

$$\int_{t_o}^{t} \int_{\Gamma_2} h_2(t,\tau,z,\xi)v(\tau,\xi)d\xi d\tau \tag{115}$$

where $u(t,z)$ is the control in Γ_1 and $v(t,z)$ is the control on the boundary $\Gamma_2 = B(\Gamma_1)$.

We now *formally* derive from Equation (115) the following stochastic equation

$$x_{tz} = x_{t_o z} + \int_{t_o}^{t} \int_{\Gamma_1} h(t-\tau,z,\xi)f(x_{\tau\xi},\xi,\tau)d\xi d\tau + \tag{116}$$

$$+ \int_{t_o}^{t} \int_{\Gamma_1} k(t-\tau,z,\xi)g(x_{\tau\xi},\xi,\tau)d\alpha_\xi(\tau)d\beta_\tau(\xi)$$

where $\alpha_z(t)$ and $\beta_t(z)$ denote the distributed Wiener-Levy processes defined above.

This model could be associated with stochastic partial differential equations as follows. Consider the deterministic equation

$$\dot{x}_{tz} = A(z,\tfrac{\partial}{\partial z})\cdot x(t,z) + f[x(t,z),z,t] \tag{117}$$

with given boundary conditions. Its Green's form is exactly Equation (115). Assume now that Equation (117) is disturbed by a distributed white noise *in the sense of Definition 12*, to yield *formally*

$$\dot{x}_{tz} = A(z, \tfrac{\partial}{\partial z})\cdot x_{tz} + f(x_{tz},z,t) + g(x_{tz},z,t)n_{tz} \tag{118}$$

with

$$n_{tz} := \left| v_2(t)w_t(z) - \frac{d\alpha_z(t)}{dt} \frac{d\beta_t(z)}{dz} \right|. \tag{119}$$

Equation (118) is not mathematically meaningful and its correct representation is given by Equation (116).

This model obviates some of the criticisms which are usually stated by abstract mathematicans against the engineering approach. □

Application to the Stability of Stochastic Distributed Systems

This model provides an approach to the study of the quantitative effects of distributed noises in the special case where the control can be resolved into the form

$$u_t(z) = \tilde{u}(t,z) + w_t(z)$$

in which $\tilde{u}(t,z)$ denotes the control which is directly applied to the system, while $w_t(z)$ is the additional disturbing input caused by a Gaussian noise. The advantage of this simple model lies in the fact that all the stochastic computation can be expanded side by side like the deterministic one, so that direct comparisons are possible.

The main conclusion is the destabilizing effects of the Gaussian noises which are quantitatively estimated by their variances. See Jumarie (1974,1975a,1975b).

CONCLUSIONS

The topic of the stability of distributed parameter systems is a large problem which can by itself be the content of a book, and unavoidably we had to restrict ourselves to some specific features of the question, in our present article. It is also an important problem as far as it is desirable that a DPS be stable before to be controlled. In this way, it is of order to remark that the famous basic coercivity assumption in the optimum control of DPS as presented by Lions (1971) is nothing else but an open loop stability prerequisite, quite similar to the open loop stability condition introduced by Popov (1962) in his by-now standard theory for lumped parameter systems.

It is quite certain that this question is appealed to be the object of extensive investigation in a very near future, due to the advent of very high dimensional systems of interest to us, together with computer distributed networks. Can we forecast the future trends of research in this way? There are number of directions which can be considered.

It seems quite likely that the hyperstability theory initiated by Popov should provide new results, mainly with regard to composition laws for the stability of DPS; what would be of obvious interest in the design of large-scale dynamic DPS. One can envisage also to extend the theory for connective stability as presently available in lumped parameter systems (Siljak, 1978) or again to consider a DPS as a large-scale system in some way to define, what makes sense since, in some instances, a DPS is viewed as an infinite dimensional system. The modern differential geometry theory popularized by the so-called "catastrophe theory" (Thom,) appears also to be straightforwardly relevant to DPS, and we can even add that the very reason for the existence of this theory is exactly the study of partial differential equations, but it appears that its practical application to DPS is very difficult, and extensive researches are performed in this way.

REFERENCES

Bellman, R. (1927). The boundedness of solutions of infinite systems of linear differential equations, *Duke Math. J.*, Vol. 14, pp 695-706.

Bellman, R. (1948). On the existence and boundedness of solutions of nonlinear partial differential equations of parabolic type. *Trans. American Math. Soc.*, Vol. 64, pp 21-44.

Blanc-Lapierre, A., and Fortet, R., (1965). *Théorie des fonctions aléatoires*, Masson et Cie, Paris, 1953. Translated by J. Gani, Gordon and Breach, N-Y.

Blodgett, R.E, and King, R.E. (1967). Absolute stability of a class of nonlinear systems containing distributed elements. *Journal of Franklin Institute*, Vol. 284, pp 153-160.

Curtain, R. (1975). A survey of infinite-dimensional filtering. *Siam Review*, Vol. 17, No 3, pp 395-411.

Doob, J.L. (1942) *American Math. Monthly*, Vol. 49, pp 648.

Freedman, M.I; Falb, P.L., and Zames, G. (1969), A Hilbert space stability theory over locally compact Abelian groups. *Siam J. Control*, Vol. 7, pp. 479-495.

Freedman, M.I., Falb, Pol., and Anton, J. (1971). Transform theoretic approach to the stability of a class of nonlinear partial differential equations, *J. Diff. Equations*, Vol. 9, pp. 25-45.

Friedman, A. (1959. Convergence of the solutions of parabolic equations to a steady state. *J. Math. and Mech.*, Vol. 2, pp 57-76.

Gelig, A.KH. (1965). Absolute stability of nonlinear distributed parameter systems. *Automation and Remote Control*, Vol. 26, No 3, pp 395-403.

Gelig, A.KH. (1966). Stability in critical cases of control systems with distributed parameters. *Automation and Remote Control*, Vol. 27, No 4, pp 525-535.

Jumarie, G. (1973). Mean square criteria for the stability of non-linear distributed systems. *Intern. J. Control*, Vol. 18, No 3, pp 469-481.

Jumarie, G. (1973). Further extension on the space mean square stability of deterministic and stochastic nonlinear distributed parameter systems. *Intern. J. Control*, Vol. 18, No 5, pp 1041-1065.

Jumarie, G. (1974). Stability criteria for distributed nonlinear sampled-data systems defined by Green's functions. *A.S.M.E. Trans; J. Dynamic Systems, Measurement and Control*, Vol. 96, Series G, No 3, pp 315-321.

Jumarie, G. (1974). Hyperstability conditions for a class of stochastic distributed systems with constant coefficients. *Ricerche di Automatica*, Vol. 5, No 2-3, pp 1-17.

Jumarie, G. (1974). A practical criterion for the input-output stability of distributed parameter systems with space-dependent coefficients.

Intern. J. Control, Vol. 21, No 2, pp 285-292.

Jumarie, G. (1975). A space function approach to the hyperstability and the average hyperstability of distributed systems with space-varying linear parts and time-varying gains. *A.S.M.E. Trans, J. Dynamic Systems, Measurement and Control*, Vol. 97, Series G, No 4, pp 345-353.

Jumarie, G. (1975). Input-output stability criteria for a broad class of stochastic nonlinear distributed systems defined by Green's functions. *A.S.M.E. Trans; Journal of Dynamic Systems, Measurement and Control*, Vol 97, Series G, No 1, pp 83-92.

Jumarie, G. (1977). Toward a practical theory of stochastic partial differential equations, *Proceeding of the IFAC International Symposium on Trends in Automatic Control Education*, 19 pps.

Jumarie, G. (1977). Further applications of the non-stationary operational calculus to the synthesis of closed-loop distributed systems. A new approach to distributed white noises. *Intern. J. Systems Sciences*, Vol. 8, N°12, pp 1337-1364.

Jumarie, G. (1980) Three models for the practical estimation of space sampling effects in nonlinear distributed system stability. *Proceedings of the 1980 Joint Automatic Control Conference, paper No WAG-C, 6pps, IEEE publications.*

Jumarie, G. (1981) Space sampling effects in nonlinear distributed parameter stability. A.S.M.E. *Trans., Journal of Dynamic Systems, Measurement and Control*, Vol. 103, Series G, No 1, pp 54-61.

Kastenberg, W.E. (1969). Stability analysis of nonlinear space development reactor kinetics. In : *Advances in Nuclear Science and Technology*, E.J. Hennelly, Ed., Vol. 5, Academic Press, New York.

Kastenberg, W.E. (1970). Comparison theorems for nonlinear multi-component diffusion systems. *J. Math. Anal. and Applic.*, Vol. 29, pp 299-305.

Kastenberg, W.E. (1973). On global stability of distributed parameter chemical reactor systems. *Chem. Engr. Sci.*, Vol. 31, pp 249-261.

Khatri, H.C. (1970). Frequency domain stability criteria for distributed parameter systems. *Journal of Basic Engineering*, ASME Trans, Ser. D, Vol. 92, No2, pp 377-384.

Lions, J.L. (1971). *Optimal Control of Systems Governed by Partial Differential Equations*, Springer-Verlag, N-Y.

Nagarajan, N. and Khatri, H.C. (1971). Frequency domain stability criteria for distributed nonlinear sampled-data systems. *Proc. I.F.A.C. Symposium on the Control of Distributed Parameter Systems*, paper No 2-4.

Narasimhan, R. (1954). On the asymptotic stability of solutions of parabolic differential equations. *J. Rational Mech. Anal.*, Vol. 3, pp 303-313.

Parzen, E. (1962). *Stochastic Processes*, Holden-Day, San Francisco.

Persidskii, K.P. (1948). On the solution stability of an infinite system of differential equations. *Adad. Nauk. SSSR Prikl. Mat. Meh.*, Vol. 12, pp 597-612.

Popov, V.M. (1962). Absolute stability of nonlinear systems of automatic control. *Automation and Remote Control*, Vol. 22, pp 857-875, 1962.

Popov, V.M. (1970). *Hyperstability of Control Systems*, 400 pps, Springer Verlag, N-Y, Berlin.

Prodi, G. (1951). Questioni di stabilita per equazioni non-lineari alle derivate parzialli di tipo parabolico. *Atti. Accad., Nazl. Lincel Rend. Cl. Sci.* F15, Mat. Nat., Vol. 10, ser. 8, pp 365-370.

Sakawa, Y. (1972) Optimal filtering in linear distributed-parameter systems. *Intern. J. Control*, Vol. 16, pp 115-127.

Siljak, D. (1969). *Nonlinear Systems*, John Wiley.

Siljak, D. (1978). *Large Scale Dynamic Systems*, North-Holland.

Thom, R. (1975). *Structural Stability and Morphogenesis*, Benjamin.

Tzafestas, S.G., and Nightingale, J.M. (1968). Optimal filtering, smoothing and prediction in linear distributed systems. *Proc. IEE*, Vol 115, pp 1207-1212.

Tzafestas, S.G., and Nightingale, J.M. (1969) Maximum likelihood approach to the optimal filtering of distributed parameter systems, *Proc. IEE*, Vol. 116, pp 1085-1093, 1969

Tzafestas, S.G. (1972. Bayesian approach to distributed parameter filtering and smoothing. *Internat. J. Control*, Vol. 15, pp 272-295, 1972

Tzafestas, S.G. (1973). On the distributed parameter least squares state estimation theory. *Internat. J. Systems Sci;* Vol 4, pp 833-853, 1973.

CHAPTER 5

STABILITY AND OPTIMAL CONTROL OF STOCHASTIC EVOLUTION EQUATIONS

A. Ichikawa

Faculty of Engineering, Shizuoka University, Hamamatsu, Japan

INTRODUCTION

For many years much effort has been devoted to the study of optimal control and system theory for partial differential equations and functional differential equations (Butkovskii, 1969; Curtain and Pritchard, 1978; Lions, 1971; Manitius, 1976; Wang, 1964). These equations are known as distributed parameter systems and can be described by several mathematical models. Here we take the approach based on the theory of semigroups and consider stability and quadratic cost control in both deterministic and stochastic cases. Semigroups naturally arise when we wish to extend exponential functions to infinite dimensions. The semigroup theory gives a unified treatment of a wide class of ordinary differential equations, parabolic and hyperbolic equations, and functional differential equations (Balakrishnan, 1976; Curtain and Pritchard, 1978; Hale, 1977; Kisyński, 1976). We can also express stochastic distributed parameter systems as infinite dimensional Ito equations. There are several ways to model noise processes in Hilbert space, and we take the Wiener process used by Curtain and Falb (1970, 1971). The semigroup formulation is helpful not only in considering the major concepts of controllability, observability and stability in system theory, but in posing various problems in optimal control and filtering. This is mainly due to the fact that the abstract notation looks exactly like that in finite dimensions and that one would naturally ask whether well-known concepts or problems are still meaningful in infinite dimensions. In fact this analogy has helped a great deal the development of the infinite dimensional theory.

There are already excellent expository books and articles which introduce the semigroup treatment of optimal control, estimation and system theory (Balakrishnan, 1976; Curtain, 1978, 1979; Curtain and Pritchard, 1978; Manitius, 1976; Pritchard and Zabczyk, 1977). So we focus our attention on stability and quadratic control of stochastic evolution equations, which is not much discussed in these references. This is an extension of Wonham (1970) to Hilbert spaces. Our mathematical framework is the same as that in (Curtain, 1978, 1979; Curtain and Pritchard, 1978), so they are of much help in reading this chapter. In the next section we collect basic results from the theory of semigroups. In the third we briefly consider optimal quadratic cost control in the deterministic case since it constitutes the preliminaries of the last section. In the fourth section we introduce Wiener processes and stochastic integrals in Hilbert space. We then consider stochastic evolution equations and define mild solutions. Another important result in this section is Ito's differential rule which is very useful in stability and optimal control.

The last two sections follow Ichikawa (1979) and consider mean square stability of stochastic evolution equations and quadratic cost control. We also introduce some results on nonlinear stability (Ichikawa, 1981). Each section contains comments in the end, where we give related topics and those which we do not touch on in this chapter.

SEMIGROUP THEORY

Strongly Continuous Semigroups

Let A be an n×n matrix and consider the differential equation

$$y'(t) = Ay(t), \quad y(0) = y_0 \tag{1}$$

in R^n. Then the unique solution of (1) is given by $y(t) = e^{At}y_0$ and the exponential matrix function e^{At} is the fundamental solution of (1). Now let Y be some Lebesgue space and take A to be the Laplacian, then A is an unbounded operator on Y. In this case (1) is a differential equation in Y and we expect it to describe the heat equation. Semigroup theory justifies this formal procedure and gives the extension of exponential functions. In other words it provides conditions which guarantee the fundamental solution of (1) in infinite dimensions. As we expect a wide class of partial differential equations and functional differential equations can be put into this abstract formulation.

We now fix the notation for the rest of this chapter. We denote by $| \ |$ norms of elements in Banach spaces and of bounded operators. If Y, Z are normed linear spaces then $L(Y,Z)$ denotes the space of bounded linear operators mapping Y into Z. In particular we write $L(Y)$ for $L(Y,Y)$. We use $< \, , >$ for inner products in Hilbert spaces. We collect basic results on semigroups (Balakrishnan, 1976; Butzer and Berens, 1967; Curtain and Pritchard, 1978; Dunford and Schwartz, 1957; Hille and Phillips, 1957; Ladas and Lakshmikantham, 1972; Yosida, 1978).

Definition 1. Let Y be a Banach space. A family of bounded linear operators $T(t) \in L(Y)$, $t \geq 0$ on Y is a semigroup if it satisfies (a) $T(0) = I$, identity on Y (b) $T(t+s) = T(t)T(s)$ for any t, $s \geq 0$, and it is strongly continuous if (c) for each $y \in Y$, $|T(t)y - y| \to 0$ as $t \to 0+$.

Useful consequences of this definition are the following.

Theorem 1. Let T(t) be a strongly continuous semigroup on a Banach space Y, then
(a) T(t)y is strongly continuous for each $y \in Y$,
(b) $$\omega_0 \triangleq \inf_{t>0}(\log|T(t)|/t) = \lim_{t\to\infty}(\log|T(t)|/t) < \infty, \tag{2}$$
(c) for any $\omega > \omega_0$ there exists a constant $M_\omega \geq 1$ such that
$$|T(t)| \leq M_\omega e^{\omega t} \quad \text{for all } t \geq 0.$$

Now we relate the semigroup T(t) to the differential equation (1). Let $D(A)$ denote the subspace of all elements y such that $(T(t)y - y)/t$ converges in Y as $t \to 0+$ and define the operator on $D(A)$: $$Ax = \lim_{t\to 0+}(T(t)y - y)/t. \tag{3}$$

Definition 2. The operator A given by (3) is the infinitesimal generator of the semigroup T(t).

Theorem 2. Let T(t) be a strongly continuous semigroup on a Banach space Y with infinitesimal generator A, then

(a) $T(t)y \in D(A)$ for all $x \in D(A)$ and $t \geq 0$,

(b) $T(t)y$ is strongly differentiable and $\frac{d}{dt}T(t)y = AT(t)y = T(t)Ay$ for all $y \in D(A)$ and $t > 0$,

(c) $T(t)y - y = \int_0^t T(r)Aydr$ for each $y \in D(A)$,

(d) A is a closed linear operator and $D(A)$ is dense in Y.

This theorem implies that if $y_0 \in D(A)$, $y(t) = T(t)y_0$ is a solution of (1). We have also the uniqueness.

Theorem 3. Let A be the infinitesimal generator of a strongly continuous semigroup $T(t)$. Then the Cauchy problem (1) with $y_0 \in D(A)$ has the unique solution $y(t) = T(t)y_0$.

Now recall that a complex number λ belongs to the resolvent set $\rho(A)$ of A if $\lambda I - A$ has a bounded inverse and the inverse is the resolvent of A, denoted usually by $R(\lambda,A)$.

Lemma 1. Let $T(t)$ be a strongly continuous semigroup on a Banach space Y with infinitesimal generator A and $|T(t)| \leq Me^{\omega t}$, then

(a) if $\omega < \text{Re}\lambda$ (the real part of λ), it follows that $\lambda \in \rho(A)$ and

$$R(\lambda,A)y = \int_0^\infty e^{-\lambda t}T(t)ydt,$$

(b) $|\lambda R(\lambda,A)| \leq 2M$ for large λ and $\lim_{\lambda\to\infty}\lambda R(\lambda,A)y = y$ for any $y \in Y$,

(c) $\lambda R(\lambda,A)Ay = \lambda^2 R(\lambda,A)y - \lambda y$, (4)

(d) $R(\lambda,A) - R(\mu,A) = (\mu-\lambda)R(\mu,A)R(\lambda,A)$ for any $\mu, \lambda \in \rho(A)$,

(e) $|R(\lambda,A)^n| \leq M/(\text{Re}\lambda-\omega)^n$ for $\text{Re}\lambda > \omega$.

The result below which is known as the Hille-Yosida Theorem provides a complete characterization of infinitesimal generators.

Theorem 4. A necessary and sufficient condition for a closed linear operator A with dense domain in a Banach space Y to generate a strongly continuous semigroup is that there exist real numbers M, ω such that for any real number $\lambda > \omega$, $\lambda \in \rho(A)$ and $|R(\lambda,A)^n| \leq M/(\lambda-\omega)^n$, $n=1,2,\ldots$. In this case $|T(t)| \leq Me^{\omega t}$. /

If Y is a Hilbert space we have a useful corollary (Balakrishnan, 1976; Kisyński, 1976).

Corollary 1. Let A be a closed densely defined linear operator on a Hilbert space Y such that for some real number β

$$\text{Re}\langle Ay,y\rangle \leq \beta|y|^2, \quad y \in D(A), \qquad \text{Re}\langle A^*y,y\rangle \leq \beta|y|^2, \quad y \in D(A^*) \tag{5}$$

where A^* is the adjoint of A. Then A generates a strongly continuous semigroup $T(t)$ and $|T(t)| \leq e^{\beta t}$.

We recall that a bounded operator mapping a Banach space into another Banach space is compact if it maps bounded sets into subsets of compact sets. A semigroup $T(t)$ is said to be compact if $T(t)$ is compact for all $t > 0$ and compact semigroups on Hilbert space have the following properties (Balakrishnan, 1976).

Theorem 5. Let $T(t)$ be a strongly continuous compact semigroup on a Hilbert space Y and let A be its infinitesimal generator. Then

(a) $T(t)$ is uniformly continuous for $t > 0$,

(b) A has a pure point spectrum consisting of at most a countable sequence of points $\{\lambda_k\}$ with corresponding eigenvectors $\{\phi_k\}$ and $\{\lambda_k\}$ cannot have an accumulation point in the finite part of the complex plane,

(c) $R(\lambda,A) = \int_0^\infty e^{-\lambda t}T(t)dt$, for $\mathrm{Re}\lambda > \omega_0$ where ω_0 is given in (2b) and the integral exists in the uniform operator topology,

(d) $T(t)\phi_k = e^{\lambda_k t}\phi_k$,

(e) $R(\lambda,A)$ exists and is compact for all $\lambda \neq \lambda_k$ and $R(\lambda,A)\phi_k = (1/\lambda-\lambda_k)\phi_k$.

Consider

$$y''(t) + By'(t) + Ay(t) = 0,\ y(0) = y_0,\ y'(0) = y_1. \tag{6}$$

Second order systems of this type are useful in applications to hyperbolic equations. Introducing a new variable z for y' we can reduce (6) to the first order equation

$$\mathcal{Y}'(t) = \mathcal{A}\mathcal{Y}(t),\ \mathcal{Y}(0) = \mathcal{Y}_0, \tag{7}$$

where $\mathcal{Y}(t) = \begin{pmatrix} y(t) \\ z(t) \end{pmatrix}$, $\mathcal{Y}_0 = \begin{pmatrix} y_0 \\ y_1 \end{pmatrix}$ and $\mathcal{A} = \begin{pmatrix} 0 & I \\ -A & -B \end{pmatrix}$. Let A be a self adjoint positive operator on a Hilbert space Y with $\langle Ay,y\rangle \geq a|y|^2$, $a > 0$, $y \in D(A)$. Then $A^{1/2}$ is well defined (Yosida, 1978) and $D(A^{1/2})$ is a Hilbert space with the graph norm. We assume $B \in L(D(A^{1/2}),Y)$, $D(B^*) \supset D(A^{1/2})$ and $\langle By,y\rangle \geq b|y|^2$, $y \in D(B)$ for some real number b. Then Corollary 1 gives the following (Ichikawa and Pritchard, 1979).

Theorem 6. Under the above conditions on A, B, the operator

$$\mathcal{A} = \begin{pmatrix} 0 & I \\ -A & -B \end{pmatrix},\ D(\mathcal{A}) = D(A)\times D(A^{1/2}) \tag{8}$$

generates a strongly continuous semigroup $\mathcal{T}(t)$ on $\mathcal{Y} = D(A^{1/2})\times Y$. If $b > 0$, there exists constants $M \geq 1$, $\omega > 0$ such that $|\mathcal{T}(t)| \leq Me^{-\omega t}$.

When we consider stability of evolution equations, semigroups with such bounds are very important. So we introduce :

Definition 3. A strongly continuous semigroup T(t) on a Banach space is exponentially stable if $|T(t)| \leq Me^{-\omega t}$ for some $M > 1$ and $\omega > 0$. The infinitesimal generator A of T(t) is said to be exponentially stable.

The following is an extension of well-known finite dimensional results (Lee and Markus, 1967) proved by Datko (1970).

Theorem 7. Let T(t) be a strongly continuous semigroup on a Hilbert space with infinitesimal generator A, then the following statements are equivalent :

(a) T(t) is exponentially stable,

(b) $\int_0^\infty |T(t)y|^2\,dt < \infty$ for all $y \in Y$,

(c) there exists a self-adjoint nonnegative operator $P \in L(Y)$ such that

$$\langle Ay,Py\rangle + \langle Py,Ay\rangle = -\langle y,y\rangle \quad \text{for all } y \in Y. \tag{9}$$

Inhomogeneous Equations

In control we need to consider the inhomogeneous equation

$$y'(t) = Ay(t) + f(t),\ y(t_0) = y_0,\ t_0 \geq 0. \tag{10}$$

Let $C^1(t_0,t_1;Y)$ be the space of strongly continuously differentiable functions on $[t_0,t_1]$ with values in Y. Then we have (Barbu, 1976; Curtain and Pritchard, 1978) :

Theorem 8. Let A generate a strongly continuous semigroup T(t) on a Banach Y and let $y_0 \in \mathcal{D}(A)$. Suppose one of the following assumptions hold :
(a) $f \in C^1(t_0,t_1;Y)$,
(b) $f(t) \in \mathcal{D}(A)$ for all $t \in [t_0,t_1]$ and $f(t), Af(t)$ are continuous on $[t_0,t_1]$.
Then the unique solution of (10) is given by

$$y(t) = T(t-t_0)y_0 + \int_{t_0}^{t} T(t-r)f(r)dr. \tag{11}$$

In applications in control we often take f(t) as a function of y(t) such as $f(t) = F(t)y(t)$, $F(t) \in \mathcal{L}(X)$. Then (11) becomes an integral equation

$$y(t) = T(t-t_0)y_0 + \int_{t_0}^{t} T(t-r)F(r)y(r)dr. \tag{12}$$

Theorem 9. Let T(t) be a strongly continuous semigroup on a Banach space Y with infinitesimal generator A and let $F(t) \in \mathcal{L}(Y)$ be strongly continuous on $[t_0,t_1]$. Then there exists a unique solution $y(t;t_0,y_0)$ of (12) which is strongly continuous. Set $U(t,t_0)y_0 = y(t;t_0,y_0)$. Then U(t,s) is well-defined on $0 \leq s \leq t \leq t_1$ and $U(t,s) \in \mathcal{L}(Y)$. Furthermore, U(t,s) is strongly continuous and uniformly bounded on $0 \leq s \leq t \leq t_1$ with properties

$$U(t,r)U(r,s) = U(t,s),\ U(t,t) = I, \quad \text{for all } 0 \leq s \leq r \leq t \leq t_1. \tag{13}$$

The operator U(t,s) with properties above is said to be an evolution operator. In general $U(t,t_0)y_0$ is not differentiable in t even for $y_0 \in \mathcal{D}(A)$ unless F(t) is strongly continuously differentiable (Phillips, 1954). However we have (Balakrishnan, 1976; Curtain and Pritchard, 1978) :

Corollary 2. Let $y \in \mathcal{D}(A)$, then U(t,s)y is strongly differentiable in s and

$$\frac{\partial}{\partial s} U(t,s)y = -U(t,s)(A+F(s))y,\ s < t. \tag{14}$$

Examples

We give some examples of infinitesimal generators.

Example 1. Let Y be a Banach space and $A \in \mathcal{L}(Y)$, then $T(t) = e^{At}$ is a group and is defined for all $-\infty < t < \infty$. In particular, if $Y = R^n$ and A is an n×n matrix, then e^{At} is a matrix exponential function.

Example 2. Let $Y = L_2(0,1)$ be the Lebesgue space of square integrable functions on [0,1]. Let $\mathcal{D}(A) = \{y \in Y |\ y, y'$ are absolutely continuous (abs. cont. for short) and $y', y'' \in Y,\ y(0) = y(1) = 0\}$ and define $Ay = \frac{d^2}{dx^2}y$. Then A is self-adjoint and $\langle Ay,y\rangle \leq -\pi^2|y|^2,\ y \in \mathcal{D}(A)$. So the assumptions of Corollary 1 are satisfied. In fact A generates the compact semigroup

$$T(t)y = \sum_{n=1}^{\infty} 2e^{-n^2\pi^2 t} \sin n\pi x \int_0^1 y(s) \sin n\pi s\, ds\ . \tag{15}$$

T(t) is exponentially stable with $|T(t)| \leq e^{-\pi^2 t}$ and solves the heat equation

$$\frac{\partial y}{\partial t} = \frac{\partial^2}{\partial x^2}y,\ 0 < x < 1,\ t > 0,\ y(0,t) = y(1,t) = 0,\ y(x,0) = y_0(x). \tag{16}$$

Example 3. More generally we take $Y = L_2(a,b)$ and define the operator A_i, $i = 1,2,3$: $A_i y = \frac{d}{dx}(p(x)\frac{dy}{dx})$ where $p(x)$ is continuous and positive on $[a,b]$ and $D(A_i) = [y \in Y|\ y,\ py'$ are abs. cont. and py', $(py')' \in Y$ with boundary condition (i)] where (1) $y(a) = y(b) = 0$, (2) $y'(a) = y'(b) = 0$, (3) $y(a) = y(b)$, $p(a)y'(a) = p(b)y'(b)$. Then $\langle A_i y,y\rangle \leq 0$, $y \in D(A_i)$ and A_i generates a contraction semigroup $T_i(t)$ i.e., $|T_i(t)| \leq 1$.

Example 4. Let $Y = L_2(0,1)$ and let A be the negative of the generator in Example 1: $Ay = -\frac{d^2}{dx^2}y$, $D(A) = \{y \in Y|\ y,\ y'$ abs. cont. and $y',\ y'' \in Y,\ y(0) = y(1) = 0\}$.

Then A is self-adjoint and positive with $\langle Ay,y\rangle \geq \pi^2|y|^2$, $y \in D(A)$. It can be shown that $D(A^{1/2}) = \{y \in Y|\ y$ abs. cont., $y(0) = y(1) = 0\}$. Now define the operator $By = b\frac{dy}{dx}$ (b real), $D(B) = \{y \in Y|\ y$ abs. cont., $y(0) = 0\}$. By Theorem 6 the operator $\mathcal{A} = \begin{pmatrix} 0 & I \\ A & B \end{pmatrix}$ with $D(\mathcal{A}) = D(A)\times D(A^{1/2})$ generates a strongly continuous semigroup on $\mathcal{Y} = D(A^{1/2})\times Y$ and solves the wave equation

$$\frac{\partial^2}{\partial t^2}y + b\frac{\partial^2}{\partial t\partial x}y - \frac{\partial^2}{\partial x^2}y = 0,\ y(0,t) = y(1,t) = 0,\ y(x,0) = y_0(x),\ \frac{\partial y}{\partial t}(x,0) = y_1(x).$$

For the operator B we may also take $B = 0$ or $By = by$. If $b > 0$ then $\mathcal{A}$ generates an exponentially stable semigroup in both cases. We may take A, B to be fourth- and second order differential operators respectively (Ichikawa and Pritchard, 1979).

Spacial domains associated with infinitesimal generators so far are one dimensional, but we can consider domains in higher dimensions. For example A can be a uniformly strongly elliptic operator on $\Omega \subset R^n$ with Dirichlet or mixed boundary condition (Friedman, 1976; Triggiani, 1980a).

Example 5. It is known (Hale, 1977) that a class of functional equations produces semigroups in Banach space. Here we introduce a product formulation of delay differential equations. Consider the linear system on a Hilbert space H (for example $H = R^m$),

$$x'(t) = A_0 x(t) + \sum_{i=1}^{n} A_i x(t+\theta_i) + \int_{-b}^{0} A_{01}(\theta)x(t+\theta)d\theta\ , \tag{17}$$
$$x(0) = x_0,\ x(\theta) = h(\theta),\ -b \leq \theta < 0,$$

where $-b < \theta_n < \theta_{n-1} \cdots < \theta_1 < 0$, $A_i \in \mathcal{L}(H)$, $i = 0,1,\cdots n$ and $A_{01} \in C(-b,0;\mathcal{L}(H))$. Taking the space $M_2(-b,0;H) = H\times L_2(-b,0;H)$ and the new state $y(t) = \begin{pmatrix} x(t) \\ x(t+\theta) \end{pmatrix}$ in $M_2(-b,0;H)$, one can show (Borisovic and Turbabin, 1969; Delfour and Mitter, 1972) that the operator

$$\mathcal{A}y = \begin{pmatrix} A_0 y(0) + \sum_{i=1}^{n} A_i y(\theta_i) + \int_{-b}^{0} A_{01}(\theta)y(\theta)d\theta \\ \frac{dy}{d\theta} \end{pmatrix} \tag{18}$$

with domain $D(\mathcal{A}) = \{\begin{pmatrix} y(0) \\ y \end{pmatrix}|\ y$, abs. cont. and $y' \in L_2(-b,0;H)\}$ generates a strongly continuous compact semigroup on $Y = M_2(-b,0;H)$. The first component of the solution of the differential equation $y'(t) = \mathcal{A}y(t)$, $y(0) = y_0 = \begin{pmatrix} x_0 \\ h \end{pmatrix}$ solves the original delay equation. It is not straightforward to show that $\mathcal{A}$ generates a semigroup. However there is an approach by Zabczyk (1978a) which is relatively easy to follow and gives a class of generators for delay equations. It is based on a certain decomposition of generators. As for stability Delfour, McCalla and Mitter (1975) have shown that for $H = R^m$ the semigroup generated by A is exponentially stable if and only if there exists an $\alpha < 0$ such that the spectrum $\sigma(A)$ of A lies entirely in $\{\lambda | \operatorname{Re}\lambda \leq \alpha\}$, where $\sigma(A) = \{\lambda | \det\Delta(\lambda) = 0\}$ and $\det\Delta(\lambda)$ is the determinant of the matrix

$$\Delta(\lambda) = \lambda I - A_0 - \sum_{i=1}^{n} A_i e^{\lambda\theta_i} - \int_b^0 A_{01}(\theta)e^{\lambda\theta}d\theta \,. \tag{19}$$

Comments

In Theorem 8 we have the differentiability of $y(t)$ for all $t \in (t_0,t_1)$, but we may relax this and replace it by the absolute continuity of $y(t)$ (Barbu, 1976). If we wish to consider the differential equation $y'(t) = A(t)y(t)$ instead of (1), we need a class of evolution operators $U(t,s)$ which are differentiable in t. The theory for such operators exists and can be found in Tanabe (1980) and Yosida (1978). Curtain and Pritchard (1976, 1978) define mild evolution operators which need only be continuous and uniformly bounded and obtain general perturbation results. We can relax the continuity of $F(t)$ in (12) and take essentially bounded operators. In this case the equality (14) is only true for almost all s. Kisyński (1976) derives semigroups from certain bilinear forms and relates them to the theory of Lions (1961) on partial differential equations. The difficult part of the proof of Theorem 7 is to show the equivalence of (a) and (b), but a rather simple proof is given in Curtain and Pritchard (1978). It is sufficient to show that (b) implies $0 > \omega_0$ (defined in (2)). There are weaker concepts of stability of semigroups such as asymptotic stability and weak stability (Pritchard and Zabczyk, 1981). A very important class of semigroups which is not discussed in this section is that of analytic semigroups (Balakrishnan, 1976; Kato, 1966). There are many useful properties for them. Semigroups in Examples 1, 2 and 3 are analytic.

QUADRATIC COST CONTROL PROBLEM

Let Y, U be real Hilbert spaces. Consider the control system

$$y'(t) = Ay(t) + Bu(t), \; y(t_0) = y_0 \,, \tag{20}$$

where A is the infinitesimal generator of a strongly continuous semigroup $T(t)$ on Y, u is a control with values in U and $B \in \mathcal{L}(U,Y)$. As we have seen in Theorem 8

$$y(t) = T(t-t_0)y_0 + \int_{t_0}^{t} T(t-r)Bu(r)dr \tag{21}$$

is the unique solution of (20) on $[t_0,t_1]$ if $y_0 \in \mathcal{D}(A)$ and $u \in C^1(t_0,t_1;U)$. However, (21) can be defined for all $y_0 \in Y$ and locally integrable functions u if the integral is understood in the sense of Bochner (Hille and Phillips, 1957). Let $L_p(t_0,t_1;U)$, $p \geq 1$ be the space of strongly mearurable functions $u(t)$ with $|u(t)|^p$ integrable. In the sequel we shall adopt the following definition which is more appropriate in control.

Definition 4. Let $y_0 \in Y$ and $u \in L_1(t_0,t_1;U)$, then (21) is said to be the mild solution of (20).

Quadratic Control on a Finite Interval

For each $u(\cdot) \in L_2(t_0,t_1;U)$ we impose the following cost :

$$J(u;t_0,y_0) = \langle Gy(t_1),y(t_1)\rangle + \int_{t_0}^{t_1}\{\langle My(t),y(t)\rangle + \langle Nu(t),u(t)\rangle\}dt \tag{22}$$

where G, M are self-adjoint nonnegative operators in $L(Y)$ (written $G \geq 0$, $M \geq 0$) and N is a self-adjoint positive operator ($N > 0$) in $L(U)$ with bounded inverse N^{-1}. Our control problem is then to minimise (22) over $L_2(t_0,t_1;U)$. In finite dimensions optimal control of this type is well-known and is given in terms of the solution of a Riccati equation (Bryson and Ho, 1975; Lee and Markus, 1967). This is one of the best examples for which dynamic programming gives a complete solution. We shall show that this is the case in infinite dimensions. Note that a feedback control is a map $K(t,y) : [t_0,t_1]\times Y \to U$ such that the integral equation given by (21) with $u(t) = K(t,y(t))$ has a unique continuous solution $y(t)$. If $u(t) = K(t,y(t)) \in L_2(t_0,t_1;U)$ then the feedback control is said to be admissible. We give a class of admissible feedback controls.

Lemma 2. Let $K(t,y)$ be strongly measurable in t for each y and satisfy

$$|K(t,y)| \leq c(1+|y|), \quad |K(t,y)-K(t,z)| \leq c|y-z|, \quad c > 0 \tag{23}$$

then $K(t,y)$ is admissible.

If $K(t,y) = K(t)y$ and $K(t) \in L(Y,U)$ is strongly continuous, then the result follows from Theorem 9. And the proof for the general case goes exactly in the same manner (Curtain and Pritchard, 1978). We give sufficient conditions for optimality.

Theorem 10. Suppose that a feedback control $\bar{u} = \bar{K}(t,y)$ and a real function $V(t,y)$ satisfy the following :

(a) $\bar{K}(t,y)$ is continuous in t and satisfies (23),

(b) $V(t,y)$ is continuous in t for each $y \in Y$ and $V(t_1,y) = \langle Gy,y\rangle$,

(c) $V(t,y)$ is Fréchet differentiable in y and the derivative $V_y(t,y)$ is continuous in t, y with $|V_y(t,y)| \leq c(1+|y|)$, $c > 0$, (24)

(d) $V(t,y)$ is differentiable in t for each $y \in D(A)$ and

$$0 = V_t(t,y) + \langle V_y(t,y),Ay+B\bar{u}\rangle + \langle My,y\rangle + \langle N\bar{u},\bar{u}\rangle$$
$$\leq V_t(t,y) + \langle V_y(t,y),Ay+Bu\rangle + \langle My,y\rangle + \langle Nu,u\rangle$$

for all $u \in U$ and $y \in D(A)$.

Then the feedback control $\bar{u} = \bar{K}(t,y)$ is optimal and $J(\bar{u};t_0,y_0) = V(t_0,y_0)$.

Let $C(t_0,t_1;Y)$ denote the Banach space of strongly continuous functions in Y. The next lemma is useful and can be proved easily.

Lemma 3. If u_n converges to u in $L_2(t_0,t_1;U)$, then $y_n(t) \to y(t)$ in $C(t_0,t_1;Y)$ where $y_n(t)$, $y(t)$ are mild solutions corresponding to u_n, u respectively.

Proof of Theorem 10. First assume $y_0 \in D(A)$ and let $\bar{y}(t)$ be the response to the feedback control $\bar{u} = \bar{K}(t,y)$. We set $\bar{u}(t) = \bar{K}(t,\bar{y}(t))$. Then there exists a sequence $u_n(t) \in C^1(t_0,t_1;U)$ which is convergent to $\bar{u}(t)$ in $L_2(t_0,t_1;U)$. Let $y_n(t)$ be the response to the control $u_n(t)$. By (24c,d), $V(t,y_n(t))$ is continuously differentiable in t. Using (24d) and integrating from t_0 to t_1 we obtain

$$V(t_0,y_0) = V(t_1,y_n(t_1)) + \int_{t_0}^{t_1}\{\langle My_n(t),y_n(t)\rangle + \langle N\bar{K}(t,y_n(t)),\bar{K}(t,y_n(t))\rangle\}dt$$
$$- \int_{t_0}^{t_1}\langle V_y(t,y_n(t)),B[u_n(t) - \bar{K}(t,y_n(t))]\rangle dt.$$

Note that $y_n(t) \to \bar{y}(t)$, $\bar{K}(t,y_n(t)) \to \bar{K}(t,\bar{y}(t)) = \bar{u}(t)$ as $n \to \infty$. Now using (24) and passing to the limit $n \to \infty$, we obtain

$$V(t_0,y_0) = \langle G\bar{y}(t_1),\bar{y}(t_1)\rangle + \int_{t_0}^{t_1}\{\langle M\bar{y}(t),\bar{y}(t)\rangle + \langle N\bar{u}(t),\bar{u}(t)\rangle\}dt$$
$$= J(\bar{u};t_0,y_0). \tag{25}$$

Let $u(t) \in L_2(t_0,t_1;U)$ be arbitrary. Then taking differentiable controls which approximates $u(t)$, we can show just as above that $V(t_0,y_0) \leq J(u;t_0,y_0)$ (26)
Since $D(A)$ is dense in Y and $J(u;t_0,y_0)$ depends continuously on y_0, we have (25) and (26) for all $y_0 \in Y$.

As in finite dimensions we seek the value function of the form $V(t,y) = \langle P(t)y,y\rangle$ for some $0 \leq P(t) \in L(Y)$. Then (24d) yields the control $\bar{u} = -N^{-1}B^*P(t)y$ and the Riccati equation

$$\frac{d}{dt}\langle P(t)y,y\rangle + 2\langle P(t)y,Ay\rangle + \langle My,y\rangle - \langle P(t)BN^{-1}B^*P(t)y,y\rangle = 0,\ y \in D(A), \tag{27}$$
$$P(t_1) = G.$$

We establish the existence and uniqueness result for it.

Theorem 11. The Riccati equation (27) has a unique solution in the class of strongly continuous self-adjoint nonnegative operators on Y.

Proof. As in the matrix case (Wonham, 1968) we iterate a sequence of nonnegative operators

$$P_n(t)y = U_n^*(t_1,t)GU_n(t_1,t)y + \int_t^{t_1} U_n^*(r,t)[M + P_{n-1}(r)BN^{-1}B^*P_{n-1}(r)]U_n(r,t)ydr,$$
$$P_0(t)y = T^*(t_1-t)GT(t_1-t)y + \int_t^{t_1} T^*(r-t)MT(r-t)ydr, \tag{28}$$

where $U_n(t,s)$ is the unique solution of the integral equation

$$U_n(t,s)y = T(t-s)y - \int_s^t T(t-r)BN^{-1}B^*P_{n-1}(r)U_n(r,s)ydr,$$

which is guaranteed by Theorem 9. Using Corollary 2 we can differentiate $\langle P_n(t)y,y\rangle$, $y \in D(A)$ to obtain

$$\frac{d}{dt}\langle P_n(t)y,y\rangle + 2\langle (A-BN^{-1}B^*P_{n-1}(t))y,P_n(t)y\rangle + \langle My,y\rangle \tag{29}$$
$$+ \langle P_{n-1}(t)BN^{-1}B^*P_{n-1}(t)y,y\rangle = 0,\ y \in D(A),\ P_n(t_1) = G,$$
$$\frac{d}{dt}\langle P_0(t)y,y\rangle + 2\langle Ay,P_0(t)y\rangle + \langle My,y\rangle = 0,\ y \in D(A),\ P_0(t_1) = G.$$

In fact $P_n(t)$ given by (28) is the unique solution of (29) since the equation

$$\frac{d}{dt}\langle Q(t)y,y\rangle + 2\langle [A-F(t)]y,Q(t)y\rangle = 0,\ y \in D(A),\ Q(t_1) = 0 \tag{30}$$

has the unique solution $Q(t) = 0$. This can be shown as follows. Let $W(t,s)$ be the solution of

$$W(t,s)y = T(t-s)y - \int_s^t T(t-r)F(r)W(r,s)ydr,$$

then it corresponds formally to the differential equation

$$y'(t) = [A - F(t)]y(t),\ y(s) = y \in D(A). \tag{31}$$

Hence using (30) and (31) we formally obtain $\frac{d}{dt}\langle Q(t)W(t,s)y,\ W(t,s)y\rangle = 0$.
Then integrating this from s to t_1 we obtain $\langle Q(s)y,y\rangle = 0$, $y \in D(A)$. This formal manipulation can be justified by introducing a sequence of approximating solutions for (31). Let $f_n(t)$ be a sequence of differentiable functions which converges to $F(t)W(t,s)y$ in $C(s,t_1;Y)$ and let $y_n(t)$ be the solution of

$$y'(t) = Ay(t) - f_n(t),\ y(s) = y. \tag{32}$$

Then we can differentiate $\langle Q(t)y_n(t),y_n(t)\rangle$. We cancel the term involving A using (30) and (32) and then integrate back from s to t_1. Finally letting $n \to \infty$ we obtain $\langle Q(s)y,y\rangle = 0$ for any $y \in D(A)$. But $D(A)$ is dense in Y so that $Q(s) = 0$ for all $t_0 \leq s \leq t_1$. Next we shall show that $P_n(t)$ is monotone increasing in n i.e., $P_n(t) \geq P_{n+1}(t) \geq 0$ for $n = 0,1,2,\cdots$ and all $0 \leq t \leq t_1$. Set $Q_n(t)=P_n(t)-P_{n-1}(t)$ then (29) yields

$$\frac{d}{dt}\langle Q_n(t)y,y\rangle + 2\langle[A - BN^{-1}B^*P_{n-1}(t)]y,Q_n(t)y\rangle - \langle Q_{n-1}(t)BN^{-1}B^*Q_{n-1}(t)y,y\rangle = 0,$$

$$Q_n(t_1) = 0,\ y \in D(A),\ n \geq 2.$$

By the uniqueness result above we have

$$\langle Q_n(t)y,y\rangle = -\int_t^{t_1} \langle U_n^*(r,t)Q_{n-1}(r)BN^{-1}B^*Q_{n-1}(r)U_n(r,t)y,y\rangle dr \leq 0 \quad \text{for all } y \in Y.$$

Thus $P_n(t) \leq P_{n-1}(t)$ for $n = 1,2,\cdots$ and all t.
Since $P_n(t)$ is a family of monotone decreasing nonnegative operators, there exists a strong limit P(t) (Riesz and SZ-Nagy, 1955) which is nonnegative. Weak continuity of P(t) follows from the uniform boundness of $P_n(t)$ and (29). Now passing to the limit $n \to \infty$ in the integrated version of (29) and then differentiating it we obtain (27). Thus we have established a solution for (27). The strong continuity of P(t) is proved in Corollary 4 below. The uniqueness follows from the fact that $\langle P(s)y,y\rangle$ is the minimal cost for the initial condition (s,y) since in Theorem 10 (t_0,y_0) is fixed, but otherwise arbitrary.

Corollary 3. The optimal control for the quadratic cost problem (21), (22) is given by the feedback law $\bar{u} = -N^{-1}B^*P(t)y$ and the minimal cost by $J(\bar{u};t_0,y_0) = \langle P(t_0)y_0,y_0\rangle$ where P(t) is the unique solution of (27).

Corollary 4. P(t) is also the unique solution in the class of strongly continuous nonnegative operators on Y of the integral equations

$$\text{(a)}\quad P(t)y = T^*(t_1-t)GT(t_1-t)y + \int_t^{t_1} T^*(r-t)[M - P(r)BN^{-1}B^*P(r)]T(r-t)y\,dr,$$

$$\text{(b)}\quad P(t)y = T^*(t_1-t)GU(t_1,t)Y + \int_t^{t_1} T^*(r-t)MU(r,t)y\,dr, \tag{33}$$

$$\text{(c)}\quad P(t)y = U^*(t_1,t)GU(t_1,t)y + \int_t^{t_1} U^*(r,t)[M + P(r)BN^{-1}B^*P(r)]U(r,t)y\,dr,$$

where U(t,s) is the unique solution of

$$U(t,s)y = T(t-s)y - \int_s^t T(t-r)BN^{-1}B^*P(r)U(r,s)y\,dr. \tag{34}$$

Proof. If P(t) satisfies one of the equations in (33), then by Corollary 2 P(t) satisfies (27). Conversely, let P(t) be the solution of (27). First calculate $\frac{\partial}{\partial s}\langle P(s-t)T(s-t)y,T(s-t)y\rangle$, $y \in D(A)$ using (27), and integrate it from t to t_1, then (33a) follows. Therefore the solution of (27) is strongly continuous. The equation (33c) is exactly the relation $V(t,y) = J(\bar{u};t,y)$. Note that $\bar{y}(t) = U(s,t)y$ is the mild solution of (20) with $t_0 = t$, $y_0 = y$ which corresponds to the optimal control $\bar{u} = -N^{-1}B^*P(s)y$. Using approximations of the form (32) and a limiting argument we can show (33b).

Quadratic Control on an Infinite Interval

Let $u \in L_2(0,\infty;U)$ and consider

$$y(t) = T(t)y_0 + \int_0^t T(t-r)Bu(r)dr \tag{35}$$

and the cost functional

$$J(u;y_0) = \int_0^\infty \{\langle My(t),y(t)\rangle + \langle Nu(t),u(t)\rangle\}dt. \tag{36}$$

A control $u(\cdot)$ is admissible if the response $y(t)$ goes to zero as $t \to \infty$. A feedback control $u = K(t,y) : [0,\infty)\times Y \to U$ is admissible if the solution $y(t)$ of the integral equation corresponding to it goes to zero as $t \to \infty$. If $K(y)\colon Y \to U$ satisfies the Lipschitz condition

$$|K(y) - K(z)| \le c|y - z|, \quad c > 0 \tag{37}$$

there exists a unique solution of the integral equation (35) with $u = K(y)$. Let V be another Hilbert space and let $C \in \mathcal{L}(Y,V)$. We introduce two useful definitions.

Definition 5. The pair (A,B) in (20) is stabilizable if there exists an operator $K \in \mathcal{L}(Y,U)$ such that $A - BK$ is exponentially stable.

Definition 6. The pair (C,A) is detectable if there exists an operator $R \in \mathcal{L}(V,Y)$ such that $A - RC$ is exponentially stable.

We now state sufficient conditions for optimality.

Theorem 12. Suppose that there exists a real continuous function $V(y)$ and an admissible feedback control $\bar{u} = \bar{K}(y)$ satisfying the following :

(a) $\bar{K}(y)$ satisfies (37),

(b) $V(y)$ is Fréchet differentiable and the derivative $V_y(s)$ is continuous with $|V_y(y)| \le c|y|$,

(c)
$$\begin{aligned} 0 &= \langle V_y(y),Ay + B\bar{u}\rangle + \langle My,y\rangle + \langle N\bar{u},\bar{u}\rangle \\ &\le \langle V_y(y),Ay + Bu\rangle + \langle My,y\rangle + \langle Nu,u\rangle \quad \text{for all } y \in \mathcal{D}(A) \text{ and } u \in U. \end{aligned} \tag{38}$$

Then the feedback control $\bar{u}$ is optimal and $J(\bar{u};y_0) = V(y_0)$.

Proof. Let $\bar{y}(t)$ be the response to the feedback control $\bar{u}$ and let $\bar{u}(t) = K(\bar{y}(t))$. Then on each finite interval $[0,t_1]$ we can approximate $u(t)$ by differentiable functions $u_n(t)$ and just as in the proof of Theorem 10 we obtain

$$V(y_0) = V(\bar{y}(t_1)) + \int_0^t \{\langle M\bar{y}(t),\bar{y}(t)\rangle + \langle N\bar{u}(t),\bar{u}(t)\rangle\}dt.$$

Since $\bar{K}(\cdot)$ is admissible, $\bar{y}(t_1)$ goes to zero as $t_1 \to \infty$ so that we obtain

$$V(y_0) = \int_0^\infty \{\langle M\bar{y}(t),\bar{y}(t)\rangle + \langle N\bar{u}(t),\bar{u}(t)\rangle\}dt = J(\bar{u};y_0).$$

Similarly using the inequality in (38c) we can show that $V(y_0) \le J(u;y_0)$ for any admissible control u.

If we set $V(y) = \langle Py,y\rangle$ for some $0 \le P \le \mathcal{L}(Y)$, then (38c) yields the control $\bar{u} = -N^{-1}B^*Py$ and the algebraic Riccati equation

$$2\langle Ay,Py\rangle + \langle My,y\rangle - \langle PBN^{-1}B^*Py,y\rangle = 0, \quad y \in \mathcal{D}(A). \tag{39}$$

We next introduce sufficient conditions due to Zabczyk (1976) that the Riccati equation (39) has a solution and that $\bar{u} = -N^{-1}B^*Py$ is admissible. Since M is nonnegative, its square root is well-defined (Riesz and SZ-Nagy, 1955), which we denote by $\bar{M}$.

Lemma 4. Suppose that $(\bar{M},A)$ is detectable and that for some operators $0 \le Q \in \mathcal{L}(Y)$ and $K \in \mathcal{L}(Y,U)$ the following inequality holds :

$$2\langle Qy,(A - BK)y\rangle + \langle My,y\rangle + \langle K^*NKy,y\rangle \le 0, \quad \text{for all } y \in \mathcal{D}(A). \tag{40}$$

Then A - BK generates an exponentially stable semigroup.

Now we show that the stabilizability of (A,B) assures the existence of a solution of the Riccati equation (39). Let (A,B) be stabilizable, then there exists an operator $K \in (Y,U)$ such that $A - BK$ is stable. We iterate the sequence P_n of nonnegative operators :

$$\text{(a) } 2\langle P_n y,(A - BN^{-1}B^*P_{n-1})y\rangle + \langle My,y\rangle + \langle P_{n-1}BN^{-1}B^*P_{n-1}y,y\rangle = 0,\ y \in \mathcal{D}(A),$$

$$\text{(b) } 2\langle P_0 y,(A - BK)y\rangle + \langle My,y\rangle + \langle NKy,Ky\rangle = 0,\ y \in \mathcal{D}(A), \tag{41}$$

which can be also written as

$$\begin{aligned}\text{(a) } & 2\langle P_n y,(A - BN^{-1}B^*P_n)y\rangle + \langle My,y\rangle + \langle P_nBN^{-1}B^*P_ny,y\rangle \\ & = -\langle (P_n - P_{n-1})BN^{-1}B^*(P_n - P_{n-1})y,y\rangle \le 0, \\ \text{(b) } & 2\langle P_0 y,(A - BN^{-1}B^*P_0)y\rangle + \langle My,y\rangle + \langle P_0BN^{-1}B^*P_0y,y\rangle \\ & = -\langle N(K - N^{-1}B^*P_0)y,(K - N^{-1}B^*P_0)Y\rangle \le 0.\end{aligned} \tag{42}$$

To show that this sequence is well-defined, the following lemma is useful.

Lemma 5. Let $S(t)$ be an exponentially stable semigroup on Y with generator $\tilde{A}$ and let $0 \le W \in \mathcal{L}(Y)$. Then the operator $Q : Qy = \int_0^\infty S^*(t)WS(t)y\,dt$ is well-defined and nonnegative and satisfies the equation $2\langle Qy,\tilde{A}y\rangle + \langle Wy,y\rangle = 0,\ y \in \mathcal{D}(\tilde{A})$. Conversely, if Q is self-adjoint and satisfies the second equation, Q is represented by the above integral.

We now define P_n by

$$\begin{aligned} P_n y &= \int_0^\infty S_n^*(t)\,[M + P_{n-1}BN^{-1}B^*P_{n-1}]S_n(t)y\,dt \\ P_0 y &= \int_0^\infty S_0^*(t)\,[M + K^*NK]S_0(t)y\,dt \end{aligned} \tag{43}$$

where $S_n(t)$ is the semigroup generated by $A - BN^{-1}B^*P_{n-1}$, $n = 1,2,\cdots$ and $S_0(t)$ by $A - BK$. Note first that P_0 is well-defined and satisfies (41b). By (42b) and Lemma 4 $A - BN^{-1}B^*P_0$ generates an exponentially stable semigroup $S_1(t)$. Hence repeating in this manner we can show by induction that (43) is well-defined and P_n satisfies (41). Now let $Q_n = P_n - P_{n-1}$, then from (41) we have

$$2\langle Q_n y,(A - BN^{-1}B^*P_{n-1})y\rangle - \langle Q_{n-1}BN^{-1}B^*Q_{n-1}y,y\rangle = 0.$$

Hence by Lemma 5 we obtain

$$\langle -Q_n y,y\rangle = \int_0^\infty \langle Q_{n-1}BN^{-1}B^*Q_{n-1}S_n(t)y,S_n(t)y\rangle dt \ge 0.$$

So P_n is monotone decreasing in n and therefore converges strongly to some $P \ge 0$. Passing to the limit $n \to \infty$ in (41) we obtain the Riccati equation (39). We can rewrite it as

$$2\langle Py,(A - BN^{-1}B^*P)y\rangle + \langle My,y\rangle + \langle PBN^{-1}B^*Py,y\rangle = 0,\ y \in \mathcal{D}(A).$$

Hence by Lemma 4 $A - BN^{-1}B^*P$ generates an exponentially stable semigroup. The uniqueness of P follows from the optimality. Summing up we have :

Theorem 13. Suppose that (A,B) is stabilizable and that $(\overline{M},A)$ is detectable. Then there exists a unique solution $P \ge 0$ of the algebraic Riccati equation (39). The feedback control $\overline{u} = -N^{-1}B^*Py$ is optimal and $J(\overline{u};y_0) = \langle Py_0,y_0\rangle$.

Stability of Semilinear Evolution Equations

Consider the semilinear equations

$$y'(t) = Ay(t) + f(y(t)),\ y(0) = y_0, \tag{44}$$

where A is the infinitesimal generator of a strongly continuous semigroup T(t) on a real Hilbert space Y and $f : Y \to Y$ satisfies the Lipschitz condition

$$|f(y) - f(z)| \le c|y - z|,\ y,\ z \in Y,\ c > 0. \tag{45}$$

The integral equation

$$y(t) = T(t)y_0 + \int_0^t T(t-r)f(y(r))dr \tag{46}$$

has a unique solution in $C(0,t_1;Y)$ for each $0 < t_1 < \infty$ and we call it the mild solution of (44). This is proved by a contraction mapping theorem on $C(0,t_1;Y)$.

Theorem 14. Let V(y) be a nonnegative function on Y with properties :
(a) V(y) is Fréchet differentiable and its derivative is continuous,
(b) $\langle V_y(y), Ay + f(y)\rangle \le -\beta V(y),\ y \in \mathcal{D}(A)$, for some real number β.
Then the mild solution of (44) satisfies $V(y(t)) \le e^{-\beta t}V(y_0)$ for all $y_0 \in Y$.

Proof. We approximate the mild solution y(t) of (44) by differentiable functions $y_n(t)$ of the form (32) and differentiate $e^{\beta t}V(y_n(t))$. Then we use the inequality in (b) and integrate from 0 to t. Finally letting $n \to \infty$ we obtain the inequality.

Corollary 5. If $\beta > 0$ and $V(y) \ge b|y|^2$ for some $b > 0$, then the mild solution of (44) is exponentially stable.

Examples.

Example 6. Consider the heat equation on [0,1]

$$\frac{\partial y}{\partial t} = \frac{\partial^2}{\partial x^2}y + b(x)u(t),\ y(x,0) = y_0(x),\ \frac{\partial y}{\partial x}(0,t) = \frac{\partial y}{\partial x}(1,t) = 0, \tag{47}$$

and the cost functional

$$J(u;y_0) = \int_0^\infty (\int_0^1 y^2(x,t)dx + Nu^2(t))dt,\ N > 0.$$

where $0 \ne b(x) \in L_2(0,1)$. In this case we take $Y = L_2(0,1)$ and

$$Ay = \frac{d^2}{dx^2}y \quad \text{with } \mathcal{D}(A) = \{y \in Y |\ y,\ y' \text{ abs. cont., } y'(0)=y'(1)=0\}.$$

Then the semigroup T(t) is given by

$$T(t)y = \sum_{n=1}^{\infty} 2e^{-n^2\pi^2 t}\cos n\pi x \int_0^1 \cos n\pi s\, y(s)ds + \int_0^1 y(s)ds,$$

and it is not exponentially stable. If we take the feedback control $u = -\langle b,y\rangle$, then

$$\langle Ay,y\rangle + \langle (b\circ b)y,y\rangle \le -\beta|y|^2,\ y \in \mathcal{D}(A),$$

where $(b\circ b)y = b\langle b,y\rangle$ and $\beta = \min\{\pi^2, \int_0^1 b^2(s)ds\}$. Using Corollary 1 we can easily show $A - b\circ b$ generates an exponentially stable semigroup. So (A,b) is stabilizable. Furthermore, the operator M in this example is the identity on Y and (I,A) is detectable (prove this). Thus all the assumptions in Theorem 12 are satisfied. Since T(t) is compact, we can represent the solution P of the Riccati equation by

$$P = \sum_{i,j=0}^{\infty} p_{ij}\phi_i \circ \phi_j, \quad \phi_i = \sqrt{2}\cos i\pi x, \quad \phi_0 = 1$$

and (39) yields

$$2\pi^2 i^2 p_{ij} + \delta_{ij} - \frac{1}{N}\sum_{k,l=0} \langle b,\phi_k\rangle\langle b,\phi_l\rangle p_{kj}p_{lj} = 0, \quad i,j = 0,1,2,\cdots, \tag{48}$$

where $p_{ij} = p_{ji}$ and $\delta_{ij} = \{\begin{smallmatrix} 0 & i \neq j \\ 1 & i = j \end{smallmatrix}$. We need to solve this infinite equations and the optimal control is then given by the feedback law

$$\overline{u} = -\frac{1}{N}\sum_{i,j=0}^{\infty} p_{ij}\langle b,\phi_j\rangle\langle\phi_j,y\rangle.$$

If in particular $b = \phi_m$ for some m, then (48) becomes

$$2\pi^2 i^2 p_{ij} + \delta_{ij} - \frac{1}{N} p_{mi}p_{mj} = 0, \quad i,j = 0,1,2,\cdots$$

which can be solved explicitly.

Example 7. Consider the nonlinear heat equation on [0,1]

$$\frac{\partial y}{\partial t} = \frac{\partial^2}{\partial x^2}y - y/(1 + |y|), \quad y(x,0) = y_0(x), \quad \frac{\partial y}{\partial x}(0,t) = \frac{\partial y}{\partial x}(1,t) = 0. \tag{49}$$

We take Y, A, T(t) as in Example 6 and $f(y) = -y/(1 + |y|)$, then f satisfies the Lipschitz condition (45) with c = 1. Now we consider $V(y) = |y|^2$, then

$$2\langle y, Ay + f(y)\rangle \leq -|y|^2 \quad \text{for all } y \in D(A).$$

Hence the mild solution corresponding to the above equation is exponentially stable.

Comments

The uniqueness of an optimal control does not appear in Theorem 10, but for the quadratic control this is an easy consequence of strict convexity of $J(u;t_0,y_0)$ in u (Balakrishnan, 1976; Lions, 1971). Quadratic control of evolution equations is considered by several authors (Curtain and Pritchard, 1976, 1978; Datko, 1971; Lukes and Russel, 1969) and the most general formulation is given by Curtain and Pritchard (1976). Our approach is different from others' in that we use dynamic programming arguments. It is justified by virtue of approximating systems. There are two possible approximations based on Theorem 8. We have used the part (a), but we may use the part (b) as well. In fact approximations based on (b) is used in the subsequent sections. We give only one example of quadratic control, but more can be found in Curtain and Pritchard (1978). See Delfour and Mitter (1972) for delay equations. The interpretation of the Riccati equation in Example 6 is in terms of eigenfunctions, but for general domains we can derive partial differential equations for the kernels of P(t) or P as in Lions (1971). See also Gibson (1979) for Riccati equations. The stability of the semilinear equation (47) is very elementary, see Ichikawa and Pritchard (1979) and Pao and Vogt (1969) for more general results. A survey of stability and stabilizability of evolution equations is found in Pritchard and Zabczyk (1981). The general theory for nonlinear evolution equations can be found in Barbu (1975) and Martin (1976). It is well known in finite dimensions (Davis, 1977) that filtering is dual to quadratic control. This is the case for evolution systems. See chapter 3 and Curtain (1978) for a survey and Vinter (1977) for filter stability. We have only discussed quadratic control, but optimal control of different types can be formulated. There are several results on maximum principle and bang-bang controls (Balakrishnan, 1965, 1976; Cole, 1968; Conti, 1968, 1970; Falb, 1964; Fattorini, 1964; Friedman, 1967a, 1967b, 1968). See also Barbu and Precupanu (1978). In our system (20) we assume that B is bounded. Then boundary control cannot be included in this formulation. But Curtain and Pritchard

(1977) has extended their earlier results for unbounded operators. Semigroup models for boundary control are also reported in Balakrishnan (1976), Lasiecka (1978, 1980a, 1980b), Nambu (1979), Triggiani (1980a, 1980b), Washburn (1979) and Zabczyk (1977, 1978). See Lions (1971) and Sorine (1978) for different approach.

STOCHASTIC EVOLUTION EQUATIONS

Stochastic Integrals in Hilbert Space

We recall definitions of random variables and Wiener processes in Hilbert space. Let $(\Omega,\mathcal{F},\mu)$ be a complete probability space and let H be a real separable Hilbert space. A map $x:\Omega \to H$ is a random variable if x is strongly measurable. If x is square integrable i.e., $x \in L_2(\Omega,\mu;H)$, then the covariance operator of x is $\mathrm{Cov}\{x\} = E\{(x - Ex)\circ(x - Ex)\}$ where E denotes the expectation and $g\circ h \in \mathcal{L}(H)$ for each $g, h \in H$ is defined by $(g\circ h)k = g\langle h,k\rangle$, $k \in H$. The covariance operator Cov{x} is a self-adjoint nonnegative trace class (or nuclear) operator (Balakrishnan, 1976; Gel'fand and Vilenkin, 1964). Let $Ex = 0$, then

$$E|x|^2 = \mathrm{tr.Cov}\{x\},\ E\langle Rx,x\rangle = \mathrm{tr.}R\ \mathrm{Cov}\{x\},\ R \in \mathcal{L}(H), \tag{50}$$

where tr. denotes the trace of an operator.
Let $x \in L_1(\Omega,\mu;H)$ and let $\overline{\mathcal{F}}$ be a sub σ-field of $\mathcal{F}$. The conditional expectation $E\{x|\ \overline{\mathcal{F}}\}$ of x relative to $\overline{\mathcal{F}}$ is such that

$$\int_F x(\omega)d\mu = \int_F E\{x|\ \overline{\mathcal{F}}\}(\omega)d\mu \text{ for all } F \in \overline{\mathcal{F}}.$$

A random variable $x \in L_2(\Omega,\mu;H)$ is Gaussian if $\langle x,e_i\rangle$ is a real Gaussian random variable for all i where $\{e_i\}$ $i = 1,2,\cdots$ is a complete orthonormal basis for H. We have the following inequality for a Gaussian random variable (Ichikawa, 1981).

Lemma 6. Let x be a Gaussian random variable in H with zero mean and covariance operator Q, then

$$E|x|^{2n} \le (2n-1)!!(\mathrm{trace}Q)^n,\ n = 1,2,\cdots \tag{51}$$

where $(2n-1)!! = (2n-1)(2n-3)\ldots3.1$ and the equality holds for $n = 1$.

We now introduce a Wiener process in H (Curtain and Pritchard, 1978).

Definition 7. A Wiener process $w(t)$, $t \in [0,t_1]$ in H is a stochastic process $w(t) \in L_2(\Omega,\mu;H)$ with properties

(a) $E\ w(t) = 0$,
(b) $\mathrm{Cov}\{w(t) - w(s)\} = (t - s)W$, where $W \in \mathcal{L}(H)$ is a nonnegative trace class operator,
(c) w(t) has independent increments, (52)
(d) w(t) has continuous sample paths on $[0,t_1]$.

Let $\sigma_t(w)$ be the σ-field generated by $w(s)$, $0 \le s \le t$, then the property (c) implies that $w(t) - w(s)$ is independent of $\sigma_s(w)$ for all $t > s$ and by (a) $E\{w(t) - w(s)|\ \sigma_s(w)\} = 0$.

Lemma 7. Let w(t) be a Wiener process in H, then

$$w(t) = \sum_{i=1}^{\infty} \beta_i(t)e_i \tag{53}$$

where $\{e_i\}$ is an orthonormal sequence consisting of eigenvectors of W with $We_i = \lambda_i e_i$, $\lambda_i > 0$ and $\beta_i(t)$ are mutually independent real Wiener processes with incremental covariance λ_i and $\mathrm{tr.}W = \sum_{i=1}^{\infty} \lambda_i < \infty$.

Corollary 6. (a) $w(t)$ is Gaussian for each $t \in [0,t_1]$, (54)

(b) $E|w(t) - w(s)|^{2n} \le (2n-1)!!(t-s)^n(\mathrm{tr.}W)^n$, where the equality holds for $n = 1$.

Let $\{F_t\}$ be an increasing family of σ-fileds $\subset F$ such that (i) $w(t)$ is measurable relative to F_t and (ii) $w(t) - w(s)$ is independent of F_s, $s < t$. (55)
Let Y be a real Hilbert space and let $B_2([0,t_1]\times\Omega;L(H,Y))$ be the space of $L(H,Y)$-valued stochastic processes $G(t)$ such that $G(t)h$ for any $h \in H$ is strongly measurable and adapted to the σ-field F_t and $E\int_0^{t_1} |G(t)|^2\,dt < \infty$. As in Curtain (1977) we shall define the stochastic integral $\int_0^{t_1} G(t)dw(t)$ using the representation (53) of $w(t)$.

Let $w_n(t) = \sum_{i=1}^{n} e_i\beta_i(t)$, then the stochastic integral

$$y_n = \int_0^{t_1} G(t)dw_n(t) = \sum_{i=1}^{n}\int_0^{t_1} G(t)e_i d\beta_i(t) \tag{56}$$

is well-defined just as in the scalar case (Arnold, 1974; Gikhman and Skorokhod, 1969, 1979). Then $\{y_n\}$ is a Cauchy sequence in $L_2(\Omega,\mu;H)$. In fact for any $m > n$ We have

$$E|y_m - y_n|^2 = \sum_{i=n+1}^{m} \lambda_i \int_0^{t_1} E|G(t)e_i|^2\,dt \le \Big(\sum_{i=n+1}^{m} \lambda_i\Big)\int_0^{t} E|G(t)|^2\,dt \to 0 \quad \text{as } n \to \infty.$$

Thus we define the stochastic integral

$$\int_0^{t_1} G(t)dw(t) = \lim_{n\to\infty}\int_0^{t_1} G(t)dw_n(t) = \sum_{i=1}^{\infty}\int_0^{t_1} G(t)e_i d\beta_i(t) \tag{57}$$

where the limit is understood in $L_2(\Omega,\mu;H)$. By an obvious modification $\int_{t_0}^{t} G(r)dw(r)$ can be defined for any $0 \le t_0 \le t \le t_1$. We now give some properties of the stochastic integral.

Lemma 8. Let $G(t) \in B_2([0,t_1]\times\Omega;L(H,Y))$. Then

(a) $E\int_0^{t_1} G(t)dw(t) = 0$,

(b) $$E\Big|\int_0^{t_1} G(t)dw(t)\Big|^2 = \begin{cases} \int_0^{t_1} E[\mathrm{tr.}G(t)WG^*(t)]dt \\ \int_0^{t_1} E[\mathrm{tr.}G^*(t)G(t)W]dt \end{cases} \le \mathrm{tr.}W\int_0^{t_1} E|G(t)|^2 dt, \tag{58}$$

(c) If $\int_0^{t_1} E|G(t)|^4\,dt < \infty$, then $\int_0^{t_1} G(t)dw(t) \in L_4(\Omega,\mu;Y)$ and

$$E\Big|\int_0^{t_1} G(t)dw(t)\Big|^4 \le 36t_1(\mathrm{tr.}W)^2\int_0^{t_1} E|G(t)|^4\,dt.$$

Proof. (a) This is an easy consequence of the equality

$$E\int_0^{t_1} G(t)dw_n(t) = \sum_{i=1}^{n} E\int_0^{t_1} G(t)e_i d\beta_i(t) = 0 \quad \text{for each n.}$$

(b) $$E\left|\int_0^{t_1} G(t)dw(t)\right|^2 = \lim_{n\to\infty} E\left|\int_0^{t_1} G(t)dw_n(t)\right|^2$$
$$= \lim_{n\to\infty} \sum_{i=1}^{n} \lambda_i \int_0^{t_1} E\langle G(t)e_i, G(t)e_i\rangle dt \tag{59}$$
$$= \begin{cases} \int_0^{t_1} E[\text{tr}.G(t)WG^*(t)]dt \\ \int_0^{t_1} E[\text{tr}.G^*(t)G(t)W]dt \end{cases} \leq \text{tr}.W\int_0^{t_1} E|G(t)|^2 dt.$$

(c) This can be proved as in the real case (Gikhman and Skorokhod, 1969).

Lemma 9. Let $G(t) \in \mathcal{B}_2([0,t_1]\times\Omega;\mathcal{L}(H,Y))$ and let $y(t) = \int_0^t G(t)dw(t)$. Then

(a) $\{y(t),\mathcal{F}_t\}$ is a martingale i.e., $E\{y(t)|\ \mathcal{F}_s\} = y(s)$ w.p.1 for $s < t$,
(b) The process y(t) has a separable version which has continuous sample paths.

This lemma can be proved as in the case of nonrandom integrants (Curtain and Pritchard, 1978). Now let G(t,s) be an $\mathcal{L}(H,Y)$-valued process defined on $[0,t_1]\times[0,t_1]$ such that G(t,s)h is strongly measurable for any $h \in H$ and $\mathcal{F}_s$-measurable for any t and h with $\int_0^{t_1}\int_0^{t_1} E|G(t,s)|^2 dsdt < \infty$.

Lemma 10. For G given above
$$\int_0^{t_1}\int_0^{t_1} G(t,s)dw(s)dt = \int_0^{t_1}\int_0^{t_1} G(t,s)dtdw(s) \quad \text{w.p.1.,} \tag{60}$$
where the right hand side is understood in the sense
$$\sum_{i=1}^{\infty}\int_0^{t_1}\int_0^{t_1} G(t,s)e_i dtd\beta_i(s).$$

Proof. The equality above is true if w is replaced by w_n. The assertion follows then taking limits.

Finally we remark that stochastic integrals can be defined for those G(t) with only $\int_0^{t_1} |G(t)|^2 dt < \infty$ w.p.1. This is done as in the scalar case (Gikhman and Skorokhod, 1969). This extension and details of stochastic integrals can be found in Ichikawa (1981).

Stochastic Evolution Equations

Let Y,H be real separable Hilbert spaces. We consider the stochastic evolution equation
$$dy(t) = [Ay(t) + f(y(t))]dt + D(y(t))dw(t),\ y(t_0) = y_0 \tag{61}$$
where A is the infinitesimal generator of a strongly continuous semigroup T(t) on Y, w(t) is a Wiener process with covariance operator W in H and $f;Y \to Y$ and $D(\cdot):Y \to \mathcal{L}(H,Y)$ satisfy Lipschitz conditions

$$\begin{aligned} |f(y) - f(z)| &\le c_1|y - z|, \quad y, z \in Y, \ c_1 > 0 \\ |D(y) - D(z)| &\le c_2|y - z|, \quad y, z \in Y, \ c_2 > 0. \end{aligned} \tag{62}$$

Let F_t be the σ-field given in (55). We introduce two concepts of a solution of (61) (Curtain, 1977, 1979).

Definition 8. $y(t)$ is a strong solution of (61) on $[t_0,t_1]$ if $y(t) \in D(A)$ with probability one for almost all t and satisfies the following :

$$\begin{aligned} &\text{(a) } y(t) \in C(t_0,t_1,L_2(\Omega,\mu;Y)) \text{ and } \int_{t_0}^{t_1} |Ay(t)|dt < \infty \text{ w.p.1,} \\ &\text{(b) } y(t) \text{ is adapted to } F_t \text{ and has continuous sample paths,} \\ &\text{(c) } y(t) = y_0 + \int_{t_0}^{t} Ay(r)dr + \int_{t_0}^{t} f(y(r))dr + \int_{t_0}^{t} D(y(r))dw(r) \text{ w.p.1.} \end{aligned} \tag{63}$$

As we see below in Theorem 16 this concept turns out to be very strong. Recall also assumptions in Theorem 8 for the inhomogeneous equation (10). So we introduce a weaker concept.

Definition 9. $y(t)$ is a mild solution on $[t_0,t_1]$ if $y(t) \in C(t_0,t_1;L_2(\Omega,\mu;Y))$ and is adapted to F_t such that

$$y(t) = T(t-t_0)y_0 + \int_{t_0}^{t} T(t-r)f(y(r))dr + \int_{t_0}^{t} T(t-r)D(y(r))dw(r). \tag{64}$$

Theorem 15. Let $t_1 > t_0 > 0$ be arbitrary and let y_0 be measurable relative to F_t with $E|y_0|^2 < \infty$. Then there exists a unique solution of (64) in $C(t_0,t_1;L_p(\Omega,\mu;Y))$, $p = 2,4$ which is adapted to F_t.

This theorem can be proved by a contraction mapping theorem (Chojnowska-Michalik, 1976; Ichikawa, 1978, 1981).

Using Lemma 10 and a Fubini theorem we can show :

Theorem 16. Suppose that the mild solution $y(t)$ of (61) satisfies the following :

$$\begin{aligned} &\text{(a) } y_0 \in D(A),\ f(y(r)),\ D(y(r))h \in D(A) \text{ w.p.1 for each } h \in H \\ &\text{(b) } \int_{t_0}^{t_1}\int_{t_0}^{t} |AT(t-r)f(y(r))|drdt < \infty \quad \text{w.p.1,} \\ &\text{(c) } \int_{t_0}^{t_1}\int_{t_0}^{t} E|AT(t-r)D(y(r))|^2 drdt < \infty. \end{aligned} \tag{65}$$

Then $y(t)$ is a strong solution of (61).

Ito's Formula

Let Y, H, $w(t)$ and F_t be given as in the previous section. Let $y(t)$ be ths stochastic process in Y given by

$$y(t) = y_0 + \int_{t_0}^{t} g(r)dr + \int_{t_0}^{t} G(r)dw(r), \quad 0 \le t_0 \le t \le t_1, \tag{66}$$

where y_0 is F_{t_0}-measurable and $g(t)$, $G(t)$ are strongly measurable and adapted to F_t with $\int_{t_0}^{t_1} |g(t)|dt < \infty$, $\int_{t_0}^{t_1} |G(t)|^2 dt < \infty$ w.p.1. Then $y(t)$ is said to have the

stochastic differential $dy = g(t)dt + G(t)dw(t)$.
The following is a generalization of Ito's formula to Hilbert spaces.

Theorem 17. Suppose that the real function $v(t,y)$ on $[t_0,t_1]\times Y$ has properties :
(a) $v(t,y)$ is continuous (on $[t_0,t_1]\times Y$),
(b) $v(t,y)$ is differentiable in t and $v_t(t,y)$ is continuous, (67)
(c) $v(t,y)$ is twice Fréchet differentiable in y, $v_y(t,y)y_1$ and $<v_{yy}(t,y)y_1,y_2>$ are continuous for each y_1, $y_2 \in Y$.
If $y(t)$ is given by (66), then $z(t) = v(t,y(t))$ has the stochastic differential

$$dz(t) = \{v_t(t,y(t)) + <v_y(t,y(t)),g(t)> + \frac{1}{2}\,\mathrm{tr}.G^*(t)v_{yy}(t,y(t))G(t)W\}dt + <v_y(t,y(t)),G(t)dw(t)>. \tag{68}$$

Proof. Let $w_n(t) = \sum_{i=1}^{n} e_i\beta_i(t)$ as in (56). Then as in the scalar case (Gikhman and Skorokhod, 1969) we obtain using Taylor's formula (Balakrishnan, 1976; Hille and Phillips, 1957)

$$dz_n(t) = \{v_t(t,y_n(t)) + <v_y(t,y_n(t)),g(t)> + \frac{1}{2}\sum_{i=1}^{n}\lambda_i<v_{yy}(t,y_n(t))G(t)e_i,G(t)e_i>\}dt + <v_y(t,y_n(t)),G(t)dw_n(t)> \tag{69}$$

where $z_n(t) = v(t,y_n(t))$ and $y_n(t) = y_0 + \int_{t_0}^{t} g(r)dr + \int_{t_0}^{t} G(r)dw_n(r)$.
Since there exists a subsequence y_{n_k} such that $y_{n_k}(t) \to y(t)$ uniformly on $[t_0,t_1]$ w.p.1 we can pass to the limit $n_k \to \infty$ to obtain (68).

In applications we need to consider a class of $v(t,y)$ which are differentiable in t only for $y \in \mathcal{D}(A)$, A the generator of a semigroup.

Corollary 7. Let A be a closed linear operator with dense domain $\mathcal{D}(A)$ in Y.
Let $v(t,y)$ satisfy (67a, c) and the following :
(b') $v(t,y)$ is differentiable in t for each $y \in \mathcal{D}(A)$ and $v_t(t,y)$ is continuous on $[t_0,t_1]\times\mathcal{D}(A)$ where $\mathcal{D}(A)$ is equiped with the graph norm of A.
Let the process $y(t)$ given by (66) have the additional property :

$$y_0 \in \mathcal{D}(A),\quad \int_{t_0}^{t_1} |Ag(t)|dt < \infty \quad\text{and}\quad \int_{t_0}^{t_1} |AG(t)|^2 dt < \infty \quad \text{w.p.1.}$$

Then $z(t) = v(t,y(t))$ has the stochastic differential (68).

The new assumptions are required to guarantee the integrability of $v_t(t,y(t))$.
The proof is similar to that of the theorem.

Comments

We have given only a few basic results on random variables and stochastic processes in infinite dimension. See for more results Bensoussan (1971), Curtain and Pritchard (1978) and Scalora (1961). The stochastic integral we use is based on Curtain and Falb (1970, 1971), however in their papers the process G(t) is assumed uniformly measurable (or strongly measurable as an $\mathcal{L}(H,Y)$ valued process (Hille and Phillips, 1957)). Other definitions can be found in Bensoussan (1971), Chow (1978) and Dawson (1972, 1975). See also Chapter 3. More general stochastic integrals can be defined with respect to a martingale (Métivier, 1975, 1977; Métivier and Pistone, 1975; Pardoux, 1976). The stochastic evolution equation (61) is not at all the most general one. We may include a term g(t)dt or we may take an unbounded D(Ichikawa, 1980). A general theory of stochastic evolution equations is given by Chojnowska-Michalik (1976) using semigroup theory and by Pardoux (1975) using Lions theory. Ito's formula in Hilbert space was first proved in Curtain and Falb (1970), but the stochastic integral there is more restricted as remarked above. As in finite dimensions it is very useful in stochastic stability and control (Fleming and Rishel, 1975). An estimate for $E|\int_0^{t_1} G(t)dw(t)|^p$, $p \geq 2$ can be calculated using Ito's formula as in the scalar case (Gihman and Skorohod, 1972, p. 26) and (58c) is obtained as a special case. Using this estimate we can establish the mild solutions of (61) in $C(t_0,t_1;L_p(\Omega,\mu;Y))$ and the stability of p-th moment $E|y(t)|^p$. See Ichikawa (1981) for details.

STOCHASTIC STABILITY

Ito's Formula and Mild Solutions

Let Y, H_i, i=1,2 be real separable Hilbert spaces. Consider the linear stochastic evolution equation

$$dy(t) = Ay(t)dt + D(y(t))dw_1(t) + Fdw_2(t),\ y(t_0) = y_0,\ t_0 \geq 0 \tag{70}$$

where A is the infinitesimal generator of a strongly continuous semigroup T(t) on Y, $D(\cdot) \in \mathcal{L}(Y,\mathcal{L}(H_1,Y))$, $F \in \mathcal{L}(H_2,Y)$, $w_i(t)$, i=1,2 are independent Wiener processes in H_i with covariance operator W_i and y_0 is measurable relative to $\mathcal{F}_{t_0}$ with $E|y_0|^2 < \infty$. By Theorem 15 we know that there exists a unique mild solution of (70) in $C(t_0,t_1;L_p(\Omega,\mu;Y))$, p=2,4, satisfying

$$y(t) = T(t-t_0)y_0 + \int_{t_0}^{t} T(t-r)D(y(r))dw_1(r) + \int_{t_0}^{t} T(t-r)Fdw_2(r). \tag{71}$$

We now show that Ito's formula enables us to calculate the quadratic functional

$$\int_{t_0}^{t_1} E\langle My(t),y(t)\rangle dt + E\langle Gy(t),y(t)\rangle,\ 0 \leq M,\ G \in \mathcal{L}(Y).$$

For this purpose we consider the linear operator differential equation

$$\frac{d}{dt}\langle P(t)y,y\rangle + 2\langle Ay,P(t)y\rangle + \langle [M + \Delta(P(t))]y,y\rangle = 0,\ y \in \mathcal{D}(A),\ P(t_1) = G, \tag{72}$$

or its integrated version

$$P(t)y = \int_{t}^{t_1} T^*(r-t)[M + \Delta(P(r))]T(r-t)y\,dr + T^*(t_1-t)GT(t_1-t)y,\ y \in Y \tag{73}$$

where $\langle\Delta(R)y,z\rangle = \text{tr.}D^*(z)RD(y)W_1$, $y,z \in Y$, $R \in \mathcal{L}(Y)$.

Lemma 11. The two equations (72) and (73) are equivalent and they have a unique solution in the class of self-adjoint nonnegative strongly continuous operators in $\mathcal{L}(Y)$.

Proof. The equivalence follows an in Corollary 4. We can construct a solution by iteration starting with $P_0(t) = 0$ and the uniqueness follows easily from the linearity of (73).

Now for a moment we assume that solution of (71) is also a strong solution of (70). Then via Corollary 7 and (67e) we can apply Ito's formula for $v(t,y) = \langle P(t)y,y\rangle$ to obtain

$$d\langle P(t)y(t),y(t)\rangle = \{-\langle My(t),y(t)\rangle + \text{tr.}F^*P(t)F\}dt$$
$$+ 2\langle P(t)y(t),D(y(t))dw_1(t) + Fdw_2(t)\rangle.$$

Integrating this from t_0 to t_1 and taking expectations, we have

$$\int_{t_0}^{t_1} E\langle My(t),y(t)\rangle dt + E\langle Gy(t_1),y(t_1)\rangle = E\langle P(t_0)y_0,y_0\rangle + \int_{t_0}^{t_1} tr.F^*P(t)FW_2 dt. \quad (74)$$

We note that (74) is a well-defined relation for the mild solution. In fact we can drop the assumption that $y(t)$ is a strong solution of (70).

Theorem 18. Let $y(t)$ be the mild solution of (70). Then the equality (74) holds, where $P(t)$ is the solution of (72).

To prove this we introduce approximations for (70). Let $\rho(A)$ be the resolvent set of A and $R(\lambda,A)$ the resolvent for $0 < \lambda \in \rho(A)$. We consider

$$dy(t) = Ay(t)dt + R(\lambda)[D(y(t))dw_1(t) + Fdw_2(t)],\ y(t_0) = y_0 \in \mathcal{D}(A), \quad (75)$$

where $R(\lambda) = \lambda R(\lambda,A)$. Then Theorem 15 guarantees the unique mild solution $y(t;\lambda)$ of (75) in $C(0,t_1;L_p(\Omega,\mu;Y))$, p=2,4. Furthermore, since $AR(\lambda) = \lambda-\lambda^2R(\lambda,A)$ is bounded (Lemma 1), the assumptions in Theorem 16 are satisfied. Thus we may apply Ito's formula for $\langle P(t)y(t;\lambda),y(t;\lambda)\rangle$ to obtain

$$\begin{aligned}&\int_{t_0}^{t_1} E\langle My(t;\lambda),y(t;\lambda)\rangle dt + E\langle Gy(t_1;\lambda),y(t_1;\lambda)\\ &= E\langle P(t_0)y_0,y_0\rangle + \int_{t_0}^{t_1} tr.F^*R(\lambda)^*P(t)R(\lambda)FWdt.\end{aligned} \quad (76)$$

The next lemma which can be proved using Lemma 1, (4b) and Lemma 8 (Ichikawa, 1979, 1981) allows us to pass to the limit $\lambda \to \infty$.

Lemma 12. $y(t;\lambda) \to y(t)$ in $C(t_0,t_1;L_2(\Omega,\mu;Y))$ as $\lambda \to \infty$, where $y(t)$ is the solution of (71).

Proof of Theorem 18. By Lemma 12 we have shown (74) for all $y_0 \in \mathcal{D}(A)$. But $\mathcal{D}(A)$ is dense in Y and $y(t)$ depends continuously on y_0; thus we obtain (74) for any $y_0 \in Y$.

Stability of Linear Equations

Let Y, A be given as in the previous section. We consider the stochastic differential equation

$$dy(t) = Ay(t)dt + D(y(t))dw(t),\quad y(0) = y_0 \quad (77)$$

and its integrated version

$$y(t) = T(t)y_0 + \int_0^t T(t-r)D(y(r))dw(r) \quad (78)$$

where $y_0 \in Y$, nonrandom, $D \in \mathcal{L}(Y,\mathcal{L}(H,Y))$ and $w(t)$ is a Wiener process in H with covariance operator W.

Definition 10. The system (77) (or (A,D)) is stable if the mild solution $y(t)$ satisfies $\int_0^\infty E|y(t)|^2 dt < \infty$ for all $y_0 \in Y$.

We have a stochastic version of Datko's result Theorem 7.

Theorem 19. The following statements are equivalent : (79)
(a) (A,D) is stable,
(b) there exists $0 \le P \in \mathcal{L}(Y)$ such that $2\langle Ay,Px\rangle + \langle\Delta(P)y,y\rangle = -\langle y,y\rangle$, $y \in \mathcal{D}(A)$,
(c) $E|y(t)|^2 \le ke^{-\alpha t}|y_0|^2$ for some $k \ge 1$, $\alpha > 0$.

Proof. Suppose (a) holds. Let $P_{t_1}(t)$ be the solution of

$$\frac{d}{dt}\langle P(t)y,y\rangle + 2\langle Ay,P(t)y\rangle + \langle[I + \Delta(P(t))]y,y\rangle = 0,\ y \in D(A),\ P(t_1) = 0. \tag{80}$$

Then specializing Theorem 18 we obtain

$$\langle P_{t_1}(0)y_0,y_0\rangle = \int_0^{t_1} E|y(t)|^2\,dt \le \int_0^{\infty} E|y(t)|^2\,dt < \infty.$$

Since $P_{t_1}(0)$ is monotone increasing in t_1 and uniformly bounded, there exists a limit $P \ge 0$ in $L(Y)$. From the integral equation of $P_{t_1}(t)$ we can easily show $P_{t_1}(t) = P_{t_1-t}(0)$. Thus from (80) we conclude that P satisfies (79). Hence (a) implies (b). Now suppose (b) holds. We apply Theorem 17 for the function $\langle Py,y\rangle$ and the approximating process of (77). Then letting $\lambda \to \infty$ just an Theorem 18 we can establish

$$\langle Py_0,y_0\rangle = E\langle Py(t),y(t)\rangle + \int_0^t E|y(r)|^2\,dr. \tag{81}$$

it follows that

$$\frac{d}{dt} E\langle Py(t),y(t)\rangle = -E|y(t)|^2 \le -|P|^{-1}E\langle Py(t),y(t)\rangle,$$

from which we obtain

$$E\langle Py(t),y(t)\rangle \le e^{-pt}\langle Py_0,y_0\rangle,\ p = |P|^{-1}. \tag{82}$$

From Lemma 4 and (79) we see that A is exponentially stable. So $|T(t)| \le ce^{-\beta t}$ for some $c \ge 1$, $\beta > 0$. From the integral equation (78) we have

$$\begin{aligned}
E|y(t)|^2 &= |T(t)y_0|^2 + \int_0^t E[tr.T(t-r)D(y(r))WD^*(y(r))T^*(t-r)]dr \\
&\le c^2e^{-2\beta t}|y_0|^2 + d\int_0^t e^{-2\beta(t-r)}E|y(r)|^2\,dr \quad \text{for some } d > 0 \\
&= c^2e^{-2\beta t}|y_0|^2 - d\int_0^t e^{-2\beta(t-r)}\frac{d}{dr}E\langle Py(r),y(r)\rangle dr \\
&\le c^2e^{-2\beta t}|y_0|^2 + de^{-2\beta t}\langle Py_0,y_0\rangle + 2\beta d\int_0^t e^{-2\beta(t-r)}E\langle Py(r),y(r)\rangle dr.
\end{aligned}$$

Using (82) we can conclude (c). Finally (c) obviously implies (a).

Corollary 8. The equation (79) has a unique solution $P \ge 0$.

Proof. From (81) we have $\langle Py_0,y_0\rangle = \int_0^{\infty} E|y(t)|^2\,dt$ which gives the uniqueness.

Stability of Semilinear Equations

Consider the stochastic differential equation

$$dy(t) = [Ay(t) + f(y(t))]dt + D(y(t))dw(t),\ y(0) = y_0 \tag{83}$$

where A, w(t) are given as in (77), $f:Y \to Y$ and $D(\cdot):Y \to L(H,Y)$ satisfy Lipschitz conditions (62) and $y_0 \in Y$ nonrandom.
Then we have the following result on exponential stability.

Theorem 20. Suppose that a real function V(y) on Y satisfies the following :

(a) V(y) is twice Fréchet differentiable and the derivatives V_y, V_{yy} are continuous in Y, $L(Y)$ respectively

(b) $|V(y)| + |y||V_y(y)| + |y|^2|V_{yy}(y)| \le c(1 + |y|^2)$

(c) $LV(y) + \alpha V(y) \le 0$. for any $y \in D(A)$, where α is a real number and

$$\mathcal{L}V(y) = \langle V_y(y), Ay + f(y)\rangle + \frac{1}{2}\langle \Delta(V_{yy}(y))y, y\rangle.$$

Then $EV(y(t)) \leq e^{-\alpha t}V(y_0)$ for the mild solution $y(t)$ of (80).

Proof. We apply Theorem 17 for the function $e^{\alpha t}V(y)$ and $y(t;\lambda)$, by Theorem 16 the strong solution of

$$dy(t) = [Ay(t) + R(\lambda)f(y(t))]dt + R(\lambda)D(y(t))dw(t),\ y(0) = y_0 \in \mathcal{D}(A).$$

Then a limiting argument as in Theorem 18 leads to the desired inequality.

Corollary 9. If $\alpha > 0$ and $V(y) \geq d|y|^2$ for some $d > 0$, then $E|y(t)|^2$ is exponentially stable.

Now consider the case $V(y) = |y|^2$. If

$$\mathcal{L}V(y) = 2\langle y, Ay + f(y)\rangle + \text{tr.}D^*(y)D(y)W \leq -\alpha|y|^2,\ \alpha > 0 \tag{84}$$

for any $y \in \mathcal{D}(A)$, then $E|y(t)|^2 \leq e^{-\alpha t}|y_0|^2$. There are many examples which satisfy

$$\langle Ay, y\rangle \leq -\beta|y|^2,\ \beta > 0,\ y \in \mathcal{D}(A) \quad \text{and } f(0) = 0,\ D(0) = 0.$$

In this case $\mathcal{L}|y|^2 \leqq -(2\beta - 2c_1 - c_2^2\ \text{tr.}W)|y|^2$ where c_1, c_2 are constants appearing in the Lipschitz conditions. Thus $2\beta > 2c_1 + c_2^2\ \text{tr.}W$ is sufficient for exponential stability of $E|y(t)|^2$.

Examples

Example 8. Consider the stochastic heat equation

$$dy(x,t) = \frac{\partial^2}{\partial x^2}y(x,t)dt + \sigma y(x,t)dw(t),\ y(0,t) = y(1,t) = 0,\ y(x,0) = y_0(x)$$

where σ is a real number and $w(t)$ is a standard Wiener process. We take Y and A as Example 2 and $D(y) = \sigma y$. In this case $\mathcal{L}|y|^2 \leqq -(2\pi^2 - \sigma^2)|y|^2$ and by Corollary 9 $E|y(t)|^2$ is exponentially stable, and the Liapunov equation (79b) has a unique solution provided $|\sigma| < \sqrt{2}\pi$. In fact we have the explicit solution

$y(t) = e^{-\frac{\sigma^2}{2} + \sigma w(t)}T(t)y_0$, where $T(t)$ is the semigroup given in Example 2 and by direct computation we have $E|y(t)|^2 \leq e^{-(2\pi^2-\sigma^2)t}|y_0|^2$.

Example 9. Consider the stochastic version of Example 7 :

$$dy(x,t) = \left[\frac{\partial^2}{\partial x^2}y(x,t) - \frac{y(x,t)}{1+|y(x,t)|}\right]dt - \frac{\sigma y(x,t)}{1+|y(x,t)|}dw(t)$$

$$\frac{\partial}{\partial x}y(0,t) = \frac{\partial}{\partial x}y(1,t) = 0,\ y(x,0) = y_0(x).$$

We take Y, A, T(t) as Example 6. Then $\mathcal{L}|y|^2 \leq -(2 - \sigma^2)|y|^2$, and by Corollary 9 $E|y(t)|^2$ is exponentially stable if $|\sigma| < \sqrt{2}$.

Comments

Stochastic stability for linear evolution equations has been studied only recently by Haussmann (1978), Ichikawa (1979) and Zabczyk (1979) and in the second order case by Curtain (1979). Stability other than in the mean square sense can be considered, for example, in terms of time-average or invariant measures (Ichikawa, 1979). Another type of stability which is not included in our presentation is sample-wise stability. We refer to Haussmann (1978), Curtain (1980a). We can extend their results to semilinear equations. Liapunov type functions V(y) in Theorem 20 can be more general and in fact further developments on nonlinear stability can be found in Ichikawa (1981). Some extension of Theorem 19 is given in Ichikawa (1980).

Stochastic stability of partial differential equations using finite dimensional approximations can be found in Yavin (1974, 1975). See Haussmann (1978) and Zabczyk (1979) for examples of delay equations and Curtain (1980a) for hyperbolic equations.

STOCHASTIC QUADRATIC CONTROL

Optimal Control on Finite Interval

Consider the controlled stochastic evolution equation

$$dy(t) = (Ay(t) + Bu(t))dt + D(y(t))dw_1(t) + Fdw_2(t) + C(u(t))dw_3(t), \; y(t_0) = y_0 \quad (85)$$

where $w_3(t)$ is a Wiener process in a real separable Hilbert space H_3 with covariance operator W_3, $u(t)$ is a control with values in a real separable Hilbert space U, $B \in L(U,Y)$, $C \in L(U,L(H_3,Y))$ and the rest are given as in (70). We say that $u(t)$ is admissible if it is adapted to F_t, the σ-field generated by $w_i(s)$, $0 \leq s \leq t$, $i=1,2,3$ with $\int_{t_0}^{t_1} E|u(t)|^2 dt < \infty$. For such $u(t)$ we can show as Theorem 15 that there exists a unique mild solution of (85) in $C(0,t_1;L_2(\Omega,\mu;Y))$. We can define the cost functional

$$J(u;t_0,y_0) = E\langle Gy(t_1),y(t_1)\rangle + \int_{t_0}^{t_1} E\{\langle My(t),y(t)\rangle + \langle Nu(t),u(t)\rangle\}dt \quad (86)$$

where $0 \leq G$, $M \in L(Y)$ and $0 < N \in L(U)$ with bounded inverse N^{-1}. Note that the feedback control $u = K(t,y)$ is admissible if $K(t,y)$: $[0,t_1]\times Y \to U$ is measurable in t and

$$|K(t,y)| \leq c(1 + |y|), \; |K(t,y)-K(t,z)| \leq c|y-z|, \; c > 0, \; y, z \in Y. \quad (87)$$

In fact (85) with $u(t) = K(t,y(t))$ has a unique mild solution $y(t)$ and $u(t) = K(t,y(t))$ is admissible.
Now define

$$\langle\Gamma(s)u,v\rangle = tr.C^*(v)SC(u)W_3, \; S \in L(U), \quad u, v \in U \quad (88)$$

$$L_uV(y) = \langle V_y(y),Ay+Bu\rangle + \frac{1}{2}\{\langle\Gamma(V_{yy}(y))u,u\rangle + \langle\Delta(V_{yy}(y))y,y\rangle + tr.F^*V_{yy}(y)FW_2\},$$

$y \in D(A)$, $u \in U$ for twice Fréchet differentiable function $V(y)$ on Y. First we give sufficient conditions for optimality.

Theorem 21. Suppose there exists a feedback control $\bar{u} = \bar{K}(t,y)$ and a real function $V(t,y)$ on $[t_0,t_1]\times Y$ with properties :
(a) $V(t,y)$ is continuous on $[t_0,t_1]\times Y$,
(b) $V(t,y)$ is twice Fréchet differentiable in y and the derivatives $V_y(t,y)y_1$, $\langle V_{yy}(t,y)y_1,y_2\rangle$ are continuous for all y_1, $y_2 \in Y$,
(c) $V(t,y)$ is differentiable in t for any $y \in D(A)$ and $V_t(t,y)$ is continuous on $[t_0,t_1]\times D(A)$,
(d) $|V(t,y)|+|y|\,|V_y(t,y)| + |y|\,|V_{yy}(t,y)| \leq c(1 + |y|^2)$, $c > 0$, (89)
(e) $V(t_1,y) = \langle Gy,y\rangle$, $y \in Y$
(f) $0 = V_t(t,y) + L_{\bar{u}}V(t,y) + \langle My,y\rangle + \langle N\bar{u},\bar{u}\rangle \leq V_y(t,y) + L_uV(t,y) + \langle My,y\rangle + \langle Nu,u\rangle$ for any $y \in D(A)$ and $u \in U$,
(g) $\bar{K}(t,y)$ is continuous and satisfies the condition (87).
Then the feedback control $\bar{u} = \bar{K}(t,y)$ is optimal and the minimal cost is $J(\bar{u};t_0,y_0) = EV(t_0,y_0)$.

Proof. Let $y(t)$ be the unique mild solution of (85) with $u = \bar{u} = \bar{K}(t,y)$. Then as Theorem 18 we can show that $J(\bar{u};t_0,y_0) = V(t_0,y_0) \leq J(u;t_0,y_0)$.

To solve the quadratic cost problem (85), (86) we need to find a function $V(t,y)$

and a control $\bar{u} = \bar{K}(t,y)$ satisfying conditions of Theorem 21. We seek a function $V(t,y)$ of the form as Wonham (1970) $V(t,y) = \langle Q(t)y,y\rangle + q(t)$, $0 \leq Q(t) \in L(Y)$. Then (89f) yields the Riccati equation

$$\frac{d}{dt}\langle Q(t)y,y\rangle + 2\langle Ay,Q(t)y\rangle + \langle My,y\rangle + \langle\Delta(Q(t))y,y\rangle$$
$$- \langle Q(t)B[N + \Gamma(Q(t)]^{-1}B^*Q(t)y,y\rangle = 0, \quad y \in D(A),\ Q(t_1) = G \tag{90}$$

$$q(t) = \int_t^{t_1} tr.F^*Q(r)FW_2dr, \tag{91}$$

and the feedback control

$$\bar{u} = -[N + \Gamma(Q(t))]^{-1}B^*Q(t)y. \tag{92}$$

Hence if we establish a solution of (90), then the control problem is completely solved. As in Theorem 11 we can prove :

Theorem 22. The Riccati equation (90) has a unique solution in the class of strongly continuous nonnegative operators in $L(Y)$.

Corollary 10. The feedback control $\bar{u} = -[N + \Gamma(Q(t)]^{-1}B^*Q(t)y$ is optimal and the minimal cost is

$$J(\bar{u};t_0,y_0) = \langle Q(t_0)y_0,y_0\rangle + \int_{t_0}^{t_1} tr.F^*Q(t)FW_2dt. \tag{93}$$

Optimal Control on an Infinite Interval

Here we consider

$$dy = (Ay + Bu)dt + D(y(t))dw_1(t) + C(u(t))dw_3(t),\ y(0) = y_0 \tag{94}$$

and the cost functional

$$J(u;y_0) = \int_0^\infty E\{\langle My(t),y(t)\rangle + \langle Nu(t),u(t)\rangle\}dt \tag{95}$$

For admissible controls we take stochastic processes $u(t)$ which is adapted to F_t, the σ-field generated by $w_i(s)$, $0 \leq s \leq t$, $i=1,3$ with $\int_0^\infty E|u(t)|^2\,dt < \infty$ such that the mild solution of (94) corresponding to it has the property : $E|y(t)|^2 \to 0$ as $t \to \infty$. The class of feedback controls $u = K(y)$ such that $|K(y) - K(z)| \leq c|y - z|$, $y, z \in Y$ is admissible if the mild solution $y(t)$ has the property above for all $y_0 \in Y$. We extend Definition 5 to the stochastic case.

Definition 11. The system (94) (or (A,B;C,D)) is stabilizable if there exists an operator $K_1 \in L(Y,U)$ such that the feedback law $u = -K_1y$ yields a stable solution $y(t)$ i.e., $\int_0^\infty E|y(t)|^2\,dt < \infty$. In this case $(A - BK_1,C,D)$ is said to be stable.

With slight modification in Theorem 19 we have :

Lemma 13. (A,B;C,D) is stabilizable if and only if there exist operators $K_1 \in L(Y,U)$, and $0 \leq P_1 \in L(Y)$ such that

$$2\langle(A - BK_1)y,P_1y\rangle + \langle[K_1^*\Gamma(P_1)K_1 + \Delta(P_1)]y,y\rangle = -\langle y,y\rangle,\ y \in D(A). \tag{96}$$

Lemma 14. If $(A - BK_1,C,D)$ is stable, then there exists $0 \leq Q_1 \in L(Y)$ such that

$$2\langle(A - BK_1)y,Q_1y\rangle + \langle(M + K_1^*NK_1)y,y\rangle + \langle(\Delta(Q_1) + K_1^*\Gamma(Q_1)K_1)y,y\rangle = 0,\ y \in D(A). \tag{97}$$

Proof. Let $Q_{t_1}(t)$ be the unique solution of

$$\frac{d}{dt}<Q(t)y,y> + <(A - BK_1)y,Q(t)y> + <M + K_1^*NK_1)y,y>$$
$$+ <[\Delta(Q(t) + K_1^*\Gamma(Q(t)K_1]y,y> = 0, \quad y \in D(A)$$
$$Q(t_1) = 0.$$

Then from a variation of Theorem 18 it follows that $Q_{t_1}(t)$ is monotone increasing in t_1 and uniformly bounded. Hence it is convergent to a solution of (97).

Now we give sufficient conditions for optimality.

Theorem 23. Suppose there exist an admissible control $\bar{u} = -\bar{K}(y)$ and a real function $V(y)$ on Y with properties :
(a) $V(y)$ is twice Fréchet differentiable and $V(y)$, $V_y(y)$, $V_{yy}(y)$ are continuous in R', Y, and $L(Y)$ respectively,
(b) $|V(y)| + |y||V_y(y)| + |y|^2|V_{yy}(y)| \leq c|y|^2$, $c > 0$ (98)
(c) $0 = L_{\bar{u}}V(y) + <My,y> + <N\bar{u},\bar{u}> \leq L_uV(y) + <My,y> + <Nu,u>$, $y \in D(A)$, $u \in U$
(d) $|\bar{K}(y) - \bar{K}(z)| \leq c|y - z|$, $y, z \in Y$,

where $L_uV(y) = <V_y(y),Ay + Bu> + \frac{1}{2}\{<\Gamma(V_{yy}(y))u,u> + <\Delta(V_{yy}(y))y,y>\}$

Then $\bar{u} = -\bar{K}(y)$ is optimal and $J(\bar{u};y_0) = V(y_0)$.

Proof. As Theorem 20 we can show via Theorem 17 that $J(\bar{u};y_0) = V(y_0) \leq J(u;y_0)$. Next we seek a function $V(y)$ of the form $V(y) = <Qy,y>$, $0 \leq Q \in L(Y)$. Then (98d) yields the algebraic Riccati equation

$$2<Ay,Qy> + <\{M + \Delta(Q) - QB[N + \Gamma(Q)]^{-1}B^*Q\}y,y> = 0, \quad y \in D(A) \tag{99}$$

and the feedback control

$$\bar{u} = -\bar{K}y, \quad \bar{K} = [N + \Gamma(Q)]^{-1}B^*Q. \tag{100}$$

We shall give sufficient conditions which guarantee a solution for (99).

Lemma 14. If (A,B;C,D) is stabilizable, then the Riccati equation has a solution $Q \geq 0$.

Proof. Let $Q_{t_1}(t)$ be the solution of the Riccati equation (90) with $G = 0$. Then $Q_{t_1}(t)$ is uniformly bounded and monotone increasing in t_1, so there exists a limit $Q \geq 0$ and Q satisfies the Riccati equation (99).

Lemma 15. Let $K \in L(Y,U)$. Suppose there exist operators $J \in L(Y)$ and $0 \leq P \in L(Y)$ such that
(a) $A - JM^{1/2}$ generates a stable semigroup $S(t)$ with $|S(t)| \leq \delta e^{-\alpha t}$, $\alpha > 0$ and $\delta^2 d^2 tr.W_1 < \alpha$, $d = |D|$, the norm of $D(\cdot)$, (101)
(b) $2<(A - BK)y,Py> + <(M + K^*NK)y,y> + <[\Delta(P) + K^*\Gamma(P)K]y,y> \leq 0$, $y \in D(A)$.

Then $(A - BK;C,D)$ is stable.

Proof. Let $y(t)$ be the solution of (94) with $u = -Ky$, then

$$y(t) = S(t)y_0 + \int_0^t S(t-r)(JM^{1/2}-BK)y(r)dr$$
$$+ \int_0^t S(t-r)D(y(r))dw_1(r) + \int_0^t S(t-r)C(Ky(r))dw_3(r)$$

It follows that

$$E|y(t)|^2 \leq 2E|\int_0^t S(t-r)D(y(r))dw_1(r)|^2 + 6|S(t)y_0|^2$$

$$+ 6E|\int_0^t S(t-r)(JM^{1/2}-BK)y(r)dr|^2$$

$$+ 6E|\int_0^t S(t-r)C(Ky(r))dw_3(r)|^2$$

$$\leq 2\delta^2 d^2 tr.W_1 \int_0^t e^{-2\alpha(t-r)} E|y(r)|^2 dr + 6\delta^2 e^{-2\alpha t}|y_0|^2$$

$$+ 6\delta^2 c \int_0^t e^{-2\alpha(t-r)} E\{<My(r),y(r)> + <Ky(r),Ky(r)>\}dr$$

$$+ 6\delta^2 c\, tr.W_3 \int_0^t e^{-2\alpha(t-r)} E<Ky(r),Ky(r)>dr$$

where $c = \max(|J|^2, |B|^2, |C|^2)$. From the assumption (b) we can show that $E<My(t),y(t)>$, $E<Ky(t),Ky(t)> \in L_1(0,\infty)$. Integrate both side of the above inequality from 0 to $+\infty$ and use (a), then we obtain $\int_0^\infty E|y(t)|^2 dt < \infty$.

Theorem 24. Suppose that (A,B;C,D) is stabilizable and that there exists an operator J satisfying (101a). Then there exists a unique solution $Q \geq 0$ of the Riccati equation (99). Moreover, the feedback control $\bar{u} = -[N + \Gamma(Q)]^{-1}B^*Qy$ is admissible and optimal with $J(\bar{u};y_0) = <Qy_0,y_0>$.

Proof. The existence of a solution of (99) follows from Lemma 14 and the admissibility of any solution of (99) from Lemma 15. The rest is a direct consequence of Theorem 23.

Examples

Example 10. Consider the scalar differential equation

$$dy(t) = [a_0 y(t) + a_1 y(t-b) + u(t)]dt + \sigma y(t)dw(t),$$
$$y(\theta) = y_0(\theta), \quad -b \leq \theta \leq 0, \quad a_0, a_1, b > 0$$

and the cost function

$$J(u) = \int_0^\infty [y^2(t) + Nu^2(t)]dt, \quad N > 0.$$

Now take the feedback control $u = -ky$. Then the $\Delta(\lambda)$ defined by (19) in this case is $\Delta(\lambda) = -(a_0 - k) - a_1 e^{-\lambda b}$ and the zero of $\Delta(\lambda)$ is negative if $k > a_0 + a_1$. Thus the system is stabilizable and the detectability condition (101a) is satisfied by the same feedback law for k sufficiently large. So all the assumptions in Theorem 24 are satisfied. Then the feedback control $\bar{u}(\cdot) = -N^{-1}[Q_{00}y(t) + \int_{-b}^0 Q_{01}(\theta)y(t+\theta)d\theta]$ is optimal where $Q = \begin{bmatrix} Q_{00} & Q_{01} \\ Q_{10} & Q_{11} \end{bmatrix}$ and kernels $Q_{01}(\theta)$, $Q_{11}(\theta,\eta)$ of Q_{01}, Q_{11} and Q_{00} satisfy the following : $2Q_{00}a_0 + Q_{10}(0) + Q_{01}(0) + 1 - N^{-1}Q_{00}^2 + \sigma^2 Q_{00} = 0$

$$\frac{dQ_{01}(\theta)}{d\theta} = (a_0 - N^{-1}Q_{00})Q_{01}(\theta) + a_0 Q_{11}(\theta,0), \quad Q_{01}(-b) = a_1 Q_{00}.$$

$$(\frac{\partial}{\partial\theta} + \frac{\partial}{\partial\eta})Q_{11}(\theta,\eta) - N^{-1}Q_{10}(\theta)Q_{01}(\eta) = 0,\ Q_{11}(-b,\eta) = a_1Q_{01}(\eta),\ Q_{11}(\theta,-b) = Q_{10}(\theta)a_1$$

For the derivation of these equations see Delfour, McCalla and Mitter (1975).

Example 11. Consider the stochastic version of Example 6.

$$dy(x,t) = [\frac{\partial^2}{\partial x^2}y(x,t) + b(x)u(t)]dt + \sigma y(t)dw(t)$$

$$\frac{\partial y}{\partial x}(0,t) = \frac{\partial y}{\partial x}(1,t) = 0, \quad y(x,0) = y_0(x)$$

$$J(u;y_0) = \int_0^\infty (\int_0^1 y^2(x,t)dx + Nu^2(t))dt, \quad N > 0$$

As we have already seen, the system satisfies both stabilizability and detectability conditions and (101a) is also met since we can achieve an arbitrary decay rate α by taking $Jy = \alpha y$, $\alpha > 0$. Thus all the assumptions of Theorem 24 are satisfied and the Riccati equation (99) has a unique solution which can be expanded

$$Q = \sum_{i,j=0}^{\infty} q_{ij}\phi_i \circ \phi_j \text{ with } 2\pi^2 i^2 q_{ij} + \delta_{ij} - N^{-1}\sum_{k,l=0}^{\infty} \langle b,\phi_k\rangle\langle b,\phi_l\rangle q_{ki}q_{lj} + \sigma^2 q_{ij} = 0.$$

Comments

This section partially extends Wonham's work (1970) to infinite dimensions. Wonham keeps the term $Fdw_2(t)$ in (94) and consider optimal stationary control in terms of invariant measures. To follow his formulation we need some preliminaries for Markov processes and invariant measures. Results in this direction are found in Ichikawa (1979). Stochastic evolution equations from Markov process point of view have been discussed in Arnold, Curtain and Kotenelez (1980) and Curtain (1980b). The quadratic problem (94), (95) is solved without using Theorem 23 in Ichikawa (1978a), where the equations (73) is established directly from (71). There the optimal control problem with average cost is also discussed. There is an important problem which is not discussed in this section, that is, the quadratic cost problem for partially observable systems. In this case admissible controls are those dependent on only observations. We need results on filtering in Chapter 3 to solve this. It is known that the problem can be decomposed into two parts, filtering and control and this fact is known as the separations principle (Curtain and Ichikawa, 1977; Curtain, 1978). See also Bensoussan and Viot (1975) for different formulation.

REFERENCES

Arnold, L. (1974). Stochastic Differential Equations: Theory and Applications. John Wiley & Sons, New York.

Arnold, L., R. F. Curtain and P. Kotelenez (1980). Nonlinear stochastic evolution equations in Hilbert space, Report no. 17, Forschungsschwerpunkt Dynamishe Systeme, Universitat Bremen.

Balakrishnan, A. V. (1965). Optimal control in Banach spaces. SIAM J. Control, 3, 152-180.

Balakrishnan, A. V. (1976). Applied Functional Analysis. Springer-Verlag, New York.

Barbu, V. (1976). Nonlinear Semigroups and Differential Equations in Banach Spaces. Noordhoff International Publishing, Leyden.

Barbu, V. and Th. Precupanu (1978). Convexity and Optimization in Banach Spaces. Sijthoff & Noordhoff, Alphen aan de Rijn.

Bensoussan, A. (1971). Filtrage Optimal des Systèmes Linéaires. Dunod, Paris.

Bensoussan, A. and M. Viot (1975). Optimal control of stochastic linear distributed parameter systems. SIAM J. Control, 13, 904-926.

Borisovic, J. G. and A. S. Turbabin. On the Cauchy problem for linear non-homogeneous differential equations with retarded arguments. Sov. Math. Dokl., 10, 401-405.

Bryson, A. E. and Y. C. Ho (1975). Applied Optimal Control. Hemisphere Publishing Corporation, Washington, D. C.

Butkovskii, A. G. (1969). Theory of Optimal Control of Distributed Parameter Systems. Elsevier, New York.

Butzer, P. L. and H. Berens (1967). Semi-Groups of Operators and Approximation. Springer-Verlag, Berlin.

Chojnowska-Michalik, A. (1976). Stochastic differential equations in Hilbert Spaces and their applications. Thesis, Institute of Mathematics, Polish Academy of Sciences.

Chow, P. L. (1978). Stochastic partial differential equations in turbulence related problems. In A. T. Bharucha-Reid (Ed.), Probabilistic Analysis and Related Topics, Vol. 1. Academic Press, New York. pp. 1-44.

Cole, J. K. (1968). Time-optimal controls of the equation of evolution in a separable and reflexive Banach space. J. Optimiz. Theory & Appl., 18, 199-205.

Conti, R. (1968). Time-optimal solution of a linear evolution equation in Banach spaces. J. Optimiz. Theory & Appl., 18, 277-285.

Conti, R. (1970). Convex programming problem in Banach spaces and applications to optimal control theory. J. Comput. & Syst. Sci., 4, 38-48.

Curtain, R. F. and P. L. Falb (1970). Ito's lemma in infinite dimensions. J. Math. Anal. Appl., 31, 434-448.

Curtain, R. F. and P. L. Falb (1971). Stochastic differential equations in Hilbert spaces. J. Diff. Eqns., 10, 412-430.

Curtain, R. F. (1977). Stochastic evolution equations with general white noise disturbance. J. Math. Anal. & Appl., 60, 570-595.

Curtain, R. F. (1978). Estimation and stochastic control for linear infinite-dimensional systems. In A. T. Bharucha-Reid (Ed.), Probabilistic Analysis and Related Topics, Vol. 1. Academic Press, New York. pp. 45-86.

Curtain, R. F. (1979). Linear stochastic Ito equations in Hilbert space. In Lecture Notes in Control and Information Sciences, 16. Springer-Verlag, Berlin.

Curtain, R. F. (1980a). Asymptotic stability of second order linear stochastic partial differential equations. J. Math. Anal. & Appl., 79, 352-369.

Curtain, R. F. (1980b). Markov processes generated by linear stochastic evolution equations. Stochastics, 4.

Curtain, R. F. and A. J. Pritchard (1976). The infinite dimensional Riccati equation for systems described by evolution operators. SIAM J. Control, 14, 951-983.

Curtain, R. F. and A. J. Pritchard (1977). An abstract theory for unbounded control action for distributed parameter systems. SIAM J. Control & Optimiz., 15, 566-611.

Curtain, R. F. and A. J. Pritchard (1978). Infinite Dimensional Linear System Theory. Lecture Notes in Control and Information Sciences, 3, Springer-Verlag, Berlin.

Curtain, R. F. and A. Ichikawa (1977). The separation principle for stochastic evolution equations. SIAM J. Control & Optimiz., 15, 367-383.

Datko, R. (1970). Extending a theorem of A. M. Liapunov to Hilbert space. J. Math. Anal. & Appl., 32, 610-616.

Datko, R. (1971). A linear control problem in abstract Hilbert space. J. Diff. Eqns., 9, 346-359.

Davis, M. H. A. (1977). Linear Estimation and Stochastic Control. Chapman and Hall, London.

Dawson, D. A. (1972). Stochastic evolution equations. Math. Biosci., 15, 287-316.

Dawson, D. A. (1975). Stochastic evolution equations and related measure processes. J. Multivari. Anal., 5, 1-52.

Delfour, M. C. and S. K. Mitter (1972). Controllability, observability and optimal feedback control of affine hereditary differential systems. *SIAM J. Control*, *10*, 298-327.

Delfour, M. C., C. McCalla and S. K. Mitter (1975). Stability and the infinite-time quadratic cost problem for linear hereditary differential systems. *SIAM J. Control*, *13*, 48-88.

Doob, J. L. (1953). *Stochastic Processes.* John Wiley and Sons, New York.

Dunford, N. and J. Schwartz (1957). *Linear operators*, Part I. Interscience, New York.

Falb, P. L. (1964). Infinite dimensional control problems I: On the closure of the set of attainable states for linear systems. *J. Math. Anal. & Appl.*, *9*, 12-22.

Fattorini, H. O. (1964). Time optimal control of solutions of operational differential equations. *SIAM J. Control*, *2*, 54-59.

Fleming, W. H. and R. W. Rishel (1975). *Deterministic and Stochastic Optimal Control.* Springer-Verlag, New York.

Friedman, A. (1967a). Optimal control for parabolic equations. *J. Math. Anal. & Appl.*, *18*, 479-491.

Friedman, A. (1967b). Optimal control in Banach spaces. *J. Math. Anal. & Appl.*, *19*, 35-55.

Friedman, A. (1968). Optimal control in Banach space with fixed end points. *J. Math. Anal. & Appl.*, *24*, 161-181.

Friedman, A. (1976). *Partial Differential Equations.* Krieger, New York.

Gel'fand, I. M. and N. Ya. Vilenkin (1964). *Generalized Functions*, Vol. 4. Academic Press, New York.

Gibson, J. S. (1979). The Riccati integral equations for optimal control problems on Hilbert spaces. *SIAM J. Control & Optimiz.*, *17*, 537-565.

Gikhman, I. I. and A. V. Skorokhod (1969). *Introduction to the Theory of Random Processes.* W. B. Saunders, Philadelphia.

Gihman, I. I. and A. V. Skorohod (1979). *The Theory of Stochastic Processes III.* Springer-Verlag, New York.

Hale, J. (1977). *Theory of Functional Differential Equations.* Springer-Verlag, New York.

Haussmann, U. G. (1978). Asymptotic stability of the linear Ito equation in infinite dimensions. *J. Math. Anal. & Appl.*, *65*, 219-235.

Hille, E. and R. S. Phillips (1957). *Functional Analysis and Semigroups*, Vol. 31. Colloq. Publ. Amer. Math. Soc.

Ichikawa, A. (1977). Dynamic programming approach to infinite dimensional systems. Control Theory Centre Report 57, University of Warwick.

Ichikawa, A. (1978a). Optimal control of a linear stochastic evolution equation with state and control dependent noise. In M. J. Gregson (Ed.), *Recent Theoretical Developments in Control*, Academic Press, London. pp. 383-402.

Ichikawa, A. (1978b). Linear stochastic evolution equations in Hilbert space. *J. Diff. Eqns.*, *28*, 266-277.

Ichikawa, A. (1979). Dynamic programming approach to stochastic evolution equations. *SIAM. J. Control & Optimiz.*, *17*, 152-174.

Ichikawa, A. (1980). Stability and control of stochastic evolution equations. In O. L. R. Jacobs (Ed.), *Analysis and Optimization of Stochastic Systems.* Academic Press, London.

Ichikawa, A. (1981). Stability of semilinear stochastic evolution equations. (to appear).

Ichikawa, A. and A. J. Pritchard (1979). Existence, uniqueness and stability of nonlinear evolution equations. *J. Math. Anal. & Appl*, *68*, 454-476.

Kato, T. (1966). *Perturbation Theory for Linear Operators.* Springer-Verlag, Berlin.

Kisyński, J. (1976). Semigroups of operators and some of their applications to partial differential equations. In *Control Theory and Topics in Functional Analysis* Vol. III. International Atomic Energy Agency, Vienna. pp. 305-405.

Kwong, R. H. and A. S. Willsky (1978). Estimation and filter stability of stochastic delay systems. SIAM J. Control & Optimiz., 16, 660-681.
Kwong, R. H. (1980). A stability theory for the linear-quadratic-Gaussian problem for systems with delays in the state, control and observations. SIAM J. Control & Optimiz., 18, 49-75.
Ladas, G. E. and V. Lakshmikantham (1972). Differential Equations in Abstract Spaces. Academic Press, New York.
Lasiecka, I. (1978). Boundary control of parabolic systems; Regularity of optimal solutions. Appl. Math. Optim., 4, 301-327.
Lasiecka, I. (1980a). Boundary control of parabolic systems; Finite-element approximations. Appl. Math. Optim., 6, 31-62.
Lasiecka, I. (1980b). Unified theory for abstract parabolic boundary problems: a semigroup approach. Appl. Math. Optim., 6, 287-333.
Lee, E. B. and L. Markus (1967). Foundations of Optimal Control Theory. John Wiley and Sons, New York.
Lions, J. L. (1961). Equations Différentielles Opérationnelles et Problèmes aux Limites. Springer-Verlag. Berlin.
Lions, J. L. (1971). Optimal Control of Systems Governed by Partial Differential Equations. Springer-Verlag, Berlin.
Lions, J. L. and E. Magenes (1972). Non-Homogeneous Boundary Value Problems and Applications I, II. Springer-Verlag, Berlin.
Lukes, D. L. and D. L. Russel (1969). The quadratic criterion for distributed systems. SIAM J. Control, 7, 101-121.
Manitius, A. (1976). Optimal control of hereditary systems. In Control Theory and Topics in Functional Analysis, Vol. III, International Atomic Energy Agency, Vienna, pp. 43-178.
Martin, R. H. (1976). Nonlinear Operators and Differential Equations in Banach Spaces. John Wiley and Sons, New York.
Métivier, M. (1975). Intégrale stochastique par rapport à des processus à valeurs dans un espace de Banach reflexit. Theory Prob. Appl., 19, 758-787.
Métivier, M. (1977). Reelle und Vektorwertige Quasimartingale und die Theorie der Stochastischen Integration. Lecture Notes in Mathematics, 607. Springer-Verlag, Berlin.
Métivier, M. and G. Pistone (1975). Une formule d'isométrie pour l' intégrale stochastique et équations d' évolution linéaires stochastiques. Z. Wahr. verw. Geb., 33, 1-18.
Nambu, T. (1979). Feedback stabilization of distributed parameter systems of parabolic type. J. Diff. Eqns., 33, 167-188.
Pao, C. V. and G. Vogt (1969). On the stability of nonlinear operator differential equations and applications. Arch. Rational Mech. Anal., 35, 30-46.
Pardoux, E. (1975). Équations aux dérivees partielles stochastiques nonlineaires monotones. These, L'Université Paris Sud, Centre D'Orsay.
Pardoux, E. (1976). Intégrales stochastiques Hilbertiennes, No. 7617. Cahiers de Mathématiques de la Décision, Université Paris-Dauphine.
Phillips, R. S. (1954). Perturbation theory for semigroups of linear operators. Trans. Amer. Math. Soc., 74, 199-221.
Pritchard, A. J. and J. Zabczyk (1981). Stability and stabilizability of infinite dimensional systems. SIAM Review., 23, 25-52.
Riesz, F. and B. SZ.-Nagy (1955). Functional Analysis. Frederick Unger, New York.
Scalora, F. S. (1961). Abstract martingale convergence theorems. Pacific J. Math., 11, 347-374.
Sorine, M. (1977). Sur les équations de Chandrasekhar associées a des opérateurs non bornés. Report No. 267. IRIA.
Tanabe, H. (1979). Equations of Evolution. Pitman, London.
Triggiani, R. (1980a). Boundary feedback stabilizability of parabolic equations. Appl. Math. Optim., 6, 201-220.
Triggiani, R. (1980b). Well-posedness and regularity of boundary feedback parabolic systems. J. Diff. Eqns., 36, 347-362.

Vinter, R. B. (1977). Filter stability for stochastic evolution equations. SIAM J. Control & Optimiz., 15, 465-485.

Wang, P. K. C. (1964). Control of distributed parameter systems. In C. T. Leondes (Ed.), Advances in Control Systems, Vol. 1. Academic Press, New York. pp. 75-172.

Washburn, D. (1979). A bound on the boundary input map for parabolic equations with applications to time optimal control. SIAM J. Control & Optimiz., 17, 652-671.

Wonham, W. M. (1967). Optimal stationary control of a linear system with state dependent noise. SIAM J. Control, 5, 486-500.

Wonham, W. M. (1968). On a matrix Riccati equation of stochastic control. SIAM J. Control, 6, 681-697.

Wonham, W. M. (1970). Random differential equations in control theory. In A. Bharucha-Reid (Ed.), Probabilistic Method in Applied Mathematics, Vol. 2 Academic Press, New York. pp. 131-212.

Yavin, Y. (1974). On the stochastic stability of a parabolic type system. Int. J. Syst. Sci., 5, 623-632.

Yavin, Y. (1975). On the modelling and stability of a stochastic distributed parameter system. Int. J. Syst. Sci., 6, 301-311.

Yosida, K. (1978). Functional Analysis. Springer-Verlag, Berlin.

Zabczyk, J. (1976). Remarks on the algebraic Riccati equation in Hilbert space. Appl. Math. Optim., 2, 251-258.

Zabczyk, J. (1977). A semigroup approach to boundary control. In proceedings IFAC Conference on Distributed Parameter Systems, Warwick, England.

Zabczyk, J. (1978a). On decomposition of generators. SIAM J. Control & Optimiz., 16, 523-534.

Zabczyk, J. (1978b). On stabilizability of boundary control systems. Report, Centre de Recherches Mathématiques, Université de Montréal.

Zabczyk, J. (1979). On stability of infinite dimensional stochastic systems. In Z. Ciesielski (Ed.), Probability Theory. Banach Center Publications, Vol. 5 Warsaw. pp. 273-281.

CHAPTER 6

ADAPTIVE AND ROBUSTNESS PROPERTIES OF CERTAIN BILINEAR DISTRIBUTED PARAMETER CONTROL SYSTEMS

D. Franke

Hochschule der Budeswenr Hamburg, Faculty of Electrotechnics, Hamburg, Federal Republic of Germany

INTRODUCTION

The theory of distributed parameter control systems has been established to a large extent within the last twenty years, and there are some fine text-books and proceedings of international conferences which summarize the state of the art. But a gap can still be observed between theory and applications in this field of automatic control. One of the reasons for this gap may be the fact that a great deal of the theory, especially in optimal control, requires exact knowledge of the plant to be controlled. Therefore, in case of parameter uncertainties or even structural uncertainties, the finest optimal control law when applied to the actual plant will yield a poor result and even may produce an unstable system. It is obviously for this reason that some attention has been focussed on the problem of parameter identification in distributed parameter systems. But a reliable identification may be time consuming, and a fast identification may not be reliable.

It seems therefore worth-wile to take a lesson from those control methods for *lumped* parameter systems which avoid parameter identification and nevertheless provide control systems with prescribed properties in the presence of plant uncertainties. Such a control system is usually called "robust". Its typical feature *is adaptivity without the need of parameter identification.*

The notion of robustness is not very sharp but the somewhat rigorous definition makes it applicable to a large variety of problems. For example, the prescribed properties may be stability, or constrained deterioration of a performance criterion, or constrained deviation from a specified nominal behaviour for a specified class of structural or parametric disturbances.

A robust controller may have a fixed structure and fixed parameters. But there is an appealing class of so-called *variable structure control systems,* whose structure is varied according to some specified strategy, dependent on the actual state of the plant. The variations of structure may be continuous or of a switching type and may refer to the controller or to the plant or to both of them.

A wide class of variable structure lumped parameter systems can be modelled by *bilinear state equations*. They are linear in state and linear in control but not jointly linear in both, and therefore take the form

$$\dot{\underline{x}}(t) = \underline{A}\,\underline{x}(t) + \sum_{i=1}^{p} u_i(t)\underline{N}_i\underline{x}(t) + \underline{B}\,\underline{u}(t),$$

where $\underline{x}$ is an n-dimensional state vector, $\underline{u}$ is a p-dimensional control vector, and $\underline{A}$, $\underline{B}$, $\underline{N}_i$, i=1, ..., p, are matrices of suitable dimensions.

From the above equation it is clear that $\underline{u}(t)$ acts as a *parametric* (or multiplicative) control, in addition to ordinary linear control. Some of the most significant benefits of multiplicative control are improvement of controllability and hence of performance in optimal control (Mohler, 1973). Some authors (Draženović, 1969; Utkin, 1977; Young, 1977) pointed out that bilinear control is a tool for the design of robust control systems. In this context, the notion of sliding modes plays a key role. In the sliding mode, the state of the system moves on a switching hypersurface towards the origin, and it is shown that in the sliding mode the system is very insensitive to even large parameter variations.

The present contribution is an attempt to generalize the concept of robust bilinear systems to certain distributed parameter systems. The chapter will be organized as follows: At first the state equations of a wide class of bilinear distributed parameter systems will be presented and illustrated by means of some examples. It will be shown that both systems with multiplicative control variables and systems with variable structure feedback control yield the same type of state equations. Then a summary of available results in the control of bilinear distributed parameter systems will be given. It seems that the theory on this class of systems stands just at its beginning, and so there is not very much to be reported. The main section of the chapter is concerned with bilinear control as a tool for the design of adaptive and robust control systems without the need of parameter identification. The considerations are essentially based on Ljapunov's direct method which avoids solution of the system's dynamical equations (Parks and Pritchard, 1969; Wang, 1966; Zubov, 1957).

An attempt has been made to present this chapter in a manner that is not too mathematical, sometimes in a more intuitive way. The author who is an engineer himself, hopes that this style will motivate the reader to further research work and applications in this new field of automatic control.

STATE EQUATIONS OF A CLASS OF BILINEAR DISTRIBUTED PARAMETER SYSTEMS

Systems with Inherent Parametric Control

There is a large variety of distributed parameter processes which can be modelled by strictly linear partial differential equations along with appropriate linear boundary conditions. But many engineering applications allow certain parameters of these linear systems to be varied within specified bounds, thus giving rise to a bilinear system.

Consider as a simple example the continuous furnace in Fig. 1. After a number of simplifications, it can be modelled by a normed equation of the type

$$\frac{\partial x(t,z)}{\partial t} + u_2(t) \cdot \frac{\partial x(t,z)}{\partial z} = a \cdot [u_1(t) - x(t,z)], \qquad 0 \leq z \leq 1, \quad t \geq t_o, \tag{1}$$

with a boundary condition

$$x(t,o) = u_3(t), \tag{2}$$

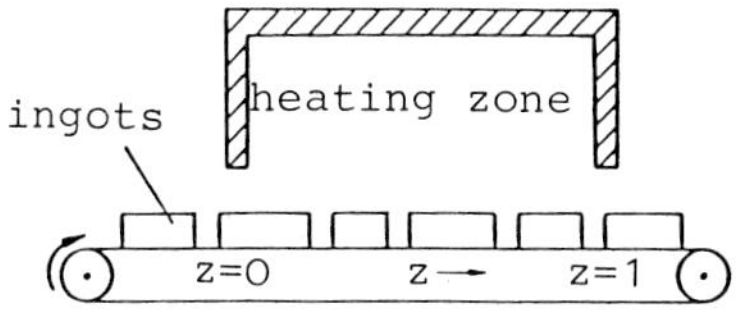

Fig. 1. Continuous heating process

and initial condition

$$x(t_o,z) = x_o(z). \tag{3}$$

An output variable of special interest will be

$$y(t) = x(t,1). \tag{4}$$

Here, $x(t,z)$ is the time and space dependent temperature inside the moving material and $u_1(t)$ is the temperature of the furnace which has been assumed to be independent of the spatial variable z. The parameter a is a constant heat transfer coefficient, and the rate of transportation, u_2, is another parameter which will be constant in any desired steady state operation. But variation of $u_2(t)$ within some prescribed technological bounds will be an efficient tool for the control of the *transient* behaviour of this plant, especially in case of boundary disturbance $u_3(t)$. The multiplicative mode of control, via $u_2(t)$, obviously offers a number of advantages when applied in addition to the linear control variable $u_1(t)$:

- controlled application of the multiplicative mode $u_2(t)$ may significantly reduce the time lag involved with strictly linear control;
- the rate of transportation, $u_2(t)$, can be varied quickly compared to the furnace temperature $u_1(t)$ which often has a considerable time delay;
- practical implementation of time-varying rate $u_2(t)$ does not impose severe problems;
- when starting up the process, u_2 has to be varied anyhow.

Another advantage of multiplicative control is adaptivity in case of plant uncertainties, and will be treated to some extent in the main section.

Consider as a second example the well-known equations of a counterflow heat-exchanger:

$$\frac{\partial x_1(t,z)}{\partial t} + u_1 \cdot \frac{\partial x_1(t,z)}{\partial z} = a_1 \cdot [x_2(t,z) - x_1(t,z)],$$

$$0 \leq z \leq 1 \tag{5}$$

$$\frac{\partial x_2(t,z)}{\partial t} - u_2(t) \cdot \frac{\partial x_2(t,z)}{\partial z} = a_2 \cdot [x_1(t,z) - x_2(t,z)],$$

with fluid temperatures x_1, x_2, flow rates u_1, u_2 and thermal coefficients a_1, a_2.

Boundary conditions are the inlet temperatures

$$x_1(t,o) = u_3(t), \quad x_2(t,1) = u_4(t). \tag{6}$$

An output variable of special interest will be

$$y(t) = x_1(t,1), \tag{7}$$

by assuming that fluid 2 is the coolant. For example, let u_1 be a constant prescribed by the technological process. Then it is obvious for physical reasons that time-dependent flow rate $u_2(t)$ is an efficient multiplicative control variable.

Multiplicative control can also arise in the *boundary conditions* of a distributed parameter system. An example is the problem of cooling a massive body from its surface:

$$\frac{\partial x(t,z)}{\partial t} = a \cdot \frac{\partial^2 x(t,z)}{\partial z^2}, \quad 0 \leq z \leq 1, \tag{8}$$

with boundary conditions

$$\left.\frac{\partial x(t,z)}{\partial z}\right|_{z=0} = u_1(t) \cdot [x(t,0) - u_o], \tag{9}$$

$$\left.\frac{\partial x(t,z)}{\partial z}\right|_{z=1} = u_2(t). \tag{10}$$

Here, $x(t,z)$ is the temperature inside the body of normed thickness 1 and u_o is the coolant temperature which can be assumed to be a constant. $u_1(t)$ is a multiplicative control variable, arising again from variations of the coolant flow rate. The heat flux $u_2(t)$ at $z=1$ can be an additive control or a disturbance.

According to Brogan (1968) and also to a paper by Stafford and Dowrick (1978), inhomogeneous boundary conditions can be made formally homogeneous by making use of an extended definition of operators. This gives rise to symbolic functions (δ-functions and their derivatives) in the partial differential equation. The application to the above problem yields

$$\frac{\partial x(t,z)}{\partial t} = a \cdot \frac{\partial^2 x(t,z)}{\partial z^2} - a \cdot u_1(t) \cdot [x(t,z) - u_o] \cdot \delta(z) + a \cdot u_2(t) \cdot \delta(z-1),$$

with *homogeneous* boundary conditions

$$\left.\frac{\partial x(t,z)}{\partial z}\right|_{z=0} = \left.\frac{\partial x(t,z)}{\partial z}\right|_{z=1} = 0.$$

The extended operator offers a unifying framework for the treatment of both homogeneous and inhomogeneous boundary value problems. Based on this concept, the above introductory examples can be generalized by defining the state equations of a wide class of bilinear distributed parameter systems as follows*):

$$\frac{\partial \underline{x}(t,z)}{\partial t} = \underline{A}(z)\underline{x}(t,z) + \sum_{i=1}^{p} u_i(t)\underline{N}_i(z)\underline{x}(t,z) + \underline{B}(z)\underline{u}(t), \tag{11}$$

$$z \in [0,1], \quad t \in [t_o, \infty),$$

with boundary conditions

$$\left.\underline{R}_o(z)\underline{x}(t,z)\right|_{z=0} \equiv \underline{0}, \quad \left.\underline{R}_1(z)\underline{x}(t,z)\right|_{z=1} \equiv 0, \tag{12}$$

*) Here subscript denotes vectors and matrices.

initial conditions

$$\underline{x}(t_o,z) = \underline{x}_o(z), \quad z\in[0,1], \tag{13}$$

and an output equation of the type

$$\underline{y}(t) = \int_o^1 \underline{C}(z)\underline{x}(t,z)dz. \tag{14}$$

In the foregoing equations, $\underline{x}(t,z)$ is an n-dimensional vector representing the state of the system, z is a spatial coordinate, and t is time. The p-dimensional input vector $\underline{u}(t)$ and the q-dimensional output vector $\underline{y}(t)$ are assumed to be lumped as in most practical situations. Moreover, let us assume for the moment that all components of $\underline{u}(t)$ are control variables and not disturbances. $\underline{A}(z)$ and $\underline{N}_i(z)$, i=1, ..., p, $\underline{R}_o(z)$ and $\underline{R}_1(z)$ are linear partial differential matrix operators with respect to z. $\underline{B}(z)$ and $\underline{C}(z)$ are matrices whose elements are functions of z. In the special case of pointwise control or measurement, $\underline{N}_i$, $\underline{B}$ and $\underline{C}$ may contain symbolic functions.

The above equations have been restricted to one spatial dimension for simplicity, but generalizations can of course be treated in a similar manner.

In some applications, additive and multiplicative control variables are strictly independent from each other. Then Eq. (11) takes the special form

$$\frac{\partial \underline{x}(t,z)}{\partial t} = \underline{A}(z)\underline{x}(t,z) + \sum_{i=1}^{p_1} v_i(t)\underline{N}_i(z)\underline{x}(t,z) + \underline{B}(z)\underline{u}(t), \tag{15}$$

$$z\in[0,1], \quad t\in[t_o,\infty),$$

with p_1-dimensional multiplicative control vector $\underline{v}(t)$, and p_2-dimensional additive control vector $\underline{u}(t)$.

By extending some definitions from lumped parameter systems theory, a system with state equation (11) will be said to be *homogeneous-in-the-state* if $\underline{B}(z) \equiv \underline{0}$, *homogeneous-in-the-input* if $\underline{A}(z) \equiv \underline{0}$, and *strictly bilinear* if both $\underline{A}(z) \equiv \underline{0}$ and $\underline{B}(z) \equiv \underline{0}$.

Variable Structure Feedback Control of Linear Distributed Parameter Systems

Consider the *linear* distributed parameter system given by Eqs. (11) - (14) with $\underline{N}_i(z) \equiv \underline{0}$, i=1, ..., p, and suppose for the control of this system a *variable structure feedback law*

$$\underline{u}(t) = \underline{w}(t) + \sum_{i=1}^{r} v_i(t) \int_o^1 \underline{E}_i(z)\underline{x}(t,z)dz. \tag{16}$$

In Eq. (16), $\underline{w}(t)$ is some reference input vector, $\underline{E}_i(z)$, i=1, ..., r, are space dependent matrices of suitable dimensions, and $v_i(t)$, i=1, ..., r, are variable coefficients. By inserting Eq. (16) into (11) with $\underline{N}_i(z) \equiv \underline{0}$, one obtains the closed-loop equation

$$\frac{\partial \underline{x}(t,z)}{\partial t} = \underline{A}(z)\underline{x}(t,z) + \sum_{i=1}^{r} v_i(t)\underline{B}(z)\int_o^1 \underline{E}_i(z)\underline{x}(t,z)dz + \underline{B}(z)\underline{w}(t) \tag{17}$$

which is a bilinear integro-differential equation with controller coefficients $v_i(t)$ acting as multiplicative control variables, whereas $\underline{w}(t)$ is an additive input vector.

A special case of Eq. (16) arises, when two alternative linear control laws are provided,

$$\underline{u}^{(1)}(t) = \underline{w}(t) + \int_0^1 \underline{E}_1(z)\underline{x}(t,z)dz, \tag{18}$$

$$\underline{u}^{(2)}(t) = \underline{w}(t) + \int_0^1 \underline{E}_2(z)\underline{x}(t,z)dz, \tag{19}$$

with specified $\underline{E}_1(z)$ and $\underline{E}_2(z)$. Then the switching from $\underline{E}_1$ to $\underline{E}_2$ and reverse can be described by a scalar binary variable v(t):

$$\underline{u}(t)=\underline{w}(t)+\frac{1}{2}\cdot\int_0^1[\underline{E}_1(z)+\underline{E}_2(z)]\underline{x}(t,z)dz+\frac{1}{2}\cdot v(t)\cdot\int_0^1[\underline{E}_1(z)-\underline{E}_2(z)]\underline{x}(t,z)dz.$$

Here, v(t) takes only values ± 1. Eqs. (18) or (19) are resulting for v = + 1 or v = - 1, respectively. The closed-loop equation then takes the bilinear form

$$\frac{\partial\underline{x}(t,z)}{\partial t} = \underline{A}(z)\underline{x}(t,z)+\underline{B}(z)\cdot\int_0^1\hat{\underline{E}}_1(z)\underline{x}(t,z)dz+\underline{B}(z)\underline{w}(t)+v(t)\underline{B}(z)\cdot\int_0^1\hat{\underline{E}}_2(z)\underline{x}(t,z)dz,$$

where

$$\hat{\underline{E}}_1(z) = \frac{1}{2}\cdot[\underline{E}_1(z) + \underline{E}_2(z)], \quad \hat{\underline{E}}_2(z) = \frac{1}{2}\cdot[\underline{E}_1(z) - \underline{E}_2(z)].$$

The analogous problem in the lumped parameter case has been studied by Becker (1979). Becker has provided near-optimal switching strategies for the binary variable basing on Ljapunov's direct method.

Another special case arises from Eq. (16) if only pointwise measurement and control are considered, hence

$$\underline{E}_i(z) = \sum_{k=1}^{m}\tilde{\underline{E}}_{ik}\cdot\delta(z-z_k), \quad i=1, \ldots, r; \quad \underline{B}(z) = \sum_{j=1}^{s}\tilde{\underline{B}}_j\cdot\delta(z-\zeta_j),$$

with constant matrices $\tilde{\underline{E}}_{ik}$ and $\tilde{\underline{B}}_j$, and specified points of measurement and control, z_k and ζ_j, respectively.

Then Eq. (17) for the variable structure feedback control system contains no integral and can be written formally as a feedforward control system very simular to Eq. (15):

$$\frac{\partial\underline{x}(t,z)}{\partial t} = \underline{A}(z)\underline{x}(t,z)+\sum_{i=1}^{r}v_i(t)\cdot[\sum_{k=1}^{m}\underline{N}_{ik}(z)\underline{x}(t,z_k)] + \underline{B}(z)\underline{w}(t), \tag{20}$$

where

$$\underline{N}_{ik}(z) = \sum_{j=1}^{s}\tilde{\underline{B}}_j\tilde{\underline{E}}_{ik}\cdot\delta(z-\zeta_j). \tag{21}$$

Again the controller coefficients, $v_i(t)$, are acting as multiplicative control variables.

A SUMMARY OF AVAILABLE RESULTS IN THE CONTROL OF BILINEAR DISTRIBUTED PARAMETER SYSTEMS

Some of the publications on bilinear distributed parameter systems actually consider lumped parameter systems, since lumped parameter approximation of the distributed plant is carried out before applying any control theory. Most authors consider deterministic systems and use mathematical models of the type of Eqs. (11) - (14).

Briggs and Shen (1969) investigated a distributed parameter thermal and stress model for a nuclear rocket. The bilinear nature of the model arises from the coolant flow rate acting as a control variable. The distributed model is discretized in the space dimension to give an arbitrary order set of ordinary differential state equations. The particular transient of interest here is that of increasing the coolant flow through the rocket from a steady-state level to some higher level as quickly as possible. Specification of a maximum tolerable thermal stress provides a constraint which is also bilinear in control and state and which makes the optimal control problem non-classical: the maximum principle does not apply directly for such constraints. Therefore, the authors apply a result derived by Berkovitz from the calculus of variations using a slack variable technique to solve the time optimal problem. It is shown that the optimal control has no switching in this case. Numerical results are given for the optimal flow program and the resultant stress distributions for various cases.

Furusho, Kanoh and Masubuchi (1978) investigate the problem of state observation of heat exchangers where the forcing functions are variations in the coefficients of the partial differential equations as a result of manipulation of flow rates. Therefore, the underlying state equations are of the type of Eq.(5). The bilinear distributed parameter system is first reduced to a bilinear lumped parameter system by using the method of weighted residuals. Then bilinear lumped observers of complete order and of reduced order are proposed as an extension of the linear Luenberger observer. The observer is driven by the inlet and outlet temperature and the flow rate and estimates the temperature at any location of the heat exchanger. Exponential stability of the error equation is assured by estimations of the norm of the error using Raleigh's quotient. The experimental result shows that the bilinear observer with only four integrators can give accurate temperature estimation even if the change of flow rate is fast and large.

An approach by Franke (1979a, 1979b, 1980a, 1980b) is based on continuous-time, deterministic and time invariant systems as modelled by Eqs. (11) - (14). Topics of special interest are structural properties, stability of feedback control structures and optimal control problems.

A reachability condition of sufficient character has been provided (Franke 1980a) for a class of scalar bilinear distributed systems whose state, $x(t,z)$, satisfies an integral equation of the following form:

$$x(t,z) = \int_{\tau=0}^{t} [g_1(t-\tau,z)u_1(\tau) + g_2(t-\tau,z)x(\tau,0)u_2(\tau)]d\tau + $$

$$+ \int_{\zeta=0}^{1} g(t,z,\zeta)x_0(\zeta)d\zeta, \quad z\in[0,1], \quad t \geq 0. \tag{22}$$

Problems with multiplicative boundary control as in Eqs. (8) - (10) can be written in such a form by means of the Green's function belonging to the linear differential operator of the system.

Let the set of admissible controls in Eq. (22), S, be given by

$$S = \{\underline{u}(t) | m_i \leq u_i(t) \leq M_i; \ i=1,2\}.$$

Then, for some $\hat{x}(z)$ to be reachable from initial state $x_0(z)$ in time T, a solution $\underline{u}(t)\in S$ of the following integral equation must exist:

$$\hat{x}(z) - \int_{\zeta=0}^{1} g(T,z,\zeta)x_0(\zeta)d\zeta = \int_{t=0}^{T} [g_1(T-t,z)u_1(t) + g_2(T-t,z)x(t,0)u_2(t)]dt. \quad (23)$$

The question of existence of solutions to (23) has been reduced to an infinite-dimensional problem of moments which can be treated in a manner provided by Butkovskiy (1969).

If only multiplicative control $u_2(t)$ is applied in (22), hence $u_1(t) \equiv 0$, then the bilinear system is homogeneous-in-the-state. If then, in addition, the zero steady state is asymptotically stable for every admissible $u_2(t)$, then the reachable set will be very restricted.

But in case of independent additive control u_1 and multiplicative control u_2 in Eq. (22), variations of u_2 offer a degree of freedom in addition to linear control which may significantly enlarge the reachable set. It is in this sense that the bilinear system is "more controllable" than the linear one.

If for a system of the type of Eq. (22) a specified $\hat{x}(z)$ is not reachable in time T, then the problem arises to steer the state of the system, $x(t,z)$, as close as possible to $\hat{x}(z)$ in time T. This is a question of optimal control, and a suitable criterion of performance is

$$J = \frac{1}{2}\int_{z=0}^{1} [\hat{x}(z)-x(T,z)]^2dz + \frac{1}{2}\cdot\sum_{i=1}^{2}\gamma_i\cdot\int_{t=0}^{T} u_i^2(t)dt,$$

with weighting coefficients $\gamma_1, \gamma_2 \geq 0$.

Optimal bilinear control problems of this type have been investigated (Franke, 1979b, 1980a) in the presence of constraints on the controls. The maximum principle (Butkovskiy, 1969) has been applied to determine the optimal feedforward control for given $x_0(z)$ and $\hat{x}(z)$.

The computation of such optimal control problems is time consuming, and therefore needs further simplification in order to allow real time implementation of optimal feedback control. A hint may be the use of Walsh-Galerkin expansions as proposed by Tzafestas (1978).

The non-optimal feedback control of bilinear distributed parameter systems has been studied to some extent by Franke (1979a, 1980a, 1980b), and has been based essentially on Ljapunov's direct method. The procedure will be repeated here in some detail, because the adaptive and robustness approach is based on some of these results.

Consider the scalar version of Eq. (11)

$$\frac{\partial x(t,z)}{\partial t} = A(z)x(t,z) + u(t) \cdot N(z)x(t,z) + B(z) \cdot u(t), \quad (24)$$

$$0 \leq z \leq 1$$

with scalar control function $u(t)$ and with homogeneous boundary conditions. $x_s(z) \equiv 0$ is a steady state of Eq. (24) corresponding to steady state input $u_s = 0$.

Let the set of admissible controls, S, be given by

$$S = \{u(t) \mid m \leq u(t) \leq M\},$$

where the interval $[m, M]$ contains the origin.

If now the linear subsystem

$$\frac{\partial x(t,z)}{\partial t} = A(z)x(t,z), \qquad 0 \leq z \leq 1,$$

is asymptotically stable with respect to some specified metric, e.g.

$$\rho(0,x) = \|x\| = \left(\int_0^1 x^2(t,z)dz\right)^{1/2},$$

then an efficient feedback control law can be constructed basing on Ljapunov's direct method. Therefore, define

$$V = \frac{1}{2} \cdot \|x\|^2 = \frac{1}{2} \cdot \int_0^1 x^2(t,z)dz$$

as a Ljapunov-functional for system (24) and make

$$\dot{V} = \int_0^1 x \cdot \frac{\partial x}{\partial t} dz = \int_0^1 x \cdot Axdz + u(t) \cdot \int_0^1 x \ (Nx+B)dz$$

as negative as possible at any instant of time, via control variable u(t), by exhausting the admissible set S, to obtain the feedback control law

$$u(t) = \frac{M+m}{2} - \frac{M-m}{2} \cdot \text{sgn} \int_0^1 x(t,z) \cdot [N(z)x(t,z) + B(z)]dz. \tag{25}$$

Obviously, the control is bang-bang type and may include sliding modes whenever

$$\int_0^1 x(t,z) \cdot [N(z)x(t,z) + B(z)]dz \equiv 0$$

during some intervals of time.

Eq. (25) will, in general, require knowledge of the total state of the system. An expedient may be the use of state observers (Köhne, 1977; Zeitz, 1977) or an approximative evaluation of the integral in (25) by means of a sufficiently large number of measurement points.

The above straight forward procedure has been applied to some elementary examples and has proved to be quite efficient.

In many practical situations bang-bang control must be excluded, because actuators and controllers are dynamical systems which impose bounded rate on the variations of u(t) in Eq. (24). Therefore, a control-loop as outlined in Fig. 2 has been investigated (Franke, 1979a, 1980a).

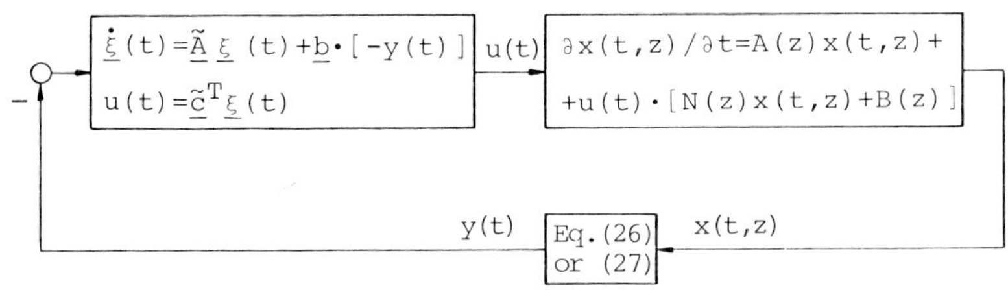

Fig. 2. Mixed lumped and distributed parameter control-loop

The output equation has been specified alternatively by

$$y(t) = \int_0^1 \delta(z,z_M)x(t,z)dz = x(t,z_M) \tag{26}$$

or by

$$y(t) = \int_0^1 x(t,z)\cdot[N(z)x(t,z) + B(z)]dz. \tag{27}$$

Eq. (26) means pointwise measurement with specified measurement point z_M. The expression in Eq. (27) coincides with the argument of the sgn-function in Eq. (25) which means that a weakened version of bang-bang law is considered.

With respect to the controller design, there will be at least one free parameter in the equations of the linear lumped subsystem in Fig. 2, e.g. the controller gain k. Therefore, the problem of stability becomes now a *parameter problem*. Design of parameter k along with assurance of a domain of asymptotic stability has been studied again via Ljapunov's direct method, using the extended Ljapunov-functional

$$V = \underline{\xi}^T(t)\underline{P}\,\underline{\xi}(t) + \frac{1}{2}\cdot\int_0^1 x^2(t,z)dz,$$

where $\underline{P}$ is a positive definite matrix.

Analytical stability conditions are obtained directly from the mathematical model without the need of approximation. On the other hand, because of the merely sufficient nature of these conditions, nothing can be said about their sharpness.

The control problem outlined in Fig. 2 has also been studied by Noldus and Derese (1980), including some extensions, e.g. to bilinear distributed systems with more than one state variable. The authors use also a Ljapunov approach. By inserting a saturation-type non-linearity between the lumped and the distributed system, they determine a range of the feedback amplifier matrices which ensure global asymptotic closed-loop stability. The algorithms for the amplifier calculation essentially require the iterative solution of an algebraic Riccati equation. A frequency domain interpretation of the stability conditions is given in terms of the frequency response matrix of the control system's linear part.

Bruni and Koch (1978) investigate a non-linear distributed parameter class of systems which is characterized by a lack of derivative in the spatial variable. The state equation of the type

$$\frac{\partial x}{\partial t} = \phi(x) + Nxu + Bu$$

is considered in a Banach space. The equation is not bilinear in the sense of the definition in section 2, but it is "nearly bilinear", since operators N and B are assumed to be linear and bounded, and operator ϕ is assumed to be locally Lipschitzian with at most a linear growth with respect to x:

$$||\phi(x)|| \leq \gamma_1 \cdot ||x|| + \gamma_2,$$

with γ_1, γ_2 nonnegative constants.

Due to the lack of spatial derivatives, investigation of theoretical properties for this class is simplified. On the other hand, this model will have some relevance in the description of important processes in biology and socio-economics as well as engineering.

So far the references are concerned with deterministic systems. Balakrishnan (1974) has investigated a very general class of stochastic bilinear partial differential equations in an abstract Hilbert space and based on semigroup theory. He proves existance and uniqueness theorems for systems with a bilinear stochastic forcing term. Both white noise and Wiener processes are considered, and the interrelationship is indicated.

BILINEAR CONTROL - A TOOL FOR THE DESIGN OF ADAPTIVE AND ROBUST DISTRIBUTED PARAMETER SYSTEMS

Motivation for Adaptive Bilinear Control

To fix ideas and to illustrate the inherent adaptive abilities of bilinear distributed parameter systems, the example of a continuous furnace is considered again as outlined in Fig. 1. The nominal behaviour is assumed to be described by the state equation

$$\frac{\partial x(t,z)}{\partial t} = - v(t) \cdot \frac{\partial x(t,z)}{\partial z} + a \cdot [u_s - x(t,z)], \qquad 0 \leqq z \leqq 1, \tag{28}$$

with boundary condition $x(t,0) \equiv 0$.

Here, the furnace temperature is assumed to be a constant, u_s, hence transportation speed $v(t)$ is the only control variable, and may be bounded by

$$(0<)m \leqq v(t) \leqq M. \tag{29}$$

The nominal steady state of Eq. (28), corresponding to some nominal value $v_s \in [m,M]$, is

$$x_s(z) = u_s \cdot [1 - \exp(-\frac{a}{v_s} z)]. \tag{30}$$

Assume now that the heat transfer coefficient, a, is an uncertain parameter. In the simplest case, the actual $a^* \neq a$ is a constant as well. As can be seen directly from Eq. (28), there exists

$$v(t) \equiv v_s^* = \frac{a^*}{a} v_s \tag{31}$$

such that (30) is preserved as a steady state solution in case of this parameter uncertainty. With respect to (29), bounds are imposed on a^*:

$$a \cdot \frac{m}{v_s} \leqq a^* \leqq a \cdot \frac{M}{v_s} .$$

Obviously if the interval $[m, M]$ of admissible $v(t)$ is sufficiently large, then so is the interval of admissible a^*.

The direct exploitation of Eq. (31) needs identification of parameter a^*. But as pointed out in the introduction, it is not in this way that the problem of adaptation will be treated in this chapter. The aim of the following section is to provide some state feedback control law of the type

$$v(t) = Fx(t,z), \tag{32}$$

where F is a non-linear functional which assures asymptotically stable steady state $x_s(z)$ of the plant even when subject to large parameter uncertainties.

This robustness property will be preserved also in case of time-varying coefficient $a^*(t)$ in the above example. Indeed, under condition $am/v_s \leq a^*(t) \leq aM/v_s$ it is evident that there does again exist an admissible multiplicative control $v^*(t) = v_s a/a^*(t)$ which provides steady state coincidence with Eq. (30). The adjustment of $v^*(t)$ according to time-varying $a^*(t)$ is carried out instantaneously, since no time delays are involved in the control law (32). It is for this reason that even fast variations of $a^*(t)$ can be admitted.

In case of space dependent $a^*(z)$, of course Eq. (30) can no longer be a steady state solution to the actual system. But it will be shown that the same control law (32) will then be able to steer the system's motion to some specified close neighbourhood of $x_s(z)$.

The above mentioned adaptive properties achieved by means of multiplicative modes of control, of course, do not liberate the designer from the task of establishing a good mathematical model of the process. Also, all available information on additive or parametric disturbances should be entered into the model. For example, a non-zero inlet temperature in the above furnace, $x(t,o) \neq 0$, would be a measurable disturbance and therefore should be considered in the model.

The additive control variable, u_s, in Eq. (28) has been a constant so far. By replacing u_s by time-dependent $u(t)$, it can be expected that very flexible control concepts open. $v(t)$ may be employed for the short-time transient behaviour and for the above mentioned adaptive properties. $u(t)$ which may be involved with considerable time delays, can be employed for the long-time behaviour. For example, steady state accuracy of the outlet temperature $x(t,1)$ can be achieved, via $u(t)$, by means of a classical PI-controller.

All these aspects will be treated in detail in the following subsections, and it is hoped that the introductory example which will be referred to throughout the text has provided some motivation for adaptive bilinear control.

Ljapunov-Design of Multiplicative Feedback Control with Adaptive and Robustness Properties

Problem formulation. Systems with independent additive and multiplicative control variables as described by a state equation of the type of Eq. (15) will be underlying the following text. But it is pointed out that all considerations can be extended in a very analogous manner to variable structure control systems described by Eq. (20) which involves spatial Dirac-functions. The additive control vector, $\underline{u}$, will be assumed to be a constant $\underline{u}_s$ in a first approach.

Now let the plant be subject to both operator disturbances and additive disturbances, thus making a time-varying system described by the state equations

$$\frac{\partial \underline{x}(t,z)}{\partial t} = \underline{A}(t,z)\underline{x}(t,z) + \sum_{i=1}^{p_1} v_i(t)\underline{N}_i(t,z)\underline{x}(t,z) + $$

$$+ \underline{B}(t,z)\underline{u}_s + \underline{d}(t,z), \quad z \in [0,1], \quad t \in [t_o, \infty), \tag{33}$$

along with homogeneous boundary conditions

$$\underline{R}_o(t,z)\underline{x}(t,z)\Big|_{z=o} \equiv \underline{0}, \qquad \underline{R}_1(t,z)\underline{x}(t,z)\Big|_{z=1} \equiv \underline{0} \tag{34}$$

and initial condition

$$\underline{x}(t_o,z) = \underline{x}_o(z). \tag{35}$$

In Eq. (33), $\underline{d}(t,z)$ means an additive disturbance. $\underline{A}(t,z)$, $\underline{B}(t,z)$, $\underline{N}_i(t,z)$, $i=1, \ldots, p_1$, $\underline{R}_o(t,z)$ and $\underline{R}_1(t,z)$ are defined as in Eqs. (11), (12), (15), but now subject to operator disturbances.

Let the set of admissible controls, S, be given by

$$S = \{\underline{v}(t)\,|\,m_i \leq v_i(t) \leq M_i,\ i=1,\ \ldots,\ p_1\}. \tag{36}$$

It will be assumed that each component $v_i(t)$ can be varied arbitrarily fast within these bounds.

The following *linear and time-invariant reference model* will be established besides the actual plant:

$$\frac{\partial \underline{x}_M(t,z)}{\partial t} = \underline{A}_M(z)\underline{x}_M(t,z) + \sum_{i=1}^{p_1} v_{iMs}\underline{N}_{iM}(z)\underline{x}_M(t,z) + \underline{B}_M(z)\underline{u}_s,$$

$$z\in[0,1], \qquad t\in[t_o,\infty), \tag{37}$$

with homogeneous boundary conditions

$$\underline{R}_{oM}(z)\underline{x}_M(t,z)\Big|_{z=o} \equiv \underline{0}, \qquad \underline{R}_{1M}(z)\underline{x}_M(t,z)\Big|_{z=1} \equiv \underline{0} \tag{38}$$

and initial condition

$$\underline{x}_M(t_o,z) = \underline{x}_{Mo}(z). \tag{39}$$

In the above equations, $\underline{x}_M(t,z)$ denotes the state of the model, with the same dimensionality as $\underline{x}(t,z)$. The time-invariant operators $\underline{A}_M(z)$, $\underline{B}_M(z)$, $\underline{N}_{iM}(z)$, $i=1, \ldots, p_1$, $\underline{R}_{oM}(z)$ and $\underline{R}_{1M}(z)$ describe some specified nominal behaviour of the plant. Of course, the fixed values of v_{iMs}, $i=1, \ldots, p_1$, must be required to satisfy

$$\underline{v}_{Ms} \in S. \tag{40}$$

Moreover, it will be assumed that the *autonomous* linear system

$$\frac{\partial \underline{x}_M(t,z)}{\partial t} = [\underline{A}_M(z) + \sum_{i=1}^{p_1} v_{iMs}\underline{N}_{iM}(z)]\underline{x}_M(t,z) \tag{41}$$

with boundary conditions (38) is asymptotically stable with respect to some metric, e.g.

$$\rho(\underline{o},\underline{x}_M) = \|\underline{x}_M\| = (\int_o^1 \underline{x}_M^T(t,z)\underline{x}_M(t,z)dz)^{1/2}, \tag{42}$$

and

$$V = \frac{1}{2}\cdot\|\underline{x}_M\|^2_{\underline{P}(z)} := \frac{1}{2}\cdot\int_o^1 \underline{x}_M^T(t,z)\underline{P}(z)\underline{x}_M(t,z)dz \tag{43}$$

is assumed to be a Ljapunov-functional for system (41) where $\underline{P}(z)$ is some symmetrical and positive definite matrix.

Therefore, the model has an asymptotically stable steady state $\underline{x}_{Ms}(z)$, corresponding to control vectors $\underline{u}_s$ and $\underline{v}_{Ms}$, which satisfies

$$[\underline{A}_M + \sum_{i=1}^{p_1} v_{iMs}\underline{N}_{iM}(z)]\underline{x}_{Ms}(z) + \underline{B}_M(z)\underline{u}_s = \underline{0}.$$

The objective is to find a feedback control law for the actual system, Eqs. (33) - (36),

$$\underline{v}(t) = \underline{F}\,\underline{x}(t,z), \tag{44}$$

with some functional $\underline{F}$ such that $\underline{v}(t) \in S$ and that, roughly speaking, the controlled system takes the following properties:

a) In case of operator coincidence with the model, hence

$$\frac{\partial \underline{x}(t,z)}{\partial t} = \underline{A}_M(z)\underline{x}(t,z) + \sum_{i=1}^{p_1} v_i(t)\underline{N}_{iM}(z)\underline{x}(t,z) + \underline{B}_M(z)\underline{u}_s, \tag{45}$$

$$\underline{R}_{oM}(z)\underline{x}(t,z)\Big|_{z=o} \equiv \underline{O}, \qquad \underline{R}_{1M}(z)\underline{x}(t,z)\Big|_{z=1} \equiv \underline{O},$$

and in the absence of additive disturbances d(t,z), control law (44) when inserted into (45) should improve the transient behaviour as compared to the known linear model. The time derivative of a Ljapunov-functional will serve as a criterion of performance in the subsequence.

b) In case of operator disturbances and additive disturbances, control law (44) should work against these disturbances such as to steer the system's state as close as possible to the predetermined steady state $\underline{x}_{Ms}(z)$ of the model. It is in this sense that we mean here adaptivity and robustness.

Process coinciding with the model. A control law $\underline{v}(t) = \underline{F}\,\underline{x}(t,z)$ will be constructed, based on Ljapunov's direct method, which assures property a) required above. Therefore, rewrite Eqs. (36), (37) in terms of deviations from the model's steady state,

$$\underline{x}(t,z) = \underline{x}_{Ms}(z) + \Delta\underline{x}(t,z), \tag{46}$$

$$\underline{v}(t) = \underline{v}_{Ms} + \Delta\underline{v}(t), \tag{47}$$

to obtain

$$\frac{\partial \Delta\underline{x}(t,z)}{\partial t} = [\underline{A}_M(z) + \sum_{i=1}^{p_1} v_{iMs}\underline{N}_{iM}(z)]\Delta\underline{x}(t,z) + \sum_{i=1}^{p_1} \Delta v_i(t)\cdot\underline{N}_{iM}(z)[\underline{x}_{Ms}(z) + \Delta\underline{x}(t,z)]. \tag{48}$$

By employing the same Ljapunov-functional as in Eq. (43), but now in terms of $\Delta\underline{x}(t,z)$, one obtains

$$\dot{V} = \int_o^1 \Delta\underline{x}^T(t,z)\underline{P}(z)[\underline{A}_M(z) + \sum_{i=1}^{p_1} v_{iMs}\underline{N}_{iM}(z)]\Delta\underline{x}(t,z)dz +$$

$$+ \sum_{i=1}^{p_1} \Delta v_i(t)\cdot\int_o^1 \Delta\underline{x}^T(t,z)\underline{P}(z)\underline{N}_{iM}(z)[\underline{x}_{Ms}(z) + \Delta\underline{x}(t,z)]dz. \tag{49}$$

The first integral on the right hand side of Eq. (49) is clearly negative definite, due to the assumption made for the autonomous system (41). Obviously, $\dot{V}$ can be made at least as negative as in the linear model by making each term of the remaining sum in Eq. (49) as negative as possible to obtain

$$\Delta v_i(t) = - v_{iMs} + \frac{M_i+m_i}{2} - \frac{M_i-m_i}{2} \cdot \mathrm{sgn}\int_0^1 \Delta \underline{x}^T(t,z)\underline{P}(z)\underline{N}_{iM}(z)[\underline{x}_{Ms}(z)+\Delta\underline{x}(t,z)]dz, \quad (50)$$

$$i=1, \ldots, p_1,$$

or, in terms of the actual $\underline{v}(t)$ and $\underline{x}(t,z)$,

$$v_i(t) = \frac{M_i+m_i}{2} - \frac{M_i-m_i}{2} \cdot \mathrm{sgn}\int_0^1 [\underline{x}(t,z)-\underline{x}_{Ms}(z)]^T\underline{P}(z)\underline{N}_{iM}(z)\underline{x}(t,z)dz, \quad (51)$$

$$i=1, \ldots, p_1.$$

Eq. (51) constitutes a suitable control law of the type of Eq. (35) which satisfies requirement a). Let us abbreviate Eq. (50), with regard to the following considerations, by

$$\Delta v_i(t) = F_i^* \Delta \underline{x}(t,z), \quad i=1, \ldots, p_1. \quad (52)$$

Example. Consider the continuous furnace already mentioned and outlined in Fig. 1, with a scalar state variable $x(t,z)$. The process of starting up the furnace will be of interest, hence

$$x_M(0,z) = x(0,z) \equiv 0, \quad z\in[0,1]. \quad (53)$$

Eq. (33) now takes the scalar form

$$\frac{\partial x(t,z)}{\partial t} = - a(t,z)x(t,z) - v(t)\frac{\partial x(t,z)}{\partial z} + a(t,z)u_s. \quad (54)$$

In terms of Eqs. (15) and (33) we have here

$$\underline{A}(t,z) \mathrel{\hat{=}} - a(t,z), \quad p_1=p_2=1, \quad \underline{N}(t,z) \mathrel{\hat{=}} - \partial/\partial z,$$

$$\underline{B}(t,z) \mathrel{\hat{=}} a(t,z), \quad \underline{d}(t,z) \mathrel{\hat{=}} 0.$$

The boundary condition may be known to be homogeneous,

$$x(t,0) \equiv 0 \text{ for all } t \geqq 0. \quad (55)$$

The normed heat transfer coefficient, $a(t,z)$, is assumed to be not known exactly. But it can be assumed that $a(t,z)$ varies within

$$0 < a_{min} \leqq a(t,z) \leqq a_{max}. \quad (56)$$

Let the set of admissible control $v(t)$ be given by

$$S = \{v(t)\,|\,0 \leqq v(t) \leqq M\}. \quad (57)$$

The following linear and time-invariant reference model will be used for the foregoing process:

$$\frac{\partial x_M(t,z)}{\partial t} = - a_M x_M(t,z) - v_{Ms}\frac{\partial x_M(t,z)}{\partial z} + a_M u_s, \quad (58)$$

with a constant coefficient a_M which satisfies $a_{min} \leqq a_M \leqq a_{max}$, and with boundary condition $x_M(t,0) \equiv 0$.

Obviously, coinciding $\underline{N} = \underline{N}_M \mathrel{\hat{=}} - \partial/\partial z$ is an inherent feature of this example which facilitates the further treatment.

It is rather elementary to show that

$$V = \frac{1}{2}\cdot \|x_M(t,z)\|^2 = \frac{1}{2}\cdot\int_0^1 x_M^2(t,z)\,dz$$

is a Ljapunov-functional for the autonomous system

$$\frac{\partial x_M}{\partial t} = - a_M x_M - v_{Ms} \cdot \frac{\partial x_M}{\partial z}, \qquad (59)$$

where $v_{Ms} \in S$. The steady state of the model is readily found to be

$$x_{Ms}(z) = u_s \cdot [1 - \exp(- \frac{a_M}{v_{Ms}} z)].$$

Process coincidence with the model requires $a(t,z) \equiv a_M$. Then the time derivative of the above Ljapunov-functional V will be made as negative as possible, according to Eq. (51), by means of the feedback control law

$$v(t) = \frac{M}{2} + \frac{M}{2} \cdot \text{sgn} \int_0^1 [x(t,z)-x_{Ms}(z)] \cdot \frac{\partial}{\partial z} x(t,z)dz. \qquad (60)$$

The undesirable spatial derivative of x(t,z) in this control law can be avoided via integration by parts to obtain

$$v(t) = \frac{M}{2} + \frac{M}{2} \cdot \text{sgn}[\frac{1}{2} x^2(t,1)-x_{Ms}(1)x(t,1)+\int_0^1 x_{Ms}'(z)x(t,z)dz]. \qquad (61)$$

Here, $x_{Ms}'(z)$ is the known spatial derivative of $x_{Ms}(z)$.

The following numerical example has been simulated:

$$a_M = 1; \quad v_{Ms} = u_s = 0{,}5; \quad M = 1.$$

The method of weighted residuals (MWR) has been employed for analog simulation, with polynomial approximation

$$x(t,z) \approx \sum_{i=1}^{5} x_i^*(t)\cdot z^i,$$

and collocation in the equidistant points $z_k = k/5$, $k = 1, \ldots, 5$. Figure 3 outlines the response of the process coinciding with the model, but with v_{Ms} being replaced by the above control law (61). The integral appearing in the control law has been evaluated approximately by means of straight lines, using the same grid points z_k as above.

Obviously, the multiplicative control variable, v(t), acts in a bang-bang manner at the beginning of the process and then keeps the required steady state in a quasi sliding mode. The sgn-function in the control law has been implemented by means of a high-gain electronic amplifier with constraints, and so it has no discontinuity in the strong mathematical sense. In the present example, a halt in the conveyor belt at the beginning of the process is intuitively advantageous.

The efficiency of control law (61) in the case of parameter variations will be demonstrated in the next two sections.

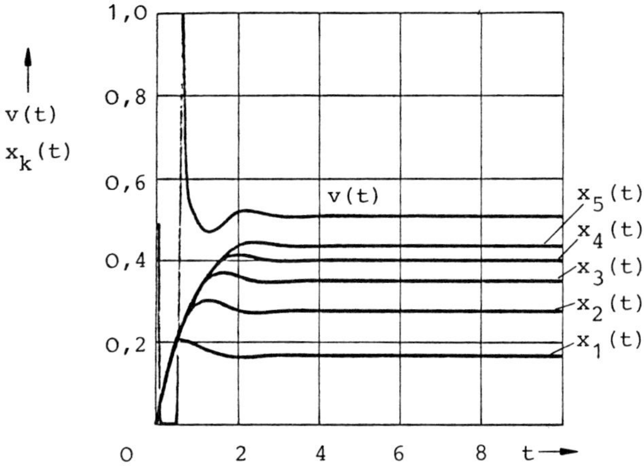

Fig. 3. Control variable v(t) and state variables $x_k(t) = x(t;k/5)$, k=1, ..., 5, with control law (61).

Disturbances which allow steady state coincidence with the model. In addition to deviations according to Eqs. (46), (47), we consider (possibly large) operator disturbances and additive disturbances:

$$\underline{A}(t,z) = \underline{A}_M(z) + \Delta\underline{A}(t,z), \qquad \underline{B}(t,z) = \underline{B}_M(z) + \Delta\underline{B}(t,z),$$

$$\underline{N}_i(t,z) = \underline{N}_{iM}(z) + \Delta\underline{N}_i(t,z), \qquad i=1, \ldots, p_1,$$

$$\underline{R}_o(t,z) = \underline{R}_{oM}(z) + \Delta\underline{R}_o(t,z), \quad \underline{R}_1(t,z) = \underline{R}_{1M}(z) + \Delta\underline{R}_1(t,z), \quad \underline{d} = \underline{d}(t,z),$$

and rewrite Eqs. (33), (34) in terms of deviations:

$$\begin{aligned}\frac{\partial\Delta\underline{x}(t,z)}{\partial t} &= [\underline{A}(t,z) + \sum_{i=1}^{p_1} v_{iMs}\underline{N}_i(t,z)]\Delta\underline{x}(t,z) + \\ &+ \Delta\underline{A}(t,z)\underline{x}_{Ms}(z) + \sum_{i=1}^{p_1} v_{iMs}\Delta\underline{N}_i(t,z)\underline{x}_{Ms}(z) + \Delta\underline{B}(t,z)\underline{u}_s + \underline{d}(t,z) + \\ &+ \sum_{i=1}^{p_1} \Delta v_i(t)\cdot\{\underline{N}_{iM}(z)[\underline{x}_{Ms}(z)+\Delta\underline{x}(t,z)]\} + \\ &+ \sum_{i=1}^{p_1} \Delta v_i(t)\cdot\{\Delta\underline{N}_i(t,z)[\underline{x}_{Ms}(z)+\Delta\underline{x}(t,z)]\},\end{aligned} \tag{62}$$

with inhomogeneous boundary conditions

$$\underline{R}_o(t,z)\Delta\underline{x}(t,z)\Big|_{z=o} = -\Delta\underline{R}_o(t,z)\underline{x}_{Ms}(z)\Big|_{z=o}, \tag{63}$$

$$\underline{R}_1(t,z)\Delta\underline{x}(t,z)\Big|_{z=o} = -\Delta\underline{R}_1(t,z)\underline{x}_{Ms}(z)\Big|_{z=1} . \tag{64}$$

But, by using the notion of extended operators (Brogan, 1968), boundary conditions (63), (64) will be made homogeneous,

$$\underline{R}_o(t,z)\Delta\underline{x}(t,z)\Big|_{z=o} \equiv \underline{o}, \qquad \underline{R}_1(t,z)\Delta\underline{x}(t,z)\Big|_{z=1} \equiv \underline{o}, \tag{65}$$

which merely modifies $\underline{d}(t,z)$ in (62) to some $\underline{d}^*(t,z)$. In the subsequence, additive and operator disturbances will be abbreviated by matrix $\underline{Z}(t,z)$, where

$$\underline{Z}(t,z) := [\Delta\underline{A}(t,z),\ \Delta\underline{B}(t,z),\ \underline{d}^*(t,z),\ \Delta\underline{N}_1(t,z),\ \ldots,\ \Delta\underline{N}_{p1}(t,z)].$$

Assume now that the initial state of the above system is $\Delta\underline{x}(t_o,z) \equiv \underline{o}$. Of course, the system will not keep this zero deviation for $t > t_o$, in general, due to time-dependent disturbances $\underline{Z}(t,z)$. But there may be a non-empty set of disturbances, say $\mathcal{Z}$, which can be eliminated by $\Delta\underline{v}(t)$, where $\underline{v}_{Ms} + \Delta\underline{v}(t)\in S$, such that $\Delta\underline{x}_s(z) \equiv \underline{o}$ becomes a steady state solution. Therefore, disturbances $\underline{Z}(t,z)\in\mathcal{Z}$ allow complete coincidence of the steady states of the model and of the disturbed real process.

From Eqs. (62), (65) it follows immediately that for $\Delta\underline{x}_s(z) \equiv \underline{o}$ to be a steady state solution the following condition must hold:

$$\underline{G}(t,z)\Delta\underline{v}(t) = \underline{h}(t,z), \tag{66}$$

where

$$\underline{G}(t,z) = [\underline{N}_1(t,z)\underline{x}_{Ms}(z),\ \ldots,\ \underline{N}_{p1}(t,z)\underline{x}_{Ms}(z)], \tag{67}$$

$$\underline{h}(t,z) = -[\Delta\underline{A}(t,z)+\sum_{i=1}^{p_1} v_{iMs}\Delta\underline{N}_i(t,z)]\underline{x}_{Ms}(z)-\Delta\underline{B}(t,z)\underline{u}_s-\underline{d}^*(t,z). \tag{68}$$

Eq. (66) is a linear algebraic equation for $\Delta\underline{v}(t)$ with generally non-quadratic coefficient matrix $\underline{G}(t,z)$. It is solvable if and only if the following rank condition holds for all $z\in[o,1]$:

$$\text{rank } \underline{G}(t,z) = \text{rank } [\underline{G}(t,z),\ \underline{h}(t,z)]. \tag{69}$$

The corresponding solution $\Delta\underline{v}(t)$ will then, in general, not be unique, but can be made to be unique by setting

$$\Delta\underline{v}(t) \equiv \underline{o} \quad \text{in case of} \quad \underline{Z}(t,z) \equiv \underline{o}$$
$$\text{and hence} \quad \underline{h}(t,z) \equiv \underline{o}. \tag{70}$$

$\underline{Z}(t,z) \equiv \underline{o}$ means coincidence between the actual process and the model, and then there is, indeed, no need for $\Delta\underline{v}(t)$ to differ from $\underline{o}$ in the steady state.

Denote the unique solution of (66) under condition (69) and with property (70) by $\Delta\underline{v}^*(t)$. So we can summarize the following *steady state result:*

> Let $\mathcal{Z}$ denote the set of those disturbances $\underline{Z}(t,z)$ which allow, for some $\Delta\underline{v}(t)$ with $\underline{v}_{Ms}+\Delta\underline{v}(t)\in S$, $\Delta\underline{x}(t,z)$ to be a solution of system (62) for all $t \geq t_o$ with boundary conditions (65) and initial state $\Delta\underline{x}_o(z) \equiv \underline{o}$.
>
> Then for a given $\underline{Z}(t,z)$ to belong to $\mathcal{Z}$ it is sufficient that it satisfies rank condition (69) and that, in addition, the unique solution of (66) with property (70), $\Delta\underline{v}^*(t)$, satisfies

$$\underline{v}_{Ms} + \Delta \underline{v}^*(t) \in S. \tag{71}$$

It should be stressed that rank condition (69) is rather restrictive. Assume for example that $\Delta\underline{A}(t,z) \equiv \underline{O}$, $\Delta\underline{B}(t,z) \equiv \underline{O}$ and $\underline{d}^*(t,z) \equiv \underline{O}$. Then (69) requires validity of

$$\sum_{i=1}^{p_1} \beta_i(t) \cdot [\underline{N}_{iM}(z) + \Delta\underline{N}_i(t,z)]\underline{x}_{Ms}(z) = \sum_{i=1}^{p_1} v_{iMs} \cdot \Delta\underline{N}_i(t,z)\underline{x}_{Ms}(z), \tag{72}$$

with certain coefficients $\beta_i(t)$. Eq. (72) can, in general, only be satisfied if

$$\Delta\underline{N}_i(t,z) = \gamma_i(t)\ \underline{N}_{iM}(z), \quad i=1, \ldots, p_1, \tag{73}$$

for all $t \geq t_o$, with certain coefficients $\gamma_i(t)$.

What about the *dynamical* behaviour of the actual process subject on the one hand to feedback control law (50) and on the other hand to disturbances $\underline{Z}(t,z) \in Z$? It will be shown in the following that for any disturbance belonging to a certain subset $\hat{Z} \subset Z$, $\Delta\underline{x}_s(z) \equiv \underline{O}$ is preserved as a global asymptotically stable steady state of the dynamical equations with control law (50).

Therefore, Eq. (43) is used again as a tentative Ljapunov-functional for the actual system to obtain now, with the aid of (62), (66), (52) and (73),

$$\dot{V} = \int_o^1 \Delta\underline{x}^T(t,z)\underline{P}(z) \cdot [\underline{A}(t,z) + \sum_{i=1}^{p_1} v_{iMs} \cdot (1+\gamma_i(t))\underline{N}_{iM}(z)]\Delta\underline{x}(t,z)dz -$$

$$- \sum_{i=1}^{p_1} (1+\gamma_i(t))\Delta v_i^*(t) \cdot \int_o^1 \Delta\underline{x}^T(t,z)\underline{P}(z)\underline{N}_{iM}(z)\underline{x}_{Ms}(z)dz +$$

$$+ \sum_{i=1}^{p_1} (1+\gamma_i(t)) \cdot F_i^*\Delta\underline{x}(t,z) \cdot \int_o^1 \Delta\underline{x}^T(t,z)\underline{P}(z)\underline{N}_{iM}(z)[\underline{x}_{Ms}(z) + \Delta\underline{x}(t,z)]dz. \tag{74}$$

It is very important to recall that $\Delta v_i^*(t)$ and $\Delta v_i(t) = F_i^*\Delta\underline{x}(t,z)$ have quite different meanings in Eq. (74). $F_i^*\Delta\underline{x}(t,z)$ is the actual bang-bang type control law which needs only state information and specification of the model. $\Delta v_i^*(t)$ are the fictitious control variables which merely assure $\Delta\underline{x}_s(z) \equiv \underline{O}$ to be a steady state solution when applied to the disturbed open-loop process.

At the first glance, the question of definiteness of $\dot{V}$ is not trivial, since $\dot{V}$ contains terms linear in $\Delta\underline{x}$ also. But a quadratic upper bound on $\dot{V}$ can be provided by exploiting the switching character of F_i^* for an estimate of the expression

$$g_i(\Delta\underline{x}) := - \Delta v_i^*(t) \cdot \int_o^1 \Delta\underline{x}^T(t,z)\underline{P}(z)\underline{N}_{iM}(z)\underline{x}_{Ms}(z)dz +$$

$$+ F_i^*\Delta\underline{x}(t,z) \cdot \int_o^1 \Delta\underline{x}^T(t,z)\underline{P}(z)\underline{N}_{iM}(z)[\underline{x}_{Ms}(z) + \Delta\underline{x}(t,z)]dz, \tag{75}$$

to obtain

$$g_i(\Delta\underline{x}) \leq \Delta v_i^*(t) \cdot \int_o^1 \Delta\underline{x}^T(t,z)\underline{P}(z)\underline{N}_{iM}(z)\Delta\underline{x}(t,z)dz$$

for any state $\underline{\Delta x}(t,z)$ and for $i=1, \ldots, p_1$.

Keeping in mind that with regard to (74) and (75)

$$\dot{V} = \int_0^1 \underline{\Delta x}^T(t,z)\underline{P}(z)[\underline{A}(t,z) + \sum_{i=1}^{p_1} v_{iMs}\cdot(1+\gamma_i(t))\underline{N}_{iM}(z)]\underline{\Delta x}(t,z)dz +$$

$$+ \sum_{i=1}^{p_1} (1+\gamma_i(t))\cdot g_i(\underline{\Delta x}),$$

and presuming that

$$\gamma_i(t) \geq -1, \quad i=1, \ldots, p_1, \tag{76}$$

we obtain an upper bound for $\dot{V}$ which is quadratic in $\underline{\Delta x}$:

$$\dot{V} \leq \int_0^1 \underline{\Delta x}^T(t,z)\underline{P}(z)\cdot\{\underline{A}(t,z)+\sum_{i=1}^{p_1} [v_{iMs}+\Delta v_i^*(\underline{Z}(t,z))]\cdot\underline{N}_i(t,z)\}\underline{\Delta x}(t,z)dz. \tag{77}$$

Here, $\underline{\Delta v}^*$ $(\underline{Z}(t,z))$ denotes the unique solution of Eq. (66) which indicates that the question of negative definiteness of $\dot{V}$ is a question of admissible $\underline{Z}(t,z)$. The investigation must be carried out individually in each concrete application, e.g. via Sylvester's criterion. So we have the following *stability result:*

Let $\hat{Z} \subseteq Z$ be the set of disturbances $\underline{Z}(t,z)$ for which

$$\int_0^1 \underline{\Delta x}^T(t,z)\underline{P}(z)\cdot\{\underline{A}(t,z)+\sum_{i=1}^{p_1} [v_{iMs}+\Delta v_i^*(\underline{Z})]\cdot\underline{N}_i(t,z)\} \underline{x}(t,z)dz \leq 0 \tag{78}$$

for all $\underline{\Delta x}(t,z)$. Then $\underline{\Delta x}_s(z) \equiv \underline{0}$ is assured to be the global asymptotically stable steady state of system (62) - (64) for any disturbance $\underline{Z}(t,z)\in\hat{Z}$.

Especially, the zero equilibrium state is immune from any disturbances $\underline{Z}(t,z)\in\hat{Z}$. This follows immediately from the property of Ljapunov-functional V.

A special subset $\tilde{Z} \subseteq \hat{Z}$ can even be obtained directly by recalling that

$$\int_0^1 \underline{\Delta x}^T(t,z)\underline{P}(z)[\underline{A}_M(z) + \sum_{i=1}^{p_1} v_{iMs}\underline{N}_{iM}(z)]\underline{\Delta x}(t,z)dz \leq 0. \tag{79}$$

Therefore if the disturbances are such that (66) can be satisfied by solution $\underline{\Delta v}^*(t)$ with property

$$\underline{\Delta A}(t,z) + \sum_{i=1}^{p_1} [(v_{iMs}+\Delta v_i^*(t))\cdot\underline{\Delta N}_i(t,z) + \Delta v_i^*(t)\underline{N}_{iM}(z)] \equiv \underline{0}, \tag{80}$$

then condition (78) reduces to (79) which provides global asymptotic stability without further calculation. Eq. (80) means, with regard to (68) that

$$\underline{h}(t,z) = \sum_{i=1}^{p_1} \Delta v_i^*(t)\underline{N}_i(t,z)\underline{x}_{Ms}(z) - \underline{\Delta B}(t,z)\underline{u}_s - \underline{d}^*(t,z)$$

and hence, by insertion into Eq. (66) that

$$\Delta \underline{B}(t,z)\underline{u}_s + \underline{d}^*(t,z) \equiv \underline{0} \tag{81}$$

for all $z\in[0,1]$ and all $t \geq t_o$. Therefore we can summarize that $\tilde{Z}$ is the subset of Z with property (81).

For disturbances $\underline{Z}(t,z)\in\tilde{Z}$, V applied to the plant decreases at least as fast as in the model, due to the sign ($\leq$) in (77).

<u>Example</u>. The continuous furnace will be considered with parameter $a(t,z) \not\equiv a_M$. According to Eqs. (67), (68) we now have

$$\underline{G}(t,z) = \underline{N}\,\underline{x}_{Ms}(z) \hat{=} - x'_{Ms}(z) = - u_s \cdot \frac{a_M}{v_{Ms}} \cdot \exp(- \frac{a_M}{v_{Ms}} z),$$

$$\underline{h}(t,z) \hat{=} \Delta a(t,z)x_{Ms}(z) - \Delta a(t,z)u_s,$$

where

$$\Delta a(t,z) = a(t,z) - a_M.$$

By means of the known $x_{Ms}(z)$, $\underline{h}(t,z)$ can be rewritten as

$$\underline{h}(t,z) \hat{=} - u_s \cdot \Delta a(t,z) \cdot \exp(- \frac{a_M}{v_{Ms}} z).$$

Therefore, rank condition (69) here requires

$$u_s \cdot \Delta a(t,z) \cdot \exp(-\frac{a_M}{v_{Ms}} z) \overset{!}{=} c(t)u_s \cdot \frac{a_M}{v_{Ms}} \cdot \exp(- \frac{a_M}{v_{Ms}} z),$$

with a coefficient c(t) depending only on time t . Hence, only space-independent $a(t) = a_M + \Delta a(t)$ can be admitted. The solution of (66) with property (70) is then $\Delta v^*(t) = \Delta a(t)v_{Ms}/a_M$. In accordance with Eq. (71), for $\Delta a(t)$ to belong to Z, it must be required that $0 \leq v_{Ms} + \Delta v^*(t) \leq M$, hence

$$(0<)\ a(t) \leq Ma_M/v_{Ms}, \tag{82}$$

which implies

$$a_{max} \leq Ma_M/v_{Ms}. \tag{83}$$

In those situations where M is not specified in advance, this condition can be satisfied by selecting M sufficiently large.

With regard to global asymptotic stability of steady state $x_{Ms}(z)$ now only condition (78) has to be tested:

$$\int_0^1 \Delta x(t,z) \cdot [-a(t)\Delta x(t,z) - (v_{Ms} + \Delta v^*(t)) \frac{\partial}{\partial z} x(t,z)]dz =$$

$$= -a(t) \cdot \int_0^1 \Delta x^2(t,z)dz - v^*(t) \cdot \frac{1}{2} \Delta x^2(t,1).$$

This expression is clearly negative definite, since $a(t) > 0$ and $v^*(t) \geq 0$, and therefore global asymptotic stability of steady state $x_{Ms}(z)$ is assured, with control law (61), for any positive time-varying coefficient a(t) satisfying

$a(t) \leq a_M \cdot M/v_{Ms}$. In the numerical example already used, $a(t)$ must satisfy $a(t) \leq 2$.

An analog simulation has been carried out for $a = 0{,}5 = a_M/2$, using again MWR-approximation. Figure 4 outlines the response of the controlled system with control law (61). It can be seen that $x(t,z)$ tends to $x_{Ms}(z)$ and that $v(t)$ tends to $v^* = v_{Ms} \cdot a/a_M = 0{,}25$ as expected, thus demonstrating the adaptive nature of the bilinear system with feedback law (61).

Once the system has reached its steady state, any variations $0 < a(t) \leq 2$ will not disturb the equilibrium state $x_{Ms}(z)$, since $v^*(t)$ will work against $a(t)$ without time delay in a sliding mode.

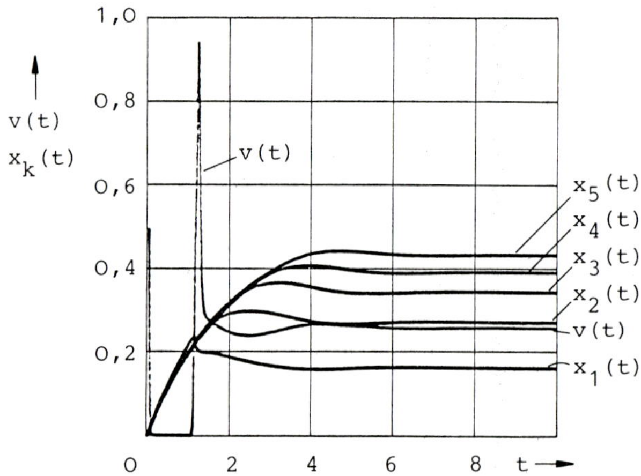

Fig. 4. Control variable $v(t)$ and state variables $x_k(t) = x(t; k/5)$, $k=1, \ldots, 5$, with control law (61) and parameter $a = a_M/2 = 0{,}5$

Disturbances which allow a steady state not coinciding with the model. In some applications the set of disturbances, $\mathcal{Z}$, may be rather small or even empty. In these situations one may put up with a deviation between the steady states of the model and the actual disturbed system, and therefore admit disturbances which exceed the set $\mathcal{Z}$.

In the following, we assume time-independent additive and operator disturbances

$$\underline{A}(z) = \underline{A}_M(z) + \Delta\underline{A}(z), \quad \underline{B}(z) = \underline{B}_M(z) + \Delta\underline{B}(z),$$

$$\underline{N}_i(z) = \underline{N}_{iM}(z) + \Delta\underline{N}_i(z), \quad i=1, \ldots, p_1,$$

$$\underline{R}_o(z) = \underline{R}_{oM}(z) + \Delta\underline{R}_o(z), \quad \underline{R}_1(z) = \underline{R}_{1M}(z) + \Delta\underline{R}_1(z), \quad \underline{d} = \underline{d}(z),$$

such that $\Delta\underline{x}(t,z) \equiv \underline{0}$ is *not* a solution to Eq. (62). But a steady state solution $\Delta\underline{x}_s(z) \not\equiv \underline{0}$ of Eq. (62) is presumed to exist for every $\Delta\underline{v}_s$, where $\underline{v}_s = \underline{v}_{Ms} + \Delta\underline{v}_s \in S$. Therefore, the following equation must hold for $\Delta\underline{x}_s(z)$:

$$[\underline{A}(z) + \sum_{i=1}^{p_1} v_{is}\underline{N}_i(z)]\Delta\underline{x}_s(z) = -\Delta\underline{A}(z)\underline{x}_{Ms}(z) - \Delta\underline{B}(z)\underline{u}_s - \underline{d}^*(z) -$$

$$- \sum_{i=1}^{p_1} v_{iMs}\Delta\underline{N}_i(z)\underline{x}_{Ms}(z) - \sum_{i=1}^{p_1} \Delta v_{is}\underline{N}_i(z)\underline{x}_{Ms}(z). \tag{84}$$

The boundary conditions of (84) are again homogeneous by means of the extended definition of operators which provides modified $\underline{d}^*(z)$ instead of $\underline{d}(z)$.

With regard to the left hand side of Eq. (84) it will be assumed that $\Delta\underline{A}(z)$ and $\Delta\underline{N}_i(z)$, i=1, ..., p_1, are such that the Green's matrix corresponding to the differential operator

$$\underline{A}(z) + \sum_{i=1}^{p_1} v_{is}\underline{N}_i(z)$$

along with the specified boundary conditions exists for any $\underline{v}_s \in S$. This provides a unique steady state solution $\Delta\underline{x}_s(z)$ of Eq. (84) dependent on $\underline{v}_s$ and on the actual disturbances.

Which is the smallest possible deviation in the sense of some norm $||\Delta\underline{x}_s(z)||$, attainable via $\Delta\underline{v}_s$? This is a non-linear problem of parameter optimization, since $\Delta\underline{v}_s$ occurs in an additive manner on the right hand side of Eq. (84) and in a multiplicative manner on the left hand side. Therefore, a simplified problem of static suboptimization will be investigated. The right hand side $\underline{r}(z)$ of Eq. (84) will be considered, and the functional

$$J^2 = \int_o^1 \underline{r}^T(z)\underline{Q}(z)\underline{r}(z)dz \tag{85}$$

with some symmetrical and positive definite matrix $\underline{Q}(z)$ will be minimized via $\Delta\underline{v}_s$, where $\underline{v}_{Ms} + \Delta\underline{v}_s \in S$. Since $\underline{r}(z)$ is linear in $\Delta\underline{v}_s$, this is a simple problem of parameter optimization and yields the solution

$$\Delta v^*_{is} = \begin{cases} m_i - v_{iMs} & \text{if } A_{i1}/A_{i2} < m_i - v_{iMs} \\ A_{i1}/A_{i2} & \text{if } m_i - v_{iMs} \leq A_{i1}/A_{i2} \leq M_i - v_{iMs} \\ M_i - v_{iMs} & \text{if } A_{i1}/A_{i2} > M_i - v_{iMs}, \end{cases} \qquad i=1, \ldots, p_1, \tag{86}$$

where

$$A_{i1} = -\int_o^1 [\underline{N}_i(z)\underline{x}_{Ms}(z)]^T\underline{Q}(z)\cdot\{[\Delta\underline{A}(z) + \sum_{i=1}^{p_1} v_{iMs}\Delta\underline{N}_i(z)]\underline{x}_{Ms}(z) +$$

$$+ \Delta\underline{B}(z)\underline{u}_s + \underline{d}^*(z)\}dz, \qquad i=1, \ldots, p_1, \tag{87}$$

$$A_{i2} = \int_o^1 [\underline{N}_i(z)\underline{x}_{Ms}(z)]^T\underline{Q}(z)\underline{N}_i(z)\underline{x}_{Ms}(z)dz, \qquad i=1, \ldots, p_1. \tag{88}$$

The corresponding steady state $\Delta\underline{x}_s^*(z)$ can be determined by means of Eq. (84) and the above mentioned Green's matrix, for any given disturbance

$$\underline{Z}(z) = [\Delta\underline{A}(z),\ \Delta\underline{B}(z),\ \underline{d}^*(z),\ \Delta\underline{N}_1(z),\ \ldots,\ \Delta\underline{N}_{p1}(z)].$$

In the special case where

$$[\Delta\underline{A}(z) + \sum_{i=1}^{p_1} v_{iMs}\Delta\underline{N}_i(z)]\underline{x}_{Ms}(z) + \Delta\underline{B}(z)\underline{u}_s + \underline{d}^*(z) \equiv \underline{O},$$

Eq. (86) yields $\Delta\underline{v}_s^* = \underline{O}$ and hence $\Delta\underline{x}_s^*(z) \equiv \underline{O}$. Indeed, in this case we have $\underline{Z}(z)\in\mathcal{Z}$.

How does the *dynamical* process behave, subject on the one hand to feedback control law (52) and on the other hand to disturbances which do not allow steady state solution $\Delta\underline{x}_s(z) \equiv \underline{O}$ of the open-loop process and hence of the closed-loop system? What one can hope is that the system's state is steered to some neighbourhood of zero from any initial state $\Delta\underline{x}_o(z)$.

In the following, an estimate will be derived for this neighbourhood.

Of course,

$$V = \frac{1}{2}\cdot\int_o^1 \Delta\underline{x}^T(t,z)\underline{P}(z)\Delta\underline{x}(t,z)\,dz$$

ceases to be a Ljapunov-functional for the system under consideration. Nevertheless, the $\Delta\underline{x}(t,z)$-notation will be kept, and $\dot{V}$ along with control law (52) will be rewritten for time-independent disturbances $\underline{Z}(z)$. By assuming, simular to Eq. (73), that $\Delta\underline{N}_i(z) = \gamma_i\underline{N}_{iM}(z)$, $\gamma_i \geq -1$, $i=1,\ \ldots,\ p_1$, the following estimate for $\dot{V}$ can be obtained:

$$\dot{V} \leq \int_o^1 \Delta\underline{x}^T(t,z)\underline{P}(z)[\underline{A}(z) + \sum_{i=1}^{p_1} v_{is}^*\underline{N}_i(z)]\Delta\underline{x}(t,z)\,dz -$$

$$- \int_o^1 \Delta\underline{x}^T(t,z)\underline{P}(z)[\underline{A}(z) + \sum_{i=1}^{p_1} v_{is}^*\underline{N}_i(z)]\Delta\underline{x}_s^*(z)\,dz. \tag{89}$$

$\Delta\underline{x}_s^*(z)$ is the "near-optimal" steady state corresponding to $\Delta\underline{v}_s^*$. Of course, $\dot{V}$ cannot be negative definite, since V is not a Ljapunov-functional. Only in the special case where $\Delta\underline{x}_s^*(z) \equiv \underline{O}$, the estimate (89) reduces to (77).

The objective must be to find a (possibly small) positive number ρ and to specify some norm $\|\Delta\underline{x}(t,z)\|$ such that $\dot{V} \leq 0$ for all $\Delta\underline{x}(t,z)$ with $\|\Delta\underline{x}(t,z)\| \geq \rho$.

Such an estimate can be found for those disturbances $\underline{Z}(z)$ which allow the first right hand integral in Eq. (89) to be estimated by

$$\int_o^1 \Delta\underline{x}^T(t,z)\underline{P}(z)[\underline{A}(z) + \sum_{i=1}^{p_1} (v_{iMs} + \Delta\dot{v}_{is}^*)\underline{N}_i(z)]\Delta\underline{x}(t,z) \leq$$

$$\leq -\ k(\underline{Z})\cdot\int_o^1 \Delta\underline{x}^T(t,z)\underline{R}(z)\Delta\underline{x}(t,z)\,dz, \tag{90}$$

where $k(\underline{Z})$ is a positive number and $\underline{R}(z)$ is a symmetrical and positive definite matrix.

By using the abbreviation

$$\int_0^1 \underline{\Delta x}^T(t,z)\underline{R}(z)\Delta\underline{x}(t,z)\,dz := ||\Delta\underline{x}(t,z)||^2_{\underline{R}(z)},$$

the second integral in (89) can be estimated by

$$-\int_0^1 \Delta\underline{x}^T(t,z)\underline{P}(z)[\underline{A}(z)+\sum_{i=1}^{p_1} v^*_{is}\underline{N}_i(z)]\Delta\underline{x}^*_s(z)\,dz \leq \int_0^1 |\Delta\underline{x}^T(t,z)\underline{R}(z)\Delta\underline{y}^*_s(z)|\,dz,$$

where

$$\Delta\underline{y}^*_s(z) := \underline{R}^{-1}(z)\underline{P}(z)[\underline{A}(z) + \sum_{i=1}^{p_1} v^*_{is}\cdot\underline{N}_i(z)]\Delta\underline{x}^*_s(z).$$

By means of Schwarz' inequality one obtains

$$\int_0^1 |\Delta\underline{x}^T(t,z)\underline{R}(z)\Delta\underline{y}^*_s(z)|\,dz \leq ||\Delta\underline{x}(t,z)||_{\underline{R}(z)} \cdot ||\Delta\underline{y}^*_s(z)||_{\underline{R}(z)}$$

and hence

$$\dot{V} \leq -k(\underline{Z})\cdot||\Delta\underline{x}(t,z)||^2_{\underline{R}(z)} + ||\Delta\underline{x}(t,z)||_{\underline{R}(z)}\cdot||\Delta\underline{y}^*_s(z)||_{\underline{R}(z)}.$$

There is a close relation between $||\Delta\underline{y}^*_s(z)||_{\underline{R}(z)}$ and J as defined by Eq. (85). Recalling $\underline{r}(z)$ as the right hand side of Eq. (84) we can express $\Delta\underline{y}^*_s(z)$ by

$$\Delta\underline{y}^*_s(z) = \underline{R}^{-1}(z)\underline{P}(z)\underline{r}^*(z),$$

where (*) indicates that $\Delta\underline{v}^*_s$ has been inserted. Therefore,

$$||\Delta\underline{y}^*_s(z)||^2_{\underline{R}(z)} = \int_0^1 \underline{r}^{*T}(z)\underline{P}(z)[\underline{R}^{-1}(z)]^T\underline{R}(z)\underline{R}^{-1}(z)\underline{P}(z)\underline{r}^*(z)\,dz =$$

$$= \int_0^1 \underline{r}^{*T}(z)\underline{P}(z)\underline{R}^{-1}(z)\underline{P}(z)\underline{r}^*(z)\,dz. \tag{91}$$

Comparison of (91) and (85) shows that

$$||\Delta\underline{y}^*_s(z)||_{\underline{R}(z)} = J_{min}$$

if the $\underline{Q}$-matrix is specified by

$$\underline{Q}(z) = \underline{P}(z)\underline{R}^{-1}(z)\underline{P}(z). \tag{92}$$

Moreover, (89) can now be written in the form

$$\dot{V} \leq -k(\underline{Z})\cdot||\Delta\underline{x}(t,z)||_{\underline{R}(z)}\cdot\{\,||\Delta\underline{x}(t,z)||_{\underline{R}(z)} - \frac{J_{min}}{k(\underline{Z})}\,\} = \sup\dot{V}, \tag{93}$$

and hence we have $\dot{V} \leq 0$ for all $\Delta\underline{x}(t,z)$ satisfying

$$||\Delta\underline{x}(t,z)||_{\underline{R}(z)} \geq \rho, \tag{94}$$

where

$$\rho = J_{min}/k(\underline{Z}). \tag{95}$$

Eqs. (93) - (95) are illustrated in Fig. 5. This guarantees that the system's motion is steered, from whatever initial state, to the domain $\|\Delta\underline{x}(t,z)\|_{\underline{R}(z)} \leq \rho$. But it should be noticed that $\Delta\underline{x}(t,z)$ is not assured to stay in this domain, since, in general, $\underline{P}(z) \neq \underline{R}(z)$.

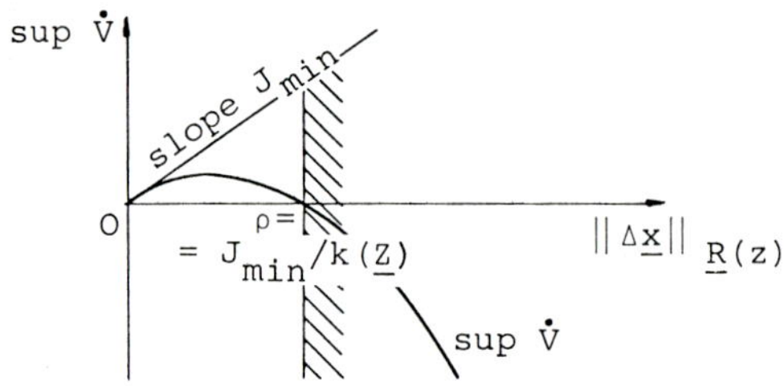

Fig. 5. Illustration of Eqs. (93) - (95)

Therefore, the Raleigh-quotient estimation

$$0 < \lambda_{min}(z) \leq \frac{\Delta\underline{x}^T\underline{P}(z)\Delta\underline{x}}{\Delta\underline{x}^T\underline{R}(z)\Delta\underline{x}} \leq \lambda_{max}(z)$$

will be employed, where $\lambda_{min}(z)$ and $\lambda_{max}(z)$ are the minimum and maximum eigenvalues of the generalized eigenvalue problem $\underline{P}(z)\underline{\Phi}(z) = \lambda(z)\underline{R}(z)\underline{\Phi}(z)$.

Now by denoting

$$\lambda_m := \min_{z\in[0,1]} \lambda_{min}(z), \qquad \lambda_M := \max_{z\in[0,1]} \lambda_{max}(z),$$

we have

$$V_{min} := \frac{\lambda_m}{2} \cdot \|\Delta\underline{x}\|^2_{\underline{R}(z)} \leq V \leq \frac{\lambda_M}{2} \cdot \|\Delta\underline{x}\|^2_{\underline{R}(z)} := V_{max}.$$

This is an estimate for V in terms of $\|\Delta\underline{x}\|_{\underline{R}(z)}$ and therefore can be entered into Fig. 5 to obtain Fig. 6. The system's motion must lie inside the tube bounded by V_{min} and V_{max}. For $\|\Delta\underline{x}\|_{\underline{R}(z)} \geq \rho$, $\dot{V}$ is sure to be negative and therefore V must decrease. Once the motion has entered the domain $\|\Delta\underline{x}\|_{\underline{R}(z)} < \rho$, it may not necessarily stay there since $\dot{V}$ may be positive. But in any case, V cannot exceed the value

$$\sigma = V_{max}(\rho) = \frac{\lambda_M}{2}\rho^2 \tag{96}$$

which can be concluded directly from the geometry of Fig. 6. Therefore, $\Delta\underline{x}(t,z)$ is sure to be steered, from whatever initial state, to the domain

$$\|\Delta\underline{x}(t,z)\|_{\underline{R}(z)} \leq \rho_1, \tag{97}$$

where ρ_1 satisfies $V_{min}(\rho_1) = \rho_1^2\lambda_m/2 = \sigma$, and hence, by means of (96), $\rho_1 = \rho\cdot\sqrt{\lambda_M/\lambda_m}$.

An example motion is sketched in Fig. 6.

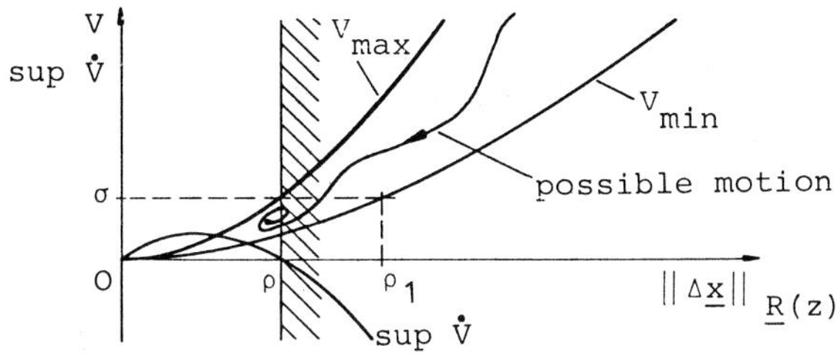

Fig. 6. V_{min} and V_{max} and a possible motion of the system

Example. The system described by Eqs. (53) - (57) is considered, but now with time-independent $a(z) = a_o + a_1 z > 0$, $a_o > 0$, $a_1 \neq 0$, and therefore does not allow steady state coincidence of the model and the real plant. (83) is supposed to be satisfied.

Eq. (84) now takes the form

$$-(a_o + a_1 z + v_s \cdot \frac{d}{dz}) \Delta x_s(z) = (a_o + a_1 z - a_M) \cdot [x_{Ms}(z) - u_s] + \Delta v_s \cdot \frac{d}{dz} x_{Ms}(z). \tag{98}$$

With boundary condition $\Delta x_s(o) = 0$, this equation has the unique solution

$$\Delta x_s(z) = u_s \cdot \left[\exp(-\frac{a_M}{v_{Ms}} z) - \exp(-\frac{a_o + a_1 z/2}{v_{Ms} + \Delta v_s} z) \right].$$

The functional J^2, Eq. (85), now takes the form

$$J^2 = \int_o^1 r^2(z) Q(z) dz,$$

where $r(z)$ is the right hand side of Eq. (98), and $Q(z)>0$ is a scalar weighting function not yet specified. Let Δv_s^*, according to Eqs. (86) - (88), be the optimum value of Δv_s.

By means of functional

$$V = \frac{1}{2} \cdot \int_o^1 \Delta x^2(t,z) dz,$$

and hence $P(z) \equiv 1$, we next find, with regard to (90), the estimate

$$\int_o^1 \Delta x(t,z) \cdot [-a(z) - (v_{Ms} + \Delta v_s^*) \frac{\partial}{\partial z}] \Delta x(t,z) dz \leq a_{min} \cdot \int_o^1 \Delta x^2(t,z) dz.$$

Therefore we have, in terms of Eq. (90), $k = a_{min} > 0$, $R(z) \equiv 1$, and hence due to (92), $Q(z) \equiv 1$.

To evaluate Eqs. (94), (95), we still need J_{min} which can be calculated since $Q(z) \equiv 1$ is specified.

One obtains, after some elementary calculations,

$$J_{min} = |a_1| \cdot \kappa,$$

where κ is a known coefficient depending only on parameters of the model. So we are sure to have

$$\dot{V} \leq 0 \text{ for all } \Delta x(t,z) \text{ satisfying}$$

$$\|\Delta x(t,z)\| = \left(\int_0^1 \Delta x^2(t,z)dz\right)^{1/2} \geq \rho := J_{min}/a_{min} = |a_1|\cdot\kappa/a_{min}.$$

Since $P(z) = Q(z) \equiv 1$ in this example, the estimate via Raleigh-quotient is dispensable ($\lambda_m = \lambda_M = 1$), and so the system's motion is sure to be steered into the ball $\|\Delta x(t,z)\| < \rho_1^M = \rho$.

For the numerical example already introduced we obtain $\rho = 0{,}0518\cdot|a_1|/a_{min}$. For $a(z) = 0{,}5 - 0{,}3z$ we have $\rho = 0{,}0776$.

Figure 7 outlines the result of analog simulation (MWR) for this $a(z)$. Obviously, $x(t,z)$ tends to a steady state which hardly differs from $x_{Ms}(z)$ as can be seen from Fig. 8.

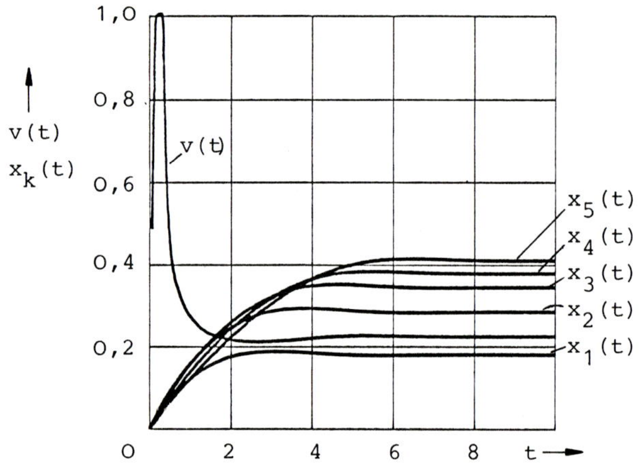

Fig. 7. Control variable $v(t)$ and state variables $x_k(t) = x(t;k/5)$, $k=1, \ldots, 5$, with control law (61) and parameter $a(z) = 0{,}5 - 0{,}3z$

A modification which avoids construction of a Ljapunov-functional. The construction of a Ljapunov-functional is dispensable if asymptotic stability of a certain linear subsystem can be established by whatever method of linear systems theory.

To start with the case of coincidence between model and plant, we assume that asymptotic stability of the autonomous linear system

$$\frac{\partial \Delta \underline{x}(t,z)}{\partial t} = [\underline{A}_M(z) + \sum_{i=1}^{p_1} v_{iMs}\underline{N}_{iM}(z)]\Delta \underline{x}(t,z), \tag{99}$$

with homogeneous boundary conditions can be assured by whatever linear method, hence

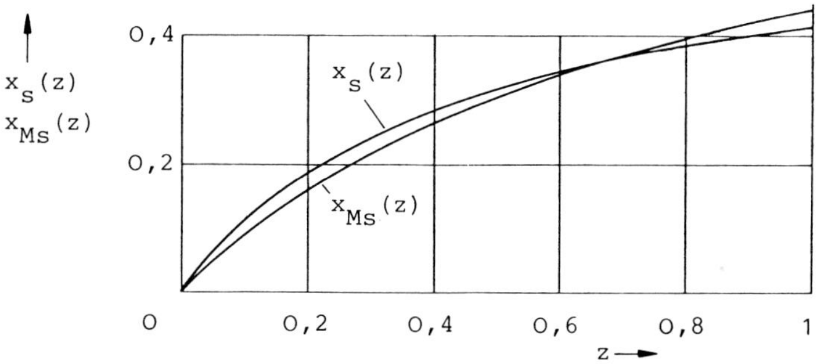

Fig. 8. Steady states of the reference model and of the controlled plant, where $a(z) = 0{,}5 - 0{,}3z$.

$$\lim_{t\to\infty} \Delta\underline{x}(t,z) = \underline{0} \tag{100}$$

in Eq. (99) for all $z\in[0,1]$ and for every $\Delta\underline{x}_o(z)$. Now let $\underline{P}(z)$ be an *arbitrary* positive definite matrix in the functional

$$V = \frac{1}{2} \cdot \int_0^1 \Delta\underline{x}^T(t,z)\underline{P}(z)\Delta\underline{x}(t,z)dz. \tag{101}$$

Of course, V now generally ceases to be a Ljapunov-functional for system (99) and therefore must not necessarily decrease monotonically with respect to time t. But because of (100), V satisfies

$$\lim_{t\to\infty} V(t) = 0.$$

Now consider the inhomogeneous system (48) along with Eq. (49) for $\dot{V}$ with the above arbitrary $\underline{P}(z)$. Then we have

$$\dot{V}_{\Delta\underline{v}\equiv\underline{0}} = \int_0^1 \Delta\underline{x}^T(t,z)\underline{P}(z)[\underline{A}_M(z) + \sum_{i=1}^{p_1} v_{iMs}\underline{N}_{iM}(z)]\Delta\underline{x}(t,z)dz$$

which need *not* be negative definite. Since control law (50), $\Delta\underline{v} = \underline{F}^*\Delta\underline{x}$, assures that the sum containing $\Delta\underline{v}(t)$ in (49) is non-positive, we obtain

$$\dot{V}_{\Delta\underline{v}=\underline{F}^*\Delta\underline{x}} \leq \dot{V}_{\Delta\underline{v}\equiv\underline{0}} \ , \tag{102}$$

and hence

$$V_{\Delta\underline{v}=\underline{F}^*\Delta\underline{x}} \leq V_{\Delta\underline{v}\equiv\underline{0}} \tag{103}$$

for all $t \geq t_o$. Eq. (103) indicates that control law (50) does not endanger global asymptotic stability. Moreover, it will improve the transient behaviour in the sense of Eq. (103).

In the example already carried out, the autonomous system (59) is clearly asymptotically stable directly from its physical background. Therefore an arbitrary weighting factor $P(z)>o$ can be inserted into the integral in control law (60).

We now proceed to plant uncertainties, and $\Delta \underline{v}^*(t)$ is supposed to exist such that $\underline{v}_{Ms}+\Delta \underline{v}^*(t)\in S$ and $\Delta \underline{x}_s(z)\equiv \underline{O}$ is a solution to Eq. (62) for disturbances $\underline{Z}(t,z)\in \mathcal{Z}$. By inserting this $\Delta \underline{v}^*(t)$ into (62) we obtain the fictitious time-varying system

$$\frac{\partial \Delta \underline{x}(t,z)}{\partial t} = [\underline{A}(t,z) + \sum_{i=1}^{p_1} (v_{iMs}+\Delta v_i^*(t))\underline{N}_i(t,z)]\Delta \underline{x}(t,z) \ , \tag{104}$$

with homogeneous boundary conditions.

It is assumed that asymptotic stability of the zero steady state of the linear system (104) can be assured for certain disturbances $\underline{Z}(t,z)\in \mathcal{Z}^*\subset \mathcal{Z}$, by whatever method, hence

$$\lim_{t\to\infty} \Delta \underline{x}(t,z) = \underline{O} \tag{105}$$

for all $z\in[0,1]$ and for every $\Delta \underline{x}_o(z)$. Again

$$\dot{V}_{\Delta \underline{v}(t)=\Delta \underline{v}^*(t)} = \int_o^1 \Delta \underline{x}^T(t,z)\underline{P}(z)[\underline{A}(t,z)+\sum_{i=1}^{p_1} (v_{iMs}+\Delta v_i^*(t))\underline{N}_i(t,z)]\Delta \underline{x}(t,z)dz \tag{106}$$

need not be negative definite with *arbitrary* $\underline{P}(z)>\underline{O}$. Nevertheless, V will tend to zero, due to (105).

It has been shown (see Eq. (77)) that

$$\dot{V}_{\Delta \underline{v}=\underline{F}^*\Delta \underline{x}} \leq \int_o^1 \Delta \underline{x}^T(t,z)\underline{P}(z)[\underline{A}(t,z) + \sum_{i=1}^{p_1} (v_{iMs}+\Delta v_i^*(t))\underline{N}_i(t,z)]\Delta \underline{x}(t,z)dz,$$

which means, by comparison with (106),

$$\dot{V}_{\Delta \underline{v}=\underline{F}^*\Delta \underline{x}} \leqq \dot{V}_{\Delta \underline{v}=\Delta \underline{v}^*(t)}, \text{ and hence } V_{\Delta \underline{v}=\underline{F}^*\Delta \underline{x}} \leqq V_{\Delta \underline{v}=\Delta \underline{v}^*(t)}.$$

This is a direct extension of (103).

In the example already discussed, the linear subsystem

$$\frac{\partial \Delta \underline{x}(t,z)}{\partial t} = - a(t)\Delta x(t,z) - [v_{Ms}+\Delta v^*(t)] \frac{\partial}{\partial z} \Delta x(t,z),$$

with boundary condition $\Delta x(t,o) = 0$, can be transformed to $D\Delta x/Dt= -a(t)\Delta x$ via the method of characteristics, and therefore it is obviously stable when subject to condition (82).

Feedback Control with Independent Additive and Multiplicative Control Variables

Statement of control problem. The additive control vector, $\underline{u}(t)$, has been kept strictly constant so far, $\underline{u}(t)\equiv \underline{u}_s$. But in many situations $\underline{u}(t)$ can act as a control as well. A typical phenomenon in many bilinear control systems is the fact that multiplicative control variables can be varied fast as compared to additive control. Therefore the feedback control structure outlined in Fig. 9 is thought to be ade-

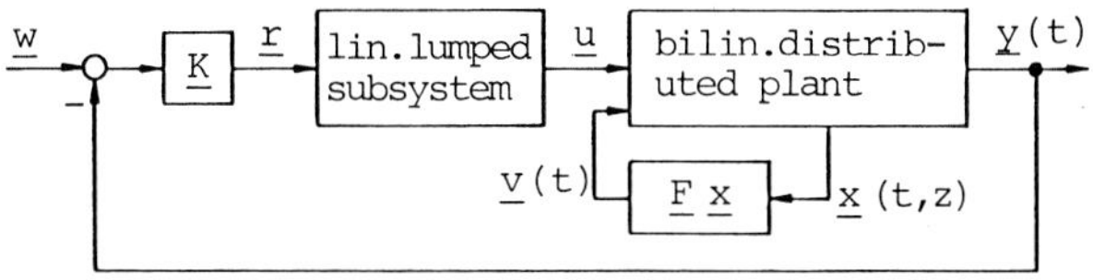

Fig. 9. Structure of the over-all control system

quate for many problems and can serve as a basis for the controller design. The bilinear distributed plant may be described by Eqs. (33) - (36), but with $\underline{u}_s$ to be replaced by $\underline{u}(t)$. The linear lumped subsystem may be given and may include actuators and dynamical elements of controllers, e.g. integrators.

The closed-loop character of the system results from feedback of some interesting output vector, $\underline{y}(t) = \underline{H}\,\underline{x}(t,z)$, via linear control law $\underline{r}(t) = \underline{K}\cdot[\underline{w}-\underline{y}(t)]$, where $\underline{H}$ is a given linear functional, $\underline{K}$ is a constant gain matrix not specified in advance, and $\underline{w}$ is a constant reference input.

In addition to linear feedback control Fig. 9 exhibits a multiplicative mode of control via the nondynamical feedback law $\underline{v}(t) = \underline{F}\,\underline{x}(t,z)$, where $\underline{F}$ is a non-linear functional not specified in advance.

Now in generalization of the problem posed for $\underline{u}(t) \equiv \underline{u}_s$, the objective is to design both matrix $\underline{K}$ and functional $\underline{F}$ and to specify a class of additive and parametric disturbances such that asymptotic stability of the desired steady state can be assured.

Outline of design procedure. A two step design procedure has been proposed by Franke (1980b) which guarantees asymptotic stability in the absence of plant uncertainties. Step 1: Open the loop $\underline{v} = \underline{F}\,\underline{x}$ in Fig. 9 and set $\underline{v}(t) \equiv \underline{v}_{Ms}$. Now apply known methods of linear control theory to specify a gain matrix $\underline{K}$ such as to provide an asymptotically stable linear multivariable control loop. Step 2: Retain $\underline{K}$ from the linear design, define a positive definite functional

$$V = \frac{1}{2} \cdot \int_0^1 \Delta\underline{x}^T(t,z)\underline{P}(z)\underline{x}(t,z)dz$$

and make $\dot{V}$ as negative as possible, via $\underline{v} = \underline{F}\,\underline{x}$, to obtain again feedback law (50)-(52).

The question of robustness in the presence of plant uncertainties can be treated in a manner very similar to the foregoing sections . Theoretical and numerical results are available and will be published elsewhere.

A Solution to the Model Following Problem

In the foregoing sections, asymptotic stability of a *steady state* has been achieved for a class of disturbances by means of multiplicative modes of control. But the same technique also is applicable when the state of the actual plant, $\underline{x}(t,z)$, is required to follow the *motion*, $x_M(t,z)$, of a reference model.

Consider the linear and time-invariant reference model (37) - (40), but with time-dependent input vector $\underline{u}(t)$ instead of $\underline{u}_s$. Hence $\underline{x}_M(t,z)$ can be considered to be a known motion corresponding to $\underline{u}(t)$, and with initial state $\underline{x}_{Mo}(z)$.

Let the actual plant again be described by Eqs. (33) - (36), and replace $\underline{u}_s$ by the same input vector, $\underline{u}(t)$, as in the model.

Then the following problem will be posed: Find a feedback control law $\underline{v}(t)=\underline{F}\,\underline{x}(t,z)$ and a set of admissible additive and operator disturbances, $\hat{Z}'$, such that for all disturbances $\underline{Z}(t,z)\in\hat{Z}'$ and for any $\underline{x}_o(z)$ the error $\Delta\underline{x}(t,z)=\underline{x}(t,z)-\underline{x}_M(t,z)$ satisfies

$$\lim_{t\to\infty} \Delta\underline{x}(t,z) = \underline{0} \tag{107}$$

for all $z\in[0,1]$.

Asymptotic stability of system (41) is again assumed in the sense of (42), (43).

Then in case of plant coincidence with the model (apart from $\underline{x}_o(z)\neq\underline{x}_{Mo}(z)$) we obtain the same equation as (48) in terms of $\Delta\underline{x}(t,z)$, merely with $\underline{x}_{Ms}(z)$ being replaced by $\underline{x}_M(t,z)$. Therefore, control law (51) with $\underline{x}_M(t,z)$ instead of $\underline{x}_{Ms}(z)$,

$$v_i(t) = \frac{M_i+m_i}{2} - \frac{M_i-m_i}{2}\cdot\operatorname{sgn}\int_0^1[\underline{x}(t,z) - \underline{x}_M(t,z)]^T\underline{P}(z)\underline{N}_{iM}(z)\underline{x}(t,z)dz =$$

$$= \underline{F}_i\underline{x}(t,z),\ i=1,\ \ldots,\ p_1, \tag{108}$$

or, in terms of deviations,

$$\Delta v_i(t) = \underline{F}_i\underline{x}(t,z) - v_{iMs} = \underline{F}_i^*\Delta\underline{x}(t,z),\ i=1,\ \ldots,\ p_1,$$

will again make $\dot{V}$ as negative as possible and thus cause $\Delta\underline{x}(t,z)$ to satisfy (107). Feedback law (108) requires the known state $\underline{x}_M(t,z)$ of the reference model in addition to the actual state $\underline{x}(t,z)$ of the plant.

In case of additive and operator disturbances, Eqs. (62) - (65) hold again, but with $\underline{u}(t)$, $\underline{x}_M(t,z)$ instead of $\underline{u}_s$, $\underline{x}_{Ms}(z)$, respectively. The set of disturbances, say $\bar{Z}'$, which allows $\Delta\underline{x}(t,z)\equiv\underline{0}$ to be a solution of these modified equations, can be obtained in a strictly analogous manner as before; all considerations remain valid when $\underline{u}_s, \underline{x}_{Ms}(z)$ are replaced by $\underline{u}(t)$, $\underline{x}_M(t,z)$. As a consequence of this replacement, $\bar{Z}'$ may of course differ from $\bar{Z}$ and hence the set $\hat{Z}'\subset\bar{Z}'$ for which (78) is satisfied may differ from $\hat{Z}$. Figure 10 shows the fundamental structure of this solution to the model following problem.

CONCLUDING COMMENTS

It has been shown that the use of multiplicative modes of control in distributed parameter systems offers a tool for the design of feedback control systems with inherent adaptive and robustness properties. The design has essentially been based on Ljapunov's direct method which avoids solution of the system's state equations. In specific situations the construction of Ljapunov-functionals can even be avoided, whenever stability of a certain linear subsystem can be assured by whatever method.

As a typical example, the control of a continuous furnace has motivated adaptive bilinear control and has been used to illustrate the material presented. In this example, the operator disturbances discussed were only parameter disturbances. But it should be pointed out that more general operator disturbances can be treated in the same manner. For example, the model state equation may be Eq. (58),

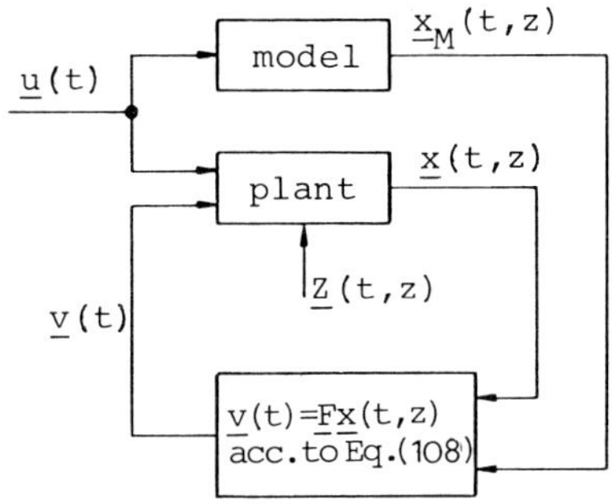

Fig. 10. Model-following control system with multiplicative modes of control, $\underline{v}(t)$.

$$\frac{\partial x_M(t,z)}{\partial t} = - a_M x_M(t,z) - v_{Ms} \cdot \frac{\partial x_M(t,z)}{\partial z} + a_M u_s,$$

and a more adequate process state equation may be

$$\frac{\partial x(t,z)}{\partial t} = - a(t,z)x(t,z) + b\cdot\frac{\partial^2 x(t,z)}{\partial z^2} - v(t)\cdot\frac{\partial x(t,z)}{\partial z} + a(t,z)u_s,$$

instead of Eq. (54). Here the operator $b\cdot\partial^2/\partial z^2$ (due to axial heat conduction) can be interpreted as an operator disturbance with regard to the model.

One should be aware of two restrictive assumptions made throughout the chapter. (1) The multiplicative modes can be varied arbitrarily fast, without any time delay, and (2), the entire state of the system is available.

As a consequence of assumption (1) sgn-functions appear in the control law, thus giving rise to sliding modes which are not always desirable. Of course, sliding modes can be avoided by replacing the sgn-functions by continuous functions, but this usually results in a reduction of robustness.

Assumption (2) is a consequence of the requirement that the *entire* state of the process should coincide with the state of the model. This requirement is needlessly restrictive in many situations.

Therefore, further investigations should be focussed on the following topics: Design of robust variable structure control systems which require only output coincidence of the model and the plant and need only output information to construct the control law. The control law should exclude sliding modes and should admit actuator dynamics for the multiplicative controls. Problems of this type are under investigation at present.

REFERENCES

Balakrishnan, A. V. (1974). Stochastic bilinear partial differential equations. In R.R. Mohler and A. Ruberti (Eds.), *Variable Structure Systems with Application to Economics and Biology*, Springer-Verlag, New York, pp. 1 - 43.

Becker, C. (1979). *Synthese strukturvariabler Regelungssysteme mit Hilfe von Ljapunow-Funktionen*, HochschulVerlag, Stuttgart.

Briggs, D.L. and C.N. Shen (1969). Switching analysis for constrained bilinear distributed parameter systems with applications. *J. Basic Eng., Trans. ASME,* 91, 277 - 283.

Brogan, W.L. (1968). Optimal control theory applied to systems described by partial differential equations. In C.T.Leondes (Ed.), *Advances in Control Systems,* Vol. 6, Academic Press, New York, pp. 221 - 321.

Bruni, C. and G. Koch (1978). A degenerate (bounded rate) class of distributed parameter systems. In A. Ruberti (Ed.), *Distributed Parameter Systems: Modelling and Identification,* Springer-Verlag, New York.

Butkovskiy, A.G. (1969). *Distributed Control Systems.* American Elesevier, New York.

Draženović, B. (1969). The invariance conditions in variable structure systems. *Automatica,* 5, 287 - 295.

Franke, D. (1979a). Ljapunow-Synthese linearer konzentrierter Regler für bilineare örtlich verteilte Strecken. *Regelungstechnik,* 27, 213 - 220.

Franke, D. (1979b). Optimal bilinear boundary control of a distributed parameter cooling process. *Preprints of the 2nd IFAC-Symposium on Optimization Methods,* Varna, paper OC.9, pp. 63 - 70.

Franke, D. (1980a). Control of bilinear distributed parameter systems. In I. Hartmann (Ed.), *Advances in Control Systems and Signal Processing,* Vol. 1, Vieweg, Braunschweig/Wiesbaden, pp. 1 - 113.

Franke, D. (1980b). Control of distributed parameter systems with independent linear and bilinear modes. *Preprints of the Third IMA-Conference on Control Theory,* Univ. of Sheffield.

Furusho, J., H. Kanoh and M. Masubuchi (1978). Observers for bilinear distributed parameter processes. In S.P. Banks and A.J. Pritchard (Eds.), *Control of Distributed Parameter Systems,* Pergamon Press, Oxford, pp. 219 - 229.

Köhne, M. (1977). *Zustandsbeobachter für Systeme mit verteilten Parametern - Theorie und Anwendung.* Fortschrittberichte der VDI-Zeitschriften, Reihe 8, Nr. 26, VDI-Verlag, Düsseldorf.

Mohler, R.R. (1973). *Bilinear Control Processes.* Academic Press, New York.

Noldus, E. and I. Derese (1980). Control of bilinear distributed parameter systems. *Preprints of the Third IMA-Conference on Control Theory,* Univ. of Sheffield.

Parks, P.C. and A.J. Pritchard (1969). On the construction and use of Ljapunov-functionals. *Proc. of the 4th IFAC-Congress,* Warszawa, paper No. 20.5.

Stafford, E.M. and G. Dowrick (1978). Applications of extended operators to diffusive systems. In S.P. Banks and A.J. Pritchard (Eds.), *Control of Distributed Parameter Systems,* Pergamon Press, Oxford, pp. 527 - 537.

Tzafestas, S.G. (1978). Design of distributed-parameter optimal controllers and filters via Walsh-Galerkin expansions. In S.P. Banks and A. J. Pritchard (Eds.), *Control of Distributed Parameter Systems,* Pergamon Press, Oxford, pp. 201 - 217.

Utkin, V.I. (1977). Variable structure systems with sliding modes. *IEEE Trans. Autom. Control,* 22, 212 - 222.

Wang, P.K.C. (1966). Stability analysis of elastic and aeroelastic systems via Ljapunov's direct method. *J. Franklin Inst.,* 281, 51 - 72.

Young, K.-K.D. (1977). Asymptotic stability of model reference systems with variable structure control. *IEEE Trans. Autom. Control,* 22, 279 - 281.

Zeitz, M. (1977). *Nichtlineare Beobachter für chemische Reaktoren.* Fortschrittberichte der VDI-Zeitschriften, Reihe 8, Nr. 27, VDI-Verlag, Düsseldorf.

Zubov, V.I. (1957). Methods of A.M. Ljapunov and their applications. *Publ. House of Leningrad Univ.*

CHAPTER 7

COORDINATED CONTROL OF DISTRIBUTED-PARAMETER SYSTEMS

K. Y. Lee

Department of Electrical Engineering, University of Houston, Houston, Texas, USA

1. INTRODUCTION

Many physical, environmental, and ecological systems are known to be described by partial differential equations due to the necessity of additional domain, spatial domain, on which variables are needed to be defined and distributed at a given instant in time (Lee, 1978a, 1978b). Thus for past decade a good measure of effort has been devoted in applying the concept of optimal control theory to such distributed parameter systems (Balakrishnan, 1965; Butkovskii, 1961; Lions, 1971; Wang, 1964). The effort, however, has been limited to providing only the theoretical guidances because of the complexities involved in computation of partial differential equations.

Toward more practical means of controlling distributed parameter systems, the ideas of sampled-data control (Lee and Barr, 1972), discrete-time control (Lee, Chow, and Barr, 1972), pointwise control (Greenberg, 1971), and combination of these (Lee, 1974) have been explored. These works have proven that distributed parameter systems can also be controlled like lumped parameter systems using realizable controllers. Nevertheless, the computational requirements again have restricted the use of these controllers.

Since these theoretical approaches provide only limited guidance to the control of distributed parameter systems, discrete approximation methods have been the subject to recent investigations (Wismer, 1971, 1973). The high dimensionality of discretized systems together with the inherent computational difficulties of optimization has led to the use of hierarchical optimization theory. The accuracy as well as the stability of the solution of the approximation problem is in general depend upon the grid size of discretization and thus the dimensionality of the approximate system. It should be noted that this hierarchical optimization approach does not reduce the dimensionality in general, rather it partitions the higher order system into several low order systems.

To reduce the dimensionality, the finite element discretization method has been applied on spatial domain (Cavin and Tandon, 1973), and the concept of reduced-order modeling technique has also been applied for adaptive control purpose (Balas and Johnson, 1979). It has been observed in large controlled flexible structures that the closed-loop controller designed for the reduced-order model cause instabilities to the original system.

Recently, the idea of coordinated control was developed to avoid the problem of high dimensionality and the instability caused by model reduction (Lee and Cooper, 1979). This approach does not make a discrete approximation of the original distributed parameter system. It partitions the system into subsystems by only making use of physical constraints.

This chapter addresses a unified approach for coordinated control of distributed parameter systems. Three different decomposition schemes, finite difference, finite element, and integration schemes, are presented in generality. In all cases, the distributed parameter system is decomposed into a number of subsystems which are coupled through interaction variables. The large-scale systems problem was then studied for the optimization of overall system in some coordinated way. It is the intent of this chapter to stress that the distributed parameter systems are infinite-dimensional and hence large-scale systems, and descentralized and coordinated control methodologies are very attractive and practical.

2. DECOMPOSITION

There can be two types of decomposition in solving the control problem for distributed parameter systems. The obvious decomposition scheme is the spatial decomposition which partitionsthe spatial domain into a number of smaller domains, and obtain corresponding mathematical descriptions. Another decomposition scheme, which is not quite common, is the functional decomposition where a vector partial differential equation is decomposed into a number of partial differential equations. In this case each subsystem is still infinite-dimensional and may further require the spatial decomposition to solve numerically. For this reason only the spatial decomposition will be considered in this section.

Three different schemes will be studied in decomposing the spatial domain. The most common approach, finite difference scheme, will be reviewed first, and followed by the finite element method. Finally a special decomposition is considered for some limited class of systems.

2.1 Basic Problem Formulation

The optimal control problem will be formulated for a very general class of systems governed by partial differential equations in this section. And the decomposition will later be applied to this problem with as much generality as the illustration permits.

The distributed parameter system is described by the following general equation

$$\frac{\partial x}{\partial t} = \mathcal{F}\,(x(z,t),u(z,t),z,t)\ , \quad z\varepsilon\Omega\ , \quad t\varepsilon T\ , \tag{1}$$

with an initial condition

$$x(z,t_0) = x_0(z)\ , \quad z\varepsilon\Omega\ , \tag{2}$$

and a very general boundary condition

$$\mathcal{G}\,(x(z,t),v(z,t),n(z,t),z,t) = 0\ , \quad z\varepsilon\Gamma\ , \quad t\varepsilon T\ , \tag{3}$$

The distributed parameter system presented above by Eqs. (1) to (3) is a vector system for the state n_x-vector function x(z,t), the distributed control n_u-vector function u(z,t), and the boundary control n_v-vector function v(z,t). The relevant dynamic behavior of the system is assumed to occur within a finite time interval $T = [t_0, t_f]$, where t_f is called the final time. Spatially, the system is confined to a closed, bounded, and simply-connected region Ω of R^m. The boundary of Ω is Γ with unit outward normal vector $n(z,t) = (n_1,\ldots,n_m)$. The operator F is a vector partial differential operator on x and is a function of x, u, z, or t. The operator G is a vector partial differential operator on x and is a function of x, u, n, z, or t.

The general performance index to be minimized in the optimal control problem is defined by

$$J = \int_{\bar{\Omega}} g_1(x(z,t_f),z,t_f)dz + \int_{t_0}^{t_f} \int_{\bar{\Omega}} g_2(x(z,t),u(z,t),v(z,t),z,t)dz\ dt\ , \tag{4}$$

where $\bar{\Omega}$ is the closure of Ω, $\bar{\Omega} = \Omega\times\Gamma$.
The integrand g_1 and g_2 are real-valued functions of class C^2 on t, and piecewise C^1 on Ω; and g_1 is C^1 on x and g_2 is C^1 on x, u, and v.

The distributed and boundary controls sometimes may be constrainted by the conditions

$$f_1(x(z,t),u(z,t),z,t)\geq 0\ ,\quad z\varepsilon\Omega\ ,\quad t\varepsilon T \tag{5}$$

$$f_2(x(z,t),v(z,t),z,t)\geq 0\ ,\quad z\varepsilon\Gamma\ ,\quad t\varepsilon T \tag{6}$$

and the state may also be constrained at the final time as

$$f_3(x(z,t_f),z,t_f) = 0,\quad z\varepsilon\Omega, \tag{7}$$

where f_1 and f_2 are respectively γ_1- and γ_2- vector functions with the same properties as g_2, and f_3 is γ_3-vector valued function with the same properties as g_1. It is assumed that the functions f_1, f_2, and f_3 are consistent with the boundary condition in Eq. (3).

The above formulation is very general and include the boundary control problem. The system described by Eq. (1) is also general to include most partial differential equations. This is because all but the highest order time derivatives can be put into state variables to form a system of first order time derivative equations. The one exception, of course, is the system of elliptic equations which may be viewed as the steady-state behavior of the system described by Eq. (1). However, this can be also a special case of the general formulation by setting the time derivative in Eq. (1) to zero and eliminate all time variables and time integrals in the above equations.

2.2 Decomposition by Finite Difference Scheme

The finite difference scheme can be applied to both spatial and time derivatives to form a large number of coupled ordinary difference equations. In this section, however, the semidiscrete approximation is employed to replace only the spatial derivatives by differences and to form a set of ordinary differential equations in time domain, resulting in a lumped parameter system. This approximation is especially useful in applying the large-scale systems theory available for lumped parameter systems.

<u>Discrete approximation by finite difference</u>. The partial domain Ω and its boundary Γ can be approximated by a finite number of mesh points of a grid Ω_h that covers the region Ω. The grid Ω_h is a collection of mesh points defined by

$$Z_i = (i_1h_1, i_2h_2, \ldots, i_mh_m) \ , \tag{8}$$

where $i = (i_1, i_2, \ldots, i_m)$ is an m-tuple of integers defined by $i_j = 0,1,\ldots,N_j$, with

$$N_j = [(z_j)_{max} - (z_j)_{min}]/h_j \ . \tag{9}$$

Here $h = (h_1, h_2, \ldots, h_m)$ is the vector of discretization intervals for the spatial coordinates.

Note that the grid Ω_h covers Ω and there may be some mesh points outside of Ω. Hence the mesh points are grouped into: interior point for a point in $\Omega \cap \Omega_h$; exterior point for a point in $C(\bar{\Omega}) \cap \Omega_h$ where $C(\cdot)$ denotes the complement; boundary point for a point in $\Gamma \cap \Omega_h$. A regular point is defined as an interior point such that all adjacent points are also interior points, and an irregular point is a point that is not regular. The irregular points can be handled by adjusting the boundary and mesh points (Young, 1962), and it is assumed here that all mesh points are regular points for simplicity.

Having defined the grid Ω_h, next step is to approximate the partial differential operator F on the grid Ω_h. This approximation will depend upon the choice of a particular finite difference scheme, and the choice will be made based upon the accuracy, numerical stability, and computational complexity. The finite difference scheme approximate a partial derivative of the state $x(z,t)$ at a mesh point Z_i in terms of differences in values of the state at neighboring mesh points. The number of mesh points required will depend upon i) the order of the partial derivative, ii) the order of the system (n_x), iii) the finite difference scheme, and iv) the number of spatial variables (m).

The approximation will affect the decomposition to be made. For our purpose, the operator F will be assumed to have at most the second-order derivatives and the simplest finite difference scheme will be used. For example if $x(z,t)$ is a single function defined on a one-dimensional z, then

$$\frac{\partial}{\partial z} x(z_i,t) = \frac{1}{h} [x(z_{i+1},t) - x(z_i,t)] \tag{10}$$

$$\frac{\partial^2}{\partial z^2} x(z_i,t) = \frac{1}{h^2} [x(z_{i+1},t) - 2x(z_i,t) + x(z_{i-1},t)] \tag{11}$$

The Eq. (10) is the forward difference with the truncation error in the order of the discretization interval h. On the other hand, the Eq. (11) represents the central difference with the truncation error in the order of h^2 (Potter, 1978).

The general expression for the approximation of the differential operator F at a mesh point Z_i can therefore be

$$F(x(Z_i,t), u(Z_i,t), Z_i, t) = F_i(x_i(t), x_{i \pm I_k}(t), u_i(t), t), \quad k = 1,2,\ldots,m \tag{12}$$

where $I_k = (0,\ldots,1,\ldots,0)$ is an m-tuple with all zeros but the kth element which is 1. The new state and control vector are defined only on mesh points as $x_i(t) = x(Z_i,t)$ and $u_i(t) = u(Z_i,t)$. The functions F_i are n_x-vector valued function and assumed to be of class C^2. Since i represents a mesh point in Ω_h, the system in Eq. (1) is now decomposed into a number of coupled ordinary vector differential equations

$$\dot{x}_i(t) = F_i(x_i(t), x_{i\pm I_k}(t), u_i(t), t), \quad k = 1,2,\ldots,m . \tag{13}$$

The number of equations is approximately equal to the number of mesh points.

For coordination control purpose, the performance index in Eq. (4) should also be decomposed into that for each subsystem. This can be achieved by performing the spatial integration on $\bar{\Omega}$ with all spatial functions approximated in terms of corresponding values on the mesh points Z_i. Among several integration schemes, the simplest one, Euler integration, decouples the Eq. (4) into the following equivalent performance index

$$J = \sum_{i\varepsilon\Omega_h} [g_{1i}(x_i(t_f)) + \int_{t_0}^{t_f} g_{2i}(x_i(t), u_i(t), v_i(t), t)dt] . \tag{14}$$

The performance index can also be partitioned into a number of indices, each corresponding to a subsystem defined.

The boundary condition can also be approximated in the similar way as for the system equation. When the system operator F contains at most the second-order partial derivatives, the boundary operator G in Eq. (3) can have up to the first-order partial derivatives. Therefore, using the finite difference scheme, Eq. (3) can be approximated as

$$G_i(x_i(t), x_{i+I_k}(t), v_i(t), t) = 0 , \qquad Z_i\varepsilon\Gamma . \tag{15}$$

Other remaining constraints in Eqs. (2), (5)-(7) are simply the evaluations at the mesh points as

$$x_i(t_0) = x_{0i} , \qquad Z_i\varepsilon\Omega , \tag{16}$$

$$f_{1i}(x_i(t), u_i(t), t) \geq 0 , \qquad Z_i\varepsilon\Omega , \tag{17}$$

$$f_{2i}(x_i(t), v_i(t), t) \geq 0 , \qquad Z_i\varepsilon\Gamma , \tag{18}$$

$$f_{3i}(x_i(t_f), t_f) = 0 , \qquad Z_i\varepsilon\Omega \tag{19}$$

Further simplification can be realized through the use of the boundary condition Eq. (15), which is assumed to be expressible as

$$x_{i+I_k}(t) = \hat{G}_i(x_i(t), v_i(t), t) , \qquad Z_i\varepsilon\Gamma . \tag{20}$$

This can be substituted into the system equation (13) for the mesh points on the boundary Γ. This will result in a modified system equation (13) with the distributed control $u_i(t)$ combined with the boundary control $v_i(t)$. Thus any boundary control problem can be transformed into the ordinary control problem. In this case the constraint (18) can also be combined to the constraint (17).

Decomposition into subsystems. The equation (13) shows some properties of the decomposed system. It describes the dynamics on one mesh point alone with coupling from neighboring points. Therefore the coordination approach can be directly applied for this system. Since the number of mesh points is very large, however, the coordination will be very difficult in practice. Thus a decision has to be made on the size of each subsystem and the number of allowable subsystems.

There can be two different approaches to group the equation (13) into a number of subsystems. When the state vector x(z,t) is high dimensional vector function, each component can be made as one subsystem and each subsystem is defined on the entire grid Ω (Wismer, 1971). Another approach is partitioning the grid Ω_h into a number of subgrids Ω_h^j which make the disjoint union of Ω_h. Usually physical constraints help in determining the boundaries for the partition. The above two approaches should be used together to find a good decomposition.

Let us assume that the system equation (13) is now decomposed into N subsystems such that each subsystem has control vector independent from other subsystems. Then the basic problem in Eqs. (13)-(20) can be described by the N coupled subsystems

$$\dot{x}_i = F_i(x_i,u_i,w_i,t) \; , \qquad i = 1,2,\ldots,N \tag{21}$$

with initial and final conditions

$$x_i(t_0) = x_{i0} \; , \tag{22}$$

$$f_i(x(t_f),t_f) = 0 \; , \tag{23}$$

the inequality constraints

$$G_i(x_i,u_i,t) \geq 0 \; , \tag{24}$$

and the performance index

$$J_i(u_i) = g_{1i}(x_i(t_f)) + \int_{t_0}^{t_f} g_{2i}(x_i,u_i,t)dt \; . \tag{25}$$

The performance index for overall system is assumed to be separable, and thus

$$J = \sum_{i=1}^{N} J_i(u_i) \; . \tag{26}$$

In the above equations (21)-(26), the same notations as before are used for convenience. The new vector w_i in Eq. (21) is called an interaction vector which represents the interactions with other subsystems, and expressed as

$$w_i = H_i(x_1,x_2,\ldots,x_k,\ldots,x_N) \; , \qquad k \neq i \; . \tag{27}$$

This concludes the decomposition of distributed parameter systems using finite-difference scheme into a number of manageable subsystems with interconnections.

2.3 Decomposition by Finite Element Scheme

The finite element method is a numerical procedure for solving the differential equations of physics and engineering. The method had its birth in the aerospace industry in the early 1950s and was first presented in the publication by Turner and others (1956). An important theoretical contribution was made by Melosh (1963) who showed that the finite element method was really a variation of the well known Raleigh-Ritz procedure.

The range of applications for the finite element method was enlarged when other investigators (Szabo and Lee, 1969; Zienkiewicz, 1971) showed that the element equations related to structural mechanics, heat transfer, and fluid mechanics could also be derived by using a weighted residual procedure such as Galerkin's method or the least-squares approach. This knowledge is a very important contribution to the theory because it allows the finite element method to be applied to any differential equations.

The fundamental concept of the finite element method is that any continuous quantity, such as temperature, pressure, or displacement, can be approximated by a discrete model composed of a set of piecewise continuous functions defined over a finite number of subdomains. The piecewise continuous functions are defined using the values of the continuous quantity at a finite number of points in its domain.

The more common problem is where the continuous quantity is unknown and is to determine the values of this quantity at certain points within the region. The construction of the discrete model is most easily explained, however, if it is assumed that the numerical value of the quantity is already known at every point within the domain. The following procedure outlines the construction of discrete model:

1. A finite number of points in the spatial domain is determined. The points are called nodal points or nodes.

2. The value of the continuous quantity at each node is denoted as a variable which is to be determined.

3. The domain is divided into a finite number of subdomains called elements.

4. The continuous quantity is approximated over each element by a polynomial that is defined using the nodal values of the continuous quantity. A different polynomial is defined for each element, but the element polynomials are selected in such a way that continuity is maintained along the element boundaries.

Discrete approximation by finite element. The spatial domain Ω with its boundary Γ are first divided into a number of elements, and the elements and nodes are numbered. The elements are usually either m-simplex or m-rectangle of different types (Ciarlet, 1978). The simplex of type (1) is the simplex with no other nodes except vertices, the type (2) is for the case when there is a node at each mid-points of the edges of the simplex. For two-dimensional space (m=2), the 2-simplex of type (1) is a triangle with 3 nodes at vertices, and the 2-simplex of type (2) is a triangle with 3 nodes at vertices and 3 nodes at the mid-points of the edges. The m-simplex of type (1) has (m+1) nodes, and the type (2) has (m+1)(m+2)/2 nodes. For simplicity, it is assumed that all elements are the m-simplex of type (1).

The labeling of the nodes should be done so that it increases the computational efficiency. The matrix equation which arises when using the finite element method has a large number of coefficients which are zero. The nonzero elements fall between the diagonal line and the line parallel to it and separated by the bandwidth $B = (R+1)n_x$, where R is the largest difference between the node numbers in each element and n_x is the number of state variables. A reduction in the bandwidth produces a reduction in the required computer memory space and a reduction in the computational time. The minimization of B depends on the minimizing of R, which can be partially achieved by labeling the nodes across the shortest dimension of the spatial domain.

Let $Z_i = (z_{1i}, z_{2i}, \ldots, z_{mi})$ be the ith node. At each node the values of state and control variables are defined as $x_i(t) = x(Z_i,t)$, $u_i(t) = u(Z_i,t)$, and $v_i(t) = v(Z_i,t)$. Then the values of all variables on each element are expressed by interpolating polynomials in terms of the values defined on nodes. When the element is an m-simplex of type (1), the interpolating polynomial is linear in spatial variables and has (m+1) coefficients. Since the number of coefficients is equal to the number of nodes on which the values of variables are defined, these coefficients can always be computed and expressed as linear polynomials in the defined variables on nodes. This results in a general representation

$$x^k(z,t) = \sum_{i\varepsilon I_k} N_i^k(z) x_i(t) = N^k(z) X^k(t) , \tag{28}$$

where k denotes for the kth element, and I_k is the set of indices for nodes belong to the kth element E_k, i.e.,

$$I_i = \{i : Z_i \varepsilon E_k\} . \tag{29}$$

The function $N_i^k(z)$ is called the shape function or interpolation function. It is a linear polynomial of spatial variables z_i, $i = 1,2,\ldots,m$. The matrix $N^k(z)$ is $n_k \times (m+1) n_x$ matrix with $n_x \times n_x$ block matrices containing $N_i^k(z)$ in the diagonals. The vector $X^k(t)$ is made of the n_x-vectors $x_i(t)$ for all $i \varepsilon I_k$. The control vector also has the same representation as the state with the same shape function.

The next major step is formulating the element equation which approximates the system equation (1). The element equations can be derived in two different ways. One approach is the variational approach which minimizes either potential energy or a functional formulation of the differential equation (Segerlind, 1976). Another approach is the use of a weighted residual procedure such as Galerkin's method or the least-squares approach (Szabo and Lee, 1969; Zienkiewicz and Parekh, 1970). The finite element equations will be formulated using the Galerkin's method here.

Galerkin's method is a means of obtaining an approximate solution to a differential equation. It does this by requiring that the error between the approximate solution and the true solution be orthogonal to the functions used in the approximation. Mathematically this is equivalent to

$$\int_{E_k} [N^k(z)^T [\, \mathcal{F}(x^k(z,t), u^k(z,t), z, t) - \frac{\partial}{\partial t} x^k(z,t)] dz = 0 \tag{30}$$

Substituting Eq. (28) into the above equation yields

$$M^k \frac{dX^k}{dt} = \int_{E_k} [N^k(z)]^T \, \mathcal{F}(N^k(z), X^k(t), U^k(t), z, t) dz , \tag{31}$$

where

$$M^k = \int_{E_k} [N^k(z)]^T N^k(z)dz \ . \tag{32}$$

Since F has partial derivatives, the volume integrals for these in Eq. (31) can be modified to corresponding volume and surface integrals by using the Green's theorem or the integration by parts.

Before presenting a general element equation, an example is given to illustrate the above procedure.

Example 1. Consider a heat transfer equation

$$\frac{\partial x}{\partial t} = \frac{\partial^2 x}{\partial z_1^2} + \frac{\partial^2 x}{\partial z_2^2} + u \ , \quad z\varepsilon\Omega \tag{33}$$

with boundary conditions

$$\frac{\partial x}{\partial n} + \alpha x = v \ , \quad z\varepsilon\Gamma \tag{34}$$

For each triangular element with (i,j,k) nodes, Eq. (28) gives the solution approximation

$$x = N_i(z)x_i(t)+N_j(z)x_j(g)+N_k(z)x_k(t) = N(z)X(t) \ . \tag{35}$$

Applying the Galerkin's method, Eq. (30) yields

$$\int_E N^T(\frac{\partial^2 x}{\partial z_1^2} + \frac{\partial^2 x}{\partial z_2^2} + u - \frac{\partial x}{\partial t})dz = 0 \tag{36}$$

The first two integrals can be expanded by using the integration by parts and the Green's theorem:

$$\int_E N^T \frac{\partial^2 x}{\partial z_1^2} dz = \int_{\partial E} N^T \frac{\partial x}{\partial z_1} n_1 dz - \int_E \frac{\partial N^T}{\partial z_1}\frac{\partial x}{\partial z_1} dz$$

$$\int_E N^T \frac{\partial^2 x}{\partial z_2^2} dz = \int_{\partial E} N^T \frac{\partial x}{\partial z_2} n_2 dz - \int_E \frac{\partial N^T}{\partial z_2}\frac{\partial x}{\partial z_2} dz$$

where ∂E is the boundary of E, and n_1, n_2 are directional cosine. Combining these with Eq. (36) yields

$$\int_{\partial E} N^T(\frac{\partial x}{\partial z_1} n_1 + \frac{\partial x}{\partial z_2} n_2)dz - \int_E (\frac{\partial N^T}{\partial z_1}\frac{\partial x}{\partial z_1} + \frac{\partial N^T}{\partial z_2}\frac{\partial x}{\partial z_2} - N^T u + N^T \frac{\partial x}{\partial t})dz = 0. \tag{37}$$

The surface integral can be rewritten in terms of the flux $\partial x/\partial n$ along the boundary where n is the outward normal to the surface. This results in

$$\int_E (\frac{\partial N^T}{\partial z_1}\frac{\partial x}{\partial z_1} + \frac{\partial N^T}{\partial z_2}\frac{\partial x}{\partial z_2})dz - \int_E N^T u dz + \int_E N^T \frac{\partial x}{\partial t} dz - \int_{\partial E} N^T \frac{\partial x}{\partial n} dz = 0 \ . \tag{38}$$

Substituting Eqs. (34) and (35) into this yields

$$M\dot{X}(t) = AX(t) + F \ , \tag{39}$$

where

$$M = \int_E N^T N \, dz \tag{40a}$$

$$A = -\int_E \left(\frac{\partial N^T}{\partial z_1}\frac{\partial N}{\partial z_1} + \frac{\partial N^T}{\partial z_2}\frac{\partial N}{\partial z_2}\right) dz - \int_{\partial E^0} N^T \alpha N \, dz \tag{40b}$$

$$F = \int_E N^T u \, dz + \int_{\partial E^0} N^T v \, dz \ . \tag{40c}$$

Here ∂E^0 refers only to elements with an external boundary on which condition (34) is specified.

When u and v in Eq. (40c) are distributed and boundary controls to be designed, then they also can be approximated as Eq. (35) using the same shape function. In this case, Eq. (39) becomes

$$M\dot{X}(t) = AX(t) + MU(t) + M^0 V(t) \ , \tag{41}$$

where

$$M^0 = \int_{\partial E^0} N^T N \, dz \ , \tag{42}$$

and it is zero for interior elements.

The matrix M in Eq. (41) is symmetric positive definite and invertible. Thus the element equation is

$$\dot{X}(t) = M^{-1}AX(t) + U(t) + M^{-1}M^0 V(t) \ . \tag{43}$$

Recall that this equation is for a single element with $X(t) = [x_i(t), x_j(t), x_k(t)]$. A neighboring element can have $X(t) = [x_i(t), x_j(t), x_\gamma(t)]$, which means that the two element equations are coupled. The equation (43) can be rearranged for all elements and form a large vector differential equation.

In general, Eq. (31) yields a general element equation

$$\dot{x}_i(t) = F_i(x_i(t), x_j(t), u_i(t), v_i(t), t) \ , \quad j \epsilon J_i \ , \tag{44}$$

where J_i is the set of indices for nodes directly connected to node i.

The performance index in Eq. (4) can also be decomposed into that for each element and then expressed as Eq. (14).

<u>Example 2</u>. For the system in Example 1, the performance index is given by

$$J = \int_{\Omega} h(z,t_f)x(z,t_f)^2 dz + \int_{t_0}^{t_f} \int_{\Omega} [q(z,t)x(z,t)^2 + r(z,t)u(z,t)^2] dz dt \ . \tag{45}$$

Using the representation in Eq. (35),

$$J = \sum_k [X_k^T(t_f)H_k X_k(t_f) + \int_{t_0}^{t_f} (X_k^T(t)Q_k X_k(t) + U_k^T(t)R_k U_k(t)dt] \ , \tag{46}$$

where k represents kth element, and

$$H_k = \int_{E_k} N^T hNdz, \ Q_k = \int_{E_k} N^T qNdz, \ R_k = \int_{E_k} N^T rNdz \ .$$

The vectors X_k and U_k can be rearranged so that the performance index is in the form

$$J = x^T(t_f)Hx(t_f) + \int_{t_0}^{t_f} [x^T(t)Qx(t) + u^T(t)Ru(t)]dt \ , \tag{47}$$

where x(t) and u(t) are vectors of nodal values.

To complete the discrete approximation, the remaining constraints, Eqs. (5)-(7), also need to be integrated over each element. This will result in constraints similar to Eqs. (17)-(19).

Decomposition into subsystems. The element equations (44) need to be grouped into a number of subsystems. This procedure is similar to the case of finite difference scheme discussed in previous section. Since the finite element method better represents the physical constraints, the decomposition might yield weaker couplings between subsystems. The mathematical representation is identical to the case of finite difference scheme, and is given by Eqs. (21)-(27).

2.4 Decomposition by Integration Scheme

A class of first-order distributed parameter systems has been the subject of many recent investigations (Johnson and Athans, 1974; Russell, 1973). The fundamental equations of electromagnetism, fluid mechanics, thermodynamics, chemistry, and relativistic kinematics may all be written directly as systems of first-order partial differential equations. Applications can also be found in environmental and ecological modeling (Lee and others, 1976). In addition, higher order systems of partial differential equations may be formally cast in first-order form by taking all but the highest order space and time derivatives as state variables, in analogy to the procedure often employed for ordinary differential equations.

For simplicity in the presentation, the first-order distributed system is assumed to be linear and controlled by (γ-2) point sources located at one dimensional spatial domain, $z_i \epsilon \Omega = [0,L]$, $0 = z_1 < z_2 < \ldots < z_{\gamma-1} < z_\gamma = L$. Then the system equation is represented by

$$\frac{\partial x}{\partial t} = \Lambda(z,t)\frac{\partial x}{\partial z} + A(z,t)x + \sum_{i=2}^{\gamma-1} B_i(t)\delta(z-z_i)u_i(t) \ , \tag{48}$$

where $\delta(\cdot)$ is the Dirac delta function, $x(z,t)$ is the n_x-vector state function, $u(t)$ is n_u-vector control function, $n_u \leq \gamma-2$. The matrix $\Lambda(z,t)$ is a diagonal matrix with negative values, and $\Lambda(z,t)$ and $A(z,t)$ are piecewise constant in z, i.e., they are constant in each interval (z_i,z_{i+1}), $i = 1,2,\ldots,\gamma-1$.

Initial and boundary conditions are given by

$$x(z,0) = x_0(z) \ , \qquad z\varepsilon[0,L] \ , \tag{49a}$$

$$x(0,t) = v(t) \ , \qquad t\varepsilon[0,T] \ , \tag{49b}$$

It is assumed that $x_0(z)$ is of class C^1 on (z_i,z_{i+1}), $i = 1,2,\ldots,\gamma-1$; $u(t)_i,v(t)$ $\varepsilon C^1[0,\infty)$; and $x_0(z)$, $u_i(t)$, and $v(t)$ satisfy the consistency conditions at $z = 0$,

$$x_0(0) = v(0),\ \Lambda(0,0)\frac{dx_0}{dz}(0) + A(0,0)x_0(0) = \frac{dv}{dt}(0) \ . \tag{50}$$

The assumption that $\Lambda(z,t)$ is diagonal is reasonable since non diagonal matrix can be diagonalized, and the case when it contains positive diagonals can also be studied with similar approach (Russell, 1973).

Partitioning into subsystems. Under the stated assumptions above, the problem in Eqs. (48)-(49) can be shown to have a unique solution $x(z,t)\varepsilon C^1\{(z,t):z\varepsilon(z_i,z_{i+1})$, $i = 1,2,\ldots,\gamma-1$, $t\varepsilon[0,T]\}$. First define $x_i(z,t)$ as the state for ith subsystem on (z_i,z_{i+1}) such that

$$x_i(z,t) = \begin{cases} x(z,t) \ , & \text{on } (z_i,z_{i+1}) \ , \quad i = 1,2,\ldots,\gamma-1 \\ x(z_i^+,t) & \text{at } z_i \\ x(z_{i+1}^-,t) & \text{at } z_{i+1} \\ x(0,t) \ , & \text{at } z_1 = 0 \ , \end{cases} \tag{51}$$

where $x(z_i^-,t)$ and $x(z_i^+,t)$ denote the left and right hand limits. In the same way matrices Λ_i, A_i, and initial conditions $x_{i0}(z)$ are defined.

At point z_i, the derivative $\partial x(z,t)/\partial z$ in Eq. (48) can be written using generalized function $\delta(z-z_i)$ and the limit values $x(z_i^-,t)$ and $x(z_i^+,t)$. Therefore Eq. (48) reduces to the form

$$[\Lambda(z_i^+,t)x(z_i^+,t)-\Lambda(z_i^-,t)x(z_i^-,t)]\delta(z-z_i)+B_i\delta(z-z_i)u_i(t) = 0 \ .$$

Partitioning x, Λ as defined in Eq. (51), yields

$$x_i(z_i,t) = C_i x_{i-1}(z_i,t)+D_i u_i(t) \ , \quad i = 1,\ldots,\gamma-1 \ , \tag{52}$$

where

$$C_i[\Lambda_i(z_i,t)]^{-1}\Lambda_{i-1}(z_i,t) \ , \quad D_i = [\Lambda_i(z_i,t)]^{-1}B_i(t) \ .$$

The set of state equations for intervals $[z_i, z_{i+1}]$ becomes

$$\frac{\partial x_i}{\partial t} = \Lambda_i(z,t)\frac{\partial x_i}{\partial z} + A_i(z,t)x_i \tag{53}$$

with the initial condition $x_i(z,0) = x_{i0}(z)$, and boundary conditions Eq. (52) and $x_1(0,t) = v(t)$. Thus the original system in Eq. (48) with pointwise controls are now converted to a homogeneous system in Eq. (53) with boundary controls in Eq. (52).

Further simplification can be achieved by normalizing the spatial intervals $[z_i, z_{i+1}]$ into $[0,1]$. This can be done by defining a new spatial variable $z'\varepsilon[0,1]$ such that

$$z = (1-z')z_i + z'z_{i+1} \quad \text{for} \quad z\varepsilon[z_i, z_{i+1}] \; . \tag{54}$$

This results in new variables

$$x_i(z',t) = x_i((1-z')z_i + z'z_{i+1}, t) \tag{55}$$

$$\Lambda_i(z',t) = \frac{1}{z_{i+1}-z_i}\Lambda_i((1-z')z_i + z'z_{i+1}, t)$$

and $A_i(z',t)$, $x_{i0}(z')$ in the same was $x_i(z',t)$ above. The previous set of state equations (53) then becomes

$$\frac{\partial x_i}{\partial t} = \Lambda_i(z,t)\frac{\partial x_i}{\partial z} + A_i(z,t)x_i \;, \qquad t\varepsilon[0,T], z\varepsilon[0,1] \tag{56a}$$

$$x_i(z,0) = x_{i0}(z) \;, \qquad z\varepsilon[0,1] \;, \qquad i = 1,\ldots,\gamma-1 \tag{56b}$$

$$x_i(0,t) = C_i x_{i-1}(1,t) + D_i u_i(t) \;, \qquad i = 2,\ldots,\gamma-1 \tag{56c}$$

$$x_1(0,t) = v(t) \;, \tag{56d}$$

where primes (') are deleted for simplicity.

The above equations show that the individual subsystem state equation (56a) is independent of other subsystem and coupled only through the boundary conditions (56c). Moreover, controls are applied only through boundaries. This makes the natural decomposition of system without any approximation or discretization. Since the system is still in the form of partial differential equations, further conversion like finite difference or finite element scheme may be necessary for computation. Due to the fact that the system is of first-order, it can be converted to ordinary differential equations by integration method (Lee and Cooper, 1979).

Integration method. By integrating the system equation (56a) with respect to the spatial variable z over each interval [0,1], an ordinary differential equation is formed. This transformation "lumps" the contribution of the state variables over each interval. This method, of course, has limitations in application. Since the state is integrated over the spatial interval, the values are not available in the interior of the interval. However, there are many applications where this method is useful. When the system (56) describes population density, we may only need to know the total number in each subsystem (Lee and others, 1976). In some other physical systems, it might be sufficient to know only the state at boundaries. In either case it is not necessary to know the distribution and thus the analysis can be greatly simplified as each subsystem is lumped in each interval.

The integrated state is defined by

$$X_i(t) = \int_0^1 x_i(z,t)dz \ . \tag{57}$$

Using the assumption that Λ_i and A_i are constant in each interval, the integration of system equation (56a) yields

$$\dot{X}_i(t) = A_i(t)X_i(t)+\Lambda_i(t)(x_i(1,t)-x_i(0,t)) \tag{58}$$

Substituting Eq. (56c), the above becomes

$$\dot{X}_i(t) = A_i(t)X_i(t)+B_i(t)u_i(t)+w_i(t) \ , \quad i = 1,\ldots,\gamma-1 \ , \tag{59a}$$

where

$$B_i(t) = -\Lambda_i(t)D_i(t) \ , \tag{59b}$$

$$w_i(t) = \Lambda_i(t)(x_i(1,t)-C_i(t)x_{i-1}(1,t)) \tag{59c}$$

with

$$D_1(t) = I \ , \qquad C_1(t) = 0 \ , \qquad u_1(t) = v(t) \tag{59d}$$

The initial condition is

$$X_i(0) = \int_0^1 x_{i0}(z)dz \ .$$

The pair $(X_i(t),x_i(1,t)$ is the solution for ith subsystem, and the interaction variable $w_i(t)$ shows the coupling with (i-1)st subsystem through boundary value at $z = 1$. Since it is not feasible to write a differential equation for the boundary values $x_i(1,t)$, an iterative procedure is developed which updates $x_i(1,t)$, $i = 1,\ldots,\gamma-1$ starting from the initial distribution $x_i(1,0) = x_{i0}(1)$ (Lee and Cooper, 1981).

To complete the formulation, the performance index for overall system is assumed to be separable, and thus

$$J = \sum_{i=1}^{N} J_i(u_i) \ , \tag{60}$$

with

$$J_i(u_i) = \frac{1}{2} X_i^T(t_f)H_iX_i(t_f) + \frac{1}{2}\int_0^{t_f} [X_i^TQ_iX_i+u_i^TR_iu_i]dt \ .$$

3. COORDINATED CONTROL

In the previous section three different approaches of decomposition were presented which decomposed the distributed parameter system into a set of coupled subsystems of ordinary differential equations. In this section optimization problem will be formulated for the general coupled subsystems developed in Eqs. (21) to (27). Following the general optimization problem, the special case of linear quadratic problem will be explored in further detail.

3.1 Formulation of Coordinated Control Problem

The optimization problem is formally formulated to minimize the performance index

$$J(u) = \sum_{i=1}^{N} J_i(u_i) , \tag{61}$$

$$J_i(u_i) = \int_{t_0}^{t_f} g_i(x_i,u_i,t)dt ,$$

subject to the N-coupled system equations

$$\dot{x}_i = F_i(x_i,u_i,w_i,t) , \qquad i = 1,\dots,N \tag{62}$$

$$w_i = H_i(x_1,x_2,\dots,x_j,\dots,x_n) , \quad j \neq i \tag{63}$$

with initial conditions

$$x_i(t_0) = x_{i0} , \tag{64}$$

and final conditions

$$f_i(x(t_f),t_f) = 0 , \tag{65}$$

and inequality constraint

$$G_i(x_i,u_i,t) \geq 0 . \tag{66}$$

This problem is a typical of so-called two-level control problem. This formulation of the problem is referred to as the global problem P_G. (Sandell and others, 1978). The global problem can be replaced by a parameterized family of N decoupled subproblems of smaller size and simpler structure. Viewing the interaction variables $w = (w_1,\dots,w_N)$ as parameters, the subproblem can be parameterized as $P_1(w_1),\dots,P_N(w_N)$. This family must be such that P_G is <u>imbedded</u> in it, that is, there is a parameter w^* at which the solution to $P_i(w_i^*)$, $i = 1,\dots,N$, yields a solution to P_G. In the language of multi-level control this imbedding is called coordinability (Mesarovic, Macko, and Takahara, 1970a, 1970b).

The subproblem $P_i(w_i)$ now becomes to minimize

$$J_i(u_i) = \int_{t_0}^{t_f} g_i(x_i,u_i,t)dt \tag{67}$$

subject to

$$\dot{x}_i = F_i(x_i,u_i,w_i,t) , \quad x_i(t_0) = x_{i0} \tag{68a}$$

$$f_i(x_i(t_f),t_f) = 0 \tag{68b}$$

$$G_i(x_i,u_i,t) \geq 0 . \tag{68c}$$

Let $x_i(w_i)$ be the state computed by the first-level, or local controller, for $P_i(w_i)$. Then the imbedding condition requires that the "error"

$$e_i(w)(t) = w_i(t)-H_i(x_j(w_j) , \quad j = 1,\dots,N, \quad j \neq i) \tag{69}$$

vanish for some w^*. The second-level controller, or coordinator, computes a new parameter w^{n+1} based upon current w^n so that the error $e(w^n)$ converges to zero. Iterative coordination process can be developed for this approach based upon the "interaction prediction principle" of Mesarovic, Macko, and Takahara (1970a, 1970b), and other investigators (Cohen, 1976; Cohen and Jolland, 1976; Singh, Hassan, and Titli, 1976).

An alternative approach is the variational approach which considers the interaction w_i as part of the control variable to the ith controller. The coordinator modifies the performance index of each local controller by augmenting the coupling constraint Eq. (63), giving the following problem in which α is a parameter:

Minimize the augmented performance index

$$J' = \sum_{i=1}^{N} \int_{t_0}^{t_f} \{g_i(x_i,u_i,t) + \alpha_i^T(t)[w_i(t)-H_i(x_j)]\}dt \tag{70}$$

subject to

$$\dot{x}_i = F_i(x_i,u_i,w_i,t) , \quad x_i(t_0) = x_{i0} \tag{71a}$$

$$f_i(x_i(t_f),t_f) = 0 \tag{71b}$$

$$G_i(x_i,u_i,t) \geq 0 \tag{71c}$$

This problem can be decomposed into N independent subproblems, $P_i(\alpha)$, when the interaction variable in Eq. (63) is separable, i.e.,

$$w_i(t) = \sum_{j=1}^{N} h_{ij}(x_j) , \quad j \neq i . \tag{72}$$

In this case the augmented performance index can be rearranged as

$$J' = \sum_{i=1}^{N} \int_{t_0}^{t_f} [g_i(x_i,u_i,t)+\alpha_i^T w_i - \sum_{\substack{j=1\\ j\neq i}}^{N} \alpha_j^T h_{ji}(x_i)]dt \tag{73}$$

By letting $x_i(\alpha),u_i(\alpha),w_i(\alpha)$ be the solution of $P_i(\alpha)$, and letting the error be

$$e_i(\alpha) = w_i(\alpha) - \sum_{j=1}^{N} h_{ij}(x_j(\alpha)), \quad j \neq i \tag{74}$$

the imbedding condition requires that there exist α^* for which $e(\alpha^*) = 0$, which is difficult without some kind of convexity restriction. Once again an iterative procedure must be developed such that α^{n+1} can be computed knowing α^n and $e(\alpha^n)$ so that $e(\alpha^{n+1})$ is smaller than $e(\alpha^n)$. For this see the "interaction balance principle" of Masarovic, Macko, and Takahara (1976), and other works (Pearson, 1971; Titli, Lefenie, and Richetin, 1973).

3.2 Computational Considerations

The second approach described in Eqs. (71) to (73) has introduced additional variable α compared to the first approach presented in Eqs. (67) to (69), and thus additional computation is necessary. However, for generality only the second approach will be presented for computational considerations.

Gauss-Seidel iterative scheme. For the global optimization problem P_G defined in Eqs. (71) to (73), the Hamiltonian is defined as

$$H = \sum_{i=1}^{N} [g_i(x_i,u_i,t) + \alpha_i^T w_i - \sum_{\substack{j=1 \\ j\neq i}}^{N} \alpha_j^T h_{ji}(x_i) + p_i^T F_i(x_i,u_i,w_i,t)] \ . \tag{75}$$

Here the constraints in Eqs. (71b) and (71c) are neglected to simplify the presentation since they do not affect the coordination problem. According to the interaction balance principle there exists an optimal parameter α^* such that P_G is imbedded in the family of subproblems $P_i(\alpha^*)$ which has the subsystem Hamiltonian

$$H_i = g_i(x_i,u_i,t) + \alpha_i^T w_i - \sum_{\substack{j=1 \\ j\neq i}}^{N} \alpha_j^T h_{ji}(x_i) + p_i^T F_i(x_i,u_i,w_i,t) \tag{76}$$

The necessary conditions for a minimum is therefore according to the maximum principle

$$\dot{x}_i = \frac{\partial H_i}{\partial p_i} = F_i(x_i,u_i,w_i,t) \tag{77a}$$

$$\dot{p}_i = -\frac{\partial H_i}{\partial x_i} = -\frac{\partial g_i}{\partial x_i} + \sum_{j=1}^{N} (\frac{\partial h_{ji}}{\partial x_i})^T \alpha_j - (\frac{\partial F_i}{\partial x_i})^T p_i \tag{77b}$$

$$\frac{\partial H_i}{\partial u_i} = \frac{\partial g_i}{\partial u_i} + (\frac{\partial F_i}{\partial u_i})^T p_i = 0 \tag{77c}$$

$$\frac{\partial H_i}{\partial w_i} = \alpha_i + (\frac{\partial F_i}{\partial w_i})^T p_i = 0 \tag{78a}$$

$$\frac{\partial H}{\partial \alpha_i} = w_i - \sum_{j=1}^{N} h_{ij}(x_j) = 0 \tag{78b}$$

For given estimates of α and w, the equation (77) is the usual first-level, or local, control algorithm. The solution x_i and p_i of subsystem is then sent to the second-level controller, or the coordinator. Using the input from all subsystems, the coordinator computes new values of α and w from the equation (78), which then are sent back to the local controllers. This iterative process is continued until no improvement is made in the error criteria in Eq. (78b). This approach is called Jacobi or Gauss-Seidel method and shown to converge by Wismer (1971).

Gradient and conjugate gradient schemes. Let $J'(x_i^*(\alpha), u_i^*(\alpha), \alpha)$ be the performance index in Eq. (73) which is minimized for a subsystem for a given parameter α. Pearson (1971) showed that this functional is concave over convex subsets of the parameters α. Therefore the global minimization problem is casted into a max-min problem, where the minimization is done by local controllers and depends on a parameter α, and the maximization is done by the coordinator over a class of parameters α.

At the local level, the problem is to minimize

$$J_i'(u_i, w_i, \alpha) = \int_{t_0}^{t_f} [g_i(x_i, u_i, t) + \alpha_i^T w_i - \sum_{\substack{j=1 \\ j \neq i}}^{N} \alpha_j^T h_{ji}(x_i)] dt \tag{79}$$

for a given parameter α, for each $i = 1,2,\ldots,N$. This can be done by any suitable method for a particular subproblem. The result is that $x_i^*(\alpha), u_i^*(\alpha), w_i^*(\alpha), p_i^*(\alpha)$, and α are all available as functions of t, where $p_i^*(\alpha)$ is the costate variable defined in Eq. (77b).

At the coordination level some method of adjusting α must be found which does not use subsystem constraint information directly. Ideally, the second level need be concerned only with violated constraints. Now, note that the variational or Fréchet derivative of $J'(\alpha)$ is given by

$$\frac{\delta J'(\alpha)}{\delta \alpha} = w^*(\alpha) - h(x^*(\alpha)) \; , \tag{80}$$

where $w = [w_1, \ldots, w_N]^T$, $h(x) = [h_1(x), \ldots, h_N(x)]^T$, and $h_i(x) = \Sigma_j h_{ij}(x_j)$. This derivative uses only information from violated constraints, or error. Since the derivative is first-order, any first-order method of maximizing $J'(\alpha)$ can be applicable. The steepest ascent, or gradient, method and the conjugate gradient method are considered here. The first one is usually slow except on exceptionally messy problems. The second method is much better and approximate the performance of second-order methods.

Steepest Ascent Method: At iteration n chose a search direction $\delta J'(\alpha^n)/\delta\alpha$, and step τ^n so that

$$\alpha^{n+1} = \alpha^n + \tau^n \frac{\delta J'(\alpha^n)}{\delta \alpha} \tag{81}$$

and then repeat for the next iteration. In general, τ^n should be a small step such that an improvement is achieved. To accelerate the convergence, τ^n may be chosen so that $J'(\alpha^{n+1})$ is maximum (Wilde, 1964). However, there is a trade-off between this additional maximization and the total computation time.

Conjugate Gradient Method: This method generates directions using the algorithm of Fletcher and Reeves (1964), and proceeds as follows:

1. Set the initial direction $d^0 = \delta J'(\alpha^0)/\delta\alpha$.

2. Find step size τ^n which maximizes $J'(\alpha^n + \tau^n d^n)$.

3. Compute $\alpha^{n+1} = \alpha^n + \tau^n d^n$.

4. Compute $g^{n+1} = \delta J'(\alpha^{n+1})/\delta\alpha$.

5. Find $\sigma^n = \langle g^{n+1}, g^{n+1}\rangle / \langle g^n, g^n\rangle$.

6. Find $d^{n+1} = \sigma^n d^n + \delta J'(\alpha^{n+1})/\delta\alpha$,

where $\langle\ ,\ \rangle$ is an inner product.

The sequence of search directions $d^0, d^1, \ldots, d^n$ is infinite in an infinite dimensional space. However in practice τ^n, σ^n, etc., cannot be found exactly and the directions lose conjugacy as the iteration proceeds. It pays then to restart with $d^m = g^m$ for $m=n+1$, say (n is the state dimension). This kind of error can be seen by checking $\langle g^{n+1}, d^i\rangle = 0$ for $i \leq n$.

3.3 Linear Quadratic Problems

The general coordination problem in the previous section involved two-point boundary value problem to solve the first level control problem for each subsystem. The state equations need to be integrated in the forward direction while the co-state equations are integrated in the backward direction. And this process is repeated as the coordinator computes new coordination parameter α in each iteration. This procedure is therefore not suited for on-line application, which is of primary concern in process controls. When the decomposed subsystems are linear, then it is possible to have real-time algorithm to be implemented for on-line application.

Consider the coupled linear systems

$$\dot{x}_i = A_i x_i + B_i u_i + w_i \ , \quad x_i(t_0) = x_{i0} \ , \tag{82}$$

with linear interactions

$$w_i = \sum_{j=1}^{N} H_{ij}\, x_j \ , \quad j \neq i \ , \tag{83}$$

and the quadratic performance index

$$J = \sum_{i=1}^{N} J_i(u_i) \ , \tag{84}$$

$$J_i(u_i) = \frac{1}{2}\int_{t_0}^{t_f} [x_i^T Q_i x_i + u_i^T R_i u_i]dt \ . \tag{85}$$

The necessary condition for a minimum of J is, from the general case in Eqs. (77) to (78),

$$\dot{x}_i = A_i x_i + B_i u_i + w_i \ , \quad x_i(t_0) = x_{i0} \ , \tag{86a}$$

$$\dot{p}_i = A_i^T p_i - Q_i x_i + \sum_{j=1}^{N} H_{ji}^T \alpha_j \ , \quad p_i(t_f) = 0 \ , \tag{86b}$$

$$u_i = -R_i^{-1}B_i^T p_i \tag{86c}$$

$$\alpha_i = -p_i \tag{87a}$$

$$w_i = \sum_{j=1}^{N} H_{ij}x_j \ , \quad j \neq i \tag{87b}$$

Noting (86b) and (87a), define a costate interaction variable

$$v_i = \sum_{j=1}^{N} H_{ji}^T \alpha_j = - \sum_{j=1}^{N} H_{ji}^T p_j \ . \tag{88}$$

Using the equations (86c) and (88), state and costate equations (86a) and (86b) can be written as

$$\begin{bmatrix} \dot{x}_i \\ \dot{p}_i \end{bmatrix} = \begin{bmatrix} A_i & -B_iR_i^{-1}B_i^T \\ -Q_i & -A_i^T \end{bmatrix} \begin{bmatrix} x_i \\ p_i \end{bmatrix} + \begin{bmatrix} w_i \\ v_i \end{bmatrix} \tag{89}$$

This is a linear system with an important property that the mapping $x_i(t_0)\to\{x_i,p_i\}$ is continuous and affine, and the mapping $\{x_i,p_i\}\to p_i(t_0)$ is also continuous. Therefore, the composite mapping $x_i(t_0)\to p_i(t_0)$ is a continuous affine mapping (Lions, 1971). This leads to a unique representation for decoupling

$$p_i(t) = K(t)x_i(t) + s(t) \quad \text{for all } t \ . \tag{90}$$

Differentiating this and using Eqs. (89) and (90) yields the following Riccati matrix differential equation

$$\dot{K}_i+K_iA_i+A_i^TK_i-K_iB_iR_i^{-1}B_i^TK_i+Q_i = 0 \ , \tag{91}$$

and a linear differential equation

$$\dot{s}_i+A_i^Ts_i-K_iB_iR_i^{-1}B_i^Ts_i+K_iw_i+v_i = 0 \ . \tag{92}$$

The local optimal control in a feedback form is

$$u_i = -R_i^{-1}B_i^T(K_ix_i+s_i) \ . \tag{93}$$

Note that the Riccati equation (91) is decoupled from other subsystems and independent of interaction variables. The coupling is done through Eq. (92) by interaction variables w_i and v_i. These equations need to be integrated backward in time, which is still not practical for on-line implementation.

It is further assumed that the problem in Eqs. (82) to (85) is time-invariant, and the final time t_f is sufficiently large so that K_i and s_i are constant functions of time. Then setting the derivatives equal to zero in Eqs. (91) and (92) yields the algebraic Riccati matrix equation

$$K_iA_i+A_i^TK_i-K_iB_iR_i^{-1}B_i^TK_i+Q_i = 0 \ , \tag{94}$$

and an expression

$$s_i = -(A_i^T - K_i B_i R_i^{-1} B_i^T)^{-1} (K_i w_i + v_i) \ . \tag{95}$$

The Riccati equation can be solved once by any well known method (Denman and Beavers, 1976), and the gains in Eq. (95) need to be computed only once initially, which qualifies the on-line application.

The above equations (93) to (95) completely determine local controller for each subsystem. The second level controller, or coordinator, is given by equations (87b) and (88):

$$w_i = \sum_{j=1}^{N} H_{ij} \, x_j \tag{96}$$

$$v_i = - \sum_{j=1}^{N} H_{ji}^T (K_j x_j + s_j) \ , \tag{97}$$

where $H_{ii} = 0, \ i = 1,\dots,N.$

At each time step, the Gauss-Seidel iterative scheme can be applied between the first and second level controllers. This method has proven to be very effective when applied to a coupled megawatt frequency and megavar voltage control problems in a generation unit (Lee and Belbachir, 1981). The computational procedure for the Gauss-Seidel iterative scheme is as follows:

1. For a fixed time, estimate x_i, s_i for $i = 1,2,\dots,N$, and set $i=1$.
2. Compute w_i, v_i from Eqs. (96) and (97) respectively.
3. Compute s_i, u_i from Eqs. (95) and (93) respectively.
4. Solve for x_i from Eq. (82) using the latest information.
5. Repeat Steps 2-4 for all $i = 1,2,\dots,N$.
6. Are Eqs. (96) and (97) satisfied to desired accuracy for all $i = 1,2,\dots,N$? If yes, stop. If no, repeat Steps 2-4 from $i=N$ to $i=1$.
7. Increase time step, and repeat Steps 2-6.

3.4 Systems Decomposed by Integration Scheme

In Section 2.4, a class of first-order system was decomposed into a set of linear ordinary differential equations for integrated state variables X_i which are coupled through the boundary values $x_i(1,t)$, $i = 1,\dots,N$ in Eq. (59):

$$\dot{X}_i = A_i X_i + B_i u_i + w_i \ , \quad i = 1,\dots,N \ , \tag{98a}$$

where

$$B_i = -\Lambda_i D_i \ , \tag{98b}$$

$$w_i = \Lambda_i(x_i(1,t)-C_i x_{i-1}(1,t)) \tag{98c}$$

with

$$D_1 = I\ , \qquad C_1 = 0\ , \qquad u_1 = v\ . \tag{98d}$$

The pair $(X_i(t),\ x_i(1,t))$ is the solution for each subsystem, and w_i is the interaction variable coupled with previous subsystem. This system coupling is different from the general case treated in the previous sections and therefore separate attention is required.

The performance index to be minimized is

$$J = \sum_{i=1}^{N} J_i(u_i)\ , \tag{99a}$$

$$J_i(u_i) = \frac{1}{2}\int_{t_0}^{t_f} (X_i^T Q_i X_i + u_i R_i u_i)dt\ . \tag{99b}$$

Since w_i is not directly related to X_i, the global problem is imbedded into subproblems to minimize J_i for each i independently. Thus the local controller is the usual feedback control which can be obtained from Eqs. (93) to (95) by setting $v_i = 0$, i.e.,

$$u_i = -R_i^{-1}B_i^T(K_i X_i + s_i)\ , \tag{100}$$

where K_i is the solution of an algebraic matrix Riccati equation

$$K_i A_i + A_i^T K_i - K_i B_i R_i^{-1} B_i^T K_i + Q_i = 0\ , \tag{101}$$

and

$$s_i = -(A_i^T - K_i B_i R_i^{-1} B_i^T)^{-1} K_i w_i\ . \tag{102}$$

The second level controller is given by (98c) as

$$w_i = \Lambda_i(x_i(1,t)-C_i x_{i-1}(1,t))\ , \tag{103}$$

which is not explicitly showing the relationship with X_i. However, since X_i is the integral of x(z,t) over [0,1] interval, it can be approximated as

$$X_i(t) \simeq \frac{x_i(0,t)+x_i(1,t)}{2}\ , \tag{104}$$

where, from (56c),

$$x_i(0,t) = C_i x_{i-1}(1,t) + D_i u_i\ . \tag{105}$$

The equation (104) shows the error

$$X_i(t) - \frac{x_i(0,t)+x_i(1,t)}{2} = \Delta X_i(t) , \tag{106}$$

and to force the error to be zero, $x_i(1,t)$ can be adjusted so that

$$X_i(t) - \frac{x_i(0,t)+x_i(1,t)+\Delta x_i(1,t)}{2} = 0 . \tag{107}$$

Comparing this with Eq. (106) yields

$$\Delta X_i(t) - \frac{\Delta x_i(1,t)}{2} = 0 ,$$

or

$$\Delta x_i(1,t) = 2\Delta X_i(t) . \tag{108}$$

The iterative scheme developed (Lee and Cooper, 1981) is summarized as follows:

1. For fixed time, estimate $x_i(1,t)$, $i = 1,\ldots,N$.
2. Compute w_i from Eq. (103).
3. Compute u_i from Eqs. (100)-(102).
4. Solve for X_i from Eq. (98a).
5. Compute $x_i(0,t)$ from Eq. (105).
6. Compute $\Delta X_i(t)$ and $\Delta x_i(1,t)$ from Eqs. (106) and (108) respectively, and update $x_i(1,t)$ by $x_i(1,t)+\Delta x_i(1,t)$ for all i.
7. Repeat Steps 2-6 until $\Delta x_i(1,t)$ is sufficiently small.
8. Increment time and repeat Steps 1-7.

The coordinated control algorithm was applied for the management of the cereal leaf. beetle insect population and shown to converge very rapidly (Lee and Cooper, 1981). This iterative scheme is very well suited for on-line applications.

4. CONCLUSION

This chapter has reviewed the coordinated control approach for distributed parameter systems. Three different methods, finite difference scheme, finite element method, and integration method, were presented to decompose the distributed parameter system into a set of coupled lumped parameter systems described by ordinary differential equations.

Finite difference scheme is rather straight-forward in application, but results in high dimensional systems and care must be made in handling irregular mesh points in boundary. Finite element method is ideal in handling irregular boundary and yields rather lower dimensional systems compared to finite difference scheme.

However, it demands the good knowledge in the approximation scheme itself. The integration scheme is limited to the first-order system and well defined boundaries to decompose. The scheme gives the integral of values over each region and values at boundary.

Some coordination control schemes were presented for the control of coupled subsystems. Two types of iterative scheme were developed. Gauss-Seidel type iterative scheme is straight-forward for application, while gradient or conjugate gradient scheme requires additional computations for speedy convergence. When the system is linear and time-invariant, a procedure of on-line coordinated control scheme is developed. This scheme has great potential in process control applications. This technique can be modified for time-varying system by using adaptive controllers.

In summary, the high dimensionality and complexity in distributed parameter systems require good decomposition and coordination schemes as necessity. This philosophy is also valid for state and parameter estimation problems. It was the purpose of this chapter to address such needs and provide a forum to direct future research areas.

5. REFERENCES

Balakrishnan, A. V. (1965). Optimal control problems in Banach spaces. SIAM J. Control, 22, 75-172.

Balas, M. J., and C. R. Johnson, Jr. (1979). Adaptive control of distributed parameter systems: The ultimate reduced-order problem. Proc. 18th IEEE Conf. on Decision & Control, 1013-1017.

Butkovskii, A. G. (1961). The maximum principle for optimum systems with distributed parameters. Autom. & Remote Control, 22, 1156-1169.

Cavin, R. K., III, and S. C. Tandon (1977). Distributed parameter system optimum control design via finite element discretization. Automatica, 13, 611-614.

Ciarlet, P. G. (1978). The Finite Element Method for Elliptic Problems. North-Holland, Amsterdam.

Cohen, G. (1976). On an algorithm of decentralized optimal control. J. Math. and Appl.

Cohen, G., and G. Jolland (1976). Coordination methods by the prediction principle in large dynamic constrained optimization problems. Proc. IFAC Symp. Large Scale Systems Theory & Appl., Udine, Italy.

Denman, E. D., and A. N. Beavers, Jr. (1976). The matrix sign function and computations in systems. Appl. Math. & Computations, 63-94.

Fletcher, R., and R. Reeves (1964). Function minimization by conjugate gradients. Computer J., 7, 149-154.

Greenberg, S. G. (1971). On quadratic optimization in distributed parameter systems. IEEE Trans. Autom. Control, 16, 153-159.

Johnson, T. L., and M. Athans (1974). A minimum principle for smooth first-order distributed systems. IEEE Trans. Autom. Control, 19, 136-139.

Lee, K. Y., and R. O. Barr (1972). Sampled-data optimization in distributed parameter systems. IEEE Trans. Autom. Control, 17, 806-809.

Lee, K. Y., S. N. Chow, and R. O. Barr (1972). On the control of distrete-time distributed parameter systems. SIAM J. Control, 10, 361-376.

Lee, K. Y. (1974). Design of optimal sampled-data pointwise controllers for distributed parameter systems. Proc. Milwaukee Symp. on Autom. Control, 359-363.

Lee, K. Y., R. O. Barr, S. H. Gage, and A. N. Kharkar (1976). Formulation of a mathematical model for insect pest ecosystem-the cereal leaf beetle problem. J. Theor. Biol., 59, 33-76.

Lee, K. Y. (1978a). Modeling and estimation of distributed parameter systems with uncertain parameters. In. A. V. Balakrishnan (Ed.), Lecture Notes in Control and Information Sciences, Vol. 1. Springer-Verlag, Berlin. pp. 325-334.

Lee, K. Y. (1978b). Bilinear distributed systems and pest management. In R. R. Mohler and A. Ruberti (Ed.), Recent Developments in Variable Structure Systems, Economics and Biology. Vol. 162. Springer-Verlag, Berlin. pp. 173-188.

Lee, K. Y., and T. G. Cooper (1979). Coordinated control of distributed parameter systems. Proc. 18th IEEE Conf. on Decision & Control, 665-607.

Lee, K. Y., and M. Belbachir (1981). A decentralized plant controller for automatic generation and voltage regulation. U. of Houston Report. Dept. of Electrical Engineering, U. of Houston, Houston, Texas.

Lee, K. Y., and T. G. Cooper (1981). Distributed parameter system decomposition and coordination. U. of Houston Report. Dept. of Electrical Engineering, U. of Houston, Houston, Texas.

Lions, J. L. (1971). Optimal Control of Systems Governed by Partial Differential Equations. Springer-Verlag, Berlin.

Melosh, R. J. (1965). Basis for derivation of matrices for the direct stiffness method. J. American Inst. for Aeronautics & Astronautics, 1, 1631-1637.

Mesarovic, M. D., D. Macko, and Y. Takahara (1970a). Theory of Hierarchical Multilevel Systems. Academic Press, New York.

Mesarovic, M. D., D. Macko, and Y. Takahara (1970b). Two coordination principles and their application in large scale systems control. Automatica, 6.

Pearson, J. D. (1971). Dynamic decomposition techniques. In D. A. Wismer (Ed.), Optimization Methods for Large-Scale Systems, McGraw-Hill, New York. Chap. 4, pp. 121-190.

Potter, M. C. (1978). Mathematical Methods in the Physical Sciences. Prentice-Hall, Englewood Cliffs.

Russell, D. L. (1973). Quadratic performance criteria in boundary control of linear symmetric hyperbolic systems. SIAM J. Control, 11, 475-509.

Sandell, N. R., Jr., P. Varaiya, M. Athans, and M. G. Safonov (1978). Survey of decentralized control methods for large scale systems. IEEE Trans. Autom. Control, 23, 108-128.

Segerlind, L. J. (1976). Applied Finite Element Analysis. John Wiley & Sons, New York.

Singh, M., M. F. Hassan, and A. Titli (1976). Multilevel feedback control for interconnected dynamical systems using the prediction principle. IEEE Trans. Syst., Man & Cybern., 6, 233-239.

Szabo, B. A., and G. C. Lee (1969). Derivation of stiffness matrices for problems in plane elasticity by Galerkin's method. Int. J. Numer. Meth. in Engineering, 1, 301-310.

Titli, A., T. Lefenie, and M. Richetin (1973). Multilevel optimization methods for the separable problems and applications. Int. J. Syst. Sci., 4.

Turner, M. J., R. W. Clough, H. C. Martin, and L. J. Topp (1956). Stiffness and deflection analysis of complex structures. J. Aeronautical Science, 23, 805-824.

Wang, P. K. C. (1964). Control of distributed parameter systems. In Leondes (Ed.), Advances in Control Systems, Vol. 1, Academic Press, New York, pp. 75-172.

Wilde, D. J. (1964). Optimum Seeking Methods. Prentice-Hall, Englewood Cliffs.

Wismer, D. A. (1971). Distributed multilevel systems. In D. A. Wismer (Ed.), Optimization Methods for Large-Scale Systems. McGraw-Hill, New York. Chap. 6, pp. 233-273.

Wismer, D. A. (1973). Decomposition and optimization of distributed parameter systems. In D. M. Himmelblau (Ed.), Decomposition of Large-Scale Systems. American Elsevier, New York.

Young, D. M. (1962). The numerical solution of elliptic and parabolic partial differential equations. In J. Todd (Ed.), Survey of Numerical Analysis. McGraw-Hill, New York. Chap. 11.

Zienkiewicz, O. C. (1971). The Finite Element Method in Engineering Science. McGraw-Hill, London.

Zienkiewicz, O. C., and C. J. Parekh (1970). Transient field problems: Two-dimensional and three-dimensional analysis by isoparametric finite elements. Int. J. Numer. Meth. in Engineering, 2, 61-71.

CHAPTER 8

THE FINITE ELEMENT METHOD IN DISTRIBUTED PARAMETER CONTROL SYSTEMS

L. Carotenuto and G. Raiconi

Department of Systems, University of Calabria, Arcavacata, Italy

INTRODUCTION

A current assertion in the literature of distributed parameter systems (d.p.s.) is that the gap between theory and practical applications is very wide: at the beginning of 1970's Robinson (1971) said that "while a substantial body of theory exists, the applications to date have been few"; in a much more recent survey Ray (1978) confirms that the situation has not substantially changed, particularly regarding such important applications as the on-line estimation and control of industrial processes. One of the principal obstacles to turning d.p.s. control theory into practice is the computational effort required in solving the partial differential equations (p.d.e.) involved: this obstacle can be overcome if sufficent attention is paid by people working in control area to the efficient methods recently provided by numerical analysis. During the past few years the finite element (f.e.) method has become extremely popular in several fields of engineering for the solving of both boundary and initial value problems and several interesting simulation studies have appeared in the literature of system engineering. At the same time numerical analysis continues to provide new theoretical results thus giving a sound basis to applications.

In the field of controls relatively few works can be found which focus attention to the f.e. method. The f.e. approximation has been used in the identification of elliptic and parabolic systems, often applied to steady state underground aquifers (Carotenuto and others, 1978, 1980a; Frind and Pinder, 1973), in the suboptimal filtering of the diffusion equation (Carotenuto and Raiconi, 1980b), in the solution of optimal design and free-boundary problems (Hang and Feng, 1977; Pironneau, 1977). The convergence of the f.e. solution for some optimal control problems has been investigated by Bosarge and Mc Knigt (1973), Winter (1978), Geveci (1979), Lasiecka (1980).

The aim of this chapter is to give a brief description of the method, of its approximation properties, of its capacity to solve a wide class of problems arising in applications of d.p.s. control theory. The exposition follows an applied mathematics approach, with particular reference to texts by Strang and Fix (1973) and

Mitchell and Wait (1977), which seem well-suited to introducing control engineers to the f.e. method; other recent texts on the subject are Zienkiewicz (1977), Norrie and De Vries (1978), Ciarlet (1978). Firstly, the construction of f.e. approximating spaces is described. Two-dimensional domains are dealt with since they enable the characteristic features of the method to be illustrated without an excessive use of notation. Then the Ritz-Galerkin f.e. method for the solution of elliptic boundary-value problems is illustrated and some results concerning the convergence of the approximate solution to the actual one are reported; the effects of some violations of the Ritz procedure, often referred to as variational crimes are also described. The following section is devoted to initial value problems, with particular emphasis on the semi-discrete Galerkin method for parabolic self-adjoint equations. Properties of the approximate eigenvalues and eigenfunctions of on elliptic operator computed by the f. e. method are outlined and then used to connect the f.e. solution to the semigroup of the distributed system. In the last section we resort to simple examples to show how it possible to build up a f.e. approximate solution to some standard problems in d.p.s. control theory.

CONSTRUCTION OF FINITE ELEMENT SPACES.

The basic idea underlying the f.e. method is to regard a domain Ω of $\mathbb{R}^m$, $m = 1,2,3$, as the union of subdomains of simple shape (elements), and to approximate a function u, defined on Ω, by a linear combination of continuous piecewise polynomial functions, each of which is nonzero on a "small number" of elements only.

In order to introduce finite element spaces let us consider a polygonal domain Ω in the plane with a triangular mesh τ on it, such that the sides of the boundary Γ of Ω belong to the mesh (mesh adapted to Ω): thus Ω is partitioned into triangles (elements) e_i, $i = 1,2,\ldots,N^e$; we assume that the elements are closed subdomains of $\bar{\Omega}$, so that any two triangles may have either a vertex or an edge in common. Let us introduce the following notations:

$P_j = (x_j, y_j)$, $j = 1,2,\ldots,N$: vertices of triangles, that is nodes of the mesh;

$P^i_{j_1}, P^i_{j_2}, P^i_{j_3}$: vertices of the element e_i;

$e^j_{i_1}, e^j_{i_2}, \ldots, e^j_{i_{n(j)}}$: elements which have a vertex at P_j;

$$\eta^j = \bigcup_{k=1}^{n(j)} e^j_{i_k}$$

h_i: the longest edge of e_i;

$h = \max_i h_i$: the mesh size.

Let us consider the space S_1 of functions $v \in C^\circ(\bar{\Omega})$ which are affine on every element:

$$v(x,y) = a^i + b^i x + c^i y \ , \ \forall (x,y) \in e_i$$

By putting

$$A_i = 2 \text{ area } (e_i),\ v_k^i = v(x_{j_k}^i, y_{j_k}^i),\ x_k^i = x_{j_k}^i,\ y_k^i = y_{j_k}^i,\quad k = 1,2,3,$$

we have

$$a_i = [v_1^i(x_2^i y_3^i - x_3^i y_2^i) + v_2^i(x_3^i y_1^i - x_1^i y_3^i) + v_3^i(x_1^i y_2^i - x_2^i y_1^i)]\, A_i^{-1}$$

$$b_i = [(v_2^i y_3^i - v_3^i y_2^i) + (v_3^i y_1^i - v_1^i y_3^i) + (v_1^i y_2^i - v_2^i y_1^i)]\, A_i^{-1}$$

$$c_i = [(v_3^i x_2^i - v_2^i x_3^i) + (v_1^i x_3^i - v_3^i x_1^i) + (v_2^i x_1^i - v_1^i x_2^i)]\, A_i^{-1}$$

which completely determine v by means of its values at the N nodes; continuity at the nodes also ensures continuity across the inter-element boundaries. S_1 is an N-dimensional vector space and v can be written as

$$v(x,y) = \sum_{j=1}^{N} q_j \phi_j(x,y),\quad (x,y) \in \bar{\Omega}$$

where $\{\phi_1,\ldots,\phi_N\}$ is a basis of S_1: a key feature of the finite element method is the choice of the basis functions, which are taken such that $\phi_k(x_j,y_j) = \delta_{kj}$, $k = 1,2,..,N$, $j = 1,2,..,N$. In the case under consideration every basis function ϕ_k is a pyramid with vertex at the node P_k, which vanishes outside η^k: we say that this basis is strictly local, in the sense that each basis function ϕ_k is nonzero only over those elements adjacent to the node P_k; moreover the coordinates of v with respect to this basis are the values of v at the nodes: $q_j = v(x_j,y_j)$, $j = 1,2,..,N$, thus justifying the denomination of nodal parameters used for them. This is in itself a valuable property, and complements the advantages obtained by choosing basis functions which are "nearly orthogonal" with respect to any weighting function: in fact $\iint_\Omega d(x,y)\phi_k(x,y)\phi_j(x,y)dxdy = 0$ when the nodes P_k, P_j are not adjacent. Where the regularity properties are concerned, the derivatives $\frac{\partial v}{\partial x}$, $\frac{\partial v}{\partial y}$ are constant on every element and discontinuous across the edges, thus giving $v \in H^1(\Omega)$.

The space S_1 defined above contains any polynomial of degree ≤ 1: on the same triangulation a finite element space S_2 can be defined, such that any polynomial of degree ≤ 2 belongs to it. This can be obtained by assuming that $v \in S_2 \subset C^\circ(\bar{\Omega})$ is a quadratic on every element e_i:

$$v(x,y) = a^i + b^i x + c^i y + d^i xy + e^i x^2 + f^i y^2,\quad (x,y) \in e_i \tag{2.1}$$

A definition of a basis for S_2 is not so straightforward: one way of constructing it is to "refine" the elements by adding to the vertices a further set of nodes, placed at the midpoints of the edges, and by associating to every node P_j (both vertex and midpoint) a function $\phi_j \in S_2$, $j = 1,2,\ldots N'$, such that $\phi_j(x_k,y_k)=\delta_{kj}$, $k = 1,2,.., N'$: function ϕ_j takes the form (2.1) on every element and the six coefficients are uniquely determined by the six nodes belonging to the element. By the definition of ϕ_j the basis $\{\phi_1,\ldots,\phi_{N'}\}$ is strictly local and the nodal parameters continues to be the values of v at the nodes:

$$v(x,y) = \sum_{j=1}^{N'} v(x_j,y_j)\ \phi_j(x,y),\quad (x,y) \in \bar{\Omega}$$

The continuity of v across the edges comes from its continuity at the nodes: in fact the restriction of v to an edge is a one-dimensional quadratic function depending only on the three nodal values at the extremes and the midpoint of the edge, which are the same for the adjacent elements.

A f.e. space is said to be of degree k-1 if it contains any polynomial of degree less than k: finite element spaces of any degree on a given triangulation can then be constructed, by adding suitably placed nodes for every element and associating to each node a basis function with the above mentioned properties: however the dimension of the space rapidly increases with k. The family of f.e. spaces which we have introduced, each endowed with the bases described, is characterized by the fact that the coordinates of $v \in S$ (nodal parameters) are the values of v at suitable points of Ω (Lagrange interpolation): an alternative choice is to assume as nodal parameters the values both of the function itself and some derivatives at the nodes. Let us consider a space of degree 3 on the above triangulation: on every element, v is cubic depending on ten parameters which can be uniquely determined by the values of v at the vertices and at seven further points; thus a Lagrangian space is obtained; alternatively we can choose as nodal parameters the values of v, $\frac{\partial v}{\partial x}$, $\frac{\partial v}{\partial y}$ at the vertices P_i, $i = 1,2,\ldots,N$, and of v at the centroids $P_{N+1},\ldots,P_{N+Ne}$: in this way a cubic on every element is uniquely determined, but a continuity constraint on $\frac{\partial v}{\partial x}$, $\frac{\partial v}{\partial y}$ at the vertices is added. Three basis functions ϕ_j, Ψ_j, θ_j per vertex P_j are defined, cubic on each element and satisfying the nodal constraints:

$$\phi_j(x_i,y_i) = \delta_{ji},\quad \frac{\partial\phi_j}{\partial y}(x_i,y_i) = \frac{\partial\phi_j}{\partial y}(x_i,y_i) = 0$$

$$\Psi_j(x_i,y_i) = 0,\quad \frac{\partial\Psi_j}{\partial x}(x_i,y_i) = \delta_{ji},\quad \frac{\partial\Psi_j}{\partial y}(x_i,y_i) = 0$$

$$\theta_j(x_i,y_i) = 0,\quad \frac{\partial\theta_j}{\partial x}(x_i,y_i) = 0,\quad \frac{\partial\theta_j}{\partial y}(x_i,y_i) = \delta_{ji}\;;$$

these vanish outside η^j and are continuous across interelement boundaries. To every centroid only ϕ_j is associated: it is nonzero over the element only, a fact which proves useful in applications. The resulting Hermitian cubic space S_{3H} whose elements are given by:

$$v(x,y) = \sum_{i=1}^{N} v(x_i,y_i)\,\phi_i(x,y) + \frac{\partial v}{\partial x}(x_i,y_i)\Psi_i(x,y) + \frac{\partial v}{\partial y}(x_i,y_i)\theta_i(x,y) + \sum_{i=N+1}^{N+N^e}(x_i,y_i)\phi_i(x,y)$$

is a subspace of the Lagrangian space of the same degree: the reduction in dimension is due to the fact that each nodal parameter is, on average, shared by more elements in the Hermitian than in the Lagrangian case. S_{3H} is still a subspace of $C^\circ(\bar{\Omega})$ and $H^1(\Omega)$ but fails to be included in $C^1(\bar{\Omega})$ since the continuity of the first derivatives at the vertices does not ensure, for a general triangulation, the continuity of derivatives across inter-element boundaries.

The procedure outlined for cubics generalizes to polynomials of odd degree $2\nu+1$ where the values of derivatives of order up to ν at the vertices are among the nodal parameters. The immediate successor to the space of the Hermite cubics is the space of piecewise quintics, continuous up to the second derivatives at the vertices of the elements: the remaining degrees of freedom can be used to impose continuity of the derivatives across interelement boundaries, thus giving a first example of a finite element space on a general triangulation which is a subspace both of $C^1(\bar{\Omega})$ and $H^2(\Omega)$

Several modifications of the basic spaces have been devised, so that the accuracy of the approximation can be improved while keeping the computational complexity acceptable (Zinkiewicz, 1977; Norrie and De Vries, 1978).

What we have said for triangular elements in $\mathbb{R}^2$ is immediately particularized to intervals of the real line. The domain $\bar{\Omega} = [a,b]$ is subdivided into closed sub-intervals (elements) and functions which are piecewise linear quadratic, .., are considered. Lagrangian spaces of degree >1 require the introduction of further nodes within the elements, and are always subspaces of $H^1(\Omega)$. Hermitian spaces are defined by taking as nodal parameters the values of the function and its derivatives at the extremes of the elements: therefore in this case Hermite cubics are $C^1(\bar{\Omega})$ unlike what happens in the case of triangular elements in two dimensions. The definition of the basis is of course simpler: for example the basis of Hermite cubics is defined by assigning, for every node, two functions which are cubic on each element, such that

$$\phi_j(x_i) = \delta_{ji} \quad , \quad \frac{d\phi_j}{dx}(x_i) = 0$$

$$\psi_j(x_i) = 0 \quad , \quad \frac{d\psi_j}{dx}(x_i) = \delta_{ji}$$

The dimension of this space is 2N, if N is the number of nodes, whilst the dimension of the Lagrangian cubic space is 3N-2.

Another class of piecewise polynomial spaces on [a,b] which are subspaces of the Hermitian space of the same degree is that of splines: for example a cubic spline on [a,b] is a piecewise cubic belonging to $C^2[a,b]$. A further reduction of dimension is obtained, but splines do not possess the typical properties of finite elements: the cubic spline with the smaller support, the B-spline, is nonzero over four elements and does not satisfy the condition $\phi_j(x_i) = \delta_{ji}$. On the other hand other properties of splines (e.g. the minimum norm property) make them rather more suited for the interpolation of known data than for the approximate solution of p.d.e. (Ahlberg, Nilson and Walsh, 1967; Shoenberg, 1969).

Coming back to plain domains, a different class of f.e. spaces arises when rectangular elements are considered. Their importance stems from the fact that in several problems the domain is itself a rectangle or a union of rectangles; furthermore rectangular elements can be used in conjunction with triangular ones to fit the boundary of the domain; finally quadrilateral elements can be easily trasformed into rectangles by a change of coordinates (isoparametric transformations). The basic approach to rectangular elements is to construct a space S whose functions on the i-th element are given by $v = \Sigma a^i_{rs} x^r y^s, 0 \leq r \leq k-1, 0 \leq s \leq k-1$; the degree of such a space is k-1. The simplest example is that of piecewise bilinear functions

$$v(x,y) = a^i_{00} + a^i_{10}x + a^i_{01}y + a^i_{11}xy \quad ,(x,y) \in e_i \ ,$$

where the coefficients are uniquely determined by the values of v at the vertices of e_i. With reference to a rectangle $\bar{\Omega} = [a,b]x[c,d]$, a partition into rectangular elements is defined by two partitions $\{x_i,\ i = 1,2,..,N\}$, $\{y_j,\ j = 1,2,..,M\}$

of [a,b] and [c,d] respectively; for every node $P_{ij} = (x_i, y_j)$ a basis function $\phi_{ij}(x,y)$ of the above space can be defined by $\phi_{ij}(x,y) = \phi_i(x)\ \phi_j(y)$, where $\phi_i(x)$ $\phi_j(y)$ are piecewise linear and such that $\phi_i(x_p) = \delta_{ip}$, $\phi_j(y_q) = \delta_{jq}$. Biquadratic and bicubic Lagrangian spaces are likewise constructed by adding suitably placed nodes. Hermitian spaces are similarly defined: an example is given by the space of bicubics on every element, parametrized by the 16 values of $v, \frac{\partial v}{\partial x}, \frac{\partial v}{\partial y}$ and $\frac{\partial^2 v}{\partial x \partial y}$ at the vertices of the element. A stricly local basis of this space is obtained by taking products of one dimensional Hermite cubic basis functions: to every node P_{ij} we associate the four functions:

$$\phi^1_{ij}(x,y) = \phi_i(x)\,\phi_j(y)\ ,\quad \phi^2_{ij}(x,y) = \phi_i(x)\psi_j(y),$$

$$\phi^3_{ij}(x,y) = \psi_i(x)\phi_j(y)\ ,\quad \phi^4_{ij}(x,y) = \psi_i(x)\psi_j(y),$$

so that v can be written:

$$v(x,y) = \sum_{i,j} v(x_i,y_j)\phi^1_{ij}(x,y) + \sum_{i,j} \frac{\partial v}{\partial y}(x_i,y_j)\phi^2_{ij}(x,y) +$$
$$+ \sum_{i,j} \frac{\partial v}{\partial x}(x_i,y_j)\phi^3_{ij}(x,y) + \sum_{i,j} \frac{\partial^2 v}{\partial x \partial y}(x_i,y_j)\phi^4_{ij}(x,y)\ , \qquad (x,y) \in \Omega$$

$v, \frac{\partial v}{\partial x}, \frac{\partial v}{\partial y}$ are continuous at the interelement boundaries: the bicubic Hermitian space on rectangles is then a subspace of $C^1(\bar{\Omega})$ and of $H^2(\Omega)$; its dimension is 4NM. A large reduction in dimension is achieved by assuming that v is a bicubic spline on $\bar{\Omega}$ (Alberg, Nilson and Walsh, 1967): a basis for the space of bicubic splines, which is a subspace of $C^2(\bar{\bar{\Omega}})$, may be obtained by taking products of one-dimensional B-splines, eccept for the case of nodes on the boundary or adjacent to the boundary, where special basis functions must be introduced: the basis is not stricly local and the remarks made for splines on $\mathbb{R}$ still hold. Rectangular elements and related spaces are important also because they can be generalized very simply to domains of $\mathbb{R}^3$ by considering exahedral elements which are much more tractable than tetrahedra, the three-dimensional elements analogous to triangles.

We cannot conclude this section without mentioning the isoparametric transformation techniques which enable quadrilaterals, with straight or curved edges, and curvilinear triangles to be dealt with: in considering domains with curved boundaries such techniques are particularly useful.

The basic idea is to change from the original (x,y) coordinates where the non-standard elements are defined, to a new coordinate system, say (ξ,η), in such a way that the corresponding elements in the (ξ,η) plane are of standard shape. On the transformed domain Ω' a finite element space S' is constructed, the original problem is formulated on Ω' and its finite element solution $v(\xi,\eta)$ is transformed back to (x,y) coordinates. A satisfactory way of choosing the coordinate transformation is to assume that the function $x(\xi,\eta)$, $y(\xi,\eta)$ mapping Ω' onto Ω are piecewise polynomials on the elements of Ω' or, in other words, that they belong to the finite element space S' itself. The classical example is that of the piecewise bilinear transformation

$$y(\xi,\eta) = \Sum x_j \phi_j(\xi,\eta)$$
$$y(\xi,\eta) = \Sum y_j \phi_j(\xi,\eta) \qquad 2.2$$

where $\{\phi_j\}$ is the standard basis of the space of piecewise bilinear functions on a square Ω' partitioned into square elements. This transformation establishes a correspondence between a polygon Ω partitioned into general quadrilateral elements and Ω': indeed to every node (ξ_j,η_j) there corresponds the node (x_j,y_j) in the (x,y) plane; along any edge of the mesh, say between (ξ_j,η_j) and (ξ_q,η_q) either ξ or η is constant and (2.2) are the equations of the segment joing (x_j,y_j) with (x_q,y_q). Moreover the Jacobian of the transformation is shown to be positive within every element when the element in the (x,y) plane has all its angles less than π: this ensures that the mapping (2.2) is invertible. Furthermore although a function belonging to S' need not be piecewise polynomial in (x,y), a polynomial of first degree $v = a_o + a_1x + a_2y$ is transformed into a function belonging to S', so that it makes sense to define the degree of S', which in this case is 1.

Using the basis of a higher degree f.e. space, quadrilatera with polynomial edges can be dealt with. Analogous arguments are used to introduce the use of isoparametric techniques for triangles with curved edges.

Approximation Properties of f.e. Spaces.

In order that a finite-dimensional space S be used fruitfully in modelling a d.p.s. it is necessary that an arbitrary function $u(x,y)$ can be approximated by a function of S with sufficient accuracy.

Firstly let us consider as before a polygonal domain $\bar{\Omega}$ and a sequence of triangular meshes τ_h adapted to Ω and parametrized by the mesh-size h. We assume that for any h all angles of the triangles are larger than $\alpha_o > 0$.

Let S^h be a f.e. space of degree k-1, with basis $\phi_1,\phi_2,\ldots,\phi_N$; for semplicity's sake S^h is assumed of Lagrangian type. Let $u \in H^k(\Omega)$: by the Sobolev lemma it follows that $H^k(\Omega) \subset C^{k-2}(\bar{\Omega})$ and then, for $k \geq 2$ it makes sense to consider the interpolate of u: $u_I^h \in S^h$,

$$u_I^h(x,y) = \sum_{j=1}^{N} u(x_j,y_j)\phi_j(x,y) \quad , (x,y) \in \bar{\Omega}$$

The fundamental result about the approximation properties of f.e. spaces is the following:

$$\| e^h \|_{H^r(\Omega)} \leq C_r h^{k-r} |u|_{H^k(\Omega)} \quad , r = 0,1 \qquad 2.3$$

where $e^h = u - u_I^h$, $e^h \in H^1(\Omega)$.

A proof of inequality (2.3) can be found in Strang and Fix (1973): we outline the main steps of the proof. Firstly it results:

$\| e^h \|_{H^r(T_1)} \leq C'_r h^{k-r} | u |_{H^k(T_1)}$, $r = 0,1$ where T_1 is the triangle with vertices

(0,0), (0,1), (1,0); in this step a key rôle is played by the property that the seminorm $|\cdot|_{H^k}$ is equivalent to the norm of the quotient space H^k/P_{k-1} *, and by the so called Bramble-Hilbert lemma, which states that if f is a continuous linear functional on $H^k(\Omega)$ such that $f(p) = 0 \ \forall u \in P_{k-1}$, then there exist γ, depending only on Ω such that

$$|f(u)| \leq \gamma \| f \|_{H^{-k}(\Omega)} \ | u |_{H^k(\Omega)} \quad \forall u \in H^k(\Omega)$$

Then, using a linear transformation which maps T_1 onto the element e_i it is shown that

$$\| e^h \|_{H^r(e_i)} \leq C''_r h_i^{k-r} |u|_{H^k(e_i)} \qquad r = 0,1$$

wher C''_r depends on the smallest angle of e_i; finally, summing over all elements and taking into account that, independently of h, all angles are bounded from below, (2.3) are obtained.

Note that the order of the interpolation error is determined by the degree k-1 of the f.e. space only if $u \in H^k(\Omega)$; when $u \in H^{k'}(\Omega)$, $k' < k$, the error is $O(h^{k'-r})$.

The convergence result can be generalized in several directions:

- the same error bound holds for Hermite spaces, provided that the nodal parameters involve derivatives of order $|\alpha| = \alpha_1 + \alpha_2 < k-1$ because otherwise the interpolate $u_I^h = \Sigma_j q_j \phi_j$, with some $q_j = \dfrac{\partial^{|\alpha|} u}{\partial x^{\alpha_1} \partial y^{\alpha_2}} (x_j, y_j)$, would not exist;
- if $S^h \subset H^q(\Omega)$, $S^h \not\subset H^{q+1}(\Omega)$, (2.3) holds for $r \leq q \leq k-1$
- the result is still valid for a rectangular mesh on a rectangle: h is the length of the longest diagonal;
- an analogous error bound holds for isoparametric elements, provided that the coordinate transformation $F: \bar{\Omega}' \to \bar{\Omega}$, $F(\xi,\eta) = [x(\xi,\eta) \ y(\xi,\eta)]^T$ is sufficiently regular and that the Jacobian of the transformation is bounded away from zero, uniformly in h: this happen if all angles are larger than $\alpha_0 > 0$ for triangles, or smaller than $\alpha_1 < \pi$ for quadrilateral elements.

In one space dimension, (2.3) holds for $k \geq 1$, $r \leq q$, where q is such that $S^h \subset H^q(\Omega)$, $S^h \not\subset H^{q+1}(\Omega)$. For domains in $\mathbb{R}^3$ the result is the same as in $\mathbb{R}^2$.

THE FINITE ELEMENT METHOD FOR ELLIPTIC BOUNDARY VALUE PROBLEMS.

Variational Equations.

Let V be a real Hilbert space, a(v,w) a symmetric, continuous and coercive ($a(v,v) \geq C \|v\|^2_V$, $C > 0$) bilinear form of V (V-elliptic), $\ell(v)$ a continuous linear functional on V: let use define the quadratic functional

$$I(v) = a(v,v) - 2\ell(v) \tag{3.1}$$

* P_{k-1} is the space of polynomials of degree less than k.

Then, (see e.g. Lions (1971)), there exists a unique $u \in V$ which minimizes $I(v)$: this is characterized by the variational equation

$$a(u,v) = \ell(v) \quad , \quad \forall v \in V \tag{3.2}$$

Example 3.1:

$$\Omega = (0,1) \quad V = H_0^1(0,1)$$

$$a(v,w) = \int_0^1 p\frac{dv}{dx}\frac{dw}{dx}dx + \int_0^1 qvw\, dx$$

$$p,\ q \in L^\infty(0,1) \quad ; \quad p(x) \geq \bar{p} > 0 \quad , \quad q(x) \geq 0 \text{, a.e. } x \in (0,1)$$

$$\ell(v) = \int_0^1 fv\, dx \quad , \quad f \in L(0,1)$$

The bilinear form is V-ellyptic and a unique $u \in H_0^1(0,1)$ satisfies (3.2) and minimizes (3.1). If the boundary value problem

$$-\frac{d}{dx}\left(p\frac{du}{dx}\right) + qu = f \tag{3.3}$$

$$u(0) = u(1) = 0$$

has a regular solution, then u satisfies (3.2), as can be verified by multiplying both sides of (3.3) by $v \in H_0^1(0,1)$ and integrating over (0,1). Furthermore u minimizes

$$I(v) = \int_0^1 [p(\frac{dv}{dx})^2 + q\ v^2]dx + 2\int_0^1 fvdx$$

In any case the solution of (3.2) (regular or not) is called weak solution of (3.3).

Example 3.2.

$$V = H_0^2(0,1)$$

$$a(v,w) = \int_0^1 [r\frac{d^2v}{dx^2}\frac{d^2w}{dx^2} + p\frac{dv}{dx}\frac{dw}{dx} + qvw]dx$$

$$r,p,q \in L^\infty(0,1) \quad , \ r(x) \geq \bar{r} > 0 \quad , \ p(x) \geq 0 \quad , \ q(x) \geq 0 \ , \text{ a.e. } x \in 0,1$$

$$\ell(v) = \int_0^1 fv\, dx \ , \quad f \in L^2(0,1)$$

The variational equation (3.2), with V, $a(v,w)$ and $\ell(v)$ as above is the weak form of the fourth order elliptic homogeneous boundary value problem

$$\frac{d^2}{dx^2}\left(r\frac{d^2u}{dx^2}\right) + \frac{d}{dx}\left(p\frac{du}{dx}\right) + qu = f$$

$$u(0) = u(1) = 0 \quad ; \quad \frac{du}{dx}(0) = \frac{du}{dx}(1) = 0$$

The above examples enable to emphasize the relationship between elliptic boundary value problems and variational equations: indeed if A is a differential elliptic operator of order 2m we can always associate to A a bilinear form on $V \subseteq H^m(\Omega)$ by taking $a(v,w) = \langle Av,w\rangle$* and performing m integrations by parts on the r.m.s. $a(v,w)$ need not be symmetric, but satisfies ($a(v,v) \geq c \|v\|^2_{H^m(\Omega)}$, $c > 0$). The equation $Au = f$ can be put into the weak form

$$a(u,v) = \langle f,v\rangle \ , \ \forall v \in V \tag{3.4}$$

and if $a(v,w)$ is symmetric, equation (3.4) qualifies as the variational equation associated to the minimum problem of a quadratic functional $I(v)$; moreover $Au = f$ is the Euler-Lagrange equation for $I(v)$.

A fundamental reference concerning elliptic problems and their variational treatment is Nečas (1967).

The Ritz-Galerkin method.

In order to approximately solve the minimum problem for (3.1) let us denote by $\{\phi_1,\phi_2,\ldots,\phi_N\}$ an N-tuple of linear independent elements of V, which we arrange in the vector $\phi^N = [\phi_1,\phi_2,\ldots,\phi_N]^T$ and by V^N the subspace of V spanned by $\{\phi_1,\phi_2,\ldots,\phi_N\}$ (trial space): $V^N = \{v^N \in V : v^N = \sum_{i=1}^{N} v_i^N \phi_i = \underline{\phi}^{NT} \underline{v}^N, \ \underline{v}^N \in \mathbb{R}^N\}$. Moreover let $A^N \in \mathbb{R}^{N\times N}$ be the matrix defined by $[A^N]_{ij} = a(\phi_i,\phi_j)$, and $\underline{b}^N \in \mathbb{R}^N$ by $[\underline{b}^N]_i = \ell(\phi_i)$. It follows from the properties of $a(v,w)$ that $\underline{A}^N$ is symmetric positive definite. The problem of minimizing $I(v)$ on V^N has a unique solution and since

$$I(v^N) = a(v^N,v^N) - 2\ell(v^N) = \underline{v}^{NT} \underline{A}^N \underline{v}^N - 2\underline{b}^{NT} \underline{v}^N$$

the minimum is attained at $u^N = \underline{\phi}^{NT} \underline{u}^N$, where $\underline{u}^N \in \mathbb{R}^N$ satisfies the equation

$$\underline{A}^N \underline{u}^N = \underline{b}^N$$

which is equivalent to the equation

$$a(u^N,v^N) = \ell(v^N) \qquad \forall v^N \in V^N \tag{3.5}$$

The procedure outlined is the classical direct (Ritz) method for obtaining an approximate solution u^N to the variational problem; u^N is the best approximation, in the "energy" norm, to the actual solution u of the variational problem:

$$a(u - u^N , u - u^N) \leq a(u - v^N , u - v^N) \quad \forall v^N \in V^N \tag{3.6}$$

inequality (3.6) is fundamental in proving the convergence of specific Ritz approximations.

A conceptually different approach (the Galerkin method) which can be regarded as a generalization of the Ritz method, is to refer directly to the variational equation (3.2) and look for that function $u^N \in V^N$ such that (3.5) holds: of course the solution u^N is the same as before if $a(v,w)$ derives from a quadratic strictly convex functional $I(v)$, but the method can be also applied to non-symmetric bilinear forms

* In order that both sides make sense $\langle .,.\rangle$ must be taken as duality pairing between V and its dual.

and to non-linear forms. Some authors still define the function $u^N \in S^N$, S^N subspace of the assumed solution space, such that

$$\langle A(u^N), v^N\rangle = \langle f, v^N\rangle \qquad \forall v^N \in W^N \tag{3.7}$$

where W^N is a subspace of a "test space" W, as Galerkin solution of the equation $A(u) = f$. On the l.h.s. of equation (3.7) it is to be understood that as many integrations by parts as necessary have been performed for the inner product to make sense, in accordance with the choice both of the solution and test spaces. The above formulation is a special case of the weighted residual method (Finlayson, 1972): a recent classification of this type of approximate methods can be found in Brebbia (1978).

An application of the Galerkin method to a non self-adjoint operator, to which a non-symmetric form is associated, is shown in the following:

example 3.3.

Let $Au = -\dfrac{d}{dx}\left(p\dfrac{du}{dx}\right) + r\dfrac{du}{dx} + qu$, p,q as in example 3.1 and $V = H^1_o(0,1)$;

the solution of the variational equation

$$\int_o^1 p\frac{du}{dx}\frac{dv}{dx}\,dx + \int_o^1 r\frac{du}{dx}\,v\,dx + \int_o^1 q\,v\,dx = \int_o^1 f\,v\,dx, \quad \forall v \in H_o(0,1)$$

is the weak solution of the boundary value problem

$$\begin{aligned} Au &= f \\ u(0) &= u(1) = 0 \end{aligned} \tag{3.8}$$

Taking $S^N = V^N = \text{span}\,\{\phi_1,\ldots,\phi_N\}$ the Galerkin approximate solution of (3.8) is $u^N(x) = \underline{\phi}^{NT}(x)\underline{u}^N$, $\underline{u}^N$ such that

$$\underline{u}^{NT}\underline{K}^N\underline{v}^N + \underline{u}^{NT}\underline{R}^N\underline{v}^N + \underline{u}^{NT}\underline{M}^N\underline{v}^N = \underline{f}^{NT}\underline{v}^N, \quad \forall\, \underline{v}^N \in \mathbb{R}^N$$

that is $\underline{u}^N = (\underline{K}^N + \underline{R}^{NT} + \underline{M}^N)^{-1}\underline{f}^N$

where

$$[\underline{K}^N]_{ij} = \int_o^1 p\frac{d\phi_i}{dx}\frac{d\phi_j}{dx}\,dx \;;\quad [\underline{R}^N]_{ij} = \int_o^1 r\frac{d\phi_i}{dx}\phi_j\,dx \;;$$

$$[\underline{M}^N]_{ij} = \int_o^1 q\phi_i\phi_j\,dx \;;\quad [\underline{f}^N]_i = \int_o^1 f\phi_i\,dx$$

The matrices $\underline{K}^N$, $\underline{M}^N$ are symmetric positive definite; we have assumed that $\underline{K}^N + \underline{R}^{NT} + \underline{M}^N$ is nonsingular, but specific r and basis functions for which it is singular can be found. A condition ensuring that $\underline{K}^N + \underline{R}^{NT} + \underline{M}^N$ is nonsingular is that $r \in C^1(0,1)$ and $dr/dx \leq 0$.

Finite Element Solution of Elliptic Equations.

In example 3.1 and 3.2 the best choice of the space V^N would be the space spanned by the first N orthonormal eigenfunctions of the operator A associated to the form

$a(v,w)$, but this requires the prior solution of a more difficult problem, at least when the coefficients are not constant. In one dimension, or in two-dimensional domains of special shape, trigonometric functions or orthogonal polynomials can be used, but in general the resulting matrices would be full. A decisive improvement is obtained by considering a finite element space with a stricly local basis. In order to clarify what this choice implies we discuss the f.e. approximation of the homogeneous boundary value problems

$$-\left[\frac{\partial}{\partial x}\left(p\frac{\partial u}{\partial x}\right)+\frac{\partial}{\partial y}\left(p\frac{\partial u}{\partial y}\right)\right]+qu=f\ ,\quad (x,y)\in\Omega \tag{3.9}$$

$$a:\ u=0 \quad \text{on } \Gamma \text{ (Dirichlet)}$$

$$b:\ \frac{\partial u}{\partial n}=0 \quad \text{on } \Gamma \text{ (Neumann)}$$

where Ω is a convex polygonal domain with boundary Γ, $p,\ q\in L^{\infty}(\Omega)$ and $p(x,y)\geq\bar{p}>0$, $q(x,y)\geq 0$ a.e. on Ω, $f\in L^2(\Omega)$

Case a). The weak solution is $u\in H^1_o(\Omega)$ such that:

$$a(u,v)=\langle f,v\rangle \quad \forall v\in H^1_o(\Omega) \tag{3.10}$$

with

$$a(u,v)=\iint_{\Omega} p\left(\frac{\partial u}{\partial x}\frac{\partial v}{\partial x}+\frac{\partial u}{\partial y}\frac{\partial v}{\partial y}\right)dxdy+\iint_{\Omega} q\ u\ v\ dxdy;$$

$a(u,v)$ is coercive on $H^1_o(\Omega)$ and u is the solution of the minimum problem: find $u\in H^1_o(\Omega)$ such that $I(u)\leq I(v)\quad \forall v\in H^1_o(\Omega)$

with

$$I(v)=\iint_{\Omega}\left\{p\left[\left(\frac{\partial v}{\partial x}\right)^2+\left(\frac{\partial v}{\partial y}\right)^2\right]+qv^2-fv\right\}dxdy \tag{3.11}$$

Case b). The weak form of (3.9) is that of case a, but u,v must be taken in $H^1(\Omega)$: indeed the function which satisfies (3.10) $\forall v\in H^1(\Omega)$ and then minimizes $I(v)$ over $H^1(\Omega)$, automatically fulfills the Neumann boundary condition: in view of this property the Neumann condition is called natural, whilst the Dirichlet condition is called essential. Note that $a(v,w)$ is coercive on $H^1(\Omega)$ only if $q(x,y)\geq\bar{q}>0$ a.e. on Ω.

Let us consider a Lagrangian finite element space $S^h\in H^1(\Omega)$ over a triangular mesh adapted to Ω, with mesh size h; let (x_j,y_j), $j=1,2,\ldots,N$ be the nodes of the mesh, numbered so that the first N^i nodes are internal, and $\phi_1,\phi_2,\ldots,\phi_N$ the associated stricly local basis functions spanning S^h. We denote by S^h_o the subspace of S^h consisting of functions vanishing on Γ: since Ω assumed to be a polygon, a function of S^h vanishes on Γ if it vanishes at the boundary nodes (x_j,y_j), $j=N_i+1,\ldots,N$, so S^h_o is spanned by $\phi_1,\phi_2,\ldots,\phi_{Ni}$.

For the application of the Ritz-Galerkin procedure to the above problem we take the approximating space to be S^h_o for case a, S^h for case b respectively: then the approximate solution of (3.9) is given by

$$u^h(x,y) = \sum_{j=1}^{M} u_j^h \, \phi_j(x,y), \quad M = N^i \text{ for case a, } M=N \text{ for case b}$$

where the vector of nodal parameters $\underline{u}^h = [u_1^h \ldots\ldots u_M^h]^T$, which directly gives the value of the approximate solution at the nodes, satisfies the equation:

$$(\underline{K}^h + \underline{M}^h)\underline{u}^h = \underline{f}^h$$

with

$$[\underline{K}^h]_{ij} = \iint_\Omega p\left(\frac{\partial\phi_i}{\partial x}\frac{\partial\phi_j}{\partial x} + \frac{\partial\phi_i}{\partial y}\frac{\partial\phi_j}{\partial y}\right)dx\,dy, \quad [\underline{M}^h]_{ij} = \iint_\Omega q\phi_i\phi_j \, dx\,dy\ ;$$

$$[\underline{f}^h]_i = \iint_\Omega f\phi_j \, dx\,dy \quad ; \; i = 1,2,\ldots,M \quad ; \quad J = 1,2,\ldots,M .$$

Note that the variational formulation enables approximating spaces to be used which would be unadmissible for the differential equation: the f.e. method fully exploits this advantage.

When a Hermitian space is used the vector of nodal parameters includes also the values of some derivatives at the nodes. Since a strictly local basis is chosen, it follows that $\underline{K}^h$, $\underline{M}^h$ are sparse: in fact $[\underline{K}^h]_{ij}$ and $[\underline{M}^h]_{ij}$ are nonzero only if the nodes P_i, P_j belong to the same element. The nonzero entries of $\underline{K}^h$, $\underline{M}^h$ can be computed by applying directly their definitions, where the integrals vanish outside $\eta^i \cap \eta^j$. An alternative procedure, which is often more efficient, consists in constructing the so-called element matrices and then assembling them into $\underline{K}^h$ and $\underline{M}^h$. Element matrices $\underline{K}_\ell$, $\ell = 1,2,\ldots,N^e$ arise from the fact that the quadratic form:

$$\underline{u}^{hT}\,\underline{K}^h\,\underline{u}^h = \iint_\Omega p\left[\left(\frac{\partial u^h}{\partial x}\right)^2 + \left(\frac{\partial u^h}{\partial y}\right)^2\right]dx\,dy = \sum_{\ell=1}^{N^e} \iint_{e_\ell} p\left[\left(\frac{\partial u^h}{\partial x}\right)^2 + \left(\frac{\partial u^h}{\partial y}\right)^2\right]dx\,dy$$

can be written as:

$$\underline{u}^{hT}\,\underline{K}^h\,\underline{u}^H = \sum_{\ell=1} \underline{u}_\ell^T\,\underline{K}_\ell\,\underline{u}_\ell$$

where $\underline{u}_\ell$ is the vector of the nodal parameters attached to the nodes belonging to e_ℓ. Several methods are proposed to evaluate the entries of $\underline{K}_\ell$. Those most commonly used involve a transformation of the element e_ℓ into a standard element e' by an appropriate coordinate change and the representation of u^h by a polynomial on e': then the connection matrix T_ℓ between the coefficients of the polynomial on e' and the original nodal parameters is evaluated, and the quadratic form is written as:

$$\underline{u}_\ell^T\,\underline{K}_\ell\underline{u}_\ell = \underline{u}_\ell^T\,\underline{T}_\ell^T\,\underline{G}_\ell\,\underline{T}_\ell\underline{u}_\ell$$

where the entries of $\underline{G}_\ell$ are of type $\iint_{e'} \xi^s\eta^r p(\xi,\eta)\,d\xi d\eta$.

When p is constant on the element, $\underline{G}_\ell$ can be computed exactly otherwise numerical integration is unavoidable. The same arguments can be repeated without change for $\underline{M}^h$.

The entries of vector $\underline{f}^h$ can also be computed by the same procedure; an alternative to numerical integration is to replace f(x,y) by its interpolation on S^h: in this case $\underline{f}^h$ is given by $\underline{f}^h = \underline{M}^h \underline{f}_I$, where $\underline{f}_I$ denotes the vector of the nodal parameters of f(x,y).

The above approach can also be applied to solving inhomogeneous boundary value problems. For equation (3.9) let us consider:

a') $u = f$ on Γ

b') $p \frac{\partial u}{\partial n} + su = b$ on Γ

In case a' assume that there exists a function $\phi_o \in H^2(\Omega)$ such that $\phi_o = g$ on Γ: then the weak solution of (3.9) is the unique u which minimizes I(v) over the linear manifold $H^1_E(\Omega) = \{v \in H^1(\Omega) : v = w + \phi_o, w \in H^1_o(\Omega)\}$, and is given by $u=\hat{u} + \phi_o$, where $\hat{u} \in H^1_o(\Omega)$ is the solution of the variational equation

$$a(\hat{u} + \phi_o, w) = \langle f,w \rangle \quad \forall w \in H^1_o(\Omega) \tag{3.12}$$

In case b') the solution u minimizes over $H^1(\Omega)$ the functional $I_b(v) = a(v,v) - 2\langle f,v \rangle + \int_\Gamma (dv^2 - 2bv)ds$ and is characterized by the variational equation

$$a(u,v) + \int_\Gamma s\ u\ v\ ds = \langle f,v \rangle + \int_\Gamma b\ v\ ds\ , \quad \forall v \in H^1(\Omega) \tag{3.13}$$

The f.e. approximation is made in the same way as before, by taking $\hat{u}$,w,v of equation (3.12) in S^h_e , and u, v of equation (3.13) in S^h. The f.e. technique applies equally to general weighted residual formulations, provided that the solution and test spaces are appropriate f.e. spaces. An often-used variant is to choose as test functions Dirac impulses suitably placed within the element, thus yielding the so called f.e. collocation method (Finlayson, 1980). The convergence of the Ritz -Galerkin f.e. solution u^h of the elliptic second order problems described above can be established by using inequality (3.6) and taking $v^h = u_I$: $\|u-u^h\|^2_{H^1(\Omega)} \leq c_1 a(u - u^h, u - u^h) \leq c_1 a(u - u_I, u - u_I) \leq c_2 \|u - u_I\|^2_{H^1(\Omega)}$. The behaviour of the approximation error then depends on the interpolation properties of S^h and is related both to the regularity of u and to the degree of S^h: if the actual solution $u \in H^k(\Omega)$ and a space of degree at least k-1 is used, from (2.5) it follows:

$$\|u - u^h\|_{H^1(\Omega)} \leq c' h^{k-1} \|u\|_{H^k(\Omega)} \tag{3.13}$$

Furthermore it can be shown that the error in L^2 norm is $O(h^k)$. In general, for an elliptic operator of order 2m and a f.e. space of degree k-1, $S^h \subset H^1(\Omega)$, the approximation error, for $s \geq 2m-k$, $u \in H^k(\Omega)$, is:

$$\|u - u^h\|_{H^s_1(\Omega)} \leq c h^{k-s} \|u\|_{H^k(\Omega)}$$

The above convergence rates are obtained when all the operations are performed exactly: when solving a specific problem this is often impossible and some further approximations must be introduced, thus committing the so-called "variational crimes".

The most important are:

a) approximate computation of $I(v^h)$ by numerical integration either over elements or along the boundary;
b) approximation of the boundary;
c) approximation of the essential boundary conditions.

All these approximations produce supplementary errors in the solution, so the error bound (3.13) becomes

$$\| u - u^h \|_{H^1(\Omega)} = ch^{k-1} \| u \|_{H^k(\Omega)} + \Delta$$

The convergence of the method is ensured when $\Delta = O(h)$ at least; furthermore an approximation is called optimal when $\Delta = O(h^{k-1})$.

Without entering into details (see e.g. Mitchell and Wait (1977)) we report some results for equation (3.9) and triangular elements. Where numerical integration is concerned the order of Δ depends on the degree of accuracy of the quadrature formula used: the approximation is convergent, for S^h of degree k-1, with formulas of degree k-2, and is optimal for formulas of degree 2(k-2) provided that p,q,f are sufficiently regular. The error depending on the approximation of a regular domain Ω by a polygon Ω_h inscribed into Ω is of order $O(h^{3/2})$ in the $H^1(\Omega)$ norm. As the boundary condition $u = g$ on Γ (Γ polygonal boundary) is concerned, the approximation made by imposing that $u^h(P_j) = g(P_j)$ at the nodes P_j of Γ does not affect the approximation error estimate.

The convergence results reported follow from inequality (3.6) which holds for the Ritz solution of an elliptic problem. Convergence in L^2 norm is also ensured for Galerkin approximations of non-symmetric coercive problems and for some specific non-coercive problems.

We conclude this section with an outline of the sequence of operations necessary to solve a given problem. The starting point is the definition of the mesh and the choice of the element to be used: it is possible to obtain problems of similar complexity by using either fine meshes with simple elements or coarse meshes with more complicated elements. A balanced choice between these alternatives requires careful consideration both of the programming and computing facility available, and of the characteristic features of the problem (shape of the domain, regularity of the coefficients and of the forcing term, boundary conditions).
In some cases the standard Ritz method would require the use of elements ensuring a high degree of continuity, which could be very difficult to be dealt with: this happens, for example, when differential operators of high order are involved, or when an accurate reconstruction of derivatives of u is necessary.

Where the solution of high order problems is concerned, convergence is still ensured when suitable "non-conforming" elemens are chosen; on the other hand the application of the so called mixed variational principles (Oden, 1972) enables to compute the unknown function and its derivatives with comparable accuracy.

Once the type of elements has been chosen, the mesh must be drown; this is often a very cumbersome task when performed manually: for this purpose computer codes are now available, which largely automatize the procedure. Their interactive capabilities are useful in performing subsequent refinements of the mesh (see e.g. Cross and Lewis, 1978).

The next step is to build up the system of equations, for example by assembling the element matrices: in order to save core occupation this operation is often performed virtually, using appropriate data structures. Finally for solving the equations several algorithms are available: still the choice among them depends on several factors such as dimension, density, bandwith of matrices, computing resources available. In the books of Bathe and Wilson (1976) and of Norrie and De Vries (1978) the main aspects of the numerical implementation of the f.e. method are treated and a wide bibliography is reported.

FINITE ELEMENT APPROXIMATION FOR INITIAL VALUE PROBLEMS

The Semi-Discrete Galerkin Approximation.

The finite element technique, initially developed for time independent problems, is extremely useful, and is finding more and more applications in solving approximately initial value problems, both for parabolic and hyperbolic sistems.

An approach similar to the Ritz method can be attempted: indeed in some cases a time dependent variational principles exists (e.g. the Hamilton principle), which ensure that the actual trajectory of a system in a time interval makes a suitable functional stationary over a set of admissible trajectories: this happens in the case of the wave equation. Even in the absence of a variational principle (such as for the heat equation), a functional can often be constructed using the adjoint formulation. To the drawbacks of this approach quoted in Mitchell and Wait (1977) we add that in the variational formulation of an initial value problem the conceptually different role played by the time and space variables is lost: on the other hand this difference is well present when distributed systems are regarded as a particular class of abstract dynamic systems with an infinite dimensional state space. Therefore we prefer to describe the so-called semi-discrete Galerkin method for the approximation of parabolic and hyperbolic initial value problems, using a f.e. trial space: this leads to represent a d.p.s. by a lumped parameter model.

In order to illustrate the method let us consider domain $\Omega \subset \mathbb{R}^2$, the space $V = H_0^1(\Omega)$, and the operator $A = -\left[\frac{\partial}{\partial x}\left(p\frac{\partial}{\partial x}\right) + \frac{\partial}{\partial y}\left(p\frac{\partial}{\partial y}\right)\right]$ with the associated symmetric coercive form:

$$a(v,w) = \iint_\Omega p\left(\frac{\partial v}{\partial x}\frac{\partial w}{\partial x} + \frac{\partial v}{\partial y}\frac{\partial w}{\partial y}\right) dx\,dy \quad ;$$

let us assume that the partial differential equation

$$\frac{\partial u}{\partial t} = \frac{\partial}{\partial x}\left(p\frac{\partial u}{\partial x}\right) + \frac{\partial}{\partial y}\left(p\frac{\partial u}{\partial y}\right) + f \quad , (x,y)\in\Omega \quad , \quad t > 0 \quad (4.1)$$

with boundary and initial conditions

$$u(x,y,t) = 0 \quad , \quad (x,y) \in \Gamma$$

$$u(x,y,0) = u_o(x,y) \quad , \quad (x,y) \in \Omega$$

has a classical solution: by mulyiplying both sides of (4.1) by $v \in H_o^1(\Omega)$ we have

$$\frac{d}{dt} \iint_\Omega u(x,y,t)v(x,y)\,dxdy = -\iint_\Omega p \left(\frac{\partial u}{\partial x}\frac{\partial v}{\partial x} + \frac{\partial u}{\partial y}\frac{\partial v}{\partial y}\right) dx\,dy \; +$$

$$+ \iint_\Omega f(x,y,t)\; v(x,y)\,dx\,dy \qquad t > 0,$$

$$\iint_\Omega u(x,y,0)\; v(x,y)\;dx\,dy = \iint_\Omega u_o(x,y)\; v(x,y)\,dx\,dy \quad t = 0$$

that is

$$\frac{d}{dt} \langle u(t)\;,\; v\rangle = -\; a(u(t),\; v) + \langle f(t),\; v\rangle \quad ; \quad t > 0; \quad v \in H_o^1(\Omega) \qquad (4.2)$$

Actually equation (4.2) holds under less severe assumptions on the regularity of u than those necessary for equation (4.1). The precise functional ambientation can be found in Lions (1971).

Now, given a finite element space $S_o^h \subset H_o^1(\Omega)$, spanned by $\phi_1, \phi_2, \ldots, \phi_{N^i}$, the Galerkin approximate solution of (4.1) is obtained by imposing the condition

$$u^h(t) \in S_o^h \;;\; \text{then}$$

$$u^h(t) = \sum_{k=1}^{N^i} u_k^h(t)\phi_k(x,y) \qquad (4.3)$$

satisfies (4.2) for every $t > 0$ and $v = v^h \in S_o^h$. The initial condition can be approximated in several ways: by the othogonal projection of u_o onto S_o^h or even, if u_o is sufficiently smooth, by the interpolate of u_o: it is required in any case that the initial state approximation be convergent.

By substituting the expression (4.3) for u and ϕ_i $(i = 1,2,\ldots N^i)$ for v into equation (4.2) we obtain the following system of ordinary differential equation for the nodal coefficients:

$$\left.\begin{array}{l} \underline{M}^h \dot{\underline{u}}^h = \underline{K}^h \underline{u}^h + \underline{f}^h \\ \underline{M}^h \underline{u}^h(0) = \underline{u}_o^h \end{array}\right. \qquad (4.4)$$

where $\underline{M}^h$, $\underline{K}^h$, $\underline{f}^h$ are defined in section 3.3 and $[\underline{u}_o^h]_i = \langle u_o, \phi_i \rangle$. The same procedure applies to homogeneous Neumann problems and to inhomogeneous Dirichlet and Neumann problems, even with time dependent boundary conditions, provided that the data are sufficently smooth.

Where hyperbolic equations are concerned the standard application example of the Galerkin finite element method is the wave equation

$$\frac{\partial^2 u}{\partial t^2} = Au + f$$

with initial conditions

$$u(0) = u_o \ , \quad \frac{\partial u}{\partial t}(0) = u'_o$$

By applying the Galerkin approximation to the weak form as before, we obtain the second order differential equation

$$\underline{M}^h \frac{d^2\underline{u}^h}{dt^2} = -\underline{K}^h\underline{u}^h + \underline{f}^h$$

$$\underline{M}^h\underline{u}^h(0) = \underline{u}_o^h \ , \quad \underline{M}^h \frac{d\underline{u}^h}{dt}(0) = \underline{u}'^h_o$$

or even:

$$\underline{\mathcal{M}}^h\dot{\underline{z}}^h = \underline{\mathcal{K}}^h\underline{z}^h + \underline{\delta}^h$$

$$\underline{\mathcal{M}}^h\underline{z}^h(0) = [\underline{u}_o^{hT} \ \underline{u}'^{hT}_o]^T$$

where

$$\underline{\mathcal{M}}^h = \begin{bmatrix} \underline{I} & \underline{0} \\ \underline{0} & \underline{M}^h \end{bmatrix} \quad \underline{\mathcal{K}}^h = \begin{bmatrix} \underline{0} & \underline{I} \\ -\underline{K}^h & \underline{0} \end{bmatrix} \quad \underline{z}^h = \begin{bmatrix} \underline{u}^h \\ \dot{\underline{u}}^h \end{bmatrix} \quad \underline{\delta}^h = \begin{bmatrix} \underline{0} \\ \underline{f}^h \end{bmatrix}$$

The Galerkin approach, in conjunction with a f.e. discretization with respect to the space variables is widely applied in the construction of finite dimensional approximations of nonlinear evolution problems. Moreover the construction of a semi-discrete Galerkin model is often the first step towards obtaining a fully-discrete model, where the ordinary differential equation is replaced by a difference equation: a popular difference-in-time scheme for first order system is that of Crank-Nicholson:

$$\underline{M}^h[\underline{u}^h(t + \Delta t) - \underline{u}^h(t)]\Delta t^{-1} = -\underline{K}^h[\underline{u}^h(t + \Delta t) + \underline{u}^h(t)]/2 + \underline{f}^h(t + \Delta t/2)$$

The finite element discretization in the time variable is not commonly used to approximate evolution problems (cf. Strang and Fix, 1973), since the direct application of the Galerkin method with the usual basis functions over a finite time interval destroys the non-anticipative character of the equation; moreover those geometrical difficulties which make the use of finite elements particularly attractive, are absent. Good results have been recently obtained by choosing non-standard basis and test functions in the numerical solution of problems where "forward propagation", is essential and standard f.e. approximation yields poor results (Ramakrishnan,1979).

Several convergence results are available for linear parabolic problems: starting from the fundamental approximation property of finite element spaces, convergence in L^2 and H^1 norms for every t is proved for suitable assumptions on the smoothness of the actual solution and its time derivative, and in the absence of variational crimes; likewise convergence is proved for some non-linear problems.

Following similar works on elliptic equations (Nitche, 1977), L^{∞} convergence results have also been obtained: see Nitche (1979), where a wide bibliography is reported.

The convergence of the semi-discrete Galerkin approximate solution of the wave equation has also been investigated: see e.g. Baker and Bramble (1979). Note that, unlike what happens for parabolic equations, the error bound is affected by a coefficient which depends on the length of the time interval.

The Approximation of Eigenfunction Expansion.

The solution of a parabolic problem can be represented in terms of the eigenvalues and eigenfunctions of the elliptic operator involved. Let us consider the equation

$$\begin{aligned} \frac{\partial u}{\partial t} &= - Au + f \qquad t > 0 \\ u(o) &= u_o \end{aligned} \tag{4.5}$$

where A is a self-adjoint elliptic operator of order 2m on V, V closed subspace of $H^m(\Omega)$ which we take to be $\overset{o}{H}{}^m(\Omega)$. The eigenvalues of A are real, positive and can be arranged in a non decreasing sequence $0 < \lambda_1 \leq \lambda_2 \leq \ldots .$, with $\lim_{i \to \infty} \lambda_i = + \infty$; moreover there exists an orthonormal (in $L^2(\Omega)$) basis of eigenfunctions v_i, i=1,2,. .. where v_i corresponds to λ_i; for any $u \in V$ the expression

$$u = \sum_{j=1}^{\infty} \langle u,v_j\rangle v_j$$

leads to the expression

$$- Au = \sum_{j=1}^{\infty} - \lambda_j \langle u,v_j\rangle v_j$$

The operator -A generates an analytic semigroup and, for $f \in L^2((0,T)x\Omega)$, the solution of (4.5) can be represented as

$$u = \sum_{j=1}^{\infty} e^{-\lambda_j t} \langle u_o,v_j\rangle v_j + \sum_{j=1}^{\infty} \left(\int_0^t e^{-\lambda_j (t-\sigma)} f_j(\sigma) d\sigma \right) v_j \tag{4.6}$$

where $f_j(\sigma) = \langle f(\sigma),v_j\rangle$.

The above formulation is the basis for extending several concepts and results from the control theory of finite dimensional linear systems to d.p.s. of type (4.5).

The f.e. approximate solution of (4.5): $u^h(t) = \sum_{i=1}^{Ni} u_i^h(t)\phi_i$, $\underline{u}^h(t) \in \mathbb{R}^{Ni}$, which satisfies

$$\begin{aligned} \underline{M}^h \dot{\underline{u}}^h &= - \underline{K}^h \underline{u}^h + \underline{f}^h \\ \underline{M}^h \underline{u}^h(o) &= \underline{u}_o^h \end{aligned} \tag{4.7}$$

with

$$[\underline{M}^h]_{ij} = \langle\phi_i,\phi_j\rangle \quad , \quad [\underline{K}^h]_{ij} = \langle A\phi_i,\phi_j\rangle = a(\phi_i,\phi_j)$$

$$[\underline{u}_o^h]_i = \langle u_o,\phi_i\rangle \quad , \quad [\underline{f}^h]_i = \langle f,\phi_i\rangle$$

can be expanded as a finite sum of terms like those in (4.6): in fact let λ_i^h, $\underline{q}_i^h$ be the eigenvalues and eigenvectors of the pencil $(\underline{M}^h,\underline{K}^h)$, that is the solutions of the problem $(\lambda^h \underline{M}^h - \underline{K}^h)\underline{q}^h = 0$. Since $\underline{K}^h$, $\underline{M}^h$ are symmetric positive definite, N^i real positive eigenvalues and N^i independent eigenvectors exist: the eigenvalues λ_i^h are arranged in non-decreasing order $0 < \lambda_1^h \leq \lambda_2^h \leq \ldots \leq \lambda_{N^i}^h$ and the eigenvectors can be normalized so that:

$$\underline{q}_i^{hT}\underline{M}^h\underline{q}_j^h = \delta_{ij}, \quad \underline{q}_i^{hT}\underline{K}^h\underline{q}_j^h = \lambda_i^h\delta_{ij}$$

Then the solution of (4.7) is given by

$$\underline{u}^h(t) = \underline{Q}\exp(-\underline{\Lambda}t)\underline{Q}^T\underline{u}_o^h + \int_o^t \underline{Q}\exp[-\underline{\Lambda}(t-\sigma)]\underline{Q}^T\underline{f}^h(\sigma)\,d\sigma$$

where $\underline{\Lambda} = \mathrm{diag}(\lambda_1^h, \ldots, \lambda_{N^i}^h)$, $\underline{Q} = [\underline{q}_1^h\,\underline{q}_2^h \ldots\ldots \underline{q}_{N^i}^h]^T$, from which, putting $v_i^h = \sum_{j=1}^{N^i} (\underline{q}_i^h)_j\phi_j$, $i = 1,2,\ldots,N^i$,

we have

$$u^h(t) = \sum_{i=1}^{N^i}\{\exp(-\lambda_i^h t)\langle u_o,v_i^h\rangle + \int_o^t \exp[-\lambda_i^h(t-\sigma)]\langle v_i^h,f(\sigma)\rangle d\sigma\}v_i^h$$

The eigenvalues λ_i^h and the functions $v_i^h \in S_i^h$ are approximations of the eigenvalues λ_i and eigenfunctions v_i of the operator A; in fact the eigenfunctions v_i of A are the stationary points (on V) of the Rayleigh quotient

$$R(v) = \frac{a(v,v)}{\langle v,v\rangle}$$

and λ_i is given by $R(v_i)$. On the other hand the eigenvectors and eigenvalues of $(\underline{M}^h, \underline{K}^h)$ are the stationary points (on $\mathbb{R}^{N^i}$) and the related values respectively, of the Rayleigh quotient:

$$R^h(\underline{q}^h) = \frac{\underline{q}^{hT}\underline{K}^h\underline{q}^h}{\underline{q}^{hT}\underline{K}^h\underline{q}^h} = \frac{a(v^h,v^h)}{\langle v^h, v^h\rangle} = R(v^h) \quad , \quad v^h \in S_o^h$$

which is just the Ritz approximation of R(v).

The main result concerning the approximation of the eigenvalues and eigenfunctions comes from the approximating properties of the f.e. spaces and is given by:

$$\begin{aligned} &\lambda_i \leq \lambda_i^h \leq \lambda_i + c\, h^{2(k-m)}\lambda_i^{k/m} \\ &a(v_i - v_i^h\, ,\, v_i - v_i^h) \leq c'\, h^{2(k-m)}\lambda_i^{k/m} \end{aligned} \qquad 1 \leq i \leq N^i \qquad (4.8)$$

where the f.e. space is taken to be of degree k-1. From (4.8) it is clear that a fixed eingenvalue can be approximated within any prescribed accuracy by refining the mesh, but the further eigenvalues thus introduced are inaccurate: in particular $\lambda^h_{N^i}$ is progressively less reliable when N^i is increased. It also follows from (4.8) that the f.e. approximate model of a parabolic system is always more stable than the corresponding d.p.s.

FINITE ELEMENT APPLICATIONS TO CONTROL SYSTEMS

Theoretical solutionsare already available for the most important problems concerning distributed parameter control systems: the work of Lions (1971) and Bensoussan (1971) are examples of an approach based on the advanced theory of partial differential equations; on the other hand Balakrishnan (1976) and Curtain and Pritchard (1978) develop an abstract theory of linear systems including d.p.s., based on the semigroup approach.

In practical applications whatever the problem may be (closed or open loop optimal control, state estimation, identification, location of sensors and/or of actuators,..) and whatever approach is used, it is necessary at some stage to introduce some approximation. It is still an open question whether approximation at the beginning or approximation at the end is preferable, and several authors have given different answers: Athans (1970) recommends that the distributed nature of the system should be retained as long as possible; more recently Pironneau and Yvon (1975) point out that in solving optimal control problems for d.p.s. it may be advantageous to the numerical stability of the optimization algorithm to discretize the problem at the beginning; the same concept is developped in Burger and Chavent (1979) with reference to parameter identification problems. The application of both approaches to the state estimation problem is outlined in Seinfeld and Koda (1978).

The approximation, either at the beginning or at the end, can be profitably performed by the f.e. method: in the approximation at the end one invariably has to solve numerically a set of partial differential equations (the system and adjoint equations to compute the gradient of a functional, the equation of the kernel corresponding to the Riccati operator and so on): for this purpose the f.e. method is a valid alternative to other methods, expecially when flexibility in dealing the geometry and the boundary conditions is important.In Lions (1977) an abstract approach is presented to the approximation of the optimality system for ellyptic problems, based on the mixed f.e. method.

On the other hand the Galerkin method has been widely used to build up finite dimensional approximate models of d.p.s., with the view of applying the techniques of lumped parameter control theory: in this framework the use of f.e. approximating spaces brings several advantages. Firstly the f.e. method enables a wide variety of systems (with non-constant coefficients, irregular domain, general boundary conditions and forcing action) to be dealt with by standard procedures; secondly the resulting system matrices have a nice structure and have been widely subjected to numerical experiments. Moreover the state variables of the lumped model are given by the nodal parameters of the approximate state of the d.p.s., thus giving to the model a direct physical significance. In many cases the direct application of the semi-discrete Galerkin approximation with a prescribed accuracy leads to models of large dimension which are practically unmanageable: for example the Riccati matrix for a parabolic system on a rectangular domain discretized by a 10x10 grid with bilinear basis, would be 100x100. Alternatively the f.e. method can be used to obtain approximate models of lower dimension by exploiting its efficiency in evaluating eigenvalues and eigenfunction of an elliptic operator. Eigenfunction expansion plays a central rôle in the development of control theory for d.p.s.,and is widely use in application to "sample problems" concerning very simple processes and geometries and where the exact eigenvalues and eigenfunctions are known; moreover a whole sector of research, namely stabilization and modal control (Triggiani, 1980),(Balas 1978) directly involve the dominant eigenvalues of the system. We also believe that problems with space varying coefficients and/or irregular domain can be treated with almost the same accuracy as "sample problems" by replacing the first n exact eigendata with the corresponding ones obtained by a f.e. approximation based on space of dimension sufficiently larger than n. For the computation of the first n eigenvalues of the pencil ($\underline{M}^h$, $\underline{K}^h$) algorithms are available which are efficient even for very large matrices (Bathe and Wilson, 1976).

EXAMPLE 5.1 - A Parametric Identification Problem.

Let us consider the problem of identifying the space varying parameter p(•) for the system:

$$\frac{\partial u}{\partial t} = \frac{\partial}{\partial x}\left(p \frac{\partial u}{\partial x}\right) + f(x,y) \qquad x \in (0,1) \ , \quad t \in (0,T) \tag{5.1}$$

$$u(0,t) = u(1,t) = 0 \quad ; \quad u(x,0) = u_o(x) \quad , \ u_o(x) \in H^1_o(0,1)$$

using the observation $\underline{z}(t)$: $z_k(t) = u(x^o_k,t)$, $k = 1,2,\dots q$, $t \geq 0$. From Tamura and Nakagiri (1977) it is known that p(•) is not identifiable with pointwise measurements. By approximating (5.1) with a Galerkin finite element model the identification of p(•) is transformed into the identification of the entries of the stiffness matrix $\underline{K}$. The dimension of the parameter space is determined, for a fixed dimension of the model, by the choice of the basis functions: the simplest choice is that of piecewise affine (figure 5.1) functions on a partition of (0,1) into N+1 elements (x_i, x_{i+1}), $i = 0,1\dots.N$, $x_o = 0$, $x_{N+1} = 1$, yielding the N-dimensional systems:

$$\underline{M}\,\dot{\underline{u}} = -\,\underline{K}\,\underline{u} + \underline{f}$$

with

$$\underline{K} = \begin{bmatrix} P_o + P_1 & -P_1 & 0 & \cdot & \cdot & 0 & 0 \\ -P_1 & P_1 + P_2 & -P_2 & \cdot & \cdot & 0 & 0 \\ 0 & -P_2 & P_2 + P_3 & \cdot & \cdot & 0 & 0 \\ \hdashline 0 & 0 & 0 & & & -P_{N-1} & P_{N-1} + P_N \end{bmatrix}$$

where the parameters P_i are proportional to the average value of p over the element

$[x_i, x_{i+1}]$:

$$P_i = \int_{x_i}^{x_{i+1}} p(x)dx \; \frac{1}{(x_{i+1}-x_i)^2} \qquad i = 0,1,\dots N$$

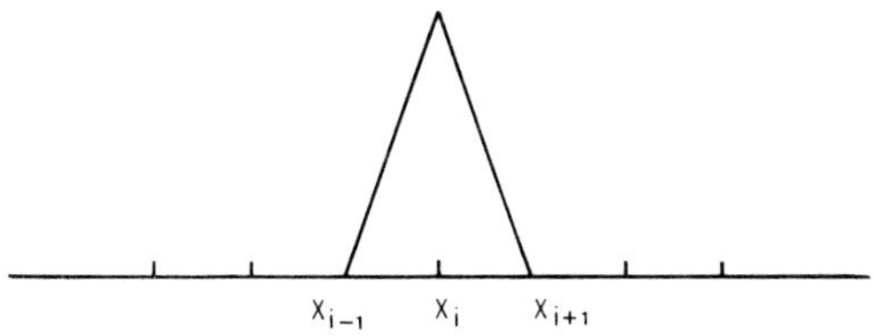

Fig. 5.1

The "mass" matrix $\underline{M}$ is also tridiagonal, with $[\underline{M}]_{ii} = \frac{x_{i+1}-x_{i-1}}{3}$, $i = 1,2,\dots N$;

$[\underline{M}]_{i,i+1} = \frac{x_{i+1}-x_i}{6}$, $i = 1,2,\dots,N-1$.

A model of the same dimension, assuming N odd, can be built up by Lagrangian quadratics defined on the N+1/2 elements $[x_i, x_{i+2}]$, $i = 0,2,\dots N-1$. A basis is constructed by considering both the function ϕ_i, quadratic on elements, such that $\phi_i(x_i) = 1$, $\phi_i(x_k) = 0, i = 2,4,\dots N-1$; $k = 0,1,\dots, N+1$, $k \neq i$, whose support is $[x_{i-2}, x_{i+2}]$, and the function Ψ_i such that $\Psi_i(x_i) = 1$, $\Psi_i(x_k) = 0, i = 1,3,\dots,N$; $k = 0,1,\dots,N+1$, $k \neq i$, whose support is $[x_{i-1}, x_{i+1}]$ (figure 5.2).

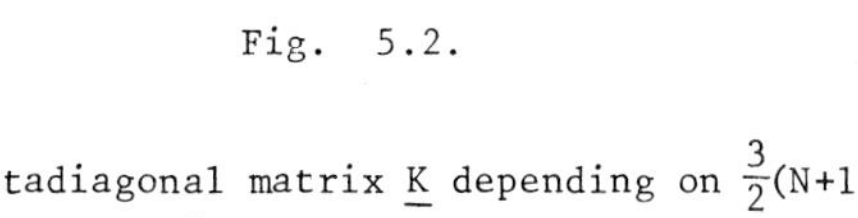

Fig. 5.2.

This choice yields a pentadiagonal matrix $\underline{K}$ depending on $\frac{3}{2}(N+1)$ parameters, which can be taken to be the moments of p:

$$\int_{x_i}^{x_{i+2}} p(x)dx \; , \; \int_{x_i}^{x_{i+2}} xp(x)dx \; , \; \int_{x_i}^{x_{i+2}} x^2 p(x)dx \qquad , \; i = 0,2,\dots, N-1$$

Then simpler elements lead to a smaller numbers of unknows: since they also give a satisfactory reconstruction of the distributed state, we use them for modelling the d.p.s.

We assume that the input f is applied at m points x_i^c, $i = 1,2,\dots,m$: $f(x,t) = \sum_{i=1}^{m} f_i(t)\delta(x-x_i^c)$. Provided that the mesh is taken in such a way that both control and observation points are nodes of the mesh the complete model is,

$$\begin{aligned} \underline{M}\,\dot{\underline{u}} &= -\underline{K}\,\underline{u} + \underline{B}\,\underline{f} \\ \underline{y} &= \underline{C}\,\underline{u} \end{aligned} \tag{5.2}$$

where $\underline{B}$ and $\underline{C}$ are given by

$$[\underline{B}]_{ij} = \begin{cases} 1 \text{ if } x_i = x_j^c \\ 0 \text{ otherwise} \end{cases}, \quad [\underline{C}]_{ij} = \begin{cases} 1 \text{ if } x_j = x_i^o \\ 0 \text{ otherwise} \end{cases}$$

and $\underline{f}$ by $\underline{f} = [f_1(t)\dots. + f_m(t)]^T$.

An investigation of the properties of system (5.2) is carried out in Carotenuto and Raiconi (1980a) where sufficient conditions for the identifiability of the parameters in $\underline{K}$ are given; such conditions are related to the location of sensors and actuators in the superimposed mesh: both from these and from the numerical experiments it proves that the accuracy in state reconstruction (which requires a fine mesh) and the identifiability of the model (which fails to be verified with a fine mesh) are conflicting.

The actual identification is performed by minimizing with respect to p_i, $0 \le i \le N$ the function

$$J = \int_0^T \| \underline{z}(t) - \underline{y}(t) \|^2 dt$$

In the previous example we assumed pointwise controls: this implies that $f \in L^2(0,T;\ H^{-1}(0,1))$ and then $u(t) \in H_0^1(0,1) \subset C^\circ(0,1)$, $\forall t \in (0,T)$, but fails to belong to $H^2(0,1)$; however convergence is still ensured (Strang and Fix 1973), even though convergence rates may be decreased. In fact the previous variational equations make sense for $f \in V'$: in the case of two space-dimensions pointwise action is mathematically much more difficult to deal with, since $\delta(x-x_o,\ y-y_o) \notin H^{-1}(\Omega)$: for example the fundamental solution $u = \log\sqrt{x+y}$ of the Poisson equation: $\nabla^2 u = \delta(x,y)$, $u = 0$ on Γ, is not in $H^1(\Omega)$.

Example 5.2 - A State Estimation Problem.

For the system (5.2) let us consider a filtering problem.

Let us assume that f(x,y) is a gaussian zero mean, white-in-time stochastic process with covariance

$$E\{f(x,t)f(z,\tau)\} = q(x,z)\,\delta(t-\tau)$$

with kernel q(x,z) symmetric, continuous on [0,1]x[0,1] and positive definite. The initial state $u_o(x)$ is assumed to be a gaussian zero mean random function with covariance kernel $p_o(x,z)$, having the same properties as $q(x,z)$. The observation is taken at the same points as in example (5.1) and is affected by a white zero mean gaussian error $\underline{\nu}(t)$ with covariance matrix R, $\underline{z}(t)$: $z_i(t) = u(x_i^o,t) + \nu_i(t)$.

By applying the Galerkin method with affine elements and basis function $\{\phi_i\}_{i=1}^N$ we obtain the N-dimensional stochastic system

$$\underline{M}\,\dot{\underline{u}} = -\underline{K}\,\underline{u} + \underline{f} \qquad (5.3)$$

$$\underline{M}\,\underline{u}(0) = \underline{u}_o$$

to which we add the output equation

$$\underline{z}(t) = c\,\underline{u} + \underline{\nu} \qquad (5.4)$$

where $\underline{f}$ is the random vector: $[\underline{f}]_i = \langle f,\phi_i\rangle$, $E\{[\underline{f}]_i\} = 0$, $E\{\underline{f}(t)\underline{f}^T(\tau)\} = \underline{Q}\delta(t-\tau)$, with the covariance matrix $\underline{Q}$ given by $[\underline{Q}]_{ij} = \int_0^1\int_0^1 q\,(x,\xi)\phi_i(x)\phi_j(\xi)dx\,d\xi$.

The initial state $\underline{u}_o$ is a zero mean random vector with covariance $\underline{P}_o$:

$$[\underline{P}_o]_{ij} = \int_0^1\int_0^1 p_o(x,\xi)\phi_i(x)\phi_j(\xi)dx\,d\xi \quad .$$

Denoting by λ_i^N , $\underline{v}_i^N$ $i = 1,2,..,N$ the eigenvalues and the normalized eigenvectors of the pencil $(\underline{M}, \underline{K})$ and by $v_i^N(x)$ the approximate eigenfunctions in S_o^N of the operator $\frac{\partial}{\partial x}(p\,\frac{\partial}{\partial x})$: $v_i^N = \sum_{k=1}^N [\underline{v}_i^N]_k\ \phi_k(x)$ $i = 1,2,...,N$, (5.3) is put into diagonal form

$$\dot{\underline{d}}^N = -\underline{\Lambda}^N\underline{d}^N + \underline{V}^{NT}\underline{f} \qquad (5.5)$$

$$\underline{d}^N(0) = \underline{V}^{NT}\underline{u}_o$$

and the piecewise affine approximation of the state of the system, is

$$u^N(x,t) = \sum_{k=1}^N [\underline{u}(t)]_k\phi_k(x) = \sum_{k=1}^N [\underline{d}^N(t)]_k v_k^N(x) \qquad (5.6)$$

The output equation (5.4) becomes $\underline{z}(t) = \underline{C}\underline{V}^N\underline{d}(t) - \underline{\nu}(t)$.

We could built up a filter for (5.2) by directly implementing the standard N dimensional Kalman filter either for (5.3) (5.4) or, equivalently, for (5.5), (5.6): but the covariance equation becomes practically intractable even for moderately large N. We propose to take as filtered estimate of u the piecewise affine function

$$\hat{u}_M^N(x,t) = \sum_{k=1}^M v_k^N(x)\ [\hat{\underline{d}}_M^N(t)]_k$$

where $\hat{\underline{d}}_M^N(t) \in \mathbb{R}^M$ is the solution of the M-th order filter algorithm ($M < N$):

$$\dot{\hat{\underline{d}}}_M^N = -\underline{\Lambda}_M^N\,\hat{\underline{d}}_M^N + \underline{G}_M^N(\underline{z} - \underline{C}_M^N\,\hat{\underline{d}}_M^N) \qquad (5.7)$$

$$\hat{\underline{d}}_M^N(0) = \underline{0}$$

$$\underline{G}_M^N = \underline{\Pi}_M^N \underline{C}_M^{NT} \underline{R}^{-1}$$

$$\frac{d}{dt}\underline{\Pi}_M^N = -\underline{\Lambda}_M^N \underline{\Pi}_M^N - \underline{\Pi}_M^N \underline{\Lambda}_M^N + \underline{V}_M^{NT} \underline{Q} \underline{V}_M^N - \underline{G}_M^N \underline{R} \underline{G}_M^{NT}$$

$$\underline{\Pi}_M^N(0) = \underline{V}_M^{NT} \underline{P}_o \underline{V}_M^N$$

and $\underline{\Lambda}_M^N = \text{diag}\,(\lambda_1^N \dots \lambda_M^N)$, $\underline{V}_M^N = [\underline{v}^N \dots \underline{v}_M^N] \in \mathbb{R}^{N \times M}$, $\underline{C}_M^N = \underline{CV}_M^N \in \mathbb{R}^{q \times M}$

Algorithm (5.7) is the reduced order filter based on the first M modes of the Galerkin N-dimensional model. In Carotenuto and Raiconi (1980b) the investigation of the estimation error is performed and it is shown that the same filter algorithm is obtained by Galerkin approximation of the optimal distributed filter, when the trial space is assumed to be $S^N = \text{span}\,\{v^N , \dots, v_M^N\}$ and the p.d.e. governing the evolution of the covariance-of-error kernel p(x,y,t) is likewise approximated.

From a qualitative point of view we can distinguish between the effect on the behaviour of the filter both of the truncation, which depends on M, and of the approximation of the first M eigenvalues and eigenfuctions, which depends on N: numerical experiments show that, for fixed M the reconstruction of the state does not improve significantly when N is increased, suggesting that the effect of the truncation is dominant. In table (5.1) the eigenvalues are reported for several N and in figures (5.3,4,5) the first three approximate eigenfunctions for the same values of N. We assumed $p(x) = e^x$.

Example 5.3 - A Linear Quadratic Open-Loop Optimal Control Problem.

Let us consider the system (5.1), with $f \in L^2(Q)$, $Q = (0,1)x(0,T)$ and the functional

$$J(f) = \int_0^T \{||u||^2_{L^2(0,1)} + \alpha ||f||^2_{L^2(0,1)}\}\,dt$$

we look for $\hat{f} \in L^2(Q)$ such that

$$J(\hat{f}) = \min_{f \in L^2(Q)} J(f)$$

with u satisfying (5.1)

The solution to the minimum problem is obtained by solving the coupled system

$$\left.\begin{aligned} &\frac{\partial u}{\partial t} = \frac{\partial}{\partial x}(p\frac{\partial u}{\partial x}) + \frac{1}{\alpha}\lambda(x,t) \\ &\frac{\partial \lambda}{\partial t} = -\frac{\partial}{\partial x}(p\frac{\partial \lambda}{\partial x}) + u(x,t) \\ &(x,t) \in Q \\ &u(x,0) = u_o(x) \quad , \lambda(x,T) = 0 \\ &u(0,t) = u(1,t) = 0 \quad ; \quad \lambda(0,t) = \lambda(1,t) = 0 \end{aligned}\right| \tag{5.8}$$

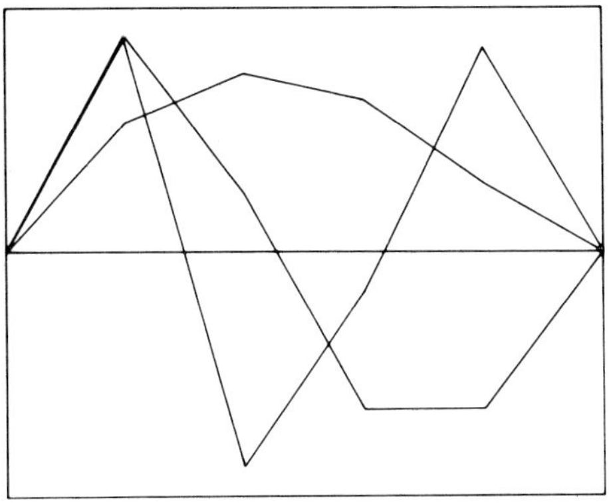

Fig. 5.3. The first three approximate eigenfunctions for N = 4

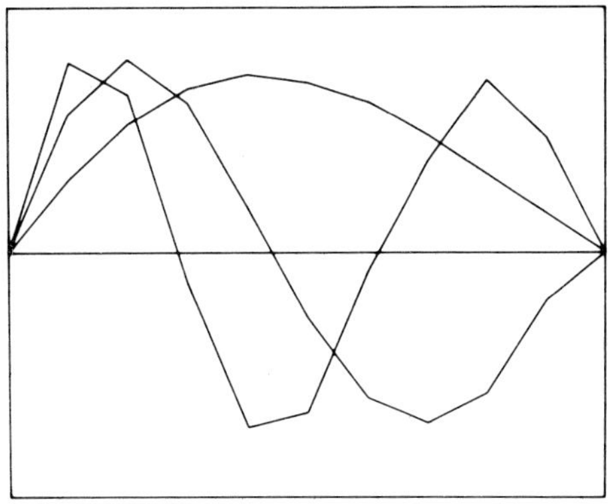

Fig. 5.4. The first three approximate eigenfunctions for N = 9

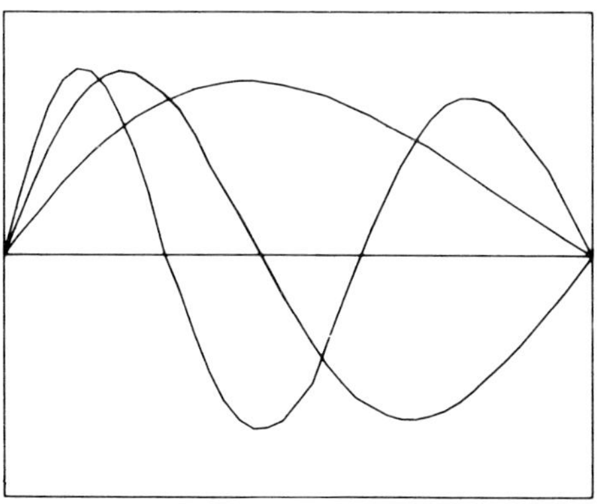

Fig. 5.5. The first three approximate eigenfunctions for N = 39.

i	N	λ_i^N	$v_i^N(0.2)$	$v_i^N(0.6)$	$v_i^N(0.8)$
1	4	16.87	1.08	1.25	.660
	9	16.39	1.05	1.22	.641
	39	16.24	1.05	1.20	.635
	99	16.24	1.05	1.20	.635
2	4	73.60	1.76	-1.28	-1.28
	9	66.38	1.57	-1.18	-1.15
	39	64.20	1.52	-1.16	-1.11
	99	64.08	1.51	-1.15	-1.11
3	4	191.0	-1.74	.314	-1.69
	9	155.4	-1.28	.162	-1.40
	39	144.5	-1.16	.105	-1.31
	99	143.9	-1.15	.102	-1.31

Table 5.1. Approximate eigenvalues and some point values of the corresponding eigenfunctions.

and is given by $\hat{f} = \frac{1}{\alpha}\lambda$; moreover $\hat{f} \in H^{1,2}(Q)$ (Lions, 1971).

In order to solve equations (5.8) let us use the semidiscrete Galerkin method assuming as trial space both for the system and the adjoint equation a f.e. space $S_o^h \subset H_o^1(0,1)$ on a partition of [0,1], spanned by $\phi_1, \phi_2, \ldots, \phi_{N^i}$; the approximate solution of (5.8) is given by $u^h(x,t) = \sum_{i=1}^{N^i} u_i^h(t)\phi_i(x)$, $\lambda^h(x,t) = \sum_{i=1}^{N^i} \lambda_i^h(t)\phi_i(x)$, where the vectors $\underline{u}^h(t) \in \mathbb{R}^{N^i}$, $\underline{\lambda}^h(t) \in \mathbb{R}^{N^i}$ satisfy the two point boundary value problem

$$\left.\begin{aligned} \underline{M}\,\dot{\underline{u}}^h &= -\underline{K}\,\underline{u}^h + \frac{1}{\alpha}\underline{M}\,\underline{\lambda}^h \\ \underline{M}\,\dot{\underline{\lambda}}^h &= \underline{K}\,\underline{\lambda}^h + \underline{M}\,\underline{u}^h \\ \underline{M}\,\underline{u}^h(0) &= \underline{u}_o^h \;,\; \underline{\lambda}^h(T) = \underline{0} \end{aligned}\right\} \quad (5.9)$$

and M, K, $\underline{u}_o^h$ are defined as usual: then $\hat{f}^h(x,t) = \frac{1}{\alpha}\lambda^h(x,t)$.

The approximation at the beginning requires that a discrete version of (5.1) and of the cost function J be built; following the semidiscrete Galerkin method, (5.1) is replaced by

$$\left.\begin{aligned} \underline{M}\,\dot{\underline{u}}^h &= -\underline{K}\,\underline{u}^h + \underline{f}^h \\ \underline{M}\,\underline{u}^h(0) &= \underline{u}_o^h \end{aligned}\right\} \quad (5.10a)$$

with $[\underline{f}^h]_i = \int_0^1 f(x,t)\phi_i(x)dx$. In the cost functional we replace $\|u\|_{L^2(0,1)}$ by $\underline{u}^{hT}\underline{M}\underline{u}^h$, whilst $\|f\|_{L^2(0,1)}$ is replaced by $\underline{f}^{hT}\underline{R}\underline{f}^h$, $\underline{R}$ being a symmetric positive definite matrix; consider the problem:

find $\hat{\underline{f}}^h \in [L^2(0,T)]^{N^i}$ such that

$$J_h(\underline{f}^h) = \int_0^T \{ \underline{u}^{hT}\,\underline{M}\,\underline{u}^h + \alpha\underline{f}^{hT}\,\underline{R}\,\underline{f}^h \}dt$$

is minimum over $[L^2(0,T)]^{N^i}$, under the constraint (5.10a). The optimal control is given by $\hat{\underline{f}}^h = \frac{1}{\alpha}\underline{R}^{-1}\underline{q}^h$, where $\underline{q}^h$ is the solution of the adjoint system

$$\left.\begin{aligned} \underline{M}\,\dot{\underline{q}}^h &= \underline{K}\,\underline{q}^h + \underline{M}\,\underline{u}^h \\ \underline{q}^h(T) &= \underline{0} \end{aligned}\right\} \quad (5.10b)$$

If we choose $\underline{R} = \underline{M}^{-1}$ the t.p.b.v.p. (5.10a,b) is exactly the same as (5.8). At this point $\hat{\underline{f}}^h(t)$ must be transformed into a distributed control $\hat{f}^h(x,t)$: considering how the elements of $\underline{f}^h$ where defined we can reasonably take $\hat{f}^h(x,t)$ such that $\hat{f}^h(t) \in S_o^h$ and

$$\int_0^1 \hat{f}^h(x,t)\phi_i(x)dx = [\hat{\underline{f}}^h(t)]_i \qquad i = 1,2,\ldots,N^i$$

thus obtaining $\hat{f}^h(x,t) = \sum_{i=1}^{N^i} \tilde{f}_i^h(t)\phi_i(x)$ with $\underline{\tilde{f}}^h = M^{-1}\underline{\hat{f}}^h = \cdot\frac{1}{\alpha}\underline{q}^h$: then, if $\underline{R} = \underline{M}^{-1}$ and $\underline{\hat{f}}^h$ is prolonged as above, the approximations at the beginning and at the end yield the same result.

An alternative approach is to build a fully discrete approximation of the control problem by using f.e. approximation both in the time and in space (Bosarge and Mc Knigt, 1973): this does not contradict what we noticed before since in this problem the state at a given time depends on the optimal input over the whole interval (0,T). The discretization can be performed either at the beginning on the state equation and cost functional, or at the end,on the system and adjoint equations, or even on the Lagrangian functional of the problem. The approach is illustrated with reference to an approximation at the beginning. Consider the domain Q partitioned by a rectangular mesh with nodes $P_{ij} = (x_i, t_j)$, $i = 0,1,..,N$, $x_o = 0$, $x_N = 1$; $j = 0,1,...,M$, $t_o = 0$, $t_M = T$; let S^h_{xo} be a f.e. subspace of $H^1_o(0,T)$ spanned by the basis functions $\phi_1,\phi_2,...,\phi_{N-1}$, associated to the nodes $x_1,x_2,...x_{N-1}$ of [0,1]; likewise let S_t be a f.e. subspace of $H^1_o(0,1)$ spanned by $\Psi_o,\Psi_1,...,\Psi_M$, associated to $t_o,t_1,...t_M$; finally let S be the f.e. space on $\bar{Q}$ spanned by $\theta_k(x,t) = \phi_i(x)\Psi_j(t)$, $k = (N-1)j+i, i = 1,2,..,N-1$, $j = 0,1,...M$;

Let u,f be approximated by u^h, $f^h \in S$:

$$u^h(x,t) = \sum_{k=1}^{k_m} u_k^h\theta_k(x,t), \quad f^h(x,t) = \sum_{k=1}^{k_m} f_k^h\theta_k(x,t) \quad , k_m = (N-1)(M+1)$$

By putting:

$$\underline{u}^h = [u_1^h\ u_2^h \ldots u_{km}^h]^T \quad , \underline{f}^h(x,t) = [f_1^h f_2^h \ldots f_{km}^h]^T \quad ,$$

$$[\underline{M}]_{k\ell} = \int_o^T\int_o^1 \theta_k(x,t)\theta_\ell(x,t)dx\,dt = \int_o^T \Psi_j(t)\Psi_r(t)\,dt \int_o^1 \phi_i(x)\phi_s(x)dx$$

$$[\underline{K}]_{k\ell} = \int_o^T\int_o^1 p\,\frac{\partial\theta_k}{\partial x}\frac{\partial\theta_\ell}{\partial x}\,dx\,dt = \int_o^T \Psi_j(t)\Psi_r(t)dt \int_o^1 p\,\frac{d\phi_i}{dx}\frac{d\phi_s}{dx}\,dx$$

$$[\underline{E}]_{k\ell} = \int_o^T\int_o^1 \theta_k\frac{\partial\theta_\ell}{\partial t}\,dx\,dt = \int_o^T \Psi_j(t)\frac{d\Psi}{dt}dt \int_o^1 \phi_i(x)\phi_s(x)dx$$

$k = (N-1)j+i$, $\ell = (N-1)r+s$

The approximate cost function is

$$J^h = \underline{u}^{hT}\,\underline{M}\,\underline{u}^h + \alpha\underline{f}^{hT}\,\underline{M}\,\underline{f}^h$$

and the fully-discrete Galerkin approximation of (5.1) is

$$(\underline{E} + \underline{K})\underline{u}^h = \underline{M}\underline{f}^h$$

The initial condition is replaced by:

$$\underline{C}\,\underline{u}^h = \underline{u}_o^h$$

where

$$\underline{C} = \begin{bmatrix} \underline{\bar{C}} & 0 \\ 0 & 0 \end{bmatrix}, \quad \underline{\bar{C}} \in \mathbb{R}^{(N+1)x(N-1)}, \quad [\underline{\bar{C}}]_{ij} = \int_0^1 \phi_i(x)\phi_j(x)dx$$

$$\underline{u}_o^h = [\underline{\bar{u}}^{hT} \vdots \underline{0}^T \dots \underline{0}^T]^T, \quad \underline{\bar{u}}^h \in \mathbb{R}^{N-1}, \quad [\underline{\bar{u}}_o^h]_i = \int_0^1 u_o(x)\phi_i(x)dx$$

Thus the control problem is reduced to a minimization of a quadratic function with linear constraints.The dimension of the problem may be very large, but the matrices involved in it have a nice structure: $\underline{M}$, $\underline{K}$ are symmetric, positive definite, $\underline{M},\underline{K},\underline{E}$ are block-banded, and each block is itself banded: for example the piecewise bilinear f.e. space, which is the simplest one over rectangles, yields block tridiagonal matrices with tridiagonal blocks.

Bosarge and Mc Knigt (1973) prove the convergence of the method for a general non--linear non-quadratic problem, under suitable regularity assumptions.

Finite elements have been appied also to the solution of the optimal control problems for lumped parameter systems by the so called Ritz-Trefftz method. The discretization is applied to the dual formulation and this proves very effective in dealing inequality constraints both on the state and on the control (Hager,1975; Hager and Strang,1975). The extension of this approach to d.p.s. appears to be very promising.

ACKNOWLEDGEMENTS.

The authors thank prof. D. Scott for revising the manuscript. This work was partially supported by the Italian National Research Council.

NOTATIONS FOR THE FUNCTION SPACES USED

Ω: bounded open set in $\mathbb{R}^n$, with boundary Γ and closure $\bar{\Omega} = \Omega \cup \Gamma$

$C^k(\bar{\Omega})$: space of k-times continuosly differentiable functions on $\bar{\Omega}$

$L^2(\Omega)$: space of square integrable functions on Ω; the symbols $<u,v>$, $||u||$ will denote inner product and norm in $L^2(\Omega)$.

$H^k(\Omega)$: space of functions u such that the derivatives up to order k belong to $L^2(\Omega)$ ($H^\circ(\Omega) = L^2(\Omega)$). $H^k(\Omega)$ is a Hilbert space with respect to the inner product $<u,v>_{H^k(\Omega)} = \sum_{0 \leq |p| \leq k} <D^p u, D^p v>$, where $p = \{p_1,\dots,p_n\}$

$|p| = p_1 + p_2 + \dots p_n$, $D^p u = \dfrac{\partial^{|p|} u}{\partial x_1^{p_1} \dots \partial x_m^{p_n}}$; $||u||_{H^k(\Omega)}$ denotes the corresponding norm; $|u|_{H^k(\Omega)}$ denotes the seminorm $|u|_{H^k(\Omega)} = (\sum_{|p|=k} ||D^p u||^2)$

$H_o^k(\Omega)$: subspace of $H^k(\Omega)$, k > 0 defined by: $H_o^k(\Omega) = \{u \in H^k(\Omega): \frac{\partial^j u}{\partial n^j} = 0,\ 0 \leq j \leq k-1$ on $\Gamma\}$, where $\frac{\partial}{\partial n}$ is the normal derivative on Γ.

REFERENCES.

Ahlberg, J.H., E.N. Nilson and J.L. Walsh (1976). The theory of splines and their applications. Academic Press, New York.

Athans, M. (1970). Towards practical theory for distributed parameter system. IEEE Trans. Autom. Control, 15, 245-247.

Baker G.A. and J.H. Bramble (1979). Semidiscrete and single step fully discrete approximation for second order hyperbolic equations. Rev. Fr. Autom. Inf. Rech. Oper. (Analyse Numerique), 13, 75-100.

Balakrishnan, A.V. (1976). Applied functional analysis. Springer Verlag, New York.

Balas, M.J. (1978). Modal control of certain flexible dynamic systems. SIAM J. Control, 16, 450-462.

Bathe, J.K. and E.L. Wilson (1976). Numerical methods in finite element analysis. Prentice Hall, Englewood Cliffs.

Bensoussan, A. (1971). Filtrage Optimal des Systeme Lineaires. Dunod, Paris.

Borsage, W.E. and R.S. Mc Knigt (1973). The Ritz-Galerkin procedure for parabolic control problems. SIAM J. Control, 11, 515-524.

Brebbia, C.A. (1978). Weighted residual classification of approximate methods. Applied Mathematical Modeling, 2, 160-164.

Burger, J. and G. Chavent (1978). Identification de paramétres repartis dans les equations aux dérivée partielles. Rev. Fr. Autom. Inf. Rech. Oper.. (Automatique), 13, 115-126.

Carotenuto, L. G.,Raiconi and G. Di Pillo (1979). A regularized solution of the identification problem for the distributed parameter model of an underground aquifer. In A. Niemi (Ed.), A Link Betwen Science and Applications of Automatic Control. Pergamon Press, Oxford. pp. 2383-2389.

Carotenuto, L. and G. Raiconi (1980a). Identifiability and identification of a Galerkin approximation for a class of distributed parameter systems. Int. J. Syst. Sci., 11, 1035-1049.

Carotenuto, L. and G. Raiconi (1980b). Error analysis of a finite element approximate filter for a class of D.P.S. In S.G. Tzafestas (Ed.), Simulation of Distributed Parameter and Large Scale Systems. North Holland, Amsterdam. pp 33-45.

Ciarlet, P.G. (1978). The Finite Element Method for Elliptic Problems. North Holland, Amsterdam.

Curtain, R.F. and A.J. Pritchard (1978). Infinite Dimensional Linear Systems. Springer-Verlag, Berlin.

Finlayson, B.A. (1972). Orthogonal collocation on finite elements: progresses and potential. Math. & Comp. Symulation, 22, 11-17.

Frind, F.O. and G.F. Pinder (1973). Galerkin solution of the inverse problem for aquifer transmissivity. Water Resources Res., 9, 1397-1410.

Geveci, T. (1979). On the approximation of the solution of an optimal control problem governed by an ellyptic equation. Rev. Fr. Autom. Inf. Res. Oper. (Analyse Numerique), 13, 313-328.

Hager, W.W.(1975). The Ritz Trefftz method for state and control constrained optimal control problems. SIAM J. Numer. Anal.,12, 854-867.

Hager, W.W. and G. Strang (1975) Free boundaries and finite elements in one dimension. Math. Comput., 29, 1020-1031.

Hang, E.J. and T.T. Feng (1977). Optimization of distributed parameter structures under dynamic loads. In C.T. Leondes (Ed.), Control and Dynamics Systems, Vol. 13, Academic Press, New York. pp.207-246.

Lasiecka, I.(1980). Boundary control of parabolic systems: finite element approximation. Applied Mathematics and Optimization, 6, 31-62.

Lewis, B.A. and M.Cross (1978). IFECS, an interactive finite element computing system. Applied Mathemetical Modelling, 2, 165-176.

Lions,J.L. (1971). Optimal Control of Systems Governed by Partial Differential Equations. Springer Verlag, Berlin.

Lions,J.L. (1977). Remarks on the theory of optimal control of distributed systems. In A.K. Aziz and others(Eds.), Control Theory of Systems Governed by Partial Differential Equations. Academic Press, New York, pp. 1-103.

Mitchell, A.R. and R.Wait (1977). The Finite Element Method in Partial Differential Equations. John Wiley & Sons, Chichester.

Nečas,J.(1967). Les méthodes directes en théorie des équations elliptiques.Masson, Paris.

Nitsche,J.A.(1977). L^{∞} convergence of finite element approximations. In Dold,A. and B. Eckmann (Eds.),Mathematical Aspects of Finite Element Methods. Springer Verlag, Berlin. pp.261-274.

Nitsche,J.A. (1979). L^{∞} convergence of finite element Galerkin approximation for parabolic problems. Rev. Fr. Autom. Inf. Res. Oper. (Analyse Numerique), 13, 31-54.

Norrie,D.H. and G.DeVries (1978). An Introduction to Finite Element Analysis. Academic Press, New York.

Oden, J.T. (1972). Generalized conjugate functions for mixed finite element approximations of boundary value problems. In A.K.Aziz (Ed.), The mathematical foundations of the finite element method with applications. Academic Press, New York, pp.629-671.

Pironneau, O. and J.P.Yvon (1975). A survey of optimization methods in optimal control for distributed parameter systems. Proc. of the IFAC 6th World Congr., 57.4.

Pironneau, O.(1977). Variational methods for the numerical solution of free boundary problems and optimum design problems. In A.K.Aziz and others (Eds.), Control Theory of Systems Governed by Partial Differential Equations. Academic Press, New York.

Ramakrishnan, C.V. (1979). An upwind finite element scheme for unsteady convective diffusive transport equation. Applied Mathematical Modelling, 3, 280-284.

Ray, W.H.(1978). Some recent application of distributed parameter system theory. A Survey. Automatica, 4, 281-287.

Robinson,A.C.(1971). A survey of optimal control of distributed parameter systems. Automatica, 7, 371-388.

Schoenberg,I.J.(1969). Approximation with Special Enphasis on Spline Functions. Academic Press, new York.

Seinfeld,J.H. and M.Koda(1976). Numerical implementation of distributed parameter filters with application to problems in air pollution. In A.Ruberti(Ed.), Distributed Parameter Systems: Modelling and Identification. Springer Verlag, Berlin, pp. 43-69.

Strang, G. and G.J.Fix(1973). An Analysis of the Finite Element Method. Prentice Hall, Englewood Cliffs.

Tamura,,S.K. and S.Nakagiri (1977). Identifiability of spatially varying and constant parameters in distributed systems of pabolic type. SIAM J. Control, 15, 785-802.

Triggiani,R. (1980). Boundary feedback stabilization of parabolic equations. Applied Mathematics and Optimization, 6, 201-220.

Winter, R. (1978). Error estimates for a Galerkin approximation of a parabolic control problem. Annali Matematica Pura ed Applicata, 17, 173-206.

Zinkiewicz,O.C. (1977). The Finite Element Method. Mc Graw Hill, New York.

CHAPTER 9

APPROXIMATION METHODS FOR DISTRIBUTED PARAMETER SYSTEMS

T. I. Seidman

Department of Mathematics, University of Maryland Baltimore County, Baltimore, MD 21228, USA

1. INTRODUCTION

The variety of distributed parameter systems which one might wish to consider is bewilderingly vast. Even the variety of those which have already received extensive treatment in the technical literature is beyond the scope of any single survey. To note only the principal types (systems governed by functional differential equations and the classical PDE's), the technical considerations are often very different for different types. These differences are, of course, compounded by the differences associated with such other distinctions as: deterministic/stochastic, linear/nonlinear, autonomous/time dependent, presence/absence of constraints on state or control, finite/infinite horizon, etc. Finally, the considerations are again quite different in treating problems of stabilization, controllability, optimal control, optimal design, observation, system identification and such combinations as adaptive control.

At best, therefore, we can only hope to suggest some major concerns and relevant approaches through consideration of a few special problems. For reasons, primarily, of my own familiarity and convenience the material of the examples presented has all been taken, with some modification, from my own work:

A convergent scheme for boundary control of the heat equation (with W.C. Chewning), SIAM J. Cont. Opt. 15 (1977) 64-72.

Parameter and state estimation for a diffusion process. In A.T. Pritchard (Ed.) Proc. IFAC 2-nd Symp. Control Distrib. Par.Syst., IFAC, Dusseldorf, 1978.

Exact boundary control of some evolution equations, SIAM J. Cont.Opt. 16 (1978) 979-999.

* This has been written while visiting at the Inst. de Math. et Sci. Phys., Université de Nice, Nice, France. Acknowledgment is due to the IMSP and to CIMPA (Centre International de Math. Pure et Appl.) for their hospitality. Acknowledgment is also due to the USARO for support under grant no. DAAG-29-77-G-0061.

Convergent Approximation Schemes for Ill-posed Problems. Control and Cybernetics 10 (1981).

Ill-posed problems arising in boundary control and observation for diffusion equations, In G. Anger, (Ed.), Inverse and Improperly Posed Problems in Differential Equations, Akademie Verlag, Berlin, 1979.

Recovery of a diffused signal, Séminaires IRIA, (1979).

Nonconvergence results for the application of least-squares estimation to ill-posed problems, JOTA 30 (1980), 535-547.

Approximation methods for distributed systems, In M.Z. Nashed (Ed.) Ill-Posed Problems:Theory and Practice, Aca.Press, N.Y., 1981.

Finite element approximation of optimal boundary nullcontrols for the heat equation, in preparation.

This is not to suggest that there has been any lack of interest in such problems:a couple of dozen names of active researchers come immediately to mind-Banks, Cannon, Céa, Chavent, Chen, Hackbusch, Lions, Malanowski, Sakawa, Seinfeld,- to name only a few. The particular selection made here has, of course, been influenced by my own particular interests. My hope, on the other hand, is that the general questions raised in these contexts can be universal, even though the technical details arising may be special.

2. GENERAL REMARKS

Before examination of any specific case, it is important to raise, briefly, the general question:Why do we approximate? since the answers suggest the nature of the approximations to be considered and the nature of the problems which concern us. Some immediate answers suggest themselves:

(A) Because we do not, in fact, know the true mathematical formulation of the problem as it arises in the real world. Because our measurements are inaccurate or our theory inadequate.

(B) Because the real world is, in its full detail, too complicated for our comprehension and we hope that analysis of a simplified model will be adequate or, at least, informative.

(C) Due to limitations of feasibility in implementation.

(D) As a tool in mathematical analysis. (All the standard concepts of calculus-series, derivative, integral, inf,...- are constructed by limit processes, implicitly involving notions of approximation, and it is hardly surprising for such constructions to arise here as well).

(E) For purposes of numerical computation.(Computers cannot, in fast, solve PDE's but only carefully chosen approximations to them of a "finitary" nature).

Needless to say, these same considerations arise in almost every area of applied mathematics. To a great extent, especially for (E), the detailed consideration of such questions has been the essential object of numerical analysis. Thus, we need only ask whether-or to what extent-the problems we are posing have already been settled by the numerical analysts rather than being, entirely or in emphasis, genuinely different and so requiring independent formulation and consideration.

It is well to begin by noting the philosophical basis of numerical analysis: *If a problem cannot be solved, as it is, "exactly", one seeks an approximating problem "close to the original" which can be solved. The (exact) solution of the approximating problem is potentially- but, unfortunately, not necessarily- an approximate solution to the exact problem.* The object of numerical analysis, then, is the construction for various problems of interest-those of distributed parameter system theory being, one hopes, among them- of solvable approximate problems for which one can, indeed, show that the solutions *are* satisfactory approximations of the solutions of the original problems.

A general principle in numerical analysis is that: If a problem is *stable* then one can hope to approximate it by a *uniformly stable* family of increasingly accurate approximating problems and can then expect that their solutions provide the desired, increasingly accurate, approximants. In this context, "stable" means *well-posed in the sense of Hadamard:* if one perturbs the data slightly (in some reasonable sense), then the perturbed problem again has a unique solution which (again in some specified sense of acceptability) differs only slightly from the unperturbed solution.

In systems theory involving only finite-dimensional state spaces one seldom encounters ill-posed problems. (In a finite-dimensional space all linear maps are continuous, all bounded sets precompact. Thus it is sufficient to verify uniqueness for $Ax=b$ in R^m to know that A has an inverse-hence, a bounded inverse-so the problem is well-posed). By contrast , in distributed parameter system theory (particularly for systems governed by FDE's or parabolic PDE's) one often works with solution operators whose range is not closed. Thus one frequently encounters ill-posed problems, especially for problems of state and system identification and for exact control. Almost by definition, the characteristic behaviour of ill-posed problems is their extreme sensitivity: arbitrarily good approximations of the problem may have no solutions at all or solutions arbitrarily far from the desired one. (This may be viewed as the limit of illconditioning). When, with care, one *does* construct an approximation scheme with "satisfactory" properties, the "price" one must pay (in terms of effort involved and precision required), as one seeks increased accuracy, *increases far more rapidly* than would be the case for well-posed problems.

3. SOME EXAMPLES

All the examples considered will relate to systems governed by parabolic partial differential equations. In this section we consider a number of examples in one space dimension while in the next section we consider some higher-dimensional optimal control problems. For such a problem the state u(t,.) might represent the spatial temperature distribution or deviation from equilibrium in some body or, alternatively, it might represent the concentration of a substance diffusing in some medium; for convenience we stick to the terminology of heat conduction.

We will use the term *"direct problem"* to refer to the problem of determining the temporal evolution of the system (i.e., the "trajectory") if one is given the initial state $u(0)=\omega$ and other relevant data. Clearly ω and $u(t,.)$ are functions of position in the body so, unless there is strong reason to believe the spatial variation can be neglected, one has a distributed problem. Such direct problems have been treated extensively-both theoretically and computationally (for a wide variety of assumptions as to the geometry of the situation, boundary conditions, etc...) since the seminal work of Fourier in the early 19-th century. For a considerable variety of notions as to what one wishes to mean by "slight perturbations" of the data and by "acceptably small" deviations of the solution (given by a corresponding variety of function spaces and norms), it is known that direct problems of this sort are well-posed.

For the direct problems, numerical computation of the solution to arbitrary accuracy is routinely accomplished by *series expansion* or, on a computer, by *discretization:* finite differences or the method of Faedo-Galerkin. By any of these methods it is possible, for example, to replace the original distributed parameter system by an approximating system of ordinary differential equations-a lumped parameter system which approximates the dynamics of the system evolution quite accurately. Our obvious thought is to treat the distributed problem by substituting such an approximation for it-at *some* stage in the analysis. This might be *initially,* in which case we formulate the corresponding system-theoretic problem for the approximating ODE system and then accept the solution obtained for *that* as our approximation of the desired result. Alternatively, one might work analytically with the distributed system to develop an iterative approximation procedure involving solution of direct problems at each step - and *then* substitute an approximating system. Several important questions arise:

Does the approach provide an acceptable approximation? Can one tell? How careful must one be?Are there "traps"? Can one do better?

Example 1

We consider the control of the 1-dimensional heat equation to a specified terminal "target state" $\hat{\omega}$ using a method of series expansion. Our domain will be $\Omega=(0,1)$ and we consider the system

$$u_t=u_{xx}+\varphi \textit{ in } Q=(0,T)\times ,$$
$$u(t,0)=0=u(t,1) \textit{ on } (0,T),\ u(0,.)=\omega \tag{1}$$

and seek a control φ such that

$$||\varphi||_Q = \min \textit{ subject to: } u(T,.)=\hat{\omega} \tag{2}$$

(Here, and subsequently, we work in an L^2 setting so norms will be L^2 norms. Thus, $||.||_Q$ represents the norm in $L^2(Q)$ given by

$$||\varphi||_Q=|\textstyle\int_Q \varphi\ |^{1/2}$$

Note that the differential of integration "dxdt" is omitted;the identifying subscript, as Q here, will also be omitted when clear from context).

Let $\{\omega_n\}$ be the normalized eigenfuctions of the spatial differential operator $Ly=-d^2y/dx^2$ with the boundary conditions $y(0)=0=y(1)$ and $\{\lambda_n\}$ the corresponding eigenvalues with $0<\lambda_1<\lambda_2<\ldots$. That is

$$L\omega_j=\lambda_j\omega_j, <\omega_j,\omega_k>_\Omega=\{1 \textit{ if } j=k; 0 \textit{ if } j\neq k\}.$$

For (1) one has, of course,

$$\lambda_j=\pi^2 j^2, \quad \omega_j(x)=\sin j\pi x,$$

but the method to be considered extends without change to more general problems (e.g., to $u_t+Lu=\varphi$ where $Ly=-\nabla.D(x)\nabla y+q(x)y$ on $\Omega\subset R^m$ with $y=0$ on the boundary of Ω); thus we use the more general notation. We write the expansions

$$\omega=\sum_{j=1}^{\infty} c_j\omega_j, \quad \varphi(t,.)=\sum_{j=1}^{\infty} f_j(t)\omega_j$$

from which it follows that the solution of (1) is given by

$$u(t,.)=\sum_{j=1}^{\infty} c_j(t)\omega_j \textit{ with } c_j(t)=e^{-\lambda_j t}c_j+\int_0^t e^{-\lambda_j(t-s)}f_j(s)ds \quad (3)$$

We wish to solve (2) and it is not hard to show that this is accomlished by taking

$$\varphi(t,.)=2\sum_{j=1}^{J}\lambda_j\left(\frac{\hat{c}_j-c_je^{-\lambda_jT}}{1-e^{-2\lambda_jT}}\right)e^{-\lambda_j(T-t)}\omega_j(.) \quad (4)$$

since the series (4) converges in $L^2(Q)$ if (and only if) the desired terminal state $\hat{\omega}$ is, indeed, attainable. So far, so good!

Adding a note of realism to this, we next observe that in order to construct (4) from $\hat{\omega},\omega$ one must first compute the coefficients $\gamma_j=\hat{c}_j-c_j\exp[-\lambda_jT]$. This, of course, will involve an additional aspect of approximation-say, computing $c_j=<\omega,\omega_j>_\Omega=\int_\Omega\hat{\omega}\,\omega_j$ by some algorithm of numerical integration. (This analysis assumes that the eigenpairs $\{(\omega_j,\omega_j)\}$ are already known exactly. This is true, of course, for the the simple model problem (1), but for more general situations a further source of error would be introduced by the preliminary approximate computation of these for $j=1,..,J$). Letting ε_j be the error in computing each γ_j and writting φ_J for the approximating control produced if $(\gamma_j+\varepsilon_j)$ is used instead of γ_j in (4), one can obtain, after some simple manipulation, the estimate:

$$\frac{1}{2}||\varphi-\varphi_J||_Q^2=\sum_{j=1}^{J}\left[\frac{\lambda_j\varepsilon_j^2}{1-e^{-2\lambda_jT}}\right]+\sum_{j>J}\left[\frac{\lambda_j\gamma_j^2}{1-e^{-2\lambda_jT}}\right]$$

For attainable $\hat{\omega}$, the last term is small for large J, corresponding to convergence of (4). However, a fixed error bound: $|\varepsilon_j|<\varepsilon$ would make the first term grow as J increased. (For the case of $\Omega\subset R^m$ one has, asymptotically, $\lambda_j\sim Cj^{2/m}$ making the first term of the order of $J^{1+2/m}\varepsilon$ for large J).

One can, of course, require that each γ_j be computed with an accuracy which improves with J. It is easily seen, however, that a controlled *relative* error ($|\varepsilon_j|<\varepsilon|\gamma_j|$) would lead to a correspondingly controlled relative error in φ_J. This seems entirely plausible for numerical integration using floating point arithmetic-but one trap remains! Neglecting ω, a plausible form of error estimate for a method of numerical integration might be:

$$|\varepsilon_j|<C_k h^k||\hat{\omega}\omega_j||_{(k)}\sim C_k' h^k \lambda_j^{k/2}||\hat{\omega}||_{(k)}$$

Here h characterizes a mesh size and the subscript on $||.||_{(k)}$ indicates consideration of derivatives up to order k with k depending on the method;we have noted that k-th derivatives of ω_j are asymptotically like $\lambda_j^{k/2}$ (i.e., like $j^{k/m}$ for the case $\Omega\subset R^m$). For a uniform (in j) estimate of the relative error, then, one might reasonably expect a need to take $h=h_j\to 0$ ($h_j\sim h_* j^{-1/m}$), continually refining the mesh as k becomes large. Again, for the special model equation (1) this consideration does not arise as there are special methods (the FFT, in particular) which avoid this difficulty.

Example 2:

We consider, next, a problem of *system identification*. We have a rod (0<x<1) in which diffusion is known to satisfy an equation of the *form:* $u_t=Du_{xx}-qu$ but the function q=q(x) and the diffusion constant D are unknown (or known with inadequate accuracy). One has u(t,1)=0 but interaction is possible at the end x=0: there one can manipulate the temperature and observe the resulting flux. Thus one considers an input/output pair ($\varphi\to\psi$) related through the diffusion equation:

$$\psi(t)=-Du_x(t,0) \textit{ with}$$

$$u_t=Du_{xx}-qu,\ u(t,1)=0,\ u(t,0)=\varphi(t),\ u(0,x)=0 \qquad (5)$$

(For convenience we have assumed 0 initial state).

By causality, linearity and the time-independence of the system, a standard argument shows that the input/output relation takes the convolution form:

$$\psi(t)=\int_0^t\sigma(t-s)\varphi(s)ds=\int_0^t\sigma(s)\varphi(t-s)ds \qquad (6)$$

where the function σ is the so-called *impulse reponse* of (5). An elementary formal calculation gives σ in terms of the eigenpairs $\{(\omega_j,\lambda_j)\}$ of the differential operator L ($Ly=-Dy''+qy$ *with* $y(0)=0=y(1)$):

$$\sigma(t)=-D^2\sum_{j=1}^{\infty}|\omega_j'(0)|^2 e^{-\lambda_j t} \tag{7}$$

One now views (6) as an integral equation (of the first kind) for σ with φ and ψ known. For fairly general (non-trivial!) inputs φ, the convolution form of (6) then enables one to show uniqueness for the solution σ. Given σ, it is clear from (7) that the sequence $\{\lambda_j\}$ of eigenvalues is uniquely determined (note that $\omega_j'(0)\neq 0$). One then has $D=\lim_j \lambda_j/\pi^2 j^2$ so D is now also unique. Suppose, finally, we add the assumption that q is known *a priori* to be smooth and symmetric (i.e., $q(x)=q(1-x)$). Then, by a classical result due to Borg, the coefficient function q is uniquely determined by the eigenvalue sequence $\{\lambda_j\}$.

In principle, this uniqueness argument for q could be used constructively for approximation: given φ and measurements of ψ, first solve the ill-posed integral equation (6) to obtain an approximation of σ; use this to determine approximately the eigenvalues λ_k appearing in the Dirichlet series (7) for σ; finally, use available numerical methods to obtain q, from this eigenvalue sequence. Each of these steps separately is numerically difficult and convergence could not be assured.

For purposes of *computation* we describe the application of a method *(history matching)* which is quite commonly used for identification problems. If one knew both q and φ (for simplicity we assume D known; say, D=1), then (5) would become a direct problem for the determination of ψ. Taking φ to be the actual input, this family of direct problems (as q varies) defines a map: $q\mapsto\bar{\Psi}(q)$. The previous argument showed that $\bar{\Psi}$ is one-to-one when restricted to the set K of admissible coefficient functions (meaning those satisfying the smoothness and symmetry assumptions above). Our problem, then, is to solve the operator equation

$$\bar{\Psi}(q)=\psi \tag{8}$$

where the ψ on the right is the observed output. For a choice of q, the value of $\bar{\Psi}(q)$ is a predicted "history" of the observable $-u_*(t,0)$ for $0<t<T$. The name of the method comes from our attempt to select q so that this *matches* (to within the measurement error) the actually observed *history* $\psi(t)$.

This possibility of error in ψ as well as the error of computation in numerical solution of the direct problem mean that we determine an approximation $q=q_\varepsilon$ by finding a coefficient function for which one has

$$q\in K,\quad ||\tilde{\Psi}(q)-\tilde{\psi}||<\varepsilon \tag{9}$$

where the tildas indicate that these are approximations to $\bar{\Psi},\psi$. Presumably ε is chosen large enough to ensure that (9) has solutions. The error estimates for $\tilde{\Psi},\tilde{\psi}$ typically permit choosing ε so the true q will satisfy (9)-although one could not expect this to be the q found nor could one recognize the fact if it were. By improving the accuracy of measurement for ψ and of computation for Ψ, one can let $\varepsilon\to 0$. We now ask whether, in this case, q_ε converges to the true q.

If we take the "smoothness" part of the specification of the set K of admissible coefficient functions to mean, say, *a priori* bounds on q and its first derivative, then $\bar{\Psi}$ is continuous from K to $L^2(0,T)$, using uniform convergence on $[0,1]$ for the notion of convergence in K. With this choice of convergence one also knows, by the Arzela-Ascoli Lemma, that K is compact. Thus, as $\varepsilon=\varepsilon(j)\to 0$ one must have a convergent subsequence q_k (corresponding to $\varepsilon(j_k)$) so $q_k\to\bar{q}$ for some $\bar{q}$ in K. Since the approximations are being improved as $\varepsilon\to 0$, one has $\tilde{\psi}\to\psi$, $\bar{\Psi}(q_k)\to\bar{\Psi}(\bar{q})$ whence $||\bar{\Psi}(\bar{q})-\psi||=0$. Since we have shown uniqueness, this means $\bar{q}$ must be the unique true solution q so every subsequence $\{q_{\varepsilon(j)}\}$ of $\{q_k\}$ contains a sub-subsequence (like $\{q_k\}$) converging to q. It follows that $q_\varepsilon\to q$ as $\varepsilon\to 0$.

Note that although the original argument for uniqueness did not provide a *computational* approach, the proof of convergence for (9) would have been impossible without it. This is typical: for well-posed problems the same techniques are often applicable both to existence/uniqueness proofs and to numerical computation while for ill-posed problems, as here, feasible approximation methods often begin with *assumptions* of existence and uniqueness whose justification is an independent problem.

Example 3:

Consider a thermal probe which we think of as a conductive wire (conductivity $D=D(x)$ for $0<x<L$; for simplicity take D constant). The end $x=L$ is in contact with the temperature of actual interest to us and there is a recording thermocouple at $x=0$ (we assume 0 heat flux at $x=0$). Thus, the temperature $\varphi(t)$ at $x=L$ is a transient, which we may view as a "signal", and this is propagated along the wire by diffusion to determine the temperature $\psi(t)$ at $x=0$, which we observe. If the transient φ is "slowly varying" compared to the characteristic time L^2/D, then the temperature distribution $u(t,.)$ along the wire will always be "close to equilibrium" *(quasi-steady state)* and so spatially almost constant. One then has $\psi(t)=u(t,0)$, $u(t,L)=\varphi(t)$ and there is no difficulty in recovering the signal φ from the observation ψ. Suppose, however, one is interested in a transient φ whose rapid variation makes this formulation inadequate. We ask: (How) can one recover the signal φ from the diffused observation ψ?

The input/output system ($\varphi\mapsto\psi$) described above takes the form:

$$\psi=u(.,0) \textit{ with}$$

$$u_t=Du_{xx},u_x(.,0)=0,\ u(.,L)=\varphi,u(0.,)=\omega \tag{10}$$

Letting A be the linear map: $\varphi\to\psi$ defined by (10) with $\omega=0$, one easily sees that A is one-to-one so φ is uniquely determined by (exact knowledge of) ψ; in the next example we will show this for a similar nonlinear problem. Since we can calculate, by a direct problem, the ψ_0 given by (10) with $\varphi=0$ (i.e., the effect of ω), we set $\psi_*=\psi-\psi_0$ and, by linearity of (10), our task is to solve the linear equation

$$A\varphi=\psi_* \tag{11}$$

Uniqueness has been noted and existence of a solution follows from our underlying assumption that we are observing an actual situation.

This is very much like (8) but, taking advantage of the linearity, we describe in somewhat more detail some methods of approach to approximate solution. In Hadamard's sense, this is an *ill-posed problem* but we see that the situation is not as hopeless as that phrase might suggest. We outline, in somewhat abstract form, three approximation procedures of quite general applicability. Two of these are *"convergent approximation schemes"*, meaning that each provides a sequence of increasingly accurate (implementable) algorithms providing a corresponding sequence of approximants known to converge to the correct solution. The "philosophy" of this is that *one can compute the true solution to any desired degree of accuracy-provided one is willing to pay the price* in terms of the accuracy and extent required of measurements and computation. The same is also true of well-posed problems, of course, but here the cost of increased accuracy increases far more rapidly.

First Method *(regularization):* this method was introduced by Tikhonov and, with variations, has been extensively applied and studied since; see the excellent book by Tikhonov and Arsenin. The strategy is to associate with the ill-posed problem a sequence of well-posed *(regularized)* approximating problems whose solutions converge correctly. Since each of the regularized problems is well-posed, it can accommodate a certain degree of perturbation (corresponding to measurement error and discretization of the direct problem) without affecting convergence. The regularization introduces stability but decreases resolution: one accepts a certain "blurring" of the recovered signal to insure against the possibly wilder effects of instability. Given a specific level of measurement accuracy, a deep problem of application is the estimation of the regularization required to give an optimal balance of stability and resolution. (This balance can sometimes be substantially improved by "filtering", using statistical properties of the measurement errors if these are available, but we will not pursue this).

Formally, solving (11) is equivalent to choosing φ to minimize the residual. This formulation is just the method of least-squares ($||A\varphi-\psi_*||^2=\min$) and, knowing this is ill-posed, we modify it, introducing a *regularization parameter* α^2. Thus we consider the well-posed *regularized* problem:

$$J_\alpha(\varphi)=\frac{1}{2}|\,||A\varphi-\psi_*||^2+\alpha^2||\varphi||^2 \;=\min \tag{12}$$

We have squared the norms because then (assuming a Hilbert space formulation) J_α is quadratic and its minimization corresponds to a linear problem:

$$[\alpha^2 I+A^*A]\varphi_\alpha=A^*\psi_* \tag{13}$$

Note that φ_α depends continuously on ψ_* since, clearly, $||\alpha^2 I+A^*A|^{-1}||<1/\alpha^2$. If the accuracy of measurement of ψ, the accuracy of the discretized computational approximations used for A and A* and the accuracy of the numerical inversion for the resulting (finite-dimensional) system of equations corresponding to (13) are each improved *faster than* α^2 then the computed approximate φ_α will converge to φ as $\alpha\to 0+$. (Actually, one would presumably *not* solve (13) as a linear system but would directly use (12) and the conjugate gradient method. See Example 7, below, for this and the convergence).

If one were to divide J_α in (12) by the constant α^2, one obviously would not affect the minimization. Writing λ for $1/\alpha^2$, this gives

$$J_\alpha(\varphi)=\frac{1}{2}\ ||\varphi||^2+\lambda||A\varphi-\psi_*||^2\ =\min \tag{12'}$$

We recognize this as a *penalty function* approach to the constrained optimization problem

$$\frac{1}{2}||\varphi||^2=\min\ \textit{subject to: } A\varphi=\psi_* \tag{14}$$

In the present case, of course, the condition uniquely determines φ: there is no genuine minimization to be done. Nevertheless, the *form* of (14) is significant since it suggests the possibility of *relaxing the constraint*, approximating (14) by

$$\frac{1}{2}||\varphi||^2=\min\ \textit{subject to: } \frac{1}{2}||A\varphi-\psi_*||^2<\varepsilon \tag{15}$$

with $\varepsilon>0$, small. Unless $\psi_*\approx 0$, the minimum in (15) is attained at the boundary of the region so < becomes = and one can use a simple Lagrange multiplier approach to (15). Except for an additive constant, this gives precisely (12')! As noted above, one sees the stabilization from (12), (13); now one sees the resolution from the equivalent form (15). Clearly $\varepsilon\to 0+$ corresponds to $\lambda\to+\infty, \alpha^2\to 0$, although the exact relation is more difficult.

Second Method *(generalized interpolation):* We select any convenient sequence $\{y_j\}$ and observe that (11) is equivalent to the infinite system of scalar equations:

$$\langle A\varphi-\psi_*, y_j\rangle=0 \quad \text{for } j=1,2,\ldots, \tag{16}$$

provided only that: $\langle z,y_j\rangle=0$ for all j implies $z=0$ (e.g., if $\{y_j\}$ were a basis). The approximation method is to choose φ_J through

$$\frac{1}{2}||\varphi||^2=\min$$

$$\textit{subject to: } \langle A\varphi-\psi_*, y_j\rangle =0 \ \textit{for}\ \ j=1,\ldots.J. \tag{17}$$

(Compare (15)). It is not difficult to see that the minimization here just means that φ_J is in the (finite-dimensional) span X_J of

$\{A^*y_1,\ldots,A^*y_J\}$. Indeed, φ_J is the *nearest point* in X_J to the unique solution of (11) (to the true solution of (14) if one were to apply this method to a problem for which the solution of (11) is not unique).
Writing

$$\varphi_J=c_1z_1+\ldots+c_Jz_J \quad (z_j=A^*y_j),$$

one can compute the coefficients $\{c_1,\ldots,c_J\}$ needed to specify φ_J by solving the JxJ linear systems:

$$\sum_{k=1}^{J} \alpha_{j,k}c_k=b_j \quad \textit{for } j=1,\ldots,J \tag{18}$$

where $\alpha_{j,k}=\langle z_j,z_k\rangle=\langle A^*y_j,A^*y_k\rangle,\ b_j=\langle\psi_*,y_j\rangle.$

If the accuracy of the determination of (b_j) (involving both the accuracy of measurement of ψ_* and of the computation of the inner product- for the present application, an integration on $[0,T]$ which, presumably, must be done numerically), the accuracy of "evaluating" A^* in determining (z_j) and then of obtaining $\alpha_{j,k}$ andthe accuracy of numerical solution of (18) are each *improved rapidly enough* as J increases then the *computed* approximate solution φ_J converges to the true solution as $J\to\infty$. For both this method and the first method one always has convergence, subject to the need for suitable improvement of accuracy of measurement and computation. However information about the *rate* of convergence can be obtained only if one has suitable *a priori* information about -say , bounds for $||\varphi||,||\varphi'||$.

For implementation of either of these methods as applied to a particular problem as (10) one must know (and be able to "evaluate") the adjoint operator A^*. For this problem, A^* is given by

$$A^*\eta=Dv_x(\cdot,L) \textit{ with} \tag{19}$$

$$-v_t=Du_{xx},\quad Dv_x(\cdot,0)=\eta,\quad v(\cdot,L)=0,\quad v(t,\cdot)=0$$

(the "initial condition" is properly given at T since the equation uses $-v_t$ and so is "backwards in time"). To see that this does, indeed, give $A^*\eta$, let $\psi=A\varphi$ as given by (10) with $u(0)=0$ and let $\zeta=Dv_x(\cdot,L)$.
Then

$$\langle\eta,A\varphi\rangle-\langle\zeta,\varphi\rangle=\int_0^T[\eta\psi-\zeta\varphi]$$

$$=\int_0^T[Du_xv-Dv_xu]\Big|_{x\ 0}^{x=L}$$

using the boundary conditions from (10), (19)

$$=\int_0^T\int_0^L[Du_{xx}v-uDv_{xx}]$$

using integration by parts

$$=\int_0^2\int_0^T (uv)_t$$

using the PDE's and interchanging integrations

$$=\int_0^L uv\Big|_{t=0}^{t=T}=0$$

using the initial conditions. Thus, $A^*\eta=\zeta$ as in (19). The evaluation of A or A* means the solution of the direct problem associated with (10) or (19). These problems are quite standard and good computational approximations to them are routinely available. (See, e.g. Example 6, below, for a treatment in a higher-dimensional context).

Third Method *(least squares):* This method somewhat resembles each of the above but differs significantly: *it may be quite dangerous!* Assume that one has a finite-dimensional subspace $Z=Z_j$ with a basis $(z_1,\ldots,z_J)$ and that one knows that φ can be approximated well in Z. The method consists of selecting the approximant φ_J through

$$\frac{1}{2}||A\varphi-\psi_*||^2=\min$$
$$\textit{subject to: } \varphi\varepsilon\ Z_J \textit{ so } \varphi=\sum_{j=1}^{J} c_j z_j \tag{20}$$

Comparing this to (17), one observes that in each case one has φ_J determined by

$$\varphi_J=c_1z_1+\ldots+c_Jz_J,$$

$$\langle A\varphi-\psi_*,y_j\rangle=0 \quad \textit{for} \quad j=1,\ldots,J.$$

But: for (17) one has the relation $z_j=A^*y_j$ whereas for (20) one has the very different relation $y_j=Az_j$.

We now consider a *sequence* $\{Z_J\}$, assuming the approximation properties are such that $||\varphi-Z_J||\to 0$ (i.e, that <u>potentially</u> one can approximate arbitrarily well in Z_J as $J\to\infty$). We ask whether one then has $\varphi_J\to\varphi$ for the approximants obtained using (20). For *well-posed problems* it is easily seen that this method ("least squares" combined with selection of accurately approximating subspaces Z_J) is quite satisfactory, giving accuracy of approximation to φ of the order of the best possible. Unfortunately, the coefficient in the relevant estimate involves $||A^{-1}||$ so the convergence argument fails for ill-posed problems such as ours. Indeed, one actually has a *non-convergence* result:For almost any such problem, there *always* exist orthonormal bases $(z_1,z_2,\ldots)$-in some sense "badly related"to the sequence of eigenfunctions of A*A- for which using this method with $Z_J=\mathrm{sp}(z_1,\ldots,z_J)$ fails: it produces a sequence (φ_J) of approximants for which $||A\varphi_J-\psi_*||\to 0$ while not only does (φ_J) not converge to φ but it need not even be a bounded sequence (the differences $||\varphi_J-\varphi||$ may actually become arbitrarily large with increasing

J). This result occurs with all computations *exact!* The method *may* work quite well for a particular choice of (z_j) but, needless to say, in the light of this result one lacks confidence: in the present state of the theory there is no general way of determining in advance whether one is in a good or a bad situation.

Example 4:

We conclude this section by considering briefly a nonlinear variant of Example 3. If the diffusion coefficient D is temperature-dependent for the range to be considered, one must replace (10) with the nonlinear input/output system:

$$\psi = u(\cdot, 0)$$

where

$$u_t = [D(u)u_x]_x, \quad u_x(\cdot,0)=0, \quad u(\cdot,L)=\varphi, \quad u(0,\cdot)=\omega \tag{21}$$

We assume that D is an *analytic* function of u. It is then known that the solution u(t,x) is an analytic function of x for each t and is infinitely differentiable in t for each x. (This holds for (t,x) in the interior of $Q=(0,T)\times(0,L)$ but we may use a trick-reflecting u as an even function across x=0, we have the same equation in $\hat{Q}=(0,T)\times(-L,L)$ and each (t,0) is now interior to $\hat{Q}$). Note that this shows that, even if the signal φ is not at all smooth, the observable output ψ must be infinitely differentiable.

For any $t>0$ we can expand u(t,x) in a power series at x=0:

$$u(t,x) = \sum_{k=0}^{\infty} c_{2k}(t)\,x^{2k}.$$

(Only even powers occur because of the symmetry of the problem across x=0; recall the condition $u_x(\cdot,0)=0$). We then substitute this into the PDE, using the (known) power series around the value $c_o = c_o(t) = u(t,0) = \psi(t)$,

$$D(u) = \sum_{j=0}^{\infty} D_j (u-c_o)^j$$

One next obtains, from the equation considered at (t,0),

$$\psi'(t) = u_t(t,0) = [D(u)u_x]_x = D(c_o)u_{xx} + D'u_x^2 = D_o \cdot 2c_2$$

since $u_x=0$ at (t,0); this determines $c_2(t)$. Differentiating the equation with respect to t and then substituting, gives

$$\psi''(t) = u_{tt} = [Du_x]_{xt} = [D_t u_x + Du_{tx}]_x$$

$$= [D'u_t u_x + Du_{tx}]_x = [D'(Du_x)_x + D(Du_x)_{xx}]_x$$

and one determines $c_4 = c_4(t)$ in terms of c_2 from this.

Repeated differentiation of the equation permits the recursive determination of all the coefficients: one need only note that the result at the k-th differentiation is an expression involving $\psi^{(k)}(t)$ on one side and $D_o^k(2k)!c_{2k}$ (with lower order terms) on the other, after evaluation at (t,0). Thus, u(t,.) is uniquely determined on (0,L) by ψ and its derivatives at t. Going to the limit $x \to L$ then gives $\varphi(t)$. This proves the invertibility of the nonlinear map: $\varphi \to \psi$.

Cannon and Ewing were actually able to *compute* u(t,x) by this power series approach in the case $D=D_o$=constant but the loss of accuracy was overwhelming for $x>(2/3)L$ so this could not be used to recover φ. As in Example 2, however, while the argument for uniqueness is not computational it is a necessary preliminary. Knowing the uniqueness, one can use methods generalizing those of Example 3. Here we only describe a variant of the generalized interpolation method.

We have proceeded so far by treating the equation as defining a map: $\varphi \to \psi$. Now we think of the differential operator and boundary evaluation as defining a quite different map

$$G: u \to [u_t - [D(u)u_x]_x, u(.,0)].$$

For a (smooth enough) function u on Q, this gives a pair of functions-one on Q and one on (0,T)-and our problem now takes the form of solving

$$G(u) = [0, \psi] \tag{22}$$

for u, after which one evaluates at x=L to get φ. Suppose one considers u in, e.g., the space $H^2(Q)$ (functions with distribution derivatives in $L^2(Q)$ up to second order; norm $||.||_{(2)}$) and uses L^2 norms on the right. Then one can choose an approximant u_k by

$$||G_k(u) - [0,\psi_k]|| + \alpha_k^2 ||u||^2 = \min \tag{23}$$

Here an approximating map G_k is, for example, constructed by the use of piecewise cubic splines on [0,L] with L/k to approximate $(Du_x)_x$ by interpolation; ψ_k is a measured (and, perhaps, interpolated) approximation to ψ and α_k^2 a (small) regularizing parameter. Knowing the uniqueness of u one can show, much as in the previous section, that if ψ_k becomes more accurate quickly enough and $\alpha_k^2 \to 0$ slowly enough, then u_k will converge to u in the sense of $H^2(Q)$ and $u_k(.,L)$ will converge to φ. The problem (23) is essentially a (nonlinear) PDE optimal control problem. (This construction is only possible if φ is assumed fairly smooth but this requirement can be reduced somewhat by using a more complcated variant, still derived from (22).)

4. FURTHER EXAMPLES

We now wish to consider some problems in higher dimensions. It is worth considering these problems if only to see that the basic considerations remain essentially the same.

Thus, we consider a bounded domain Ω in m-dimensional space R^m (for physically significant problems m=1,2,3) with smooth boundary surface $\partial\Omega$. For a fixed time interval $[0,T]$, let $Q=(0,T)\times\Omega$ and let S be the lateral "surface" $(0,T)\times\partial\Omega$; the "initial" and "terminal" boundaries of Q are $\Omega_o=\{0\}\times\Omega$ and $\Omega_T=\{T\}\times\Omega$

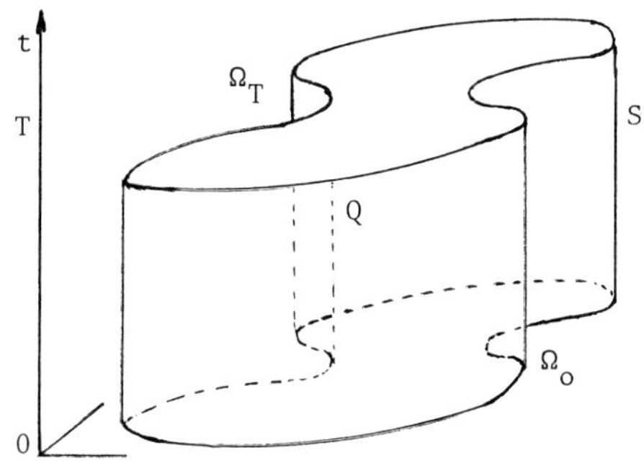

We now adopt the point of view that our only possible direct interaction with the evolution of the system (diffusion of heat in Ω) is at the boundary. Certainly this is plausible for a solid body in R^3. In this case the problem takes a form with no direct parallel in the ODE case, since the control will not appear in the differential equation but will be in the inhomogeneous boundary conditions.

Example 5

We consider the Dirichlet nullcontrol problem for an inhomogeneous body. This means that the system evolution is governed by

$$u_t=\nabla.D\nabla u,\ u|_S=\varphi,\ u(0,.)=\omega \tag{24}$$

where ω is the initial state and we wish to select/manipulate the Dirichlet boundary data φ in such a way that the terminal state u(T) will be exactly 0. Subject to some technical conditions, it is known that this is possible for every ω in $L^2(\Omega)$ for a wide variety of coefficient functions D(.). We will actually choose φ through

$$\tfrac{1}{2}||\varphi||^2=\min \quad \textit{subject to: } u(T,.)=0, \tag{25}$$

since the terminal condition does not uniquely determine φ. (The norm is that of $L^2(S)$). We note that the *optimal* φ has the characterization

$$\varphi=v_\nu \quad \textit{for some solution of} \quad -v_t=\nabla.D\nabla v,\ v|_S=0 \tag{26}$$

($v_\nu=D\bar{n}.\nabla v$ where $\bar{n}$ is the unit outward normal to Γ).

In this example our interest is not in computation but in a "stability" problem. We might well know the form (24) and, by measurement (or by inference, comparable to the system identification of Example 2), have obtained an *approximation* $\hat{D}=\hat{D}(x)$ to the actual diffusion coefficient $D=D(x)$. Suppose we were able to analyse the system (24) with $\hat{D}$ used in place of D and to determine the corresponding optimal control $\hat{\varphi}$. It would be rather unpleasant to find that although one has obtained a very good approximation $\hat{D}$ for D, nevertheless $\hat{\varphi}$ is far from φ. We ask whether convergence of D_k to D, for an increasingly accurate sequence of approximations, implies convergence of φ_k to φ for the corresponding (exactly determined) optimal controls.

The answer is "moderately satisfactory". If we suppose the convergence of D_k to D is in the sense of uniform convergence on $\bar{\Omega}$ with some number K>1 of derivatives, then for any $\varepsilon>0$ one has L^2 convergence of φ_k to φ (perhaps with some number of derivatives, depending on K) when restricted to $[0,T-\varepsilon]\times\partial\Omega$, i.e., when staying slightly away from the terminal time. On the other hand, the best available *global* result is *weak* convergence in $L^2(S)$. One cannot even be confident of approximating the optimal cost $\frac{1}{2}||\varphi||^2$ since it is not known whether $\lim||\varphi_k||$ can be strictly larger than $||\varphi||$ (indeed, if $\lim||\varphi_k||\leq||\varphi||$ one would necessarily have *strong* convergence $\varphi_k\to\varphi$ in $L^2(S)$).

A final difficulty is that if one were to use φ_k for (24), producing u_k, then one need not have $u_k(T)\to 0$ in $L^2(\Omega)$. This last is not due to the weakness of the convergence but to the use of Dirichlet controls, for which one does not have continuity to $L^2(\Omega)$ of the map: $\varphi\mapsto u(T)$. (Had we considered the corresponding Neumann control problem.

$$u_t=\nabla.D\nabla u,\quad u_\nu=\varphi,\quad u(0,.)=\omega \tag{27}$$

with (25), then weak convergence $\varphi_k\rightharpoonup\varphi$ would be adequate to ensure $u_k(T)\to 0$). One *does* obtain uniform convergence to 0 for $u_k(T)$ and derivatives when restricted to any interior subset $\Omega_*\subset\Omega$ maintaining a positive distance δ from Γ; the difficulties near the terminal time induce corresponding difficulties for u(T) only near Γ.

The proofs of these results are far from simple but we may briefly describe the *structure* of the argument. A careful examination of the proof of existence of φ_k gives a uniform bound in $L^2(S)$ so (using weak compactness and passing to a subsequence) φ_k converges weakly to a limit φ_*. Letting φ_* give u_* by (24), one shows $\int u_*(T)\zeta=0$ for each smooth enough ζ so φ_* is a nullcontrol. The most difficult part is to show that φ_* is the unique *optimal* nullcontrol φ. This is done by showing that optimality of each φ_k -given characterizations *like* (26)- implies, in the limit, (26) for φ_*. The details of this part of the argument give the stronger form of convergence on $[0,T-\varepsilon]\times\partial\Omega$ while the convergence of $u_k(T)$ on φ_* comes from interior regularity theorems.

Example 6:

We consider the same problem (24), (25) and now seek to *compute* φ. For this purpose one must, of course, reduce the distributed problem (at some stage) to a finite-dimensional problem and we formulate the *Faedo-Galerkin semi-discretization* of (24).

Introduce the notation: $X=L^2(\Omega)$, $V=H^1(\Omega)=\{$functions in $L^2(\Omega)$ with L^2 distributional derivatives of first order$\}$, $V_o=H^1_o(\Omega)=\{u\epsilon V: u|_\Gamma=0\}$ which is just the nullspace of the (continuous) *trace* map $\underline{\tau}: V\to X: u\to u|_\Gamma$. A *weak formulation* of (24) is now

$$\int_\Omega [\dot{u}v+D\nabla u.\nabla u]=0,$$

$$\int_\Omega [u(0)-\omega]v=0 \quad (\text{all } v\epsilon V_o) \tag{24'}$$

$$\underline{\tau}u=\varphi$$

To approximate this we use a finite-dimensional subspace V^h of V and seek $u^h=u^h(t)$ taking values in V^h such that

$$\int_\Omega [\dot{u}^h v+D\nabla u^h.\nabla v]=0,$$

$$\int_\Omega [u^h(0)-\omega]v=0 \quad (\text{all } v\epsilon V^h_o), \tag{28}$$

$$\underline{\tau}^h u^h=\varphi^h$$

Here $V^h_o=\{u\epsilon V^h:\underline{\tau}u=0\}$ is the nullspace of the restriction $\underline{\tau}^h$ of $\underline{\tau}$ to V^h and φ^h is the orthogonal projection of φ into the range X^h of $\underline{\tau}^h$.

We see that this reduces to a system of ordinary differential equations. Write $u^h=v^h+w^h$ where v^h takes values in V^h_o and w^h takes values in W^h, the orthogonal complement of V^h_o in V^h. One immediately has $w^h=\underline{\sigma}^h\varphi$ where $\underline{\sigma}^h: H\to W^h\ V^h$ is the *pseudo-inverse* of $\underline{\tau}^h$; the rest of (28) becomes conditions on v^h. Let $(v_1,\ldots,v_j)$ and $(w_1,\ldots,w_k)$ be convenient bases for V^h_o and W^h and expand v^h, w^h:

$$v^h(t)=y_1(t)v_1+\ldots+y_J(t)v_J,$$

$$w^h(t)=\psi_1(t)w_1+\ldots+\psi_k(t)w_k$$

writing $Y=Y(t)=(y_1,\ldots,y_J)$, $\Psi=\Psi(t):(\psi_1,\ldots,\psi_K)$. One now lets $\underline{G}$ be the Gramian matrix of $\{v_j\}$ (entries:$\langle v_j,v_k\rangle=\int_\Omega v_jv_k$), $-\underline{A}$ the "weighted" Gramian matrix of $\{\nabla v_j\}$ (entries; $\langle\nabla v_j,\nabla v_k\rangle_D=\int_\Omega D\nabla_j\nabla v_k$), $\underline{Q}$ the Gramian matrix of $\{\underline{\tau}w_k\}$ (entries: $\langle(\underline{\tau}w_j)(\underline{\tau}w_k)\rangle$), $\underline{B}$ the matrix with entries $\langle\nabla u_j,\nabla w_k\rangle_D=\int D\nabla u_j\cdot\nabla v_k$. Then (28) becomes

$$\underline{G}\dot{Y}=\underline{A}Y+\underline{B}\ ,\quad Y(0)=Y_o \tag{29}$$

where $\underline{G}Y_o$ has entries $<\omega,v_j>=\int_\Omega \omega v_j$. One also has

$$||\varphi^h||^2=\int_o^T \Psi.\underline{Q}\Psi.$$

The particular choices of the spaces V^h and of the bases are, of course, critical for the implementation of this approach. A major concern of the *finite element method* is the construction of these so the matrices involved will be sparse. We only note here that a variety of results are available as to the convergence of u^h to u (for the direct problem:given φ).

An obvious approximation method for (24), (25) is to select φ^h through

$$\frac{1}{2}\int_o^T \Psi.\underline{Q}\Psi=\min \quad \textit{subject to:}\ Y(T)=0 \tag{30}$$

with (29). (This notation is consistent with that of (28) since minimization of $||\varphi||$ there implies φ in X^h). The problem (29), (30) is a standard linear optimal control problem with quadratic cost and is routinely solvable by standard methods-subject to the controllability rank condition, which will be satisfied if W^h is not too small.

The approximation by discretization is analogous to Example 5 and it is reasonable to hope to justify this by an analogous argument. That is, one tries, to show that φ^h converges to φ as $h\to 0+$, corresponding to refinement of the discretization. Under suitable technical assumptions on the spaces V^h, this is possible. One obtains essentially the same sort of "moderately satisfactory" results as in Example 5.

The methods of *regularization* and *generalized interpolation* (first two methods of Example 3) can also be applied here. Corresponding to (11), one now takes A to be the map: $\varphi\to u(T)$ given by (24) with $u(0)=0$ and takes $\psi_*=-u(T)$ given by (24) with $\varphi=0$. (Without further elaboration, we note that the method of generalized interpolation was originally invented for precisely this problem). Regularization by relaxation of the terminal constraint will be considered in the context of Example 7.

Example 7:

We consider, a final example the optimal approximate control of (27) to (a neighborhood of) an arbitrary terminal target state $\bar{\omega}$. We may consider this in either of the equivalent forms (compare (15), (12')):

$$\frac{1}{2}||\varphi||^2=\min \ \textit{subject to:}\quad ||u(T,.)-\bar{\omega}||<\varepsilon \tag{31}$$

$$J(\varphi)=\frac{1}{2}||\varphi||^2+\lambda||u(T.,)-\bar{\omega}||^2=\min \tag{31'}$$

with ε,λ suitably related. For computation it is more convenient to work with the form (31') and we assume λ is given, ignoring the question of determining the equivalent λ during the computation if one were given ε.

For the moment we do not take $\lambda\to\infty$ but ask only how to proceed with the minimization in (31') for fixed λ, using the *conjugate gradient method*. (For simplicity we ignore the question of discretization and assume that we can solve direct problems, as (27), "exactly"- i.e., with arbitrarily good accuracy by methods such as that described in the preceding example. This causes no significant problems in the present context).

Let A be the linear map: $\varphi\to u(T)$ given by (27) with $u(0)=0$ and let $\hat{\omega}=u(T)-\bar{\omega}$ with $u(T)$ here given by (27) with $\varphi=0$. Then

$$J(\varphi)=\frac{1}{2}\left[||\varphi||^2+\lambda||A\varphi-\hat{\omega}||\right]$$

The method proceeds, at each stage after initialization, to use a *one-dimensional search* to obtain a new estimate φ_{k+1} for the minimizer of J and then to select a new search direction φ_{k+1} for the next stage. We write $\bar{\varphi}$, $\bar{\delta}$, $\bar{F}$ for the "old" values φ_k, δ_k, F_k (note: F_k will be $J'(\varphi_k)$) and φ, δ, F for φ_{k+1}, δ_{k+1}, F_{k+1}. The sequence of steps at each stage is then:

$$\begin{aligned} &\text{(i)} && J(\varphi)=\min \textit{ subject to: } \varphi=\bar{\varphi}+s\bar{\delta}\ (s\varepsilon R),\\ &\text{(ii)} && F=J'(\varphi)=\varphi+\lambda A^*A\varphi+\lambda A^*\hat{\omega},\\ &\text{(iii)} && \delta=F-c\bar{\delta} \text{ with } c=\langle F,F-\bar{F}\rangle/||\bar{F}||^2. \end{aligned} \tag{32}$$

For initialization, one guesses an approximation φ_o and obtains $\delta_o=F_o$ as in (ii). For our problem (27) we note that the adjoint A* is given by

$$\begin{aligned} &A^*\upsilon=-w|_S\\ &-w_t=\nabla.D\nabla w, \quad w_\nu=0, \quad w(T)=\upsilon \end{aligned} \tag{33}$$

The argument is essentially the same as for (19). Since J is quadratic, one easily sees that (i) becomes

$$\text{(i')}\ \varphi=\bar{\varphi}+s\bar{\delta} \qquad \textit{with}$$

$$s=\frac{\langle\bar{\varphi},\bar{\delta}\rangle+\lambda\langle A\bar{\varphi},A\bar{\delta}\rangle}{\langle\bar{\delta},\bar{\delta}\rangle+\lambda\langle A\bar{\delta},A\bar{\delta}\rangle}=\frac{\langle\bar{F},\bar{\delta}\rangle}{||\bar{\delta}||^2+\lambda||A\bar{\delta}||^2}$$

Examination of these steps shows that at each stage one applies A to φ and then applies A* to the result while computing F in step (ii) ($A^*\hat{\omega}$ need be computed only at initialization); one also requires $A\bar{\delta}$ so as to compute s in step (i'). . Thus, each stage of the conjugate gradient method involves solving three *direct* problems. At the price of a possible build-up of round-off error, one can compute $A\varphi_o$ initially and then recursively use the fact that (i') gives $A\varphi=A\bar{\varphi}+sA\delta$ to reduce to *two* the number of direct problems to be solved at each iterative stage. (An interesting alternative procedure would be to observe that the solution φ of (31') is given by $\varphi=-\lambda A^*\upsilon$ with $\upsilon=A\varphi+\hat{\omega}$. One then sees that υ is given by minimization of

$$\tilde{J}(\upsilon)=\frac{1}{2}[||\upsilon-\hat{\omega}||^2+\lambda||A^*\upsilon||^2] \quad (=J(\varphi)/\lambda)$$

so that one could apply the conjugate gradient method to $\tilde{J}$ instead).

We have considered (27), rather than (24), for this example so: $A:L^2(S)\to L^2(\Omega)$ is continuous (for (24) one could proceed by working with norms in $H^{-1}(\Omega)$). A standard argument then shows convergence of the iteration. There is no difficulty in combining this with the use of a discretization to compute A,A^* approximately.

If one were to think of ε as small ($\varepsilon\to 0+,\lambda\to\infty$), then (31) is an obvious relaxation of

$$\frac{1}{2}||\varphi_o||^2=\min \textit{ subject to: } u(T,.)=\bar{\omega}. \tag{34}$$

For an "arbitrary" $\bar{\omega}$ (34) is impossible but, *assuming* $\bar{\omega}$ *is attainable* (e.g. for $\bar{\omega}=0$), one might denote the solution of (31) by φ_ε and ask whether φ_ε converges to the solution φ_o of (34) as $\varepsilon\to 0+$. The proof is a simple functional analysis argument: Take $\varphi_j=\varphi_\varepsilon$ for $\varepsilon=\varepsilon_j\to 0+$. Since φ_o satisfies the constraint in (31) for each ε, one must have $||\varphi_j||\leq||\varphi_o||$ so the sequence $\{\varphi_j\}$ is bounded and one can extract a subsequence converging weakly to some φ_*. As the corresponding $\{u_j(T)\}$ converge, by the constraint, to $\bar{\omega}$ one finds that φ_* is a control exactly to $\bar{\omega}$ with $||\varphi_*||\leq \mathit{lim}||\varphi_k||\leq||\varphi_o||$. Thus one has $\varphi_*=\varphi_o$ by (34) and also $||\varphi_j||\to||\varphi_o||$ which implies that the weak convergence is actually $L^2(S)$ convergence. (If one were to discretize, as in the preceding example, and use increasingly good subspaces (V^h depending on ε), then the same argument can be used to show $\tilde{\varphi}_\varepsilon\to\varphi_o$ for the computed approximants $\tilde{\varphi}_\varepsilon$).

5. CONCLUDING REMARKS

The discussion above has only been able to touch briefly on some principal themes and considerations in the application of approximation methods to some distributed parameter systems and to indicate a few such approaches. Let us summarize these principal themes:

[A] Most lumped parameter systems used in standard engineering practice are, in fact approximations to distributed parameter systems. A 1000-mile transmission line is obviously a distributed system; a $\frac{1}{2}$inch piece of wire in a circuit is also-but the fact can here be successfully ignored. (It is worth the mental effort to recognize these approximations-if only to determine their ranges of validity. Most approximations occur in modeling and have never been explicitly analyzed as such). The theory of distributed parameter systems can only be of practical value to the extent that one can recognize situations of economic/practical interest for which (a) ODE models are inadequate and (b) feasible computational procedures for a distributed model make it possible to do enough better to warrant the effort.

[B] All considerations of approximations for distributed parameter systems begin with the theory, representations and computational algorithms associated with the *direct problems* which express the basic dynamics of the systems. These direct problems define functions which one wishes to invert or minimize. Thus, the general body of theory for methods of approximate solution of abstract operator equations and of computational optimization can serve as a source of powerful and useful techniques applicable to problems of distributed system theory once the problems have been suitably formulated.

[C] Many of the problems which arise in distributed system theory are "ill-posed". Such problems are extremely sensitive to apparently minor points of interpretation or implementation which might, in other sorts of problems, make little difference. The term "ill-posed" need not mean that one has already made an error of formulation (Hadamard's original attitude) but does mean that one must train one's intuition properly and exercise extreme care not to make such errors!

CHAPTER 10

SELECTION OF MEASUREMENT AND CONTROL LOCATIONS IN DISTRIBUTED-PARAMETER SYSTEMS

M. Amouroux* and J. P. Babary**

*Laboratoire de Physique Appliquée, University of Perpignan, France
**Laboratoire d'Automatique et d'Analyse des Systèmes, Toulouse, France

INTRODUCTION

The computation of the optimal control law with respect to a given performance criterion for a dynamical system requires, in general, an estimate of the system state from output measurements.
In most distributed processes, measurements are taken, in practice, at a finite number of spatial locations and the control inputs are applied only at a finite number of zones of the spatial domain. This raises the problem of defining suitable criteria for the selection of the transducer and control locations.
This problem can be treated from a theoretical or an engineering point of view. It is the latter that we have choosen in what follows.
In fact, every control study must be followed by a study of the effective realization of the algorithm; so, rather than searching for an optimal solution of the theoretical problem, it is better to try to be optimal vis a vis this realization. This implies that one has to insert into the initial criterion some parameters characterizing the possibility of realization, the optimal position of actuators and sensors being then obtained from a compromise between the "simplicity of the realization" and the realized optimum.
So, at first, it is possible to define a class of admissible control laws or estimators (realizable) and to determine in this class, the control (filter) which minimizes the quality criterion. Such a class must be defined from technological elements of the realization. If we want, for instance, to implement a real-time algorithm, one has to consider the constraints imposed by the device elaborating the control law. For instance, if the transition operator has to be computed on line, the dimensions of the matrices must be compatible with a relatively small computation time.
In many cases the realization must be easily modifiable in order to adapt itself to any eventual modifications of parameters, or initial or final conditions. (If there are modifications, ideally, the parametric optimization problem should not be solved each time).
The selection of measurement and control locations is made using criteria methods developped subject to the above considerations.
This chapter has three main subjects. The problem of the optimization of actuator positions is presented in section 2, after a rapid survey of the various papers published over the last few years on this subject; section 3 concerns optimal measurement locations and the final section treats the problem globally.

Since there are as many solutions as criteria for control problems, it is difficult to cover the whole field in each paragraph. We have decided to treat the case of minimum energy control with constraints on the final state for the actuator problem. We do not consider the classical regulator problem since in the treatement of the sensor problem we use Riccati's equation to obtain the optimal filter. However both criteria are developed in the final paragraph.

CONTROL LOCATION PROBLEM

The pointwise (or zone) control of distributed parameter dynamical systems has grown in importance over the past few years. This control deals with a particular form of action for which, contrary to the distributed control, the actuators have a mathematical representation under separated form (space-time). This assumption which might seem at first sight restrictive, is on the contrary, very near to the physical reality. Moreover, concerning the theoretical aspect, Vidyasagar and Higgins (I973) have obtained conditions assuring the existence of a control which acts in a finite number of points, approximating in the best way a distributed control which appears to be the best one possible, since it means that one can act in an independent way on each point of the space domain. From this assumption it might be considered as well their position, in order to obtain a given approximation. This problem can be included in a more general study which consists of optimizing a set of geometrical elements for distributed parameter systems. There are as many different ways to solve this problem as there are different formulations of the control problem.

This optimization can be made, as said before, in order to determine a "lumped" control which is the nearest possible to the optimal distributed control, or, the action mode (pointwise or zone control) being already chosen, these geometrical elements can be obtained from the direct optimization of the initial quality criterion. This last point is the one we present in this paper.

Most of the studies in this field are connected with the choice of a quadratic criterion on the state and the control. Lions (I972) for example, states the overall problem of the simultaneous optimization of the position of pointwise actuators, and of the control law for parabolic problems. He gives existence and uniqueness conditions of the solution. Numerical examples, solved using discretization methods, and with one or two actuators, are presented by Saguez (I974). Courdesses (I977), Greenberg (I969) and Kessler (I972) propose a solution of this same problem in two steps : first, they set the partial differential Riccati equation, which they solve from an expansion of the solution on the eigenfunction basis of the operator which appears in the model; then they minimize an approximated criterion, obtained from this solution, with respect to the position of the pointwise actuators. A similar expansion is proposed by Aidarous (I976) and Wong and Soudack (I975) for discrete stochastic systems. They also proposed numerical algorithms based on the computation of the gradient, the optimization of the location of the actuators and on the solution of the Riccati equation, these two last computations being alternated.

In addition to these studies with the same objective, one can mention the following, connected with different criteria :

Foster (I969) tackles the optimal position problem of action zones (with given structures) in connection with open-loop control laws. He determines this position in such a way that, for a class of given inputs, the output of the system is maximal. Burger (I976) considers a geometrical approach in which the optimization of the pointwise location of the actuators is done from the minimization of the difference between the desired state and the nearest reachable state; therefore, the chosen criterion concerns the accuracy of the steady state.

Wang (1969) is interested in the influence of the number and position of pointwise actuators on the controllability of an approximated model and on the dynamics of this model.
Amouroux and Babary (1973) have also proposed a criterion that enables one to know the controllability degree of the system with respect to the location of the pointwise actuators. The advantage of these two last criteria lies in the fact that they are independent of the nature of the considered problem.
The problem of determining the best shape and location of the actuators in a parabolic distributed parameter system, having regard to the minimization of an energy criterion is proposed in this paragraph. The precise control problem is to find a minimal energy control law which enables one to reach a desired state, considering the shape and location of the actuators as parameters.

Problem Formulation

We consider linear time-invariant distributed systems, defined over an open set $\Omega \subset R^s$ and described by a parabolic equation of the type:

$$\frac{dy(t)}{dt} = My(t) + \sum_{j=1}^{p} u_j(t) g_j(x, \eta_j) \tag{1}$$

with $y(0) = y_0 \in L_2(\Omega)$

where $- \; y \in W(0,t) = \{y : y \in L_2[0,t;L_2(\Omega)] \; ; \; \frac{dy}{dt} \in L_2[0,t;L_2(\Omega)]\}$

$- \; u = (u_1, \ldots, u_p)^T \in L_2[0,t;R^p]$

- M is the infinitesimal generator of a strongly continuous semigroup S(t)

- the domain of M, $\mathcal{D}(M)$, is dense in $L_2(\Omega)$ and is described by:

$$\mathcal{D}(M) = \{y : My(t) \in L_2(\Omega), \; Ly=0 \text{ on } \partial\Omega\}$$

L being a linear operator defined on the boundary $\partial\Omega$ of the spatial domain

- $g_j(x,\eta_j)$ characterises the jth control zone and has compact support; η_j $(j=1,\ldots,p)$ is a vector defining the location and the shape of the jth actuator and are the parameters to be optimized.

Optimality Criterion

Let $y_d(x)$ be the desired profile to be reached at a fixed time t_f. The problem is then the following:

$$\min_{u,\eta} ||u||^2_{(L_2[t_0,t_f])^p}$$

with constraint (2)

$$y(x,t_f) = y_d(x)$$

The solution of eqn. (1) can be written as:

$$y(t) = S(t-t_0)y_0 + \int_{t_0}^{t} S(t-\tau)\left[\sum_{i=1}^{p} g_i(\eta_i)u_i(\tau)\right] d\tau \tag{3}$$

Let $\tilde{y} = y_d - S(t_f - t_0)y_0$ and $\eta = [\eta_1, \ldots, \eta_p]^T$

Then problem (2) becomes:

$$\min_{u,\eta} ||u||^2_{[L_2(t_0,t_f)]^p} \triangleq \min_{u,\eta} J(u,\eta)$$
$$\tilde{y} = H(u) \tag{4}$$

Where H is the following operator:

$$H = \int_{t_0}^{t_f} S(t_f-\tau)g^T(\eta)(.)d\tau \tag{5}$$

With
$$g^T(\eta) = [g_1(\eta_1),\dots,g_p(\eta_p)]$$

Porter (1966) states that if η is known and such that the system is controllable and if $\tilde{y} \in \mathrm{Im}(HH^*)$ an optimal control exists and

$$||u_0||^2 = \langle \tilde{y},(HH^*)^{-1}\tilde{y} \rangle \tag{6}$$

Problem (4) can thus be changed into a parametric optimization problem over the different controllability domains of the system.

Solution Method

Let $\{\phi_i(x)\}$ be the setof eigenfunctions of the operator M. We may expand the state variable y on this basis

$$y(x,t) = \sum_{i=1}^{\infty} a^i(t)\phi_i(x) = a(t)^T\phi(x) \tag{7}$$

The application of the eigenfunction method leads to an equivalent solution of (3)

$$a(t) = \Phi(t-t_0)a(t_0) + \int_{t_0}^{t} \Phi(t-\tau)B(\eta)u(\tau)d\tau \tag{8}$$

(8) is the solution of the equivalent model of system

$$\begin{aligned}\dot{a}(t) &= Aa(t) + B(\eta)u(t)\\ a(t_0) &= a_0\\ y(x,t) &= C(x)a(t) = [\phi_1(x),\ \phi_2(x),\dots]\ a(t)\end{aligned} \tag{9}$$

where
$$a(t) = [a^1(t),\dots,a^i(t),\dots]^T$$

$$\Phi(t) = [\Phi_{ij}(t)]_{\substack{i=1,\dots,\infty\\ j=1,\dots,\infty}} \quad ; \quad \Phi_{ij}(t) = \langle S(t)\phi_j,\ \psi_i\rangle_{L_2(\Omega)}$$
$$B(\eta) = [b_{ij}]_{\substack{i=1,\dots,\infty\\ j=1,\dots,p}} \quad ; \quad b_{ij}(t) = \langle g_j(x,\eta_j),\phi_i\rangle_{L_2(\Omega)} \tag{10}$$

$$a(t_0) = [a^1(t_0),\dots,a^i(t_0),\dots]^T \ ; \ a^i(t_0) = \langle y_0(x),\ \psi_i\rangle_{L_2(\Omega)} \tag{11}$$

$\{\psi_i\}$ are the eigenfunctions of the adjoint operator M^*

The optimal control law (for η fixed) takes the form:

$$u_0(t) = H^*(HH^*)^{-1}\left[a_d - \phi(t_f - t_0)a_0\right]$$

$$= B^T(\eta)\ \phi^T(t_f - t)\left[\int_{t_0}^{t} \phi(t_f-\tau)B(\eta)B^T(\eta)\ \phi^T(t_f-\tau)d\tau\right]^{-1}\left[a_d - \phi(t_f-t_0)a_0\right]$$

and the optimum criterion:

$$||u_0||^2 = \tilde{a}^T(t_0)\left[\int_{t_0}^{t} \phi(t_f-\tau)B(\eta)B^T(\eta)\phi^T(t_f-\tau)d\tau\right]^{-1}\tilde{a}(t_0)$$

where $\tilde{a}(t_0) = a_d - \phi(t_f - t_0)$

Rather than calculating η using a numerical method to minimize (6), we propose an implementable control law from which it is possible to define a new criterion to determine η.

Suboptimal Control And Optimization Of Actuators

If we consider an approximation of y (N-order truncation) the approximated state vector is the solution of the truncated model:

$$\begin{aligned} \dot{a}_N(t) &= A_N a_N(t) + B_N(\eta)u(t) \\ a_N(t_0) &= a_{N0} \\ y_N(x,t) &= C_N(x)a_N(t) \end{aligned} \tag{12}$$

where:

$$(A_N)_{ij} = \begin{cases} \lambda_i & \text{if } i=j \quad (\lambda_i \text{ eigenvalues of operator M}) \\ 0 & \text{if } i\neq j \end{cases}$$

$$(B_N)_{ij} = (B)_{ij} \quad \text{if } i=1,\ldots,N \text{ and } j=1,\ldots,p$$

$$a_N = \left[a^1,\ldots,a^N\right]^T$$

$$C_N = \left[\phi_1,\ldots,\phi_N\right]$$

Then the optimal control for our optimization problem is given by

$$u_{N0}(t) = B_N^T(\eta)\ \Phi_N^T(t_f-t)\ U_N^{-1}(\eta)\tilde{a}_N \tag{13}$$

with

$$\Phi_N(t) = \exp(A_N t), \quad \tilde{a}_N = a_{Nd} - \Phi_N(t_f-t_0)a_{N0}$$

(a_d being the desired state corresponding to $y_d(x)$), and

$$U_N(\eta) = \int_{t_0}^{t_f} \Phi_N(t_f-\tau)B_N(\eta)B_N^T(\eta)\ \Phi_N^T(t_f-\tau)d\tau \tag{14}$$

Two propositions concerning this control law have already been proven (Amouroux and Babary 1975).

Proposition I:

$\forall N$, the control law (13) applied on system (8) enables an optimal control of the subsystem obtained by the projection of the state vector in an N-dimensional space.

Proposition II:

$$y_N(x,t_f) \to y_d(x) \quad \text{if } N \to \infty$$

Definition of a modified criterion: The application of $u_{NO}(t)$ to original system (8) leads to an error at the final instant t_f which can be important if the location of the actuators is badly chosen.
In order to avoid this situation, and to be nearer the optimality conditions, we insert in the initial energy criterion one penalty term characterizing the error at time t_f, and we consider the following problem:

$$\min_\eta I_m(\eta) \triangleq \min_\eta \{I(u_{NO})+q|| \varepsilon(x,t_f)||^2_{L_2(\Omega)}\} \tag{15}$$

where

$$\varepsilon(x,t_f) = y_d(x) - y(x,t_f) \quad ; \quad q > 0$$

Calculation of $I_m(\eta)$: Using equality $y(x,t_f) = C(x)a(t_f)$, we get

$$||\varepsilon(x,t_f)||^2 = \langle C(x)(a_d-a(t_f)),C(x)(a_d-a(t_f))\rangle_{L_2(\Omega)}$$

$$= [a_d-a(t_f)]^T C'[a_d-a(t_f)]$$

where

$$C' = \int_\Omega C(x)C^T(x)dx \;;\quad C'_{ij} = \int_\Omega \phi_i(x)\phi_j(x)dx$$

According to the proposition I, we have

$$a_{Nd} = a_N(t_f)$$

Therefore

$$||\varepsilon(x,t_f)||^2 = [0^T \vdots (a_{rd}-a_r(t_f))^T] \begin{bmatrix} C'_N & \vdots & C'_a \\ \cdots & \cdots & \cdots \\ C'^T_a & \vdots & C'_r \end{bmatrix} \begin{bmatrix} 0 \\ \cdots \\ a_{rd}-a_r(t_f) \end{bmatrix}$$

$$= [a_{rd}-a_r(t_f)]^T C'_r [a_{rd}-a_r(t_f)]$$

Let $\qquad \tilde{a}_r \triangleq a_{rd} - \Phi_r(t_f-t_0)a_{r0}$

Then

$$||\varepsilon(x,t_f)||^2 = [\tilde{a}_N^T \vdots \tilde{a}_r^T] \underbrace{\begin{bmatrix} Z^T(x^c)C'Z(\eta) & \vdots & Z^T(\eta) \quad C' \\ \cdots\cdots & \vdots & \cdots\cdots \\ C'^T Z(\eta) & \vdots & C' \end{bmatrix}}_{\triangleq P} \begin{bmatrix} \tilde{a}_N \\ \cdots \\ \tilde{a}_r \end{bmatrix}$$

in which

$$Z(\eta) \triangleq \int_{t_0}^{tf} \Phi_r(t_f-\tau)B_r(\eta)\, B_N^T(\eta)\phi_N^T(t_f-\tau)\, d\tau.\ U_N^{-1}(\eta)$$

For η belonging to the controllability domain of (12), P is a non-negative definite symmetrical and bounded operator. Besides this :

$$I(u_{NO}) = ||u_{NO}||^2\ \tilde{a}_N^T\ U_N^{-1}(\eta^c)\ \tilde{a}_N$$

Therefore

$$I_m(\eta) = [\tilde{a}_N^T \vdots \tilde{a}_r^T]\Delta(\eta) \begin{bmatrix} \tilde{a}_N \\ \cdots \\ \tilde{a}_r \end{bmatrix} \tag{16}$$

where

$$\Delta(\eta) = q \begin{bmatrix} Z^T(\eta)\, C'Z(\eta) + \frac{1}{q} U_N^{-1}(\eta) & \vdots & Z^T(\eta)\, C' \\ \cdots\cdots & \vdots & \cdots\cdots \\ C'^T Z(\eta) & \vdots & C' \end{bmatrix} \tag{17}$$

For η belonging to the controllability domain of (12), $\Delta(\eta)$ is a positive definite bounded operator (symmetrical and defined in l_2).
The numerical computation of the criterion is achieved by truncation using gradient method.

<u>Expression of grad (I_m)</u>

If we consider criterion I_m truncated at order $M > N$, we get :

$$I_m^M(\eta) = \tilde{a}_N^T\ U_N^{-1}\tilde{a}_N + q\tilde{a}_N Z^T C' Z\tilde{a}_N + 2qa_N^T C'^T a_M + qa_M^T\ C'\ a_M$$

$$\frac{\partial I_m^M}{\partial \eta_i} = \text{trace}\ \tilde{a}_N\tilde{a}_N^T\ U_N^{-1}\left\{\frac{\partial U_N}{\partial \eta_i}\left[1 + U_N^{-1}Z_2C'Z_2\right] + Z_1C'Z_1\right\}\ U_N^{-1}$$

$$+\ 2\ \text{trace}\ \tilde{a}_M\tilde{a}_N^T\ \left\{C'Z_1 + Z_2U_N^{-1}\frac{\partial U_N}{\partial \eta_i}\right\} U_N^{-1}$$

where

$$Z_1 = \int_{t_0}^{t_f} \phi_M \frac{\partial B_M}{\partial \eta_i} B_N^T + B_M\frac{\partial B_N}{\partial \eta_i}\, d\tau\ U_N^{-1}$$

$$Z_2 = \int_{t_0}^{t_f} \Phi_M B_M B_N^T \Phi_N^T d\tau \frac{\partial U_N^{-1}}{\partial \eta_i}$$

Criterion independent of the initial and final states

The solution of the previous problem depends upon the considered initial and final states. For a concrete application the implementation of actuators must be considered for a series of experiments in which these states can (be) change(d) from one experiment to another. So, it could be interesting to modify the previous criterion, for instance :

$$\min_{\eta} \sup_{||\tilde{a}||< K} \{\tilde{a}^T \Delta(\eta) \tilde{a}\} \tag{18}$$

$$= \min_{\eta} ||\Delta(\eta)|| = \min_{\eta} K \sqrt{\max \text{ eigenvalue } (\overline{\Delta^T \Delta(\eta)})}$$

giving valid results for a class of initial and final states.
However, the use of such a criterion presents some negative aspects :
- the error at final instant could be important for some values of initial and final states;
- difficult implementation of an optimization method based on computation of the gradient.

Note: We can slightly modify the criterion in the following manner :

$$\min_{\eta \in \mathcal{D}} \left\{ \sup_{||\tilde{a}||< K} I(u_o) \right\}$$

where

$$\mathcal{D} \triangleq \{ x^c : x^c \in \text{controllability domain and} \sup_{||a||<K} ||\varepsilon(x,t_f)||^2 \leqslant \varepsilon_0 \}$$

ε_0 is the upper bound of the error, compatible with the chosen N-order approximation.

The main advantage of the eigenfunction method is the control law expression itself. Indeed, all essential parameters which could be modified appear in this expression, avoiding therefore another solution of the parametric optimization problem. This fact is very interesting, for example, for the application of control schemes in 'adaptive open-loop control' problems.

Example 1 : Pointwise control.

Let us consider the system described by the equation

$$\frac{\partial y(x\ t)}{\partial t} = \alpha \frac{\partial^2 y(x,t)}{\partial x^2} + \gamma y(x,t) + \delta (x-x^c) u(t)$$

$$y(x,0) = y_0(x),\ x \in [0,1] \tag{19}$$

$$y(0,t) = y(1,t) = 0$$

y_d given

The operator $M = (\alpha(\partial^2/\partial x^2) + \gamma)$ with the given boundary conditions is self adjoint. Its eigenvalues and eigenfunctions are:

$$\lambda_i = \gamma - \alpha i^2 \pi^2 \qquad i=1,2,\ldots$$

$$\phi_i(x) = \psi_i(x) = \sqrt{2}\,\sin(i\pi x)$$

The numerical values of the model parameters are the following:

$$\alpha = 0.2, \qquad \gamma=1$$

Then

$$\lambda_i = 1 - 0.2 i^2 \pi^2$$

The initial state is assumed to be nought: $y_0(x)=0, (t_0=0)$. The desired state is given by the function

$$y_d(x) = x(1-x^3)$$

This profile is to be reached at $t_f=5s$

(a) Criterion $I_m(x^c)$: Let us now consider the control law (13). This law can be written as

$$u_{NO}(t) = \sum_{i=1}^{N} \sum_{j=1}^{N} \exp\left[\lambda_i(t_f-t)\right] \frac{\gamma_{ij}}{\psi_i(x^c)} a_{jd} \tag{20}$$

where γ_{ij} is the (i,j) term of the inverse matrix of U_N defined by

$$(U_N)_{i,j} = \frac{\exp\left[(\lambda_i+\lambda_j)t_f\right]-1}{\lambda_i+\lambda_j}$$

The error at the final instant is written as

$$\begin{aligned} ||y_d(x) - y(x,t_f)|| &\triangleq ||\varepsilon|| \\ &= \left[\sum_{i=N+1}^{\infty} \left(a_d^{\,i} - \sum_{k=1}^{N} \frac{\psi_{N+i}(x^c)}{\psi_k(x^c)} \right.\right. \\ &\qquad \left.\left. \times \sum_{j=1}^{N} \frac{\exp\left[(\lambda_i+\lambda_j)t_f\right]-1}{\lambda_i+\lambda_j} \gamma_{jk} a_d^{\,k}\right)^2\right]^{1/2} \end{aligned} \tag{21}$$

Note. This error can tend towards infinity if x^c is chosen such that $\psi_k(x^c) = 0$

(b) results: The minimization problem (15) has been numurically solved using the gradient method. Obviously, the results depend on the parameter q of the criterion. If $q \to 0$, with N=4 and MO=30, the optimal values are

$$x_0^c = 0.79 \quad \text{and} \quad ||u_{NO}|| = 0.70$$

The final state error is then $||\varepsilon|| = 0.5 \times 10^{-2}$.
Figure 1 represents the energy value with respect to the location x^c of the actuator for three values of N (N = 2, 3, 4).

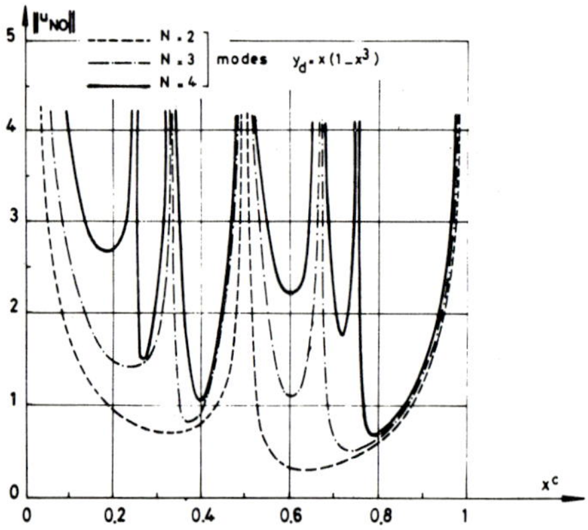

Figure 1

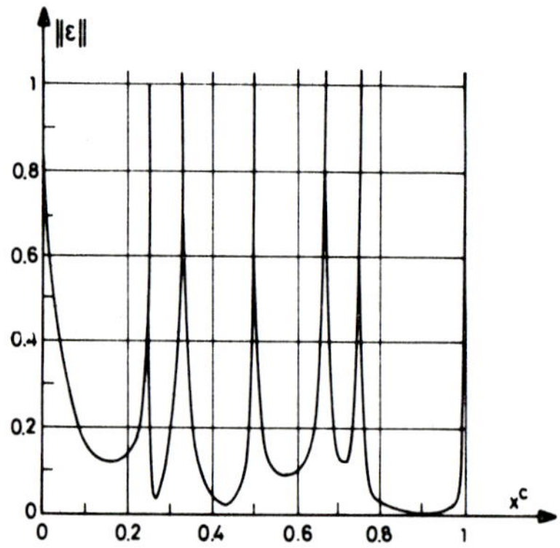

Figure 2

If $q \to \infty$, we get:

$$x_0^c = 0.90 \quad ; \quad ||u_{NO}|| = 1.15 \quad \text{and} \quad ||\varepsilon|| = 0.43 \times 10^{-2}$$

We can see from Fig.2 that, in the domain [0.79-0.90] the final state error is almost constant. The optimal location for q = 100 is $x_0^c = 0.85$.

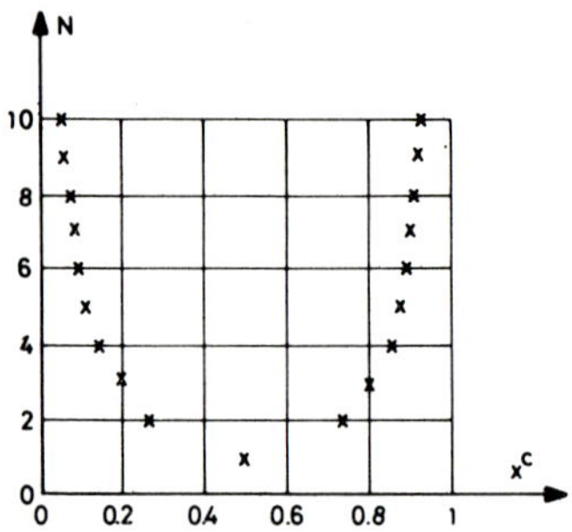

Figure 3

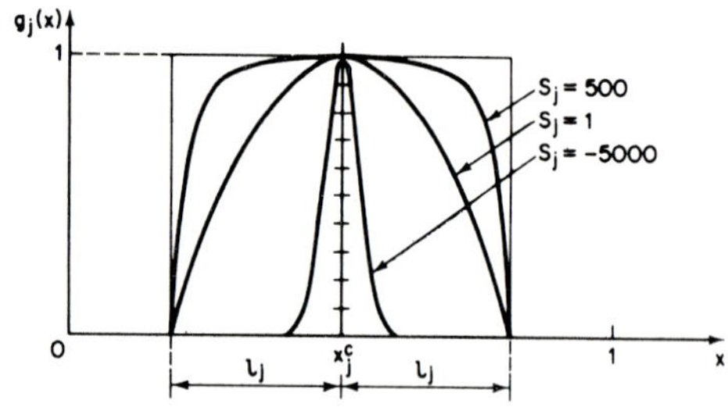

Figure 4

If we choose a criterion independent of initial and final states, for instance:

$$\sqrt{|\text{max eigenvalue } (\Delta(x^c)^T \, \Delta(x^c))|}$$

we get two symmetrical optimal positions, depending only on the truncation order, such as it is seen in Fig. 3 (q = 100). The symmetry of results is due to the symmetry of operator and boundary conditions.

Summing up the results.

Criterion	x^c	$\|\|u_0\|\|$	$\|\|\varepsilon\|\|$
Energy	[0.79-0.90] (according to q)	Between 0.70 and 1.15	Between $0.43x10^{-2}$ and $0.5x10^{-2}$
Energy independent of y_0 and y_d	0.15 .... 0.85	.... 1.80 .... 0.85	.. $11x10^{-2}$. $0.48x10^{-2}$

Example 2: Zone of action.

Let us consider the same system controlled over one zone, the action is characterized by the function:

$$g(x,\eta) = \frac{\exp(s\ell^2)}{1-\exp(s\ell^2)} \{\exp[s(x-x^c)^2 - s\ell^2] - 1\} \tag{22}$$

$$x \in \omega \overset{\Delta}{=} [x^c - \ell \; ; \; x^c + \ell]$$

x^c is the position of the center of the zone; 2ℓ is the length of the zone; s characterizes the distribution of the magnitude of the control. So, η is composed of three parameters

$$\eta = [x^c, \ell, s]^T \tag{23}$$

The minimization problem has been numerically solved using the same gradient method. The optimal values are: $x_0^c = 0.85$ and $\ell_0 = 0.15$ for N = 4, q = 100;

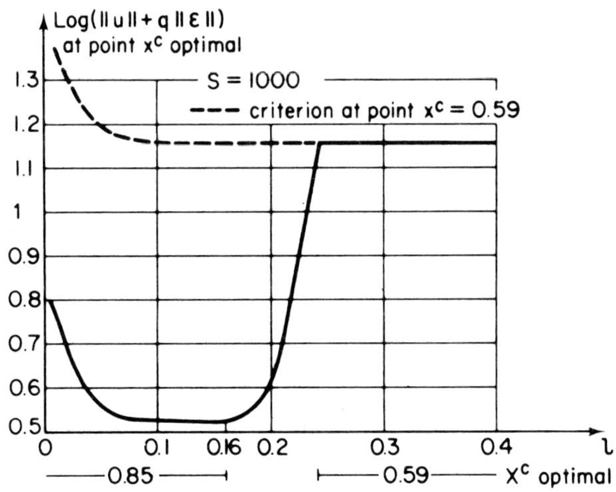

Figure 5

Figure 5 shows the evolution of I_m with respect to ℓ, values of which being computed for the optimal x_0^c corresponding to the chosen ℓ_0. We notice that when ℓ is greater than 0.15 the optimal location x_0^c varies continously up to x^c=0.75 and for each pair (x^c,ℓ) such that $x^c+\ell=1$. For $\ell>0.25$ the optimal location is $x_0^c=0.59$ and the criterion remains constant until $x_0^c+\ell_0=1$.

<u>Influence of initial and final states</u>: If we interchange the preceding initial and final states with the same dynamics we obtain:

pointwise action case: $x_0^c = 0.505$

zone action case: $x_0^c = 0.505, \quad \ell_0 = 0.495, \quad s \to \infty$

In fact, $x_0^c = 0.505$ is a point very close to $x^c = 0.5$ which is a point of uncontrollability of the system. This does not matter here since the weakly controlled modes are stable and decay rapidly (the desired state being y_d=0). Moreover, the amplitude of the first mode is maximal at x^c=0.5.
The choice of criterion greatly affects the results and the efficiency of the control In the paper (Amouroux and Babary 1978), we emphasize the caution (from a practical viewpoint) required in choosing a criterion which is independent of initial and final states.

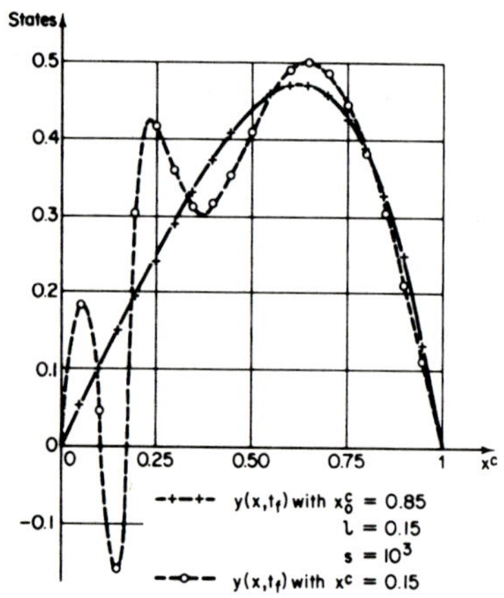

Figure 6

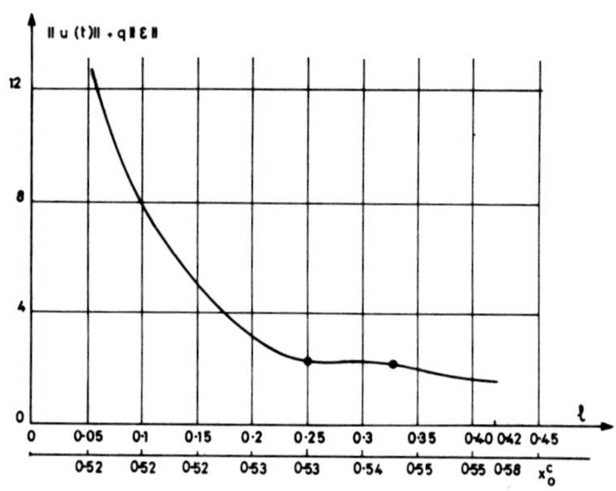

Figure 7

Figure 6 shows that, with this type of criterion, results obtained on the final state $y(x,t_f)$ can be poor. Results improve when the number of actuators is increased (El Jai 1977).

<u>Influence of the dynamics (unstable system)</u>: If α=0.2 and γ=10 for example, the two first eigenvalues of M are positive; (λ_1=8.02 and λ_2=2.10).
In this case the optimal values are

$$x_0^c = 0.58, \qquad \ell_0 = 0.42, \qquad (x_0^c + \ell_0 = 1)$$

The distance between x_0^c and the point of uncontrollability is now greater.

Numerical runs show that for any ℓ, the optimal value of x_0^c is always in the range 0.52-0.58. Fig 7 shows the criterion I_m with respect to ℓ corresponding to the optimal value of x^c for the considered length of the zone.

SENSOR LOCATION PROBLEM

This problem has been more frequently studied than the other. It has mostly been associated with the state estimation od D.P.S. and in a few cases with the identification problem (Nakagiri et al 1977 and Courdesses et al 1981).

The accuracy of the state estimates depends on the number and locations of the sensors. Generally, this number is governed by economy considerations and it is desirable to select each sensor point in order to give the best estimate of the system.

The sensor location problem for D.P.S. was treated from different viewpoints by several authors. Bensoussan (1972), Omatu etal (1978), Curtain et al (1976), Chen and Seinfeld (1975) proved the existence theorem concerning a solution of this problem using variational inequalities, functional analysis or the matrix minimum principle. Yu and Seinfeld (1973) developed a suboptimal technique for sequentially located sensors by using the eigenfunction method. Amouroux et al (1976) propose a modified conjugate gradient algorithm in the case of continuous time measurements. The discrete time problem was treated by Aidarous et al (1975).

In the above works, it has been necessary to solve the Riccati differential equation to determine the filter equation and optimal sensor locations based in general on the minimization of the scalar measure of the error covariance. Kumar and Seinfeld (1978) simplify the numerical solution by defining a criterion based on the calculation of an upper bound for the error covariance matrix which reduces the problem to the solution of a linear differential equation.

Other criteria have been proposed in the litterature. Cannon and Klein (1970) suggest an algorithm which minimizes the upper bound of the error function between the true solution and its approximation for a linear heat conduction system. Caravani et al (1975) consider the sensor location problem with an unknown initial state; they define a difference between this state and an approximation to it,which they calculate.

In this paragraph we formulate the optimum measurement location problem for linear time invariant D.P.S. subject to noise disturbances in the dynamics and in the observation. Measurements are assumed to be available continuously in time. As in the preceding paragraph we develop the solution in two steps:

- assuming that the sensor locations are fixed,the optimal filter is used to construct the best estimate of the state.
- the optimality criterion for the sensor s location is chosen to be the spatial integral of the trace of the steady state error covariance matrix.

The effects of changing the noise statistics are discussed in an example.

Problem Formulation

We consider a linear stochastic distributed parameter defined over Ω:

$$\frac{dy(t)}{dt} = My(t) + w(t) \tag{24}$$

M, $\mathcal{D}(M)$ are given as previously.

y_0 is a gaussian stochastic function whose mean and covariance are given as:

$$E(y_0) = y_0(x),\ E(||y_0||^2) < \infty \qquad x \in \Omega$$
$$E\left[y_0(x)\ y_0(r)\right] = P_0(x,\ r) \tag{25}$$

P_0 is a non negative symmetric operator
w(t) is a white Gaussian process with:

$$E\left[w(t)\right] = 0$$
$$E\left[w(x,t)w(r,\tau)\right] = Q(x,r)\ \delta(t-\tau) \tag{26}$$

Q(x,r) is symmetric and non negative

We assume that observations of the state are made at q points of Ω.
Let us define $y_m(t) = \left[y(x_1^m,t),\ldots,y(x_q^m,t)\right]^T$
The observation equation is:

$$z_m(t) = y_m(t) + v(t) \tag{27}$$

v(t) is a white gaussian process defined at the q points x^j

$$E(v(t)) = 0$$
$$E\left[v(x^m,t)\ v^T(x^m,\tau)\right] = R(x^m)\ \delta(t-\tau)\mathbb{1} \tag{28}$$

and w, v and y_0 are assumed statistically independent of each other.
More details about the well posedness of the problem can be found in Curtain (1976)

Optimal Estimation

The optimal filter for linear D.P.S. has been derived by many authors for both type of measurements, i.e continuous in time and space (Tzafestas 1968, Curtain 1976) and continuous in time , discrete in space (Sakawa 1972)
Let us denote $\hat{y}(t)$ the optimal estimate of the system's state based on measurements $z_m(t)$, $t_0<\tau<t$.
The optimal estimate, in the sense of the minimization of $E\{||y(t)-\hat{y}(t)||\}$ is

$$\begin{cases} \dfrac{d\hat{y}}{dt} = M\hat{y}(t) + P_m(t)\ \left\{ R^{-1}\ (x^m)\ \left[z_m(t) - \hat{y}_m(t)\right]\right\} \\ \hat{y}(0) = \bar{y}_0 \\ L\hat{y} = 0 \qquad \text{on } \partial\Omega. \end{cases} \tag{29}$$

where: $P_m(t) = \left[P(x_1^m,t),\ldots.P(x_q^m,t)\right]$

$$P(t) = E\left[(y(t)-\hat{y}(t))\cdot(y(t)-\hat{y}(t))\right]$$

The covariance equation is:

$$\frac{d}{dt} \langle P(t)z,z \rangle = 2 \langle M^* z, P(t)z \rangle - \langle R^{-1}(x^m) CP_m(t)z,\ CP_m(t)z \rangle - \langle Qz,z \rangle \tag{30}$$

$$P(0) = P_0 \quad ; \quad z \in \mathcal{D}(M)$$

In view of equations (29) and (30) it is clear that the optimal estimate and the error covariance are affected by the choice of the measurement points x^m through the terms $R(x^m)$ and $P_m(t)$. Results concerning the existence of an optimal location of sensors can be found in Curtain and Ichikawa (1976). In what follows we shall consider the solutions of the equations (29) and (30) which are necessary to optimize the sensors' location.

Solution To The Filter Equations

The most common approach to solve these equations is the eigenfunction method. Let $\{\phi_i\}$ an $\{\lambda_i\}$ the set of eigenfunctions and eigenvalues of the operator M. We may expand y, $\hat{y}$, P and Q on this basis:

$$y(t) = \sum_{i=1}^{\infty} a_i(t)\ \phi_i$$

$$\hat{y}(t) = \sum_{i=1}^{\infty} \hat{a}_i(t)\ \phi_i$$

$$P(t)z = \sum_{i=1}^{\infty} \sum_{j=1}^{\infty} p_{ij}(t)\ \phi_j \langle z,\phi_i \rangle \quad ; \quad p_{ij}(t) = p_{ji}(t)$$

$$Q(t)z = \sum_{i=1}^{\infty} \sum_{j=1}^{\infty} q_{ij}(t)\ \phi_j \langle z,\phi_i \rangle \quad ; \quad q_{ij}(t) = q_{ji}(t)$$

Substituing these expressions in (29) and (30) yields:

$$\frac{d\hat{a}_i(t)}{dt} = \lambda_i \hat{a}_i(t) + \Big[\sum_{j=1}^{\infty} p_{ij}(t)\ \phi_j(x^m)\Big] \cdot R^{-1}(x^m)\ \Big[z_m(t) - \sum_{j=1}^{\infty} \hat{a}_k(t)\ \phi_k(x^m)\Big] \tag{31}$$

$$\hat{a}_i(0) = \langle \bar{\mathbf{y}}_0, \psi_i \rangle$$

$$\frac{dp_{ij}(t)}{dt} = (\lambda_i + \lambda_j)\ p_{ij}(t) - \sum_{k=1}^{\infty} \sum_{\ell=1}^{\infty} p_{ik}(t)\ \phi_k(x^m) \cdot R^{-1}(x^m) \cdot \phi_\ell(x^m)\ p_{\ell j}(t) + q_{ij} \tag{32}$$

$$p_{ij}(0) = \mu_{ij}$$

finally we have:

$$p_{ij}(t) = E\ \Big[(a_i(t) - \hat{a}_i(t))\ (a_j(t) - \hat{a}_j(t))^T\Big]$$

Suboptimal Solution and the Optimal Sensor Locations Algorithm

For computational needs, one can approximate the infinite dimensional solution by the first N terms and solve the corresponding truncated versions of (31) and (32). Such a finite dimensional approximation converges to the infinite dimensional solution as $N \to \infty$, if the system is N-mode observable we define the following matrices:

$$\hat{a}_N(t) = \text{col}\ (\hat{a}_i(t)) \quad ; \quad \Lambda = \text{diag}(\lambda_i) \quad i=1,\dots,N$$

$$P_N(t) = \begin{bmatrix} p_{11}(t)\dots p_{1N}(t) \\ \\ p_{N1}(t)\dots p_{NN}(t) \end{bmatrix} \quad \Phi^m = \begin{bmatrix} \phi_1(x_1^m)\dots & \phi_N(x_1^m) \\ \\ \phi_1(x_q^m)\dots & \phi_N(x_q^m) \end{bmatrix} \quad Q = \begin{bmatrix} q_{11}\dots q_{1N} \\ \\ q_{N1}\dots q_{NN} \end{bmatrix}$$

Then the system of equations (31) (32) can be approximated by:

$$\frac{d\hat{a}_N(t)}{dt} = \Lambda\,\hat{a}_N(t) + P_N(t)\,\Phi^T \left\{ R^{-1} \left| z_m(t) - C_N\Phi\,\hat{a}_N(t) \right| \right\} \tag{33}$$

$$\frac{dP_N(t)}{dt} = \Lambda\,P_N(t) + P_N(t)\,\Lambda - P_N(t)\,\Phi^{m^T} R^{-1}\Phi^m P_N(t) + Q \tag{34}$$

In order to determine the optimal locations of the sensors we define a criterion of the accuracy of estimation:

$$J = \text{trace}\ \overline{P_N} = \sum_{i=1}^{N} \overline{p_{ii}} \tag{35}$$

where $\bar{P}_N$ is the steady state covariance matrix. Chen and Seinfeld (1975) present two different criteria:

$$\min \{ \text{ trace } P_N(t_f) \} \qquad \text{for } t_f \text{ fixed}$$

$$\min \left\{ \int_0^{t_f} (\text{trace } P_N(t))\ dt \right\}$$

They develop an algorithm based on the necessary conditions for optimality. An iterative procedure based on the calculation of the gradient is used below.

Gradient calculation: Let x^j be a point of measurement.

$$\frac{\partial J}{\partial x_j} = \text{trace}\ \frac{\partial \overline{P_N}}{\partial x_j} \quad , \quad j=1,\dots,q.$$

Differentiating (34) with respect to x_j we obtain:

$$\left[\Lambda - \bar{P}_N(t)\,\Phi^{m^T} R^{-1}\,\Phi^m\right]\frac{\partial \bar{P}_N(t)}{\partial x^j} + \frac{\partial \bar{P}_N(t)}{\partial x^j}\left[\Lambda - \Phi^{m^T} R^{-1}\Phi^m\,\bar{P}_N(t)\right] = \tag{36}$$

.....

$$= \bar{P}_N(t) \frac{\partial \Phi^{m^T}}{\partial x^j} R^{-1} \Phi^m \bar{P}_N(t) + \bar{P}_N(t) \Phi^{m^T} R^{-1} \frac{\partial \Phi^m}{\partial x^j} \bar{P}_N(t) -$$

$$\bar{P}_N(t) \Phi^{m^T} R^{-1} \frac{\partial R}{\partial x^j} R^{-1} \Phi^m \bar{P}_N(t)$$

(36) is an algebric lyapunov equation for the matrix $\frac{\partial \bar{P}_N}{\partial x^j}$ and can be solved numerically at each step of the algorithm

<u>Example</u>:

Let us consider the system described by:

$$\frac{\partial y(x,t)}{\partial t} = \alpha \frac{\partial^2 y(x,t)}{\partial x^2} + \gamma y(x,t) + w(x,t)$$

$$\left.\frac{\partial y}{\partial x}\right|_{x=0} = \left.\frac{\partial y}{\partial x}\right|_{x=1} = 0$$

The eigenvalues and eigenfunctions are:

$$\lambda_1 = \gamma \ ; \ \lambda_i = \alpha - \gamma(i-1)^2 \pi^2 \ ; \ i=2,3\ldots$$

$$\phi_1 = 1 \ ; \ \phi_i = \sqrt{2} \cos(i-1) \pi x \ ; \ i=2,3\ldots$$

For $\alpha = 0.1$, $\gamma = 1.1$, N=5, q=1 and 2 we summarize the results in table 2.

$Q(x)$	$2+2x(1-x)$	$2+2x(1-x^3)$	$2+2x(1-x)$	$2+2x(1-x^3)$
$R(x)$	$2+2x(1-x)$	$2+2x(1-x)$	$2+2x(1-x^3)$	$2+2x(1-x^3)$
x_1^m	0.34 0.66	0.34	0.340	0.340
J_{opt}	33.10	34.26	34.10	35.30
x_1^m	0.16	0.16	0.16	0.17
x_2^m	0.84	1.00	0.84	1.00
J_{opt}	5.58	5.64	5.82	5.92

Table 2

Fig. 8 and 9 show the convergence of the algorithm and the evolution of trace $\bar{P}_N$.

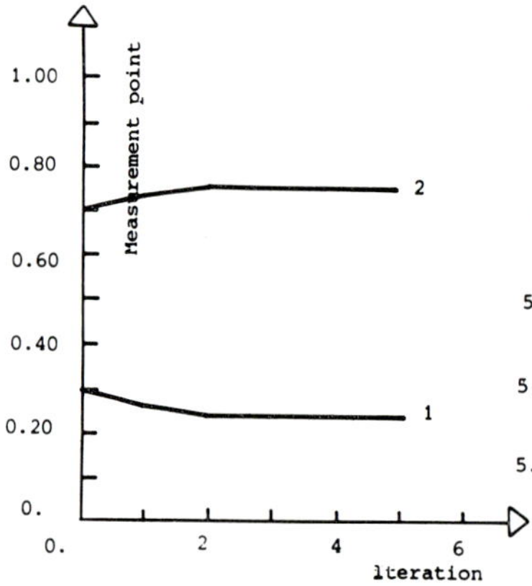

Figure 8

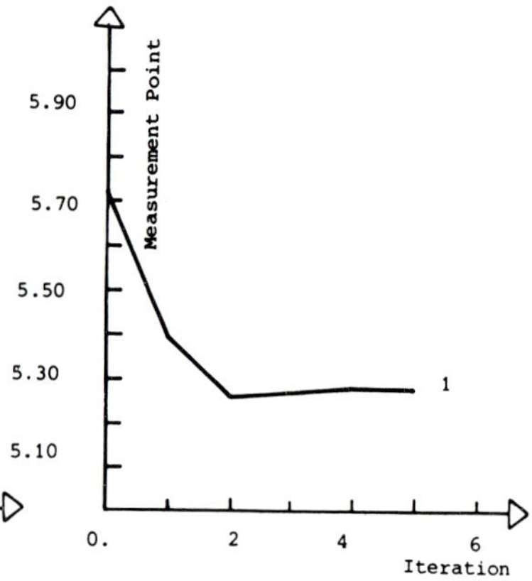

Figure 9

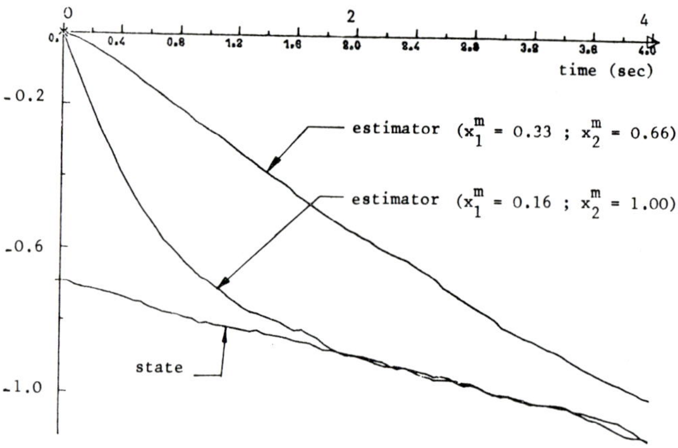

Figure 10

Fig. 10 shows the evolution of the second mode of the system in the case of optimal location of the sensors and in the case assuring the observability condition. A general comment we can make is that symmetrical noise covariances yield a symmetrical solution of measurement points.

Remark: We have taken the example studied by Yu and Seinfeld (1973 p 769). From the same covariances the results are as follows.

- Sequential optimization (Yu and Seinfeld)

$x_1^m = 0.36 \qquad x_2^m = 1 \qquad \text{trace } \bar{P}_N = 4.65$

- Global optimization

$x_1^m = 0.20$ $x_2^m = 0.8$ trace $\bar{P}_N = 3.95$

Figure 11 shows the evolution of the trajectories in each case.

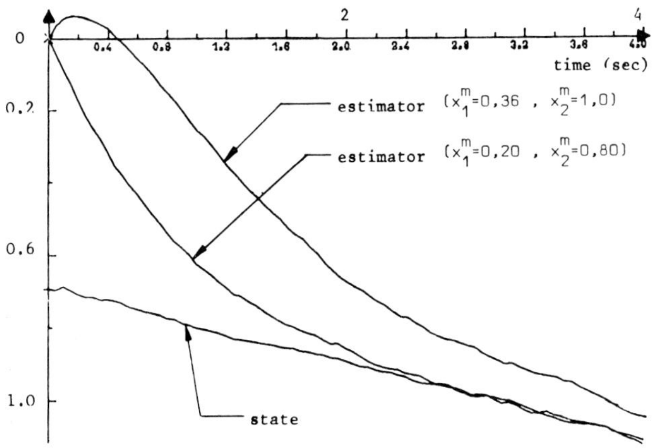

Figure 11

SENSOR AND CONTROL LOCATION PROBLEM

This problem arises whenever it is necessary to control a system whose state is not known.

The separation theorem allows construction of a linear controller in two stages. First the best possible estimate of the state variables is obtained by an estimator and then an optimal controller operates on this state estimate.

One might think of applying the same approach to determine the positions of the sensors and actuators. This, however, leads to nonoptimal results as the problem is strongly non linear; the best estimate thus obtained takes account of neither the control criteria nor the implementation of the various elements of the feedback loop. It is interesting to note that further mathematical analysis show a direct relationship between the sensor positions and the control law and also between the actuator parameters and the filter.

The global optimization of sensors and actuators has been very little studied up to now. In general the following approach is used: first the control structure is determined independently of the sensor-actuator problem; then a new criterion, constructed on the basis of the initial criterion, is used to optimize the variables. Kaiser(1971), Lee and Koppel (1973) were among the first to consider this problem in thecases of temperature distribution control and Counter current processes.

More recently the open loop control problem with unknown initial state and the stochastic regulator problem have been developped Amouroux et al (1977), Ichikawa and Ryan (1979), Malandrakis (1979). It is these two problems that we present below.

Minimum Energy Problem With Unknown Initial State

We consider the deterministic system (1); we assume that measurements are taken at q points $\{x_i^m\}$ of Ω ; These measurements are corrupted by an additive noise:

$$z_m(t) = y_m(t) + \varepsilon(t)$$

where $\varepsilon(t) \in R^q$ is a stationary gaussian zero mean white process with

$$E\{\varepsilon(t)\,\varepsilon^T(\tau)\} = \sigma^2\,\delta(t-\tau)\,\mathbb{I} \qquad (37)$$

We consider the problem of determining the control u(t) which drives the system from the (unknown) initial state y(0) to a required final state $\tilde{y}(t_f)$ while minimizing the energy criterion. We assume, for convenience, that $\tilde{y}(t_f) = 0$. If this is not the case, obvious changes are required in the subsequent developments.
The control law is computed on the basis of an estimate of the system state at time t_m, with $0 < t_m < t_f$.
We use the same eigenfunction method for solving the problem
The estimation of $a(t_m)$ is performed from output measurements $z_m(t)$ by minimizing the error functional:

$$J_m = \int_0^{t_m} \| z_m(t) - C(x^m)\, a(t) \|^2 dt$$

Optimal estimation and control: Under the assumption that system (9) is observable, the optimal estimate of $a(t_m)$ is given by:

$$\hat{a}(t_m) = \Phi(t_m)\Big[\int_0^{t_m} \Phi(t) C(x^m) C(x^m)\, \Phi(t)\, dt\Big]^{-1} \cdot \int_0^{t_m} \Phi(t) C(x^m)\, z_m(t)\, dt \qquad (38)$$

The control which drives the state from $\hat{a}(t_m)$ to $\tilde{a}(t_f)=0$ and minimizes the energy criterion:

$$J_c(u) = \int_{t_m}^{t_f} \| u(t) \|^2 dt$$

is given by (first paragraph)

$$u_0(t) = -B(\eta)\Phi(t_f - t) \cdot \Big[\int_{t_m}^{t_f} \Phi(t_f-\tau) B(\eta) B(\eta) \Phi(t_f-\tau)\, d\tau\Big]^{-1} \Phi(t_f - t_m)\, \hat{a}(t_m) \qquad (39)$$

provided that system (9) is controllable.
From (38) (39) is appears that the computation of the control law $u_0(t)$ requires the inversion of two infinite dimensional operators.To overcome this difficulty, we admit (as in the first paragraph) that the system evolution a(t) can be approximated by means of the state evolution $a_N(t)$ of a finite dimensional system described by the state transition equation:

$$a_N(t) = \Phi_N(t)\, a_N(0) + \int_0^t \Phi_N(t-\tau) B_N(\eta)\, u(\tau)\, d\tau \qquad (40)$$

and by the output equation:

$$s(t) = C_N(x^m)\, a_N(t)$$

where $a_N(t) \in R^N$, $s(t) \in R^q$

It is now possible to obtain an approximate estimate $\hat{a}_N(t_m)$ of $a(t_m)$ and a control law $u_{ON}(t)$ which approximates $u_O(t)$.

Let

$$M_N(\eta) = \int_{t_m}^{t_f} \Phi_N(t_f-t)\ B_N(\eta)B_N\ (\eta)\ \Phi_N\ (t_f-t)dt \tag{41}$$

$$N_N(x^m) = \int_0^{t_m} \Phi_N(t)\ C_N\ (x^m)\ C_N(x^m)\ \Phi_N(t)dt \tag{42}$$

and assume that $M_N(x^c)$, $N_N(x^m)$ are non singular. Then, the optimal state estimation and the optimal control law for the approximating finite-dimensional system are expressed by:

$$\hat{a}_N(t_m) = \Phi_N(t_m)\ N_N^{-1}(x^m) \int_0^{t_m} \Phi_N\ (t)C_N\ (x^m)z_m(t)dt \tag{43}$$

$$u_{ON}(t) = -B_N\ (\eta)\ \Phi_N\ (t_f-t)\ M_N^{-1}(\eta)\ \Phi_N(t_f-t_m)\hat{a}_N(t_m) \tag{44}$$

Denote by $y(t_f)$ the state of system (1) which is attained by applying the control $u_O(t)$, $t_m<t<t_f$

Then the square error on $L_2(\Omega)$ is given by:

$$||e||^2_{L_2(\Omega)} = ||y(t_f)||^2_{L_2(\Omega)} = ||a(t_f)||^2_{\ell_2}$$

It will be shown in thenext section that $E\{||e||^2_{L_2(\Omega)}\}$ and $E\{||\bar{u}_O||^2_{L_2(t_m,t_f)}\}$ depend on the spatial locations of sensors and controls x^m,η, on the noise variance σ^2 and on the true initial state $a(0)$.

<u>Sensor and control location criterion</u>: We define an optimality criterion for the selection of x^m,η which consists in the minimization of the functional:

$$I(x^m,\eta) = \sup_{||a(0)||<k} |\ E\{||e||^2_{L_2(\Omega)}\} + \lambda E\{||u_{ON}||^2_{L_2(t_m,t_f)}\}| \tag{45}$$

where $\lambda> 0$ is a weighting coefficient and

$$\eta \in F_{N^c} \triangleq \{\eta,\ M_N^{-1}(\eta)\ \text{exists}\} \tag{46}$$

$$x^m \in F_{N^m} \triangleq \{x,\ N_N^{-1}(x)\ \text{exists}\} \tag{47}$$

<u>Explicit expression of $I(x^m,\eta)$</u>: We first derive an explicit expression for $E\{||e||^2_{L_2(\Omega)}\}$. Taking (44) into account, and letting:

$$M(\eta) \triangleq \int_{t_m}^{t_f} \Phi(t_f-\tau)B(\eta)\ B_N\ (\eta)\ \Phi_N\ (t_f-\tau)d\tau \tag{48}$$

we obtain:

$$a(t_f) = \Phi(t_f)\ a(0) + \int_{t_m}^{t_f} \Phi(t_f-\tau)\ B(\eta)\ \bar{u}(\tau)d\tau$$

$$= \Phi(t_f)\ a(0) - M(\eta)\ M_N^{-1}(\eta)\ \Phi_N(t_f-t_m)\ \hat{a}_N(t_m) \qquad (49)$$

Substituting (43) into (49) and defining

$$N(x^m) \triangleq \int_0^{t_m} \Phi_N\ (t)\ C_N\ (x^m)\ C(x^m)\ \Phi(t)dt \qquad (50)$$

it results:

$$a(t_f)=\Phi(t_f)a(0)-M(\eta)M_N^{-1}(\eta)\Phi_N(t_f)N_N^{-1}(x^m)N(x^m)a(0) -$$

$$M(\eta)M_N^{-1}(\eta)\Phi_N(t_f)N_N^{-1}(x^m) \int_0^{t_m} \Phi_N(t)C_N(x^m)\varepsilon(t)dt \qquad (51)$$

Letting:

$$W(x^m,\eta) \triangleq \Phi(t_f)-M(\eta)\ M_N^{-1}(\eta)\ \Phi_N(t_f)\ N_N^{-1}(x^m)\ N(x^m) \qquad (52)$$

$$V(x^m,\eta,t) \triangleq M(\eta)M_N^{-1}(\eta)\ \Phi_N(t_f)\ N_N^{-1}(x^m)\ \Phi_N(t)\ C_N(x^m) \qquad (53)$$

eq. (51) can be put in the form:

$$a(t_f) = W(x^m,\eta)\ a(0) - \int_0^{t_m} V(x^m,\eta,t)\ \varepsilon(t)dt \qquad (54)$$

Taking expected values and recalling (37) it results:

$$E\{||e||^2_{L_2(\Omega)}\}=||W\ (x^m,\eta)a(0)||^2_{L_2(\Omega)} + \sigma^2\int_0^{t_m}\text{trace}\{V\ (x^m,\eta,t)V(x^m,\eta,t)\ dt \qquad (55)$$

We obtain now an explicit expression for

$$E\ \{||u_{ON}||^2_{L_2(t_m,t_f)}\}.$$

By (43) (44) it can be written:

$$||u_{ON}||^2_{L_2(t_m,t_f)}= \int_{t_m}^{t_f}||B_N(\eta)\Phi_N(t_f-t)M_N^{-1}(x^c)\Phi_N(t_f-t_m)\hat{a}_N(t_m)||^2dt =$$

$$=\int_{t_m}^{t_f}||B_N(\eta)\Phi_N(t_f-t)M_N^{-1}(\eta)\Phi_N(t_f)N_N^{-1}(x^m)|N(x^m)a(0)+\int_0^{t_m} \Phi_N(\tau)C_N(x^m)\varepsilon(\tau)d\tau|\ ||^2dt$$

Taking expected values and letting:

$$Q(x^m,\eta,t) = B_N(\eta)\ \Phi_N(t_f-t)\ M_N^{-1}(\eta)\ \Phi_N(t_f)\ N_N^{-1}(x_m)\ N(x^m) \qquad (56)$$

$$S(x^m,\eta,t) = M_N^{-1}(\eta)\ \Phi_N(t)\ N_N^{-1}(x^m)\ \Phi_N(t)\ C_N(x^m) \qquad (57)$$

We obtain:

$$E\{||u_{ON}||^2_{L_2(t_m,t_f)}\}=\int_{t_m}^{t_f}||Q(x^m,\eta,t)a(0)||^2dt+\sigma^2\int_{t_m}^{t_f}\text{trace}\left[S\ (x^m,\eta,t)M_N(\eta)S(x^m,\eta,t)\right]dt \tag{58}$$

Therefore, by (55) (58) the functional $I(x^m,\eta)$ becomes:

$$I(x^m,\eta)=\sup_{||a(0)||<k}\{||W\ (x^m,\eta)a(0)||^2+\lambda\int_{t_m}^{t_f}||Q(x^m,\eta,t)a(0)||^2dt\}+$$

$$+\sigma^2\{\int_0^{t_m}\text{trace}\{V\ (x^m,\eta,t)V(x^m,\eta,t)\}dt+\lambda\int_{t_m}^{t_f}\text{trace}\{S\ (x^m,\eta,t)M_N(\eta)S(x^m,\eta,t)\}dt\} \tag{59}$$

where $\lambda>0$ is a weighting coefficient.

Observe now that, for fixed $x^m\in F_{N^m}$, $\eta\in F_{N^c}$, the linear operator

$$W\ (x^m,\eta)\ W(x^m,\eta)\ +\ \lambda\int_{t_m}^{t_f}Q\ (x^m,\eta,t)\ Q(x^m,\eta,t)dt$$

is a symmetric operator, defined on ℓ_2 and therefore is bounded. It follows, from (59)

$$I(x^m,\eta)=k^2||W\ (x^m,\eta)W(x^m,\eta)\ +\ \lambda\int_{t_m}^{t_f}Q\ (x^m,\eta,t)Q(x^m,\eta,t)dt||\ +$$

$$+\sigma^2\left[\int_0^{t_f}\text{trace}\{V\ (x^m,\eta,t)V(x^m,\eta,t)\}dt+\lambda\int_{t_m}^{t_f}\text{trace}\{S\ (x^m,\eta,t)M_N(\eta)S(x^m,\eta,t)\}dt\right] \tag{60}$$

In the rhs of eq. (60) there appear infinite-dimensional matrices; an approximate procedure for the computation of $I(x^m,\eta)$ will be considered in the next section.

<u>Approximate evaluation of $I(x^m,\eta)$</u>: Let us partition the infinite matrices $M(\eta)$ $N(x^m)$, $\Phi(t_f)$ as:

$$M(\eta)=\begin{bmatrix}M_M(\eta)\\ \cdots\cdots\\ M_r(\eta)\end{bmatrix},\quad N(x^m)=\left[N_M(x^m)\ :\ N_r(x^m)\right]$$

$$\Phi(t_f)=\begin{bmatrix}\Phi_M(t_f) & : & \Phi_a(t_f)\\ \cdots & \cdots & \cdots\\ \Phi_b(t_f) & : & \Phi_c(t_f)\end{bmatrix} \tag{61}$$

where $M_M(\eta)$ is (M x p), $N_M(x^m)$ is (N x M) and $\Phi_M(t_f)$ is M x M, with M>N.

By (52), (61) it can be written:

$$W\ (x^m,\eta)=\begin{bmatrix}W_M(x^m,\eta) & : & W_a(x^m,\eta)\\ \cdots & \cdots & \cdots\\ W_b(x^m,\eta) & : & W_c(x^m,\eta)\end{bmatrix}$$

where

$$W_M(x^m,\eta)=\Phi_M(t_f)-N_M(x^m)N_N^{-1}(x^m)\ \Phi_N(t_f)M_N^{-1}(\eta)M_M(\eta) \tag{62}$$

Similarly, by (56) (61) we can set

$$Q(x^m,\eta,t) = [\,Q_M(x^m,\eta,t) : Q_r(x^m,\eta,t)\,]$$

being

$$Q_M(x^m,\eta,t) = B_N(\eta)\ \Phi_N(t_f - t)\ M_N^{-1}(\eta)\ \Phi_N(t_f)\ N_N^{-1}(x^m)\ N_M(x^m) \tag{63}$$

Finally, from (53) (61) we obtain

$$V(x^m,\eta,t) = \begin{bmatrix} V^M(x^m,\eta,t) \\ \cdots\cdots \\ V_r(x^m,\eta,t) \end{bmatrix}$$

with

$$V_M(x^m,\eta,t) = M_M(\eta)\ M_N^{-1}(\eta)\ \Phi_N(t_f)\ N_N^{-1}(x^m)\ \Phi_N(t)\ C_N(x^m) \tag{64}$$

Letting

$$a(0) = [\,a^T{}_M : a^T{}_r\,]^T \quad \text{with} \quad a_M \in R^M$$

It is now possible to define the functional:

$$I_M(x^m,\eta) = \sup_{||a_M||_{R^M} < k} \{||W_M(x^m,\eta)a_M||^2 + \lambda \int_{t_m}^{t_f} ||\Phi_M(x^m,\eta,t)a_M||^2 dt\} +$$

$$+\sigma^2 [\int_0^{t_m} \text{trace}\{V_M(x^m,\eta,t)V_M(x^m,\eta,t)\}dt + \lambda\int_{t_m}^{t_f} \text{trace}\{S\ (x^m,\eta,t)M_N(\eta)S(x^m,\eta,t)\}dt] =$$

$$k^2||W_M(x^m,\eta)W_M(x^m,\eta) + \lambda\int_{t_m}^{t_f} Q_M(x^m,\eta,t)Q_M(x^m,\eta,t)dt|| + \sigma^2[\int_0^{t_m} \text{trace}\{V_M(x^m,\eta,t)V_M(x^m,\eta,t)\}dt$$

$$+ \lambda\int_{t_m}^{t_f} \text{trace}\{S\ (x^m,\eta,t)M_N(\eta)S(x^m,\eta,t)\}dt] \tag{65}$$

By construction, it results that the functional $I_M(x^m,\eta)$ approximates $I(x^m,\eta)$ for $M \to \infty$, that is

$$\lim_{M\to\infty} I_M(x^m,\eta) = I(x^m,\eta) \tag{66}$$

Moreover, it results, for any given M > N:

$$I(x^m,\eta) = \sup_{||a(0)||<k} \{||W_M a_M + W_a a_r||^2 + ||W_b a_M + W_c a_r||^2 +$$

$$+\lambda \int_{t_m}^{t_f} ||Q_M a_M + Q_r a_r||^2 dt + \sigma^2\int_0^{t_m} \text{trace}\{V_M V_M\}dt + \sigma^2\int_0^{t_m} \text{trace}\{V_r V_r\}dt$$

$$+\lambda\sigma^2\int_{t_m}^{t_f} \text{trace}\{S\ M_N S\}dt \geqslant \sup_{||a_M||<k} \{||W_M a_M||^2 + \lambda\int_{t_m}^{t_f} ||Q_M a_M||^2 dt\}$$

$$+\sigma^2\int_0^{t_m} \text{trace}\{V_M V_M\}dt + \lambda\sigma^2\int_{t_m}^{t_f} \text{trace}\{S\ M_N S\}dt = I_M(x^m,\eta) \tag{67}$$

Similarly, it can be easily verified that

$$I_{M+1}(x^m,\eta) > I_M(x^m,\eta) \tag{68}$$

Inqs. (67) (68) indicate that, as M increases, we get a better and better approximation of $I(x^m,\eta)$.

Example:

Consider a one-dimensional diffusion process described by the parabolic equation:

$$\frac{\partial y(x,t)}{\partial t} = \beta \frac{\partial^2 y(x,t)}{\partial x^2} + u(t)g(x^c,x) \tag{69}$$

with $\quad x \in]0,1[\;,\quad t \in [0,t_f]$

$$y(0,t)=y(1,t)=0 \qquad 0 \leq t \leq t_f \tag{70}$$

$$g(x^c,x) = \begin{cases} 1 & x^c < x < x^c + \ell \\ 0 & \text{otherwise} \end{cases} \tag{71}$$

where ℓ is the width of the control zone.

Let

$$y(x,0)=y_0(x)$$

be the unknown initial state and assume that measurements are taken at a point $x^m \in]0,1[$:

$$z_m(t) = y(x^m,t) + \varepsilon(t)$$

Assume as an orthonormal basis on $L_2(\Omega)$ the set $\{\phi_i\}$ of the eigenfunctions of the spatial differential operator $\frac{\partial^2}{\partial x^2}$, with boundary conditions (70). It results

$$\phi_i(x) = \sqrt{2}\,\sin(i\pi x) \qquad i=1,\ldots,\infty$$
$$\lambda_i = -\beta(i\pi)^2 \tag{72}$$

Then, the operator $\Phi(t)$ in (8) becomes:

$$\Phi(t) = \text{diag}\,\{e^{\lambda_1 t}, e^{\lambda_2 t},\ldots\} \tag{73}$$

and the components of $B(x^c)$ are given by

$$b_i(x^c) = \frac{2\sqrt{2}}{i\pi}\sin\left(\frac{i\pi\ell}{2}\right)\sin\left(i\pi\left(x^c+\frac{\ell}{2}\right)\right) \qquad i=1,\ldots,\infty \tag{74}$$

Consider an approximating finite dimensional system of order N, obtained by neglecting the higher order modes. The matrix $\Phi_N(t)$ in (40) becomes:

$$\Phi_N(t) = \text{diag}(e^{\lambda_1 t}, e^{\lambda_2 t},\ldots, e^{\lambda_N t}) \tag{75}$$

and the infinite matrices defined in (48) (50) are given by:

$$M(x^c) = (b_i(x^c) b_j(x^c) \frac{e^{(\lambda_i+\lambda_j)(t_f-t_m)}-1}{\lambda_i+\lambda_j}) \qquad \begin{matrix} i=1,\dots,\infty \\ j=1,\dots,N \end{matrix} \tag{76}$$

$$N(x^m) = (\phi_i(x^m)\phi_j(x^m) \frac{(e^{(\lambda_i+\lambda_j)t_m}-1}{\lambda_i+\lambda_j}) \qquad \begin{matrix} i=1,\dots,N \\ j=1,\dots,\infty \end{matrix} \tag{77}$$

The matrices $M_N(x^c)$, $N_N(x^m)$, $M_M(x^c)$, $N_M(x^m)$ which appear in (65) are obvious cut-off matrices of $M(x^c)$ and $N(x^m)$.

The sets F_{N^c}, F_{N^m} defined in (46) (47) become:

$$F_{N^c} = \{x:\ 0<x<1,\ x+1<1,\ x+\frac{1}{2} \neq \frac{k}{i},\ k,\ i=1,\dots,N\}$$

$$F_{N^m} = \{x:\ 0<x<1,\quad x\neq \frac{k}{i}\ ,\ k,\ i=1,\dots,N\}$$

The location for the minimum has been evaluated by direct search on the domain $x^c \in F_{N^c}$, $x^m \in F_{N^m} \cap [0,0.5]$, since the symmetry of the problem ensures that.

$$I(x^m,x^c) = I(1-x^m,\ 1-x^c-\ell)$$

In the case $N=4$, $\beta=0.05$, $\ell=0.2$, $\sigma=0.02$, $t_m=1$, $t_f=2$

The optimal locations are

$$x^m=0.22 \qquad x^c=0.50$$

The optimal values in the case of separate optimization is $x^m=0.15$ $x^c=0.18$

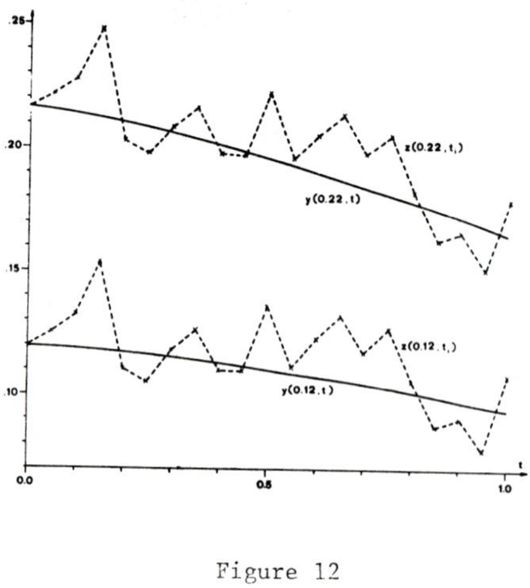

Figure 12

Fig. 12 shows the noiseless output $y(0.22,t)$ and the samples of the measurements $z(0.22,t_i)$ for $t_i=0.5i$, $i=0, 1,...,20$.
In fig. 13 is drawn the control $u_{ON}(t)$, $t_m<t<t_f$
In Fig. 14 are drawn $y_0(x)$, $y(x,t_m)$, the estimated state $\hat{y}(x,t_m)=\sum_{i=1}^{N}\hat{a}_i(t_m)\phi_i(x)$, the state $y_\ell(x,t_f)$ of the unforced system and the state $y(x,t_f)$ attained at time t_f by applying the control $u_{ON}(t)$

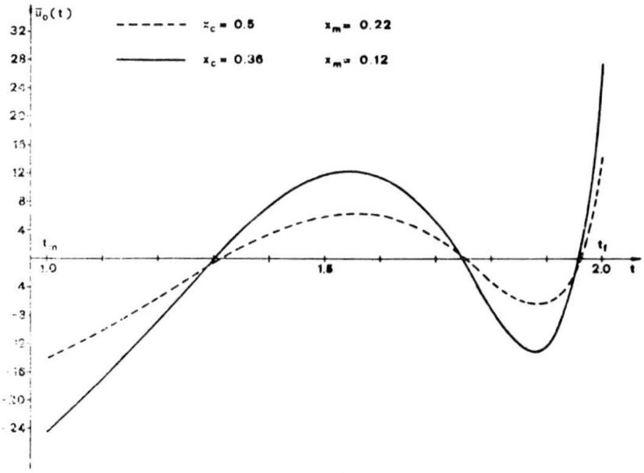

Figure 13

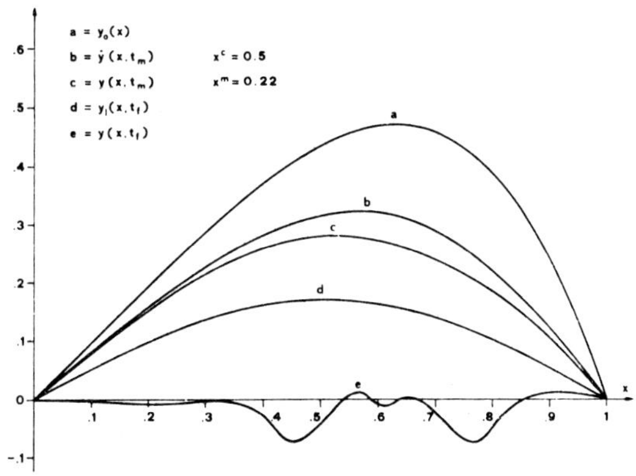

Figure 14

Stochastic regulator problem

We consider the system (24) with an additive pointwise control term:

$$Bu = \sum_{j=1}^{p} u_j(t) \,.\, \delta(x-x_j^c)$$

A control u is sought as to minimize the functional:

$$J_c(u)=E\{<y(t_f),S_c y(t_f)>\} + \int_0^{t_f} E\{<y(t),Q_c y(t)>+<u(t),R_c u(t)>\}dt \qquad (78)$$

where $S_c \geqslant 0$, $Q_c \geqslant 0$, $R_c > 0$ with bounded inverse R^{-1}.
The expression of the solution is well known (separation principle)

$$u_0(t) = -R_c^{-1} P_c^c (t)\hat{y}(t) \qquad (79)$$

where $\quad P_c^c(t) = [P_c(x_1^c,t),\ldots, P_c(x_p^c,t)]^T$

and $P_c(t)$ is a solution of the Riccati equation:

$$\frac{d}{dt} <P_c y,y> + 2<My,P_c y> + <y,Q_c y> - <R_c^{-1}P_c^c y,P_c^c y> = 0 \qquad (80)$$

$$y\in \mathcal{D}(M) \;;\; P_c(t_f)=S_c$$

$\hat{y}(t)$ is the filtered estimate of y(t) given by the solution of:

$$\frac{d\hat{y}(t)}{dt} = M\hat{y}(t) + \sum_{j=1}^{p} u_j(t)\, \delta(x-x_j^c) + P_m(t)\, R^{-1}[z_m(t)-\hat{y}_m(t)] \qquad (81)$$

$$\hat{y}(0) = E(y_0) = \bar{y}_0$$

where $P_m(t)$ is a solution of equation (30)

Setting $\quad (P_c(t)y)(x) = \int_\Omega K_c(x,\xi,t)y(\xi)d\xi$

and $\quad (P(t)y)(x) = \int_\Omega K_m(x,\xi,t)y(\xi)d\xi$

Then the optimal cost for the original problem (78) is:

$$J_c^0=\int_0^{t_f}\{\int_\Omega\int_\Omega[K_c(x,r,t)K_m(x^m,x,t)R^{-1}K_m^T(x^m,r,t)+Q_c(x,r)K_m(x,r,t)]dxdr\}dt +$$

$$+\int_\Omega\int_\Omega S_c(x,r)K_m(x,r,t_f)dxdr \qquad (82)$$

where $K_m(x^m,x,t) = [K_m(x_1^m,x,t),\ldots,K_m(x_q^m,x,t)]$
Obviously this optimal cost (82) depends on the control and measurements points. The optimal problem for the determination of these parameters becomes:

$$\min_{x^c,x^m} J_c^0(x^c,x^m) \qquad (83)$$

Note: For more details concerning this problem the reader is refered to(Ichikawa and Ryan 1979) and Malandrakis (1979). An interesting case is the one concerning the infinite time regulator. If the deterministic system associated to the initial system is controllable and observable, the solution has the same structure, the feedback gains are constant and equations (30) (80) become algebraic equations; we note by $\bar{K}_c$ and $\bar{K}_m$ their solution.

The optimal cost rate is given by:

$$J_\infty^0 = \lim_{t_f \to \infty} \frac{1}{t_f} J_c^0 = \int_\Omega \int_\Omega [\bar{K}_c(x,r)\bar{K}_m(x^m,x)R^{-1}(x^m)\bar{K}_m^T(x^m,r)+Q_c(x,r)\bar{K}_m(x,r)]\,dx\,dr \tag{84}$$

As noted $\bar{K}_m$ and $\bar{K}_c$ are independent of the estimate $\hat{y}(t)$ and so the optimization proposed can be realized off-line.

Solution of the problem: We use the eigenfunction method, assuming that a complete orthonormal basis $\{\phi_i\}$ exist.

We expand $\bar{K}_m, \bar{K}_c$, Q_c and Q on this basis:

$$\bar{K}_c(x,r) = \sum_{i=1}^{\infty}\sum_{j=1}^{\infty} p_{ij}^c\, \phi_i(x)\, \phi_j(r) \quad ; \quad k_{ij}^c = k_{ji}^c$$

$$\bar{K}_m(x,r) = \sum_{i=1}^{\infty}\sum_{j=1}^{\infty} p_{ij}^m\, \phi_i(x)\, \phi_j(r) \quad ; \quad k_{ij}^m = k_{ji}^m$$

$$Q_c(x,r) = \sum_{i=1}^{\infty}\sum_{j=1}^{\infty} q_{ij}^c\, \phi_i(x)\, \phi_j(r) \quad ; \quad q_{ij}^c = q_{ji}^c$$

$$Q(x,r) = \sum_{i=1}^{\infty}\sum_{j=1}^{\infty} q_{ij}\, \phi_i(x)\, \phi_j(r) \quad ; \quad q_{ij} = q_{ji}$$

We define the following matrices:

$$\Phi^c = \begin{bmatrix} \phi_1(x_1^c) & \dots & \phi_n(x_1^c) & \dots \\ \\ \phi_1(x_p^c) & \dots & \phi_n(x_p^c) & \dots \end{bmatrix} \qquad \Phi^m = \begin{bmatrix} \phi_1(x_1^m) & \dots & \phi_n(x_1^m) & \dots \\ \\ \phi_1(x_q^m) & \dots & \phi_n(x_q^m) & \dots \end{bmatrix}$$

Substituting in (30) (80) we obtain the algebraic Riccati equations:

$$A\bar{P} + \bar{P}A - \bar{P}\Phi^{m^T} R^{-1}\, \Phi^m\, \bar{P} + \bar{Q} = 0 \tag{85}$$

$$A\bar{\Sigma} + \bar{\Sigma}A - \bar{\Sigma}\Phi^{c^T} R_c^{-1}\, \Phi^c\, \bar{\Sigma} + Q_c = 0 \tag{86}$$

where $A = \mathrm{diag}(\lambda_i)$, $\bar{P} = [p_{ij}^m]$, $\bar{\Sigma} = [p_{ij}^c]$, $\bar{Q} = [q_{ij}]$, $\bar{Q}_c = [q_{ij}^c]$

Then, the cost rate (84) and the optimal problem are:

$$J_\infty^0 = \mathrm{trace}\ [\bar{P} + \bar{\Sigma}\ \bar{P}\ \Phi^{m^T}\ R^{-1}\ \Phi^m\ \bar{P}] \tag{87}$$

$$\min_{x^c, x^m}\ J_\infty^0(x^c, x^m) \tag{88}$$

For computational needs, one can approximate the infinite matrices in (87) by the first N terms.

Example:

We consider the example of section 2:

$$\frac{\partial y(x,t)}{\partial t} = \alpha \frac{\partial^2 y(x,t)}{\partial x^2} + \gamma y(x,t) + \sum_{j=1}^{p} \delta(x-x_j^c)\, u_j(t) + w(x,t) \;; \quad x \in]0\ 1[$$

$$\left.\frac{\partial y}{\partial x}\right|_{x=0} = \left.\frac{\partial y}{\partial x}\right|_{x=1} = 0$$

$$z_m = y_m(t) + v(t)$$

The criterion to be minimized is:

$$J_c = \int_0^\infty \Big[\int_\Omega Q_c(x)\, y^2(x,t)\, dx + \sum_{j=1}^{p} R_{cj}\, u_j^2(t) \Big]\, dt$$

We take $\alpha=0.1$, $\gamma=1.1$, $N=5$, $p=1$, $R_c=2$

Table 3 summarizes the results obtained for one controller and two sensors, and for different combinations of $R(x)$, $Q(x)$, $Q_c(x)$.

$Q(x)$	$2+2x(1-x)$	$2+2x(1-x)$	$2+2x(1-x^3)$
$R(x)$	$2+2x(1-x)$	$2+2x(1-x^3)$	$2+2x(1-x^3)$
$Q_c(x)$	$4+2x(1-x)$	$4+2x(1-x)$	$4+2x(1-x)$
x_1^c	0.41	0.58	0.41
x_1^m	0.18	1.00	0.22
x_2^m	0.88	0.88	0.78
$J^0_{,5}$	214.16	224.76	243.31

Table 3

By comparing tables 2 and 3 we see for this example that does not exist a separation property for the optimal position of the sensors and controllers.

CONCLUSION

The problem of optimal location of sensors and actuators that has been presented in this chapter is an important one for D.P.S. since it falls directly into practical realization studies. On the other hand, it can be said that are as many problems as there exists control and estimation criteria.

Throughthe presentation of some classical control problems and examples, we have tried to establish a phylosophy and a starting point that will leads us towards possible solutions. The adopted procedure has consisted in each case in choosing

a solution in two stages:
- firstly, the measure and control structures are supposed to be known, so the equations of the optimal controls and filters are determined from classical theory.
- secondly a criterion that takes into account the necessary approximations in the obtention of implementable algorithms is defined and optimized as function of sensors and actuators position.
The problem of determining the number of sensors (actuators) has not been considered. Evidently, the estimation (control) quality increases if more measuring (control) points are added. The questions arises then at what value of q, the reduction of criterion does not present any more interest? The question can be ansered satisfactoraly only if a term characterizing the cost of the different elements is included in the criterion.
When examining the results obtained in the examples, the importance of good sensor (actuator) location must be underlined as their number is reduced (less than two); this problem desappears a q surpasses three (El Jai 1977)
The numerical methods of non linear programming converge rapidly but it must be bear in mind that the minima obtained is a local minima, therefore numerous initializations must be performed. The strong non convexity of the criterion becomes particularly troublesome as the number of parameters that must be optimized increases.

REFERENCES

Aidarous, E. (1976). Optimal allocation strategies in stochastic D.S. Ph.D. Thesis Université catholique de Louvain (Belgium)

Aidarous, E., Gevers, M.R., Installe, M.J. (1975) Optimal sensors allocation strategies for a class of stochastic D.S. Int. J. Control, vol 22, n°2.

Amouroux, M. (1977). Localisation optimale de capteurs et actionneurs pour la commande d'une classe de systèmes à paramètres répartis Dr. Etat, Université de Toulouse (France)

Amouroux, M., Babary, J.P. (1975). Optimal pointwise control for a class of D.P.S. 6th IFAC Cong. Boston (USA)

Amouroux, M., Babary, J.P. (1978). On the optimal pointwise control and parametric optimization of D.P.S. Int. J. Control, vol 28, n°5, 789-807

Amouroux, M., Babary, J.P. (1979). On optimization of zones of action for an optimal control problem for D.P.S. Int. J. Control, vol 29, n°5, 861-869

Amouroux, M., Babary, J.P., El Jai, A. (1978). Adaptive open loop control for a class of D.P.S. Automatica, vol 14, 603-607

Amouroux, M., Babary, J.P., Malandrakis, C.G. (1976). Optimal location of sensors for linear stochastic D.P.S. Proc. IFIP W. Conf. on D.P.S. Rome. Lect. Notes in control and Inf. Sciences. n°1, 92-113

Amouroux, M., Di Pillo, G., Grippo, L. (1976) Optimal selection of sensor and control locations for a class of D.P.S. Ricerche di automatica, vol 7, n°1 92-112

Bensoussan, A. (1972). Optimization of sensors location in a distributed filtering problem. Int. Symposium on Stability of Stochastic dynamical systems. Lect. Notes in Math. Springer Verlag, n° 294

Burger, J. (1976). Emplacement optimal des actionneurs ponctuels ou par zones dans un système gouverné par une équation linéaire d'évolution du premier ordre. Rev. Fr. Autom. Inf. Rech. Oper (France) Avril, 87-102

Cannon, J.R., Klein, R.E. (1970). Optimal selection of measurement location in a conductor for approximate determination of temperature distribution. Proc. J.A.C.C. Atlanta (USA)

Caravani, P., Di Pillo, G., Grippo, L. (1975) Optimal location of a measurement point in a diffusion process. 6th IFAC Cong. Boston (USA)

Chen, W.H., Seinfeld, J.H. (1975). Optimal location of process measurements Int. J. Control, vol 21, n°6, 1003-1014

Colantuoni, G., Padmanabhan, L. (1978). Optimal sensor selection in sequential estimation problems. Int. J. Control, vol. 28, n°6, 821-845

Courdesses, M. (1977) .Sur la commande ponctuelle des systèmes à paramètres répartis avec critère quadratique. Thèse de 3e cycle, Université de Toulouse (France)

Courdesses, M., Polis, M.P., Amouroux, M. (1981). On identifiability of parameters in a class of parabolic D.S. IEEE Trans .on Autom. Control, vol 26, n°2, 474-477.

Curtain, R.F., Ichikawa, A., Ryan, E P. (1976). Optimal location of sensors for filtering of D.P.S. Lect .Notes in Control and Inf. Sciences, n°1 Springer Verlag.

Davis, J M., Perkins, W.R. (1971) .Optimal control of D.P.S. with separable controller .Proc. IFAC Symp. on the Control of D.P.S., Banff (Canada)

El Jai, A. (1977) Sur la commande avec estimation de l'état initial d'une classe de systèmes à paramètres répartis. Rev. Fr. Autom. Inf. Rech. Oper (France) vol 11, 313.

Ewing, D.J., Higgins, T.J. (1971) .A sensor algorithm for D.P.S Proc. Ninth. Annual Allerton Conf. on Circuit and System Theory.

Foster, A.M. (1969) A design procedure for a class of D.P.S. Ph .D. Syst .Eng. Div. Case Western Reserve Univ. (USA)

Goodson, R.E., Klein, R.E. (1970) A definition and some results for distributed system observability. IEEE Trans. Autom. Control vol 15, n°2.

Greenberg, S .G. (1969). Optimal pointwise feedback control for D.P.S. Ph. D. M.I.T.

Hamidi, M. (1975) Optimal control location for partial differential equations Ph.D. Univ. of California Los Angeles

Howland, H.R. (1968). On observing the state of a D.P.S. Ph. D. Univ. of Pensylvania

Ichikawa, A., Ryan, E.P. (1979). Sensor and controller location problems for D.P.S. Automatica, vol 15, 347-352

Kaiser, K.W. (1971). A method of determining the heater sensor configuration in temperature distribution controllers. Proc. IFAC Symp. on the control of D.P.S. Banff (Canada)

Kessler, K. (1972). Finite control and observation of D.P.S. Doct.of Sciences, Washington Univ.

Kumar, S., Seinfeld, J.H. (1978). Optimal location of measurement for D.P.S. estimation. IEEE Trans. Autom. Control, 23, n°4, 690-698

Lee, H.H., Koppel, L.B., Lim, H.C. (1973) Optimal sensor locations and controller settings for a class of contercurrent processes. Ind. Eng. Chem. Process Dev. vol 12, n°1, 1973.

Le Letty et al (1976) Systèmes à paramètres répartis et à commande distribuée. Rapport ATP CNRS n°1001

Lions,J.L. (1972) Some aspects of the optimal control of D.P.S. Reg. Conf. series in Appl. Math. Philadelphia, Pennsylvania.(USA)

Malandrakis, C. (1979) Optimal sensor and controller allocation for a class of D.P.S. Int. J. Systems Sci. vol 10,n°11, 1283-1299

Martin, J.C. (1979). Dynamic selection of actuators for lumped an D.P.S. IEEE Trans. on Autom. Control vol 24, n°1, 70-78

Martin, J.C. (1979) Scanning observability of linear systems. IEEE Trans. on Autom. Control, vol 24, n°3, 489-491

Matsumoto, J., Ito, K. (1970). Feedback control of D.P.S. with spatially concentrated controls. Int. J. Control vol12, 401-419

Mehra, R.K. (1976) Optimization of measurement schedules and sensor designs for linear dynamic systems. IEEE Trans. on Autom. Control, vol 21, 55-64

Nakagiri, S., Kitamura, S. (1977) Identifiability of spatially varying and constant parameters in distributed systems of parabolic type. SIAM. J. Control Optimiz. vol 15, 785-802

Omatu, S., Koide, S., Soeda, T. (1978). Optimal sensor location problem for a linear D.P.S. IEEE. Trans. on Autom. Control, vol 23, n°4, 665-675

Porter, W.A. (1966) Modern foundations of systems engineering. Mac Millan Company New York.

Ray, W.H. (1977). Some recent applications of D.P. control theory: A survey. Proc. 2nd IFAC Sympo. Control of D.P.S. Coventry (England)

Saguez, C. (1974). Contrôle ponctuel et contrôle en nombres entiers des systèmes distribués. Thèse de Doct. Ing. Université de Paris VI (France)

Sakawa, Y. (1975) Observability an related problems for partial differential equations of parabolic systems. SIAM. J. Control vol 13, n°1

Sakawa, Y. (1972) Optimal filtering in linear D.P.S. Int. J. Control, vol 18, 117-127

Seinfeld, J.H. (1972) Optimal location of pollutant monitoring stations in an airshed. Atmospheric Environment. vol 6, 847-858

Tzafestas, G. (1972). On optimum distributed parameter filtering and fixed-interval smoothing for colored noise. IEEE Trans. on Autom. Control, vol 17, n°4

Vidyasagar, A., Higgins, T.J. (1973) A basic theorem on distributed control and point control. Trans. ASME. J. Dyn. Systems, Meas. and Control. vol 95, 64-67.

Wang, J. (1969) The analysis an synthesis of a class of linear D.P.S. Ph. D. Univ. of California

Wong , J.K., Soudack, A.C. (1975). Stochastic optimal pointwise control for linear discrete time D.P.S. Int. J. Control, vol 22, n°5, 593-609

Yu, T.K., Seinfeld, J.H. (1973). Observability and optimal measurement location in linear D.P.S. Int. J. Control vol 18, n°4, 785

Part II

Applications

CHAPTER 11

EXPERIMENTAL APPLICATION OF DISTRIBUTED-PARAMETER CONTROL THEORY

G. K. Lausterer* and E. Eitelberg**

*Siemens AG, Systemtechnische Entwicklung, Karlsruhe, Federal Republic of Germany
**Institut für Regelungs- und Steuerungssysteme, University of Karlsruhe, Federal Republic of Germany

1. INTRODUCTION

Distributed-parameter control theory is not developed for the mere beauty of its equations, but to solve complicated industrial control problems. In times where energy and natural resources become scarce and enviromental concern is growing, while the only thing that is going down in price is computing power, the application of modern control theory becomes increasingly important. But before theoretical results are directly applied to an industrial process, their feasibility should be field-tested. Usually first practical experience is gained from laboratory processes.

For further clarification it is helpful to divide the reported applications of distributed-parameter control theory into the following classes:

a) Pure computer simulations, where the theory was applied to a model of some physical relevance (no field data available).
b) Off-line simulations using measured process data as inputs or for comparison (field data from laboratory or industrial process available).
c) Actual real-time implementation to an existing laboratory or industrial process (on-line real-time data transfer between process and process computer/microprocessor).

Surveys on distributed-parameter state estimation algorithm and applications and on some recent applications of distributed-parameter control theory have been given by Ray (1975,1977). He notes that although there have been an enormous number of reported applications, most of them are purely computational and are thus of class a, where the final step of actual implementation and comparison with experiments has not been made. While a fair number of class b applications have been described, there are very few reports on class c applications, all of which are only concerning laboratory processes. Thus, even though the first papers on distributed parameter control theory appeared as early as 1964 (Wang and Tung; Wang), until today there was no report in the literature of an actual application of distributed-parameter control theory to an industrial process.

Any application involving a digital computer requires the use of a numerical procedure in order to obtain the desired numerical results. For this purpose the partial differential equations have to be discretized or "lumped" appropriately. This lumping may be done immediately after the distributed-parameter model is established (early lumping) and subsequently the well known control theory for lumped systems may be applied. Alternatively, the design of estimators (filters, observers) and controllers may be carried out using the full power of the distributed-parameter control theory and only for the purpose of numerical treatment the transfer to ordinary differential equations is made (late lumping). There are several advantages to late lumping as has been pointed out by Lausterer and Ray (1979) and other authors (e.g. Koda, 1979), even though for linear systems the resulting equations are, at least in principle, identical. However, the choice of the free design parameters such as sensor and actuator locations and number, controller and estimator weighting parameters, and covariance initial conditions is difficult to justify if one uses early lumping.

This chapter presents a one-dimensional and a two-dimensional linear example, where all relevant problems of distributed-parameter control theory are discussed:
- modelling and parameter estimation;
- controllability, observability, stability;
- number and location of sensors and actuators;
- choice of the weighting functions;
- numerical and real-time requirements;
- sensitivity to parameters and noise.

Furthermore, a moving boundary problem is included to complete the discussion. It is not our intention to dwell on theoretical derivations, but rather we want to emphasize the practical relevance of the results obtained from the theory. Thus mathematical derivations are reduced to a minimum in favour of a more extensive description and discussion of practical aspects.

2. A ONE-DIMENSIONAL LINEAR EXAMPLE

One of the most thoroughly studied distributed parameter systems (DPS) is the aluminium slab of Mäder (1975a). The fact, that this particular system is rather well described by a very simple spatially one-dimensional linear partial differential equation (PDE) with homogeneous boundary conditions, has greatly facilitated the synthesis of various control and state estimation algorithms. The real-time computations have been feasible, since the heat conduction dynamics are not only simple but also relatively slow.

(i) Description of the Process

The laboratory-size heat conduction system is a horizontal aluminium slab 1 m long, 2 cm thick and 25 cm wide as depicted in Fig. 2.1.

The slab has 20 heating lamps oriented transversely both above and below the slab for providing temperature profile control in z direction. In addition, cooling water flows through 20 holes drilled transversely through the middle of the slab. There are 21 thermocouples located in the center of the slab along the z direction. One is at each end and the remainder are at the midpoints between the cooling water channels. The faces at $z = 0$ and $z = 1$ m are insulated.

Mäder (1975a,b) has made a thorough modelling and control study of this apparatus and has found that a very good model of the center line temperature is given by

$$\rho C_p \frac{\partial T(z',t)}{\partial t} = k \frac{\partial^2 T(z',t)}{\partial z'^2} - \alpha_w (T - T_w) + C\, u(z',t) \;, \tag{2.1}$$

$$t > 0 \;, \qquad 0 \leq z' \leq 1 \;,$$

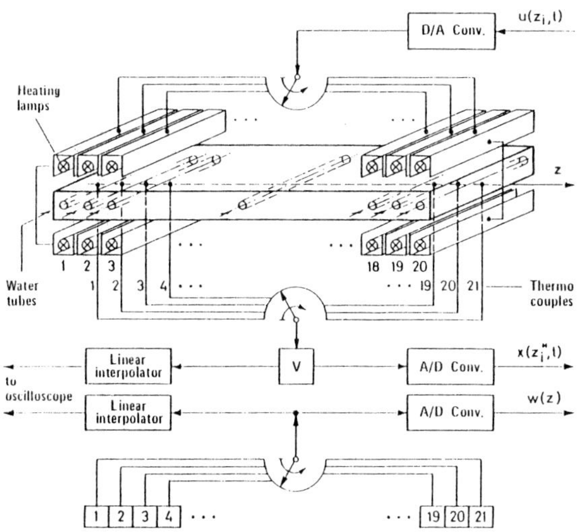

Fig. 2.1. Schematic view of the experimental apparatus.

where ρ, C_p, k are the density, heat capacity and effective thermal conductivity of the aluminium slab, α_w is an overall heat transfer parameter determined experimentally for the particular system, T_w is the average coolant temperature which is not used for any control purpose, C is an empirical function of the power input $u(z',t)$ to the heating lamps but due to an additional nonlinear compensating characteristic may be considered roughly as a constant. The boundary conditions become

$$\frac{\partial T}{\partial z'} = 0 \quad \text{at } z' = 0 \text{ and } z' = 1 \ . \qquad (2.2)$$

If the quantities

$$z = z'/l \ , \qquad x = T - T_w \ ,$$

$$\alpha = \frac{k}{\rho C_p l^2} \ , \qquad \beta = \frac{\alpha_w}{\rho C_p} \ , \qquad (2.3)$$

$$\gamma = \frac{C}{\rho C_p}$$

are defined, then the model takes the following simpler form

$$\frac{\partial x(z,t)}{\partial t} = \alpha \frac{\partial^2 x(z,t)}{\partial z^2} - \beta x(z,t) + \gamma u(z,t) \ , \qquad t > 0 \ , \qquad 0 \le z \le 1 \ , \qquad (2.4)$$

$$\frac{\partial x}{\partial z} = 0 \ , \qquad z = 0,\ 1 \ , \qquad (2.5)$$

where the modelling parameters have been determined by Mäder (1975a) and are given as follows:

$$T_w = 296.5 \text{ K } (= 23\ ^oC) \ , \qquad \alpha_w = 1.53\cdot 10^4 \ \frac{W}{m^3\cdot K} \ , \qquad l = 1 \text{ m} \ ,$$
$$\rho C_p = 2.58\cdot 10^6 \ \frac{W\cdot s}{m^3\cdot K} \ , \qquad k = 134 \ \frac{W}{m\cdot K} \ , \qquad C = 1.19\cdot 10^4 \ . \qquad (2.6)$$

To implement various control and state estimation algorithms flexibly, the EAI Pacer 100 digital computer of the Institut für Systemdynamik und Regelungstechnik of the University of Stuttgart has been used, although as Mäder (1973) has shown, some control algorithms can be realized also with analog elements. In order to link the computer with the experimental equipment, the A/D and D/A converters of the hybrid section of the EAI 680 analog computer have been used. In addition, the analog hardware sketched in Fig. 2.1, had to be designed for scanning the 21 thermocouples, the temperature profile set point, and administering the power to the heating lamps. The data sampling interval t_s can be chosen equal to 0.4, 4 or 40 seconds. However, for the following experiments 4 s have been chosen. It must be pointed out, that the serial profile scanning and transfer leads to a time delay t_d. For the data sampling interval of 4 s the delay is equal to 9 seconds (Mäder, 1975a), which, however, can normally be neglected, since the dominant time constant of the slab is between 2 and 3 minutes. The whole equipment works in a repetitive manner. The duration of a repetition t_r is the sum of the data sampling interval (usually t_s = 4 s) and the numerical processing time t_n,

$$t_r = t_s + t_n , \tag{2.7}$$

which are independent of each other. In the following t_n depends on the specific experiment, although it is usually much less than the sampling time t_s.

(ii) Modal Control

Mäder (1975a and 1976) has experimented with two kinds of modal control. One of them was a special kind of spatial discretization with an additional state transformation to a diagonal form. The other was a true distributed system modal transformation and the latter will be described in the following.

Algorithm. The experimental apparatus of Mäder (1975a,b) allows the realization of the modal control with the eigenfunctions of the system (2.4, 2.5), because the temperature profile x as well as the input u are accessible in almost a continuous way. The control loop has been structured so that the modes of the heat conduction process remain decoupled also in closed loop operation. In the case of the given system the Laplace-transformed control variable reads:

$$u(z,s) = \sum_{i=o}^{n} \phi_i(z)\cdot G_{Ri}(s)\cdot\omega_i \int_o^1 \phi_i(\zeta)\ \{w(\zeta) - x(\zeta,s)\}\ d\zeta \tag{2.8}$$

where s is the complex frequency variable. Here the first n+1 spatial eigenfunctions of the system (2.4) with boundary conditions (2.5)

$$\phi_i(z) = \cos\ (i\,\pi\, z)\ , \qquad i = 0,\ 1,\ 2,\ \ldots,\ n \tag{2.9}$$

and the corresponding factors

$$\omega_o = 1\ , \quad \omega_i = 2\ , \qquad i = 1,\ 2,\ 3,\ \ldots,\ n \tag{2.10}$$

are used. The structure of the controller (2.8) is depicted in Fig. 2.2.

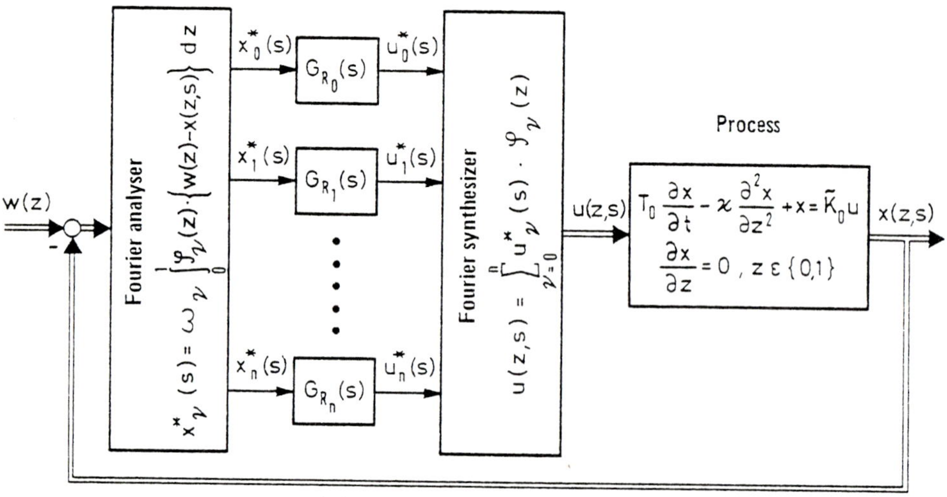

Fig. 2.2. Modal control structure.

The implementation of the Fourier analysis and synthesis is described by Mäder (1975a,b). The controller transfer functions $G_{Ri}(s)$ can be chosen arbitrarily. According to Mäder a standard PI-structure, identical for all n+1 modes, was chosen:

$$G_{Ri}(s) = G_R(s) = K_R\left(1 + \frac{1}{T_I s}\right) , \qquad i = 0, \ldots, n , \tag{2.11}$$

which lead to reasonable results.

Experimental results. In the following one of the measured temperature profile histories is shown. The described PI-control is applied to the first six eigenfunctions ($n = 5$). The controller parameters were chosen as $K_R = 17$ and $T_I = 31$ s. However, for practical reasons it was necessary to use a controller with variable structure. In order to avoid overshooting the integral channel of the controller is activated only after 150 s of the transient time. Figure 2.3 shows the transient temperature profiles approaching the desired profile. One can remark that for dynamical purposes we have only a proportional controller.

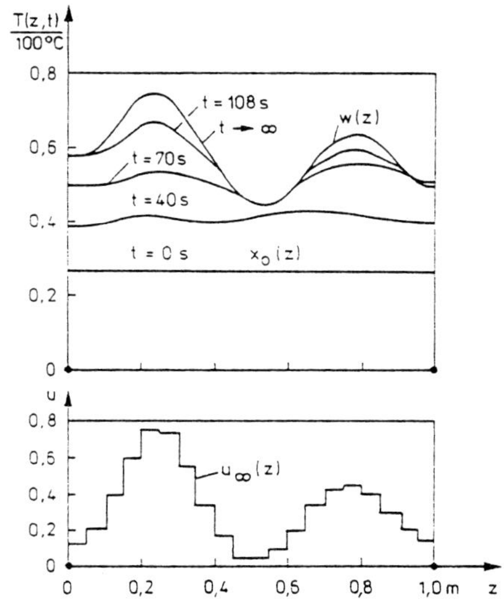

Fig. 2.3. Modal Control.

(iii) Control with Filter

It is unusual that the temperature profile of a slab is measured in almost a continuous way. In general, only few measurements will be available at discrete points of the slab. For various reasons (such as control) one might then have to estimate the entire temperature profile along the slab, which can be achieved by a distributed-parameter state estimator (filter or observer). In the following a distributed-parameter filter is described, which has been tested in both open-loop experiments as well as in the previously described proportional (plus integral) feedback control loop (Ajinkya, Köhne, Mäder, Ray, 1975).

Filter algorithm. The deterministic form of the model is given by equations (2.4) and (2.5) and the measuring device takes the general form

$$y_i(t) = x(z_i^*,t) + \eta_i(t) , \qquad i = 1, 2, \ldots, m , \tag{2.12}$$

where the $x(z_i^*,t)$ are the true values of the slab temperature at the measurement points $z = z_i^*$, and $\eta(i)$ represents a measurement error. Ajinkya and coauthors (1975) have formulated the problem in the following way: given measurements $y_i(t)$ it is desired to estimate $x(z,t)$ such that the least-squares criterion is minimized:

$$I = \int_0^{t_f}\int_0^1\int_0^1 \left\{ \frac{\partial x(z,t)}{\partial t} - \alpha \frac{\partial^2 x(z,t)}{\partial z^2} + \beta x(z,t) - \gamma u(z,t)\right\} R(z,z',t) \left\{ \frac{\partial x(z',t)}{\partial t} - \alpha \frac{\partial^2 x(z't)}{\partial z'^2} + \beta x(z',t) - \gamma u(z',t)\right\} dzdz'dt + \int_0^{t_f} R_o(t) \left|\frac{\partial x(0,t)}{\partial z}\right|^2 dt + \int_0^{t_f} R_1(t) \left|\frac{\partial x(1,t)}{\partial z}\right|^2 dt + \int_0^{t_f} \sum_{i=1}^{m}\sum_{j=1}^{m} [y_i(t) - x(z_i^*,t)]\, \underline{Q}_{ij}(t)\, [y_j(t) - x(z_j^*,t)]\, dt \,, \tag{2.13}$$

The weighting coefficients R_o, R_1 and $Q(t)$ are positive definite and $R(z,z',t)$ is defined by

$$\int_0^1 R^+(z,\rho,t)\, R(\rho,z',t)\, d\rho = \delta(z - z') \tag{2.14}$$

where $R^+ > 0$ and $\delta(\cdot)$ is the Dirac delta function. The filter equations take the following form as indicated by Ajinkya and coauthors (1975)

$$\frac{\partial \hat{x}(z,t)}{\partial t} = \alpha \frac{\partial^2 \hat{x}(z,t)}{\partial z^2} - \beta\hat{x}(z,t) + \gamma u(z,t) + \sum_{i=1}^{m}\sum_{j=1}^{m} P(z,z_i^*,t)\, Q_{ij}(t)\, [y_j(t) - \hat{x}(z_j^*,t)] \,, \tag{2.15}$$

$$\frac{\partial \hat{x}(0,t)}{\partial z} = \frac{\partial \hat{x}(1,t)}{\partial z} = 0 \,, \tag{2.16}$$

where $\hat{x}(z,t)$ is the optimal estimate of the state $x(z,t)$ and the covariance of this estimate can be found from the solution

$$\frac{\partial P(z,z',t)}{\partial t} = -2\beta P(z,z',t) + \alpha \left\{ \frac{\partial^2 P(z,z',t)}{\partial z^2} + \frac{\partial^2 P(z,z',t)}{\partial z'^2} \right\} - \sum_{i=1}^{m}\sum_{j=1}^{m} P(z,z_i^*,t)\, Q_{ij}(t)\, P(z_j^*,z',t) + R^+(z,z',t) \tag{2.17}$$

for $t > 0$ and $z,z' \in (0,1)$. The boundary conditions take the form

$$\frac{\partial P(z,z',t)}{\partial z} + \alpha R_o^{-1}(t)\, \delta(z') = 0 \,, \qquad z = 0 \,, \tag{2.18}$$

$$\frac{\partial P(z,z',t)}{\partial z} - \alpha R_1^{-1}(t)\, \delta(z'-1) = 0 \,, \qquad z = 1 \,, \tag{2.19}$$

$$\frac{\partial P(z,z',t)}{\partial z'} + \alpha R_o^{-1}(t)\, \delta(z) = 0 \,, \qquad z' = 0 \,, \tag{2.20}$$

$$\frac{\partial P(z,z',t)}{\partial z'} - \alpha R_1^{-1}(t)\, \delta(z-1) = 0 \,, \qquad z' = 1 \,. \tag{2.21}$$

At this stage Ajinkya and coauthors (1975) have decided to introduce modal approximation, namely all distributed variables are approximately expressed in terms of the first three eigenfunctions of the system. In comparison, Mäder had used six of them.

Experimental results. In the implementation of the above filter the covariance values have been computed in real time. The set of 9 ordinary differential equations was solved by a Runge-Kutta method with a data sampling interval of approximately 4 seconds and a corresponding integration time step of about 8 seconds so that the values for the Runge-Kutta method at $\Delta t/2$ were available. The filter calculations required something less than 0.02 seconds per integration time step.

Only a much simplified version of the above filter was actually applied by Ajinkya and coworkers (1975). They considered only the measurement noise and set R_0^{-1}, R_1^{-1}, and R^+ to zero. As a consequence the covariance equations yield a zero steady state solution and thus the filter gradually becomes decoupled from the physical process and measurement noise after the initial phase of the experiment. This means that the filter then almost operates in an open-loop fashion and accordingly the final control is more or less in a feed-forward mode. The only remaining weighting, the matrix $\underline{Q}(t)$ was chosen as a diagonal matrix with all non-zero elements taking the value 0.3 regardless of the actual level of experimental errors encountered. Because the natural error in the thermocouples was quite small, for some runs additional noise was synthesized and added to the data. The information flow for the open-loop experiments is shown in Fig. 2.4.

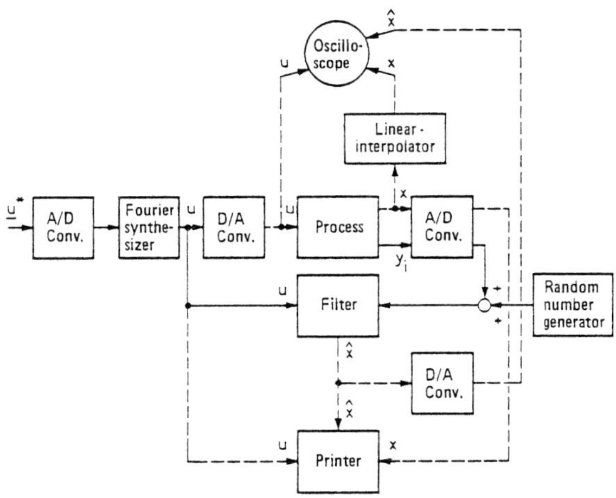

Fig. 2.4. The flow of information during the open-loop experiments ($u_0^* = 50$ °C, $u_1^* = 20$ °C, $u_2^* = 0$ °C).

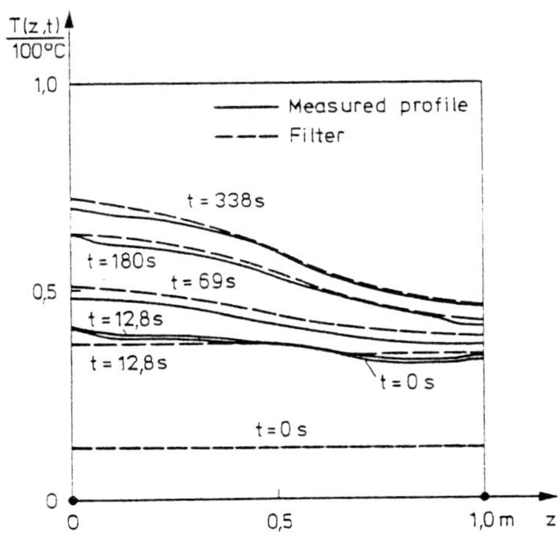

Fig. 2.5. Open-loop filter performance with one measurement.

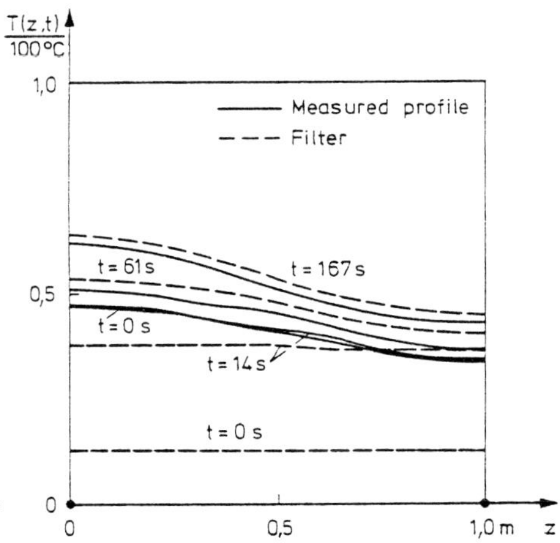

Fig. 2.6. As Fig. 2.5, with added measurement noise.

First only one measurement at z = 0 was used. Figure 2.5 shows the result of an experiment without any additional measurement noise. P(z,z',0) = 500 was assumed with no explanations.

Figure 2.6 shows practically the same experiment but with an added Gaussian random measurement error with zero mean and 8 °C standard deviation, where P(z,z',0) = 800 was assumed.

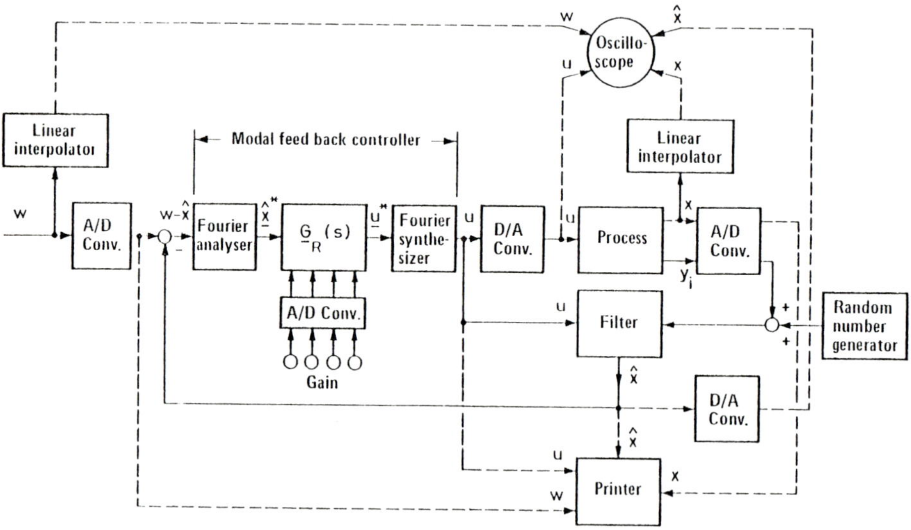

Fig. 2.7. Structure for the closed-loop experiments.

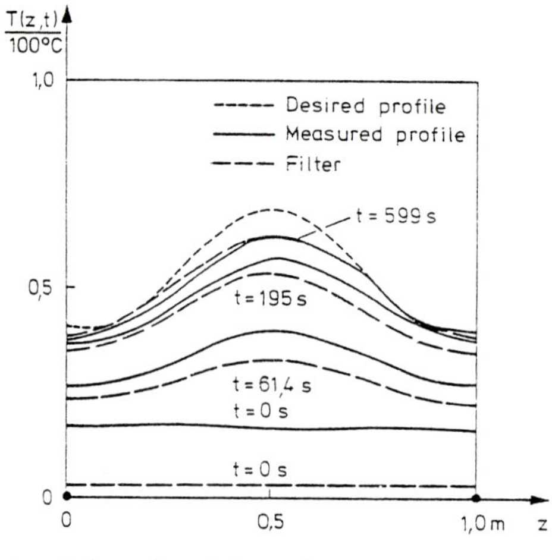

Fig. 2.8. Closed-loop (asymptotically open-loop) filter performance with one measurement.

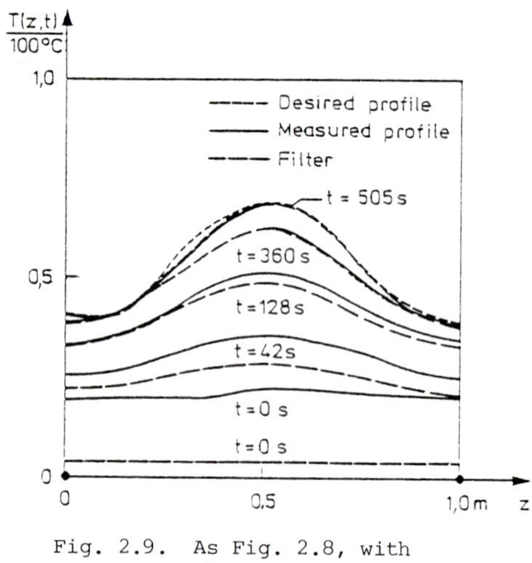

Fig. 2.9. As Fig. 2.8, with three measurements.

Also three measurements at $z = 0.25, 0.5, 0.75$ were used. The results were very similar to those with one measurement. The information flow for the closed-loop operation is shown in Fig. 2.7.

Notice, that with the experimental conditions here the link from y_i and the random number generator to the filter is relevant only during the initial stage of the experiment. As a controller the earlier described modal controller of Mäder (1975a,b) was used, although in view of only three eigenfunctions in the filter one should have reduced also the controller from six to three eigenfunctions. Otherwise appearance of at least one of the three higher modes in the desired profile w (Fig. 2.7) would bring the respective controller integrators to saturation. Figure 2.8 shows the performance with one measurement at $z = 0$, 8 °C standard deviation added measurement noise and $P(z,z',0) = 400$. Figure 2.9 shows the performance with three measurements at $z = 0.25, 0.5, 0.75$ with the same added measurement noise.

(iv) Control with Observer

Observers for distributed-parameter systems have been derived by a number of authors (e.g. Johnson, 1969; Orner and Foster, 1971; Lin and Lapidus, 1976; Köhne, 1977). For obvious reasons the following is restricted to the heat conduction system of Mäder (1975a), i.e. eq. (2.4).

Observer algorithm. Since both the previously described filter and the presently described observer of Köhne (1977) serve an identical purpose, i.e. estimating the states of a process, they have also an essentially identical structure:

$$\frac{\partial \hat{x}(z,t)}{\partial t} = \alpha \frac{\partial^2 \hat{x}(z,t)}{\partial z^2} - \beta \hat{x}(z,t) + \gamma u(z,t) - \underline{g}^T(z)\ \underline{Q}(t)\ [\hat{\underline{y}}(t) - \underline{y}(t)]\ , \tag{2.22}$$

$$\frac{\partial \hat{x}(0,t)}{\partial z} = \frac{\partial \hat{x}(1,t)}{\partial z} = 0\ . \tag{2.23}$$

The observer output $\hat{y}$ is defined according to the measurements at z_j^*:

$$\hat{y}_j(t) = \hat{x}(z_j^*,t)\ . \tag{2.24}$$

Although the matrix $\underline{Q}(t)$ in the filter equation (2.15) and in the observer eq. (2.22) looks the same, its meaning is different. For the filter synthesis the weighting matrix $\underline{Q}(t)$ apart from being positive definite (Ajinkya and coauthors, 1975) can be chosen arbitrarily, whereas here one must proceed much more carefully if a stable estimation is desired. The choice of the spatial weighting functions $g_j(z)$ for the present example seems to have been manageable (Köhne, 1977) (due to the very good understanding of the physical processes), however this need not be so just for any general system.

Open and closed-loop performance. Köhne (1977) also described the actual implementation of the above observer. Open or closed-loop operation can be chosen by a simple switching as shown in Fig. 2.10.

In the open-loop operation mainly the influence of $\underline{g}(z)$ has been studied by Köhne (1977). He has found that, in the case of the examined heat conduction process, it is advantageous to use spatially linear weighting functions $g_j(z)$ for the error feedback. In the case of two measurement points at $z_i^* = 0, 1$ Köhne has chosen two weighting functions as shown in Fig. 2.11.

He has not specified the matrix $\underline{Q}$. The experimental results are presented in Fig. 2.12. Contrary to the filter experiments by Ajinkya and others (1975), where Köhne was a coauthor, in his own work Köhne (1977) has examined the observer also

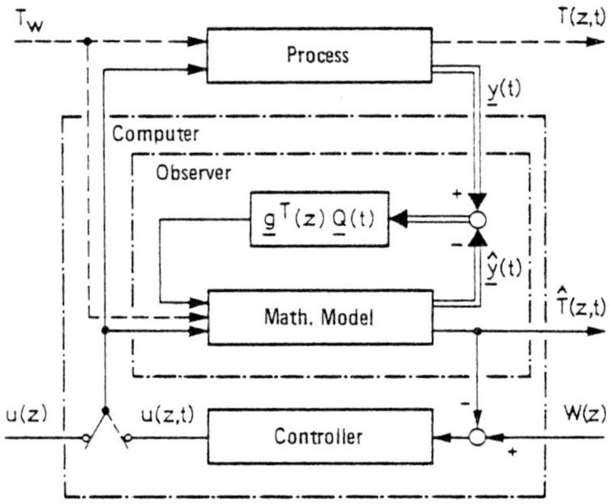

Fig. 2.10. Information flow in the experiments with observer and controller.

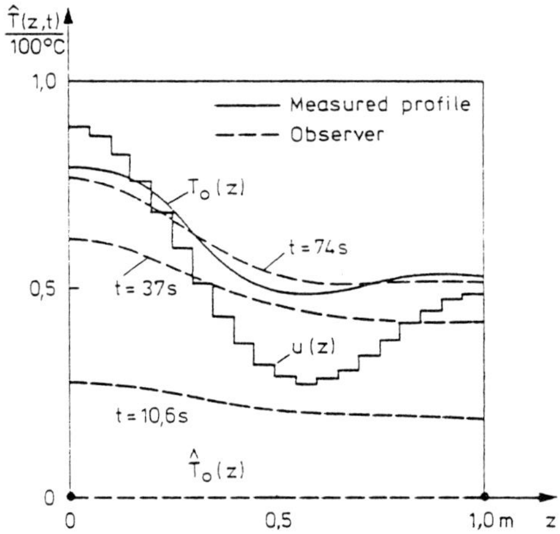

Fig. 2.12. Open-loop observer behaviour with linear weighting and two measurements.

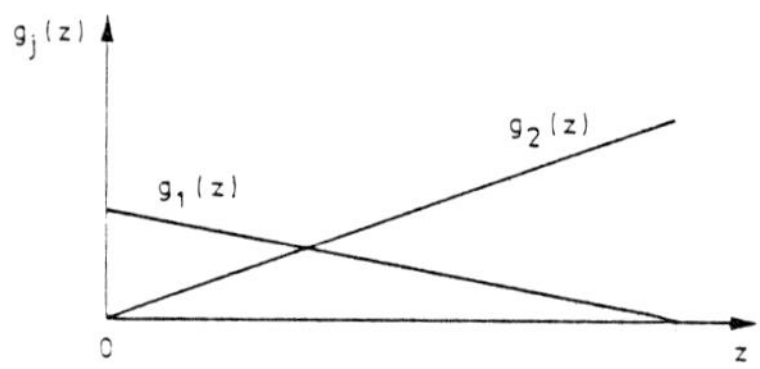

Fig. 2.11. Linear weighting.

under truly closed-loop conditions. Instead of the P(I)-controllers of Mäder (1975a,b), an optimal controller approximated by six eigenfunctions (2.9) was used by him. The above observer with two boundary measurements was tested. The results are displayed in Fig. 2.13. It takes about 90 seconds until the observed and actual profiles are almost equal. After additional 53 seconds the desired profile $w_1(z)$ is achieved. To achieve a new desired profile $w_2(z)$ it takes now only 90 seconds.

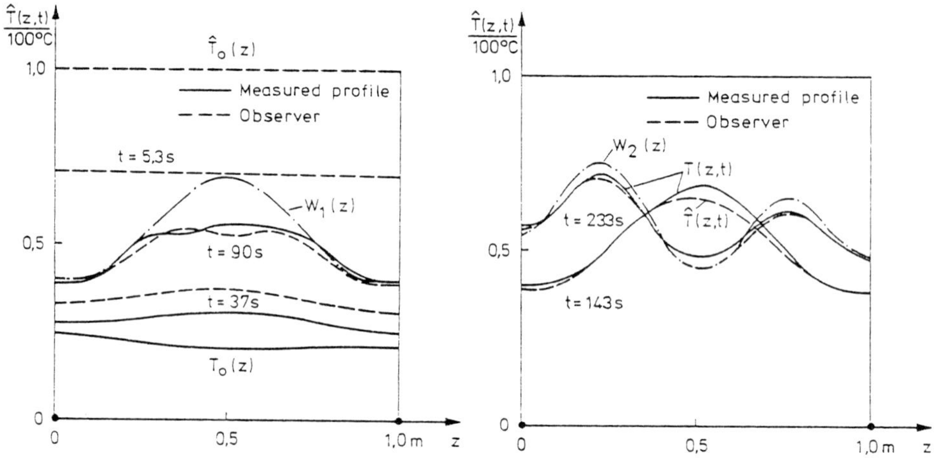

Fig. 2.13. Temperature control with two measurements and observer.

(v) Conclusions

In this section a one-dimensional laboratory heat conduction process has been used to test the application of distributed-parameter system theory. For the given process the filter, observer and feedback algorithms worked well, however, a few critical remarks have to be made. The process itself is very well-behaved, has homogeneous boundary conditions and the desired profiles are rather smooth. Thus only few eigenfunctions are required. It is not obvious that these results will carry over to other systems. Especially with boundary control one can suspect that more eigenfunctions will be necessary to adequately represent the transient profiles. The question whether the use of a filter or observer is advantageous is not readily answered from a scientific standpoint. It is more a matter of personal preference. The authors feel that a filter is somewhat superior to the observer for the following reasons:

a) The physical relevance of the weighting functions in the performance index is more obvious than the relevance of the observer weights.
b) The stability of the filter is always guaranteed.
c) The solution of the matrix Riccati equation is generally rather well behaved and not very sensitive to the choice of the design parameters. It can be obtained off-line and thus does not increase the on-line computational burden in comparison to the observer.

It has to be stressed that the example described above was in fact the first real application of distributed-parameter theory to a physical process. The success of these experiments has certainly encouraged experimental work, some of which is reported below.

3. A TWO-DIMENSIONAL LINEAR EXAMPLE

The heating of steel ingots in soaking pits prior to rolling represents a major operation in the steel manufacturing process. Soaking pits require vast amounts of energy. However, no attempt has so far been made to apply modern distributed-parameter control theory to this problem. Even though in some plants very modern process control and instrumentation equipment is used, only conventional control algorithms are implemented. The object of soaking pit control is to raise the ingot temperature to a desired level determined by the rolling requirements, with greatest possible temperature uniformity inside the ingot while minimizing energy expenditure for the heating process. This problem can well be handled using the existing theory of distributed-parameter systems. In practice, however, the problem is somewhat complicated due to constraints on the control and the state variables, i.e. maximum fuel flow cannot be exceeded and surface melting of the ingots must be avoided. As has been pointed out by Lausterer and Ray (1979), these constraints may indirectly be considered by choosing the control weights appropriately. The project reported below was thought to be an initial study of the feasibility of the application of distributed-parameter control theory to soaking pit control. Some of the experimental results have been described by Lausterer, Ray and Martens (1978), Ray, Greiss and Lausterer (1979), and in more details by Lausterer (1977).

(i) Description of the Process

The experimental apparatus studied consisted of the three subsystems shown in Fig. 3.1: a) the model ingot, b) the electronic interface, c) the process computer. Considering the rapid advancement of the technology of electronic components, we shall only describe the model ingot in details here. The heart of the apparatus, the model ingot, is shown in Fig. 3.2a. It consists of a stainless steel cylinder with a 1/2 in. hole drilled through the center allowing the passage of cooling water. Even though this arrangement does not directly correspond to real ingots, it has the decisive advantage of allowing rapid cooling and thus small time constants permitting shorter experiments and easier repetition. The hollow cylinder is enclosed by ceramic heaters (Fig. 3.2a,b), representing a three-zone furnace so that temperature profiles may be controlled both axially and radially. At four radial positions r_j and eight axial locations z_i (measuring planes) thermocouples were placed. The locations z_i of the eight measuring planes were chosen to ensure observability (cf. section 4) i.e. the zeros of the first axial eigenfunctions were avoided.

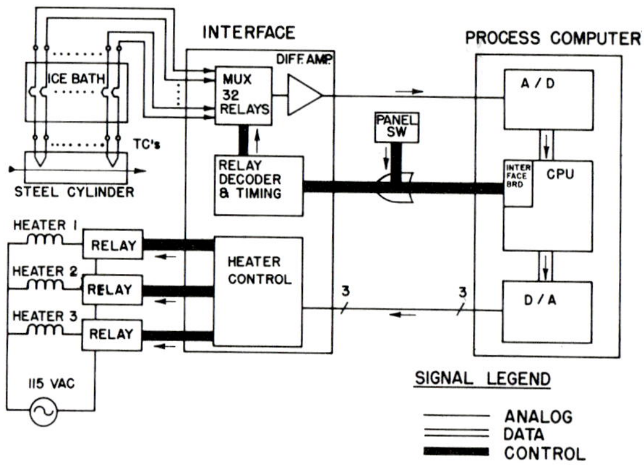

Fig. 3.1. System block diagram.

Figure 3.2b shows a radial cross section through a measuring plane. The particular direction of the holes for the thermocouples was chosen so that it would be perpendicular at the tip to the radial heat flow and thus minimize flow disturbances. Table 3.1 lists the geometric data. It has to be stressed that the design of the ingot model and the choice of the material was primarily based on dynamic simulation. Figures 3.3a and 3.3b show photographs of the experimental setup and the three-zone furnace, where the hose supplying cooling water and the thermocouple leads may be seen.

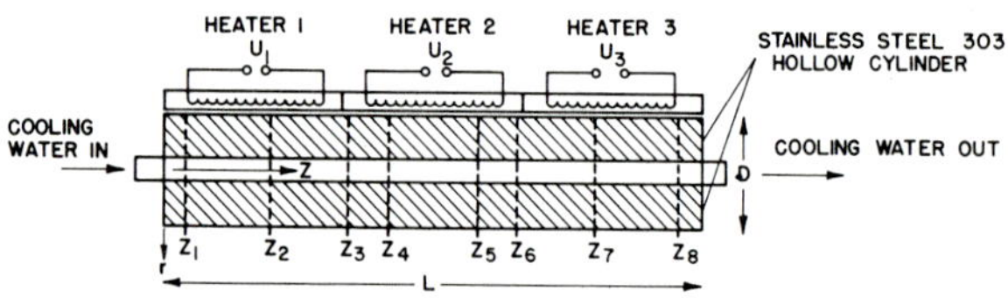

Fig. 3.2a. Axial cross section of the experimental apparatus.

In modelling this system, the following assumptions were made:

a) there is heat flow only in the r and z direction so that there is angular uniformity of temperature in the cylinder;

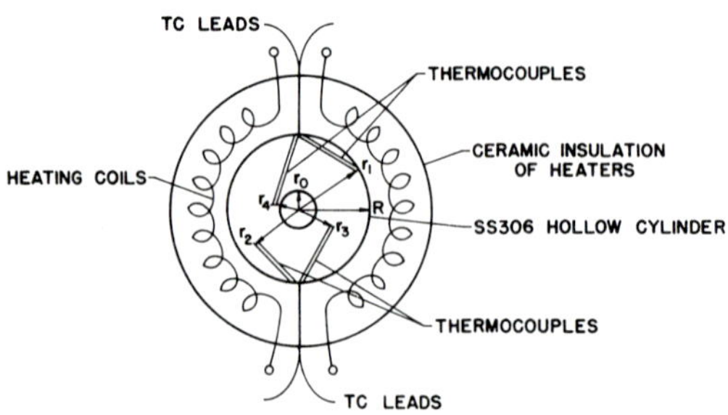

Fig. 3.2b. Radial cross section of the experimental apparatus.

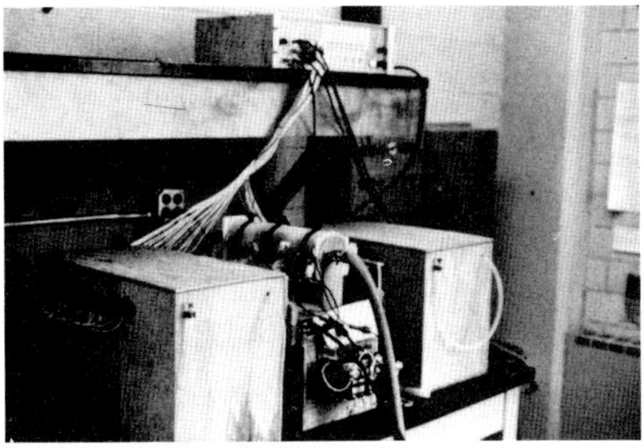

Fig. 3.3a. Experimental setup.

Fig. 3.3b. Water-cooled three-zone furnace.

TABLE 3.1 Geometric Data

Length $L = 12''$
Diameter $D = 2\text{-}3/8''$
Outer radius $R_c = 1\text{-}3/16''$
Inner radius $r_0 = 1/4''$
Locations of the thermocouples
Axial distance of the measuring planes from entrance ($z = 0$):
$z_1 = 0.5''$
$z_2 = 2.4''$
$z_3 = 4.1''$
$z_4 = 5.0''$
$z_5 = 7.0''$
$z_6 = 7.9''$
$z_7 = 9.6''$
$z_8 = 11.5''$
Radial distance of thermocouple tip from centerline ($r = 0$):
$r_1 = 35/32''$
$r_2 = 28/32''$
$r_3 = 18/32''$
$r_4 = 11/32''$

TABLE 3.3 System Parameters

$T_w = 22°C$
$\rho C_p = 4.0267 \times 10^6$ W-sec/m^3°C
$k = 15.3$ W/m°C
$h = 5100$ W/m^2°C
$\alpha = 3.8 \times 10^{-6}$ sec/m^2 $= k/\rho C_p$

b) the insulation at z' = 0 and z' = L is perfect, thus the heat flux is zero at that point;

c) the heat input from the ceramic heaters are correlated linearly with the power input;

d) the water cooling can be represented by Newton's law of cooling applied uniformly over the total inner surface;

e) the density, specific heat and especially the thermal conductivity and the heat transfer coefficient at the inner wall do not vary with temperature and thus can be represented by constants.

If these assumptions hold we can formulate the mathematical model of the cylinder as follows:

$$\frac{\partial T}{\partial t'} = \alpha\left[\frac{\partial^2 T}{\partial r'^2} + \frac{1}{r'}\frac{\partial T}{\partial r'} + \frac{\partial^2 T}{\partial z'^2}\right] \qquad (3.1)$$

$$0 \le z' \le L\ , \quad r'_o \le r \le R\ , \quad t' > 0$$

where T is the temperature in the cylinder in °C, r' is the radius and z' is

TABLE 3.2 Heater Parameters

$p_{10} = 0.8509$	$p_{20} = -0.2104$	$p_{30} = -1.8314$	$c_1 = 0.9$	$a = 26.4\,\frac{W}{V}$
$p_{11} = 1.8117$	$p_{21} = 5.1046$	$p_{31} = 7.1662$	$c_2 = 0.96$	$b = 77.\,W$
$p_{12} = -4.5261$	$p_{22} = -5.0726$	$p_{32} = -4.464$	$c_3 = 1.0$	$d = -12.32\,\frac{W}{V^2}$
				$e = 88\,\frac{W}{V}$

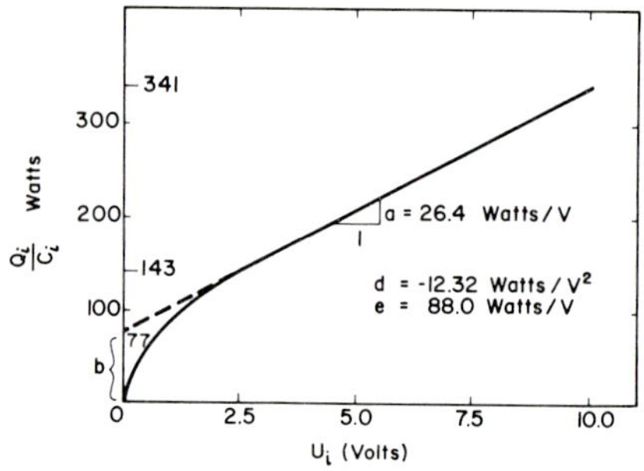

Fig. 3.4. Correlation of heat input

the length of the cylinder, and $\alpha = k/\rho C_p$ is the thermal diffusivity.

The boundary conditions can be stated as

$$\frac{\partial T}{\partial z'} = 0 \qquad \text{at} \qquad z' = 0 \tag{3.2}$$

$$\frac{\partial T}{\partial z'} = 0 \qquad \text{at} \qquad z' = L \tag{3.3}$$

$$k\,\frac{\partial T}{\partial r'} = h(T - T_w) \qquad \text{at} \qquad r' = r'_o \tag{3.4}$$

$$k\,\frac{\partial T}{\partial r'} = q'(z',t) \qquad \text{at} \qquad r' = R \tag{3.5}$$

where k is the thermal conductivity and h is the overall heat transfer coefficient. T_W is the mean water temperature and $q'(z',t)$ is the heat flux from the heaters at the outer surface into the cylinder, determined by

$$q'(z',t) = \sum_{i=1}^{3} q_i(t)\; f_i(z')\; \sigma_i(z') \tag{3.6}$$

where

$$q_i(t) = Q_i'/2\pi R L \tag{3.7}$$

$$f_i(z') = P_{io} + P_{i1}z' + P_{i2}z'^2 , \qquad \bar{z}_i' \le z' \le \bar{z}_{i+1}' , \qquad i=1,2,3 \tag{3.8}$$

and

$$\sigma_i(z') = H(z'-\bar{z}_i') - H(z'-\bar{z}_{i+1}') . \tag{3.9}$$

Here H is the Heaviside function and

$$\bar{z}_1' = 0 , \qquad \bar{z}_2' = L/3 , \qquad \bar{z}_3' = 2L/3 , \qquad \bar{z}_4' = L$$

define the heater sections.

The actual heat output of each heater in Watts was determined from steady state experiments using the relationship

$$Q' = F \rho_w C_w \Delta T \tag{3.10}$$

where Q' is the amount of heat taken up by the cooling water per second, F is the measured flowrate, ρ_W and C_W are the density and specific heat of water and ΔT is the temperature difference between the inlet and outlet of the cylinder, also determined experimentally. It was found that the heater output could be approximated by

$$Q_i'(t) = c_i\left(au_i(t) + b\right), \qquad 2.5 \le u_i \le 10 , \qquad i=1,2,3 , \tag{3.11}$$

and

$$Q_i'(t) = c_i\left(du_i^2(t) + cu_i(t)\right) , \qquad 0 \le u_i \le 2.5 , \qquad i=1,2,3 , \tag{3.12}$$

which is shown in Fig. 3.4. The parameters p_{ij} defined in (3.8) are also determined experimentally by fitting parabola to the axial steady-state temperature profiles obtained from measurements. The heater parameters are summarized in table 3.2, while the system parameters are listed in table 3.3. The only parameter that could not readily be obtained from direct measurements or physical tables is the global heat transfer coefficient h. Even though the thermal conductivity k may also be found from tables and h can be estimated from correlations, it is preferred to determine these values experimentally from measured data as outlined below. The steady-state temperature profile in a hollow cylinder with inner radius r_o and outer radius R is given as

$$T(r) = \frac{qR}{k} \ln\left(\frac{r}{r_o}\right) + \frac{qR}{hr_o} + T_w . \tag{3.13}$$

From this formula both k and h may be independently determined. Subtracting the measured temperature at two different radial locations in one measuring plane we can write

$$\Delta T_{ij} = T(r_i) - T(r_j) = \frac{qR}{k} \ln\left(\frac{r_i}{r_j}\right) , \qquad i,j=1,\ldots,4 \tag{3.14}$$

yielding

$$k_{ij} = \frac{qR}{T_{ij}} \ln\left(\frac{r_i}{r_j}\right) , \qquad i,j=1,\ldots,4 . \tag{3.15}$$

With $q = Q'/2\pi RL$ known from (3.10), k was determined using a least squares fit. Similarly, the global heat transfer coefficient h is determined from

$$h_{ijk} = \frac{qR}{r_o[\left(T(r_k) - T_w\right) - \frac{qR}{k_{ij}} \ln\left(\frac{r_k}{r_o}\right)]} , \qquad i,j,k=1,\ldots,4 \tag{3.16}$$

again using a least squares regression. It is remarkable that this value for h was also predicted by the simple correlation

$$h(t) = 150(1 + .001t) \frac{V_s^{0.8}}{D^{0.2}} \tag{3.17}$$

with an error of less than .6% (V_s is the flow velocity and D the inner tube diameter). During the course of the experiments it was found necessary to include heater dynamics into the model which can be approximately represented by

$$\dot{\underline{v}} = \underline{A}\ \underline{v} + \underline{B}\ \underline{u} \tag{3.18}$$

where $\underline{A}$ and $\underline{B}$ are diagonal matrices, $\underline{v}(t)$ is the dimensionless amplitude vector of the heat flux of the heaters, $\underline{u}(t)$ represents the power input into the heaters coming from the A/D converters. $\underline{A}$ and $\underline{B}$ contain the eigenvalues of the heaters, i.e. the reciprocal time constants which were found to be identical for each zone with a value of approximately 180 s.

Actually the overall system given by (3.1) and (3.18) would state a coupled lumped and distributed-parameter system, the theory of which is treated, for instance, by Ajinkya, Ray, Yu and Seinfeld (1975) and Lausterer (1977). However, it was subsequently assumed that the initial state of the heaters were known exactly, and therefore, since there are practically no input disturbances, the state of the heaters can be predicted with reasonable accuracy just by simulation. Similarly, in the derivation of the optimal linear-quadratic control law below we have neglected the heater dynamics and assumed that the desired heat flux is imposed directly to the outer surface of the cylinder. This certainly introduces some error in the control due to the actual heat flux in our laboratory process lagging the theoretical value assumed by the controller (not by the filter since the heater dynamics were accurately modelled). However, for an industrial system like a soaking pit the heater dynamics would essentially be negligible, because gas or oil fired furnaces respond very fast (within seconds) in comparison to the slow dynamics of real ingots (minutes to hours). Thus the heater dynamics may be neglected without loss of practical relevance.

Let us now define the dimensionless quantities:

$$\Theta = (T - T_w)/T_w\,, \qquad z = z'/L\,, \qquad r = r'/R\,, \qquad r_o = r_o'/R\,,$$

$$t = \alpha t'/R^2\,, \qquad \alpha' = R^2/L^2\,, \qquad Bi = hR/k\,, \tag{3.19}$$

$$q(z,t) = q'(z',t)R/kT_w\,, \qquad \underline{g}^T(z)\,\underline{v}(t) = q(z,t)\,.$$

Then we obtain the dimensionless partial differential equation:

$$\frac{\partial\Theta}{\partial t} = \frac{\partial^2\Theta}{\partial r^2} + \frac{1}{r}\frac{\partial\Theta}{\partial r} + \alpha'\frac{\partial^2\Theta}{\partial z^2} + \delta(r-1)\,\underline{g}^T(z)\,\underline{v}(t)\,, \tag{3.20}$$

$$r_o \le r \le 1\,, \qquad 0 \le z \le 1\,, \qquad t \ge 0$$

$$\frac{\partial\Theta}{\partial z} = 0\,, \qquad \text{at} \quad z=0\,, \quad z=1 \tag{3.21}$$

$$\frac{\partial\Theta}{\partial r} = Bi\,\Theta\,, \qquad \text{at} \quad r = r_o \tag{3.22}$$

$$\frac{\partial\Theta}{\partial r} = 0\,, \qquad \text{at} \quad r = 1 \tag{3.23}$$

where the boundary conditions were made homogenous by employing the δ-function $\delta(r-1)$ in the system partial differential equation (3.20).

The spatially discrete temperature measurements are given by

$$y_{ik}(t) = \Theta(r_i, z_k, t) + \eta_{ik}(t) \qquad i=1,2,\ldots N_r\,, \qquad k=1,2,\ldots M_z \tag{3.24}$$

where $\Theta(r_i,z_k,t)$ is the true dimensionless temperature at location $r = r_i$ and $z = z_k$ and $\eta_{ik}(t)$ represents the measurement error. N_r and M_z denote the maximum number of radial and axial measurement positions used by the estimator.

(ii) Optimal State Estimation

Optimal state estimation equations for distributed-parameter systems have been derived by many authors (cf., for instance, Tzafestas and Nightingale, 1968; Meditch, 1971; Sakawa, 1972; Atre and Lamba, 1972). Thus only the results are stated here. They have been derived (Lausterer, 1977; Lausterer and Ray, 1979) such that the following least squares performance index is minimized:

$$I_F = \int_0^{t_f} \left\{ \int_0^1\int_0^1\int_{r_o}^1\int_{r_o}^1 \left[\frac{\partial\Theta(r,z,t)}{\partial t} - f(r,z,t)\right] R(r,r',z,z',t) \left[\frac{\partial\Theta(r',z',t)}{t} - f(r',z',t)\right]\cdot\right.$$

$$\left.\cdot\, dr\,dr'\,dz\,dz' + \sum_{i=1}^{N_r}\sum_{j=1}^{N_r}\sum_{k=1}^{M_z}\sum_{l=1}^{M_z} [y_{ik}(t) - \Theta(r_i,z_k,t)]\, Q_{ijkl}(t)\, [y_{jl}(t) - \Theta(r_j,z,t)]\right\} dt \tag{3.25}$$

where $\hat{\Theta}(r,z,t)$ denotes the optimal estimate of Θ and $f(r,z,t)$ is the right-hand side (RHS) of equation (3.20). The first term minimizes the model error, while the second minimizes the measurement error. $Q_{ijkl}(t)$ is a positive definite weighting function and R has a generalized inverse defined by

$$\int_0^1\int_0^1 R^+(r,\rho,z,\xi,t)\, R(\rho,r',\xi,z',t)\, d\rho\, d\xi = \delta(r-r')\,\delta(z-z') \tag{3.26}$$

where R^+ is also positive definite and symmetric. Such a generalized inverse was introduced by Tzafestas and Nightingale (1968a) and must not be confused with the generalized matrix inverse. The filter takes the form

$$\frac{\partial\hat{\Theta}}{\partial t} = \frac{\partial^2\hat{\Theta}}{\partial r^2} + \frac{1}{r}\frac{\partial\hat{\Theta}}{\partial r} + \alpha'\frac{\partial^2\hat{\Theta}}{\partial z^2} + \delta(r-1)\ \underline{g}^T(z)\ \underline{v}(t) + $$

$$+ \sum_{i=1}^{N_r}\sum_{j=1}^{N_r}\sum_{k=1}^{M_z}\sum_{l=1}^{M_z} P(r,r_i,z,z_k,t)\ Q_{ijkl}(t)\ [y_{jl}(t) - \hat{\Theta}(r_j,z_l,t)] \ . \qquad (3.27)$$

The original system equations (3.20) have been supplemented by a feedback term, where the difference between measurements and estimates is weighted with appropriate gain factors. $P(r,r',z,z',t)$ is determined from

$$\frac{\partial P}{\partial t} = \frac{\partial^2 P}{\partial r^2} + \frac{1}{r}\frac{\partial P}{\partial r} + \frac{\partial^2 P}{\partial r'^2} + \frac{1}{r'}\frac{\partial P}{\partial r'} + \alpha'\left(\frac{\partial^2 P}{\partial z^2} + \frac{\partial^2 P}{\partial z'^2}\right) -$$

$$- \sum_{i=1}^{N_r}\sum_{j=1}^{N_r}\sum_{k=1}^{M_z}\sum_{l=1}^{M_z} P(r,r_i,z,z_k,t)\ Q_{ijkl}(t)\ P(r_j,r',z_l,z',t) + R^+(r,r',z,z',t) \qquad (3.28)$$

with boundary conditions

$$\frac{\partial P}{\partial z} = 0 \qquad \text{at} \quad z=0 \quad \text{and} \quad z=1 \qquad (3.29)$$

$$\frac{\partial P}{\partial r} = \mathrm{Bi}\, P \qquad \text{at} \quad r = r_o \qquad (3.30)$$

$$\frac{\partial P}{\partial r} = 0 \qquad \text{at} \quad r = 1 \qquad (3.31)$$

and similar conditions at $z' = 0$, $z' = 1$, $r' = r_o$ and $r' = 1$. We have to point out that in the case where both process and measurement noise are Gaussian white noise processes with covariances R^+ and Q^{-1}_{ijkl} respectively, P is a true estimate covariance, otherwise it must be considered a filter parameter without precise statistical significance.

(iii) Optimal Feedback Control

The optimal feedback control law is derived minimizing the quadratic cost functional

$$I_C = \int_0^1\int_0^1\int_{r_o}^1\int_{r_o}^1 [\Theta_d(r,z,t_f) - \Theta(r,z,t_f)]\ \gamma_f(r,r',z,z')\ [\Theta_d(r',z',t_f) - \Theta(r',z',t_f)]\ \cdot$$

$$\cdot\ dr\, dr'\, dz\, dz' + \int_0^{t_f}\left\{\int_0^1\int_0^1\int_{r_o}^1\int_{r_o}^1 [\Theta_d(r,z,t) - \Theta(r,z,t)]\ \gamma_d(r,r',z,z',t)\ \cdot\right.$$

$$\left.\cdot\ [\Theta_d(r',z',t) - \Theta(r',z',t)]\, dr\, dr'\, dz\, dz' + [\underline{u}^*(t) - \underline{u}(t)]^T\ \underline{\Gamma}_u(t)\ [\underline{u}^*(t) - \underline{u}(t)]\right\} dt \qquad (3.32)$$

where $\Theta_d(r,z,t)$ is the desired temperature profile and $\underline{u}^*(t)$ is the control setting corresponding to this desired profile. γ_f and γ_d are positive semidefinite weighting functions of the deviations of the state variables from the desired state, while $\underline{\Gamma}_u(t)$ is a positive definite matrix weighting the control expense.

Now the optimal control law reads

$$\underline{u}(t) = \underline{u}^*(t) + \underline{\Gamma}_u^{-1} \int_0^1\int_0^1\int_{r_o}^1\int_{r_o}^1 R\,(r,r',z,z',t)\,[\Theta_d(r',z',t)-\Theta(r',z',t)]\,\underline{g}(z)\,\delta(r-1)\,dr\,dr'\,dz\,dz' \tag{3.33}$$

where $R_c(r,r',z,z',t)$ is the solution of

$$-\frac{\partial R_c}{\partial t} = \frac{\partial^2 R_c}{\partial r^2} - \frac{\partial}{\partial r}\left(\frac{1}{r} R_c\right) + \frac{\partial^2 R_c}{\partial r'^2} - \frac{\partial}{\partial r'}\left(\frac{1}{r'} R_c\right) + \alpha'\left(\frac{\partial^2 R_c}{\partial z^2} + \frac{\partial^2 R_c}{\partial z'^2}\right) + \gamma_d(r,r',z,z',t) -$$

$$- \int_0^1\int_0^1\int_{r_o}^1\int_{r_o}^1 R_c(r,\rho,z,\xi,t)\,\delta(\rho-1)\,\underline{g}^T(\xi)\,\underline{\Gamma}_u^{-1}\,\delta(\rho'-1)\,\underline{g}(\xi')\,R_c(\rho',r',\xi',z',t)\,d\rho\,d\rho'\,d\xi\,d\xi' \tag{3.34}$$

Note the correspondence to the filter equations above. However, the boundary conditions are the adjoint of those for $P(r,r',z,z',t)$. Derivations of the optimal controller equations as stated above have been given by numerous authors. The reader is referred to some early and survey papers (Wang, 1964; Brogan, 1968; Robinson, 1971) and books (Sage, 1968; Butkowskii, 1969; Denn, 1969; Lions, 1971; Ray and Lainiotis, 1978).

(iv) Modal Decomposition

For the purpose of obtaining numerical solutions of the filter and controller equations stated above it is convenient to approximate the partial differential equations by a finite series of ordinary differential equations and then solve the resulting set of equations by a suitable numerical integration scheme. Since our system (3.20) - (3.23) is linear and homogeneous, an obvious choice is to use modal approximation. Thus we expand $\Theta(r,z,t)$ in terms of the eigenfunctions of the system

$$\Theta(r,z,t) \cong \sum_{n=1}^{N}\sum_{m=1}^{M} a_{nm}(t)\,\phi_n(r)\,\psi_m(z) \tag{3.35}$$

where N and M are the total number of radial and axial eigenfunctions $\phi_n(r)$ and $\psi_m(z)$ used in the approximation and $a_{nm}(t)$ is their amplitude. The eigenfunctions are determined as

$$\phi_n(r) = A_n\left[J_o(\sqrt{\mu_n}\,r) - \frac{J_1(\sqrt{\mu_n})\,Y_o(\sqrt{\mu_n}\,r)}{Y_1(\sqrt{\mu_n})}\right] \tag{3.36}$$

and

$$\psi_m(z) = \begin{cases} 1 & m = 1 \\ \sqrt{2}\cos(m-1)\pi z & m > 1 \end{cases} \tag{3.37}$$

where A_n is calculated from

$$A_n^2 \int_{r_o}^1 r\left[J_o(\sqrt{\mu_n}\,r) + \frac{J_1(\sqrt{\mu_n})}{Y_1(\sqrt{\mu_n})}\,Y_o(\sqrt{\mu_n}\,r)\right]^2 dr = 1 \tag{3.38}$$

and μ_n satisfies the transcendental equation

$$Bi\,Y_o(\sqrt{\mu_n}\,r_o) + \sqrt{\mu_n}\,Y_1(\sqrt{\mu_n}\,r_o) - [Bi\,J_o(\sqrt{\mu_n}\,r_o) + \sqrt{\mu_n}\,J_1(\sqrt{\mu_n}\,r_o)]\,\frac{Y_1(\sqrt{\mu_n})}{J_1(\sqrt{\mu_n})} = 0 \tag{3.39}$$

the roots of which were computed using a simple interval halving method. The dynamics of the modal system may be written as

$$\dot{\underline{a}} = \underline{\Lambda}\ \underline{a} + \underline{B}_D \underline{v}(t) \tag{3.40}$$

where $\underline{\Lambda}$ is the diagonal system matrix comprised of the eigenvalues

$$\lambda_{nm} = \mu_n + \alpha' [(m-1)\pi]^2 \tag{3.41}$$

and $\underline{B}_D$ is an (NM×3) input matrix given by

$$\underline{B}_D = \{\underline{b}_{nmi}\}, \qquad n=1,2,\ldots N \quad m=1,2,\ldots M \quad i=1,2,3 \tag{3.42}$$

with

$$\underline{b}_{nmi} = \int_o^1 \int_{r_o}^1 \underline{g}_i(z)\ \delta(r-1)\ \phi_n^*(r)\ \psi_m^*(z)\ dr\ dz \tag{3.43}$$

where the asterisk denotes the adjoint eigenfunction. Applying the eigenfunction expansion to the optimal filter (3.27) and covariance equations (3.28) we obtain

$$\frac{d\hat{a}_{nm}(t)}{dt} = -\lambda_{nm}\hat{a}_{nm}(t) + \sum_{i=1}^{3} \underline{b}_{nmi}^T(t)\ \underline{v}_i(t) +$$

$$+ \sum_{k'=1}^{N_C} \sum_{p=1}^{M_C} \sum_{i=1}^{N_r} \sum_{k=1}^{M_z} P_{nk'mp}(t)\phi_{k'}(r_i)\psi_p(z_k)Q_{ik}\left[y_{ik}(t) - \sum_{n'=1}^{N} \sum_{m'=1}^{M} \hat{a}_{n'm'}(t)\phi_{n'}(r_i)\psi_{m'}(z_k)\right] \tag{3.44}$$

and

$$\frac{dp_{nk'mp}(t)}{dt} = -\gamma_{nk'mp}\, p_{nk'mp}(t) + r_{nk'mp}(t) -$$

$$- \sum_{n'=1}^{N_C} \sum_{l'=1}^{N_C} \sum_{m'=1}^{M_C} \sum_{p'=1}^{M_C} \sum_{i=1}^{N_r} \sum_{k=1}^{M_z} p_{nl'mp'}(t)\ \phi_{l'}(r_i)\ \psi_{p'}(z_k)Q_{ik}\ \phi_{n'}(r_i)\ \psi_{m'}(z_k)\ p_{n'k'm'p}(t) \tag{3.45}$$

where

$$\gamma_{nk'mp} = \lambda_{nm} + \lambda_{k'p} \tag{3.46}$$

are the eigenvalues of the covariance equation and

$$r_{nk'mp}(t) = \int_o^1 \int_o^1 \int_{r_o}^1 \int_{r_o}^1 R^+(r,r',z,z',t)\ \phi_n(r)\ \phi_{k'}(r')\ \psi_m(z)\ \psi_p(z')\ r\,r'\,dr\,dr'\ dz\,dz'\ . \tag{3.47}$$

Here we have assumed $Q_{ijkl}(t)$ to be diagonal and time independent (i.e. $Q_{ijkl}(t) = Q_{ik}$). N_C and M_C are the total numbers of radial and axial eigenfunctions used in the covariance equations. Similarly the optimal controller equations are expanded in terms of the adjoint eigenfunctions yielding the optimal control law

$$\underline{u}(t) = \underline{u}^*(t) + \underline{\Gamma}_u^{-1} \sum_{n=1}^{N_R} \sum_{k'=1}^{N_R} \sum_{m=1}^{M_R} \sum_{p=1}^{M_R} r^C_{nk'mp}(t)\ \underline{b}_{nm}\left(a^d_{k'p}(t) - a_{k'p}(t)\right) \tag{3.48}$$

where

$$\underline{b}_{nm} = \int_o^1 \int_{r_o}^1 r\phi_n(r)\ \psi_m(z)\ \underline{g}(z)\ \delta(r-1)\ dr\ dz \tag{3.49}$$

and $r^C_{nk'mp}(t)$ is determined by

$$-\frac{dr^C_{nk'mp}}{dt} = -\gamma_{nk'mp}\, r^C_{nk'mp}(t) + \gamma^d_{nk'mp} - \sum_{n'=1}^{N_R}\sum_{m'=1}^{N_R}\sum_{l'=1}^{M_R}\sum_{p'=1}^{M_R} r^C_{n'k'm'p}(t)\, \underline{b}^T_{n'm'}\, \underline{\Gamma}_u^{-1}\, \underline{b}_{l'p'}\, r^C_{nl'mp'}(t) \quad (3.50)$$

with

$$\gamma^d_{nk'mp}(t) = \int_o^1\int_o^1\int_{r_o}^1\int_{r_o}^1 \gamma_d(r,r',z,z',t)\,\phi_n(r)\,\phi_{k'}(r')\,\psi_m(z)\,\psi_p(z')\,dr\,dr'\,dz\,dz' \quad (3.51)$$

Here N_R and M_R denote the number of radial and axial eigenfunctions used for the controller Riccati equations.

(v) Suboptimal Riccati Equations

Equations (3.45) and (3.49) represent large systems of nonlinear ordinary differential equations even with only few eigenfunctions used and require time-consuming numerical integration. Thus it is desirable to simplify these equations. If the initial conditions and weighting functions are chosen to be diagonal, it is easy to observe that the only off-diagonal elements appear in the quadratic term. Computing these elements we find a strong diagonal dominance suggesting that the off-diagonal elements may be neglected without introducing an intolerable error. The suboptimal Riccati equations then read

$$\frac{dp_{nnmm}(t)}{dt} = -\gamma_{nnmm}\, p_{nnmm}(t) - p^2_{nnmm}(t)\sum_{i=1}^{N_r}\sum_{k=1}^{N_z} Q_{ik}\,\phi_n(r_i)^2\,\psi_m(z)^2 + r_{nnmm} \quad (3.52)$$

$$-\frac{dr^C_{nnmm}(t)}{dt} = -\gamma_{nnmm}\, r^C_{nnmm}(t) - r^{C2}_{nnmm}(t)\,\underline{b}^T_{nm}\,\underline{\Gamma}_u^{-1}\,\underline{b}_{nm} + \gamma^d_{nnmm} \quad (3.53)$$

yielding an analytical solution in the form

$$y(t) = \frac{a_1 + a_2 e^{rt}}{a(1+a_3 e^{rt})} \quad . \quad (3.54)$$

Parameter changes can now easily be made without requiring costly off-line calculations. It must be stressed, however, that his decoupling is only justified after diagonal dominance has been proved. The suboptimal filter and suboptimal feedback controller equations now read

$$\frac{d\hat{a}_{nm}(t)}{dt} = -\lambda_{nm}\,\hat{a}_{nm}(t) + \sum_{i=1}^{N_r}\sum_{k=1}^{M_z} p_{nnmm}(t)\,\phi_n(r_i)\,\psi_m(z_k)\,Q_{ik}\;\cdot$$

$$\cdot\left[y_{ik}(t) - \sum_{n'=1}^{N}\sum_{m'=1}^{M}\hat{a}_{n'm'}(t)\,\phi_{n'}(r_i)\,\psi_{m'}(z_k)\right] + \sum_{i=1}^{3}\underline{b}^T_{nmi}\,\underline{v}_i(t) \quad (3.55)$$

$$\underline{u}(t) = \underline{u}^*(t) + \underline{\Gamma}_u^{-1}\sum_{n=1}^{N_R}\sum_{m=1}^{M_R} r^C_{nnmm}(t)\,\underline{b}_{nm}\left(a^d_{nm}(t) - a_{nm}(t)\right) \quad (3.56)$$

with considerable savings in real-time requirements.

(vi) Choice of the Weighting Functions

The dynamic performance of the filter is, of course, depending on the eigenvalues of the system, but to a large extend also on the proper choice of weighting functions in the filter performance index as given in (3.25). Similarly the control performance is strongly influenced by an appropriate choice of the controller weighting functions in (3.32). In fact, it turns out that a proper specification of these weighting functions is crucial to the dynamic behaviour of filter and controller. The choice of these functions is often already difficult for lumped systems, but with distributed systems one must specify not only time dependency, but also the spatial shape of these functions. It requires a good deal of physical insight and engineering intuition to specify these weights and it is essential to build in as much physical knowledge as possible. It is precisely at this point, where the advantages of late lumping become apparent and therefore we shall describe the selection procedure in detail. For a linear system with additive Gaussian white noise as formulated above, it can be shown that

$$P(r,r',z,z',0) = E\{[\Theta(r,z,0) - \hat{\Theta}(r,z,0)][\Theta(r',z',0) - \hat{\Theta}(r',z',0)]\} \tag{3.57}$$

is the covariance of the initial estimation error. Clearly, no technical process contains true Gaussian white noise, but for our system it may safely be assumed that the natural noise contains much higher frequences than the dominant eigenvalues would cover. Thus it may be regarded as a low-pass filter and hence the idealization of Gaussian white noise is expected to be a valid assumption. Thus we can say that $R^+(r,r',z,z',t)$ should be chosen proportional to the modelling error covariance and $Q_{ijkl}(t)$ inversely proportional to the measurement error covariance. Of course, now the question arises, where do we obtain the values of these covariances from. The measurement error covariance may, in fact, be obtained experimentally from measurements, but the modelling error must be estimated from physical plausibility. Let us assume that the initial error covariance can be written in the form

$$P(r,r',z,z',0) = S_o^2 \; f^r(r,r') \; f^z(z,z') \tag{3.58}$$

where S_o^2 represents the mean square initial error and f^r and f^z are spatial functions. In order to determine a meaningful shape of the spatial weighting we consider a typical initial temperature distribution $\Theta(r,z,0)$ and compare it to the initial condition we used for our filter in our experiments, which is a spatially uniform constant temperature distribution $\Theta(r,z,0) = \Theta_o$, where Θ_o was usually taken to be corresponding to the cooling water temperature. Since the radial temperature gradient in the ingot is necessarily always positive (heat applied at outer surface, cooling from inner surface) it makes sense to choose the simplest radially increasing function symmetric in r and r'. Thus we set

$$f^r(r,r') = r\,r' \; . \tag{3.59}$$

This choice is also of particular mathematical convenience permitting analytical solutions of the integrals resulting from eigenfunction expansion. A reasonable choice for $f^z(z,z')$ seems to be

$$f^z(z,z') = \cos\,[\tfrac{\pi}{2}\,(z - z')] \; . \tag{3.60}$$

Another possibility is

$$f^z(z,z') = \delta(z - z') \; . \tag{3.61}$$

Similarly, we can select the modelling error as

$$R^+(r,r',z,z',n\Delta t)_{dis} = T^2_{mod} \cos\,[\tfrac{\pi}{2}\,(z - z')] \tag{3.62}$$

or

$$R^+(r,r',z,z',n\Delta t)_{dis} = T^2_{mod}\,\delta(z - z') \; . \tag{3.63}$$

Note that in both cases we assumed that the process error is fully correlated radially. The cosine function implies that the error is correlated decreasing with axial distance, while the delta function reflects totally uncorrelated spacial error. This choice can be defended by the small ratio of radius to length ($r_o^2/L^2 < 10^{-2}$). The constants S_o^2 and T_{mod}^2 (the mean-square amplitude of the errors) can be chosen reasonably close from physical considerations as a first rough guess and subsequently be improved during the course of the experiments (experimental tuning). It has to be stressed, however, that even without tuning all our constants chosen with some physical background gave acceptable performance at the first instant, and the subsequent tuning brought only marginal improvements. Note that T_{mod}^2 could be a function of time. This would be appropriate, for instance, if the parameters were time-varying, whereas the model parameters are kept constant, and thus the modelling error becomes a function of time. If we assume time independence and no spatial correlation of the measurement error, we can define

$$Q_{ijkl}(n\Delta t)_{dis} = \frac{1}{T_{meas}^2} \delta_{ij} \delta_{kl} \tag{3.64}$$

where δ_{ij} and δ_{kl} are Kronecker-symbols and T_{meas}^2 represents the variance of the measurement error. These assumptions are certainly reasonable, since the error of an individual sensor should in general not depend on time nor on its location. If, however, we have knowledge about the individual quality of a particular sensor, it would be easy to include this knowledge by defining an individual weighting factor $T_{meas,ik}^2$. Since in our example all sensors were of identical type and, moreover, their small differences were largely compensated by careful calibration, equation (3.64) is well justified.

Let us now discuss the weighting functions for the controller as stated in the performance index I_C. The choice of these functions is not as obvious as in the case of the filter. The situation is further complicated because in most practical cases the controls and often also the state variables are constrained. This means that the weights of the optimal linear-quadratic feedback controller have to be chosen so that the control variables remain unconstrained. In practice, it was found advantageous to select the controller weighting functions by a trial and error procedure in such a way that the controller did not saturate, but for the maximum deviation occuring the full control range is used. This means that the control action is as strong as possible and the deviation is corrected along the fastest trajectory.

Recognizing that off-diagonal elements in $\underline{\Gamma}_u$ bear no physical significance and that each heater section is almost identical, we define

$$\underline{\Gamma}_u = k_u^{-1} \underline{I} \tag{3.65}$$

where $\underline{I}$ is the identity matrix and k_u is chosen by trial and error such that the maximum permissible power output of the controllers is not exceeded with the largest deviation occuring (worst case tuning). It was found that the controller Riccati equations tend to their steady-state values very fast. This suggested that it is reasonable to let t_f approach infinity and use the steady-state solution. In this case γ_f has to vanish. For the state variable weighting function, we define

$$\gamma_d(r,r',z,z',t) = \gamma^r(r,r')\,\gamma^z(z,z') \tag{3.66}$$

Clearly we want to steer our system as close to the desired state as possible with no axial preference. Furthermore there is no real physical significance of correlating the deviation at location (r,z) with the deviation at (r',z'). Thus we define

$$\gamma^r(r,r') = r\delta(r-r') \ , \tag{3.67}$$

$$\gamma^z(r,z') = \delta(z-z') \ . \tag{3.68}$$

It is necessary to point out that the factor r in (3.67) was introduced not only for mathematical convenience. It means that we give more weight to the control closer to the outer surface, which reflects the fact that the filter receives only surface measurements. Thus we are also more certain about estimates closer to the surface. Moreover, in actual ingot heating we are most concerned about surface temperature control in order to avoid surface melting, whereas the internal temperature distribution is only of interest at the final time when the ingots are to be drawn. It is also obvious that in order to minimize soaking time (thereby maximizing pit throughout), the heat flux must be kept at a maximum, which means that we have to keep the surface temperature as close to the melting temperature as possible. All of these reasons suggest then that we should increase the weighting function radially.

(vii) Experimental Results

More than 100 experiments were carried out to investigate open and closed-loop performance of the filter controller system. The relevant parameters for the experiments are given in Tables 3.4 and 3.5 respectively. They include the heating program u_1, u_2, u_3 (0 V = no heating, 10 V = maximum power) for the open-loop experiments and the values u_i used as set points for the closed-loop experiments. The type of filter and controller used is defined as follows:

OA = optimal filter with axial correlation given by (3.60), (3.62), (3.44);
SA = suboptimal filter with axial correlation given by (3.60), (3.62), (3.55);
SN = suboptimal filter with no axial correlation given by (3.61), (3.63), (3.55);
OC = optimal feedback controller given by (3.48);
SC = suboptimal feedback controller given by (3.56);
FS = fixed step manual control without feedback.

In order to present the large number of experiments in condensed form and to facilitate comparison, a statistical analysis was performed using the following definitions:

$$\varepsilon_{ik}(t_n) = T(r_i,z_k,t_n) - T_E(r_i,z_k,t_n) \tag{3.69}$$

$$\varepsilon_F(t_n) = \frac{1}{N_r M_z} \sum_{i=1}^{N_r} \sum_{k=1}^{M_z} |\varepsilon_{ik}(t_n)| \tag{3.70}$$

$$S_F(t_n) = \frac{1}{N_r M_z} \sum_{l=0}^{n} \sum_{i=1}^{N_r} \sum_{k=1}^{M_z} \varepsilon_{ik}^2(t_l) \tag{3.71}$$

$$\bar{\varepsilon}_{ik}(t_n) = T_D(r_i,z_k,t_n) - T(r_i,z_k,t_n) \tag{3.72}$$

$$\varepsilon_C(t_n) = \frac{1}{N_r M_z} \sum_{i=1}^{N_r} \sum_{k=1}^{M_z} |\bar{\varepsilon}_{ik}(t_n)| \tag{3.73}$$

$$S_C(t_n) = \frac{1}{N_r M_z} \sum_{l=0}^{n} \sum_{i=1}^{N_r} \sum_{k=1}^{M_z} \bar{\varepsilon}_{ik}^2(t_l) \tag{3.74}$$

where T is the actual, T_E is the estimated, and T_D is the desired temperature, while t_n is defined as t_n = nt. It is possible to attach some physical significance to these quantities: ε_F represents the average absolute error of the filter per thermocouple in °C at time $t = t_n$, whereas S_F is the sum of the squared error per thermocouple accumulated up to t_n. Thus, S_F may be viewed as a quadratic performance index describing the convergence history of the filter. Similarly ε_C represents the average absolute deviation from the desired state at t_n and S_C in turn is a quadratic measure of controller performance. Since we do not know the actual temperature $T(r_i,z_k,t_n)$, we assumed that T can effectively be replaced by T_M, the measured temperature. An error analysis in several steady-states proved that his assumption is well justified. The average absolute measurement error due to natural noise is typically in the order of $\varepsilon_F(t_n) \approx .1\ ^oC$ per thermocouple and $S_F(t_f)$, the sum of the mean squared error per thermocouple over a period of t_f = 320 s is typically about 2 $^oC^2$. Thus it is obvious from a comparison with the values in Tables 3.4 and 3.5 that the error introduced due to natural noise is in fact negligible.

Table 3.4 shows the sum of the squared error S_F and the absolute error ε_F at t = 0 and t_f = 320 s for various combinations of axial and radial eigenfunctions and heating programs, variation of covariance parameters, types of filters, various noise levels and numbers of sensors used. Since it is not possible to reproduce the initial conditions exactly for every group of experiments, we can only compare two experimental runs by considering also their initial errors $S_F(0)$ and $\varepsilon_F(0)$.

Accuracy of the modal approximation. The most important property of any numerical solution algorithm besides stability is its accuracy. But in order to appreciate the accuracy of the algorithms used in our experiments let us recall the approximations that have been made so far: a linear model for a nonlinear process (especially the thermal conductivity is a function of temperature), the system parameters are estimated, the true state cannot be measured, calibration error of the thermocouples, finite resolution of D/A and A/D converters, finite word length of the computer, heaters modelled with first order dynamics only, heater dynamics are neglected in the control law.

The number of eigenfunctions very much determines the accuracy of the numerical solution. Even though the program provided for different numbers for filter, covariance, and controller Riccati equations, in all experiments we choose $N = N_C = N_R$ and $M = M_C = M_R$. Unfortunately, greater accuracy with more eigenfunctions requires more real-time. Thus, a trade-off is necessary in cases where real-time limitations are crucial. In the experiments of group A the influence of the number of eigenfunctions was investigated. It is evident that the accuracy of the approximation depends very much on the heating program. Whereas for program L = (10., 7., 4.) it would seem sufficient to employ only three axial modes, for program B = (10., 5., 10.) we definitely need more. Another interesting aspect is that for the symmetric profile with program L the addition of the fourth and sixth (anti-symmetric) axial modes does not make much difference, but there is a marked improvement by adding the fifth and seventh (symmetric) axial modes. The reverse is true with program B. This suggests that in special cases it will be advantageous to use more even (or odd) eigenfunctions depending on the symmetry of the profiles to be approximated. Figure 3.5 shows ε_F and S_F versus time for run # 1 and # 7. Both runs show initially the same convergence, but then ε_F of run # 1 increases steadily because with only three axial modes the filter is not able to approximate the profiles well as they become more and more curved. This situation is more clearly seen in Figs. 3.6, 3.7 and 3.8, where Fig. 3.6 shows the typical initial condition for the filter, which was chosen erroneously low in order to demonstrate its convergence properties. Remarkable is the good convergence after only 40 seconds. This means that the filter is able to track the dynamic behaviour of the system closely after a very short initial convergence period.

Influence of the covariance parameters. The influence of the choice of S_o, T_{meas} and T_{mod} is investigated in group B experiments (Table 3.4). The system started from the standard initial condition and was driven to state A = (5., 10., 5.), the filter was given a zero initial condition $(\Theta(r,z,0) = 0$ corresponds to the cooling water temperature as initial temperatures$)$ and all eight outer sensors were used

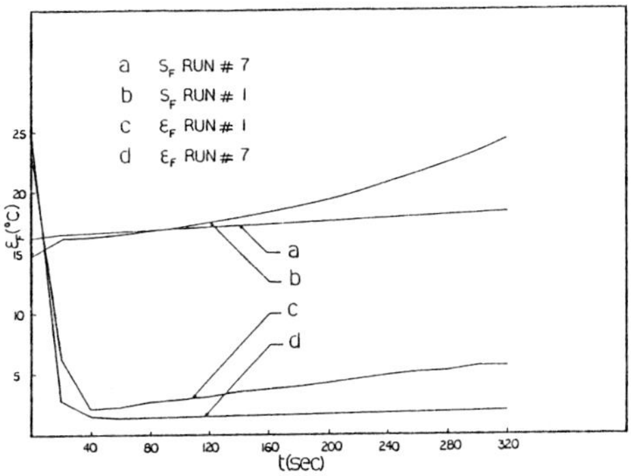

Fig. 3.5. Influence of the number of eigenfunctions.

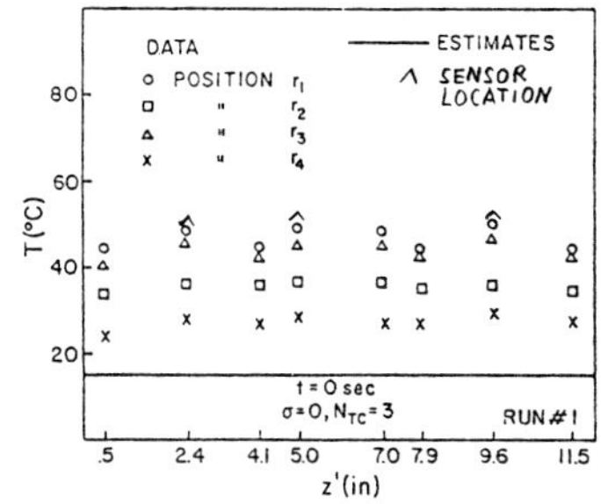

Fig. 3.6. Axial temperature distribution, typical initial condition for filter.

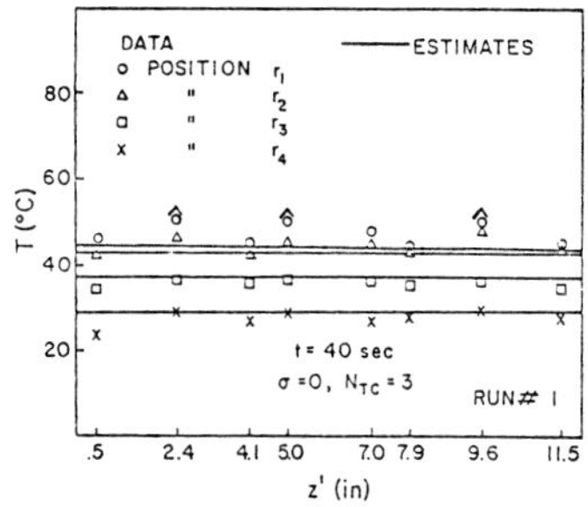

Fig. 3.7. Axial temperature distribution. Filter performance with only 3 axial eigenfunctions after 40 s.

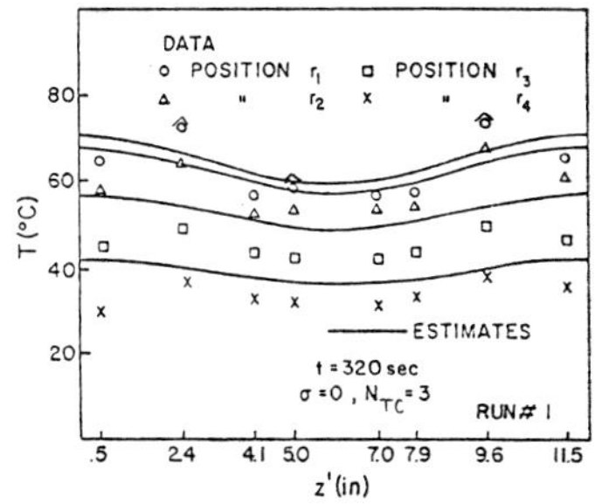

Fig. 3.8. Axial temperature distribution. Filter performance with only 3 axial eigenfunctions after 320 s.

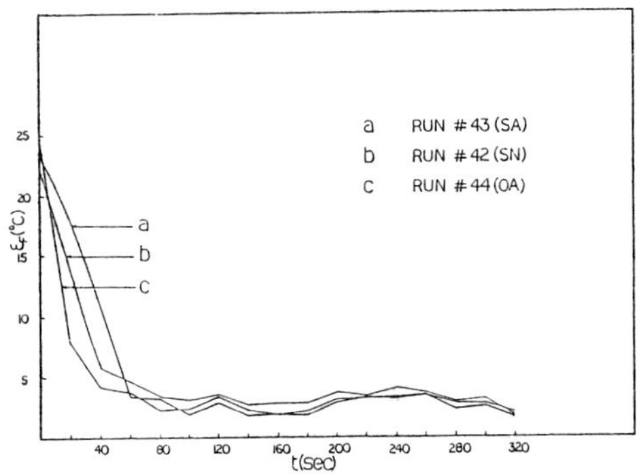

Fig. 3.9. Comparison of filter performance with eight sensors used; OA: full optimal filter; SA: suboptimal filter with axial correlation; SN: suboptimal filter (no axial correlation).

with no noise added. We can learn from Table 3.4 that a) decreasing T_{meas} improves $\varepsilon_F(t_f)$, b) increasing T_{mod} improves both $\varepsilon_F(t_f)$ and $S_F(t_f)$, c) $T_{meas} = 2$ appears to be a critical lower bound for T_{meas} after which the numerical algorithm becomes almost unstable, d) decreasing S_o improves $S_F(t_f)$. To explain this behaviour we view these parameters as factors determining the feedback gain that amplifies the estimation error of the filter. This gain is an increasing function of S_o and T_{mod} and a decreasing function of T_{meas}. Thus the "optimal" weighting parameters will be the highest values of S_o and T_{mod} and the lowest value of T_{meas} that our algorithm can tolerate. Our experimental

TABLE 3.4 Conditions and Summary of Results for Open-Loop Filter Experiments

Group	Figure	Run	Heating Program			No. of Eigenf.		Filter Parameters			Type of Filter	Filter Performance					
		#	u_1(V)	u_2(V)	u_3(V)	N N_C	M M_C	$-S_o$ °C	$-T_{meas}$ °C	$-T_{mod}$ °C/sec		σ	N_{TC}	$S_F(0)$	$S_F(t_f)$	$\varepsilon_F(0)$	$\varepsilon_F(t_f)$
A	3.5	1	10.	5.	10.	1	3	31.6	2.	.02	SA	0.	3	591.5	975.6	23.05	5.79
A		2	"	"	"	1	7	"	"	"	"	"	"	603.7	899.7	23.30	4.37
A		3	"	"	"	2	3	"	"	"	"	"	"	623.2	796.2	23.67	3.93
A		4	"	"	"	2	4	"	"	"	"	"	"	597.1	767.3	23.18	4.00
A		5	"	"	"	2	5	"	"	"	"	"	"	600.1	714.7	23.24	2.79
A		6	"	"	"	2	6	"	"	"	"	"	"	612.0	726.4	23.50	2.80
A,C	3.5,3.10	7	10.	5.	10.	2	7	31.6	2.	.02	SA	0.	3	651.0	738.5	24.35	2.20
A		8	10.	7.	4.	2	3	31.6	2.	.02	SA	0.	3	612.2	723.1	23.48	2.42
A		9	"	"	"	2	4	"	"	"	"	"	"	604.1	709.4	23.33	2.35
A		10	"	"	"	2	5	"	"	"	"	"	"	656.4	762.3	24.37	2.29
A		11	"	"	"	2	6	"	"	"	"	"	"	604.0	692.6	23.16	2.35
A		12	10.	7.	4.	2	7	31.6	2.	.02	SA	0.	3	585.0	664.2	23.00	2.00
B		20	5.	10.	5.	2	7	47.4	5.	.012	SA	0.	8	693.3	1057.2	25.19	5.14
B		22	"	"	"	"	"	47.4	3.	.012	"	"	"	667.6	1397.7	24.72	2.71
B		24	"	"	"	"	"	31.6	3.	.012	"	"	"	652.2	830.7	24.42	2.82
B		26	"	"	"	"	"	31.6	2.	.012	"	"	"	661.6	1201.9	24.76	2.11
B,C	3.11	28	"	"	"	"	"	31.6	2.	.02	"	"	"	629.8	1053.0	23.91	1.74
B		30	5.	10.	5.	2	7	31.6	1.5	.02	SA	0.	8	648.7	26518.5	24.32	1.77
C		31	5.	10.	5.	2	7	31.6	2.	.02	SN	0.	8	588.3	1000.3	23.00	1.40
C		32	5.	10.	5.	2	7	31.6	2.	.02	OA	0.	8	616.4	784.4	23.64	1.80
D		33	5.	10.	5.	2	7	31.6	2.	.02	SN	10.	8	580.7	1149.6	22.83	1.56
D	3.11	34	"	"	"	"	"	"	"	"	SA	10.	8	664.0	1366.8	24.55	1.80
D		35	"	"	"	"	"	"	"	"	OA	10.	8	611.4	888.8	23.53	1.93
D		36	5.	10.	5.	2	7	31.6	2.	.02	SN	20.	8	625.6	1854.0	23.71	2.33
D	3.11,3.12	37	"	"	"	"	"	"	"	"	SA	20.	8	633.2	1818.8	23.80	1.71
D		38	"	"	"	"	"	"	"	"	OA	20.	8	675.8	1226.5	24.87	2.39
C		39	10.	5.	10.	2	7	31.6	2.	.02	SN	0.	8	599.1	1047.2	23.26	1.65
C		40	"	"	"	"	"	"	"	"	SA	0.	8	566.5	970.1	22.53	1.89
C		41	"	"	"	"	"	"	"	"	OA	0.	8	663.0	869.3	24.66	2.08
D	3.9	42	10.	5.	10.	2	7	31.6	2.	.02	SN	10.	8	523.6	1099.5	22.01	1.90
D	3.9	43	"	"	"	"	"	"	"	"	SA	10.	8	606.1	1271.8	23.24	1.86
D	3.9	44	"	"	"	"	"	"	"	"	OA	10.	8	626.0	905.7	23.81	2.00
C		45	5.	10.	5.	2	7	31.6	2.	.02	SN	0.	3	616.6	2521.0	23.62	3.34
C		46	"	"	"	"	"	"	"	"	SA	0.	3	590.4	669.2	23.00	1.97
C		47	"	"	"	"	"	"	"	"	OA	0.	3	599.9	797.0	23.30	1.74
D		48	5.	10.	5.	2	7	31.6	2.	.02	SN	10.	3	606.6	2894.5	23.41	3.70
D		49	"	"	"	"	"	"	"	"	SA	10.	3	658.5	831.9	24.48	1.94
D		50	"	"	"	"	"	"	"	"	OA	10.	3	645.8	900.4	24.23	1.75
C	3.10,3.13	51	10.	5.	10.	2	7	31.6	2.	.02	SN	0.	3	680.0	2861.4	24.86	3.95
C	3.10	52	"	"	"	"	"	"	"	"	OA	0.	3	597.6	757.3	23.35	1.88
D		53	10.	5.	10.	2	7	31.6	2.	.02	SN	10.	3	634.8	3059.0	24.00	3.86
D		54	"	"	"	"	"	"	"	"	SA	10.	3	654.5	840.7	24.41	2.26
D		55	"	"	"	"	"	"	"	"	OA	10.	3	629.8	883.4	23.93	1.99
C		56	5.	10.	5.	2	7	31.6	2.	.02	SN	0.	1	588.1	4523.0	23.03	7.82
C		57	"	"	"	"	"	"	"	"	SA	0.	1	600.9	1211.8	23.35	2.56
C		58	"	"	"	"	"	"	"	"	OA	0.	1	594.4	1442.9	23.20	2.40
D		59	5.	10.	5.	2	7	31.6	2.	.02	SN	10.	1	593.9	4561.9	23.14	7.98
D		60	"	"	"	"	"	"	"	"	SA	10.	1	656.5	1333.7	24.46	3.50
D		61	"	"	"	"	"	"	"	"	OA	10.	1	656.3	1574.1	24.45	3.21
C		62	10.	5.	10.	2	7	31.6	2.	.02	SN	0.	1	632.2	5122.5	23.94	8.62
C		63	"	"	"	"	"	"	"	"	SA	0.	1	651.3	1449.0	24.35	3.50
C		64	"	"	"	"	"	"	"	"	OA	0.	1	621.2	1609.9	23.69	3.21
D		65	10.	5.	10.	2	7	31.6	2.	.02	SN	10.	1	650.0	5270.2	24.30	9.22
D	3.14	66	"	"	"	"	"	"	"	"	SA	10.	1	635.9	1461.6	24.09	4.07
D		67	"	"	"	"	"	"	"	"	OA	10.	1	623.0	1574.9	23.81	4.29

experience with a wide range of physically reasonable parameter showed that the filter was remarkably stable. This may be attributed to the fact that the solution of the matrix Riccati equation resulting from the linear-quadratic optimization problem usually has a flat optimum, i.e. is not very sensitive to the weighting parameters, and furthermore our system is rather well behaved (parabolic p.d.e.).

Comparison of filter performance. Here we investigated the influence of the type of filter (SN = suboptimal, no axial correlation, SA = suboptimal with axial correlation, OA = optimal with axial correlation) without added noise (group C experiments, and with added noise (group D experiments). The system was again started from the standard initial condition, the filter was started from zero and heating program A = (5., 10., 5.) was used. We conclude from Table 3.4 and Figs. 3.9 and 3.10:

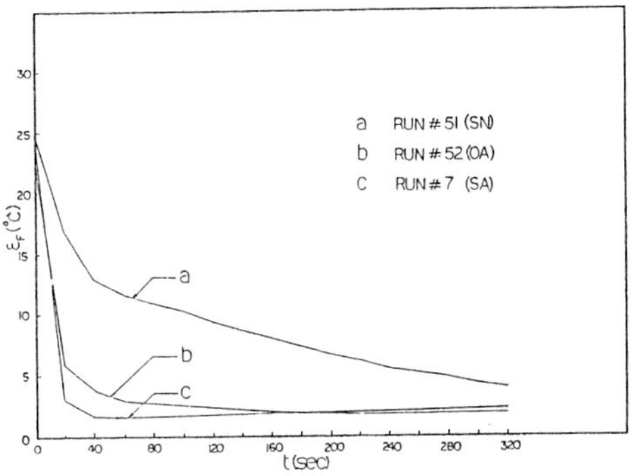

Fig. 3.10. Influence of shape of axial correlation with three sensors used.

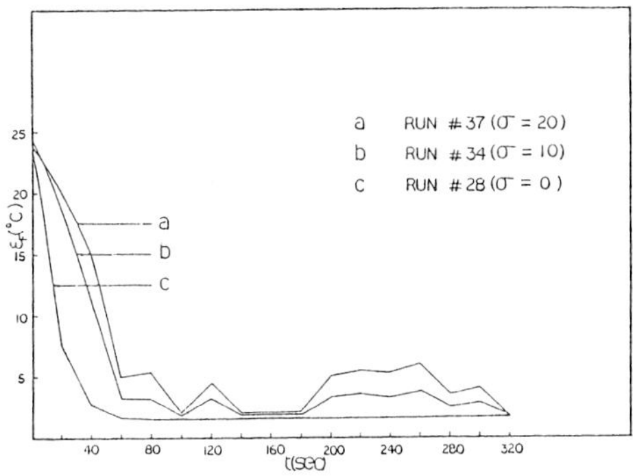

Fig. 3.11. Influence of added noise (σ in ^{O}C), eight sensors used.

a) Filters SA and OA provide very good final tracking as well as good convergence in all cases, even with high noise and only one sensor.
b) Filter SN yields only good estimates with eight sensors, however these have the best values for $\varepsilon_F(t_f)$. Apparently the delta functions for the axial error covariance allow a better adjustment to the actual profiles.
c) It is surprising that there is almost no difference between SA and OA, in fact with 3 and with one sensor the suboptimal filter has better initial convergence $\left(\text{lower } S_F(t_f)\right)$ but a somewhat higher final error

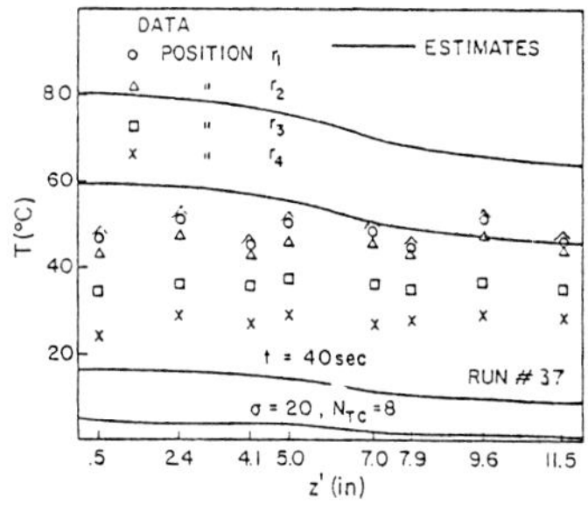

Fig. 3.12a. Axial temperature distribution. Filter performance with added noise at 40 s.

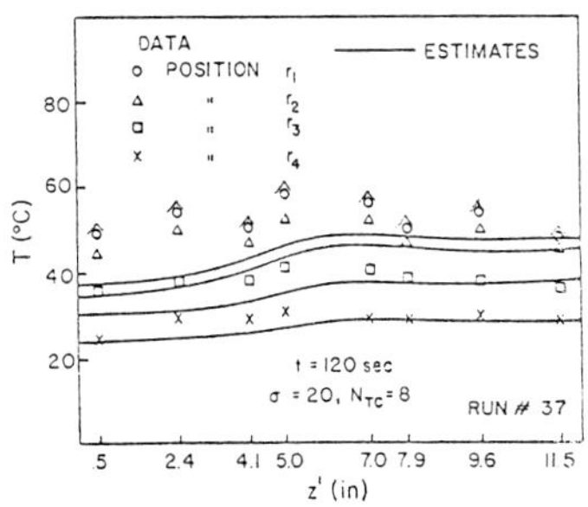

Fig. 3.12b. Axial temperature distribution. Filter performance with added noise at 120 s.

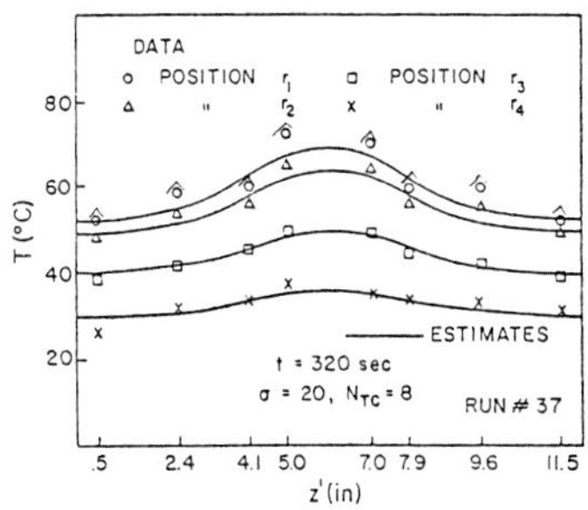

Fig. 3.12c. Axial temperature distribution. Filter performance with added noise at 320 s.

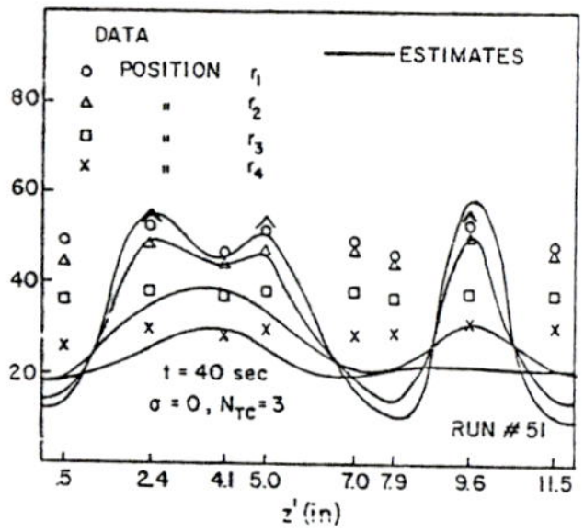

Fig. 3.13a.

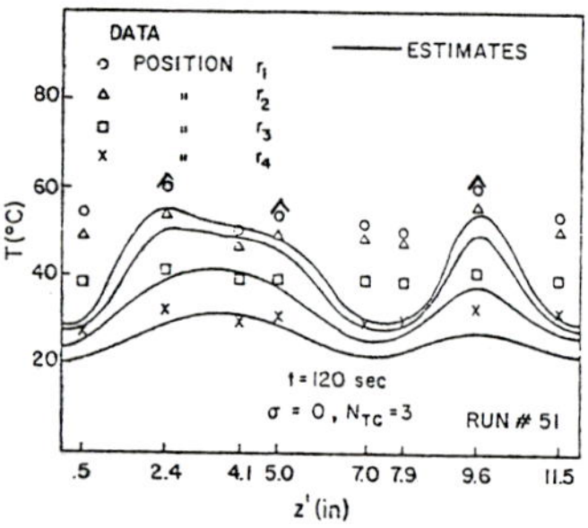

Fig. 3.13b.

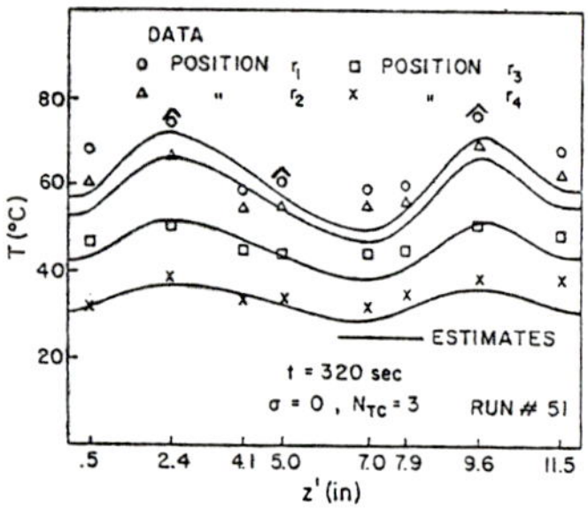

Fig. 3.13c.

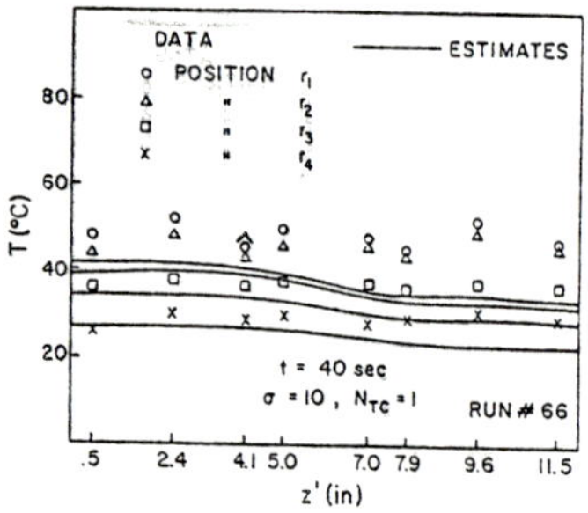

Fig. 3.14a.

$\varepsilon_F(t_f)$, which can be attributed to the approximation of the scalar Riccati solution by an exponential function yielding higher initial feedback gain.

Figure 3.9 shows that there is hardly a difference between the filters if all 8 outer sensors are used, however Fig. 3.10 demonstrates the difference between the uncorrelated and the correlated filters with only 3 sensors. The influence of added zero mean pseudo-random noise with Gaussian distribution is shown in Fig. 3.11 and also in Fig. 3.12, where the filter overshoots considerably due to noise (Fig. 3.12a), while the later profiles (Fig. 3.12b,c) show good convergence. We can state that all filters have surprisingly low sensitivity to noise and that even extreme noise can be tolerated (runs # 36 to # 38).

Influence of number and location of sensors. The influence of the number of sensors was already revealed in Table 3.4 and in Fig. 3.9 to 3.12. It is obvious that with more sensors used the filter will in general produce better estimates. The influence of the actual sensor locations is illustrated best in Fig. 3.13, where the uncorrelated filter (SN) with three sensors is presented. While the approximation is fairly good at locations 2, 4 and 7, where the actual sensors used for the filter are placed (denoted by ^), the estimates remain fairly poor at intermediate locations. Due to the δ-functions in the weights (3.64) this filter has no knowledge about the state at any other point except where the sensors are actually located. Thus the other points have transients only dictated by their initial condition and the heating program with no feedback (open-loop simulation). Here the advantage of the axial error correlation becomes obvious. As a matter of fact, only with this

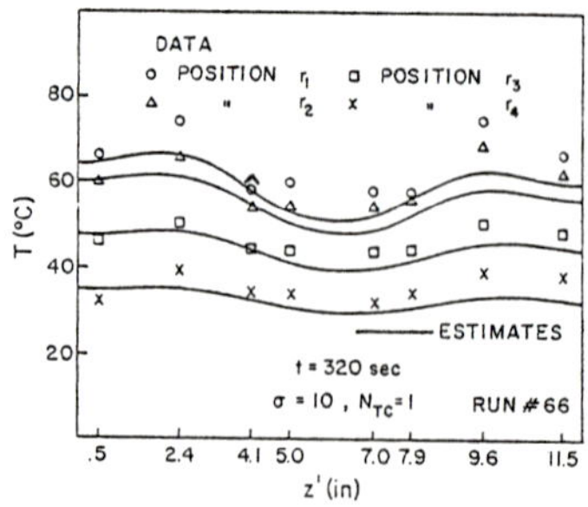

Fig. 3.14b.

TABLE 3.5 Conditions and Summary of Results for Feedback Control Experiments

Figure	Run #	Desired Profile u_1(V)	u_2(V)	u_3(V)	N N_C N_R	M M_C M_R	$\sim S_o$ °C	$\sim T_{meas}$ °C	$\sim T_{mod}$ °C/sec	Type of Filt.	Cont.	Control System Performance σ	N_{TC}	k_u	$S_F(0)$	$S_F(t_f)$	$\varepsilon_F(0)$	$\varepsilon_F(t_f)$	$S_C(0)$	$S_C(t_f)$	$\varepsilon_C(0)$	$\varepsilon_C(t_f)$
	68	6.	6.	6.	-	-	-	-	-	-	FS	-	-	-	-	-	-	-	349.3	3089.9	17.76	6.04
	69	6.	6.	6.	2	7	31.6	2.	.02	SA	SC	0.	8	4.	6.0	53.3	2.24	1.60	323.1	2328.1	16.98	2.58
	70	6.	6.	6.	2	7	31.6	2.	.02	OA	OC	0.	8	4.	11.0	62.1	3.08	1.62	291.8	2030.4	16.74	2.22
	71	4.	7.	4.	-	-	-	-	-	-	FS	-	-	-	-	-	-	-	234.5	2203.2	13.96	5.36
	72	4.	7.	4.	2	7	31.6	2.	.02	SA	SC	0.	8	4.	2.3	40.7	2.39	1.46	252.6	1720.1	14.53	2.15
	73	4.	7.	4.	2	7	31.6	2.	.02	OA	OC	0.	8	4.	7.2	52.7	2.35	1.49	214.7	1540.0	13.37	2.23
3.15	74	7.	4.	7.	-	-	-	-	-	-	FS	-	-	-	-	-	-	-	328.5	3158.8	16.96	7.02
	75	7.	4.	7.	2	7	31.6	2.	.02	SA	SC	0.	8	4.	2.2	53.1	1.23	1.83	366.1	2476.8	17.79	2.55
	76	7.	4.	7.	2	7	31.6	2.	.02	OA	OC	0.	8	4.	4.5	60.9	1.76	1.77	336.9	2302.1	17.22	2.60
	77	7.	5.	3.	-	-	-	-	-	-	FS	-	-	-	-	-	-	-	234.2	2236.7	13.62	5.37
	78	7.	5.	3.	2	7	31.6	2.	.02	SA	SC	0.	8	4.	3.2	45.7	1.57	1.59	242.4	1716.3	13.82	2.24
	79	7.	5.	3.	2	7	31.6	2.	.02	OA	OC	0.	8	4.	6.0	50.7	2.23	1.55	232.2	1603.2	13.26	2.24
	98	7.	4.	7.	2	7	31.6	2.	.02	SA	SC	0.	3	4.	330.6	415.6	27.30	2.21	326.9	2441.9	16.85	3.45
	99	7.	4.	7.	2	7	31.6	2.	.02	SA	SC	10.	3	4.	-319.3	462.4	17.00	2.19	334.7	2257.6	17.15	2.65
3.15	100	7.	4.	7.	2	7	31.6	2.	.02	SA	SC	0.	3	4.	200.6	180.8	9.44	2.22	326.1	2499.2	16.85	3.61
3.16	101	7.	4.	7.	2	7	31.6	2.	.02	SA	SC	0.	1	4.	329.6	527.2	27.29	1.90	326.4	2166.0	16.86	2.52
3.17	102	7.	4.	7.	2	7	31.6	2.	.02	SA	SC	10.	1	4.	328.7	522.2	17.31	1.80	326.6	2204.0	16.84	2.50
3.15	103	7.	4.	7.	2	7	31.6	2.	.02	SA	SC	0.	1	4.	100.9	238.3	9.55	1.50	329.5	2582.3	16.96	3.62
3.16	104	7.	4.	7.	2	7	31.6	2.	.02	SA	SC	10.	1	4.	102.8	230.8	9.63	1.54	330.1	2599.6	17.04	3.61
3.18																						

kind of correlation the filter becomes practically feasible, since in most applications one has to rely on very few sensors with noisy measurements. The dynamic performance of the suboptimal filter with axial correlation (SA), only one sensor and added random noise is shown in Figs. 3.14a, 3.14b.

Analysis of combined filter and controller performance. The influence of the filter performance on the feedback controller under various conditions is demonstrated in Table 3.5. The controller supplied with state estimates by the filter is used to drive the system from the initial state to the desired state. The filter is given an erroneous initial condition, in some experiments artificial noise was added and the influence of the number of sensors used is studied. In two cases optimal and suboptimal algorithms are compared. We can conclude that a) the initial condition given to the filter has a strong influence on the control; b) the number of sensors used and the addition of noise has a noticeable influence on the control; c) but the difference between the optimal and the suboptimal algorithms is not significant. The latter statement was also verified in a number of experiments, where the filter was given a correct initial condition with no added measurement error and eight sensors so that it tracked the temperature profile very closely. No appreciable difference between the optimal and the suboptimal controller was found. In order to judge the relative performance of the feedback controllers, manual fixed step (FS) control was also excerted. It is obvious that a low initial condition speeds up the control due to a higher feedback (the filter pretends a larger deviation than there is in actual case). This fact is demonstrated in Fig. 3.15, where the control $u_2(t)$ for the middle heater is shown. While run # 102 starts with a very high initial control output, run # 104 starts very low and as the filter converges from above, actually increases its control output until the filter has converged. Since the system is slower in run # 104 due to lower controls initially and thus further away from the desired profile, final controls must necessarily be higher than for run # 102. Thus the controller tries to make up later for what it lost initially, and the final difference in the state of the system at t_f is therefore not very large. In the steady state, both curves a and b will approach c, of course. The actual convergence of the filter in run # 102 and # 104 is shown in Fig. 3.16. Notice that $\varepsilon_F(t)$ is the absolute error and thus the direction of the approach is not indicated in this figure. The profiles for these runs are shown in Figs. 3.17 and 3.18. That the difference at 320 sec is not very large demonstrates that even a wrong guess for the initial condition of the filter has actually very little effect on the closed-loop performance of the control system.

Real-time and memory requirements. All experiments were performed using a PDP 11/55 real-time process computer. The full optimal filter equations with seven axial and two radial eigenfunctions were solved in less than 5 s per 20 s time step. The execution time includes the actual sampling of all 32 thermocouples at about 8 ms each, writing data on disk for storage, reading optimal covariance and controller data from disk and displaying data on a terminal screen. Considering that the time constants for real ingots in steel-making processes are typically two orders of magnitude larger, there is ample time for real-time implementation. Memory requirements for the filter and controller algorithms were 24 k, while 32 k were used for the off-line covariance program, which after slight modifications was also used to solve the controller Riccati equations. Execution times for the covariance equations (84 coupled o.d. es.) were typically 3 to 5 hours for 320 s of system time. Thus these equations could certainly never be solved on-line with the available minicomputer, whereas the suboptimal covariance equations required only fraction of a second of real-time per time step.

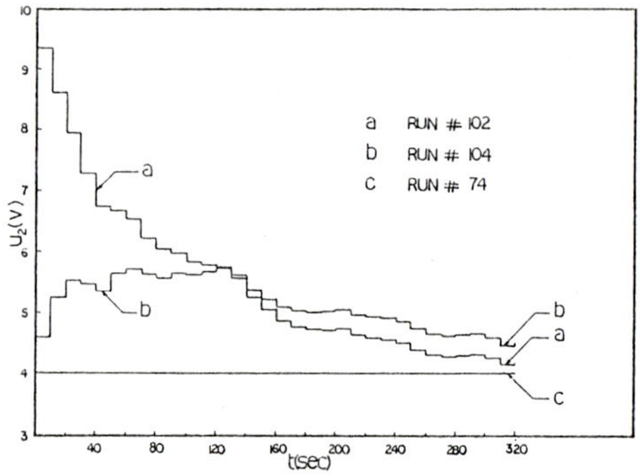

Fig. 3.15. Influence of filter on controller; controller output for middle heater; run # 102: low initial condition for filter; run # 104: high initial condition for filter.

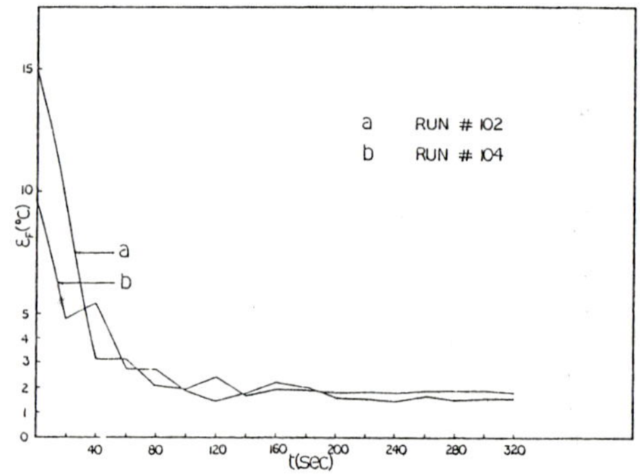

Fig. 3.16. Convergence of filter with controller; absolute error of filter estimates; run # 102: low initial condition for filter; run # 104: high initial condition for filter.

(viii) Conclusions

From the results of our experiments we can draw the following conclusions:

a) On-line filtering and feedback control of a two-dimensional heat conduction system is feasible with reasonable accuracy.
b) Suboptimal algorithms can be used with little loss in performance but great savings in off-line computations and programming effort.
c) If only few sensors are available, it is essential to assume axial error correlation for the initial covariance and the modelling error.
d) The filter algorithm is very robust. The filter is stable even with a poor choice of filter parameters and high noise levels and shows good convergence.
e) The choice of the filter parameters is assisted by physical considerations.
f) The influence of the filter on the controller is noticeable but comparatively small. The performance of the combined filter and controller system is excellent.

Given these results, we hope that the potential industrial users of real-time distributed-parameter control systems feel encouraged to actually implement state estimators and controllers based on distributed-parameter system theory.

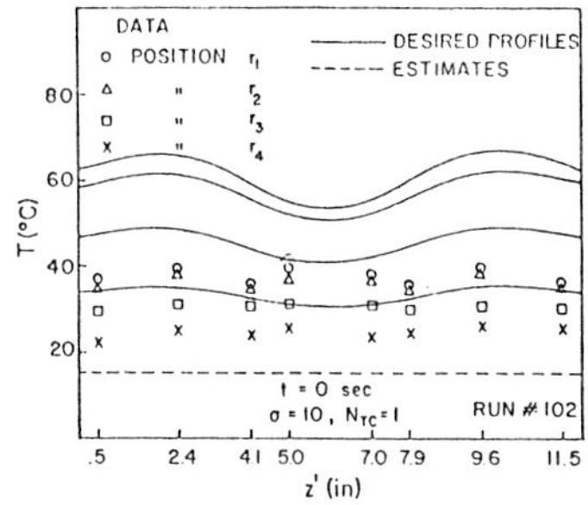

Fig. 3.17a. Combined filter and controller performance; low initial condition for filter.

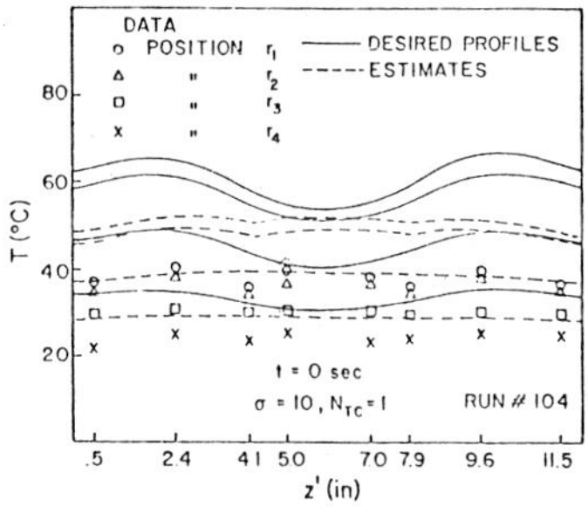

Fig. 3.18a. Combined filter and controller performance; high initial condition for filter.

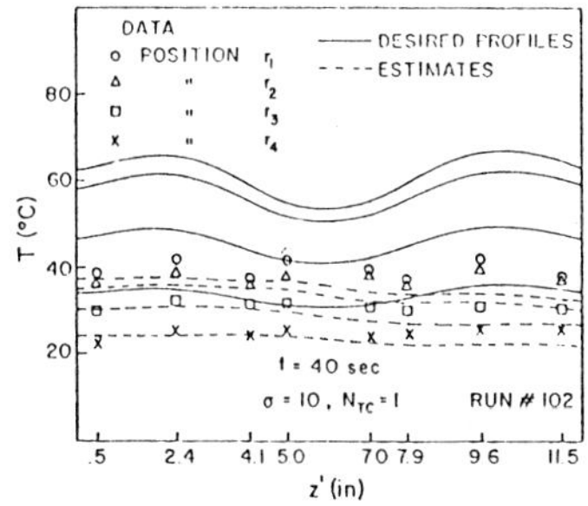

Fig. 3.17b.

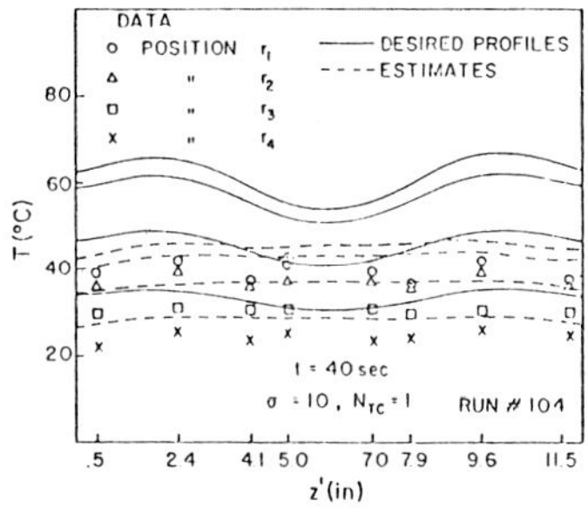

Fig. 3.18b.

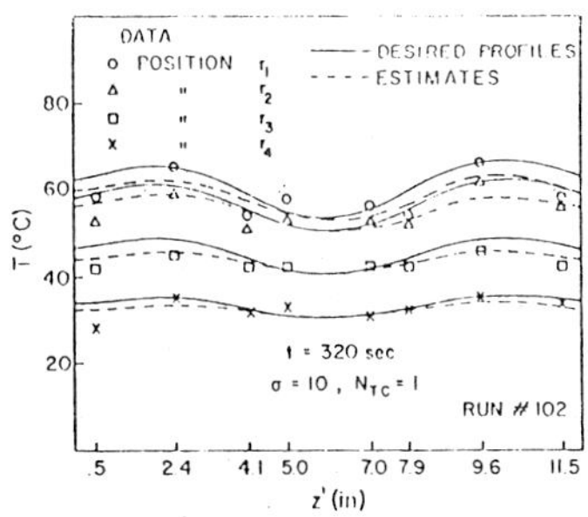

Fig. 3.17c.

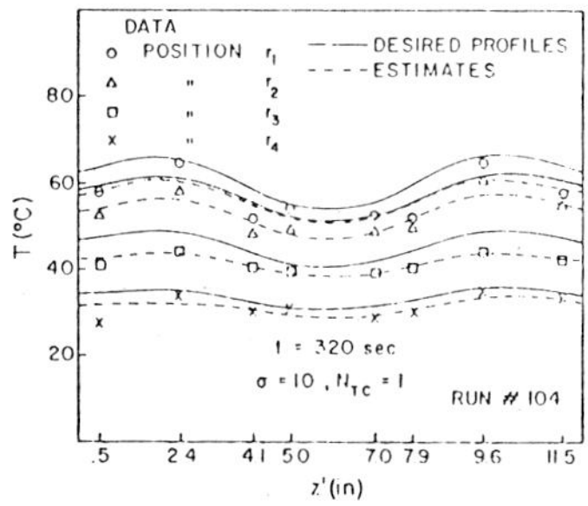

Fig. 3.18c.

4. A MOVING BOUNDARY PROBLEM

Greiss and Ray (1980) report an application of a nonlinear state estimator with discrete time measurements to a laboratory process with moving boundaries. Also an optimal feedback controller was designed, the performance of which in conjunction with the above state estimator was acceptable, but did not seem to offer any special advantages.

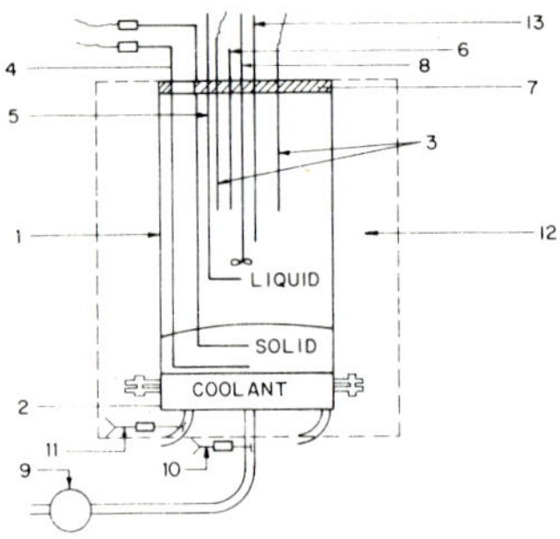

Fig. 4.1. The experimental apparatus. 1. glass cylinder; 2. cooling chamber; 3. heaters; 4,5,10,11. thermocouples; 6. thermometer; 7. cover; 8. stirrer; 9. flow meter; 12. insulation; 13. calibrated rod for solid thickness measurement.

(i) Description of the Process

The apparatus, shown in Fig. 4.1, consists of an insulated glass cylinder with an aluminium heat exchanger attached to its bottom, two heaters attached to the top and a stirrer to ensure uniformity of the liquid temperature. Thermocouples are placed along the cylinder axis at thirteen different positions and at the coolant water inlet and exit of the heat exchanger. The material to be cast (stearic acid in this case) is maintained somewhat above its melting point by the electric heaters and solidified material grows on the cooled bottom of the chamber. To be consistent with industrial metal casting conditions, only the signal from the bottom thermocouple (i.e. at the mold wall) was made available to the state estimator (Greiss, Ray, 1980). The coolant chamber inlet temperature is controlled by adjusting the flow of hot water while flow regulators keep the total flow constant. The actually observed and controlled process in the above apparatus is the solidification problem, characterized by the simple geometric picture shown in Fig. 4.2. The heated stearic acid liquid at temperature T_l is brought in contact with the cooled mold wall where heat is removed at a rate $q_c(t)$. A solid phase begins to form and the solid-liquid boundary, $b'(t)$, as well as the solid temperature profile $T_s(z,t)$ will change with time. A mathematical model of this process may be formulated as follows:

$$\frac{\partial T_s}{\partial t} = \alpha_s \frac{\partial^2 T_s}{\partial z^2} , \qquad 0 \leq z \leq b'(t) , \tag{4.1}$$

$$\rho_s \Lambda \frac{db'}{dt} = k_s \left. \frac{\partial T_s}{\partial z} \right|_{z=b'} - h_l \left(T_l(t) - T_{mp}\right) , \tag{4.2}$$

where the boundary conditions are

$$k_s \frac{\partial T_s}{\partial z} = q_c(t) , \qquad z = 0 , \tag{4.3}$$

$$T_s = T_{mp} , \qquad z = b'(t) . \tag{4.4}$$

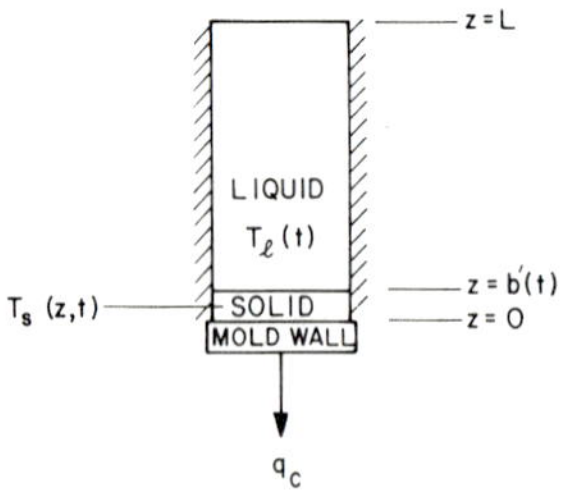

Fig. 4.2. Schematic representation of a solidification problem.

Actually the cooling at the mold wall has been accomplished by water cooling so that the heat removal rate in eq. (4.3) is given by

$$q_c(t) = h\left(T_s\Big|_{z=o} - T_w(t)\right) . \tag{4.5}$$

Thus instead of the boundary condition (4.3) we have

$$T_s - \frac{k_s}{h}\frac{\partial T_s}{\partial z} = T_w(t) , \qquad z = 0 . \tag{4.6}$$

Greiss and Ray (1980), with no explanation, have based the observer and controller synthesis on the condition (4.3) with the supposedly independent $q_c(t)$. The parameters of the experimental system take the following values:

$$T_{mp} = 69\ ^{o}C , \quad \Lambda = 47.54\ \frac{cal}{g} , \quad k_s = 0.000412\ \frac{cal}{cm\ s\ ^{o}C} , \quad L = 25.4\ cm ,$$

$$C_{ps} = 0.40\ \frac{cal}{g\ ^{o}C} , \quad \rho_s = 0.941\ \frac{g}{cm^3} , \quad \alpha_s = \frac{k_s}{\rho_s C_{ps}} = 0.001095\ \frac{cm^2}{s} , \tag{4.7}$$

$$h_1 = 0.00322\ \frac{cal}{cm^2\ s\ ^{o}C} , \quad h = 0.0904\ \frac{cal}{s\ cm^2\ ^{o}C} .$$

(ii) State Estimation

Algorithm. The crucial step in controlling the location of the moving boundary is to develop a method of estimating its position in real time. For the present example the state estimation problem is to provide estimates of $b'(t)$ $\left(\text{and } T_s(z,t)\right)$ based on a discrete time temperature measurement at the mold wall. The solution of this nonlinear problem is so "... formal ... and ... unrigorous ..." (Greiss and Ray, 1980) that it is not presented in this chapter. However a short description of the basic ideas can be useful. Greiss and Ray (1980) have considered it most convenient to convert the moving boundary problem (4.1) to one over a fixed domain by transforming variables. Furthermore, the quadratic criterion (2.13) (applied to the present problem) is minimized approximately (see Greiss, Ray (1980) and for more details Greiss (1978)). The solution of the resulting nonlinear partial and ordinary differential equations was carried out (between the sampling time points) by use of a Galerkin procedure. Instead of the common use of the continuous or quasi-continuous estimation error feedback into the model equations Greiss and Ray (1980) allow rather large sampling intervals and use the error $\{T_s(0,t_k) - \hat{T}_s(0,t_k)\}$ for the updating of $\hat{T}_s(z,t)$ and $\hat{b}'(t)$ at only these sampling points t_k.

Open and closed loop performance. For the actual implementation four terms in the Galerkin expansion procedure were found to be sufficient (Greiss and Ray, 1980). In order to test the performance of the state estimation algorithm, a rather large number of open-loop runs have been made. In Greiss and Ray (1980) "... a representative sample of these ..." was presented. In all of the runs for some unexplained reason the state estimation was begun at t_o = 40 s. The presented open-loop tests can be characterized with the adjoining data.

Run No.	Initial error in $\hat{b}'(0)$	in $\hat{T}(0,0)$	Noise level σ
9	+ 0.3 cm	+ 34 ^{o}C	0
17	+ 0.3 cm	+ 34 ^{o}C	2 ^{o}C
20	+ 0.3 cm	+ 34 ^{o}C	0
21	+ 0.3 cm	+ 34 ^{o}C	0

Figure 4.3 shows the results of Run 17 where random errors are added to the mold surface temperature measurements. These random errors were taken from a Gaussian distribution with 2 ^{o}C standard deviation.

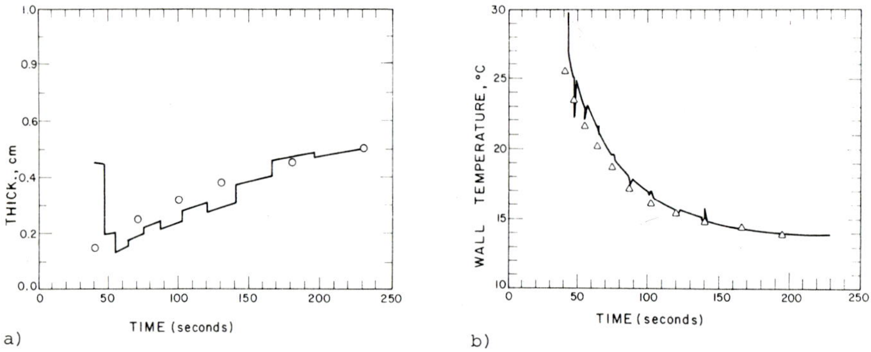

Fig. 4.3. Solid thickness and mold wall temperature for Run 17.

For the closed-loop control of the solid thickness Greiss and Ray (1980) have designed a "linear-quadratic" feedback law, which uses the estimates $\hat{T}_s$ and $\hat{b}'$ for control purposes. Results of two runs are presented on Figs. 4.4 and 4.5. For both runs it was desired to bring the solid to a thickness of 1 cm. The first of these runs had no measurement errors and is shown on Fig. 4.4. When Gaussian random measurement error with $\sigma = 2\ ^{O}C$ was added, the estimator converged much worse, although the actually controlled process remained unchanged! Hence Greiss and Ray (1980) themselves ask "... whether a simple proportional-integral controller would not have done just as well or better for this application."

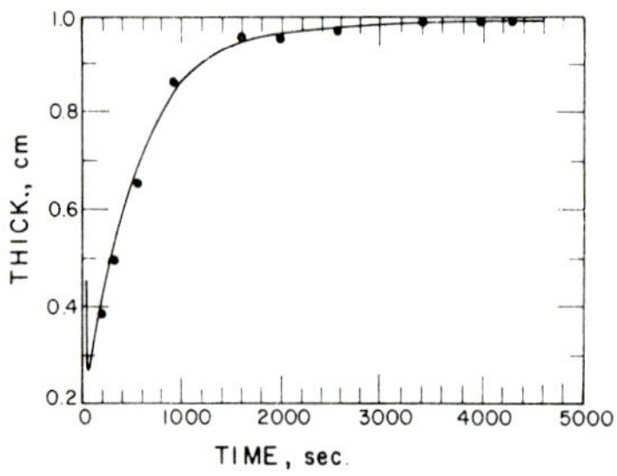

Fig. 4.4. Controlled solid thickness without measurement errors.

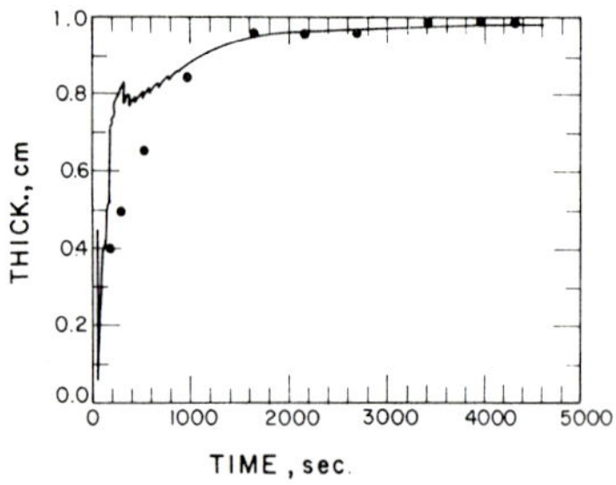

Fig. 4.5. Controlled solid thickness with measurement errors.

(iii) Conclusions

The presented experiments of the state estimation for this nonlinear process seem to be encouraging, even more so if one considers that the on-line computations, carried out on a PDP 11/55 computer, required maximally about 8 % of real time. This, as Greiss and Ray (1980) cautiously put it, "... would seem to make the proposed algorithm attractive for industrial applications." We cannot share this optimism, because a number of questions are still open.

The estimated thickness $\hat{b}'(t)$ was restricted between 0.1 and 1.6 cm in all presented experiments, although the available length L was 25.4 cm (see Fig. 4.2).

For small $\hat{b}'(t)$ the estimation algorithm becomes very stiff leading to numerical difficulties. Furthermore the limit $b'(t) \to 0$ can not be investigated due to the introduction of $\hat{b}^{-1}(t)$ into the differential equation. For large $b'(t)$ the thickness depends very sensitively on the model parameters and on $T_l(t)$, which can be seen already from the steady-state solutions. Due to the not exactly known influence of the liquid phase stirring and the rather complex cooling mechanism the heat transfer coefficients h_l and h can not be determined exactly only from theoretical correlations. They must have been tuned to match the measured experimental data (parameter identification). It might have been easier under these circumstances to directly identify a lumped parameter model, e.g. of the first order (with $b'(t)$ as state variable and $T_l(t)$, $T_w(t)$, and $T_s(0,t)$ as inputs) or of the second order (with $b'(t)$ and $T_s(0,t)$ as state variables and $T_l(t)$ and $T_w(t)$ as inputs). The example presented above was the first real application of distributed-parameter theory to a moving boundary problem. The experiments have demonstrated the applicability of the theory and posed some relevant questions for further experimental work. Thus any potential experimental investigator should feel encouraged to enter this very interesting field.

5. DISCUSSION OF PRACTICAL PROBLEMS

In this chapter we concentrate on the discussion of practical problems, because we feel that this was greatly neglected in the past. The practicing engineer, who is not used to handle large sets of complicated equations routinely, might well loose the courage to carry on in treating a distributed-parameter problem, if he is not assisted by some carefully worked out example. There are three fields of practical relevance that need special attention: the physical nature of the problem given, the careful design of all free parameters, and the practical implementation. These fields are strongly interdependent and should never be viewed in an isolated manner. Very often the problems associated with practical implementation are underestimated and then make the presented solutions unpractical. The most important question that should always be asked first is whether a distributed-parameter model is really required for adequate representation of the studied system. In many cases simple lumped models will suffice for practical purposes. A nice example, where a model of very low order is used to represent a multi-stage extraction column is presented by Gilles (1980). Eitelberg (1979,1980) employed a model reduction technique to a distributed-parameter system with a fourth-order spatial operator (deep-sea mining pipe). It must also be questioned whether a lumped model would not have been adequate for the moving boundary problem described above, since the controlled variable is the boundary itself, i.e. a lumped state variable. Although we have argued in favor of late lumping whenever a distributed-parameter model is used, we have to stress that distributed-parameter models should only be used where they are unavoidable.

(i) Physical Investigations

It cannot be overemphasized that the crucial step in the application of distributed-parameter theory is the modelling phase. Without an adequate model all later steps become more or less meaningless. Of course, modelling is always a compromise between the required accuracy and the computing effort. A discussion of the problems associated with mathematical modelling of a special distributed-parameter system (once-through steam generator) was given by Lausterer, Franke and Eitelberg (1980).

(ii) Design Considerations

The design of an experimental apparatus or of an industrial plant, i.e. the specification of its relevant parameters like size, shape, functions etc., may be greatly assisted by the application of distributed-parameter systems theory. Not only the design of feedback loops, but also the position of sensors and actuators may be determined by theoretical considerations. Of course, theory can never replace practical experience and physical insight. A discussion of the problem of optimal sensor location in the presence of nonstationary noise was given by Morari and O'Dowd (1980). They demonstrated the validity of their results using the experimental apparatus of Lausterer that was described in section 4. Especially the choice of the weighting parameters in the performance index should be based on physical considerations. This will in many cases simplify or even obviate the subsequent tuning procedure. Numerical simulation can assist the process of system design and parameter selection and in general provides improved understanding of the nature and behaviour of the process studied. Thus numerical simulation should in general precede the practical implementation. It is also necessary in most cases for model validation.

(iii) Practical Implementation

In the past, practical implementation of modern control theory, especially of the lengthy algorithms resulting from distributed-parameter theory, was strongly influenced by hardware restrictions. With the introduction of microprocessors, the hardware situation has considerably improved. The distributed-process automation systems* now available offer excellent means to implement modern control concepts even in large scale industrial plants (Borsi and Pavlik, 1980).

The major difficulties in the implementation of distributed-parameter estimation or control algorithms arise in the numerical treatment of the equations. Problems of numerical stability are often encountered. They are in general due to stiffness problems, i.e. the range of the eigenvalues is very large. Sometimes numerical problems also arise because the digital word length is insufficient and thus round-off errors accumulate. The latter problems can be solved by increasing the word length (at the cost of increasing real-time requirements), while the former require special precausions. Numerical instabilities occur, if explicit methods are used and the time step is chosen too large. Reducing the time step may lead to impractical computing times rendering real-time implementation impossible. The numerical difficulties may be avoided using implicit algorithms. For nonlinear systems, these algorithms require time-consuming iterations, however, unless the equations are linearized for each timestep. In this case a linear system of equations has to be solved, which for distributed-parameter-systems usually consists of sparse matrices. Computational requirements may then greatly be reduced by the use of appropriate numerical techniques. An account of implicit numerical methods was given by Eitelberg (1979b).

Another common method to reduce the computational burden is the use of suboptimal algorithms. As has been demonstrated with the two-dimensional example, suboptimal algorithms may greatly simplify the algorithm with little loss in performance. Thus the use of simplifications should always be considered before practical implementation. Some caution is necessary, however, not to loose essential properties of the system in the simplification process. Thus all simplifying assumptions must be carefully checked against the performance of the full system. Physical insight into the problem at hand can greatly assist the selection of suboptimal algorithms.

*Note that distributed here means that there are several processors in the system, i.e. the computing power is distributed over the control system. Another word for distributed would be decentralized in this context.

6. SUMMARY AND OUTLOOK

In this chapter we presented experimental results gained with the practical implementation of distributed-parameter algorithms for a one and a two-dimensional linear as well as a moving boundary example. Even though the experimental evidence stands on a narrow base, it can be concluded that real-time implementation of distributed-parameter algorithms is feasible. The most serious obstacle to a widespread application of distributed-parameter theory is the "mathematical apparatus" required, which might be discouraging to many practicing control engineers. Thus it is imperative to present many well worked-out examples of successful applications of the theory in order to encourage engineers not so familiar with partial differential equations, filtering techniques etc. Unfortunately, the only real application of distributed-parameter theory that was reported since Ray's survey in 1977, is that of Morari and O'Dowd (1980).

Even though the reported applications were all successful, a word of caution is necessary. All processes studied, except perhaps the moving boundary problem, are very good-natured and well-behaved (parabolic p.d.es.). For other types of problems, e.g. tubular chemical reactors (hyperbolic p.d.es.), simulations also indicate the feasibility of practical implementations, however it is expected that more problems will be encountered relating to sensitivity, stability etc. Furthermore many questions still need satisfactory answers:
- the problems associated with boundary control,
- improved modelling and model reduction techniques,
- off-line or on-line identification of distributed-parameter systems.

At present, the theory of distributed-parameter systems is well developed, but practical examples are very few in number. It is hoped that the discussion above will stimulate practical research in this interesting and rewarding field.

7. REFERENCES

Ajinkya, M.B., M. Köhne, H.F. Mäder, and W.H. Ray (1975). The experimental implementation of a distributed-parameter filter. Automatica, 11, pp. 571-575.

Atre, S.R., and S.S. Lamba (1972). Derivation of an Optimal Estimate for Distributed Parameter Systems via Maximum Principle, IEEE Trans. Autom. Contr. AC-17, 388-390.

Brogan, W.L. (1968). Optimal Control Theory Applied to Systems Described by Partial Differential Equations, Adv. in Contr. Syst. V, 221-317.

Butkovskii, A.G. (1969). Distributed Control Systems, American Elsevier, N.Y.

Denn, M.M. (1969). Optimization by Variational Methods, McGraw-Hill, N.Y.

Eitelberg, Ed. (1979a). Modellreduktion linearer zeitinvarianter Systeme durch Minimieren des Gleichungsfehlers, pp. 66-91. HochschulVerlag, Freiburg.

Eitelberg, Ed. (1979b). Numerical Simulation of Stiff Systems with a Diagonal Splitting Method. Mathematics and Computers in Simulation 21, pp. 109-115.

Eitelberg, Ed. (1980). Modellreduktion für ein Tiefseeförderrohr. Regelungstechnik, 28, pp. 409-419.

Greiss, F.K. (1976). MS Thesis. State University of New York at Buffalo.

Greiss, F.K., and W.H. Ray (1976). The application of distributed-parameter state estimation theory to a metallurgical casting operation. Proceedings International Symposium on New Trends in Systems Theory. Paris. Springer, Berlin.

Greiss, F.K. (1978). PhD Thesis. University of Wisconsin. Madison.

Greiss, F.K., and W.H. Ray (1980). Stochastic Control of Processes Having Moving Boundaries - An Experimental Study. Automatica, 16, pp. 157-166.

Gilles, E.D. (1980). Modelling - a Measurement Aid. Process Automation, 1, pp. 14-23.

Johnson, G.W. (1969). A Deterministic Theory of Estimation and Control, IEEE Trans. Autom. Control AC-14, 380-384.

Köhne, M. (1976). Theorie der Zustandsbeobachter mit verteilten Parametern. Regelungstechnik, 24, 277-282, 314-316.

Köhne, M. (1977). Zustandsbeobachter für Systeme mit verteilten Parametern-Theorie und Anwendung. Fortschritt-Berichte der VDI, Reihe 8, Nr. 26.

Koda, M. (1979). Finite difference implementation of distributed-parameter filters. Automatica, 15, 687-692.

Lausterer, G.K. (1977). Ph.D. Thesis. State University of New York, Bufallo.

Lausterer, G.K. W.H. Ray, and H.R. Martens (1978). Real time distributed-parameter state estimation applied to a two dimensional heated ingot. Automatica, 14, 335-344.

Lausterer, G.K., and W.H. Ray (1979). Distributed-parameter state estimation and optimal feedback control - an experimental study in two space dimensions. IEEE Trans. Autom. Control AC-24, 179-190.

Lausterer, G.K., J. Franke, and Ed. Eitelberg (1980). Mathematical Modelling of a Steam Generator. Digital Computer Applications to Process Control. Proc. 6th IFAC/IFIP International Conference, Dusseldorf.

Lin, Y.A., and L. Lapidus (1976). Observer Theory for Distributed Parameter Systems. Int. J. Syst. Sci., 7, 731-742.

Lions, J.L. (1971). Optimal Control of Systems Governed by Partial Differential Equations, Springer-Verlag, N.Y.

Mäder, H.F. (1975a). Zeitoptimale Steuerung und modale Regelung eines technisch realisierten Wärmeleitsystems. Doctoral dissertation, Universität Stutgart.

Mäder, H.F. (1975b). Experimenteller Aufbau einer Temperatur-Profilregelung. Regelungstechnische Praxis, 17, pp. 247-251.

Mäder, H.F. (1976). Modell und modale Regelung eines technisch realisierten Wärmeleitsystems. Regelungstechnik, 24, pp. 347-354.

Meditch. J.S. (1971). Least-Squares Filtering and Smoothing for Linear Distributed Parameter Systems. Automatica, 7, 315-322.

Morari, M., and M.J. O'Dowd (1980). Optimal Sensor Location in the Presence of Nonstationary Noise. Automatica, 16, pp. 463-480.

Orner, P.A., and A.M. Foster (1971). A Design Procedure for a Class of Distributed Parameter Control Systems. Trans. ASME. J.Dyn. Syst., Measrmts. Control, 93, 86-93.

Ray, W.H. andD.G. Lainiotis, Eds., (1978). Identification, Estimation, and Control of Distributed Parameter Systems. New York: Dekker.

Ray, W.H., (1977). Some recent applications of distributed parameter control theory - A survey, in Proc. 2nd IFAC Symp. Distributed Parameter Contr. Theory, Coventry.

Ray, W.H. (1975). Distributed parameter state estimation algorithms and applications - a survey. Proceedings 6th World Congress, IFAC, Boston.

Ray, W.H. (1978). Applications of distributed parameter systems theory. Automatica, 14, 281.

Ray, W.H., F.K. Greiss, and G.K. Lausterer (1979). Some Potential Applications of Distributed Parameter Systems Theory to Metallurgical Heating and Casting Operations. Metallurgical Transactions 3, 10B, 533-537.

Robinson, A.C. (1971). A Survey of Optimal Control of Distributed-Parameter Systems. Automatica, 7, 371.

Sage, A.P. (1968). Optimum Systems Control, Prentice-Hall, Englewood Cliffs, N.J.

Sakawa, Y. (1972). Optimal Filtering in Linear Distributed-Parameter Systems. Int. J. Control, 16, 115-127.

Tzafestas, S.G., and J.M. Nightingale (1968a). Concerning Optimal Filtering Theory of Linear Distributed-Parameter Systems. Proc. IEE, 115, 1737-1742.

Tzafestas, S.G., and J.M. Nightingale (1968b). Optimal Filtering, Smoothing and Prediction in Linear Distributed Parameter Systems. Proc. IEE, 115, 1207-1212.

Wang, P.K.C., and F. Tung (1964). Optimum Control of Distributed-Parameter Systems. Transact. ASME, J. Basic Eng., 78, 67-79.

Wang, P.K.C. (1964). Control of Distributed Parameter Systems. Advances in Control Systems I, 75-172.

CHAPTER 12

DISTRIBUTED PARAMETER IDENTIFICATION IN GEOPHYSICS — PETROLEUM RESERVOIRS AND AQUIFERS

J. H. Seinfeld and C. Kravaris

Department of Chemical Engineering, California Institute of Technology, Pasadena, California 91125, USA

INTRODUCTION

The classic identification problem arising in the modeling of petroleum reservoirs and subsurface aquifers is to estimate the parameter $\alpha(x,y)$ in the parabolic partial differential equation,

$$\frac{\partial u}{\partial t} = \frac{\partial}{\partial x}\left(\alpha(x,y)\frac{\partial u}{\partial x}\right) + \frac{\partial}{\partial y}\left(\alpha(x,y)\frac{\partial u}{\partial y}\right) + q(x,y,t) \tag{1}$$

based on noisy measurements of $u(x,y,t)$, $u^{obs}(x_i,y_i,t)$, at a set of discrete spatial locations, $i = 1,2,\ldots, m$. In such applications the dependent variable u represents pressure, q accounts for the withdrawal or injection of fluid in the region, and α is the transmissivity that determines the ease with which fluid flows in the reservoir. The initial condition to (1) is $u(x,y,0) = u_o(x,y)$, and a typical boundary condition is no fluid flow across the boundary of the region, i.e. $\partial u/\partial\omega = 0$, where ω represents the normal to the boundary. This identification problem is the subject of this chapter.

Consider first the problem of solving (1) given $\alpha(x,y)$. It is well known that there exists an operator A such that the solution of (1) can be expressed as $u = A\alpha$. It is also known that (i) the solution is unique, i.e. A is unique, and (ii) the solution is stable with respect to small changes in α, i.e. A is continuous. The identification problem is the inverse of that just stated, namely given data on u, determine α. For this reason, identification problems of this sort are frequently referred to as *inverse problems.* In this chapter we will refer to the basic problem of estimating $\alpha(x,y)$ as the identification or inverse problem.

Assume that u is known and that the desired solution α belongs to a function space F, that, at present, need not be specified. Clearly, for all $u \in AF$, $A\alpha = u$ will have at least one solution. It is natural to inquire: (i) Does A have a unique inverse, and (ii) Is the (unique or not) inverse operator A^{-1} continuous? Alternatively stated, will small changes in u result in arbitrarily small changes in α? If the answer to both questions is affirmative, then our problem is *well-posed* for all $u \in AF$. It is worth stressing that for the inverse problem to be well-posed it is not enough to insure the uniqueness of α. In fact, the crucial property will turn out to be the stability of the solution with respect to the observation data. We will now show that, even when the solution for α is guaranteed to be unique, the inverse problem is generally an unstable one.

Consider for simplicity the steady state, homogeneous form of (1),

$$\frac{\partial}{\partial x}\left(\alpha(x,y)\,\frac{\partial u}{\partial x}\right) + \frac{\partial}{\partial y}\left(\alpha(x,y)\,\frac{\partial u}{\partial y}\right) = 0 \tag{2}$$

which can be considered as a first-order, linear partial differential equation for $\alpha(x,y)$. The solution of (2) can be obtained easily by choosing an orthogonal coordinate system (s,ψ), such that (2) becomes (Emsellem and de Marsily, 1971)

$$\frac{\partial}{\partial s}\left(\alpha(s,\psi)\,\frac{du}{ds}\right) = 0 \tag{3}$$

The solution of (3) is

$$\alpha(s,\psi) = \frac{c(\psi)}{du/ds} \tag{4}$$

where $c(\psi)$ is an arbitrary function of ψ. If α is known along a line Γ crossing the streamlines of the flow, then $c(\psi)$ can be determined. Because, in general, data on α are not available along such a line, $c(\psi)$ is indeterminate and (4) is non-unique, although if such data are prescribed, uniqueness is obtained.

Now consider the question of continuous dependence of the solution α on u, one that reduces to that of the continuity of the operator,

$$K[\cdot] = \frac{1}{\frac{d}{ds}[\cdot]}$$

We will now show that this operator defined on the metric space F with metric, $\rho_F(u_1,u_2) = \sup_{s\in\mathscr{R}}|u_1(s)-u_2(s)|$, is not continuous. Consider the function

$$u_\delta(s) = \begin{cases} s\cos\left(\frac{\delta}{s}\right) & s \neq 0 \\ 0 & s = 0 \end{cases} \tag{5}$$

$u_\delta(s)$ possesses the following properties: (i) It is continuous in $\mathscr{R}$, (ii) for $\delta = 0$, $u_\delta(s) = s$, (iii) $|s - u_\delta(s)| < \delta$ for all $s \in \mathscr{R}$ and $\delta > 0$. Hence $u_\delta(s)$ belongs to the δ-neighborhood of the function $\hat{u}(s) = s$, in the sense of the topology induced by the metric of F. Now, $K[\hat{u}(s)] = 1$ and

$$K[u_\delta(s)] = \left[\cos\left(\frac{\delta}{s}\right) + \frac{\delta}{s}\sin\left(\frac{\delta}{s}\right)\right]^{-1} \tag{6}$$

The right hand side of (6) becomes infinite for $s = \delta/\xi_n$, where ξ_n is a root of $\xi + \cot\xi = 0$. Thus, for all $\delta > 0$ we have found a $u_\delta \in N_\delta(\hat{u})$ such that $K[u_\delta(s)]$ does not belong to any neighborhood of $K[\hat{u}(s)] = 1$, and consequently the operator K is not continuous. The inverse problem of determining α given u is therefore *ill-posed*. A similar demonstration of ill-posedness of the basic problem has been given by Yakowitz and Duckstein (1980).

The above example was selected only for its expository convenience. The basic problem is that α depends on u only through its derivatives, and it is well known that given any function, a neighboring function can be selected that has its derivative arbitrarily far from that of the given function. This ill-posedness manifests itself even in numerical approximation of derivatives. Even in the hypothetical case in which the data are exact, the solution of the identification problem may nevertheless be nonunique because of the lack of complete information about u. Both nonuniqueness and instability are therefore present, and they manifest themselves by similar symptoms.

The essential difficulty associated with estimating α is to alleviate the inherent ill-posedness of the problem. Consequently, this chapter is devoted to a study of the identification problem with special attention to its ill-posedness or ill-conditioning. We begin with a summary of prior literature on the problem and proceed to describe the basic approaches that are available. Because of our focus on the ill-posed nature of the current problem, we then present a brief review of methods for the solution of general ill-posed problems. Finally, drawing on the theory of solution of ill-posed problems, we propose a new approach to the current problem that offers promise from both theoretical and computational points of view.

SURVEY OF LITERATURE IN RESERVOIR AND AQUIFER IDENTIFICATION

The identification problem of interest has been defined in the previous section, namely to estimate $\alpha(x,y)$ in (1) based on spatially discrete, noisy measurements of $u(x,y,t)$, $u^{obs}(x_i,y_i,t)$, $i = 1,2,\ldots, m$. It appears that two principal lines of attack have been followed with regard to this problem. One consists of regarding (1) as a first-order partial differential equation in α, where u and its spatial and temporal derivatives are assumed to be known over the entire region based on the available data. Several authors have referred to this as the *direct approach*. The uniqueness of the equation for α can be insured by prescribing the value of α along a curve Γ that cuts all the streamlines of the flow or by prescribing the flow, rather than α, at every streamline. With either condition, the problem reduces simply to an initial-value problem for α. The second main line of attack, often referred to as the *indirect approach*, consists of a trial and error procedure that seeks to improve an existing estimate of α in an iterative manner until the predicted u is sufficiently close to the observed u. Specifically, the identification problem is formulated as the minimization of a suitable measure of the difference between observed and predicted $u(x_i,y_i,t)$. The so-called indirect approach seems to offer several advantages over the direct approach in that it is far less demanding in terms of the necessary observation data, specifically, the spatial and temporal derivatives of u over the region need not be known, only the temporal histories of u at the observation locations. The indirect approach is, in fact, the line of attack of most distributed parameter identification problems (Kubrusly, 1977), and is the approach on which we focus our attention later in this chapter.

Table 1 presents a list of many of the prior studies on the current identification problem. Upon a survey of the literature it becomes apparent that two more or less independent lines of research on this problem have existed, one associated with petroleum reservoirs and one directed toward subsurface hydrological systems, although from time to time one finds cross references between the two bodies of work. The papers listed in the table are divided into three major topics, direct methods, indirect methods, and those devoted to specific consideration of the ill-posed nature of the problem and the statistical aspects of the parameter estimation. These two aspects are treated together because a great deal of insight into the ill-conditioning of the problem is available from a statistical analysis of the estimates. In the papers adopting the indirect approach a large variety of minimization methods have been employed. Because such methods have received extensive coverage in several textbooks and review papers they need not be reviewed here (Eykhoff, 1974; Seinfeld and Lapidus, 1974; Beck and Arnold, 1977; Kubrusly, 1977). In the next section methods for solution of the identification problem are discussed, with emphasis on dealing with the ill-conditioning.

TABLE 1 Available Literature on the Estimation of Transmissivity $\alpha(x,y)$ in

$$\frac{\partial u}{\partial t} = \frac{\partial}{\partial x}\left(\alpha \frac{\partial u}{\partial x}\right) + \frac{\partial}{\partial y}\left(\alpha \frac{\partial u}{\partial y}\right) + q$$

Reference	Approach
Stallman (1963) Sammel (1964) Nelson (1968) Nelson and McCollum (1969) Deininger (1969) Kleinecke (1971) Frind and Pinder (1973) Sagar et al. (1975) Yakowitz and Noren (1976)	*Direct approach* in which (1) is regarded as a first-order p.d.e. in $\alpha(x,y)$ where u and its spatial and temporal derivatives are assumed known over the entire region. Later papers have employed finite-element approximations for u and α, e.g. Frind and Pinder (1973)
Jacquard and Jain (1965) Jahns (1966) Haimes et al. (1968) Vermuri and Karplus (1969) Vemuri et al. (1969) Coats et al. (1970) Slater and Durrer (1971) Yeh and Tauxe (1971ab) Chen and Seinfeld (1972) Marino and Yeh (1973) Carter et al. (1974) Chen et al. (1974) Chavent et al. (1975) Distefano and Rath (1975) Wasserman et al. (1975) Chang and Yeh (1976) Gavalas et al. (1976) Cooley (1977, 1979) Shah et al. (1978) Seinfeld and Chen (1978)	*Indirect approach* in which $\alpha(x,y)$ is determined to minimize a performance index such as $J = \int_0^T \sum_{i=1}^m [u^{obs}(x_i,y_i,t) - u(x_i,y_i,t)]^2\, dt$ subject to (1). Many of the papers employ *zonation* in which the region is divided into N subregions (zones) having uniform α. Minimization methods used include direct gradient methods, steepest descent algorithms derived from optimal control theory, and linear programming.
Emsellem and de Marsily (1971) Lovell et al. (1972) Neuman (1973,1980ab) Gavalas et al. (1976) Yeh and Yoon (1976) Shah et al. (1978) Neuman and Yakowitz (1979) Yakowitz and Duckstein (1980)	Treatments of the ill-posed nature of the problem based on concepts of Bayesian estimation and incorporation of prior information about α into the problem. Statistical analysis used to determine optimum number of parameters, e.g. Yeh and Yoon (1976) and Shah et al. (1978).

METHODS FOR SOLUTION OF THE IDENTIFICATION PROBLEM

The solution of the identification problem can be divided into two steps: (i) formulation of the problem, including the performance index and the manner of representing $\alpha(x,y)$ and (ii) development of a method of actually carrying out the estimation. As noted above, the key difficulty is dealing with the ill-conditioning of the basic identification problem, and that aspect must generally be faced when formulating the problem. As noted, methods of actual parameter estimation have been discussed widely and need not be reviewed here. Therefore, in this section we focus on methods for solution of the identification problem that specifically address the matter of ill-conditioning.

The only way to improve the ill-conditioning of the identification problem is to incorporate additional a priori information into the problem beyond that available in the model (1) and the observational data. Such a priori information may take the form of assumptions concerning the nature of α, such as its spatial variation or an assumed functional form to which α is to adhere, or the a priori information may be introduced through terms added to the identification performance index. All of these approaches act to restrict the space of functions within which the estimate of α may lie. Thus, the essence of how one alleviates the ill-conditioning of the identification problem is how one incorporates a priori information into it.

Zonation

Since (1) is frequently solved numerically on a grid lying over the region, an obvious first approach is to assume that $\alpha(x,y)$ takes on different values in each grid cell. The number of such grid cells for even a modest sized reservoir can be quite large (the order of 10^3), leading to an equally large number of unknown parameters. It is well known that when one attempts to estimate such a large number of parameters, even though the optimization routine may lead to an "optimal" set of estimates, the accuracy associated with the estimates is poor (Gavalas et al., 1976; Shah et al., 1978).

The use of an array of grid cells to characterize the spatial heterogeneity of a reservoir is, of course, a computational convenience. That the spatial heterogeneity of the parameters should conform to the same grid is also a computational convenience. When this approach is used, however, not only do the estimated parameters lack uniqueness, but the estimated distributions do not preserve the smoothness features inherent in the geologists' concept of the reservoir. A common way to alleviate the problem of nonuniqueness is what is sometimes referred to as *zonation,* that is simply requiring that the parameters be uniform over regions of the reservoir that contain multiple numbers of grid cells (Jacquard and Jain, 1965; Jahns, 1966; Coats et al., 1970; Carter et al., 1974; Gavalas et al., 1976; Cooley, 1977, 1979)

When prior information about the transmissivities is not taken into account, the zonation approach makes intuitive sense, and, if the number N of zones is sufficiently small, should lead to a well-behaved solution. It is important to note, however, that when N is small, the modeling error becomes large since the true α distribution cannot be represented even if the solution is well-behaved. The question is how large N can become before ill-conditioning sets in, and what should be the optimum size and shape of each zone. The determination of an optimal zoning pattern was first addressed by Emsellem and de Marsily (1971), who proposed solving a sequence of inverse problems starting with a small number of zones and gradually increasing the number until there is no improvement in model fit. Distefano and Rath (1975) adopted a criterion for deciding on the optimal number of zones based on visual observation of the spatial behavior of the estimated α field. Yeh and Yoon (1976) were apparently the first to introduce statistical notions into the stepwise zoning process. They used the variance of the estimation error as a criterion for deciding whether a particular zone should be subdivided or kept intact at each step. Shah et al. (1978) developed a detailed theoretical basis for the selection of an optimum number of zones. The trace of the covariance matrix used as a measure of the over-all accuracy of estimation was studied as a function of N, and a procedure for selecting the optimum parameterization was developed. A brief summary of that analysis is presented later in this section.

Neuman (1973) suggested that incorporating prior information about the transmissivities directly into the statement of the identification problem could alleviate ill-conditioning. A similar idea was advanced by Gavalas et al. (1976) who formulated the identification problem in a Bayesian framework. By using a priori statistical information on the unknown parameters, the problem becomes statistically better determined. Shah et al. (1978) showed that when reliable a priori statistics are available, the Bayesian approach is superior in that it leads to a smaller variance of the estimation error. Neuman and Yakowitz (1979) employed statistical theory in estimating transmissivities on the basis of steady state data on u. They formulated the identification problem as one of minimizing a performance index consisting of two terms, one, the mismatch between predictions and observations, and second, a Bayesian-like term accounting for deviations from a priori estimate of α, similarly to Gavalas et al. (1976). The selection of the arbitrary scaling parameter that multiplies the second term of the performance index was studied in some detail. When reliable a priori statistics are available, the Bayesian approach has been shown to be effective in reducing the total estimation error. Such statistics may not be available however.

An alternative to zonation is to specify the functional form obeyed by α up to a set of unknown coefficients. For example, Distefano and Rath (1975) used zonation and then assumed that α is a linear function of position in each zone.

In either instance, the original problem of estimating $\alpha(x,y)$, a continuous function of spatial position, is converted into one of estimating a finite number of parameters, $\alpha_1, \alpha_2, \ldots, \alpha_N$. The ill-conditioned nature of the problem is, of course, not circumvented if N is sufficiently large. The underlying nature of the problem can be nicely illustrated by considering its statistical aspects.

Statistical Aspects of the Identification Process

We have expressed the solution of (1) as $u = A\alpha$, and have discussed the problems arising because of the nonuniqueness and lack of continuity of the inverse operator A^{-1}. For the purpose of illustrating the ill-conditioning of the basic identification problem, let us assume that (1) has been discretized both spatially and temporally, so that u and α are vectors and A is a matrix, all of appropriate dimension. Shah et al. (1978) and Padmanabhan (1980) have developed the statistical aspects of the identification problem, and we present a synopsis of those analyses here.

When $\dim \alpha > \dim u$, $u = A\alpha$ has infinitely many solutions given by

$$\hat{\alpha} = A^{+}u + (I - A^{+}A)\kappa \tag{7}$$

where A^{+} is the pseudo-inverse

$$A^{+} = (A^{*}A)^{-1}A^{*} \tag{8}$$

and where A^{*} is the adjoint of A. κ is an arbitrary vector with dimension equal to that of α. The first term in (7) is a particular solution and the second represents any element from the null space of A. Since κ is arbitrary, the estimate $\hat{\alpha}$ may also be arbitrarily different from the true value α. In other words, the estimation error has no upper bound. The dimension of the null space is the number of zero eigenvalues of A^*A. In this sense, A is singular. In a practical sense, overparametrization can occur even without A being strictly singular. It is enough if A^*A has one or more of its eigenvalues much smaller than the rest. (A would then be ill-conditioned.) The parameters along the associated eigenspace are virtually indeterminable. As these eigenvalues approach zero, in the limit the eigenspace becomes the null space. In the eigenspace a slight variation in the data u would be interpreted as being induced by a large change in the parameters, since we have $\delta\hat{\alpha} = A^+(\delta u)$. Here A^+ has large eigenvalues in the eigenspace so that $||\delta\hat{\alpha}||$ is large even though $||\delta u||$ is small. The estimation is ill-posed even though A is invertible in the strict sense.

We have noted that one approach to achieving better-conditioned estimates is to reduce the number of unknown parameters. This process can be represented as replacing the parameter vector α by its restriction to a smaller subspace of the parameter space F. If we characterize this subspace by its basis vectors M, the parameter space can be resolved into M and its orthogonal complement M' in F. Correspondingly, we can resolve α into α_M and $\alpha_{M'}$ by $\alpha = \alpha_M + \alpha_{M'}$, where $\alpha_M = MM^+\alpha$, the projection of α into M. The replacement of α by α_M leads to so-called *parameterization error.* Since α_M lies in a smaller space, we may characterize it as $\alpha_M = M\zeta$, in terms of a vector ζ with fewer components. Thus, the estimation of α is replaced as that of ζ. If the data are represented as $u^o = u + \eta$, where η is the error vector, then

$$\begin{aligned} u^o &= u + \eta = A\alpha + \eta \\ &= A(M\zeta + \alpha_{M'}) + \eta \\ &= [(AM)\zeta] \quad + \quad [A\alpha_{M'}] \quad + \quad \eta \qquad (9) \\ &= \underset{\text{Model}}{\text{Approximate}} \quad \underset{\text{Error}}{\text{Parameterization}} \quad \underset{\text{Error}}{\text{Measurement}} \end{aligned}$$

which shows the composition of the observed quantity u^o. The estimate is $\hat{\zeta} = (AM)^+u^o$, or, in terms of the original parameter vector, $\hat{\alpha} = M(AM)^+u^o$.

The error in matching observed and predicted performance can be defined as

$$J(\hat{\alpha}) = E\{||\, u^o - A\hat{\alpha}||^2\} \tag{10}$$

where $E\{\cdot\}$ is the expectation. Eq. (10) thus becomes

$$\begin{aligned} J(\hat{\alpha}) &= E\{||[I - (AM)(AM)^+]u^o||^2\} \\ &= \text{trace}\{\rho_{(AM)}\, E(u^o u^{o*})\rho_{(AM)}\} \end{aligned} \tag{11}$$

where $\rho_{(AM)} = I - (AM)(AM)^+$ is the projection into the null space of $(AM)^+$. The expectation $E(u^o u^{o*})$ is $E\{(A\alpha + \eta)(A\alpha + \eta)^*\} = A\alpha\alpha^*A^* + E(\eta\eta^*)$. We see that the estimation error can be reduced if $\rho_{(AM)} \to 0$, i.e., as (AM) assumes full rank.
In other words, by increasing the number of parameters, it is possible to produce a perfect match of observations and predictions, regardless of measurement errors.

The parameter estimation error is defined as

$$J_\alpha(\hat{\alpha}) = E\{||\hat{\alpha} - \alpha||^2\} \tag{12}$$

The estimate $\hat{\alpha}$ is given by

$$\begin{aligned}\hat{\alpha} &= M(AM)^+u^o \\ &= M(AM)^+[A(\alpha_M + \alpha_{M'}) + \eta]\end{aligned} \tag{13}$$

Thus,

$$J_\alpha(\hat{\alpha}) = ||\alpha_{M'}||^2 + \text{trace}\ \{M(AM)^+\ E(\eta\eta^*)(M(AM)^+)^*\} \tag{14}$$

which consists of two terms, the parameterization error and the measurement error. The parameterization error, $||\alpha_{M'}||^2$, approaches zero as ν = rank M approaches dim x, i.e., as the number of parameters is increased. The second term contains $M(AM)^+$, which has the eigenvalues $1/\mu_1, 1/\mu_2, \ldots, 1/\mu_\nu$, where $\{\mu_i^2\}$ are the largest ν eigenvalues of A*A. As ν increases, the degree of ill-conditioning of $M(AM)^+$ also worsens because $1/\mu_\nu$ increases with ν. Thus, the effect of measurement error is amplified as the number of parameters is increased. For every problem there is an optimum level of parameterization measured by ν at which $J_\alpha(\hat{\alpha})$ reaches a minimum. Padmanabhan (1980) has included modeling error in the above analysis. Although we have not considered modeling error here, i.e., the inexactness of A, it is frequently of more importance than measurement error in the identification problem.

REGULARIZATION METHODS FOR THE SOLUTION OF ILL-POSED PROBLEMS

Introduction

In the previous sections we have discussed the ill-posed nature of the identification problem arising from nonuniqueness and instability of solutions. No mathematical method can really alleviate nonuniqueness unless it selects some solution that has certain physically desirable properties (e.g. smoothness). However, as far as decreasing instability is concerned, mathematical methods do have a great potential.

The regularization approach is a stability-oriented approach, with which one obtains stable approximate solutions to ill-posed problems that are in accordance with the level of error in the data. The regularization method has two crucial properties:

i) In problems with inherent nonuniqueness, it selects (for a given level of error) the smoothest solution, in the sense that it corresponds to the least value of an appropriately chosen functional (called the stabilizing functional).

ii) In problems where no nonuniqueness is present, the approximate solution it yields converges to the true solution in the limit as the level of error in the data tends to zero.

The Concepts of Well-Posed and Ill-Posed Problems

Consider the operator equation

$$Az = u \tag{15}$$

where u and z are elements of the metric spaces U and F with metrics $\rho_U(u_1,u_2)$ for $u_1,u_2 \in U$ and $\rho_F(z_1,z_2)$ for $z_1,z_2 \in F$. We say that a solution $z = R(u)$ of (15) is *stable* on the spaces (F,U) if for every positive number ε, there exists a positive number $\delta(\varepsilon)$ such that the inequality $\rho_U(u_1,u_2) < \delta(\varepsilon)$ implies $\rho_F(z_1,z_2) \leq \varepsilon$, where $z_1 = R(u_1)$ and $z_2 = R(u_2)$ with u_1 and u_2 in U and z_1 and z_2 in F.

The problem of solving (15) is said to be *well-posed* on the pair of metric spaces (F,U) if the following three conditions are satisfied:

i) for every $u \in U$ there exists a solution $z \in F$

ii) the solution is unique

iii) the solution is stable on the spaces (F,U).

Problems that do not satisfy all the above conditions are said to be *ill-posed*. Note that existence of a solution is assured if and only if $U \subset AF$; uniqueness if and only if A is injective; stability if and only if A^{-1} is continuous. The (exact) solution of a well-posed problem of the form (15) is simply

$$z = A^{-1}u \tag{16}$$

For a long time, it was the accepted point of view that any mathematical problem corresponding to any physical problem has to be well-posed. This point of view, while natural as applied to certain phenomena that have been studied over the years, is not valid for all problems. It is well-established now that a great number of physically meaningful problems are mathematically ill-posed.

In most practical problems, the RHS of (15) is not known exactly. Rather, we have an approximation u_δ of the true u, u_T, with accuracy δ, i.e. $\rho_U(u_\delta,u_T) \leq \delta$. It is natural then to seek an approximate solution to (15) in the class Q_δ of elements z for which $\rho_U(Az,u_\delta) \leq \delta$. The question is which element of Q_δ to choose. If the problem is well-posed, we may choose

$$z = A^{-1}u_\delta \tag{17}$$

If the problem is ill-posed, on the other hand, the result (17) is of little value.

The Regularization Method

Consider again the solution of the operator equation (15) with approximately known right hand side $u_\delta(\rho_U(u_\delta,u_T) < \delta)$ and suppose that A^{-1} is unique but not continuous. Then, an approximate solution z_δ of (15) cannot be obtained by (17).

Since the numerical parameter δ characterizes the error in the right-hand side of (15), it is natural to define z_δ with the aid of an operator depending on a parameter having a value chosen in accordance with the error δ in the data u_δ. Specifically, as $\delta \to 0$, i.e. as u_δ approaches (in the metric of the space U) the exact value u_T, the approximate solution z_δ must approach (in the metric of the space F) the exact solution z_T.

Suppose that the elements $z_T \in F$ and $u_T \in U$ are related by $Az_T = u_T$. An operator $R(u,\beta)$ depending on a parameter β is said to be a *regularizing operator* for the equation $Az = u$ in a neighborhood of $u = u_T$ if:

(i) There exists $\delta_1 > 0$ such that $R(u,\beta)$ is defined for every $\beta > 0$ and every $u \in U$ for which $\rho_U(u,u) \leq \delta \leq \delta_1$, and

(ii) There exists a function $\beta = \beta(\delta)$ such that, for every $\varepsilon > 0$, there exists a number $\delta(\varepsilon) \leq \delta_1$ such that $u_\delta \in U$ and $\rho_U(u_T,u_\delta) \leq \delta(\varepsilon)$ imply that $\rho_F(z_T,z_\beta) \leq \varepsilon$, where $z_\beta = R(u_\delta,\beta(\delta))$.

If $\rho_U(u_T,u_\delta) \leq \delta$, we can take for an approximate solution of (15) with approximate RHS u_δ the element $z_\beta = R(u_\delta,\beta)$ obtained with the aid of the regularizing operator $R(u,\beta)$, where $\beta = \beta(\delta,u_\delta)$ in accordance with the error in the initial data u_δ. This solution is called a *regularized solution* of (15), and the numerical parameter β is called the *regularization parameter*. Thus, the problem of finding an approximate solution of (15) that is stable under small changes in the right-hand member reduces to (i) finding regularizing operators and (ii) determining the regularization parameter β from supplementary information pertaining to the problem, such as the size of the error in u_δ. This method of constructing approximate solutions is called the *regularization method*.

Theorem 1. (Tikhonov and Arsenin, 1977, p. 49) Let A denote an operator from F into U and let $\bar{R}(u,\beta)$ denote an operator from U into F that is defined for every element $u \in U$ and every $\beta > 0$ and that is continuous with respect to u. If

$$\lim_{\beta \to 0} \bar{R}(Az,\beta) = z$$

for every $z \in F$, then the operator $\bar{R}(u,\beta)$ is a regularizing operator for (15).

Let $\Omega(z)$ denote a continuous, non-negative functional defined on a subset F_1 of F that is everywhere dense in F. $\Omega(z)$ is said to be a *stabilizing functional* if it possesses the following properties:

(i) z_T belongs to the domain of definition of $\Omega(z)$

(ii) For every $d > 0$, the set of elements z of F_1 for which $\Omega(z) \leq d$ is a compact subset of F_1.

Suppose we know that the deviation of the right-hand member of (15), u_δ, from the exact value, u_T, does not exceed δ, i.e. $\rho_U(u_\delta,u_T) \leq \delta$. It is then natural to seek an approximate solution in the class $Q_\delta \triangleq \{z \in F \mid \rho_U(Az,u_\delta) \leq \delta\}$. We cannot, however, take an arbitrary element of z_δ of Q_δ as an approximate solution of (15) with approximate right-hand member u_δ because such a solution will not in general be continuous with respect to δ. We need a principle for selecting the possible solutions that insures continuous dependence on δ.

Suppose that $\Omega(z)$ is a stabilizing functional defined on $F_1 \subset F$. We will consider only those elements of Q_δ on which $\Omega(z)$ is defined, i.e. the elements of the set $F_{1,\delta} = F_1 \cap Q_\delta$. Among the elements of this set, we seek the one (or ones) that will minimize $\Omega(z)$ on $F_{1,\delta}$. Let z_δ denote such an element, that can be regarded as the result of applying to u_δ an operator $\tilde{R}$ depending on a parameter δ, $z_\delta = \tilde{R}(u_\delta,\delta)$.

Theorem 2. (Tikhonov and Arsenin, 1977, p. 52) An operator $\tilde{R}(u,\delta)$ into F_1, defined on the set of pairs (u,δ) with $u \in U$ and $\delta > 0$, so that $z_\delta = \tilde{R}(u_\delta,\delta)$ minimizes the stabilizing functional $\Omega(z)$ on the set $F_{1,\delta} = F_1 \cap \{z \in F | \rho_U^\delta(Az,u_\delta) \leq \delta\}$, is a regularizing operator for (15), and hence z_δ can be taken as an approximate solution of (15).

With Theorem 2, the problem of finding an approximate solution to (15) with an approximate right-hand member consists of minimizing a stabilizing functional $\Omega(z)$ on the set $F_{1,\delta}$. It should be pointed out, however, that the numerical solution of this problem, as it stands, is often difficult. Hence it is desirable to reformulate it somewhat. Before doing so, we define:

(i) $\Omega_0 = \inf_{z \in F_1} \Omega(z)$

(ii) $M_0 = \{z \in F_1 | \Omega(z) = \Omega_0\}$

(iii) $\Omega(z)$ is said to be *quasimonotonic* if, for every $z_0 \in F_1$ that does not belong to M_0, every neighborhood of z_0 includes a $z_1 \in F_1$ such that $\Omega(z_1) < \Omega(z_0)$.

Theorem 3. (Tikhonov and Arsenin, 1977, p. 56) The greatest lower bound of a functional $\Omega(z)$ on a set $F_{1,\delta}$ for which $M_0 \cap F_{1,\delta} \neq \emptyset$ is attained for every element of $M_0 \cap F_{1,\delta}$. The greatest lower bound of a quasimonotonic functional $\Omega(z)$ on a set $F_{1,\delta}$ for which $M_0 \cap F_{1,\delta} = \emptyset$ is attained for an element z_δ for which $\rho_U(Az_\delta,u_\delta) = \delta$.

From Theorem 3 we conclude that the problem of minimizing a quasimonotonic stabilizing functional $\Omega(z)$ on the set $F_{1,\delta}$ is equivalent to minimizing $\Omega(z)$ on the set F_1 subject to the constraint $\rho_U(Az,\hat{u}_\delta) = \delta$, provided that $\rho_U(Az_0,u_\delta) > \delta$ for every $z_0 \in M_0$. This is a conditional extremum problem that can be solved by use of Lagrange multipliers. That is, let us minimize the so-called *smoothing functional*,

$$M^\beta(z,u_\delta) = \rho_U^2(Az,u_\delta) + \beta\Omega(z) \tag{18}$$

where the numerical parameter β is determined from the condition $\rho_U(Az_\beta,u_\delta) = \delta$, and where z_β is an element at which the functional $M^\beta(z,u_\delta)$ attains its greatest lower bound. Clearly this approach is equivalent to the previous one if and only if there exists a β such that $\rho_U(Az_\beta,u_\delta) = \delta$.

Theorem 4. (Tikhonov and Arsenin, 1977, p. 63) Let A denote a continuous operator from F into U. For every $u \in U$ and $\beta > 0$ there exists a $z_\beta \in F_1$ for which the functional

$$M^\beta(z,u) = \rho_U^2(Az,u) + \beta\Omega(z)$$

attains its greatest lower bound,

$$\inf_{z \in F_1} M^\beta(z,u) = M^\beta(z_\beta,u)$$

Thus, an operator $R_1(u,\beta)$ into F_1 is defined on the set of pairs (u,β), where $u \in U$ and $\beta > 0$ so that $z_\beta = R_1(u,\beta)$ minimizes $M^\beta(z,u)$. Sufficient conditions for uniqueness of z_β can be exhibited for example if the operator A is linear, the set F is a Hilbert space, and $\Omega(z)$ is a quadratic stabilizing functional. For a nonlinear operator A, z_β may not be unique.

Theorem 5 establishes that $R_1(u,\beta)$ is a regularizing operator for (15). Let us denote by T_{δ_1} the class of functions that are non-negative, non-decreasing and continuous on an interval $[0,\delta_1]$.

Theorem 5. (Tikhonov and Arsenin, 1977, p. 65) Let z_T denote a solution of (15) with right-hand member u_T. Then for any $\varepsilon > 0$ and any $\beta_1(\delta)$ and $\beta_2(\delta)$ in the class T_{δ_1} such that $\beta_2(0) = 0$ and $\delta^2/\beta_1(\delta) \leq \beta_2(\delta)$, there exists a number $\delta_0(\varepsilon, \beta_1,\beta_2) \leq \delta_1$ such that for $\tilde{u} \in U$ and $\delta \leq \delta_0$, $\rho_U(\tilde{u},u_T) \leq \delta$ implies that $\rho_F(z_T,\tilde{z}_\beta) \leq \varepsilon$, where $\tilde{z}_\beta = R_1(\tilde{u},\beta)$, for all β satisfying

$$\frac{\delta^2}{\beta_1(\delta)} \leq \beta \leq \beta_2(\delta)$$

Suppose that a subset Φ of a metric space F admits a metrization $\rho_\Phi(z_1,z_2)$ such that $\rho_F(z_1,z_2) \leq \rho_\Phi(z_1,z_2)$. If the sphere $\rho_\Phi(z,z_0) \leq d$ (with center z_0) is compact in F (in the metric of F), then Theorem 4 is valid, i.e. there exists an element $z_\beta \in \Phi$ minimizing the functional

$$M^\beta(z,\tilde{u}) = \rho_U^2(Az,\tilde{u}) + \beta\Omega(z)$$

where $\Omega(z) = \rho_\Phi^2(z,0)$. If the exact solution z_T of (15) belongs to the set Φ, then the operator $R_2(\tilde{u},\beta)$ that provides, for every $\beta > 0$ and $\tilde{u} \in U$, an element $\tilde{z}_\beta$ minimizing $M^\beta(z,\tilde{u})$ is a regularizing operator.

Example 1. Consider the case in which F is the space of continuous functions $z(\zeta)$ in the interval $[a,b]$ with the C-metric,

$$\rho_F(z_1,z_2) = \sup_{a \leq \zeta \leq b} |z_1(\zeta) - z_2(\zeta)|$$

and Φ is the space of functions that have square-integrable generalized derivatives up to p-th order, i.e. Φ is the Sobolev space W_2^p. The metric in W_2^p is defined by

$$\rho_\Phi(z_1,z_2) = \left\{ \int_a^b \sum_{r=0}^{p} q_r(\zeta) \left[\frac{d^r(z_1-z_2)}{d\zeta^r} \right]^2 d\zeta \right\}^{1/2} \tag{19}$$

where $q_0(\zeta), q_1(\zeta), \ldots, q_{p-1}(\zeta)$ are given non-negative continuous functions, and $q_p(\zeta)$ is a given positive continuous function. For every p, W_2^p is a Hilbert space, and a closed ball in it is compact in C. Hence $\rho_\Phi^2(z,0)$ defines a stabilizing functional on the set $F_1 = W_2^p \subset F$.

The functionals defined by*

$$\Omega(z) = \int_a^b \sum_{r=0}^{p} q_r(\zeta)\left(\frac{d^r z}{d\zeta^r}\right)^2 d\zeta \tag{20}$$

where the $q_r(\zeta)$ satisfy the conditions noted above, are called *stabilizers of the p-th order*. If all $q_r(\zeta)$ are constants, they are called *stabilizers of the p-th order with constant coefficients* or *Tikhonov stabilizers*. Stabilizers of p-th order are quasi-monotonic and have the property that their greatest lower bound $\Omega_0 = \Omega(0) = 0$.

On the question of the existence of a β such that $\rho_U(Az_\beta, u_\delta) = \delta$, we have:

Theorem 6. (Tikhonov and Arsenin, 1977, p. 91) If $\rho_U^2(Az_\beta, u)$ is a single-valued function of β, then for every positive number $\delta < \rho_U(Az_0, u)$, where

$$z_0 \in \{z \mid \Omega(z) = \inf_{y \in F_1} \Omega(y)\}$$

there exists a $\beta(\delta)$ such that $\rho_U(Az_{\beta(\delta)}, u) = \delta$.

Some Remarks on the Regularization Method. The fact that F and U are metric spaces is used in the regularization method to express the closeness of elements. Ivanov (1969) has shown that the main results of the regularization theory remain valid for topological spaces F and U.

In this section it has been established that regularized solutions to ill-posed problems of the form Az = u are stable with respect to small changes in u and converge to the true solution z_T in the limit as the level of error in the data tends to zero. The regularization theory does not provide the rate of convergence of the regularized solution to the true one. A study of this issue has been made by Franklin (1974). Franklin has shown that for mildly ill-posed problems, such as numerical differentiation, the rate of convergence is fast, e.g. $O(\delta^{1/2})$. However, for the severely ill-posed problem of the backward heat equation, the convergence rate goes as a power of $1/\log(1/\delta)$. It can go even slower for more severely ill-posed problems. It is noteworthy that the present identification problem is mildly ill-posed in this sense.

The Approach of Miller

Consider again the problem of solving an operator equation of the form (15). Miller (1970) reformulated the problem as follows:

Given the Hilbert spaces F, U and W, the bounded linear operators A: F $\to$ U and B: F $\to$ W, and an element $u \in U$, and assuming that B^{-1} exists and is bounded, suppose that $z_T \in F$ satisfies

$$||Az_T - u|| \leq \delta \tag{21}$$

*When z is a function of several variables, $\zeta_1, \zeta_2, \ldots, \zeta_m$, we can define stabilizers of the p-th order as

$$\Omega(z) = \int \sum_{r=0}^{p} q_r(\zeta_1,\ldots,\zeta_n)\left[\sum \frac{r!}{k_1!\ldots k_n!}\left(\frac{\partial^r z}{\partial\zeta_1^{k_1}\ldots\partial\zeta_n^{k_n}}\right)^2\right] d\zeta_1\ldots d\zeta_n$$

$$||Bz_T|| \leq \Delta \tag{22}$$

The problem is to find an element $\tilde{z} \in F$ that approximates z_T in the sense that the seminorm $<\tilde{z}-z_T>$ is small for δ small.

Consider first the case where both δ and Δ are known. Miller defines

$$\mu(\delta,\Delta) = \sup\{<z>:z \in F, ||Az|| \leq \delta \ , \ ||Bz|| \leq \Delta\}$$

$$\mu_1(\delta,\Delta) = \sup\{<z>:z \in F \ , \ ||Az||^2 + \left(\frac{\delta}{\Delta}\right)^2 ||Bz||^2 \leq 2\delta^2\}$$

and shows that if z_T satisfies (21) and (22) then it also satisfies

$$||Az_T - u||^2 + \left(\frac{\delta}{\Delta}\right)^2 ||Bz_T||^2 \leq 2\delta^2 \tag{23}$$

Conversely, any z_T satisfying (23) also satisfies (21) and (22), except for a factor of at most $\sqrt{2}$. Furthermore,

$$\mu(\delta,\Delta) \leq \mu_1(\delta,\Delta) \leq \sqrt{2}\ \mu(\delta,\Delta)$$

This result suggests that we may combine the two constraints (21) and (22) and sacrifice at most a factor of $\sqrt{2}$.

<u>Method 1</u>. Miller proposed to take as an approximate solution the element $\tilde{z} \in F$ that minimizes

$$||Az-u||^2 + \left(\frac{\delta}{\Delta}\right)^2 ||Bz||^2$$

This element is given by

$$\tilde{z} = \left(A^*A + \left(\frac{\delta}{\Delta}\right)^2 B^*B\right)^{-1} A^*u$$

where superscript * denotes the adjoint of an operator. Note that the operator $(A^*A + \left(\frac{\delta}{\Delta}\right)^2 B^*B)$ has a bounded inverse under the assumptions made for the operators A and B. This guarantees stability of the approximate solution $\tilde{z}$.

Miller also gives an error estimate

$$<\tilde{z} - z_T> \leq \mu_1(\delta,\Delta)$$

provided that z_T satisfies (23).

Next Miller presents a more general approach for which it is not necessary that both δ and Δ are specified. A pair (δ,Δ) is said to be *permissible* if there exists a $z \in F$ satisfying (21) and (22).

Now let z_β be a minimizer of

$$||Az-u||^2 + \beta||Bz||^2$$

and let $\delta_\beta = ||Az_\beta-u||$ and $\Delta_\beta = ||Bz_\beta||$. Clearly z_β minimizes $||Az-u||$ with respect to the constraint $||Bz|| \leq \Delta_\beta$. Likewise z_β minimizes $||Bz||$ with respect to the constraint $||Az-u|| \leq \delta_\beta$. It is also easily seen that δ_β and Δ_β are continuously increasing and decreasing functions of β, respectively. Thus, the set E of permissible pairs is exactly the set of points that are above and to the right of the curve $(\delta_\beta,\Delta_\beta)$, $0 < \beta < \infty$ (Figure 1). Here the case $\beta = 0$ corresponds to the minimization of $||A\bar{z}-u||_T$ alone, in which case Δ_β may be ∞, and the case $\beta = \infty$ corresponds to minimization of $||Bz||$ alone, in which case $\delta_\beta = ||u||$ and $\Delta_\beta = 0$. Moreover, it is easily shown that E is a convex set, hence computation of β only a finite number of points on its boundary curve is needed to ascertain the curve.

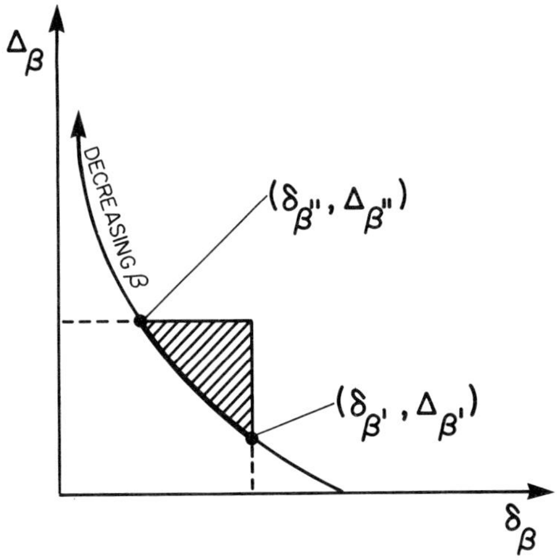

Fig. 1

Method 2. If δ and Δ are known, the unknown solution must lie in the shaded area of Figure 1. Furthermore, any z_β whose corresponding $(\delta_\beta,\Delta_\beta)$ touches the shaded area can be taken as an approximate solution. Note that $<z_\beta - z_T> \leq 2\mu(\delta,\Delta)$.

Method 3. Assume that only δ is known. Miller proposes to take as an approximate solution $\tilde{z}$ that element of F that minimizes $||Bz||$ subject to $||Az-u|| \leq \delta$. Then $\tilde{z} = z_{\beta'}$, where β' is the value of β such that $\delta_{\beta'} = \delta$, as shown in Figure 1.

Method 4. Assume that only Δ is known. Miller proposes to take as an approximate solution $\tilde{z}$ that element of F that minimizes $||Az-u||$ subject to $||Bz|| \leq \Delta$. Then $\tilde{z} = z_{\beta''}$, where β'' is the value of β such that $\Delta_{\beta''} = \Delta$, as shown in Figure 1.

Discussion. Upon examining Method 1, we note that Miller's theory relies heavily on the fact that A is a linear operator, whereas the regularization method of the previous section includes nonlinear operators. $||Bz||^2$ clearly defines a stabilizing functional, but the assumption that B is bounded is too restrictive for the problems of interest. The operator B corresponding to Tikhonov stabilizers is, in general, not bounded (Franklin, 1978). The regularization parameter $(\delta/\Delta)^2$ is a quadratic function of the error in the data, which is consistent with Theorem 5 ($\beta_1 = \Delta^2$ and $\beta_2 = \delta^2/\Delta^2$). However, such an a priori selection of the regularization parameter seems unnecessary.

In methods 2-4, $||Az-u||^2 + \beta||Bz||^2$ is clearly a smoothing functional with β the regularization parameter. Methods 2-4 suggest an interest generalization of the method of selection of the regularization parameter of the last section. Note however that if the operator A is nonlinear, the set E will no longer in general be convex and the functions $\delta_\beta(\beta)$ and $\Delta_\beta(\beta)$ no longer single-valued. Hence, for a nonlinear operator A it is possible that β' and β'' do not exist for given δ and Δ. This fact creates a serious complication when dealing with nonlinear problems, in which case the local convexity of E must be investigated either theoretically or numerically.

Wahba's Version of the Regularization Method - Weighted Cross Validation

Wahba's approach to regularization consists of constructing regularized approximate solutions by minimizing a smoothing functional. The difference of this approach from the previously described ones is the method of selection of the regularization parameter. The method, termed *weighted cross-validation* does not require any a priori statistical information concerning measurement error. The weighted cross-validation technique seems to have originated from problems of interpolation, and has been applied to solution of Fredholm integral equations of the first kind. The idea is quite general, however, so that extension to general nonlinear ill-posed problems seems possible but far from trivial (Neuman and Yakowitz, 1979; Wahba, 1980).

Consider for the moment the problem of solving the operator equation

$$Az(x) = u(x) \tag{24}$$

from discrete noisy measurements u_1^{obs}, u_2^{obs}, ..., u_n^{obs}, of u(x) at the points $x_1, x_2, \ldots, x_n$. If A is the identity operator, then this is the standard interpolation problem. (Note that although the identity operator is obviously continuous, the interpolation problem is ill-posed since solutions are not unique). The interpolation problem can be approached by the regularization method. Assuming that the true z(x) has continuous second derivatives and choosing

$$\int_{x_1}^{x_n} [z''(x)]^2 \, dx$$

as a stabilizing functional, one can obtain regularized solutions to the interpolation problem by minimizing the smoothing functional

$$M^\beta(z,u^{obs}) = \frac{1}{n}\sum_{i=1}^{n}\left[u_i^{obs} - z(x_i)\right]^2 + \beta \int_{x_1}^{x_n}\left[z''(x)\right]^2 dx \tag{25}$$

It can be shown that the solution of the minimization problem leads to a natural cubic spline,

$$s(x) = a_i + b_i(x-x_i) + c_i(x-x_i)^2 + d_i(x-x_i)^3$$
$$x \in [x_i, x_{i+1}] \tag{26}$$

Expressions for the coefficients are given in the literature (Reinsch, 1967; Merz, 1978). For this discussion we are interested only in the expression for $a = (a_1, a_2, \ldots, a_n)$.

Let P be a symmetric, tridiagonal, (n-2)x(n-2) dimensional matrix with elements,

$$p_{ii} = \frac{2}{3}(x_{i+2}-x_i) \qquad p_{i,i+1} = p_{i+1,i} = \frac{1}{3}(x_{i+2}-x_{i+1})$$

and let Q be the tridiagonal nx(n-2) dimensional matrix with elements

$$q_{ii} = \frac{1}{x_{i+1}-x_i}, \quad q_{i+1,i} = -\left[\frac{1}{x_{i+1}-x_i} + \frac{1}{x_{i+2}-x_{i+1}}\right], \quad q_{i+2,i} = \frac{1}{x_{i+2}-x_{i+1}}$$

then (Merz, 1978)

$$a = Lu^{obs} \tag{27}$$

where $L = I - Q(Q^TQ + n\beta P)Q^T$

Reinsch (1967) suggested that if the standard deviation σ of the errors in the data is approximately known, β should be chosen such that

$$\frac{1}{n}\sum_{i=1}^{n}(u_i^{obs} - S(x_i))^2 = \sigma^2$$

or

$$\frac{1}{n}\,||(I-L(\beta))u^{obs}||^2 = \sigma^2 \tag{28}$$

where $||\cdot||$ denotes Euclidian norm. This method for determining β gives excellent results for most practical purposes, but breaks down whenever error statistics are unavailable.

The determination of β by cross-validation proceeds as follows (Wahba and Wold, 1975; Craven and Wahba, 1979). Let $s_\beta^{[k]}$ be the spline using all data points except the kth. We take the ability of $s_\beta^{[k]}$ to predict the missing data point u_k^{obs} as a measure of the effectiveness of β. An overall measure of this effectiveness is the value of

$$V_0(\beta) = \frac{1}{n}\sum_{k=1}^{n}\left[s_\beta^{[k]}(x_k)-u_k^{obs}\right]^2 \tag{29}$$

It can be shown from (27) and some algebra that

$$V_0(\beta) = \frac{1}{n}\sum_{k=1}^{n}\frac{[u_k^{obs}-a_k(\beta)]^2}{[1-L_{kk}(\beta)]^2} \tag{30}$$

a weighted sum of squared residuals of the fit to the data.

One could use the minimizer of $V_0(\beta)$ as the optimum β, a procedure known as ordinary cross-validation. Using statistical arguments, Wahba suggested that ordinary cross-validation be used only for periodic splines with equally-spaced data points. Note that in this case the weights $(1-L_{kk}(\beta))^{-2}$ of squared residuals are all equal. In the general case one should minimize

$$V(\beta) = \frac{1}{n}\sum_{k=1}^{n}\left[s^{[k]}(x_k) - u_k^{obs}\right]^2 w_k(\beta) \tag{31}$$

where w_k account for non-equally spaced data and possible non-periodicity of the spline. Equation (31) can be written as

$$V(\beta) = \frac{1}{n}\sum_{k=1}^{n} \frac{w_k(\beta)}{[1-L_{kk}(\beta)]^2} [u_k^{obs}-a_k(\beta)]^2 \tag{32}$$

According to Wahba, $w_k(\beta)$ have to be chosen in such way that the weights $w_k(\beta)/[1-L_{kk}(\beta)]^2$ are all equal and $V(\beta) = V_0(\beta)$ in the special case of equally-spaced data and periodic spline. Then

$$w_k(\beta) = \left[\frac{1 - L_{kk}(\beta)}{\frac{1}{n}\,\mathrm{tr}(I-L(\beta))}\right]^2 \tag{33}$$

and

$$V(\beta) = \frac{\frac{1}{n}||(I-L(\beta))u^{obs}||^2}{\left[\frac{1}{n}\,\mathrm{tr}(I-L(\beta))\right]^2} \tag{34}$$

where $||\cdot||$ denotes Euclidian norm. The minimization of $V(\beta)$ given by (34) yields the weighted cross-validation estimate of β.

Consider now the problem of approximately solving the integral equation

$$\int_0^1 K(x,\zeta)z(\zeta)\, d\zeta = u(x) \tag{35}$$

given discrete noisy measurements $u_1^{obs}, u_2^{obs}, \ldots, u_n^{obs}$ at the points $x_1, x_2, \ldots, x_n$.

Assuming that $z \in L_2[0,1]$ and choosing a Tikhonov stabilizer of zeroth order as a stabilizing functional, one can obtain regularized solutions $z_\beta(x)$ by minimizing the smoothing functional,

$$M^\beta(z,u^{obs}) = \frac{1}{n}\sum_{i=1}^{n}\left[\int_0^1 K(x_i,\zeta)z(\zeta)\, d\zeta - u_i^{obs}\right]^2 + \beta\int_0^1 [z(\zeta)]^2 d\zeta \tag{36}$$

Wahba (1977) has shown that

$$z_\beta(x) = \big(K(x_1,x), K(x_2,x), \ldots, K(x_n,x)\big)\ (\tilde{Q} + n\beta I)^{-1}(u_1^{obs}, u_2^{obs}, \ldots, u_n^{obs})^T \tag{37}$$

where

$$\tilde{Q}_{jk} = \int_0^1 K(x_j,\zeta)K(x_k,\zeta)\, d\zeta$$

Now the weighted cross-validation function is given by

$$V(\beta) = \frac{1}{n}\sum_{k=1}^{n}\left[\int_0^1 K(x_k,\zeta)z_\beta^{[k]}(\zeta)\, d\zeta - u_k^{obs}\right]^2 w_k(\beta) \tag{38}$$

where $z_{\beta}^{[k]}(x)$ is the regularized solution obtained by omitting the k-th data point, u_k^{obs}.

Wahba (1977,1980), using a similar argument as in the interpolation problem, has shown that the appropriate choice of $w_k(\beta)$ is

$$w_k(\beta) = \left[\frac{1-L_{kk}(\beta)}{\frac{1}{n}\,\mathrm{tr}(I-L(\beta))} \right]^2 \tag{39}$$

where

$$L(\beta) = \tilde{Q}(\tilde{Q} + n\beta I)^{-1}$$

Then

$$V(\beta) = \frac{\frac{1}{n}\,||(I-L(\beta))u^{obs}||^2}{\left[\frac{1}{n}\,\mathrm{tr}(I-L(\beta))\right]^2} \tag{40}$$

The idea of cross-validation is quite general and is applicable nonlinear problems. The key problem is how to select the weights $w_k(\beta)$ in the cross-validation function $V(\beta)$. As pointed out by Wahba (1980), the problem of choosing the weights $w_k(\beta)$ for nonlinear problems has not yet been solved. However, one may use ordinary cross-validation, or even (40) when the operator is "mildly" nonlinear.

REGULARIZATION APPROACH TO THE IDENTIFICATION PROBLEM

Consider again (1) in which it is desired to estimate $\alpha(x,y)$ on the basis of $u^{obs}(x_i,y_i,t)$ that contains normally distributed errors with standard deviation σ.

Define the space U to be the semimetric space of real-valued functions $u(x,y,t)$ with semimetric

$$\rho_U(u_1,u_2) = \left[\frac{1}{T} \int_0^T \frac{1}{m} \sum_{i=1}^{m} \left[u_1(x_i,y_i,t) - u_2(x_i,y_i,t)\right]^2 dt \right]^{1/2} \tag{41}$$

Also define the space F to be the metric space of continuous, real-valued functions (x,y) defined on the set $D \subset \mathcal{R}^2$ with the metric

$$\rho_F(\alpha_1,\alpha_2) = \sup_{(x,y)\in D} |\alpha_1(x,y) - \alpha_2(x,y)| \tag{42}$$

Following our discussion in the Introduction, the identification problem can be posed as finding an approximate solution in F of the operator equation $A\alpha = u$ from error-corrupted measurements of u. For its solution, we propose a regularization method based on the general regularization theory presented previously. Our method is also in agreement with Miller's Method 3.

We first define a stabilizing functional. Assuming that the desired solution belongs to the Sobolev space W_2^2, i.e. the function space of square-integrable generalized derivatives up to second order, we can use as stabilizing functional a Tikhonov stabilizer of second order,

$$\Omega(\alpha) = \iint_D \left[\left(\frac{\partial^2\alpha}{\partial x^2}\right)^2 + 2\left(\frac{\partial^2\alpha}{\partial x \partial y}\right)^2 + \left(\frac{\partial^2\alpha}{\partial y^2}\right)^2 \right] dxdy \tag{43}$$

Then we can define a smoothing functional

$$M^{\beta}(\alpha,u^{obs}) = \rho_U^2(A\alpha,u^{obs}) + \beta\Omega(\alpha) \tag{44}$$

The minimizer of $M^{\beta}(\alpha,u^{obs})$, $\alpha_{\beta}(x,y)$, is a regularized solution of the identification problem.

Next we need to determine the regularization parameter β so that the two terms in (44) are matched in accordance with the level of error in the data. Since the standard deviation σ of the measurement error is assumed to be known, we can use it to estimate the optimum deviation (in the sense of the semimetric of U) of $A\alpha_{\beta}$ from the measurements, i.e.

$$\rho_U^2(A\alpha_{\beta},u^{obs}) = \sigma^2 \tag{45}$$

As pointed out previously, a value of β such that (45) is satisfied might not exist. Sufficient conditions for existence are given by Theorem 6. We can illustrate this point as follows:

Consider the function $\mathcal{T}$: $W_2^2 \to \mathcal{R}^2$ defined by $\mathcal{T}(\alpha) = (\rho_U^2(A\alpha,u^{obs}),\Omega(\alpha))$. The minimizers α_{β} of the functional $\rho_U^2(A\alpha,u^{obs}) + \beta\Omega(\alpha)$ will generate the boundary of the range of $\mathcal{T}$, as β scans $\mathcal{R}^+$, provided that the range of $\mathcal{T}$ is convex (Figure 2a). If it is not convex, then α_{β} will generate only a part of the boundary (Figure 2b). Hence, if $\sigma^2 \in (a,b)$ (Figure 2b), then no β exists such that (45) holds. In this case, one can generate the missing part of the boundary and obtain regularized solution with desired deviation from the data by minimizing $\Omega(\alpha)$ subject to $\rho_U^2(A\alpha,u^{obs}) = \sigma^2 (\sigma \in (a,b))$, as suggested by Theorem 3.

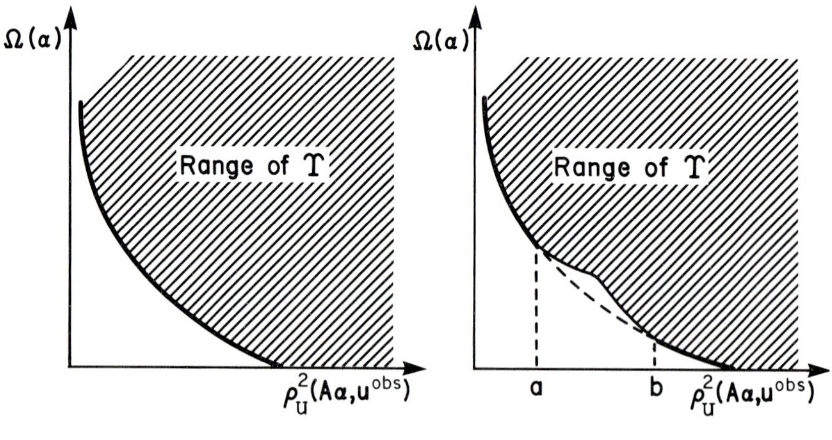

Fig. 2a Fig. 2b

If the operator A were linear, it can be shown readily that the range of $\mathcal{T}$ would convex (see Miller's approach, Methods 2-4). For nonlinear operators, the range of $\mathcal{T}$ will not in general be convex. Numerical experiments seem to be called for this regard.

The results of the approach suggested in this section will be presented elsewhere.

REFERENCES

Beck, J. V. and K. J. Arnold (1977). Parameter Estimation in Engineering and Science. John Wiley, New York.

Carter, R. D., L. F. Kemp, Jr., A. C. Pierce, and D. L. Williams (1974). Performance matching with constraints. Soc. Pet. Eng. J., 14, 187-196.

Chang, S. and W. Yeh (1976). A proposed algorithm for the solution of the large-scale inverse problem in ground water. Water Resources Research, 12, 365-374.

Chavent, C., M. Dupuy, and P. Lemonnier (1975). History matching by use of optimal control theory. Soc. Pet. Eng. J., 15, 74-86.

Coats, K. H., J. R. Dempsey, and J. H. Henderson (1970). A new technique for determining reservoir description from field performance data. Soc. Pet. Eng. J., 10, 66-74.

Chen, W. H. and J. H. Seinfeld (1972). Estimation of spatially varying parameters in partial differential equations. Int. J. Control, 15, 487-495.

Chen, W. H., G. R. Gavalas, J. H. Seinfeld, and M. L. Wasserman (1974). A new algorithm for automatic history matching. Soc. Pet. Eng. J., 14, 593-608.

Cooley, R. L. (1977). A method of estimating parameters and assessing reliability for models of steady state groundwater flow. 1. Theory and numerical properties. Water Resour. Res., 13, 318-324.

Cooley, R. L. (1979). A method of estimating parameters and assessing reliability for models of steady state groundwater flow. 2. Application and statistical analysis. Water Resour. Res., 15, 603-617.

Craven, P. and G. Wahba (1979). Smoothing noisy data with spline functions. Numerische Mathematik, 31, 377-403.

Deininger, R. A. (1969). Linear programing for hydrologic analyses. Water Resour. Res., 5, 1105-1109.

Distefano, N. and A. Rath (1975). An identification approach to subsurface hydrological systems. Water Resour. Res., 11, 1005-1012.

Emsellem, Y. and G. de Marsily (1971). An automatic solution for the inverse problem. Water Resour. Res., 7, 1264-1283.

Eykhoff, P. (1974). System Identification. John Wiley, New York.

Franklin, J. N. (1974). On Tikhonov's method for ill-posed problems. Mathematics of Computation, 28, 889-907.

Franklin, J. N. (1978). Minimum principles for ill-posed problems. SIAM J. on Math. Anal., 9, 638-650.

Frind, E. O. and G. F. Pinder (1973). Gelerkin solution of the inverse problem for aquifer transmissivity. Water Resour. Res., 9, 1397-1410.

Gavalas, G. R., P. C. Shah, and J. H. Seinfeld (1976). Reservoir history matching by Bayesian estimation. Soc. Pet. Eng. J., 16, 337-350.

Haimes, Y. Y., R. L. Perrine, and D. A. Wismer (1968). Identification of aquifer parameters by decomposition in multi-level optimization. Israel J. of Tech., 6, 322.

Ivanov, V. K. (1969). Ill-posed problems in topological spaces. Sibirskiy Matematicheskiy Zhurnal, 10, 1065-1074.

Jacquard, P. and C. Jain (1965). Permeability distribution from field pressure data. Soc. Pet. Eng. J., 5, 281-294.

Jahns, H. O. (1966). A rapid method for obtaining a two-dimensional reservoir description from well pressure response data. Soc. Pet. Eng. J., 6, 315-327.

Kleinecke, D. (1971). Use of linear programing for estimating geohydrologic parameters of groundwater basins. Water Resour. Res., 7, 367-375.

Kubrusly, C. (1977). Distributed parameter system identification: A survey. Int. J. Control, 26, 509-535.

Lovell, R. E., L. Duckstein, and C. C. Kisiel (1972). Use of subjective information in estimation of aquifer parameters. Water Resour. Res., 8, 680-690.

Marino, M.A. and W-G. Yeh (1973). Identificationof parameters in finite leaky aquifer system. J. Hydraul. Div. Amer. Soc. Civil Eng., 99, 319-336.

Merz, P. (1978). Spline smoothing by generalized cross-validation. Chevron Research Company, Richmond, California.

Miller, K. (1970). Least-squares methods for ill-posed problems with a prescribed bound. SIAM J. on Math. Anal., 1, 52-74.

Nelson, R. N. (1968). In-place determination of permeability distribution for heterogeneous porous media through analysis of energy dissipation. Soc. Pet. Eng. J., 8, 33-42.

Nelson, R. N. and W. L. McCollum (1969). Transient energy dissipation methods of measuring permeability distributions in heterogeneous porous materials. Rep. CSC 691229, Water Resour. Div., U.S. Geol. Surv., Washington, D. C.

Neuman, S. P. (1973). Calibration of distributed parameter groundwater flow models viewed as a multiple-objective decision process under uncertainty. Water Resour. Res., 9, 1006-1021.

Neuman, S. P. (1980b). A statistical approach to the inverse problem of aquifer hydrology. 3. Improved solution method and added perspective. Water Resour. Res., 16, 331-346.

Neuman, S. P. and S. Yakowitz (1979). A statistical approach to the inverse problem of aquifer hydrology. 1. Theory. Water Resour Res., 15, 845-860.

Newman, S. P., G. Fogg, and E. Jacobson (1980a). A statistical approach to the inverse problem of aquifer hydrology. 2. Case study. Water Resour. Res., 16, 33-58.

Padmanabhan, L. (1980). Private communication.

Reinsch, C. H. (1967). Smoothing by spline functions. Numerische Mathematik, 10, 177-183.

Sagar, B., S. Yakowitz, and L. Duckstein (1975). A direct method for the identification of the parameters of dynamic nonhomogeneous aquifers, Water Resour. Res., 11, 563-570.

Sammel, E. A. (1964). Evaluation of numerical analysis methods for determining variations in transmissibilities, Int. Ass. Sci. Hydrol. Publ., 64, 239-251.

Seinfeld, J. H. and W. H. Chen (1978). Identification of petroleum reservoir properties in Distributed Parameter Systems. Identification, Estimation and Control, edited by W. H. Ray and D. G. Lainiotis, Marcel Dekker, 497-554.

Seinfeld, J. H. and L. Lapidus (1974). Mathematical Methods in Chemical Engineering, Vol. 3. Process modeling estimation and identification. Prentice-Hall, Englewood Cliffs.

Shah, P. C., G. R. Gavalas, and J. H. Seinfeld (1978). Error analysis in history matching: The optimum level of parameterization. Soc. Pet. Eng. J., 18, 219-228.

Slater, G. E. and E. J. Durrer (1971). Adjustment of reservoir simulation models to match field performance. Soc. Pet. Eng. J., 11, 295-305.

Stallman, R. W. (1963). Calculation of resistance and error in an electric analog of steady flow through nonhonogeneous aquifers. U. S. Geol. Surv. Water Supply Pap. 20, 1544-1546.

Tikhonov, A. N. and V. Ya. Arsenin (1977). Solutions of Ill-Posed Problems. Winston-Wiley.

Vermuri, V. and W. J. Karplus (1969). Identification of nonlinear parameters of groundwater basins by hybrid computation. Water Resour. Res., 5, 172-185.

Vermuri, V., J. A. Dracup, and R. C. Erdmann (1969). Sensitivity analysis method of system identificationand its potential in hydrologic research. Water Resour. Res., 5, 341-349.

Wahba, G. (1975). Smoothing noisy data with spline functions. Numerische Mathematik, 24, 383-393.

Wahba, G. and S. Wold (1975). A completely automatic french curve: Fitting spline functions by cross-validation. Communications in Statistics, 4, 1-17.

Wahba, G. (1977). Practical approximate solutions to linear operator equations when data are noisy. SIAM J. on Numer. Anal., 14, 651-667.

Wahba, G. (1980). Ill-posed problems: Numerical and statistical methods for mildly, moderately, and severely ill-posed problems with noisy data. Technical Report No. 595, University of Wisconsin.
Wasserman, M. L., A. S. Emanuel, and J. H. Seinfeld (1975). Practical application of optimal control theory for history-matching multiphase simulator models. Soc. Pet. Eng. J., 15, 347-355.
Yakowitz, S. and L. Duckstein (1980). Instability in aquifer identification: Theory and case studies. Water Resour. Res., 16, 1054-1064.
Yakowitz, S. and P. Noren (1976). On the identification of space variable parameters in dynamic linear partial differential equations. J. Math. Anal. Appl., 53, 521-538.
Yeh, W. W. and Y. S. Yoon (1976). A systematic optimization procedure for the identification of inhomogeneous aquifer parameters, in Advances in Groundwater Hydrology. American Water Resources Association, Minneapolis, Minnesota, 72-82.
Yeh, W-G. and G. W. Tauxe (1971a). Optimal identificationof aquifer diffusivity using quasi-linearization. Water Resour. Res., 7, 955-962.
Yeh, W-G. and G. W. Tauxe (1971b). A proposed technique for identification of unconfined aquifer parameters. J. Hydrol., 12, 117-128.

CHAPTER 13

IDENTIFICATION FOR DISTRIBUTED PARAMETER CHEMICAL ENGINEERING SYSTEMS — FIXED-BED TUBULAR REACTORS

W. F. Ramirez

Department of Chemical Engineering, University of Colorado, Boulder, Colorado, USA

INTRODUCTION

A number of important chemical engineering problems have been treated using distributed parameter filtering and identification theory. These include problems dealing with fixed-bed catalytic tubular reactors, heat transfer, oil recovery, water pollution, and air pollution. In this chapter we review the scope of applications reported in the literature. We then focus on fixed-bed tubular reactors in order to illustrate the concepts needed to successfully implement dynamic and steady state distributed parameter filtering to chemical engineering problems.

SCOPE OF CHEMICAL ENGINEERING APPLICATIONS

Besides using distributed identification theory to estimate temperature and composition profiles in fixed-bed catalytic tubular reactors, a number of other important problems have been studied. These include heat transfer processes, oil recovery, water pollution, and air pollution.

The applications in heat transfer include the work of Ajinkya et al. (1975) on the heating of a slab in a furnace, Amouroux et al. (1975) who considered a multizone furnace problem, Beck (1970) who treated the heating of metal blocks, and Greiss and Ray's (1976) treatment of the casting process. Ingot heating has been considered by Lausterer et al. (1977), and heat exchanger identification by Wright and Bacon (1974).

A water pollution problem was considered by Yih and Davidson (1975) and air pollution by Bankoff and Hanzevach (1975) as well as Seinfeld and co-workers (Lamb, 1973; Kyan, 1973).

The identification of state variables and model parameters for oil recovery processes has proven to be another area where identification theory has yielded impressive results. Work includes that of Seinfeld (1980), Gavalas and Seinfeld (1973), Chen et al. (1974), Chavent (1976), Chavent and workers (1975), and Wasserman et al. (1975).

FIXED-BED CATALYTIC REACTORS

From the beginning, chemical engineers enterested in distributed parameter filtering and identification have realized the importance of applying theoretical results to the identification and estimation of temperature profiles, composition profiles, and catalyst activities for fixed-bed catalytic tubular reactor systems. Such reactors are at the heart of many commercial processes and improved operation and knowledge of these systems results in direct economic gains.

Early work included that of Seinfeld (1969) who considered the determination of the reaction velocity constant for the first-order decomposition of acetic anhydride in an isothermal laminar-flow tubular reactor with radial dispersion operated at steady-state conditions. Distributed parameter filtering theory was not used but instead a steepest descent algorithm was developed to minimize a least squares error fit of experimental data and model output. This sensitivity coefficient approach is not a true example of sequential identification but illustrates the traditional approach to identification of model parameters.

The on-line sequential estimation of state variables and kinetic parameters for plug flow reactors with slow catalyst decay was treated by Gavalas and Seinfeld (1969). In this work they reduced the distributed parameter estimation problem to a traditional Kalman Filter for lumped systems described by ordinary differential equations. By assuming a quasi-steady state, namely, that the compositions and the temperatures are always given by their steady state values and in addition that there are no major model uncertainties in these parameters, they formulated a Kalman filter for the catalyst decay parameter k given by the model

$$\frac{dk}{dt} = -\alpha k^m \tag{1}$$

where α and m are constants. They applied these concepts to a single irreversible isothermal dehydrogenation reaction.

$$A \rightarrow B + C \ . \tag{2}$$

Results showed this filter efficient in estimating the composition and the slowly varying rate constant, k, but not the rate decay parameters α and m.

A packed bed reactor in which n-pentane is isomerized to isopentane and cracked to gas and in which the catalyst is fouled by coke formed as a by-product was treated by Ajinkya et al. (1974). The experimental apparatus was designed so that samples of the flowing gaseous components could be withdrawn at discrete axial positions. At the end of a run, the catalyst was unloaded and the carbon content was determined by combustion. These authors developed a sequential dynamic filter for the single state of the carbon content on the catalyst using "measurements" of the coke profile computed indirectly from gas-phase concentration data. A complete distributed parameter dynamic filter was applied to this single state variable problem. The filter reconciled significant discrepancies between the model and measurements.

A complete distributed parameter sequential dynamic filter was also used by Seinfelt et al. (1971) to estimate the time dependent concentration for an isothermal, plug-flow chemical reactor with a second-order irreversible reaction,

$$2A \rightarrow B + C \tag{3}$$

This system illustrates the computational complexities associated with implementing a dynamic distributed parameter filter. The mathematical model for the composition profile is,

$$\frac{\partial c}{\partial t} + \frac{\partial c}{\partial z} = -\beta c^2 \tag{4}$$

with boundary conditions

$$c\,(z,o) = c_{ss}(z) \tag{5}$$

$$c\,(o,t) = \alpha_o(t) \tag{6}$$

$$\frac{d\alpha_o}{dt}(t) = \xi(t) \; ; \; \alpha_o(o) = 1 \tag{7}$$

where c = composition
ξ = random variable excitation

The measurements are assumed to be at discrete spatial locations, z_i and corrupted with noise $\eta_i(t)$,

$$y(z_i,t) = x(z_i,t) + \eta_i(t) \qquad i = 1, z, \ldots m$$

where m = number of measurement locations along the reactor.

The steady-state solution to the deterministic model of equation (4) is

$$c_{ss}(z) = \frac{1}{(\frac{1}{\alpha_o} + \beta z)} \tag{8}$$

Optimal estimates $\hat{c}(z,t)$ and $\hat{\alpha}_o(t)$ for this distributed parameter system with correlated noise given by a covariance $P(z,z',t)$ are,

$$\frac{\partial \hat{c}(z,t)}{\partial t} + \frac{\partial \hat{c}(z,t)}{\partial z} = -\beta \hat{c}^2(z,t) + \sum_{i=1}^{m} P(z,z_i,t)[y(z_i,t) - \hat{c}(z_i,t)] \tag{9}$$

with $\hat{c}(s,t) = \hat{\alpha}_o(t)$ (10)

$$\frac{d\hat{\alpha}_o(t)}{dt} = \sum_{i=1}^{m} P(z_i,o,t)[y(z_i,t) - \hat{c}(z_i,t)] \tag{11}$$

The covariance $P(z,z',t)$ is computed from

$$\frac{\partial P(z,z',t)}{\partial t} + \frac{\partial P(z,z',t)}{\partial z} + \frac{\partial P(z,z',t)}{\partial z'} = -2\beta[\hat{c}(z,t)P(z,z',t)$$

$$+ \hat{c}(z',t)P(z,z',t)] - \sum_{i=1}^{m} P(z,z_i,t)P(z_i,z',t) \tag{12}$$

with

$$\frac{\partial P(z,o,t)}{\partial t} + \frac{\partial P(z,o,t)}{\partial z} = -2\beta\hat{c}(z,t)P(z,o,t) - \sum_{i=1}^{m} P(z_i,o,t)P(z,z_i,t) \tag{13}$$

$$\frac{dP(o,o,t)}{dt} = -\sum_{i=1}^{m} P^2(z_i,o,t) + Q \tag{14}$$

where Q is the covariance of the random variable excitation, ξ, and the initial condition is

$$P(z,z',o) = a\ \exp(-|z-z'|) \tag{15}$$

Numerical methods are needed to solve even this single state variable identification set of equations. A forward explicit finite difference method can be employed. The general idea is to replace the partial derivatives by first order finite difference approximations so that an explicit, marching-in-time and distance solution is obtained. For this problem we have three independent variables: time, t, and the spatial variables z and z'. Two spatial variables are required due to the correlated nature of spatial noise. For a variable u, first order finite difference formulas are

$$\frac{\partial u}{\partial t}(z,z',t) = \frac{u_{i,j,n+1} - u_{i,j,n}}{\Delta t} \tag{16}$$

$$\frac{\partial u}{\partial z}(z,z',t) = \frac{u_{i,j,n} - u_{i-1,j,n}}{\Delta z} \tag{17}$$

$$\frac{\partial u}{\partial z'}(z,z',t) = \frac{u_{i,j,n} - u_{i,j-1,n}}{\Delta z'} \tag{18}$$

Using these first order finite difference algorithms, a simple explicit solution scheme results. For $\Delta z = \Delta z' = .005$ and $\Delta t = .005$ when we have 3 equally spaced measurement locations, rapid convergence to steady state (20 time steps)was obtained in the face of noisy measurements. The entire filter calculations required about 1 second of CPU time on the University of Colorado's Cyber 172 system.

Soliman and Ray (1979) discuss the development of a distributed filter for systems with two different time scales. The system they considered was a packed bed tubular reactor with fast composition dynamics and a slow thermal wave or dynamic response. They derived a fast estimator for estimating composition dynamics and a separate slow estimator for both temperature and composition. This later filter used a quasi-steady state assumption for the fast concentration effect. Three continuous-in-time temperature measurements and a single continuous-in-time exit composition measurement were assumed. Even with this data their fast filter for composition dynamics did not converge and therefore was useless. The slow filter gave reasonable results for both temperature and composition over the slower time scale. The estimation of fast dynamic effects coupled with slow dynamic terms remained a problem area.

Kuruoglu, Ramirez, and Clough (1981) consider the identification of composition and temperature profiles down a fixed-bed catalytic styrene reactor along with the identification of the catalytic activity for the main dehydrogenation reaction. Since the process time constants for their pilot scale reactor were in the order of one second while the reaction activity time constant was of the order of days, a quasi-steady-state assumption was used where the process state dynamics were neglected. A new steady-state estimator assuming uncorrelated spatial noise was developed to estimate state profiles and a dynamic lumped parameter Kalman Filter was used to estimate the catalyst activity. Both filters were shown to be efficient and usable for on-line implementation. Using optimally estimated composition and temperature data in the dynamic filter significantly improved the estimation of catalyst activity. Raw measurement errors are of the order of 2% which leads to a filtered catalyst activity response shown in Figure 1. The true

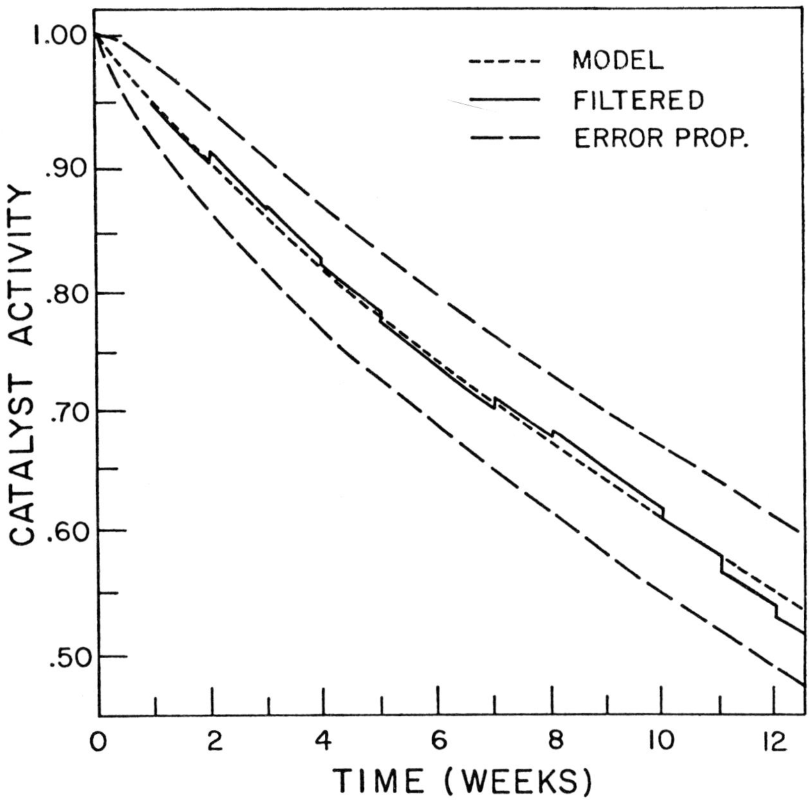

Fig. 1 Identification of catalytic activity

activity is shown as the dotted curve. One standard deviation error bounds on the mathematical model are given by the dashed curves. The activity estimates are updated sequentially every week and are shown by the solid curve. Figure 2 shows the catalyst identification when steady-state filtered composition and temperature data are used. The improvement in the identification of catalyst activity is obvious. A 2% measurement error has essentially been reduced to the level of 0.5%

STEADY-STATE SEQUENTIAL DISTRIBUTED FILTERING

Recent results from our laboratory allow for the inclusion of spatially correlated noise for the estimation of steady-state distributed processes. In this work we take advantage of previously developed distributed filter theory (Tau, 1969) and a new concept of steady-state estimation for distributed systems.

The steady-state form of the system to be considered can be represented as

$$\underline{o} = \underline{f}(\underline{x}, \frac{\partial \underline{x}}{\partial z}) + \underline{\nu}(z) \tag{19}$$

where $\underline{\nu}(z)$ is a random variable added to right-hand side to represent model uncertainties. It is assumed to be white noise with a covariance,

$$E\{\underline{\nu}(z)\underline{\nu}^T(z')\} = \underline{\underline{Q}}(z,z') \text{ and a mean } E\{\underline{\nu}(z)\} = \underline{0} \tag{20}$$

The measurements are linear functions of the state and taken at a finite number of discrete points along the spatial domain:

$$\underline{y}(t_j, z_i) = \underline{\underline{H}}_i \underline{x}(z_i) + \underline{n}_{ji} \quad i = 1, \ldots m, \quad j = 1, \ldots \tag{21}$$

where $\underline{n}_{ji}$ is a white measurement noise characterized by

$$E\{\underline{n}_{ji}\underline{n}^T_{lk}\} = \underline{\underline{R}}_{ji}\delta_{jl}\delta_{ik} \quad \text{and} \quad E\{\underline{n}_{ji}\} = \underline{0} \tag{22}$$

and m = number of measurement points.

The basic computational procedure for steady-state sequential filtering is as follows: First, the deterministic form of the model given by Equation (19) is solved for an a priori estimate of the states, $\hat{\underline{x}}^-(z)$. Then using a set of measurements, updated optimal estimates, $\hat{\underline{x}}^+(z)$ are obtained using the updating algorithm derived below. Improved estimates are used as a basis for processing the next data set. For the sake of simplicity, the subscripts representing time in Equations (21) and (22) will be dropped during the derivation of the filter equations since one set of measurements is taken at a time.

We assume that the form of the optimal estimates immediately after a measurement set is given by

$$\hat{\underline{x}}^+(z) = \hat{\underline{x}}^-(z) + \sum_{i=1}^{m} \underline{\underline{K}}(z, z_i)[\underline{y}(z_i) - \underline{\underline{H}}_i\hat{\underline{x}}^-(z_i)] \tag{23}$$

Following Tau (1969), the gain $\underline{\underline{K}}(z, z_i)$ is determined by minimizing the covariance of the a posteriori estimate

$$\underline{\underline{P}}^+(z,z') = E\left\{\tilde{\underline{x}}^+(z)\left(\tilde{\underline{x}}^+(z')\right)^T\right\} \tag{24}$$

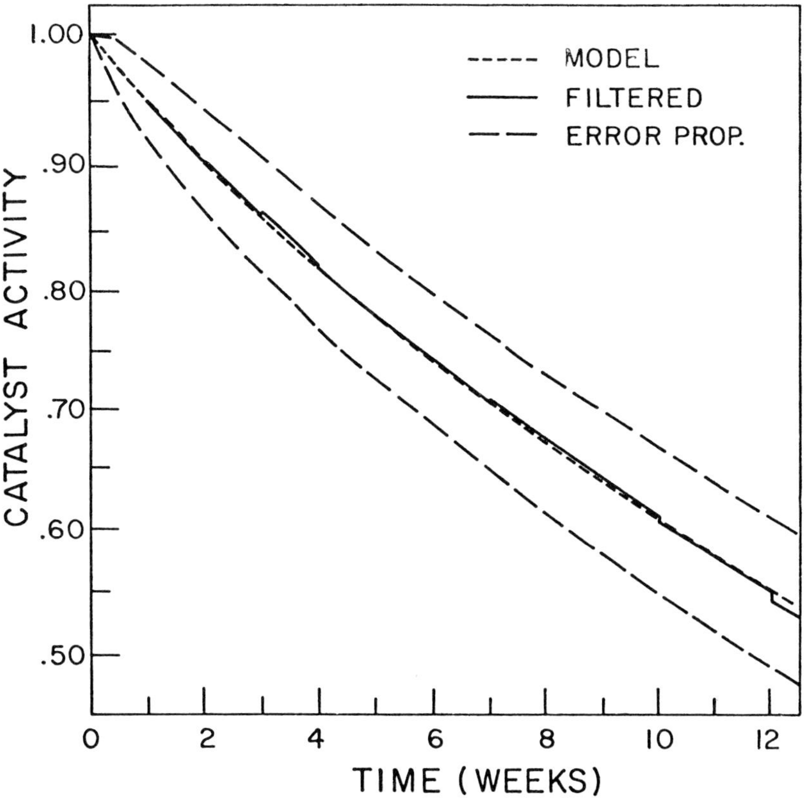

Fig. 2 Identification of catalytic activity

where $$\underline{\tilde{x}}^{+}(z) = \underline{x}(z) - \underline{\hat{x}}^{+}(z) \tag{25}$$

In Equation (25), $\underline{x}(z)$ = true value and $\underline{\hat{x}}^{+}(z)$ = a posteriori estimate.

Rearranging Equation (23) using Equations (21) and (25) gives

$$\underline{\tilde{x}}^{+}(z) = \underline{\tilde{x}}^{-}(z) - \sum_{i=1}^{m} \underline{\underline{K}}(z, z_i)[\underline{n}_i + \underline{\underline{H}}_i \tilde{x}^{-}(z_i)] \tag{26}$$

where $$\underline{\tilde{x}}^{-}(z_i) = \underline{x}(z_i) - \underline{\hat{x}}^{-}(z_i) \tag{27}$$

Multiplying Equation (26) by its transpose at z' gives

$$\begin{aligned}
\tilde{x}^{+}(z)\left(\tilde{x}^{+}(z')\right)^{T} &= \underline{\tilde{x}}(z')\left(\underline{\tilde{x}}^{-}(z')\right)^{T} \\
&- \sum_{i=1}^{m} \underline{\underline{K}}(z, z_i)\underline{n}_i\left(\underline{\tilde{x}}^{-}(z')\right)^{T} + \underline{\underline{H}}_i\underline{\tilde{x}}^{-}(z_i)\left(\underline{\tilde{x}}^{-}(z')\right)^{T} \\
&- \sum_{i=1}^{m}\left[\underline{\tilde{x}}^{-}(z)\underline{n}_i^{T} + \underline{\tilde{x}}^{-}(z)\left(\underline{\tilde{x}}^{-}(z_i)\right)^{T}\underline{\underline{H}}_i^{T}\right]\underline{\underline{K}}^{T}(z',z_i) \\
&+ \sum_{i=1}^{m}\sum_{j=1}^{m} \underline{\underline{K}}(z, z_i)\underline{n}_i\underline{n}_j^{T}\underline{\underline{K}}^{T}(z',z_j) \\
&+ \sum_{i=1}^{m}\sum_{j=1}^{m} \underline{\underline{K}}(z,z_i)\underline{\underline{H}}_i\tilde{x}^{-}(z_i)\underline{n}_j^{T}\underline{\underline{K}}^{T}(z',z_j) \\
&+ \sum_{i=1}^{m}\sum_{j=1}^{m} \underline{\underline{K}}(z,z_i)\underline{\underline{H}}_i\underline{\tilde{x}}^{-}(z_i)\left(\underline{\tilde{x}}^{-}(z_j)\right)^{T}\underline{\underline{H}}_j^{T}\underline{\underline{K}}^{T}(z',z_j)
\end{aligned} \tag{28}$$

We take the expected value of both sides of Equation (28) using the fact that the measurements are uncorrelated and the definition of Equations (22) and (24) to obtain

$$\begin{aligned}
P^{+}(z,z') &= P^{-}(z,z') - \sum_{i=1}^{m} \underline{\underline{K}}(z,z_i)\underline{\underline{H}}_i\underline{\underline{P}}^{-}(z_i,z') \\
&- \sum_{i=1}^{m} P^{-}(z,z_i)\underline{\underline{H}}_i^{T}\underline{\underline{K}}(z',z_i) + \sum_{i=1}^{m} \underline{\underline{K}}(z,z_i)\underline{\underline{R}}_i\underline{\underline{K}}^{T}(z',z_i) \\
&+ \sum_{i=1}^{m} \underline{\underline{K}}(z,z_i)\underline{\underline{H}}_i\underline{\underline{P}}^{-}(z_i,z_i)\underline{\underline{H}}_i^{T}\underline{\underline{K}}^{T}(z',z_i)
\end{aligned} \tag{29}$$

Differentiating Equation (29) with respect to

$\underline{\underline{K}}(\cdot,z_i)$, i = 1,2, ..., m and setting the result to zero gives the optimum gain,

$$\underline{\underline{K}}(z,z_i) = \underline{\underline{P}}^{-}(z,z_i)\underline{\underline{H}}_i^{T}\left(\underline{\underline{H}}_i P^{-}(z_i,z_i)\underline{\underline{H}}_i^{T} + \underline{\underline{R}}_i\right)^{-1} \tag{30}$$

With this result for the optimal gain, the covariance matrix after measurement becomes:

$$\underline{\underline{P}}^+(z,z') = \underline{\underline{P}}^-(z,z') - \sum_{i=1}^{m} \underline{\underline{P}}(z,z_i)\underline{\underline{H}}_i^T\left(\underline{\underline{H}}_i P^-(z_i,z_i)\underline{\underline{H}}_i^T + \underline{\underline{R}}_i\right)^{-1}\underline{\underline{H}}_i\underline{\underline{P}}^-(z_i,z') \quad (31)$$

As each set of data is obtained, Equations (23) and (31) are solved with $\hat{x}^-(z)$ and $\underline{\underline{P}}^-(z,z')$ equal to the previously updated values $\hat{\underline{x}}^+(z)$ and $\underline{\underline{P}}^+(z,z')$. To start the process initial values for the state and error covariance $\hat{\underline{x}}^-(z)$ and $\underline{\underline{P}}^-(z,z')$ are needed. State values, $\hat{x}^-(z)$, are obtained by the solution of the steady-state deterministic model equations. The error covariance, $\underline{\underline{P}}^-(z,z')$ is calculated solving the steady-state form of the dynamic model error propagation prediction equation

$$\frac{\partial \underline{\underline{P}}}{\partial t} = 0 = \left(\frac{\partial \underline{f}}{\partial \underline{x}}\right)_z \underline{\underline{P}}(z,z') + \underline{\underline{P}}(z,z')\left(\frac{\partial \underline{f}}{\partial \underline{x}}\right)^T_{z'} + \left(\frac{\partial \underline{f}}{\partial \underline{x}_z}\right)\left(\frac{\partial \underline{\underline{P}}(z,z')}{\partial z}\right)$$

$$+ \left(\frac{\partial \underline{\underline{P}}(z,z')}{\partial z'}\right)\left(\frac{\partial \underline{f}}{\partial \underline{x}_{z'}}\right) + \underline{\underline{Q}}(z,z') \quad (32)$$

where $\underline{x}_z = \frac{\partial \underline{x}}{\partial z}$ and $\underline{x}_{z'} = \frac{\partial \underline{x}}{\partial z'}$.

The general unsteady-state form of Equation (32) is presented by Tzafestas (1978).

The sequential steady-state distributed filter derived is tested using the mathematical model for the fixed-bed catalytic reactor used for the thermal dehydrogenation of ethylbenzene to form styrene monomer presented by Clough and Ramirez (1976). Measurements are obtained by adding random noise generated by a Gaussian white noise generator to the deterministic values. Since there are ten distinct chemical components plus temperature, a complete treatment would necessitate the calculation of covariance and gain matrices, $\underline{\underline{P}}$ and $\underline{\underline{K}}$, dimensioned eleven by eleven. To ease the computational burden, a simplified model using only temperature and the three major components, i.e., ethylbenzene, styrene, and benzene, is considered. Other components can be evaluated based on the filtered values of these four states using deterministic mass balances.

Throughout the simulation studies, a 1% model uncertainty and a 2% measurement noise are assumed for each of the states. Also an exponential distribution function was used for the spatial correlation of the model uncertainty, namely,

$$\underline{\underline{Q}}_{diag}(z,z') = 0.0001 \exp(-|z - z'|) \quad (33)$$

and

$$\underline{\underline{R}}_{diag}(z_i, z_i) = 0.0004 \qquad , i = 1, \ldots, 5 \quad (34)$$

Also, $\underline{\underline{P}}_{diag}(0, z') = 0.0004 \exp(-z')$ (35)

was used when solving Equation (32). This condition means that a 2% level of uncertainty was assumed at the inlet boundary which was also distributed into the spatial domain with an exponential decay function. The solution to the partial differential equations in z and z' of this problem was carried out using a simple first-order explicit finite-difference technique. Computations were performed on a CYBER 172 system at the University of Colorado.

Assuming that the true values are given by the model values, three different cases were studied. In case number 1, there are no temperature measurements and concentrations are measured at the exit. For case number 2, temperature measurements are taken at z = 0.5 and concentrations are measured at the exit. And finally, in case number 3, there are four temperature measurements at z = 0.2, 0.4, 0.6, and 0.8 and concentration measurements at the exit. For all these three cases, the temperature estimates and error propagations are presented in Figures 3-5. These results show that with only one exit concentration measurement, the internal level of uncertainty remains high, close to the level without measurement. However, with internal temperature measurements this uncertainty level is reduced significantly so that the confidence placed on the filtered states is much larger.

For the results presented in Figures 3-5, only one iteration or set of data is processed. As the number of measurement locations is increased, elements of covariance matrix, $\underline{\underline{P}}$, go to zero immediately showing that convergence is extremely fast. This behavior was also observed for discrete Kalman filters for lumped parameter systems (Jo and Bankoff, 1976) when the initial model errors are large compared to measurement errors.

This extremely rapid convergence is not very desirable since the converged filter value is based on very little actual data. To decrease the convergence rate of the covariance matrix, $\underline{\underline{P}}$, Equations (23), (29), and (30) were modified based on the number of measurement locations as follows:

$$\underline{\hat{x}}^{+}(z) = \underline{\hat{x}}^{-}(z) + \frac{1}{m}\sum_{i=1}^{m}\underline{\underline{K}}(z,z_i)\left\{\underline{y}(z_i) - \underline{\underline{H}}_i\underline{\hat{x}}^{-}(z_i)\right\} \tag{36}$$

$$\underline{\underline{P}}^{+}(z,z') = \underline{\underline{P}}^{-}(z,z') - \frac{1}{m}\sum_{i=1}^{m}\underline{\underline{K}}(z,z_i)\underline{\underline{H}}_i^T\underline{\underline{P}}^{-}(z_i,z') \tag{37}$$

$$\underline{\underline{K}}(z,z_i) = \underline{\underline{P}}^{-}(z,z_i)\underline{\underline{H}}_i\left[\underline{\underline{H}}_i\underline{\underline{P}}^{-}(z_i,z_i)\underline{\underline{H}}_i^T + m\underline{\underline{R}}(z_i,z_i)\right]^{-1} \tag{38}$$

In Equation (38), $\underline{\underline{R}}\,(z_i,z_i)$, the measurement noise level, is multiplied by the number of measurement locations, m, to reflect an increase in the uncertainty, as the number of measurement locations increase. In Equations (36) and (37), the summed quantities are multiplied by $\frac{1}{m}$ which gives us an "average" over all measurement locations. By making these modifications the $\underline{\underline{P}}$ decay is rather slow allowing more measurements to be processed and effect the converged filtered values. The concepts used in these modifications are similar to those used to develop limited memory filters for lumped systems (Jo and Bankoff, 1976). We have essentially proposed a limited memory distributed parameter filter.

Two more cases using the modified version of the steady-state distributed filter are considered. The $\underline{\underline{R}}(z_i,z_i)$, $\underline{\underline{Q}}(z_i,z')$, and $\underline{\underline{P}}(0,z')$ values are unchanged and given by Equations (33)-(35). The results for these last two cases were obtained after processing 20 measurement sets instead of the single set used previously. Measurements taken were for internal temperature measurements and concentration measurements as described in case 3. For case 4, the true values correspond to model values. The results of the estimated temperature, styrene, and benzene profiles are shown in Figures 6, 7, and 8, respectively. Comparison of Figures 5 and 6 gives a clear indication of the advantage of the modified filter and processing more measurements. The temperature and concentration estimates are very close to the true values despite the high level of measurement noise.

Next, in case 5, the model was simplified by disregarding the water-gas shift reactions which are considered to be of secondary importance. This simplification

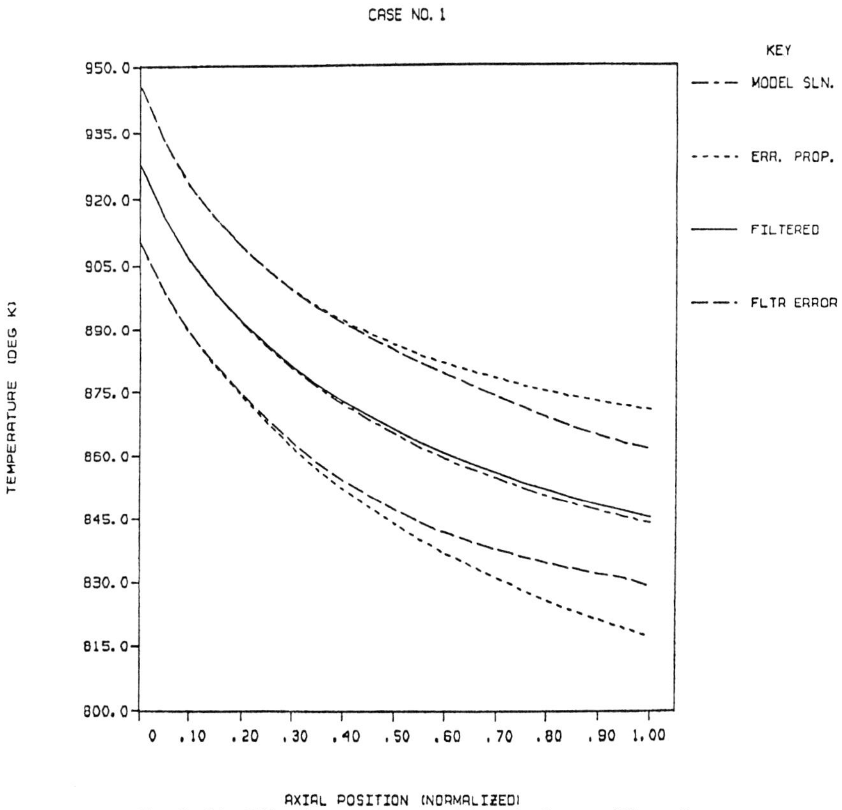

Fig. 3 Identification ased on one data set of compositions at reactor exit

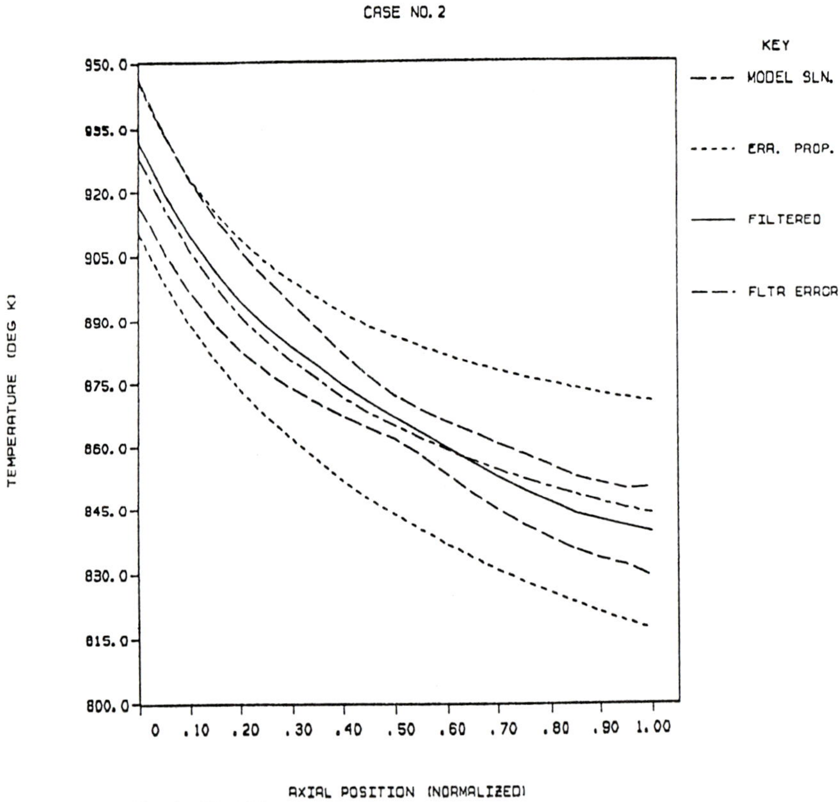

Fig. 4 Identification based on one data set of compositions at reactor exit and temperature at z = .5

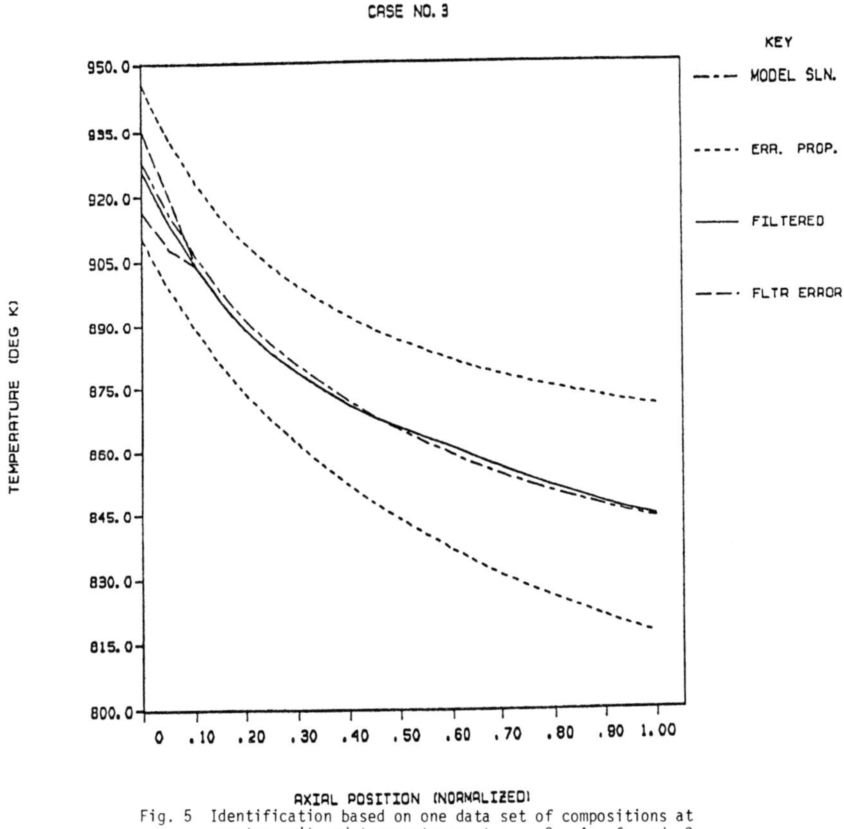

Fig. 5 Identification based on one data set of compositions at reactor exit and temperatures at z = .2, .4, .6, and .8

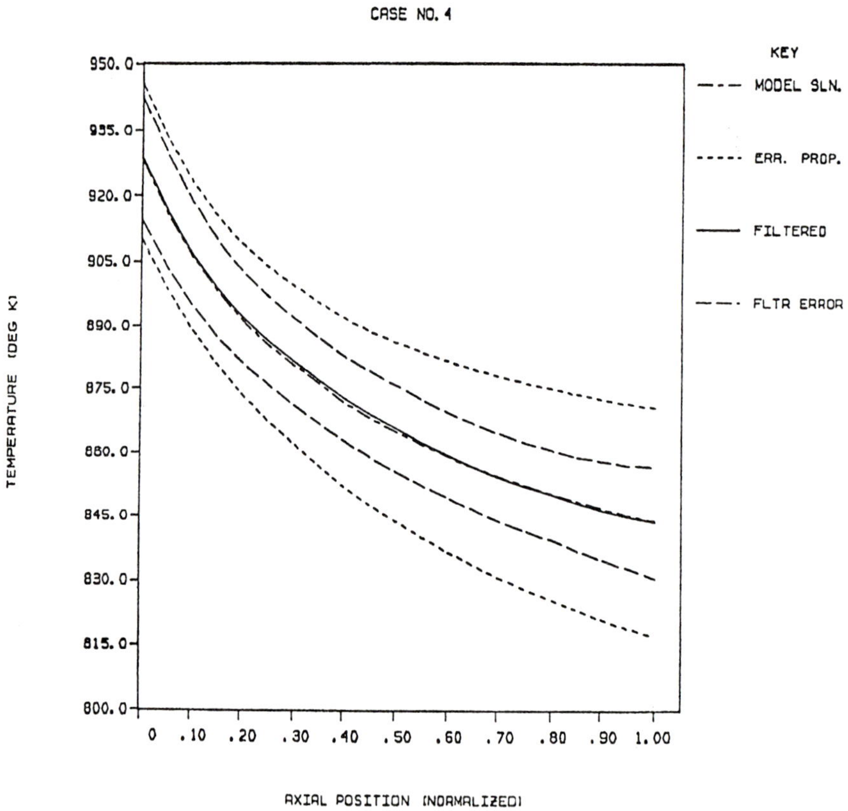

Fig. 6 Identification based on limited memory filter and twenty data sets

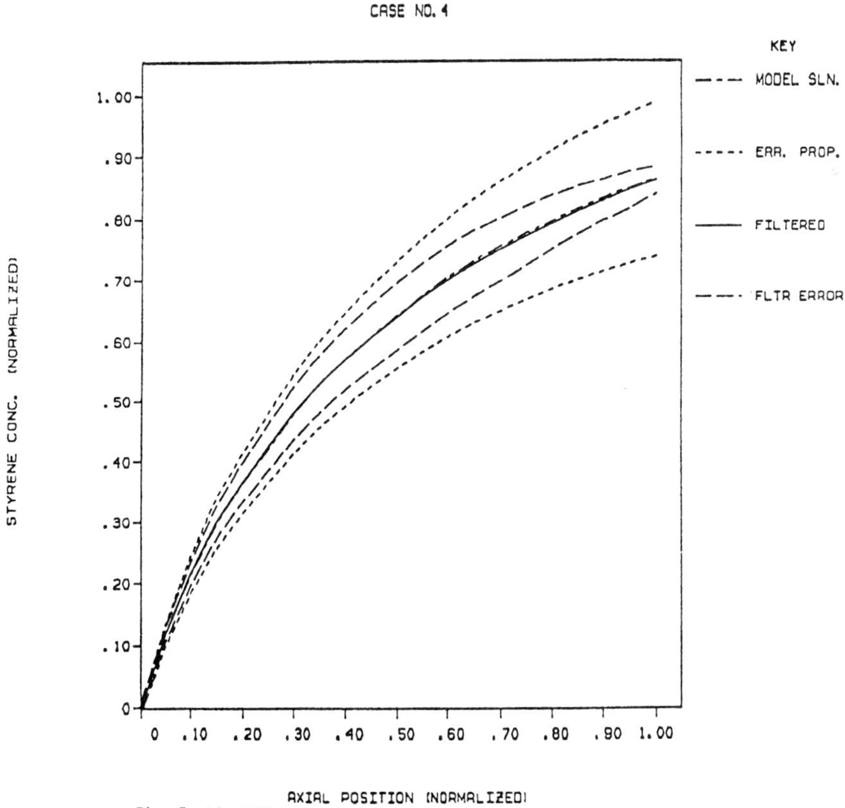

Fig. 7 Identification based on limited memory filter and twenty data sets

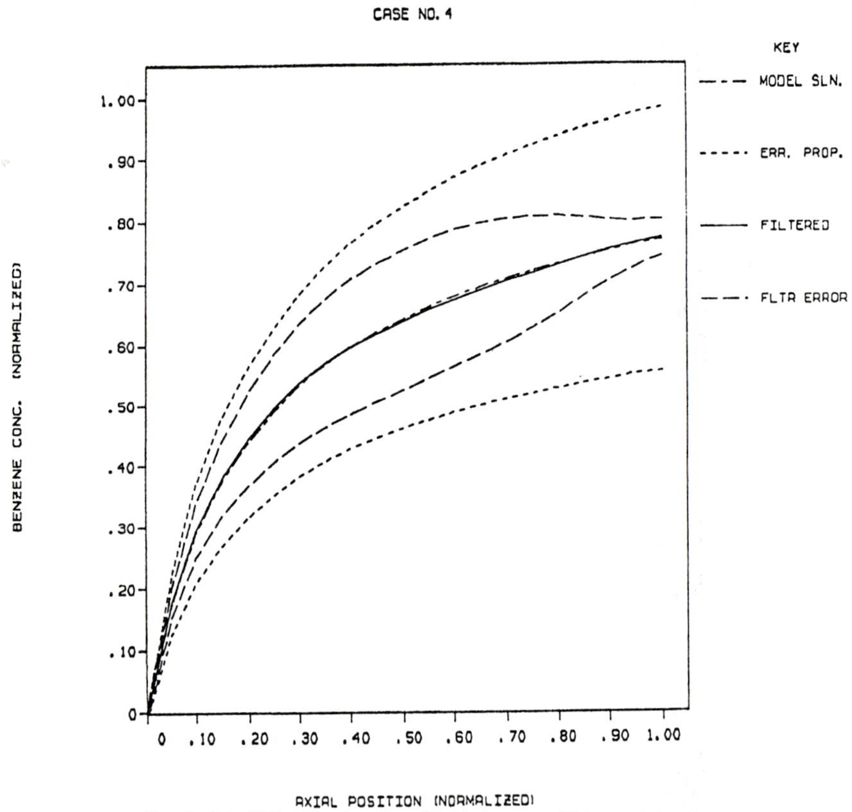

Fig. 8 Identification based on limited memory filter and twenty data sets

reduces the complexity of the mathematical model easing the computational burden and therefore making it more feasible for on-line usage. The true values correspond to the model values of the complete model. Figures 9, 10, and 11 show the results for temperature, styrene, and benzene profiles. The estimated profiles for the temperature and styrene concentration are very close to the true profiles. For the low concentration by-product benzene, filtered concentration profiles are best at the entrance and exit sections of the reactor. However, in the middle part of the reactor, there is a negative drift in the estimated profile. This is due to the extremely high nonlinearity of the system and the fact that our distributed filter is based on linear theory. Also, as seen in Figure 6, the confidence limits for benzene are rather broad in the center sections with just internal temperature measurements. Such a drift is, therefore, not surprising for a trace by-product component such as benzene. Figure 12 shows the actual convergence rate of the filtered temperature at a measurement location. The modified limited memory filter converges around the tenth measurement to the actual, rather than the approximate, model value in the face of rather large measurement noise.

Finally, the modified sequential steady-state distributed filter was tested using actual data obtained by Clough (1975) during his experimental program. The data used in this computation consist of 15 sets of four internal temperature measurements taken at $z = 0.2$, 0.4, 0.6, and 0.8, and compositions at the exit. These data were taken during a near steady-state operation of a laboratory-scale styrene Pilot Plant over a period of 11 hours. The temperature measurements had a standard deviation of less than 2% whereas the concentration measurements were found to have standard deviations of about 5%. The temperature and composition measurements were averaged and compared with the model predictions and it was found that the temperature predictions deviated by about 30%. Based on these initial analyses of data, the following parameters are defined for the steady-state distributed filtering:

For temperature

$$Q_{11}(z,z') = 0.001 \exp(-|z - z'|) \tag{39}$$

$$R_{11}(z_i,z_i) = 0.0004 \quad , i = 1, \ldots 4 \tag{40}$$

$$P_{11}(0,z') = 0.0004 \exp(-z') \tag{41}$$

For compositions:

$$Q_{nn}(z,z') = 0.0225 \exp(-|z - z'|) \tag{42}$$

$$R_{nn}(z_i,z_i) = 0.0025 \exp(-|z - z'|) \quad , i = 5 \tag{43}$$

$$P_{nn}(0,z') = 0.0025 \exp(-|z - z'|) \tag{44}$$

where $n = 2, 3, 4$.

In Equations (39)-(44) the subscripts represent the corresponding diagonal elements of the related matrices. Equations (39)-(44) mean that we have defined a 1% model uncertainty and a 2% measurement error for temperature, and a 15% model uncertainty and a 5% measurement error for the compositions. Some of the results obtained from the distributed filtering of actual data are presented in Figures 13 and 15. Figures 13 and 14 show the temperature and styrene composition profiles, respectively. When compared with the averaged values of the data, the filtered profiles are very good and represent the actual reactor profiles better than either the model predictions or any single set of measurements. Figure 15 shows the temperature measurements and the filtered values at the measurement location $z = 0.8$.

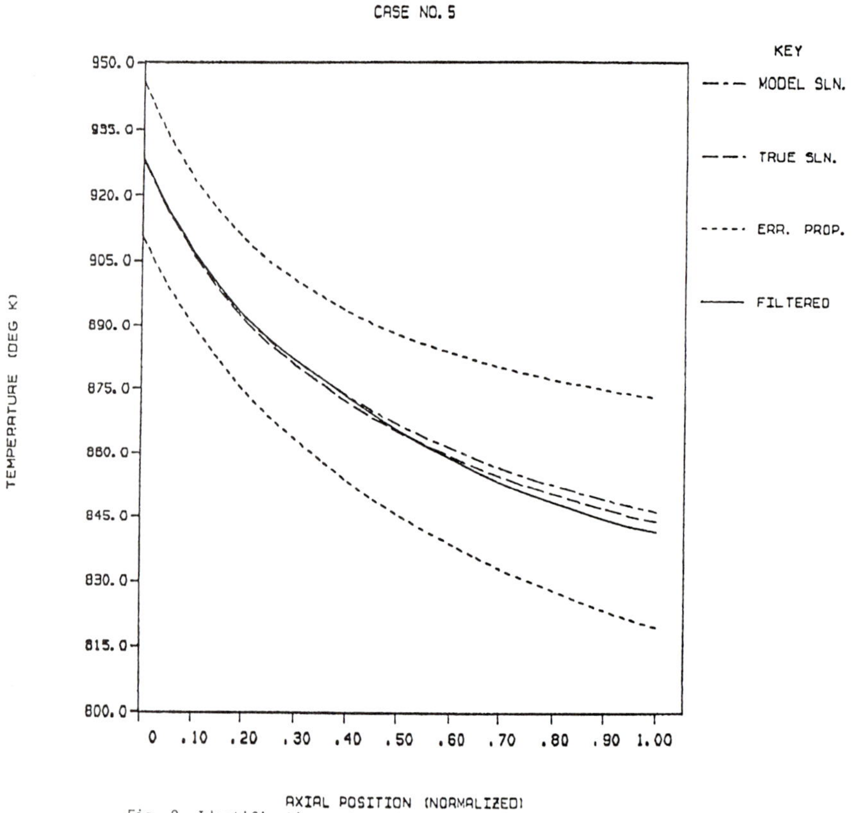

Fig. 9 Identification using simplified mathematical model and limited memory filter

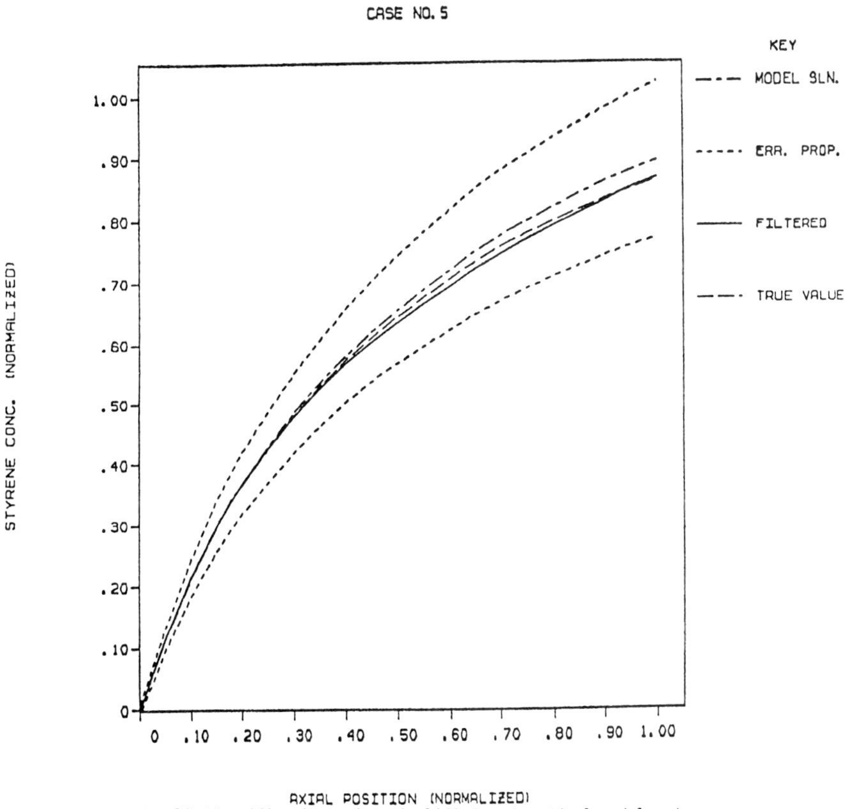

Fig. 10 Identification using simplified mathematical model and limited memory filter

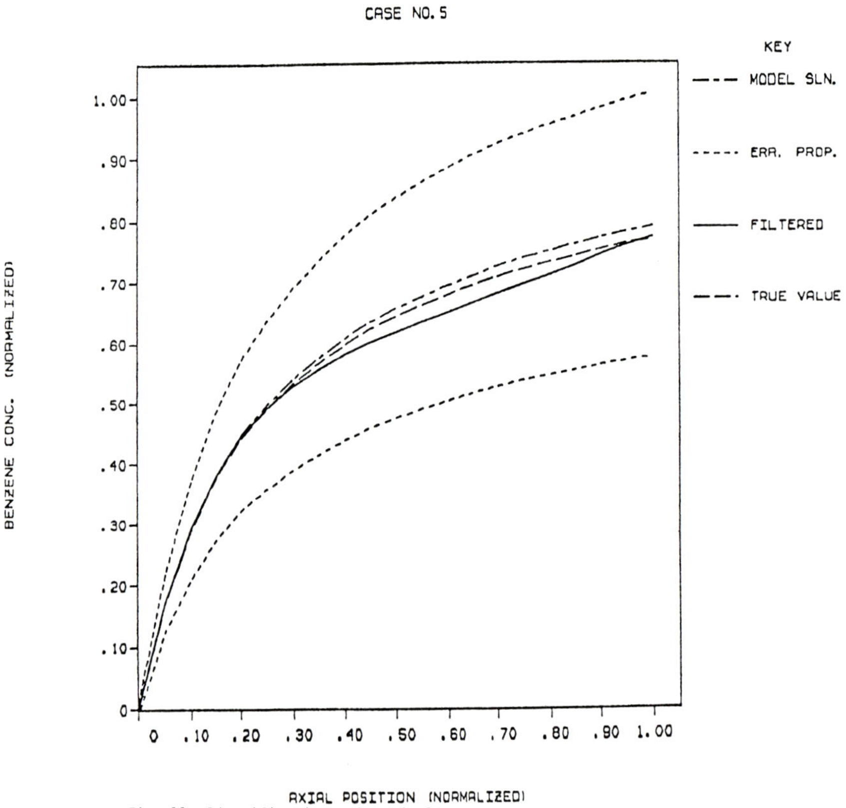

Fig. 11 Identification using simplified mathematical model and limited memory filter

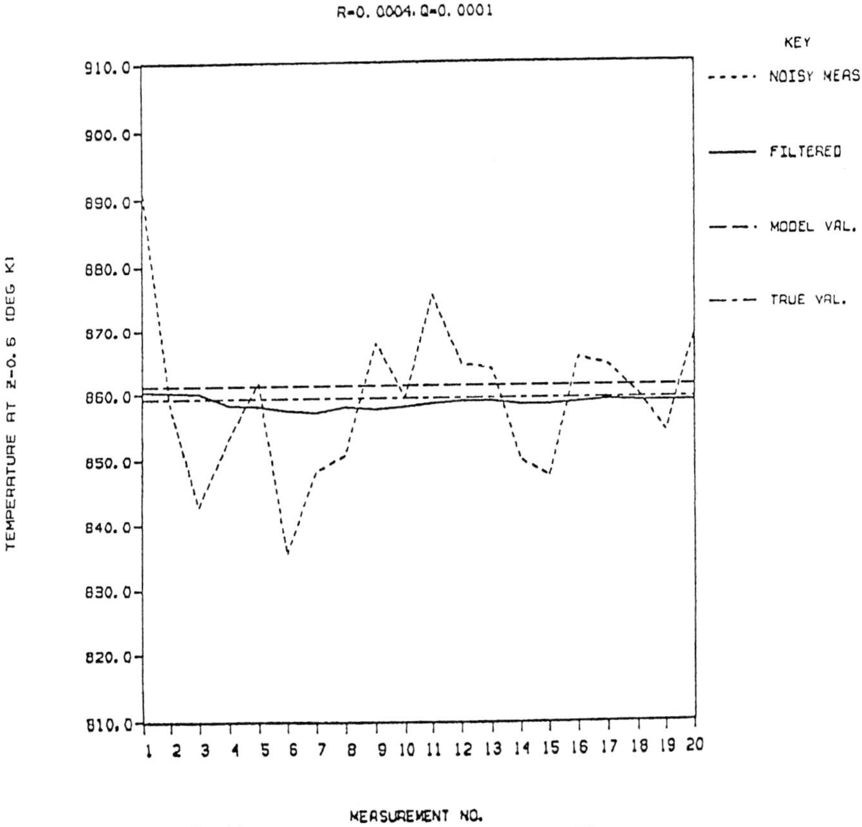

Fig. 12 Convergence rate of limited memory filter

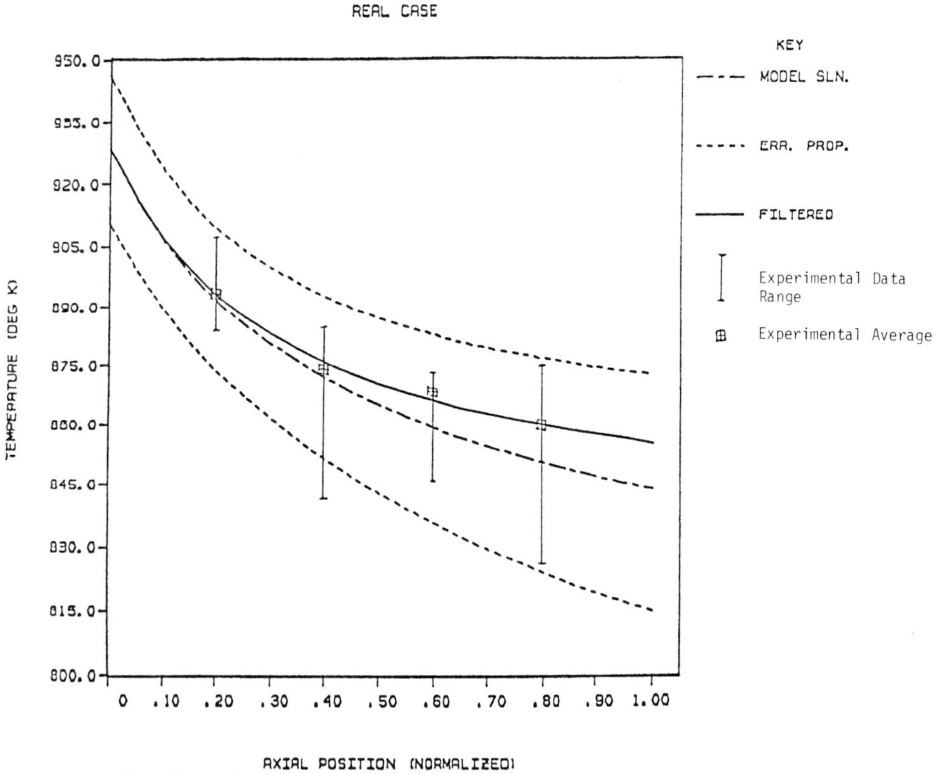

Fig. 13 Identification using pilot-plant experimental data

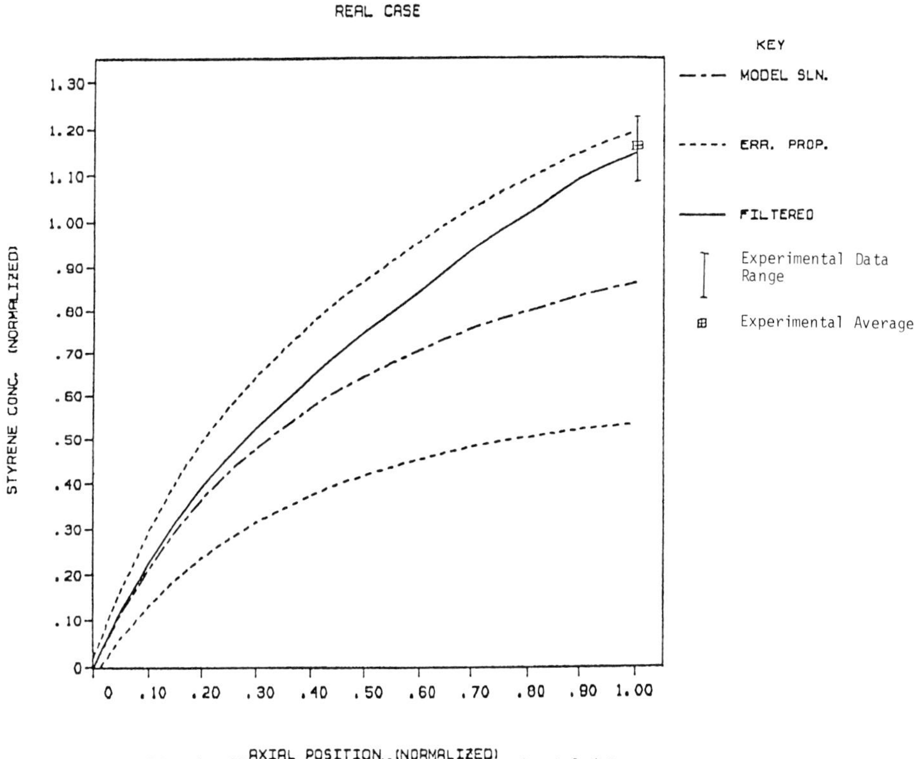

Fig. 14 Identification using pilot-plant experimental data

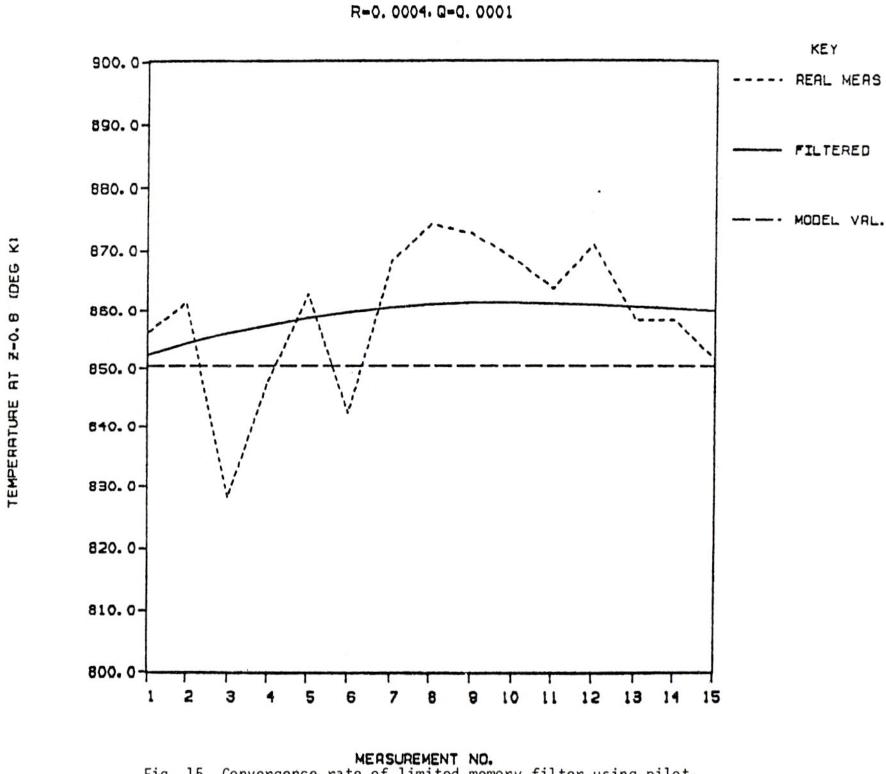

Fig. 15 Convergence rate of limited memory filter using pilot-plant experimental data

REFERENCES

Ajinkya, M. B., M. Kohne, H. F. Mader, and W. H. Ray (1975). The experimental implementation of a distributed parameter filter. Automatica, 11, 571.
Ajinkya, M. B., G. F. Froment, and W. H. Ray (1974). On-line estimation of catalyst activity in packed-bed reactors having catalyst decay. I&EC Proc. Des. Dev., 13, No. 2, 107.
Amouroux, M., J. P. Babary, and Y. Queminer (1975). Controle par calculateur numerique de la temperature d'un four d'elaboration de materiaux et composants electronique. Proc. 3rd Cong. Nat. Inf. y Auto., Madrid, Oct.
Bankoff, S. G. and E. L. Hanzevack (1975). The adaptive filtering transport model for prediction and control of pollutant concentrations in an urban airshed. Atmos. Env., 9, 793.
Beck, J. V. (1970). Nonlinear estimation applied to the nonlinear inverse heat conduction problem. Int. J. Heat and Mass Transfer, 13, 703.
Chavent, G. (1976). Diphasic incompressible flow in porous media. Lecture Notes in Math. No. 503, Springer-Verlag.
Chavent, G., M. Duprey, and P. Lemonnier (1975). History matching by use of optimal control theory. S.P.E. J., 15, 74.
Chen, W. H., G. R. Gavalas, J. H. Seinfeld, and M. O. Wasserman (1974). A new algorithm for automatic history matching. S.P.E. J., 14, 593.
Clough, D. E. (1975). Optimization and control of the dehydrogenation of ethylbenzene to form styrene. Ph.D. Thesis, University of Colorado.
Clough, D. E. and W. F. Ramirez (1976). Mathematical modeling and optimization of the dyhydrogenation of ethylbenzene to form styrene. AIChE J., 22, 1097.
Gavalas, G. R. and J. H. Seinfeld (1969). Sequential estimation of states and kinetic parameters in tubular reactors with catalyst decay. Chem. Eng. Sci., 24, 625.
Gavalas, G. R. and J. H. Seinfeld (1973). Reservoirs with spatially varying properties. S.P.E. J., 13, 335.
Greiss, F. K. and W. H. Ray (1976). The application of distributed parameter state estimation theory to a metallurgical casting operation. Proc. IFAC/IFIP Symposium on New Trends in Systems Theory, Paris.
Jo, J. H. and S. G. Bankoff (1976). Digital monitoring and estimation of polymerization reactors. AIChE J., 22, 361.
Kuruoglu, N., D. E. Clough, and W. F. Ramirez (1981). Distributed parameter estimation and identification for systems with fast and slow dynamics. Chem. Eng. Sci., 36, 1357.
Kyan, C. P. and J. H. Seinfeld (1973). Real-time control of air pollution. AIChE J., 19, No. 3, 579.
Lamb, R. G. and J. H. Seinfeld (1973). Mathematical modeling of urban air pollution. Env. Sci. Tech., 7, 253.
Lausterer, G. K., H. R. Martens, and W. H. Ray (1977). The real time application of distributed parameter state estimation theory to a two-dimensional heated ingot. Proc. 2nd IFAC Symp. on Distributed Parameter Control Theory, Coventry.
Seinfeld, J. H. (1969). Identification of parameters in partial differential equations. Chem. Eng. Sci., 24, 65.
Seinfeld, J. H., G. R. Gavalas, and M. Hwang (1971). Nonlinear filtering in distributed parameter systems. J. Dyn.Syst., Meas. and Control, 157, Sept.
Soliman, M. S. and W. H. Ray (1979). Nonlinear state estimation of packed-bed tubular reactors. AIChE J., 25, 718.
Tau, F. E. (1969). An optimum filtering for a class of linear distributed-parameter systems. J. Basic Engr., 91, 173.
Tzafestas, S. G. (1978). In W. H. Ray and D. G. Lainiotis (Eds.), Distributed Parameter Systems--Identification, Estimation, and Control. Dekker, New York.

Wasserman, M. L., A. E. Emanuel, and J. H. Seinfeld (1975). Practical applications of optimal control theory to history matching multiphase simulator models. S.P.E. J., 15, 347.

Watson, A. T., G. R. Gavalas, J. H. Seinfeld, and P. T. Woo (1980). History matching in two-phase petroleum reservoirs. S.P.E. J., 20, 521.

Wright, J. D. and D. W. Bacon (1974). Statistical identification of the dynamic behaviour of a dual input heat exchanger network. Can. J. Ch.E., 52, 405.

Yih, S. M. and B. Davidson (1975). Identification in nonlinear distributed parameter water quality models. Water Resources Res., 11, 696.

CHAPTER 14

DISTRIBUTED PARAMETER HEAT EXCHANGERS — MODELING, DYNAMICS AND CONTROL

H. Kanoh

Mechanical Engineering Department, Faculty of Engineering, Osaka University, Suita, Osaka, Japan

1. INTRODUCTION

Heat exchangers are used extensively in nearly every industrial process such as power plant, internal-combustion engine, gas turbines, air-conditioning systems, refineries and myriad chemical plants. With the evolution of many processes, precise and fast operations for startup, shutdown, emergency situation and load change were required. The traditional design based on steady-state data became inadequate, and attention has been focused on the understanding and evaluation of the dynamic behavior of heat exchangers. Since P.Profos [1] made the first dynamic model for a simple heat exchanger in 1943 and Y.Takahashi [2] presented the first transfer functions for the common heat exchangers in 1951, many studies have been done. This chapter is intended to be a guide to mathematical modeling and fundamental properties of dynamics of heat exchangers. As no attempt has been made to survey all parts of research in this field recently, a historical review of modeling and dynamics is given in Section 2. The basic framework for modeling is constructed and various mathematical models are developed in Section 3. Section 4 is devoted to a discussion of calculation of frequency responses in general manner by which transfer functions of typical heat exchangers are derived. A new result on general transfer functions of one-shell multi-pass heat exchangers is given here. Interpretation of some aspects of dynamic behavior of heat exchangers is discussed in Section 5. Lumped parameter-simulation technique for heat exchangers is touched upon in Section 6. Various control problems concerning heat exchangers are surveyed in Section 7. The publications in this field are collected extensively in the bibliography section [1-206]. The literature is classified and cited in chronological order. Although we cannot refer to all of them, it is hoped that the readers can rely on the title and the data in the Bibliography to know the activities in each subject area. (Numbers in brackets designate literature in the Bibliography, and references without number will be found in the References.)

2. HISTORY OF RESEARCH ON MODELING AND DYNAMICS OF HEAT EXCHANGERS

(i) Dynamics of one-dimensional-flow heat exchangers

The first modeling of the dynamic behavior of the heat exchanger was made by Profos (1943)[1] for a percolation-type heat exchanger in his Dissertation. But the boom in modeling and investigation of dynamics came in the 1950s. The pioneer was Takahashi (1951)[2,3] who presented transfer functions for common heat exchangers and investigated controllability by means of stability limits. In 1954, Dusinberre

[5] presented a numerical method to get the transient response of heat exchangers, and Rizika [7] considered a special case of parallel-flow and counter-flow heat exchangers in which the relatively thick intermediate wall dominates the dynamics and the dynamics of the fluids can be neglected. Cohen and Johnson (1956)[9] studied the dynamics of a double-pipe heat exchanger, and they found the resonance phenomenon in the frequency response which had been predicted by Debolt (1954)[6]. Clark and his co-workers made a series of studies on the dynamic responce of a single-tube heat exchanger (1958)[14] [16] [22] with internal heat sources. They confirmed experimentally the existance of the resonance in their reply to discussers for their paper [22]. Masubuchi (1958)[17], (1960)[20] has presented mathematical models for various types of multipass heat exchangers and their transfer functions, and found that the characteristic equations for the one-shell multipass heat exchanger are essentially third order.

In the 1960s, the emphasis was gredually switched from studying the dynamic response with inlet temperature change or that of one-dimensional-flow heat exchangers to the more practical and more difficult problem of flow-forced dynamics or the dynamics of cross-flow heat exchangers. These will be discussed under separate items. In 1967, Noldus [36] studied analytically the dynamics of parallel-flow and counter-flow heat exchangers and he showed that counter-flow dynamics corresponds essentially to the behavior of a second-order system, and that parallel-flow dynamics correaponds to that of a second-order system with a dead time. Plate-heat exchanger systems were studied extensively by Ito and Masubuchi [103, 104, 105] in 1975-77. Bayonet-heat exchangers were studied by Todo [48, 54] in 1976. Muller (1977)[100] investigated the dynamic behavior of the finned surface in heat exchangers. The overall dynamics can be composed under the special conditions of the transfer function of the heat exchanger without the fin effect and the transfer function of the finned surface. Experimental results of frequency response verified theoretical results. Furusho, Kanoh & Masubuchi (1979)[101] made a distributed parameter model of the plate-fin-type heat exchanger. Transfer functions were derived from three-dimensional analysis, the validity of which is verified by experemental results. Kanoh (1979)[98] studied the type of dynamics of heat exchangers and defined the type of dynamics and showed that some heat exchangers switch their dynamics from one type to the other type as the parameter changes.

(ii) Flow forced dynamics

In the early 1960s, the study of the dynamics of flow-forced heat exchangers began. Models of such processes are bilinear, so called "nonlinear" from the process-dynamics point of view, because the heat-transfer coefficients are dependent on flow rate. (Strictly speaking, the heat-transfer coefficients are also dependent on temperature, but this dependence is often neglected.) Although a simple first-order model has been solved exactly, almost all studies use linearized models which are derived from a perturbation analysis. In 1961, Hempel [59] presented the first linearized equation assuming that the heat-transfer coefficient is constant with respect to temperature, but dependent on the fluid velocity. His experimental results show the existence of resonance in flow-forced dynamics. An exact treatment was shown by Koppel (1962)[60] for a nonlinear first-order partial differential equation describing flow-forced percolator dynamics. He compared the exact solution with the approximate linearized solution for a step change in velocity and found that significant error can result in the approximate solution for large magnitudes of step change. But Stermole and Larson (1963)[62] showed that the linearized model was in good agreement with experimental data over the full range checked for small flow rate change. Privott and Ferrell (1966)[67] made mathematical models of flow-forced liquid-liquid heat exchangers by taking account of heat-transfer coefficients dependent on both fluid velocity and temperature but they did not derive transfer function expressions. IN 1967, Todo [69] made an investigation of parallel and counter-heat exchanger dynamics with flow-rate change and obtained transfer functions as in extended form of Takahashi's results. Gilles (1974)[73] presented transfer functions of the counter-flow heat exchanger with flow-rate change by taking into account variations of the heat transfer coefficients with respect to fluid flow-rate and temperatures.

(iii) Cross-flow heat exchangers

The dynamic behavior of cross-flow heat exchangers is described by a partial differential equation with two independent spatial variables as well as with an independent time variable. Therefore, the analysis of them is more complex problem. There are two types of cross-flow heat exchangers; a tube cross-flow heat exchanger and a plate cross-flow heat exchanger. The former is used in building heating and air conditioning. A primary fluid flows inside the tube and a secondary fluid flows normally to the tube over the outer surface. In the analysis of this type, the primary fluid is assumed to be mixed, that is, it is at a uniform temperature across its flow passage which contains the entire flow. Both fluids are considered to be unmixed in the plate cross-flow heat exchanger.

In 1959, Dusinberre [75] showed a method of numerical calculation for a special case of a cross-flow heat exchanger with both fluids unmixed. Gartner and Harrison [76] presented a basic equation describing the tube cross-flow heat exchanger and obtained a transfer function in 1963. They modified this model in 1965 [77] to a more simple one in which the secondary fluid dynamics is neglected. Myer and Mitchell [79] made a fundamental equation of the plate cross-flow heat exchanger, but they did not find the transfer function. Masubuchi [83] treated analytically three types of cross-flow heat exchangers; both fluids unmixed, one fluid mixed the other unmixed, and both fluids mixed. He made exact transfer functions by using the two-dimensional Laplace Transformation. In 1969, Gartner and Danne [81] developed a mathematical model for the serpentine cross-flow heat exchanger with fluid velocity change. Their experimental results obtained by employing a periodic binary perturbation method correlated well with their model computations. Tamm [82] extended the model of Gatner and Danne to multirow cross-flow heat exchangers, and Shekar and Green (1970) [41] treated the same problem as Tamm which has a slight difference in the surface condition of coil. Bender [86] studied analytically the plate cross-flow heat exchanger with fluid velocity change and applied its result to the tube cross-flow heat exchanger. Terasaka, Kanoh & Masubuchi (1979) [145] considered the dynamics of a rotary regenerative heat exchanger which has a two-dimensional temperature distribution. The heat exchanger was analyzed as a problem of coupling two cross-flow heat exchangers. The Galerkin's method was used to approximate the temperature profiles of the rotating matrix along the boundaries.

(iv) Simulation of heat exchangers

(a) Numerical simulation of heat exchanger dynamics

It is necessary to use numerical computation methods for solving the equations of the mathematical models. Many numerical computation methods and approximation methods to facilitate the computation have been developed for solving the equations. Schmit [140] has surveyed this field extensively.

There are four evolutionary phases according to the development of the computation facilities over the past 30 years: electrical circuits analog, analog computer, hybrid analog computer, and digital computer. In the first three phases, the partial differential equations are reduced to a set of ordinary differential equations using finite difference techniques. Usually the spatial derivative is replaced by a finite difference expression. This procedure is called the parallel method, which guarantees the computational stability but requires a huge amount of equipment to secure accuracy.

Electrical circuits simulation In 1956, Mozley [112] used a passive electrical network to predict the dynamics of a concentric pipe heat exchanger. He approximated the distributed parameter heat exchanger with five-lumped sections in the longitudinal direction. Each section is simulated by R-C networks. The experimantal results obtained indicated that the analogue transfer functions were sufficiently accurate for control purposes. Kourim (1957)[114] simulated the parallel-flow and counter-flow heat exchangers by electrical circuits. Cima & London (1958)[12] used the four-lumped electrical analogue for the two-fluid exchanger to simulate a gas-turbine regenerator. Kummel (1969)[127] adopted a variable conductor with high-speed switch and a buffer amplifier to simulate a heat exchanger controlled by manipulation of the flow-rate.

Analog computer Friske, Morris, Otto & Williams (1960)[116] studied dynamics and control of a one-shell four-pass heat exchanger using the six-section model. This six-section model required 121 operational amplifiers, 3 servo multipliers and 199 coefficient potentiometers. Privott & Ferrell (1967)[124] simulated a flow-forced, concentric-tube, liquid-liquid heat exchanger by analog computer, taking account of the nonlinear dependence of heat-transfer coefficients on temperature and flow rate.

Hybrid computer and digital computer In the use of the hybrid computer, the finite difference is taken in time instead of in distance. This procedure is called the series method or CSDT (Continuous Space Discrete Time) method. In the CSDT method, the equations derived are integrated continuously with respect to the spatial variable and this integration is repeated for successive time steps. Although economy of equipment can be achieved, variables in one-time step have to be stored for use at the next step, so that a function-store playback device is required. Schmidt & Clark (1969)[128] and Dunnett & Hutchinson [138] discussed CSDT analog simulation of parallel-flow and counter-flow heat exchangers. The transient response of heat exchangers can be calculated through the use of the digital computer by finite differencing the derivatives with respect to time and distance. Wood & Sastry (1972)[135] introduced a digital simulation technique known as CSMP (Continuous System Modelling Program). Bender (1972)[137] presented a finite difference model of crossflow heat exchanger which is suitable to digital calculation. He observed that for 50×50 spacial subdivisions no difference exists between the exact solution and the approximate solution. Choug & Schmidt (1976)[141] also discussed digital algorithms for dynamic analysis of parallel-flow and counter-flow heat exchangers and showed that the algorithms are fully verified against the experimantal results.

(b) Approximation of heat exchanger dynamics

Approximation of transfer functions Exact transfer functions of heat exchangers have, in general, complex transcendental forms and are inconvenient to treat with. The most popular method of approximating the transfer functions is replacement of them with a dead time and 1st- or 2nd-order lag, the parameters of which are determined by the curve fitting in the frequency response diagram. Although many of the papers used this method, this approach suffers from no relationship between the parameters determined by fitting and the original ones; each application will require a separate fitting of the parameters.

Paynter & Takahashi (1956) [110] developed a method of evaluating an approximate transfer function from the exact one by expanding the logarithm of the exact transfer function. The coefficients in the expansion (called "cumulants") are used for approximate transfer functions.

The alternative approach is the successive-approximation method used by Schöne (1966) [120]. Schöne's method assumes an approximate-transfer function of double-pipe heat exchangers to be

$$G(s) = \frac{m + Ts}{1 + Ts} \frac{1 - e^{-(a_1 + b_1 s)}}{a_2 + b_2 s}$$

where a_1, a_2, b_1, b_2 are constants which are determined analytically from the high frequency limit of the exact transfer function and m is determined from the low frequency limit, and T is arbitrary. Friedly (1967) [122] modified Schöne's method and made comparison of the approximation with the exact dynamics for a double-pipe heat exchanger with several conditions, multi-pass heat exchangers and a packed-bed reactor. He concluded that it is possible to approximate the dynamics of a variety of plug-flow processes and that this method guarantees good agreements between approximate and exact transfer functions at both the high and low frequencies.

Lumped dynamical model of heat exchangers As a model for control system design, a lumped-parameter dynamical model is frequently required. Besides the obvious finite difference method, there are two kinds of methods to obtain the dynamical model; the cell method, which partitions a distributed process into a sequence of cells by finite difference concept and physical interpretation, and the method of weighted residuals (MWR) which uses a modal expansion.

Fricke, Morris, Otto & Williams (1960)[116] made a cell model (sectionalized model) of an industrial one-shell four-passes heat exchanger. The cell-model method is initiated by considering the process as a series of perfectly mixed cells with each representing an increment along the distance. They showed that the cell model gives good agreement with experimental deta and that the cell model required at least six cells to simulate the industrial multipass heat exchanger. Proksa (1962)[118] and Ball (1963)[119] also used cell model for calculating the frequency response or parallel-flow and counter-flow heat exchangers. The nonlinear cell model was developed by Ruedrich, Chongooranrd & Durbin (1970)[130], (1973)[72].

The MWR can be used to reduce the partial differential equation to an ordinary differential equation with respect to the time, which determines the approximate lumped-parameter dynamical model of the heat exchangers. This approach is usually quite superior to the procedure of finite differencing the space variables. In the latter case, the resulting state model might have more than 50 variables, which is too large a number to be used for the control system design. While MWR usually leads to only 4-5 variables; thus the control system of the heat exchanger can be synthesized. By using this approach which converts the problem into an equivalent lumped one, we can treat the problem through recent detailed results of the lumped parameter (especially linear) system. For applications of the MWR to various fields, the reader should refer to an inspiring book by Finlayson (1972).

Gordon-Clark (1966) developed a modified Galerkin method to analyze the connective system which contains heat exchangers. Stafford and Nightingale (1970)[132] used the generalised Fourier method to obtain causal time-domain descriptions of the dynamics of counter-flow heat exchangers, but their numerical results seemed not to be very good. Kanoh (1977)[143] studied MWR approximation of heat exchangers (monotubular, parallel-flow, counter-flow and multipass) and showed that for the simulation of these heat exchangers the MWR method requires 3-6 variables to describe a temperature profile. Terasaka, Kanoh & Masubuchi (1979)[146] made an approximate lumped parameter model of a cross flow heat exchanger by Galerkin's method and the collocation method. Several numerical examples show that the finite difference method requires considerably larger numbers of subdivisions (more than 50×50) to secure the same accuracy as the MWR model does (36th order) to simulate the static and dynamic characteristics.

3. MODELING OF HEAT EXCHANGERS

(i) Deriviation of basic equation
Consider short sections of fluid and wall exchanging heat mutually as shown in Fig.1. The following assumptions are imposed on the derivation of basic equation.

(1) The fluid temperature, velocity, and the wall temperature are uniform across the flow cross section and wall cross section respectively.

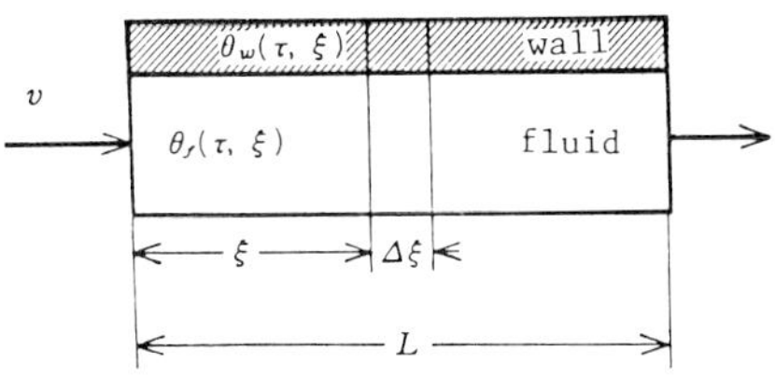

Fig.1 Section of a heat exchanger

(2) Axial heat conduction is negligible in both fluid and wall.
(3) The heat transfer coefficient is constant with length and time.
(4) The fluid is incompressible and all properties of fluid and wall are constant.
(5) There is no internal heat generation.
(6) The fluid is incompressible.

According to the first law of thermodynamics, the energy transferred in time interval $\Delta\tau$ to the section $(\xi, \xi+\Delta\xi)$ of the fluid must be equal to the change in the enthalpy of the fluid minus the change of pressure:

$$\Delta Q = \rho_f S_f \Delta\xi \left(\Delta i - \frac{1}{\rho_f} \Delta p \right) \tag{1}$$

where ρ_f: Density [kg/m³], S_f: Cross sectional area [m²], Q: Transferred energy [kJ], i: Enthalpy [kJ/kg], p: Pressure [kPa], τ: Time [h], C: Specific heat [kJ/kg. K°].

Neglecting the friction loss and heat conductivity in fluid, we can assume that ΔQ can be written using Newton's law of cooling as follows:

$$\Delta Q = \frac{\alpha_1 A_1}{L} (\theta_w - \theta_f) \Delta\xi \Delta\tau \tag{2}$$

As the fluid is incompressible, the enthalpy change can be written as follows:

$$\Delta i = C_f \Delta\theta_f + \frac{1}{\rho_f} \Delta p. \quad C_f = C_p = C_v \tag{3}$$

Substituting Eqs.(2), (3) into (1) and taking a limit yields

$$\rho_f C_f S_f \frac{D\theta_f}{D\tau} = \frac{\alpha_1 A_1}{L} (\theta_w - \theta_f) \tag{4}$$

where $D/D\tau$ is a total differential which is composed of the sum of partial differentials and convective differentials,

$$\frac{D}{D\tau} = \frac{\partial}{\partial\tau} + v\frac{\partial}{\partial\xi} \tag{5}$$

Then Eq.(4) becomes

$$\frac{\partial\theta_f}{\partial\tau} + v\frac{\partial\theta_f}{\partial\xi} = \frac{\alpha_1 A_1}{\rho_f C_f S_f L} (\theta_w - \theta_f) \tag{6}$$

Similarly, by applying the balance for an elemental section of wall, we obtain

$$\frac{\partial\theta_w}{\partial\tau} = \frac{\alpha_1 A_1}{\rho_w C_w S_w L} (\theta_f - \theta_w) + \frac{\alpha_2 A_2}{\rho_w C_w S_w L} (\theta_a - \theta_w) \tag{7}$$

Making system parameters dimensionless forms, we obtain the following equations

$$\frac{\partial\theta_f}{\partial t} + u\frac{\partial\theta_f}{\partial x} = a_1 (\theta_w - \theta_f) \tag{8}$$

$$\frac{\partial\theta_w}{\partial t} = b_1 (\theta_f - \theta_w) + b_2 (\theta_a - \theta_w) \tag{9}$$

Boundary condition:

$$\theta_f(t, 0) = \theta_{fi}(t) \tag{10}$$

where

$$a_1 = \alpha_1 A_1 / \rho_f C_f S_f v_s, \quad b_1 = \alpha_1 A_1 / \rho_w C_w S_w v_s, \quad b_2 = \alpha_2 A_2 / \rho_w C_w S_w v_s, \quad t = \tau v_s / L, \quad x = \xi / L, \quad u = v / v_s$$

If the wall is very thin and its heat capacity is very small, Eq.(9) can be rewritten as follows:

$$\varepsilon\frac{\partial\theta_w}{\partial t} = \varepsilon b_1 (\theta_f - \theta_w) + \varepsilon b_2 (\theta_a - \theta_w)$$

where ε is a small number. Singular perturbating Eq.(9-1), we obtain

$$b_1(\theta_f - \theta_w) + b_2(\theta_a - \theta_w) = 0$$

or

$$\theta_w = (b_1\theta_f + b_2\theta_a)/(b_1 + b_2) \tag{9-1}$$

Substituting this expression into Eq.(8), we get

$$\frac{\partial \theta_f}{\partial t} + u\frac{\partial \theta_f}{\partial x} = a_1'(\theta_a - \theta_f),\quad a_1' = a_1 b_2/(b_1 + b_2) \tag{8-1}$$

Eqs.(8) and (9) are the basic equations from which many heat exchanger's equations can be yielded as shown in the following.

(ii) <u>One dimensional flow heat exchangers</u>

Eqs.(8), (9) (or (8-1)) are the basic equations from which many heat exchanger's equations can be yielded. Single-tube heat exchanger's equation is just the same as Eqs.(8), (9) (or (8-1)).

<u>One-shell multi-pass heat exchangers</u>

For simplicity, it is assumed that the effects of walls can be neglected. There are four cases as shown in Fig.2 where n indicates the number of passes. Applying Eq.(8-1) to both fluids we obtain

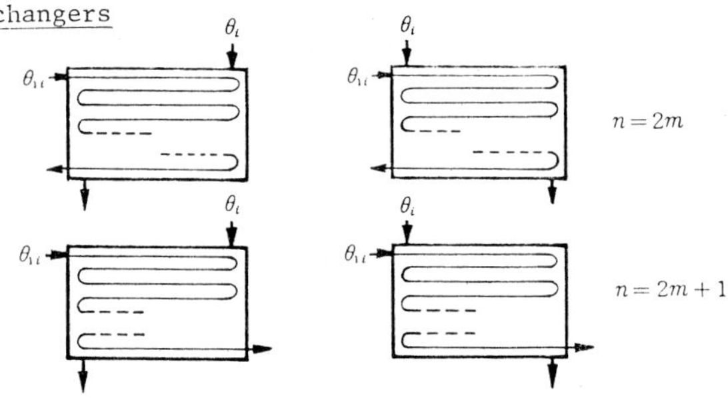

Fig.2 Flow types of multi-pass heat exchangers

$$\begin{cases} \dfrac{\partial \theta_k}{\partial t} + (-1)^{k-1} u_1 \dfrac{\partial \theta_k}{\partial x} = a_1(\theta - \theta_k) & k = 1, 2, \cdots, n \\ r\dfrac{\partial \theta}{\partial t} \pm u_2 \dfrac{\partial \theta}{\partial x} = \sum_{k=1}^{n} a(\theta_k - \theta) & \begin{array}{l} + \text{ for parallel start} \\ - \text{ for counter start} \end{array} \end{cases} \tag{11}$$

where θ_k represents the temperature of kth pass fluid from the entrance and θ represents the temperature of the shell fluid. Boundary conditions are given as follows:

$$\begin{cases} \theta_1(t, 0) = \theta_{1i}(t) & \\ \theta_{2k}(t, 0) = \theta_{2k+1}(t, 0), & k = 1, 2, \cdots, \left[\dfrac{n-1}{2}\right] \\ \theta_{2k-1}(t, 1) = \theta_{2k}(t, 1), & k = 1, 2, \cdots \left[\dfrac{n}{2}\right] \\ \theta(t, \{{}^{0}_{1}) = \theta_i(t) & \end{cases} \tag{12}$$

where the upper term corresponds to the parallel start and the lower term to the counter start and $[x]$ indicates the maximum integer not exceeding x.

(iii) <u>Flow forced heat exchangers</u>

The heat transfer coefficients are dependent of both the fluid temperature and the fluid velocity. But the temperature dependence is often neglected. In order to evaluate the heat transfer coefficient, numerous experiments have been made by many people and from these the value of the exponent for the fluid velocity has been established as 0.8 for turbulent flow in horizontal clean pipe. For other conditions, the exponent has various values according to circurmstances. In the following, a_1 and b_1 in Eqs.(8), (9) are treated as functions of the fluid velocity u.

<u>Single-tube heat exchanger</u>

The basic equations are given by Eqs.(8), (9). To obtain a feasible model, consider the behaviour about a steady state. For steady state, Eqs.(8), (9) are reduced to the following equations:

$$\frac{d\theta_{fs}}{dx} = a_1(\theta_{ws} - \theta_{fs}), \quad \theta_{fs}(0) = \theta_{fsi} \tag{8-s}$$

$$b_1(\theta_{fs} - \theta_{ws}) + b_2(\theta_{as} - \theta_{ws}) = 0 \tag{9-s}$$

where the subscript s indicates a steady state. (Note that $v_s = 1$)

Letting θ_{as} be uniform through the distance and solving Eqs.(8-s), (9-s), we obtain

$$\theta_{ws}(x) - \theta_{fs}(x) = \frac{b_2(\theta_{as} - \theta_{fsi})}{b_1 + b_2} e^{-a_1' x} = \mu e^{-a_1' x} \tag{13}$$

Then the variables are split into the steady state and a deviation from it as follows:

$$u = 1 + u(t), \quad \theta_f = \theta_{fs}(x) + \theta_f(x, t), \quad \theta_w = \theta_{ws}(x) + \theta_w(x, t) \tag{14}$$

where the deviations have the same names as the original ones not to increase the number of notations. Using the expression (14), we rewrite Eqs.(8), (9) as

$$\frac{\partial \theta_f}{\partial t} + (1+u)\frac{\partial \theta_f}{\partial x} = a_1(u)(\theta_w - \theta_f) + (a_1(u) - a_1 - a_1 u)(\theta_{ws}(x) - \theta_{fs}(x)) \tag{8-N}$$

$$\frac{\partial \theta_w}{\partial t} = b_1(u)(\theta_f - \theta_w) + b_2(\theta_a - \theta_w) + (b_1(u) - b_1)(\theta_{fs}(x) - \theta_{ws}(x)) \tag{9-N}$$

Using the Taylor's expansion, we can approximate the coefficients as follows:

$$a_1(u) = a_1 + c_1 u, \quad b_1(u) = b_1 + c_2 u \tag{15}$$

Substituting these relations into Eqs.(8-N), (9-N) yields

$$\frac{\partial \theta_f}{\partial t} + (1+u)\frac{\partial \theta_f}{\partial x} = (a_1 + c_1 u)(\theta_w - \theta_f) + (c_1 - a_1)\mu e^{-a_1' x} u \tag{8-B}$$

$$\frac{\partial \theta_w}{\partial t} = (b_1 + c_2 u)(\theta_f - \theta_w) + b_2(\theta_a - \theta_w) - c_2 \mu e^{-a_1' x} u \tag{9-B}$$

These equations represent <u>a bilinear model</u> of flow-forced heat exchanger which is used for a considerably large range of fluid velocity change. Dropping the product terms from bilinear equations, we obtain the following <u>linearlized equations.</u>

$$\frac{\partial \theta_f}{\partial t} + \frac{\partial \theta_f}{\partial x} = a_1(\theta_w - \theta_f) + (c_1 - a_1)\mu e^{-a_1' x} u(t) \tag{8-L}$$

$$\frac{\partial \theta_w}{\partial t} = b_1(\theta_f - \theta_w) + b_2(\theta_s - \theta_w) - c_2 \mu e^{-a_1' x} u(t) \tag{9-L}$$

<u>One-shell-multipass heat exchangers</u>

Let the variables in Eq.(11) be split into steady states and deviations from them as follows:

$$\begin{cases} u_1 = 1 + u_1(t), \quad u_2 = 1 + u_2(t), \\ \theta = \theta_s(x) + \theta(x, t), \quad \theta_k = \theta_{ks}(x) + \theta_k(x, t), \quad k = 1, 2, \cdots, n \end{cases} \tag{16}$$

where the deviations have the same names as the original ones not to increase the number of notations and the steady states $\theta_s(x)$ and $\theta_{ks}(x)$ satisfy the following equations and boundary conditions (17) with steady state inputs.

$$\begin{cases} (-1)^{k-1} \dfrac{d\theta_{ks}}{dx} = a_1(\theta_s - \theta_{ks}), \quad k = 1, 2, \cdots, n \\ \pm \dfrac{d\theta_s}{dx} = \displaystyle\sum_{k=1}^{n} a(\theta_{ks} - \theta_s) \end{cases} \tag{17}$$

Approximating the coefficients as follows:

$$\begin{cases} a_1 = a_1 + c_1 u_1(t) + c_3 u_2(t) \\ a = a + c_2 u_1(t) + c_4 u_2(t) \end{cases} \tag{18}$$

and substituting Eqs.(16), (18) into Eqs.(11), we obtain the bilinear model

$$\begin{cases} \dfrac{\partial \theta_k}{\partial t} + (-1)^{k-1}(1+u_1)\dfrac{\partial \theta_k}{\partial x} = (a_1 + c_1 u_1 + c_3 u_2)(\theta - \theta_k) + ((c_1 - a_1)u_1 + c_3 u_2)(\theta_s(x) - \theta_{ks}(x)), \\ \qquad\qquad k = 1, 2, \cdots, n \end{cases}$$

$$r\frac{\partial \theta}{\partial t} \pm (1+u_2)\frac{\partial \theta}{\partial x} = \sum_{k=1}^{n}(a+c_2u_1+c_4u_2)(\theta_k-\theta) + \sum_{k=1}^{n}(c_2u_1+(c_4-a)u_2)(\theta_{ks}(x)-\theta_s(x)) \tag{11-B}$$

Dropping the product terms from the bilinear equations, we obtain the following linearized equations:

$$\begin{cases} \dfrac{\partial \theta_k}{\partial t} + (-1)^{k-1}\dfrac{\partial \theta_k}{\partial x} = a_1(\theta-\theta_k) + (\theta_s(x)-\theta_{ks}(x))((c_1-a_1)u_1(t)+c_3u_2(t)), \\ \qquad\qquad k=1, 2, \cdots, n \\ r\dfrac{\partial \theta}{\partial t} \pm \dfrac{\partial \theta}{\partial x} = \sum_{k=1}^{n}a(\theta_k-\theta) + \sum_{k=1}^{n}a(\theta_k-\theta) + \sum_{k=1}^{n}(\theta_{ks}(x)-\theta_s(x))(c_2u_1(t)+(c_4-a)u_2(t)) \end{cases} \tag{11-L}$$

(iv) Crossflow heat exchanger

The crossflow heat exchanger has two types as shown in Fig.3.

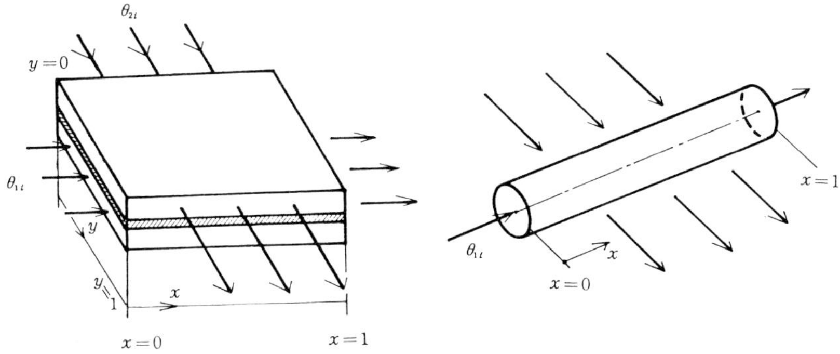

Fig.3(a) Plate type Fig.3(b) Tube type

(a) Plate-type-crossflow heat exchanger

The heat balance equations can be obtained by applying Eqs.(8), (9) to respective fluids as follows:

$$\begin{cases} \dfrac{\partial \theta_1}{\partial t} + u_1\dfrac{\partial \theta_1}{\partial x} = a_1(\theta_w-\theta_1) \\ \dfrac{\partial \theta_2}{\partial t} + u_2\dfrac{\partial \theta_2}{\partial y} = a_2(\theta_w-\theta_2) \\ \dfrac{\partial \theta_w}{\partial t} = b_1(\theta_1-\theta_w) + b_2(\theta_2-\theta_w) \end{cases} \tag{19}$$

$$\theta_1(t, 0, y) = \theta_{1i}(t), \quad \theta_2(t, x, 0) = \theta_{2i}(t) \tag{20}$$

(b) Tube-type-crossflow heat exchanger

In this type of heat exchanger, since the secondary fluid has a relatively short time to contact with the tube in which the primary fluid flows, the secondary fluid is considered to have an effective temperature $\bar{\theta}_2$ for heat transfer which is the average of the incoming temperature and the outgoing temperature. Considering the heat balance in the small element stirred completely, we obtain the following equation for $\bar{\theta}_2$ instead of the second equation in (19). (Gartner and Daane (1965)[77])

$$\begin{cases} \dfrac{\partial \theta_1}{\partial t} + u_1\dfrac{\partial \theta_1}{\partial x} = a_1(\theta_w-\theta_1) \\ \dfrac{\partial \bar{\theta}_2}{\partial t} = u_2(\theta_{2i}-\bar{\theta}_2) + a_2(\theta_w-\bar{\theta}_2) \\ \dfrac{\partial \theta_w}{\partial t} = b_1(\theta_1-\theta_w) + b_2(\bar{\theta}_2-\theta_w) \end{cases} \tag{21}$$

$$\bar{\theta}_2 = (\theta_{2i} + \theta_{2o})/2 \tag{22}$$

(v) Spiral heat exchanger

The spiral heat exchanger consists of a pair of concentric spiral passages whose cross section for flow is rectangular. There are four types of flow patterns as shown in Fig.4. Almost all commercial heat exchangers are used as the counter-flow type.

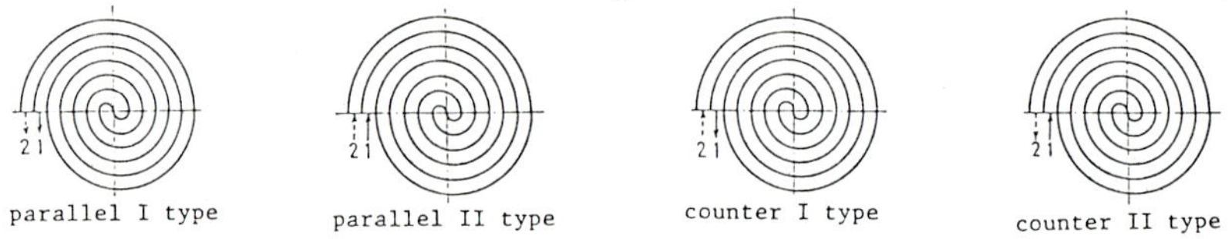

Fig.4 Flow patterns

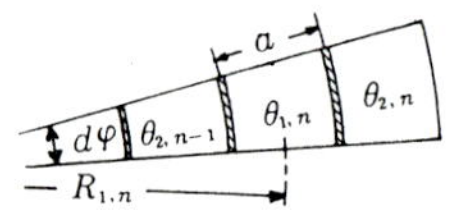

Fig.5 A small wedge element

By considering a heat balance on a small wedge of any "turn" as shown in Fig. 5, a series of n differential equations may be developed as follows:

$$\begin{cases} \dfrac{\partial \theta_{1,n}}{\partial \tau} \pm \dfrac{1}{2\pi N P_{1,n}} \dfrac{\partial \theta_{1,n}}{\partial \varphi} = \dfrac{a_1}{P_{1,n}} \{(\theta_{2,n-1} - \theta_{1,n}) P_{1,n}^i + (\theta_{2,n} - \theta_{1,n}) P_{1,n}^o\} & (n = 2, 3, \cdots, N) \\ r\dfrac{\partial \theta_{2,n}}{\partial \tau} \pm \dfrac{1}{2\pi N P_{2,n}} \dfrac{\partial \theta_{2,n}}{\partial \varphi} = \dfrac{a_2}{P_{2,n}} \{(\theta_{1,n} - \theta_{2,n}) P_{2,n}^i + (\theta_{1,n+1} - \theta_{2,n}) P_{2,n}^o\} & (n = 1, 2, \cdots, N-1) \end{cases}$$

$$\begin{cases} \dfrac{\partial \theta_{1,1}}{\partial \tau} \pm \dfrac{1}{2\pi N P_{1,1}} \dfrac{\partial \theta_{1,1}}{\partial \varphi} = \dfrac{a_1}{P_{1,1}} (\theta_{2,1} - \theta_{1,1}) P_{1,1}^o \\ r\dfrac{\partial \theta_{2,N}}{\partial \tau} \pm \dfrac{1}{2\pi N P_{2,N}} \dfrac{\partial \theta_{1,1}}{\partial \varphi} = \dfrac{a_2}{P_{2,N}} (\theta_{1,N} - \theta_{2,N}) P_{2,N}^i \end{cases} \tag{23}$$

where

$$R_{1,n} = (2n - 3/2)a + a\Phi/\pi \qquad P_{j,n} = R_{j,n}/L_1, \quad P_{j,n}^o = R_{j,n}^o/L_1, \quad P_{j,n}^i = R_{j,n}^i/L_1 \; (j = 1, 2)$$

$$R_{2,n} = (2n - 1/2)a + a\Phi/\pi$$

$$R_{1,n}^o = (2n - 1)a + a\Phi/\pi \qquad a_1 = \frac{kbL_1}{w_1 v_1}, \quad a_2 = \frac{kbL_1}{w_2 v_2}, \quad r = \frac{v_1}{v_2}$$

$$R_{1,n}^i = (2n - 2)a + a\Phi/\pi$$

$$R_{2,n}^o = (2na) + a\Phi/\pi \qquad \tau = \frac{t}{L_1/v_1}, \quad \varphi = \frac{\Phi}{2\pi N}$$

$$R_{2,n}^i = (2n - 1)a + a\Phi/\pi$$

a is the spacing between walls, L_1 is the total length of the passage.

The boundary conditions for each flow pattern are given at $\varphi = 0$ and at $\varphi = 1/N$ as follows:

(i) parallel I type

$$\left.\begin{aligned} &\theta_{1,n}(\tau, \varphi_1) = \theta_{1,n+1}(\tau, 0) \\ &\theta_{2,n}(\tau, \varphi_1) = \theta_{2,n+1}(\tau, 0) \\ &\theta_{1,1}(\tau, 0) = \theta_{1i}(\tau) \\ &\theta_{2,1}(\tau, 0) = \theta_{2i}(\tau) \end{aligned}\right\} (n = 1, 2, \cdots, N-1)$$

(ii) parallel II type

$$\left.\begin{aligned} &\theta_{1,n}(\tau, \varphi_1) = \theta_{1,n+1}(\tau, 0) \\ &\theta_{2,n}(\tau, \varphi_1) = \theta_{2,n+1}(\tau, 0) \\ &\theta_{1,N}(\tau, \varphi_1) = \theta_{1i}(\tau) \\ &\theta_{2,N}(\tau, \varphi_1) = \theta_{2i}(\tau) \end{aligned}\right\} (n = 1, 2, \cdots, N-1)$$

(iii) counter I type

$$\left.\begin{aligned} &\theta_{1,n}(\tau, \varphi_1) = \theta_{1,n+1}(\tau, 0) \\ &\theta_{2,n}(\tau, \varphi_1) = \theta_{2,n+1}(\tau, 0) \\ &\theta_{1,1}(\tau, 0) = \theta_{1i}(\tau) \\ &\theta_{2,N}(\tau, \varphi_1) = \theta_{2i}(\tau) \end{aligned}\right\} (n = 1, 2, \cdots, N-1)$$

(iv) counter II type

$$\left.\begin{aligned} &\theta_{1,n}(\tau, \varphi_1) = \theta_{1,n+1}(\tau, 0) \\ &\theta_{2,n}(\tau, \varphi_1) = \theta_{2,n+1}(\tau, 0) \\ &\theta_{1,n}(\tau, \varphi_1) = \theta_{1i}(\tau) \\ &\theta_{2,1}(\tau, 0) = \theta_{2i}(\tau) \end{aligned}\right\} (n = 1, 2, \cdots, N-1) \tag{24}$$

(vi) Rotary regenerative heat exchanger

The periodic-flow heat exchanger or rotary regenerator is used extensively in such a field as the gas turbine cycle of the exhaust-gas preheater for a boiler plant. The heat exchanger is described schematically as shown in Fig.6. The rotating matrix provides a flow of metal, from the hot gas to the cold gas, for the purpose of transporting the otherwise wasted thermal energy from exhaust gas to the feed gas.

The following assumptions are used to facilitate the analysis;

1) All matrix elements have uniform physical properties which are constant in time.
2) Temperatures of the fluids and matrix are uniform in the normal direction to the flow and the thermal conductivity of the matrix zero in the direction of fluid flow and matrix metal flow.
3) Entering fluid temperatures and flow rates are uniform over the flow cross section.
4) No mixing of the fluids occurs during the switch from hot to cold flow.
5) There is no gap between hot and cold sectors.

Consider a small wedge of the rotor shown in Fig.7, then

$$\text{mass of fluid} = R^2 \rho_f \delta \Delta\psi \Delta x/2$$

$$\text{mass of matrix} = R^2 (1-\delta) \rho_w \Delta\psi \Delta x/2$$

$$\text{heat transfer area} = A^* \Delta\psi \Delta x/P$$

where δ is matrix fraction voids to volume and A^* is heat transfer area per unit length along the flow. Energy balances in this element yield

Hot side: $0 \leqq \psi \leqq P_h$

$$\begin{cases} \dfrac{\partial\theta_{fh}}{\partial t} + u_h \dfrac{\partial\theta_{fh}}{\partial x} = \dfrac{k_h A_h^*}{C_{fh}\rho_{fh}(R^2/2)\delta P_h}(\theta_{wh}-\theta_{fh}) \\ \dfrac{\partial\theta_{wh}}{\partial t} + u \dfrac{\partial\theta_{wh}}{\partial\psi} = \dfrac{k_h A_h^*}{C_{wh}\rho_{wh}(R^2/2)(1-\delta)P_h}(\theta_{fh}-\theta_{wh}) \end{cases} \tag{25}$$

Cold side: $0 \leqq \psi \leqq P_c$

$$\begin{cases} \dfrac{\partial\theta_{wc}}{\partial t} + u \dfrac{\partial\theta_{wc}}{\partial\psi} = \dfrac{k_c A_c^*}{C_{wc}\rho_{wc}(R^2/2)(1-\delta)P_c}(\theta_{fc}-\theta_{wc}) \\ \dfrac{\partial\theta_{fc}}{\partial t} + u_c \dfrac{\partial\theta_{fc}}{\partial x} = \dfrac{k_c A_c^*}{C_{fc}\rho_{fc}(R^2/2)\delta P_c}(\theta_{wc}-\theta_{fc}) \end{cases} \tag{26}$$

where "+" in the double symbol is used for parallel flow and "-" is used for counterflow, subscripts f, w, h and c mean fluid, matrix, hot-side and cold-side respectively. Boundary conditions are as follows:

$$\begin{cases} \theta_{fh}(t, 0, y) = \theta_{fhin}(t) \\ \theta_{wh}(t, x, 0) = \theta_{wc}\left(t, \left\{ {x \atop 1-x} \right., 1\right) \\ \theta_{wh}(t, x, 1) = \theta_{wc}\left(t, \left\{ {x \atop 1-x} \right., 0\right) \\ \theta_{fc}(t, 0, y) = \theta_{fcin}(t) \end{cases} \tag{27}$$

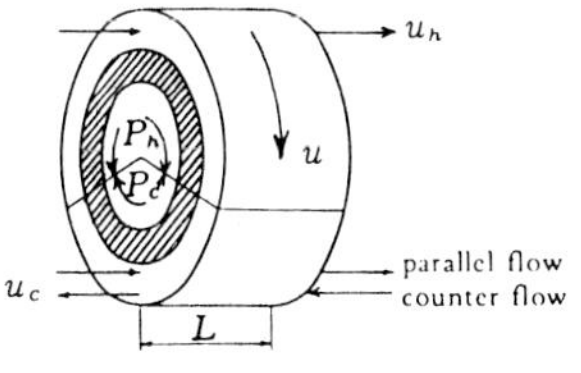

Fig.6 Rotary regenerative heat exchanger

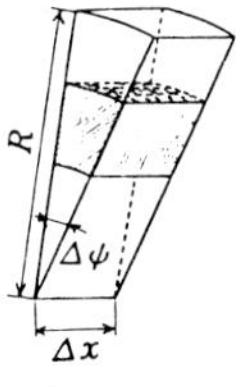

Fig.7 A small element of rotor

where the upper terms correspond to parallel-flow and the lower terms correspond to the counter-flow.

Since both equations for the hot side and the cold side take the same form of the basic equation describing the cross-flow heat exchanger, this heat exchanger can be analyzed as a problem of coupling two cross-flow heat exchangers. Hence, a mathematical model of the heat exchanger can be represented schematically as shown in Fig.8.

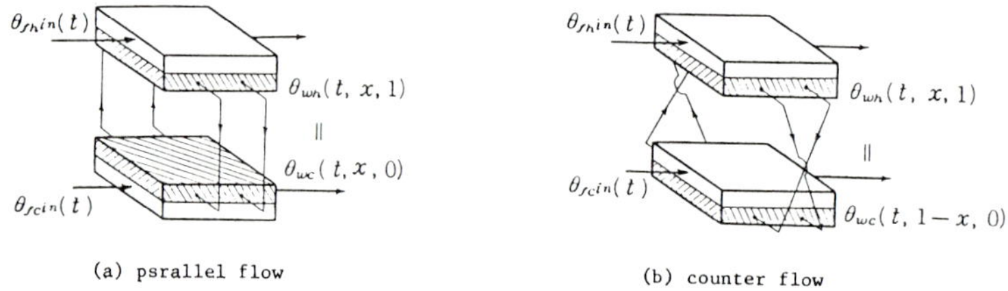

Fig.8 Mathematical model of the rotary heat exchanger.

4. ANALYSIS OF DYNAMIC BEHAVIOUR OF HEAT EXCHANGERS

(i) General method to calculate response

As shown in the previous section, many heat exchangers are described by the following equations:

$$\begin{cases} \frac{\partial \theta}{\partial t} + V\frac{\partial \theta}{\partial x} = A_{11}(x)\theta + A_{12}(x)\theta_w + B_1(x)u_1(t) \\ \frac{\partial \theta_w}{\partial t} = A_{21}(x)\theta + A_{22}(x)\theta_w + B_2(x)u_2(t) \end{cases} \tag{28}$$

with boundary condition

$$P_1\theta(t, 0) + P_2\theta(t, 1) = r(t) \tag{29}$$

where

$$\theta = \mathrm{col}(\theta_1, \theta_2, \cdots, \theta_n), \quad V = \mathrm{diag}(v_1, v_2, \cdots, v_n), \quad r = \mathrm{col}(r_1, r_2, \cdots, r_n)$$

P_1, P_2 are $n\times n$ constant matrices and A_{11}, A_{12}, A_{21}, A_{22} are $n\times n$ matrices dependent on x, $r(t)$ represents the inlet temperature inputs and $u_1(t)$, $u_2(t)$ represent the flow rate inputs or the wall temperature inputs. Laplace transforming t to s with zero initial condition gives

$$\frac{d\theta(s, x)}{dx} = A(s, x)\theta(s, x) + B(x)u(s) \tag{30}$$

$$P_1\theta(s, 0) + P_2\theta(s, 1) = r(s) \tag{31}$$

where

$$A(s, x) = V^{-1}(A_{11}(x) - sI) + V^{-1}A_{12}(sI - A_{22}(x))^{-1}A_{21}(x)$$

$$B(x)u(s) = V^{-1}B_1(x)u_1(s) + V^{-1}(sI - A_{22}(x))^{-1}B_2(x)u_2(s)$$

$$V^{-1} = \mathrm{diag}(1/v_1, 1/v_2, \cdots, 1/v_n)$$

Let $Y(s, x)$ be a fundamental system of solutions which can be expressed in the following equation:

$$\frac{dY(s, x)}{dx} = A(s, x)Y(s, x), \quad Y(s, 0) = 0 \tag{32}$$

Solving Eq.(32) with initial value $\theta(s, 0)$ yields

$$\theta(s, x) = Y(s, x)[\theta(s, 0) + \int_0^x Y^{-1}(s, z)B(z)dz \cdot u(s)] \tag{33}$$

Therefore,

$$\theta(s, x) = (P_1 + P_2 Y(s, 1))^{-1}[r(s) - P_2 Y(s, 1)\int_0^1 Y^{-1}(s, z)B(z)dz \cdot u(s)] + Y(s, x)\int_0^x Y^{-1}(s, z)B(z)dz \cdot u(s) \tag{34}$$

is the transfer function to the system (28), (29).

Numerical computation of the frequency response from (34) can be obtained by solving Eq.(32) for fixed value $s = jw$ using the Runge-Kutter method.

(ii) A special case: Matrix A does not depend on x.
If the matrix A does not depend on x and its Jordan form is diagonal, we can avoid the numecal integration. Laplace transformation x to p of Eq.(30) yields

$$\theta(s, p) = (pI - A)^{-1}\theta(s, 0) + (pI - A)^{-1}B(p)u(s) \tag{35}$$

Then we obtain

$$\theta(s, x) = \exp(Ax)\theta(s, 0) + e^{Ax}\int_0^x e^{-Az}B(z)dz \cdot u(s)$$

and from Eq.(31),

$$\theta(s, x) = e^{Ax}(P_1 + P_2 e^{A})^{-1}r(s) + e^{Ax}[Y(x) - (P_1 + P_2 e^{A})^{-1}P_2 e^{A})^{-1}P_2 e^{A}Y(1)]u(s) \tag{36}$$

where

$$Y(x) = \int_0^x e^{-Az}B(z)dz \tag{37}$$

Let the eigenvalues and eigenvectors of the matrix A be $p_1, p_2, \cdots, p_n$ and $v_1, v_2, \cdots, v_n$ respectively and define a matrix

$$T = [v_1, v_2, \cdots, v_n] \tag{38}$$

then

$$T^{-1}AT = \Lambda = \mathrm{diag}[p_1, p_2, \cdots, p_n] \tag{39}$$

and

$$T^{-1}e^{Ax}T = e^{\Lambda x} = \mathrm{diag}[e^{p_1 x}, e^{p_2 x}, \cdots, e^{p_n x}]$$

Therefore, from Eqs.(36), (37)

$$\begin{cases} \theta(s, x) = G(s, x)r(s) + [H(s, x) - G(s, x)P_2 H(s, 1)]u(s) \\ G(s, x) = Te^{\Lambda x}(P_1 T + P_2 Te^{\Lambda})^{-1} \end{cases} \tag{40}$$

$$H(s, x) = Te^{\Lambda x}Y_1(s, x), \qquad Y_1(s, x) = \int_0^x e^{-\Lambda z}T^{-1}B(z)dz. \tag{41}$$

The frequency responses for input $r(t)$ can be obtained easily by simple algorithm which consists of the calculations of eigenvalues and eigenvectors of A and Eq.(41).

The frequency responses for input $u(t)$ can be calculated without numerical integration of Eq.(41), since in many case $B(x)$ has the form

$$B_{ij}(x) = \sum_{k=1}^{m_{ij}} a(i, j, k)x^{k-1}\exp(b(i, j, k)x) \tag{42}$$

where $a(i, j, k)$ and $b(i, j, k)$ are constant. Therefore the integration in Eq.(41) can be calculated analytically. Many heat exchangers are involved in this case ; as an example one-shell multipass heat exchangers are analyzed as follows.

One-shell multipass heat exchangers

The basic equation and the boundary condition are given by Eqs. (11-L), (12), the Laplace transformation of which makes

$$A(s,x)=\left[\begin{array}{ccccc:c} -f_1 & & & & & a_1 \\ & f_1 & & 0 & & -a_1 \\ & & -f_1 & & & a_1 \\ & 0 & & \ddots & & \vdots \\ & & & & (-1)^n f_1 & (-1)^{n-1}a_1 \\ \hdashline \pm a & \pm a & \cdots & & \pm a & \mp f \end{array}\right], \quad \begin{array}{l} f_1 = s + a_1 \\ f = rs + na \end{array} \tag{43}$$

then

$$det[pI-A]=\begin{cases}(p^2-f_1^2)^{m-1}[(p^2-f_1^2)(p+f)+a_1 anf_1], & \text{for } n=2m \\ (p+f_1)(p^2-f_1^2)^{m-1}[(p+f)(p^2-f_1^2)+a_1 anf_1+a_1 ap], & \text{for } n=2m+1\end{cases} \tag{44}$$

The characteristic equations (44) have multiple roots $\pm f_1$ for $n \geqq 5$, and three roots p_1, p_2 and p_3 are determined by

$$\begin{cases} p^3+fp^2-f_1^2 p+(na_1 a-ff_1)f_1=0, & \text{for } n=2m \\ p^3+fp^2-(f_1^2+a_1 a)p+(na_1 a-ff_1)f_1=0, & \text{for } n=2m+1 \end{cases} \tag{45}$$

The eigenvectors corresponding to the multiple roots $\pm f_1$ span a subspace with dimension equal to its multiplicity; hence the Jordan form of A becomes diagonal matrix. It is easy to show this by checking that the rank reductions of matrix $(pI-A)$ for $p=\pm f_1$ are equal to the multiplicities. The transform matrix T can easily be found to be

$$T=\left[\begin{array}{cccc:ccc} & & & & \alpha_{11} & \alpha_{12} & \alpha_{13} \\ & I_{n-2} & & & \alpha_{21} & \alpha_{22} & \alpha_{23} \\ & & & & \vdots & \vdots & \vdots \\ \hdashline \alpha & \beta & \alpha & \cdots & \alpha_{n-1,1} & \alpha_{n-1,2} & \alpha_{n-1,3} \\ \beta & \alpha & \beta & \cdots & \alpha_{n,1} & \alpha_{n,2} & \alpha_{n,3} \\ 0 & 0 & 0 & \cdots & 1 & 1 & 1 \end{array}\right] \tag{46}$$

where

$$\alpha_{ij}=\frac{(-1)^{i-1}a_1}{p_j+(-1)^{i-1}f_1}, \quad \begin{array}{l} i=1,2,\cdots,n. \\ j=1,2,3. \end{array} \qquad \alpha=-\frac{1+(-1)^n}{2}, \quad \beta=-\frac{1-(-1)^n}{2}$$

and

$$\Lambda=diag[-f_1, f_1, -f_1, \cdots, (-1)^{n-2}f_1, p_1, p_2, p_3]$$

From (11-L),

$$B(x)u(t)=(u_1(t)B_1+u_2(t)B_2)\,\theta_s(x) \tag{47}$$

where

$$B_2=\left[\begin{array}{ccc:c} -c_3 & & 0 & c_3 \\ & \ddots & & \vdots \\ 0 & & -c_3 & c_3 \\ \hdashline -(c_4-a) & \cdots & -(c_4-a) & c_4-a \end{array}\right], \quad B_1=\left[\begin{array}{ccc:c} -(c_1-a_1) & & 0 & c_1-a_1 \\ & \ddots & & \vdots \\ 0 & & -(c_1-a_1) & c_1-a_1 \\ \hdashline -c_2 & & & c_2 \end{array}\right] \tag{48}$$

And from Eq.(23) and boundary condition (12), we obtain

$$\theta_s(x) = e^{A_0 x}(P_1 + P_2 e^{A_0})^{-1} r_s \tag{49}$$

where $A_0 = A|_{s=0}$ and r_s indicates the steady state inputs. Hence, $B(x)$ can be expressed in the form of Eq.(42).

(iii) Transfer function of typical heat exchangers

Single tube heat exchanger

The basic equation and the boundary condition are given by Eqs.(8), (9) and (10). Two dimensional Laplace transformation $(t \to s,\ x \to p)$ of them yields

$$\begin{cases}(p + s + a_1)\theta_f(s, p) = a_1\theta_w(s, p) + \theta_{fi}(s) \\ (s + b_1 + b_2)\theta_w(s, p) = b_1\theta_f(s, p) + b_2 \dfrac{\theta_a(s)}{p}\end{cases}$$

Eliminating θ_w and inverse Laplace transformation respect to p make the following transfer functions

$$\theta_f(s, x) = e^{-(s+a_1'')x}\theta_{fi}(s) + \frac{a_1 b_2(1 - e^{-(s+a_1'')x})}{(s + a_1'')(s + b_1 + b_2)}\theta_a(s) \tag{50}$$

$$= e^{-(s+a_1')x}\theta_{fi}(s) + \frac{a_1'(1 - e^{-(s+a_1')x})}{s + a_1'}\theta_a(s) \tag{50-1}$$

where

$$a_1'' = a_1 \left| \frac{s + b_2}{s + b_1 + b_2} \right|$$

Parallel-flow and counter-flow heat exchangers

The basic equations are given by (11-L) of $n=1$, where the steady state solution of Eq.(23) is given by

$$\theta_s(x) - \theta_{1s}(x) = \mu e^{-(a_1 \pm a)x} \tag{51}$$

where

$$\mu = (\theta_{is} - \theta_{1is})$$
$$= (\theta_{is} - \theta_{1is})(a_1 - a)/(a_1 - a\ \exp(a_1 - a))$$

Hence, we can calculate the transfer functions by using the method mentioned in (i). System matrices are given by

$$A(s) = \begin{bmatrix} -f_1 & a_1 \\ \pm a & \mp f \end{bmatrix},\quad B(x) = \begin{bmatrix} \mu(c_1 - a_1)e^{p_0 x} \\ \mp \mu c_2 e^{p_0 x} \end{bmatrix},\quad p_0 = -(a_1 + a)$$

$$P_1 = \begin{bmatrix} 1 & 0 \\ 0 & \begin{smallmatrix}1\\0\end{smallmatrix} \end{bmatrix},\quad P_2 = \begin{bmatrix} 0 & 0 \\ 0 & \begin{smallmatrix}0\\1\end{smallmatrix} \end{bmatrix},\quad r(s) = \begin{bmatrix} \theta_{1i}(s) \\ \theta_i(s) \end{bmatrix}$$

The eigenvalues of $A(s)$ are the roots p_1, p_2 of the following characteristic equation of A :

$$det[pI - A] = p^2 + (f_1 \pm f)p \pm ff_1 \mp aa_1 \tag{52}$$

The transformation matrix T and $Y_1(x)$ are given as follows:

$$T = \begin{bmatrix} \dfrac{a_1}{p_1 + f_1} & \dfrac{a_1}{p_2 + f_1} \\ 1 & 1 \end{bmatrix}$$

$$Y_1(x) = \frac{\begin{bmatrix} \mu(\mp a(c_1 - a_1) - (p_1 + f_1)c_2)(e^{(p_0 - p_1)x} - 1)(p_0 - p_2) \\ \mu(\pm a(c_1 - a_1) + (p_2 + f_1)c_2)(e^{(p_0 - p_2)x} - 1)(p_0 - p_1) \end{bmatrix}}{(p_0 - p_1)(p_1 - p_2)(p_2 - p_0)}$$

Then from Eq.(40), we obtain the transfer functions as follows:
Parallel flow heat exchanger:

$$\begin{bmatrix} \theta_1(s,\ x) \\ \theta_2(s,\ x) \end{bmatrix} = \frac{\begin{bmatrix} (p_2+f_1)e^{p_1 x}-(p_1+f_1)e^{p_2 x} & a_1[e^{p_2 x}-e^{p_1 x}] \\ a[e^{p_2 x}-e^{p_1 x}] & (p_2+f_1)e^{p_2 x}-(p_1+f_1)e^{p_1 x} \end{bmatrix}}{p_2-p_1} \begin{bmatrix} \theta_{1i}(s) \\ \theta_{2i}(s) \end{bmatrix}$$

$$+ \frac{\begin{bmatrix} \sum_{i=0}^{2} J_1(p_i)e^{p_i x}(p_{[i+1]}-p_{[i+2]}) \\ \sum_{i=0}^{2} J_2(p_i)e^{p_i x}(p_{[i+1]}-p_{[i+2]}) \end{bmatrix}}{(p_0-p_1)(p_1-p_2)(p_2-p_0)} u_1(s) \qquad (54)$$

where

$$J_1(p)=(c_1-a_1)(p+f)+a_1 c_2$$
$$J_2(p)=(p+f_1)c_2+a(c_1-a_1)$$

Counter flow heat exchanger:

$$\begin{bmatrix} \theta_1(s,\ x) \\ \theta_2(s,\ x) \end{bmatrix} = \frac{\begin{bmatrix} (p_2+f_1)e^{p_1 x+p_2}-(p_1+f_1)e^{p_2 x+p_1} & a_1(e^{p_2 x}-e^{p_1 x}) \\ a(e^{p_1 x+p_2}-e^{p_2 x+p_1}) & (p_2+f_1)e^{p_2 x}-(p_1+f_1)e^{p_1 x} \end{bmatrix}}{(p_2+f_1)e^{p_2}-(p_1+f_1)e^{p_1}} \begin{bmatrix} \theta_{1i}(s) \\ \theta_{2i}(s) \end{bmatrix}$$

$$+ \begin{bmatrix} H_1(s,\ x)-G_1(s,\ x)H_1(s,\ 1) \\ H_2(s,\ x)-G_2(s,\ x)H_2(s,\ 1) \end{bmatrix} u_1(s) \qquad (55)$$

where

$$G_1(s,\ x)= a_1(e^{p_1 x}-e^{p_2 x})/[(p_2+f_1)e^{p_2}-(p_1+f_1)e^{p_1}]$$
$$G_2(s,\ x)=[(p_2+f_1)e^{p_2 x}-(p_1+f_1)e^{p_1 x}]/[(p_2+f_1)e^{p_2}-(p_1+f_1)e^{p_1}]$$
$$H_1(s,\ x)=\mu \sum_{i=0}^{2} J_3(p_i)e^{p_i x}(p_{[i+1]}-p_{[i+2]})/(p_0-p_1)(p_1-p_2)(p_2-p_0)$$
$$H_2(s,\ x)=\mu \sum_{i=0}^{2} J_4(p_i)e^{p_i x}(p_{[i+1]}-p_{[i+2]})/(p_0-p_1)(p_1-p_2)(p_2-p_0)$$
$$J_3(p)=(c_1-a_1)(p-f)+a_1 c_2$$
$$J_4(p)=(p+f_1)c_2-a(c_1-a_1)$$

In above equations, the brackets [ ] in the subspripts indicate the residuals mod 3.

<u>One-shell two-pass heat exchangers</u> (See Fig.11 in the next section)
In this case, the parallel-start type is called P-C type and the counter-start type is called C-P type. For simplicity, only transfer functions corresponding to the inlet temperature inputs are calculated. From Eqs.(11-L), (12), we have

$$A(s,\ x)=\begin{bmatrix} -f_1 & 0 & a_1 \\ 0 & f_1 & -a_1 \\ \pm a & \pm a & \mp f \end{bmatrix},\quad r(s)=\begin{bmatrix} \theta_{1i}(s) \\ 0 \\ \theta_i(s) \end{bmatrix} \quad P_1=\begin{bmatrix} 1 & 0 & 0 \\ 0 & 0 & 0 \\ 0 & 0 & \mid \substack{1 \\ 0} \end{bmatrix},\quad P_2=\begin{bmatrix} 0 & 0 & 0 \\ 1 & -1 & 0 \\ 0 & 0 & \mid \substack{0 \\ 1} \end{bmatrix}$$

Let p_1, p_2 and p_3 be the roots of the characteristic equation

$$p^3 \pm fp^2 - f_1^2 p \pm (2aa_1 - ff_1)f_1 = 0$$

then from Eq.(46),

$$T=[\alpha_{ij}],\quad Te^{Ax}=[\alpha_{ij}e^{p_j x}],\quad i,\ j=1,2,3.$$

where

$$\alpha_{1j}=a_1/(p_j+f_1),\quad \alpha_{2j}=-a_1/(p_j-f_1),\quad \alpha_{3j}=1 \qquad j=1,2,3.$$

and

$$(P_1 T + P_2 T e^{\Lambda}) = \begin{bmatrix} \alpha_{11} & \alpha_{12} & \alpha_{13} \\ (\alpha_{11} - \alpha_{21}) e^{p_1} & (\alpha_{12} - \alpha_{22}) e^{p_2} & (\alpha_{13} - \alpha_{23}) e^{p_3} \\ \begin{cases} 1 \\ e^{p_1} \end{cases} & \begin{cases} 1 \\ e^{p_2} \end{cases} & \begin{cases} 1 \\ e^{p_3} \end{cases} \end{bmatrix}$$

Transfer functions $\theta(s, x)/\theta_{1i}(s)$ $\theta(s, x)/\theta_{3i}(s)$ are equal to the first column and the third column of a matrix $T \exp(\Lambda x)(P_1 T + P_2 T \exp(\Lambda))^{-1}$ respectively. Applying a Laplace expansion to the matrix, we obtain

$$\frac{\theta_j(s, x)}{\theta_{1i}(s)} = \frac{det \begin{bmatrix} \alpha_{j1} e^{p_1 x} & \alpha_{j2} e^{p_2 x} & \alpha_{j3} e^{p_3 x} \\ (\alpha_{11} - \alpha_{21}) e^{p_1} & (\alpha_{12} - \alpha_{22}) e^{p_2} & (\alpha_{13} - \alpha_{23}) e^{p_3} \\ \begin{cases} 1 \\ e^{p_1} \end{cases} & \begin{cases} 1 \\ e^{p_2} \end{cases} & \begin{cases} 1 \\ e^{p_3} \end{cases} \end{bmatrix}}{det[P_1 T + P_2 T e^{\Lambda}]} \quad j = 1, 2, 3$$

$$\frac{\theta_j(s, x)}{\theta_{3i}(s)} = \frac{det \begin{bmatrix} \alpha_{11} & \alpha_{12} & \alpha_{13} \\ (\alpha_{11} - \alpha_{21}) e^{p_1} & (\alpha_{12} - \alpha_{22}) e^{p_2} & (\alpha_{13} - \alpha_{23}) e^{p_3} \\ \alpha_{j1} e^{p_1 x} & \alpha_{j2} e^{p_2 x} & \alpha_{j3} e^{p_3 x} \end{bmatrix}}{det[P_1 T + P_2 T e^{\Lambda}]} \quad j = 1, 2, 3 \tag{56}$$

Cross-flow heat exchanger

The basic equations and the boundary conditions are given by Eqs.(19), (20). For simplicity, the velocities of both fluids are assumed to be constant. [See Bender (1972)[86] for the case of flow change] Three dimensional Laplace transformation $(t \to s,\ x \to p,\ y \to q)$ of the equations yields

$$\begin{bmatrix} p + f_1 & -g_1 \\ -g_2 & q + f_2 \end{bmatrix} \begin{bmatrix} \theta_1(s, p, q) \\ \theta_2(s, p, q) \end{bmatrix} = \begin{bmatrix} \theta_{1i}(s, q) \\ \theta_{2i}(s, p) \end{bmatrix}$$

where

$$f_1 = (s + a_1 - a_1 b_1/f_3)/u_1, \quad f_2 = (s + a_2 - a_2 b_2/f_3)/u_2, \quad f_3 = s + b_1 + b_2$$
$$g_1 = a_1 b_2/(f_3 u_1), \qquad g_2 = a_2 b_1/(f_3 u_2)$$

Then

$$\begin{bmatrix} \theta_1(s, p, q) \\ \theta_2(s, p, q) \end{bmatrix} = \frac{\begin{bmatrix} q + f_2 & g_1 \\ g_2 & p + f_1 \end{bmatrix} \begin{bmatrix} \theta_{1i}(s, q) \\ \theta_{2i}(s, p) \end{bmatrix}}{(q + f_2)(p + f_1) - g_1 g_2}$$

Inverse Laplace transformation with respect to p, q makes (See Voelker and Doetsch, 1950)

$$\begin{cases} \theta_1(s, x, y) = e^{-f_1 x} \theta_{1i}(s, y) + \int_0^y \sqrt{\frac{g_1 g_2 x}{\eta}}\, h_1(x, \eta)\, \theta_{1i}(s, y - \eta)\, d\eta + g_1 \int_0^x h_0(\xi, y)\, \theta_{2i}(s, x - \xi)\, d\xi \\ \theta_2(s, x, y) = e^{-f_2 y} \theta_{2i}(s, x) + \int_0^x \sqrt{\frac{g_1 g_2 y}{\xi}}\, h_1(\xi, y)\, \theta_{2i}(s, x - \xi)\, d\xi + g_2 \int_0^y h_0(x, \eta)\, \theta_{1i}(s, y - \eta)\, d\eta \end{cases} \tag{57}$$

where

$$h_0 = \exp(-f_1 x - f_2 y) I_0(2\sqrt{g_1 g_2 x y}), \qquad h_1 = \exp(-f_1 x - f_2 y) I_1(2\sqrt{g_1 g_2 x y})$$

When inputs θ_{1i}, θ_{2i} are uniform along the distance, Eq.(57) becomes

$$\begin{bmatrix} \theta_1(s, x, y) \\ \theta_2(s, x, y) \end{bmatrix} = \begin{bmatrix} h_0(x, y) + f_2 \int_0^y h_0(x, \eta) d\eta & g_1 \int_0^x h_0(\xi, y) d\xi \\ g_2 \int_0^y h_0(x, \eta) d\eta & h_0(x, y) + f_1 \int_0^x h_0(\xi, y) d\xi \end{bmatrix} \begin{bmatrix} \theta_{1i}(s) \\ \theta_{2i}(s) \end{bmatrix} \tag{58}$$

The transfer function of cross-flow heat exchanger obtained is represented by using an integration term. This is an inevitable feature of the two dimensional process, because the transfer function of an outlet temperature at one point $(x, 1)$ is a sum of influences from the inlet temperature of every point on interval $[0, x]$.

5. PROPERTIES OF DYNAMICS OF HEAT EXCHANGERS

(i) Resonance phenomena of frequency response of heat exchangers
The term "resonance" has been used in a somewhat misleading manner in the literature. The resonance of heat exchangers means a periodic fluctuation in the magnitude and phase lag of the frequency response and does not mean the resonance used in the mechanics.

The resonance phenomena was predicted by DeBolt [6] and Cohen and Johnson [9] in shell- and tube-heat exchangers, and by Masubuchi [19] in multi-pass heat exchangers, and also has been investigated in detail by Yang [21], and Clark and Arpaci [22] for single-tube heat exchangers with an internal heat source. Its existence has been confirmed in experimental studies [22].

Harriott explained the resonance by using a single-tube heat exchanger where the fluid is heated through the wall. From equation (50-1), we get the following transfer function:

$$G(s, 1) = \theta_f(s, 1)/\theta_a(s) = a[1 - \exp(-(s+a))]/(s+a)$$

As this transfer function consists of two functions $1-\exp(-(s+a))$ and $1/(s+a)$ in cascade, it is necessary only to add the respective magnitude in decibels and phases for each frequency for $s = jw$. The vector locus of $1-\exp(-(s+a))$ and its magnitude and phase are plotted in Fig.9 where the fluctuation in the frequency response occurs.

Fig.9

Vector locus, magnitude and phase lag of $1-k\exp(-j\omega)$, $k<1$

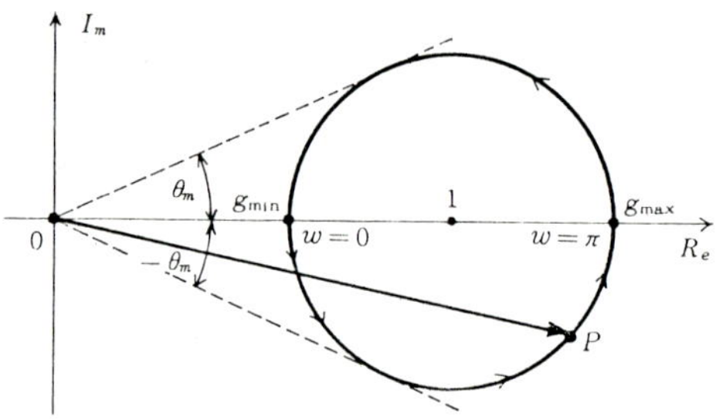

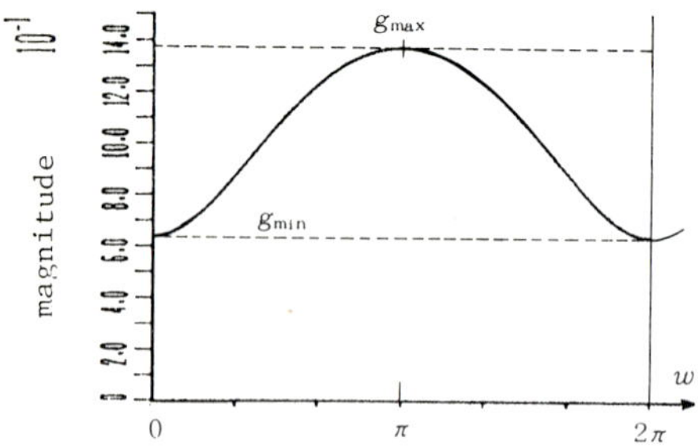

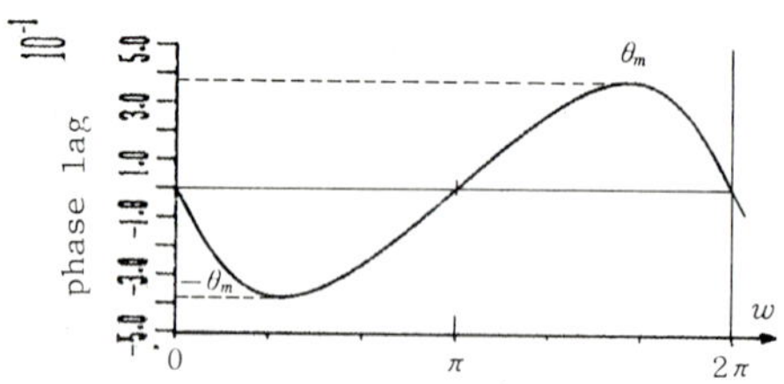

However, alternative interpretation reflecting the distributed parameter structure of heat exchangers is possible. Since the transfer function consists of the sum of two transfer functions $1/(s+a)$ and $k\exp(-s)/(s+a)$, we can consider that the resonance arises because of interference between a quick response and a delayed response as shown in Fig.10.

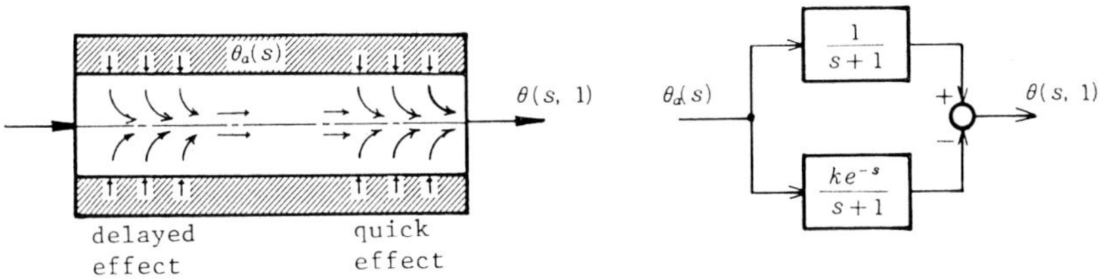

Fig. 10 Interference of two effects in a single tube heat exchanger.

This idea is useful to interpret more complicated heat exchanger dynamics, for instance, that of one-shell two-pass heat exchangers. As mentioned before, there are two flow-types, the P-C type and C-P type, as illustrated in Fig.11.

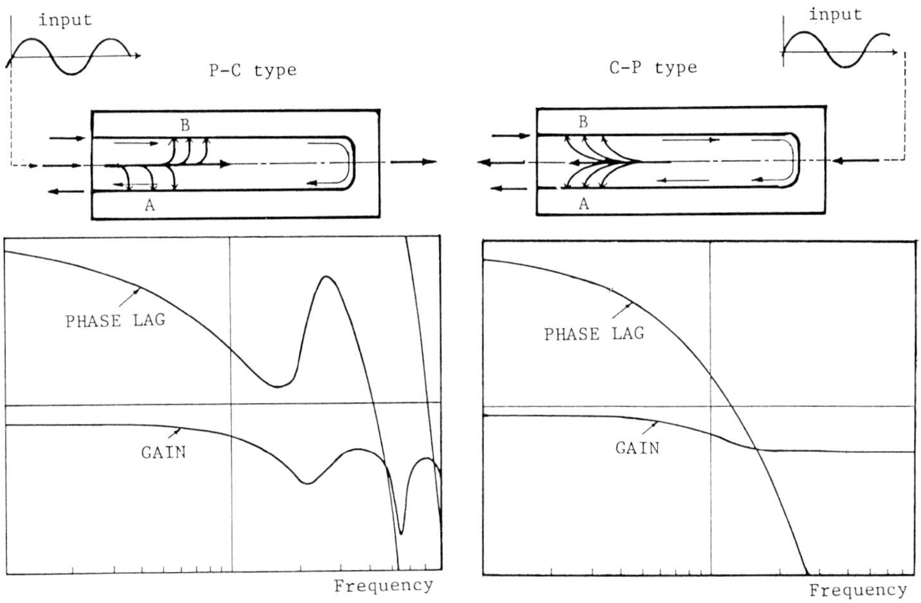

Fig.11 One-shell two-pass heat exchangers.

In the P-C type, the resonance appears clearly as a result of the interference effect between an immediate response through pass A and a delayed response through pass B. In the C-P type, the resonance is not clear because the pure delay dynamics is more dominant than the resonance effect. However, note that appearances of the resonance depend on the ratio of flow rates u_1/u_2 and on the heat capacities of both fluids.

(ii) Resonance and controllability

A relationship between resonance and controllability is shown by illustrating an indented stability limit in the feedback controlled single-tube heat exchanger whose dynamics has the resonance. (Fig.12) The stability limit reflects the perildic fluctuation in the magnitude and phase lag of the resonance frequency response and shows a conditional stability; for example, when K_I increases with fixed $K_P=0$, the controlled system is stable at A and unstable at B but becomes stable again at C then turns into unstable at D.

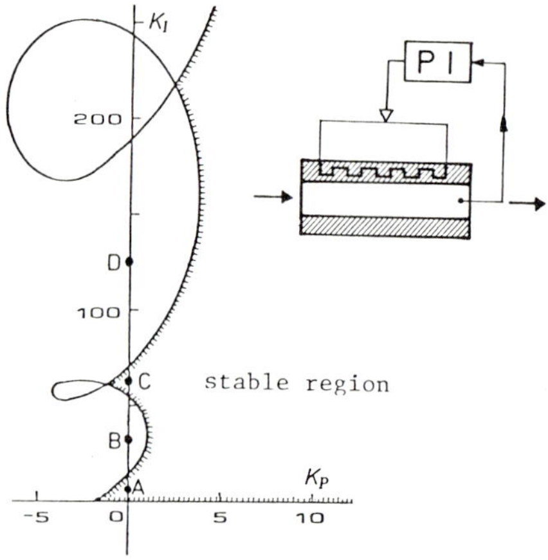

Fig.12 Conditional stability of a single tube heat exchanger.

(iii) Types of heat exchanger dynamics
Many investigations of the dynamics of parallel-flow, counter-flow and multi-pass heat exchangers have presented analytical transfer functions and shown several numerical or experimental results. There are, however, few works studying what kind of heat exchanger has what kind of dynamics.

Noldus [36] studied the transfer functions of parallel-flow and counter-flow heat exchangers and showed that the parallel-flow heat exchanger can be approximated by second-order lag plus dead time, and that the counter-flow heat exchanger can be approximated by second-order lag. Kanoh [98] defined the types of dynamics of heat exchangers and showed that the one-shell two-pass heat exchanger shows transilience from the parallel-flow type to the counter-flow type between a very narrow parameter range.

In this section, we define the type of heat exchanger dynamics and investigate it in detail. First of all, we define two types of heat exchanger dynamics: the parallel-flow type in which the phase lag exceeds -180° line, and the counter-flow type in which the phase lag does not exceed -180° line. From this point of view, we examine the heat exchanger dynamics in the following:

Parallel-flow heat exchanger
The transfer function $G(s)=\theta_1(s,1)/\theta_{2i}(s)$ in Eq.(53) can be modified as follows:

$$G(s)=a_1\exp\left(-\frac{g-R}{2}\right)\frac{1-\exp(-R)}{R} \tag{66}$$

where

$$g=f_1+f,\quad R=\sqrt{(f_1-f)^2+4a_1a},\quad f_1=s+a_1,\quad f=rs+a$$

In the high frequency range,

$$R\doteqdot f_1-f=(1-r)s+a_1-a$$

therefore, the transfer function $G(s)$ consists of pure-delay element and resonance element $(1-\exp(-R))/R$. Then the phase-lag of parallel-flow heat exchanger always exceeds -180° line.

Counter flow heat exchanger
The transfer function $G(s)=\theta_1(s,1)/\theta_{2i}(s)$ in Eq.(54) can be modified as follows:

$$G(s)=\frac{2a_1}{g+R\dfrac{1+\exp(-R)}{1-\exp(-R)}} \tag{67}$$

where

$$g=f_1+f,\quad R=\sqrt{g^2-4a_1a}$$

In the high frequency range,

$$R \doteqdot f_1 + f = g = (1 + r)s + a_1 + a$$

then,

$$G(s) \doteqdot \frac{1 - \exp(-g)}{g} \tag{68}$$

which represents the same resonance transfer function as the single-tube heat exchanger. Hence, the counter-flow heat exchanger has the counter-flow type dynamics.

(iv) Stability limit and types of dynamics

There is a close relationship between the type and the controllability. From the definitions, it will be found easily that when controlled by a proportional feedback, the counter-flow type cannot be unstable for any large gain, but the parallel-flow type can be unstable at a finite gain. Fig.13 shows the stability limits for a parallel-flow and counter-flow heat exchangers controlled by a proportional plus an integral controller.

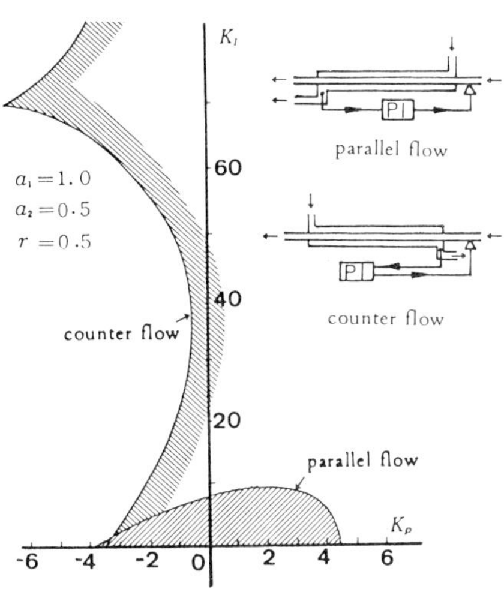

Fig.13
Stability limits of PI feedback control systems of parallel-flow and counter-flow heat exchangers.

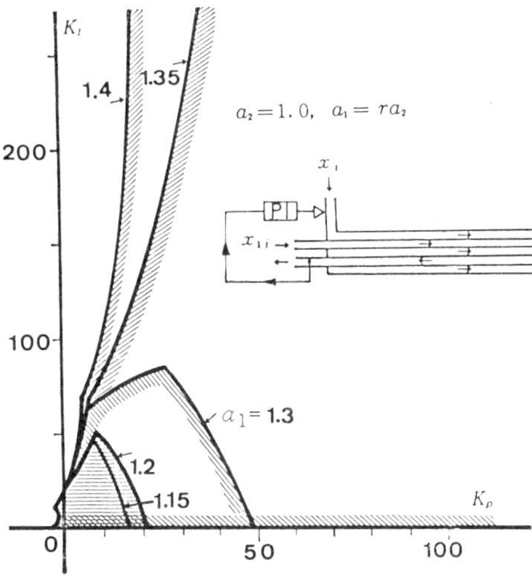

Fig.14
Stability limits of PI feedback system of P-C type one-shell two-pass heat exchanger.

In some heat exchangers which have both a parallel-flow part and a counter-flow part, such as multi-shell and multi-pass heat exchangers, the type of dynamics depends on the layout of shell and tube and operation condition.

One-shell two-pass heat exchanger

The transfer functions are given by Eq.(55) or Eq.(56). The C-P type shows always parallel-flow type, while the dynamics of the P-C type changes according to parameters a_1, a_2, r. Fig.14 shows stability limits corresponding to the change of the value of a_1 for a special case : $a_2 = 1.0$, $a_1 = ra_2$ which represents the change of the flow rate. The transilience from the parallel-flow type to the counter-flow type takes place in a very narrow range $a_1 = 1.3 \sim 1.35$. Fig.14 shows the possibility that a small fluctuation of parameter may cause instability to stable system.

This switching of dynamics types can be seen in other heat exchangers, such as a two-shell four-pass heat exchanger or a plate heat exchanger [106].

(v) Effect of wall heat capacity

Another important factor which influences the dynamics of heat exchangers is wall capacity. To show the fundamental feature of it, let us consider the single-tube heat exchanger facing an inlet-temperature change. The transfer function is given in Eq.(50) as

$$G(s)=\theta_f(s,1)/\theta_{fi}(s)=e^{-(s+a)+\frac{a_1 b_1}{s+b_1+b_1}}$$

Then the effect of wall capacity can be represented by a transfer function which has a phase lag as following:

$$\psi=\frac{-a_1 b_1 w}{w^2+(b_1+b_2)^2},\quad s=jw$$

which takes the minimum value $-0.5a_1 b_1/(b_1+b_2)$ at $w=b_1+b_2$. This makes a depression in the phase lag of $G(s)$ and becomes dangerous for large a_1 and relatively small b_1 and b_2. This is the case when superheaters in boiler plant is modeled by a single-tube heat exchanger. See P.Profos, 1962, p.163. Fig.15 shows the effect of the wall capacity on the frequency response.

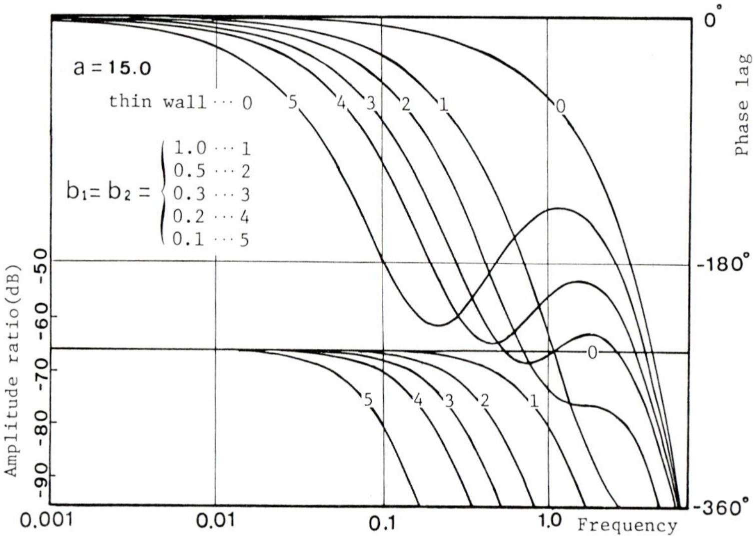

Fig. 15 Effect of wall capacity on the single tube heat exchanger dynamics.

As the large wall capacity increases the phase lag, the previous discussion on the type of heat exchanger dynamics is not valid for heat exchangers with large heat capacity of a wall. To see this, let us consider a counter-flow heat exchanger described by the following equations:

$$\begin{cases}\dfrac{\partial\theta_1}{\partial t}+u_1\dfrac{\partial\theta_1}{\partial x}=a_1(\theta_w-\theta_1)\\ r\dfrac{\partial\theta_2}{\partial t}-u_2\dfrac{\partial\theta_2}{\partial x}=a_2(\theta_w-\theta_2)\\ \dfrac{\partial\theta_w}{\partial t}=b_1(\theta_1-\theta_w)+b_2(\theta_2-\theta_w)\end{cases}$$

The effect of heat capacity of wall on the frequency response is shown in Fig.16. The phase lag becomes more negative than -180°, as the heat capacity increases.

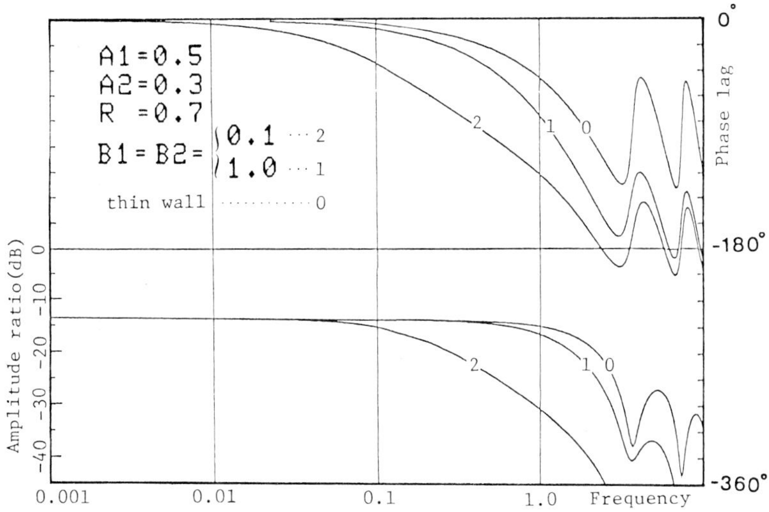

Fig.16 Effect of wall heat capacity on a counter-flow heat exchanger.

6. LUMPED PARAMETER SIMULATION OF HEAT EXCHANGERS BY METHOD OF WEIGHTED RESIDUALS

(i) Outline of the method

Consider a partial differential equation

$$\frac{\partial\theta}{\partial t}=\mathscr{L}\left(\theta,\ \frac{\partial\theta}{\partial x},\ x,\ t\right),\qquad x\in\Omega \tag{69}$$

where x denotes a one- or two-dimensional spatial variable. We assume an approximate solution $\hat{\theta}(t,x)$ for this equation under some boundary condition of the form

$$\hat{\theta}_i(t,x)=h_i(t)+\sum_{j=1}^{N}x_{ij}(t)\phi_j(x),\quad (i=1,2,\cdots,n) \tag{70}$$

where $\phi_j(x)\,(j=1,2,\cdots,N)$ are linearly independent functions called trial functions, $x_{ij}(t)$ are unknown time functions to be determined and $h_i(t)$ are such that $\hat{\theta}(t,x)$ satisfies the boundary condition exactly. N is a number of terms necessary in the approximation. Let the differential equation residual $R(\hat{\theta})$ be defined as

$$R(\hat{\theta})=\frac{\partial\hat{\theta}}{\partial t}-\mathscr{L}\left(\hat{\theta},\ \frac{\partial\hat{\theta}}{\partial x},\ x,\ t\right) \tag{71}$$

Now, the residuals $R(\hat{\theta})$ is made orthogonal to the subspace spanned by a set of functions $\psi_i(x)$ called weighting function. Thus,

$$<\psi_j,\ R(\hat{\theta})>=0\qquad j=1,2,\cdots,N \tag{72}$$

where $<\cdot,\cdot>$ is the inner-product notation defined as

$$<u,\ v>=\int_{\Omega}uv\,d\Omega \tag{73}$$

The set of equations (62) yields ordinary differential equations for the unknown $x_{ij}(t)$.

(ii) One-shell multi-pass heat exchangers

To use orthonormalized Legendre polynomials as trial function, we transform the spatial variable x in Eq.(11-L) into $2l+1$. Then we have

$$\begin{cases} \dfrac{\partial\theta_k}{\partial t}+2(-1)^{k-1}\dfrac{\partial\theta_k}{\partial l}=a_1(\theta-\theta_k)+\zeta_k(l)v_1(t), \quad k=1,2,\cdots,n \\ r\dfrac{\partial\theta}{\partial t}+2\dfrac{\partial\theta}{\partial l}=\sum_{k=1}^{n}a(\theta_k-\theta)-\sum_{k=1}^{n}\zeta_k(l)v_2(t) \end{cases}$$

$$-1\leqq l\leqq 1$$

where

$$v_1(t)=((c_{11}-a_{1s})u_1(t)+c_{12}u_2(t)), \quad v_2(t)=(c_{21}u_1(t)+(c_{22}-a_s)u_2(t))$$

$$\zeta_k(l)=\theta_s(l)-\theta_{ks}(l), \quad k=1,2,\cdots,n$$

Approximate solutions satisfying the boundary condition (12) are assumed to be

$$\begin{cases} \hat{\theta}_0(t,l)=\theta_{1i}(t) \\ \hat{\theta}_k(t,l)=(l+(-1)^{k-1})u'(l)x_k(t)+\hat{\theta}_{k-1}(t,(-1)^k) \quad k=1,2,\cdots,n \\ \hat{\theta}(t,l)=(l\pm 1)u'(l)x(t)+\theta_i(t) \end{cases} \tag{74}$$

where $u(l)=[u_{i+1}(l)=\sqrt{i+1/2}\,P_i(l),\ i=1,2,\cdots,N-1\ ]$ and $P_i(l)$ is the Legendre polynomial. Evaluating the differential equation residual $R(\hat{\theta})$, making the residual orthogonal to the weighting function $u(t)$, we obtain the following ordinary differential equation

$$\begin{cases} S_k\dot{x}_k(t)+2\sum_{i=1}^{k-1}E_i\dot{x}_i(t)=2(-1)^k T_k x_k(t)+[a_1\theta_i(t)-a_1\theta_{1i}(t)-\dot{\theta}_{1i}(t)]p \\ \qquad +a_1[S_{+1}x(t)-S_k x_k(t)-2\sum_{i=1}^{k-1}(-1)^{i-1}E_i x_i(t)]+p_k v_1(t) \\ \qquad\qquad k=1,2,\cdots,n \\ rS_{\pm 1}\dot{x}(t)=\sum_{k=1}^{n}a[S_k x_k(t)+2\sum_{i=1}^{k-1}(-1)^{i-1}E_i x_i(t)-S_{\pm 1}x(t)]+2T_{\pm 1}x(t) \\ \qquad +[na\theta_{1i}(t)-na\theta_i(t)-r\dot{\theta}_i(t)]p-\sum_{k=1}^{n}p_k v_2(t) \end{cases} \tag{75}$$

where

$$S_k=S_{-1)^{k-1}}\cdots, \quad T_k=T_{(-1)^{k-1}}, \quad E_k=E_{(-1)^{k-1}}, \quad k=1,2,\cdots,n$$

$$S_{\pm 1}=\int_{-1}^{1}(l\pm 1)u'(l)\,dl, \qquad T_{\pm 1}=\int_{-1}^{1}u(l)\frac{d}{dl}[(l\pm 1)u'(l)]dl$$

$$E_{\pm 1}=\int_{-1}^{1}u(l)u'(\pm 1)\,dl, \qquad p=\int_{-1}^{1}u(l)\,dl=[\sqrt{2},\ 0,\ 0,\ \cdots,\ 0]'$$

$$p_k=\int_{-1}^{1}u(l)\zeta_k(l)\,dl=\int_{-1}^{1}u(l)[\theta_s(l)-\theta_{ks}(l)]\,dl, \quad k=1,2,\cdots,n$$

We only need the calculations of inverse of two matrices $S_{\pm 1}$ to transform Eq.(75) into the standard-state space expression $\dot{x}=Ax+Bu$. The integrations of $S_{\pm 1}$, $T_{\pm 1}$ can be easily done by using the property of Legendre polynomials ; the first row in $E_{\pm 1}$ is $\sqrt{2}u'(\pm 1)$ and the remaining rows are all zeros. If this procedure is applied to Eq.(11-B), a bilinear lumped parameter model of one-shell multi-pass heat exchangers is obtained.

The approximate frequency responses of the outlet temperature θ_1 for the change of the inlet temperature $\theta_{2i}(s)$ calculated from Equs.(75) are compared with the exact solution and the finite difference solutions in Figs. 17, 18 for parallel-flow and counter-flow heat exchangers respectively. It is shown that the MWR results for N=4,8 agree very well with the exact frequency response but the finite difference results have a large error. A further advantage of the MWR result is that the MWR approximation with only two terms agrees well with the exact frequency response in the low frequency range and the range of good agreement reaches towards the high frequency range as N increases.

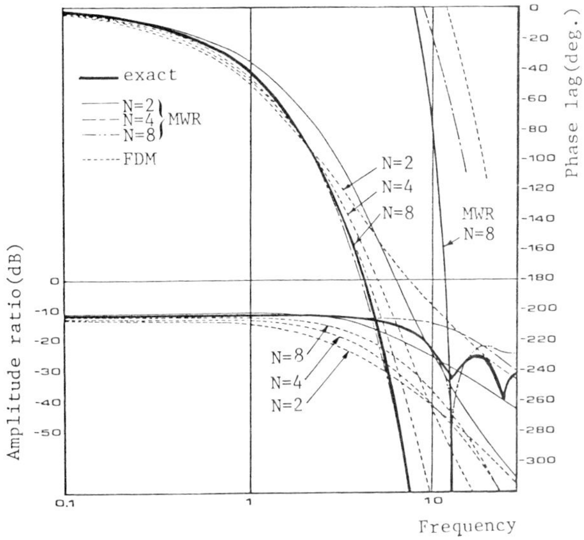

Fig.17
Approximate frequency responses of a parallel flow heat exchanger.

a_1=1.0
a_2=0.5
r=0.5

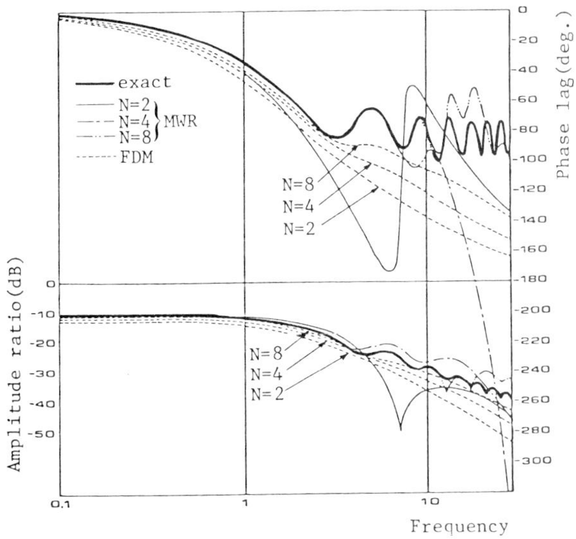

Fig.18
Approximate frequency responses of a counter flow heat exchanger.

a_1=1.0
a_2=0.5
r=0.5

(iii) Cross-flow heat exchanger
The basic equations and boundary conditions are given by Eqs.(19),(20). An approximate solution is assumed to be

$$\begin{cases} \bar{\theta}_1(t, x, y) = xu^t(x, y)\boldsymbol{\alpha}(t) + \theta_{1i}(t, y) \\ \bar{\theta}_2(t, x, y) = yu^t(x, y)\boldsymbol{\beta}(t) + \theta_{2i}(t, x) \\ \bar{\theta}_w(t, x, y) = u^t(x, y)\boldsymbol{\gamma}(t) \end{cases} \tag{76}$$

where

$$u^t(x, y) = [1, x, y, x^2, xy, y^2, \cdots, x^M, x^{M-1}y, \cdots, y^M] \tag{77}$$

which has $N = (M+1)(M+2)/2$ elements.

Taking $u(x, y)$ itself as the weighting function and using MWR procedure, we obtain the following ordinary differential equations.

$$\begin{cases} A_x\dot{\boldsymbol{\alpha}} + \dot{\theta}_{1i}(t) + u_1 B_x\boldsymbol{\alpha} = a_1(C\boldsymbol{\gamma} - A_x\boldsymbol{\alpha}) - a_1\boldsymbol{\theta}_{1i}(t) \\ A_y\dot{\boldsymbol{\beta}} + \dot{\theta}_{2i}(t) + u_2 B_y\boldsymbol{\beta} = a_2(C\boldsymbol{\gamma} - A_y\boldsymbol{\beta}) - a_2\boldsymbol{\theta}_{2i}(t) \\ C\dot{\boldsymbol{\gamma}} = b_1(A_x\boldsymbol{\alpha} - C\boldsymbol{\gamma}) + b_2(A_y\boldsymbol{\beta} - C\boldsymbol{\gamma}) + b_1\boldsymbol{\theta}_{1i}(t) + b_2\boldsymbol{\theta}_{2i}(t) \end{cases} \tag{78}$$

where

$$A_x = \langle u, xu^t \rangle, \quad A_y = \langle u, yu^t \rangle, \quad C = \langle u, u^t \rangle,$$
$$B_x = \langle u, \frac{\partial}{\partial x}(xu^t) \rangle, \quad B_y = \langle u, \frac{\partial}{\partial y}(yu^t) \rangle,$$
$$\boldsymbol{\theta}_{1i}(t) = \langle u, \theta_{1i} \rangle, \quad \boldsymbol{\theta}_{2i}(t) = \langle u, \theta_{2i} \rangle$$
$$\langle p(x, y), q(x, y) \rangle = \int_0^1 \int_0^1 p(x, y)q(x, y)\,dxdy$$

Numerical examinations of the above mentioned MWR procedure have been done for one-dimensional flow heat exchangers (single-tube, parallel-flow, counter-flow, one-shell two-pass and plate-type heat exchangers) by Kanoh (1977)[143], Ito & Masubuchi [144] and Ito & Masubuchi [106]; for cross-flow heat exchanger by Terasaka, Kanoh & Masubuchi [146]. The results of these investigations show that the MWR model compares favorably the finite difference model, in which the MWR procedure is more accurate and makes a more reduced-order lumped parameter model than the finite difference one. The reduction of order is remarkable in cross flow heat exchanger simulation. The dynamics of such a complicated heat exchanger as the rotary regenerator can be analyzed through the MWR technique by Terasaka, Kanoh & Masubuchi [145].

7. A SURVEY OF RESEARCH ON CONTROL PROBLEMS OF HEAT EXCHANGERS

The basic conventional control schemes have been introduced by many authors; Sanders (1959)[150], Harriott (1964), Buckley (1964), Forman (1966) [155] and Gould (1970). The performance of the PI-control system for heat exchangers was investigated by Takahashi (1951)[2] and Catheron, Goodhye & Hansen (1959)[148]. Koppel (1966)[154] treated a heat exchanger having a transfer function $(1-\exp(-s))/s$ controled by piecewise constant manipulation of flow rate. Switching times minimizing various performance criteria for the exact and linearized models were obtained.

Feedforward or cascade control has been studied as a method of improving the performance of the conventional control system of heat exchangers. The parameters and/or the positions of the feedforward sensors were adjusted to minimize a performance-index such as the integral of the error squared. (Koppel (1966)[183], Koppel, Kamman & Woodward (1970)[157]; feedback computer control of a flow-forced heat exchanger.) Ball (1977)[186] considered an optimum feed-forward controller for counter heat exchanger formulated in Hilbert space as a minimum norm problem.

Koppel (1967)[170] studied the optimal control of a tubular heat exchanger. He converted the basic partial differential equation into an equivalent ordinary differential equation with delay term and obtained time optimal control law using the maximum principle. Koppel & Shin (1968)[171] presented a strong minimum principle for a class of heat-exchangers with performance criteria such as minimum integral-square-error. The distributed parameter Riccati equation was derived for the linear case. Seinfeld & Lapidus (1968)[172] examined several techniques for optimal control of the hyperbolic distributed parameter system which included a direct search on the performance index and a method of steepest ascent based on a maximum principle Seinfeld & Lapidus (1968)[173] pointed out the existence of singular solution in the optimal control. Weigand (1970)[175] presented an optimization technique considered in the context of functional analysis and necessary conditions were obtained by variational techniques in Hilbert space. Both analytical and numerical solutions and a method for a good suboptimal control are presented. The optimal control techniques frequently used for heat exchangers are included in the maximum principle derived by Sirazetdinov (1964) for distributed paramater system which is described by hyperbolic partial differential equation. Okamoto et al (1975)[181] discussed the optimal flow-rate control problem of the counter-flow heat exchanger by using Sirazetdinov's theorem. The pattern of optimal control was bang-bang control with singular control.

Frequency domain synthesis of optimal control law was also considered. The spectrum factorization procedure was applied to the optimal control problem for the heat exchanger by Knudsen & Kummel (1974)[180].

The optimal programs obtained are too sophisticated for implementation compared with conventional controllers, because they require measurement of the entire profile of the temperature and unfeasible computation time to solve the two point boundary value problem of partial diffrrential equation. Hence it remains in a theoretical stage at the present time.

Mutharasan & Coughanour (1974)[162] developed a digital control algorithm for a flow-forced and a wall temperature forced tubular heat exchanger. Their algorithm consists of estimation of inlet state from the past values of the manipulated variable values and calculation of the present manipulated variable values. Mutharasan & Luus (1976)[164] presented a DDC algorithms based on the direct searching for the optimal parameters. Unbehauen, Schmid & Bottiger (1976)[142] applied seven DDC algorithms to the temperature control of a tubular heat exchanger. These algorithms are: PID, cascade, compensation (pole assignment), deadbeat, halfproportional, adaptive and optimal state feedback controller using an observer. The experimental results of their investigation showed that a PID controller with anti-windup is the best means to control the heat exchanger under different operating conditions. Tomizuka & Rosenthal (1978)[169] developed a digital state vector feedback controller with integral and preview actions for a flow-forced shell and tube-heat exchanger.

Model reference adaptive control was applied by Davidon & Durbin (1974)[187] to a finned tube heat exchanger and was shown to be effective in improving the dynamics. Ito, Kanoh & Masubuchi (1977)[144] applied the modal control technique to parallel-flow and counter-flow heat exchangers. Numerical results indicate that the controlled systems have widened-band width and larger stable regions in the control parameter plane. The bilinear observer was developed by Furusho, Kanoh & Masubuchi (1977) [199] to estimate intermediate temperatures of flow-forced heat exchangers.

Ansari (1970)[189] presented a sufficient condition for the stability of a mono-tubular heat exchanger controlled by flow rate. Clough & Ramirey (1970)[188] analyzed the stability of counter-current and parallel-flow heat exchangers with and without diffusion. The conclusion was naturally that all cases are stable. Kanoh (1974)[161] studied PI-feedback control of parallel-flow and counter-flow heat exchangers and found the limit cycle phenomenon caused by the nonlinearity of the manipulation of flow-rate. The existence of the limit cycle was verified by experimental results. Kitamura, Kanoh, Koshiro & Nishimura (1975)[191] studied the stability of the P- and PI-controlled heat exchangers. They showed the conditions for the steady states to be unique and that the unique steady state is globally asymptotically stable and that a stable limit cycle exists in the PI-control system if the steady state is not asymptotically stable. Ansari (1976)[192] considered the bypass controlled heat exchanger and he found that stability requires a possitive feedback for disturbance and that if the control is designed on the basis of minimizing the error in the final temperature alone, the optimal negative feedback law leads to sustained oscillations on the intermediate variables.

8. REFERENCES

Buckley,p. (1964). Techniques of process control, Wiley, New York.
Finlayson, B. A. (1972). The method of weighted residuals and variational principles, Academic Press, New York.
Gordon-Clark, M. R. (1966). Dynamic models for convective systems, Ph.D. thesis, Dept. of Elec. Eng. M.I.T., Cambridge, Mass.
Gould, L. A. (1969). Chemical process control, Addison-Wesley.
Harriott, P. (1964). Process control, McGraw-Hill, New York.
Profos, P. (1962). Die Regelung von Dampfanlagen, Springer-Verlag, Berlin.
Sirazetdinov, T. K. (1964). On the theory of optimal processes with distributed parameters, Avtomatika i Telemekhanika,25-4, 463-472.
Voelker, D. and Doetsch, G. (1950). Die zweidimensionale Laplace-Transformation, Verlag Birkhauser AG., Basel.

9. BIBLIOGRAPHY OF DYNAMICS AND CONTROL OF HEAT EXCHANGERS

One dimensional flow heat exchangers

[1] Profos, P. (1943). Die behandlung von regelproblemen vermittels des frequenzganges des regelkreises. Dissertation Zürich.

[2] Takahashi, Y. (1951). Automatic control of heat exchange. Bull. JSME, 54-393, 426-. (in Japanese)

[3] Takahashi, Y. (1952). Transfer function analysis of heat exchange process. In A. Tustin (Ed.), Automation & Manual Control, 235-247.

[4] Takahashi, Y. (1953). Regeltechnische eigenschaften gleich-und gegenstrom-warmeaustauschern. Regelungstechnik, 2-1, 32-35.

[5] Dusinberre, G. M. (1954). Calculation of transient temperatures in pipes and heat exchangers by numerical methods. Trans. ASME, 76, 421-426.

[6] Debolt, R. R. (1954). Dynamic characteristics of a steam-water heat exchangers. MS Thesis Univ. of California.

[7] Rizika, J. W. (1954). Thermal lags in flowing systems containing heat capacitors. Trans. ASME, 76, 411-420.

[8] Rizika, J.W. (1956). Thermal lags in flowing incompressible fluid systems containing heat capacitors. Trans. ASME, 78, 1407-1413.

[9] Cohen, W. C. & E. F. Johnson (1956). Dynamic characteristics of double-pipe heat exchangers. I & EC, 48-6, 1031-1034.

[10] Masubuchi, M. (1958). Control of multi-pass heat exchangers, Part 1 & 2. Bull. JSME, 24-139 & 147, 209-213, 934-940. (in Japanese)

[11] Profos, P. (1958). Dynamik der uberhitzerregelung. Regelungstechnik, 7-6, 239-246.

[12] Cima, R. M. & A. L. London (1958). The transient response of a two-fluid counterflow heat exchanger — The gas-turbine regenerator Trans. ASME, 1169-1179.

[13] Clark, J. A., V. S. Arpaci & K. M. Treadwell (1958). Dynamic response of heat exchangers having internal heat sources-Part I & II. Trans. ASME, 80-3, 612-624, 625-634.

[14] Morris, H. J. (1959). The dynamic response of shell and tube heat exchangers to temperature disturbances. ASME Paper.

[15] Iscol, L. & R. J. Altpeter (1959). Frequency response of multipass shell and tube heat exchanger. ASME Paper, 59-IRD-4.

[16] Arpaci, V. S. & J. A. Clark (1959). Dynamic response of heat exchangers having internal heat sources, Part 3. Trans. ASME Ser. C, 81, 253-266.

[17] Masubuchi, M. (1959). Control of multi-pass heat exchangers, Part 3 & 4. Bull. JSME, 25-155, 703-708, 708-711. (in Japanese)

[18] London, A. L., F. R. Biancardi & J. W. Mitchell (1959). The transient response of gas-turbine-plant heat exchangers—Regenerators, intercoolers, Precoolers, and ducting. Trans. ASME Ser. A, 81, 433-448.

[19] Masubuchi, M. (1959). Study of dynamics of coupled heat exchangers. Automatic Control, 6-3, 126-132. (in Japanese)

[20] Masubuchi, M. (1960). Dynamic response and control of multipass heat exchangers. Trans. ASME Ser. D, 51-65.

[21] Yang, W. J. (1961-8). Dynamic response and resonance phenomena of a single-solid, single-

fluid heat exchangers, Part 1 & 2. Bull. JSME, 27-180, 27-183. (in Japanese)

[22] Yang,W. J., J. A. Clark & V. Š. Arpaci (1961) Dynamic response of heat exchangers having internal heat sources, Part 4. ASME Paper, Ser. C, 83-3, 321-338.

[23] Hsu,J. P. & G. Nathan (1962-11). Transfer functions of heat exchangers. AIChE J., 8-5, 593-598.

[24] Enns,M. (1962). Comparison of dynamic models of a superheater. Trans. ASME Ser.C, 375-385. Delvaux L. (1963). Experimental study of the dynamic behaviour of a heat exchanger and of a mixing process. IFAC, 275-289.

[26] Harriott, P. & H. Weigandt, (1964-9). Counter-current heat exchange with vaporizing immiscible transfer agent. AIChE J., 10-5, 755-758.

[27] Chen, T. N. & K. R. Galle (1964). Transfer function for steam-generating heat exchangers. ISA Trans., 3-2, 175-183.

[28] Thal-Larsen, H. & W. V. Loscutoff (1964). Fluid-temperature transients in a dual-heat-exchanger system. Trans. ASME, Ser. D, 3, 23-31.

[29] Hotta, K. & M. Imaeda (1965). Simplifed dynamic model of steam-liquid shell and tube heat exchanger. Kagaku Kogaku, 29-2, 980-987.

[30] Letan R. & E. Kehat (1965). Mixing effects in a spray-column heat rxchanger. AIChE J., 11-5 804-808.

[31] Stainthorp, F. P. & A. G. Axon (1965). The dynamic behaviour of a multipass steam-heated exchanger. Chem. Eng. Sci., 20, 107-119.

[32] Larsen, H. T. & W. V. Loscutoff (1966). Fluid center-line-velocity effect on fluid-temperature transient in a heat exchanger. Trans. ASME Ser. D, 408-414.

[33] Antoine, M. J. (1966). Etude et essais de la regulation d'un echangeur thermique. Automatisme XI-3, 168-177.

[34] Isermann, V. R. (1966-10). Vorausbestimung der regelgute der dampf-temperatur-regelung von trommekesseln mit dem anakogrechner, Teil 1. Regelungstechnik, 14-10, 469-480.

[35] Kawai, S., T. Machiyama & M. Koizumi (1967). Experimental study on the transient performance of shell tube type heat exchangers. Trans. Soc. Instrum. & Control Eng., 3-1, 1-10. (in Japanese)

[36] Noldus, E. (1967-3). Analytisch untersuchungen des ubertragungs-verhaltens von gleich-und genestrom warme austauschern. Regelungstechnik 15-3, 112-117.

[37] Isermann, R. (1969). Mathematical models for the dynamic behaviour of steam heat exchangers for steam flow or condensate flow regulation. IFAC, Sess.47-2, 18-32.

[38] Davoust, G. (1969). Determining of heat exchangers dynamics basing upon their structural characteristics. IFAC Warszawa, Sess.47.5, 46-63.

[39] Schöne, A. (1969-4). Zum dynamischen verhalten von warmeaustauschern mit konzentrischen rohren und rohrbundeln. Regelungstechnik, 17-4, 154-161.

[40] Takeda, Y., K. Hotta & K.Iinoya (1969). Fluid mixing of the shell side steam in multitube heat exchanger. Kagaku Kogaku, 33-10, 981-987. (in Japanese)

[41] Shekar, S. C. & G. H. Green (1970-6), Dynamic study of a chill water cooling and dehumidifying coil. ASHRA Trans., 76-2141, 36-51.

[42] Myers, G. E., J. W. Mitchell & C. F. Lindeman, Jr. (1970). The transient response of heat exchangers having an infinite capacitance rate fluid. ASME Ser. C, 92-2, 269-275.

[43] Beckman, L. V. & V. J. Law (1971). Unsteady state simulation of the counterflow double pipe heat exchanger. Proc. of 1st IFAC Symp. on D.P.S., Paper 6-3.

[44] McNeill, W. A. (1973). Method-of-characteristics of transient single-pass heat-exchange process. ASME Ser. C, 95-2, 286-287.

[45] Unbehauen, H., Chr. Schmid, F. Bottiger & G. Lausterer, (1974). Anwindung von DDC-algorithmen zur regelung eines warmetauschers. Regelungstechnik, 22-8, 226-235.

[46] Vendrov, I. B. & I. I. Kostanji (1974). Non-linear mathematical model of system of jointly operating evaporators and heat-exchangers. Sov. Autom. Control, 6, 44-49.

[47] Ioseliani, A. N., A. A. Mikhalevich, V. B. Nesterenko & M. E. Salukvadze (1975). Problem of optimization of parameter for heat exchangers. IFAC 6th World, Ser.59-3, 1-5.

[48] Todo, I. (1976). Dynamic response of Bayonet-type heat exchanger (Part 1). Bull. JSME, 19-136, 1135-1140.

[49] Todo, I. (1976). Distributed temperature response of multipass heat exchanger. Trans. Soc. Instrum. & Control Eng., 12-5, 581-587.(in Japanese)

[50] Singh, K. P. (1976). Predicting flow induced vibration in U-bendregions of heat exchangers: an engineering solution. J. Franklin Inst., 302-2, 195-206.

[51] Vendrov, I. B. & I. I. Kostanzhi (1977). Dynamic characteristics of jointly operating evaporators and heat-exchangers. Sov. Autom. Control, 5, 34-36.

[52] Schult, M. & F. Mayinger (1977). Dynamic characteristics of double-pipe heat exchangers with double phase change. Paper ASME, No.77-WA/HT-26, 13.

[53] Vendrov, I. B. & I. I. Kostanghi (1977). Dynamic characteristics of jointly operating evaporators and heat-exchangers. Sov. Autom. Control, 5, 34-36.

[54] Todo, I. (1978). Dynamic response of Bayonet-type heat exchanger. Bull. JSME, 21-154, 644-651.

[55] Hashizume, T. (1978). Characteristics of heat exchanger with condensing. Trans. Soc. Instrum. & Control Eng., 14-4, 467-472.(in Japanese)

[56] Saraf, G. R. & P. L. Dhar (1979-3). Simulation of shell-and-tube condenser. Simulation, 32-2, 83-92.

[57] Earl, W. B. & T. K. Tan (1975). Process dynamics of simple heat exchanger. New Zealand Engin., 30-10, 298-299.

Flow forced dynamics

[58] Takahashi, Y. (1959). Response of heat exchangers with flow rate change. JIDO SEIGYO, 6-1, 2-7.(in Japanese)

[59] Hempel, A. (1961-6). On the dynamics of steam liquid heat exchangers. Trans. ASME Aer. D, 244-252.

[60] Koppel, L. B. (1962). Dynamics of a flow-forced heat exchanger. I & EC Fund., 1-2, 131-134.

[61] Stermole, F. J. & M. A. Larson (1963-6). Dynamic response of heat exchangers to flow rate changes. I & EC Fund., 2-1, 62-67.

[62] Stermole, F. J. & M. A. Larson (1964-9). The dynamics of flow forced distributed parameter heat exchangers. AIChE J., 10-5, 688-694.
[63] Hanus, B. (1965-1). Vereinfachte nachbildung des regelverhaltens eines dampfuberhitzers am analogrechner. Regelungstechnik, 13-1, 14-20.
[64] Kamman, D. T. & L. B. Koppel (1966-5). Dynamics of flow-forced heat exchanger. I & EC Fund., 5-2, 208-211.
[65] Penrod, D. D. & E. D. Crandall (1966). Dynamics of flow-forced systems. I & EC Fund., 5-4, 581-583.
[66] Ray, W. H. (1966). Forced-flow heat exchanger dynamics. I & EC Fund., 5-1, 138-139.
[67] Privott, W. J. & J. K. Ferrell (1966). Dynamic analysis of a flow-forced, concentric tube heat exchanger. Chem. Eng. Progr. Sympo. Ser.62-64, 200-208.
[68] Smith, J. M. (1966). Dynamics of flow-forced systems. I & EC Fund., 5-4, 581-584.
[69] Todo, I. (1967). Dynamic response of parallel-flow and counter-flow heat exchangers to the change of flow rate. Trans. JSME, 32-252, 1215-1226. (in Japanese)
[70] Stermole, F. (1967). Correspondence "dynamics of a flow-forced heat exchanger". I & EC Fund., 6-2, 318-320.
[71] Imaeda, M., S. Honda & S. Sugiyama (1970). Dynamics of heat exchangers subject to simultaneous changes in flow rate and inlet fluid temperatures. Int. Chem. J., 10-2, 303-308.
[72] Ruedrich, R. A., G. M. Rana & L. D. Durbin (1973). Flow forced dynamics and steam temperature control in a pinned tube exchanger with multi-cell model simulation. JACC, 375.
[73] Gilles, G. (1974-9). New results in modeling heat exchanger dynamics. ASME Ser. G, 96-3, 277.
[74] Todo, I. (1975). Dynamic response of parallel flow and counter flow heat exchangers to flow rate change. Trans. Soc. Instrum. & Control Eng., 11-5, 600-606.

Cross flow heat exchangers

[75] Dusinberre, G. M. (1959-2). Calculation of transient in a cross-flow heat exchanger. J. Heat Trans., 61-67.
[76] Gartner, J. R. & H. L. Harrison (1963-6). Frequency response transfer functions for a tube in crossflow. ASHRAE Trans., 69-1841, 323-330.
[77] Gartner, J. R. & H. L. Harrison (1965). Dynamic characteristics of water-to air crossflow heat exchangers. ASHRAE Trans., 71-1934, 212-224.
[78] Myers, G. E., J. W. Mitchell & B. Nagaoka (1965-1). A method of estimating cross-flow heat exchanger transients. ASHRAE Trans., 71-1933. 225-230.
[79] Myers, G. E., J. W. Mitchell & R. F. Norman (1967-2). The transient response of crossflow heat exchangers, evaporators and condensers. Trans. ASME Ser. C, 75-80.
[80] McNamara, R. T. & H. L. Harrison (1967-6). A lumped parameter approach to crossflow heat exchanger dynamics. ASHRAE Trans., 2052, IV1.1-IV1.9.
[81] Gartner, J. R. & L. E. Danne (1969). Dynamic response relations for a serpentive crossflow heat exchanger with water velocity disturbance ASHRAE Trans., 75-1, 53-68.
[82] Tamm, H. (1969). Dynamic response relations for multi-row crossflow heat exchangers. ASHRAE Trans., 75-1, 69-80.
[83] Masubuchi, M. (1969). Dynamic response of crossflow heat exchangers. IFAC Warszawa, Sess. 47.
[84] Isermann, V. R. & H. Jantschke (1970-1). Das regeldynamische verhalten wasser-und dampfbeheizten kreugstromwarmebertragen in klimaanlagen. Regelungstechnik und Proze -Datenverbeitung, 3, 115-122.
[85] Schulz, R. (1970). Das dynamische verhalten eines kreuzstromwarmetauschers. Warme Band, 76-2, 13-18.
[86] Bender, E. (1972). Dynamic response of cross-flow heat exchangers for variations in temperature and massflow. IFAC 5th World, Sess.8.3, 1-10.
[87] Sindelar, R. (1973). Mathematisches modell fur die dynamik des kreuzstromwarmeastauschers. Regelungstechnik und Prozess-Datenverarbeitung 21-8, 261-266.
[88] Brasz, J., J. H. Voskamp & O. Rademaker (1973) The dynamic behaviour of a gas-solid cross-flow heat exchanger. J. Measure. Control.
[89] Tobias, J. R. (1973). Simplified transfer function for temperature response of fluids flowing through coils, pipes or ducts. ASHRAE Trans., 79, Part II, 19-22.
[90] Tamm, H. & G. H. Green (1973). Experimental multi-row crossflow heat exchanger dynamics. ASHRAE Trans., 79-2276, 9-18.
[91] Pearson, J. T., R. G. Leonard & R. D. McCutchan (1974). Gain and time constant for finned serpentine crossflow heat exchangers. ASHRAE Trans., 80, Part II, No.2321, 255-267.
[92] Bhargava, S. C., F. C. McQuiston & L. D. Zirkle (1975). Transfer functions for crossflow multi-row heat exchangers. ASHRAE Trans., No.2380, 294-314.
[93] Boot, J. L., J. T. Pearson & R. G. Leonard (1977). An inproved dynamic response model for finne serpentine cross-flow heat exchangers. ASHRAE Trans., 83-1, 218-239.
[94] Yamashita, H., S. Yamaguchi & R. Izumi (1977). Analysis of the dynamic characteristics of crossflow heat exchangers with both fluids unmixed (on the transiente response to a step change in the inlet temperature). Trans. JSME, 43-371, 2687-2694. (in Japanese)

Spline heat exchangers

[95] Takashima, S., H. Kanoh & M. Masubuchi (1976) Dynamics of spiral-plate heat exchanger for flow rate change. Memoirs of the Faculty Education, Kumamoto Univ., No.25, Natl. Sci., 17-28. (in Japanese)
[96] Takashima, S., H. Kanoh & M. Masubuchi (1976). Dynamics of spiral heat exchanger. Trans. JSME, 42-362, 3152. (in Japanese)

Survey

[97] Williams, T. J. & H. J. Morris (1961). A survey of the literature on heat exchanger dynamics and cotrol. Chem. Eng. Progr. Symp. Ser. 57-36, 20-33.
[98] Kanoh, H. (1979). Some aspects of dynamics and control of heat exchangers. Proc. IMACS Europ. Simul. Meeting (Patras, Greese).

Finned heat exchangers

[99] Stoecker, W. F., J. E. Shahan & S. A. Mumma (1971). Dynamic response of a finned-coil refrigerant evaporator to step changes in refrigerant flow rate. ASHRAE Trans., 2199, Pt.2, 80-87.

[100] Muller, K. (1977). Das dynamische verhalten von rippenrohren in warmeubertragen. Regelungstechnik, 25-11, 350-357.

[101] Furusho, J., H. Kanoh & M. Masubuchi (1979). Distributed-parameter modeling of plate-fin type heat exchanger. Proc. IMACS Europ. Simul. Meeting (Patras, Greece).

[102] Furusho, J., K. Uematsu, H. Kanoh & M. Masubuchi (1979). Dynamics of plate-fin type heat exchangers. Trans. Soc. Instrum. & Control Eng., 15-1, 125-130. (in Japanese)

Plate heat exchangers

[103] Masubuchi, M. & A. Ito (1975). Dynamic analysis of plate heat exchangers. Trans. Soc. Instrum. & Control Eng., 11-5, 613-621. (in Japanese)

[104] Ito, A. & M. Masubuchi (1977). Experimental frequency analysis of plate heat exchange system. Tech. Rep. Osaka Univ., 27-1361, 261-269.

[105] Masubuchi, M. & A. Ito (1977-4). Dynamic analysis of a plate heat exchanger system. Bull. JSME, 20-142, 434-441.

[106] Ito, A. & M. Masubuchi (1978). Dynamic analysis and modal control of plate heat exchanger systems. IFAC 7th World, Sess.6.3, 343-349.

Approximation of heat exchanger dynamics

[107] Takahashi, Y. (1955). Graphically determine the dynamics of heat percolation. Control Eng., 2-5, 46-50.

[108] Juhasz, S. & J. Clark (1956-9). Hydraulic analogy for transient conditions in heat exchangers. 4th Int. Inst. & Meas. Conf., Paper GMP 2C.

[109] Liebemann, G. (1956). A new electrical analog method for the solution of transient heat-conduction problems. Trnas. ASME, 655-665.

[110] Paynter, H. M. & Y. Takahashi (1956). A new method of evaluating dynamic response of counterflow and parallel-flow heat exchangers. Trnas. ASME, 749-758.

[111] Paynter, H. M. (1956). On an analogy between stochastic processes and monotone dynamic systems. Preprint Conf. Control Tech. Heidelberg.

[112] Mozley, J. M. (1956). Predicting dynamics of concentric pipe heat exchangers. I & EC, 48-6, 1035-1041.

[113] Ford, R. L. (1956-1). Electrical analogue for heat exchangers. Proc. IEE Paper, 103, No.1934 65-82.

[114] Kourim, G. (1957). Die electrische nachbildung der instationaren thermischen vorgange beim warmeaustauscher. Regelungstechnik, 5, 163-167.

[115] Hainsworth, B. D., V. V. Tivy & H. M. Paynter, (1957). Dynamic analysis of heat exchanger control. Preprint of ISA J..

[116] Fricke, L. M., H. J. Morris, R. E. Otto & T. J. Williams (1960). Process dynamics and analogue-computer simulation of shell-and-tube heat exchangers. Chem. Eng. Progr. Symp. Ser. 56-31, 80-85.

[117] Thal-Larsen, H. (1960). Dynamics of heat exchangers and their models. Trans. ASME Ser.D, 489-504.

[118] Prooksa, F. (1962). Verfahren zur berechnumg der frequenz-gange von gleich-und gegenstromwarmeaustauschern. Regelungstechnik, 5, 206-210.

[119] Ball, S. J. (1963). Approximate models for distributed-parameter heat-transfer systems. IFAC, 131-139.

[120] Schone, A. (1966-6). A special class transfer functions for the description of the dynamic behaviour of heat exchangers. IFAC 3th Proc., 10B, 1-7.

[121] Dorri, M. K. (1966). Approximate solution of an analog computer of certain partial differential equations of heat exchange. Autom. & Remote Control, 27, 1445-1451.

[122] Friedly, J. C. (1967). Asymptotic approximations to plug flow process dynamics. JACC, 216-228.

[123] Filipovich, V. (1967-7). Simulation of transient processes in heat exchangers with application of generalized differential-difference equations. Autom. & Remote Control, 28-7, 1126-1131.

[124] Privott, W. J. Jr. & J. K. Ferrell (1967-10). Simulating a flow-forced heat exchanger. Instrum. & Control Syst., 40, 121-123.

[125] Imaeda, M., S. Hayashi & S. Sugiyama (1967). Dynamic response of shell and tube heat exchangers. KAGAKU KOGAKU, 31-8, 812-818.

[126] Chi, T. & S. Seshadri (1969-1). An Approximate solution for countercurrent heat exchangers. AIChE J., 15-1, 39-46.

[127] Kummel, M. (1969). Analog simulation and control of a distributed parameter system. Chem. Eng. Sci., 24, 1055-1062.

[128] Schmidt, J. R. & D. R. Clark (1969). Analog simulation techniques for modeling parallel-flow heat exchangers. Simulation, 12, 15-21.

[129] Silvey, T. I. & J. R. Barker (1970). Hybrid computing techniques for solving parabolic and hyperbolic partial differential equations. Comput. J., 13-2, 164-170.

[130] Ruedrich, R. A., S. Chongvoranond & L. D. Durbin (1970). Non-linear network model for the dynamics and control of a condensing vapor heat exchanger. JACC, Sess.23-D, 621-632.

[131] Gilles, E. D. & M. Zeitz (1970). Modal simulation method for distributed parameter system JACC, Sess.29-B, 779-787.

[132] Stafford, E. M. & J. M. Nightingale (1970). Generalised fourier methods for 1st-order systems. Proc. IEE, 117-9, 1864-1868.

[133] Heidemann, R. A., C. E. Huckaba, F. S. Eisen, L. I. Weissman & G. M. Gallatig. (1971). Dynamics of convection heat exchangers. Canadian J. Chem. Eng., 49, 147-153.

[134] Otte, G. (1972). Modellierung einer klasse von systemen mit verteilten parametern auf digitalrechnern. Mess. Steuern Regeln, 15-7, 259-263.

[135] Wood, R. K. & V. A. Sastry (1972-3). Simulation studies of a heat exchanger. Simulation, 18-3, 105-113.

[136] Otte, G. (1972). Algorithmic model formation of systems with distributed parameters. IFAC 5th world, Part 4, 36.2, 1-9.

[137] Bender, Von E. (1972). Das dynamische verhalten von kreuzstromwarmeaustuschen fur massenstromvariationen. Regelungstechnik und pro-

zess-daterverarbait, 20-1, 13-20.
[138]Dunnett, R. M. & N. Hutchinson (1974). Serial solution of heat-exchanger equations without a hybrid computer. Proc. IEE, 121-7, 730-734.
[139]Matsuyama, H. & E. Oshima (1974). Representation of dynamics of multipass heat exchangers. KAGAKU KOGAKU, 38-7, 526-533.(in Japanese)
[140]Schmidt, F. W. (1974). Numerical simulation of the thermal behaviour of convective heat transfer equipment. Afgan, Schlunder Ed. "Heat exchanger: Design and Theory Soucebook", McGraw-Hill, 491-524.
[141]Choung, K. C. & J. R. Schmidt (1976). Digital algorithms for dynamic analysis of a heat exchanger. Proc. Heat Transfer Fluid Mech. Inst., 483-494.
[142]Unbehauen, H., Chr. Schmid & F. Bottiger (1976). Comparison and application of DDC algorithms for heat exchangers. Automatica, 12-5, 393-402.
[143]Kanoh, H. (1977). Approximation of the dynamics of heat exchangers by the method of weighted residuals. Tech. Reports Osaka Univ., 27-1386, 507-515.
[144]Ito, A., H. Kanoh & M. Masubuchi (1977). MWR approximation and modal control of parallel- and counterflow heat exchangers. IFAC Symp. DPS Proc., 12-24.
[145]Terasaka, H., H. Kanoh & M. Masubuchi (1979). Dynamic characteristics of rotary regenerative heat exchanger by Galerkin's method. Bull. JSME, 22-170, 1067-1073.
[146]Terasaka, H., H. Kanoh & M. Masubuchi (1980). Approximate dynamic analysis of crossflow heat exchanger by method of weighted residuals. Bull. JSME, 23-177, 432-438.
[147]Kanoh, H., J. Furusho & M. Masubuchi (1980). Reduced order lumped parameter model of heat exchangers. J. Soc. Instrum. & Control Eng., 19-7, 636-639.(in Japanese)

Conventional control of heat exchangers

[148]Catheron, A. R., S. H. Goodhue & P. D. Hansen (1959). Control of shell and tube heat exchangers. ASME Paper, 59-IRD-14.
[149]Fricke, L. H., H. J. Morris & R. E. Otto(1959) Process dynamics automatic control and analog computer simulation of shell and tube heat exchanger. ASME Paper.
[150]Sanders, C. W. (1959). Better contorl of heat exchangers. Chem. Eng., 66-19, 145-148.
[151]Douglas, J. M., J. C. Orcutt & P. W. Berthiaume (1962). Design and control of feed-eddluent, exchanger-reactor systems. I & EC Fund., 1-4, 253-257.
[152]Berger, R. L., E. C. Hoelscher & C. Kippenhan (1964). Temperature control and steam demand characteristics of a load leveling heat exchanger. ASHRAE Trans., 70-1906, 434-439.
[153]Koppel, L. B. (1965). Dynamics of a class of nonlinear, distributed-parameter, chemical reactors. I & EC Fund., 4-3, 269-275.
[154]Koppel, L. B. (1966-8). Control of tubular heat exchangers and chemical reactors by piecewise constant manipulation of flow rate. I & EC Fund., 5-3, 413-422.
[155]Forman, E. R. (1966). Control dynamics in heat transfer. Chem. Eng., 1-3, 91-96.
[156]Zelnski, R. E., R. A. Lund & R. E. Sowla (1968) An investigation of a closed-loop system for duct air temperature control. ASHRAE Trans., 74, VI 1.1-VI 1.9.
[157]Koppel, L. B., D. T. Kamman & J. L. Woodward (1970). Two-point practical contorl of a class of distributed parameter processes. Ind. Eng. Chem. Fund., 9-2. 198-205.
[158]Kirsch, L. W. & C. K. Sanathanan (1971). Contrrol of a distributed parameter system from its boundary. IFAC 1st Symp. on DPS, 6-2.
[159]Messa, C. J., G. W. Poehlein & A. S. Foust (1971). Heat exchanger modeling by conservative scalar pulse testing. I & EC Process Des. Develop., 10-4, 466-472.
[160]Lee, H. H., L. B. Koppel & H. C. Lim (1972). Integrated approach to design and control of a class of countercurrent processes. I & EC Process Des. Develop., 11-3, 376-382.
[161]Kanoh, H. (1974). Control of stirred tank heat exchanger and parallel-counter heat exchanger by manipulation of flow rate. Tech. Reports Osaka Univ., 24-1180, 257-277.
[162]Mutharasan, R. & D. R. Coughanowr (1974). Feedback direct digital control algorithms for a class of distributed-parameter systems. Ind. Eng. Chem. Process Des. Develop., 13-2, 168-176.
[163]Masubuchi, M. & A. Ito (1974). Incomplete feedback control in a heated tube flow process. Trans. JSME, 40-329, 176-180. (in Japanese)
[164]Mutharasan, R. & R. Luus (1976). Linear direct digital control algorithms for a class of distributed processes. Ind. Eng. Chem. Process Des. Develop., 15-1, 137-141.
[165]Mutharasan, R. & D. R. Coughanowr (1976). Sampled-data proportional control of a flow-forced tubular reactor. Ind. Eng. Chem. Process Des. Develop., 15-1, 141-144.
[166]Mutharasan, R. & D. R. Coughanowr (1976). Sampled-data control of a distributed-parameter process. Ind. Eng. Chem. Process Des. Develop., 15-2, 378-381.
[167]Furusho, J., H. Kanoh & M. Masubuchi (1977) Feedback controllers for systems linear in control. Tech. Reports Osaka Univ., 27-1387, 517-525.
[168]Yang, C. H. & T. J. Ward (1977). Structual analysis control of a heat excahnger. JACC, 2, 893-898.
[169]Tomizuka, M. & D. E. Rosenthal (1978). On the optimal digital state vector feedback controller with integral and preview actions. JACC, 3, 155-168.

Optimal control of heat exchangers

[170]Koppel, L. B. (1967). Optimum control of a class of distributed-parameter processes. I & EC Fund., 6-2, 299-303.
[171]Koppel, L. B. (1968). Optimal control of a class of distributed-parameter systems with distributed controls. I & EC Fund., 7-3, 414-422.
[172]Seinfeld, J. H. & L. Lapidus (1968). Computational aspects of the optimal control of distributed-parameter systems. Chem. Eng. Sci., 23, 1461-1483.
[173]Seinfeld, J. H. & L. Lapidus (1968). Singular solutions in the optimal control of lumped- and distributed-parameter systems. Chem. Eng. Sci., 23, 1485-1499.
[174]Baldwin, J. T., R. A. Ruedrich & L. D.Durbin (1969). Optimal boundary control signal synthesis for the ackflow cell-network model of

a nonlinear heat exchange reactor with axial mixing. JACC, 803-814.

[175] Weigand, W. A. (1970). Optimal control of plug-flow heat exchanger with control produced by wall flux or wall temperature. I & EC Fund., 9-4, 641-651.

[176] Huang, H. S. & W. J. Yang (1970). Optimum control of heat exchangers with internal heat generation. Appl. Sci. Res., 23, 95-112.

[177] Gilles, G. & M. Nougaret (1972). Optimal Control of a pilot heat-exchanger. IFAC, 5th, pt.4 36.3.

[178] Lee, H. H., L. B. Koppel & H. C. Lim (1973). Optimal sensor locations and controller settings for class of countercurrent processes. Ind. Eng. Chem. Process Des. Develop., 12-1, 36-41.

[179] Iddan, G. J. (1974). Control of some distributed parameter systems with disturbances. JACC, 436-442.

[180] Knudsen, J. K. H. & M. Kummel (1974). Frequency domain solution of the optimal control problem for linear distributed system . Chem. Eng. Sci., 29, 521-531.

[181] Okamoto, M., M. SHima, S. Ikeda & H. Tokumaru (1975). Optimal control of a bilinear hyperbolic system. Trans. Soc. Instrum. & Control Eng., 11-2, 214-220. (in Japnese)

[182] Trybus, L. A. (1979). Pointwise control of a counterflow process. Int. J. Syst. Sci., 10-7, 807-815.

Feedforward control of heat exchangers

[183] Koppel, L. B. (1966-8). Dynamics and control of a class of nonlinear, tubular, parametrically forced heat exchangers and chemical reactors. I & EC Fund., 5-3, 403-413.

[184] Paraskos, J. A. & T. J. Mcavoy (1970). Feedforward computer control of a class of distributed parameter processes. AIChE J., 16-5, 754-761.

[185] Goford, P., J. F. MacGregor & J. D. Wright (1976). Predictive stochastic feedforward-feedback control of a heat exchanger-stirred tank system. JACC, THRY 2-5, 394-398.

[186] Ball, D. J. (1977). Optimum feedforward control of distributed parameter systems. Int. J. Sys. Sci., 8-1, 81-85.

Adaptive control of a heat exchanger

[187] Davidson, J. M. & L. D. Durbin (1974). Model reference adaptive control specification for steam heeted fined tube heat exchanger. JACC, 474-483.

Stability problems of heat exchanger control

[188] Clough, D. E. & W. F. Ramirez (1970). Stability of counter-current and parallel flow heat exchangers with and without diffusive effects. JACC, Sess.23-A, 596-601.

[189] Ansari, J. S. (1970). Stability of a class of distributed parameter feedback systems. JACC, 112-115.

[190] Ansari, J. S. (1973). Stability of differential-difference equations representing heat exchanger and certain other systems. Int. J. Control, 17-1, 193-198.

[191] Kitamura, S., H. Kanoh, M. Koshiro & S. Nishimura (1975). Stability analysis of a flow rate controlled heat exchangers. Trans. Soc. Instrum. & Control Eng., 11-2, 239-244.

[192] Ansari, J. S. (1976-6). Stability analysis and optimization of by-pass controlled heat exchanger. ASME Ser.G, 98-2, 161-166.

Identification problems of heat exchangers

[193] Corlis, R. G. & R. Luus (1969). Use of residuals in the identification and control two-input, single-output system. I & EC Fund., 8-2, 246-252.

[194] Carpenter, W. T., M. J. Wozny & R. E. Goodson (1970). Distributed parameter identification using the method of characteristics. JACC, Sess. B, 565-572.

[195] Pedersen, C. O. & C. D. Mouen (1973). Application of system identification techniques to the determination of thermal response factors from experimental data. ASHRAE Trans., No.2287 RP-102, 127-136.

[196] Wright, J. D. & D. W. Bacon (1974). Statistical identificaiton of the dynamic behaviour of a dual input heat exchanger network. Canadian J. Chem. Eng., 52, 405-.

[197] Baur, U. & R. Isermann (1977). On-line identification of a heat exchanger with a process computer—a case study—. Automation, 13-5, 487-496.

[198] Bauer, B. & H. Unbehauer (1978). On-line identification of a load-dependent heat exchanger in closed loop using a modefied instrumental variable method. IFAC 7th World, Sess.6.4, 351-359.

Bilinear observer for heat exchanger

[199] Furusho, J., H. Kanoh & M. Masubuchi (1977). Observers for bilinear distributed parameter process. IFAC Sympo. Warwick Univ..

Testing of heat exchanger dynamics

[200] Lees, S. & J. O. Hougen (1956). Pulse testing a model heat exchange process. I & EC, 48-6, 1064-1068.

[201] Izawa K., K. Furuta, H. Oono, T. Ito & T. Akiyama (1966). Measurement of dynamic characteristics of heat exchanger using binary random signal. Trans. Soc. Instrum. & Control Eng., 2-2, 101-112. (in Japanese)

[202] Ito, K. & T. Akiyama (1966). Dynamic response of shell and tube heat exchanger. Trans. Soc. Instrum. & Control Eng., 2-2, 94-100.(in Japanese)

[203] Messa, C. J., G. W. Poehlein & A. S. Foust (1971). Heat exchanger modeling by conservative scalar pulse testing. I & EC, Process Des. Develop., 10-4, 466-472.

[204] Davoust, G. (1969). Determination of heat-exchanger dynamics on basis of their structual characteristics. IFAC Warszawa. Sess.47.

Supplements

[205] Gutierrez R.H.J & A.R.Cooper (1979). Heat exchanger process dynamics review. The Chemical Engineering J.,17-, 13-18.

[206] Cooper A.R. & H.J.Gutierrez R. (1979) Superposition of liquid flow rate disturbances in a heat exchanger at low Reynolds number. The Chemical Engineering J., 17-, 19-40.

CHAPTER 15

MODELING AND CONTROL OF DISTRIBUTED PARAMETER NUCLEAR REACTORS

D. M. Wiberg

Department of System Science, University of California, Los Angeles, USA

INTRODUCTION

Nuclear reactors are excellent examples of processes in which spatial effects are important. As such, they can furnish applications for distributed parameter control systems. The heat power produced by the nuclear reaction depends not only on time but on the spatial position in the reactor. Therefore, the error signal formed by the deviation of the actual power production from that desired is distributed in space and changes with time. The control rods act at various points within the reactor, and their position influences the control. Some of the techniques discussed by Amouroux and Babary in this volume can be used to select the positions of the control rods and also those of the various measuring instruments. Flux meters, thermocouples, and many other instruments are used to measure neutron population, temperature, and other pertinent variables at different proximities to the reactor.

As in most physical processes, the control and measurement points for nuclear reactors are discrete. Here it will be shown that better results are obtained by keeping the distributed formulation to find the control, and then discretizing in space rather than first discretizing and then applying discrete parameter control theory. Therefore the nuclear reactor application leads to a true distributed parameter control problem when the shape of the power distribution in space is considered. An example of this type of problem is the control of spatially unstable xenon oscillations as discussed in Wiberg (1967a). This is in contradistinction to the problem of controlling only the spatially averaged power, which is not a true distributed parameter control problem.

This chapter is not a comprehensive survey of the application of distributed parameter techniques to nuclear reactor systems. Rather, it represents a unification and updating of the author's viewpoint, which is more academic and theoretical than that of a practicing nuclear reactor control engineer. A spatial control system has not been implemented on a reactor to date, because reactors with the spatial dimensions and power levels great enough to require a spatial control system have not yet been built. However, the insights gained from considering the distributed parameter viewpoint do have applications even in today's reactors.

---

*Also, Consultant, The Aerospace Corporation. This manuscript was prepared with the assistance of The Aerospace Corporation.

The chapter is divided into sections. The next section of the chapter reviews the development of the general distributed state equations. Linearization, incorporation of the power control system, noise modeling, and controllability of the reactor are discussed in Section 3. The fourth section is general exposition of non-interacting modal expansions, Section 5 incorporates this approach into filtering and control, and Section 6 then applies these methods to a liquid metal fast breeder reactor. The chapter concludes with Section 7 which shows how the premature discretization of the delay equation found in the previous section leads to erroneous results.

NUCLEAR REACTOR DISTRIBUTED STATE EQUATIONS

The heat power produced by a nuclear reactor results mainly from the speed of the atomic fission fragments, which is eventually transferred to the thermal motion of the molecules of the surrounding medium. The heat power is therefore proportional to the number of fissions, which is $\Sigma_f\Phi$ where Σ_f is the fission cross section and Φ is the neutron flux. The neutron flux Φ equals vn, where n is the neutron density and v is the velocity of the neutrons. Therefore the power distribution and other properties of the reactor can be computed either from the flux Φ or the neutron density n, assuming v is known.

The distribution of neutrons is found by writing a neutron balance equation. The neutron balance in a power reactor in which the reactor materials are not moving is most accurately described by the Boltzmann transport theory model. In this model, the neutron flux is dependent not only on space and time, but also on neutron velocity and the direction of neutron motion. However, for reactor regions in which there are no sharp flux gradients, the Boltzmann equation may be adequately approximated by the multigroup diffusion equations. This set of equations considers the neutron balance at different velocity groups, averaging over neutron direction and over the particular velocity interval for the group under consideration.

Furthermore, the model can incorporate groups of concentrations C of delayed neutron precursors. A delayed neutron is a neutron born from a fission fragment decay chain that produces an atom called a delayed neutron precursor. With a time constant specific to the group of delayed neutrons, the precursor decays to other fragments including the delayed neutron. The averaged effect of the delayed neutrons upon all the neutrons is that the time from one generation of neutrons to the next is about five or ten seconds. This averaged effect permits mechanical positioning of the control rods. However, if the fraction of excess neutrons created per generation is greater than the fraction β of delayed neutrons, then the neutrons created directly from fission are sufficient to sustain the chain reaction. Then the time from one generation of neutrons to the next becomes much less than one second. In such circumstances, mechanical control is impossible. This situation is called "prompt-critical" and must be avoided by any control system.

The multigroup diffusion equations are written in terms of velocity group fluxes $\Phi(r, t)$ and the delayed neutron precursor groups $C(r, t)$, both of which are functionally dependent on the space variable r (of one, two, or three dimensions) and the time variable t. Thus they form a set of coupled partial differential equations, which are the distributed parameter formulation considered in this chapter. For example, if the velocity group fluxes are divided into one fast group Φ_F and one slow group Φ_S, and all the delayed neutron precursors are averaged into one group C, the multigroup equations are

$$\begin{Bmatrix} \frac{1}{v_S} & 0 & 0 \\ 0 & \frac{1}{v_S} & 0 \\ 0 & 0 & 1 \end{Bmatrix} \frac{\partial}{\partial t} \begin{Bmatrix} \Phi_F \\ \Phi_S \\ C \end{Bmatrix}$$

$$= \begin{Bmatrix} \{\nabla \cdot D_F \nabla - (\Sigma_{aF} + \Sigma_R) + (1 - \beta)\nu\beta_{fF}\}\Phi_F + (1 - \beta)\nu\Sigma_{fS}\Phi_S + \lambda_c C \\ \Sigma_R \Phi_F + (\nabla \cdot D_S \nabla - \Sigma_{aS})\Phi_S \\ \beta\nu\Sigma_{fF}\Phi_F + \beta\nu\Sigma_{fS}\Phi_S - \lambda_c C \end{Bmatrix} \tag{1}$$

Here the ∇ represents the gradient operator in r, so that these are partial differential equations. The boundary conditions are that Φ_F, Φ_S, $D_F\nabla\Phi_F$, and $D_S\nabla\Phi_S$ are continuous at all reactor region interfaces, and that Φ_F, Φ_S, and C vanish at the boundaries of the reactor (extrapolated to account for some return from the surrounding medium).

This multigroup model is derived in detail in standard nuclear engineering texts such as Glasstone and Edlund (1952), and only a brief explanation is given here. The rate of cnange of neutrons is expressed as $\partial n/\partial t$ and equals $(\partial\Phi/\partial t)/v$. This derivative equals neutrons gained minus neutrons lost. The neutrons lost from the reactor by diffusion can be represented by $\nabla \cdot D\nabla\Phi$, where D is the diffusion coefficient. Also the neutrons lost by absorption are equal to $\Sigma_a\Phi$. Fast neutrons are gained by fission, where the number of fissions is $\Sigma_{fF}\Phi_F + \Sigma_{fS}\Phi_S$, with average number of neutrons per fission ν, and the delayed neutron fraction is β. Fast neutrons are also gained by decay of neutron precursor concentration C with time constant λ_C. Fast neutrons are slowed to slow neutrons at a rate $\Sigma_R\Phi_F$. The balance of these effects gives the multigroup diffusion equations, which can easily be extended to many velocity groups of flux and many delayed neutron precursor groups. If the various coefficients such as D or Σ depended only on r and t, the multigroup model would consist of linear partial differential equations. However, these coefficients usually depend on temperature, fission produce poisons, fuel burn up, control rods, etc., which in turn depend on the flux and therefore these equations become nonlinear.

Temperature effects can be described only very generally because they are quite dependent on reactor type. The effect of temperature on each of the various coefficients must be estimated by physical considerations and/or material measurements. This estimation has been made for many nuclear reactors. Furthermore, because the multigroup equations will be linearized, there is need only for the first term in the Taylor series, e.g. the first partial derivative of any cross section with respect to temperature, $\partial\Sigma/\partial T$. Of course this term must be evaluated as a function of space r and operating temperature $T_0(r)$. This term models the effects of temperature in the multigroup equations.

To obtain an equation for the temperature, the solid core and the coolant are considered separately. Within the solid core of the reactor, the heat conduction balance is

$$\rho(r)C_T(r)\frac{\partial T(r,t)}{\partial t} = \nabla\cdot\kappa(r)\nabla T(r,t) + e\Sigma_{fF}(T,r)\Phi_F(r,t) + e\Sigma_{fS}(t,r)\Phi_S(r,t) \tag{2}$$

where ρ is the density, C_T the thermal capacity, κ the conduction coefficient, and e the heat produced per fission. The boundary conditions for this equation are similar to those for the multigroup equations, in such that between solid bodies the temperature and the heat flux are both continuous at solid interfaces. If T is the temperature difference from the surroundings at the reactor extrapolated boundaries, T vanishes also.

Coolant is usually pumped through the reactor core to remove the heat. Consider an average coolant bulk temperature T(x,t) that varies along the coolant flow direction x. Assume the coolant has a constant velocity v, density ρ, and thermal capacity C_T. Then

$$\rho C_T\frac{\partial T(x,t)}{\partial t} = -\rho C_T v\frac{\partial T(x,t)}{\partial x} + h\left[T_{wall}(x,t) - T(x,t)\right] \tag{3}$$

where h is the heat transfer coefficient and $T_{wall}(x,t)$ is the coolant channel wall temperature. The known (measured) coolant inlet temperature is the boundary condition T(0,t).

Fission product poison effects are discussed in detail in Wiberg (1967a). Here only a brief review is given. Fission products give rise to decay chains that produce fission produce poisons. Examples of poisons are xenon and samarium isotopes (Xe^{135} and Sm^{149}) that have a very large slow neutron absorption cross section. Here only the xenon process is considered because the samarium process is similar and has less effect. Denote the concentration of xenon atoms by N_x and the microscopic xenon slow neutron cross section by σ^x_{as}. Then the effect on the multigroup equations is

$$\Sigma_{as} = \Sigma_{as0}(r) + \sigma^x_{as}N_x(r,t) \tag{4}$$

The xenon concentration is found from an xenon balance equation.

$$\frac{\partial N_x(r,t)}{\partial t} = \lambda_I N_I(r,t)+\gamma_x\Sigma_{fF}(r)\Phi_F(r,t)+\gamma_x\Sigma_{fS}(r)\Phi_S(r,t)-\lambda_x N_x(r,t)-\sigma^x_{as}N_x(r,t)\Phi_S(r,t) \tag{5}$$

Here the xenon concentration gains from decay of xenon precursor (iodine 135) concentration N_I with time constant λ_I and also from a small fraction γ_x that comes directly from the total number of fissions $\Sigma_{fF}\Phi_F + \Sigma_{fS}\Phi_S$. Xenon concentration is decreased by radioactive decay to cesium 135 with time constant λ_X and by neutron capture to xenon 136.
The iodine concentration is found from an iodine balance equation, in which γ_I is the fraction of the iodine concentration produced by fission.

$$\frac{\partial N_I(r,t)}{\partial t} = \gamma_I\Sigma_{fF}(r,t)\Phi_F(r,t)+\gamma_I\Sigma_{fS}(r,t)\Phi_S(r,t)-\lambda_I N_I(r,t) \tag{6}$$

There is no xenon or iodine outside the reactor, so N_x and N_I vanish at the extrapolated boundaries.

Although fuel burn up is an important spacial process in the nuclear reactor, very little attention has been given to it as a process that can be controlled by feedback. Fuel burn up occurs very slowly, generally over a period of months, and is compensated for by the reactor operators adjusting the shim rods. But the actions of the reactor operators can be considered as the implementation of part of a spatial control scheme. Although this consideration may not yet be practical, some type of feedback scheme might bear further investigation with the advent of very large reactors.

The effect of fuel burn up on the fission cross section is

$$\Sigma_f(r,t) = \Sigma_{f0}(r) - f\int_0^t \Sigma_f(r,t)\Phi(r,t)dt \tag{7}$$

where f is the fraction of fuel burned up per unit time and Σ_{f0} is the initial fission cross section. Differentation of this gives

$$\frac{\partial \Sigma_f(r,t)}{\partial t} = -f\Sigma_f(r,t)\Phi(r,t) \tag{8}$$

Because there is no fuel outside the reactor, Σ_f vanishes at the extrapolated boundaries.

The effect of control rods is discussed in detail in Wiberg (1967a). If the insertion or extraction of the control rods is considered as an absorption cross section sink or source at the interior end of the control rod, there results

$$\Sigma_{as}(r,t) = \Sigma_{as0}(r) + \sum_{n=1}^{N} u_n(t)\delta(r-r_n) \tag{9}$$

where there are N control rods considered with interior ends at r_n. The strength of the absorption cross section sink or source is $u_n(t)$, which is the control function of time to be determined.

This method is a reasonable approximation when the control rods absorb only some of the neutrons in the control rod region. However, often the region which contains the control rods is almost completely devoid of neutrons because the control rods have such a large absorption cross section. In this case the control problem becomes in essence a moving boundary control problem. An approximation to this moving boundary can be made by experimentally determining an appropriate effective control rod end. However, methods for handling the moving boundary control problem in a nuclear reactor setting have not yet been developed.

In the preceding paragraphs, equations have been developed that describe the following processes in the nuclear reactor as a function of time: the neutron flux, the heat conduction, the fission product poisoning, the fuel burn-up, and the control rod action. These equations are common to many nuclear reactor problems and serve as examples for the subsequent development of the theory in this chapter. In some cases other equations describing different processes in time may be needed for a complete description of the reactor. All the equations relevant to the particular problem at hand can be gathered together into one vector-valued partial differential equation for the state of the nuclear reactor.

The choice of an appropriate state vector for the distributed parameter nuclear reactor control problem depends not only on the particular reactor and problem considered but also on the time frame of the problem and the goals of the modeling. For instance, power control is a matter of seconds, fission product poisoning is a matter of hours, and fuel burn up is a matter of months. In some cases, control may not be the primary goal of the modeling. On-line estimation of the state vector may be used to give the reactor operators a picture of processes within the reactor. Parameter estimation may be used to find design values of the reactor parameters.

When the equations that are both necessary and sufficient to describe the time behavior of the physical process considered, when an appropriate time frame in which to cast these equations had been determined, and when an appropriate goal for modeling the process has been selected, then the equations can be combined into the state space partial differential equation

$$\mathbf{U} \frac{\partial Z}{\partial t} = f \tag{10}$$

where the state vector is $Z(r,t)$ whose components are the state variables and $\mathbf{U}$ and f are partial differential operators in r on Z. At the extrapolated boundaries, Z vanishes.

THE LINEAR STOCHASTIC MODEL WITH CONTROL

The appearance of the spatial Laplacian operator ∇^2 in the multigroup diffusion equations (1) and consequently in the state equation (10), together with the vanishing of the state vector Z at the smooth, closed, and bounded exprapolated boundaries implies a modal form of solution. In particular, the steady state position of the control rods is adjusted so that there exists a fundamental mode $Z_o(r)$ satisfying (10), independent of time t. If the multigroup equations (1) were linear, then the control rods would be adjusted in such a way that $Z_o(r)$ would be the fundamental eigenfunction. Even in the nonlinear case, the physics of the situation demand the existence of such a fundamental mode, $Z_o(r)$. In such a state the reactor is called "critical". Physically, criticality implies that exactly as many neutrons are born as are lost, so that the chain reaction is self-sustaining with no excess. It is assumed in this mathematical development that the control rod steady state positions are such that the reactor is critical in the presence of poisons, etc., at the desired steady state power level. This assumption implies that $Z_o(r)$ is the equilibrium state about which all Taylor series expansions are taken. Furthermore, the purpose of the spatial control system is then to return any deviations from this equilibrium state to zero.

The deviations of the state variables from their steady state are defined as

$$\phi_S(r,t) = \Phi_S(r,t) - \Phi_{SO}(r), \quad \phi_F(r,t) = \Phi_F(r,t) - \Phi_{FO}(r), \quad i(r,t) = N_1(r,t) - N_{IO}(r)$$

$$c(r,t) = C(r,t) - C_0(r), \quad z(r,t) = Z(r,t) - Z_o(r), \quad \theta(r,t) = T(r,t) - T_0(r)$$

$$x(r,t) = N_x(r,t) - N_{xO}(r) \tag{11}$$

If the deviation $z(r,t)$ of the state vector $Z(r,t)$ is assumed to be small, then higher order terms can be omitted to obtain the set of linear partial differential equations

$$\mathbf{U}(r) \frac{\partial}{\partial t} z(r,t) = \mathbf{L}(r) z(r,t) + \mathbf{G}(r) u(t) \tag{12}$$

where $\mathbf{U}(r)$, $\mathbf{L}(r)$ and $\mathbf{G}(r)$ are matrices containing functions of r (including symbolic functions such as a Dirac delta) and linear operatiors in r such as ∇. Since Z vanishes at the extrapolated boundaries of the reactor, so does z. This represents the distributed parameter state equations of the linearized system under the control of all control rods, with no noise.

To include the noise terms in these equations, the physical processes of the reactor itself must be considered. For example, each fission produces a random number of neutrons, ususlly two or three. The stochastic description of the chain reaction is a branching process. Every term in the multigroup equations (1) has some random feature in its nature. In fact, the definition of the term cross-section is the probability of an interaction. Other physical processes associated with the reactor usually have some random feature also. For example, the temperature equation (3) considers the bulk coolant temperature T. But only local coolant temperature, in the swirls and eddies of the flow, can be measured. The random temperature effects of these local flow variations introduce noise into the measurement of the bulk fluid temperature no matter how accurately the local temperature is measured.

To find an appropriate noise model in the distributed parameter case, the lumped parameter case is considered first. The stochastic representation of the overall reactor power can be found from standard derivations such as Borgwaldt (1968) to be

$$\langle P^2 \rangle = E_f P_o D |H(j\omega)|^2 + E_f P_o / W \tag{13}$$

where $\langle P^2 \rangle$ is the auto power spectral density, in megawatts squared seconds, E_f is the energy per fission, usually about 3.2×10^{-17} megawatt seconds, W is the detector efficiency, usually about 10^{-9}, and D is the Diven factor, usually about 0.8, equal to ν^{-2} times the expected value of $\nu(\nu-1)$. P_o is the steady state power, equal to $E_f \Sigma_f \Phi_o$ integrated over the reactor volume. Here $H(j\omega)$ is the reactivity transfer function {Borgwaldt (1968)}, for the lumped reactor.

To lump the distributed model (10), multiply (10) by the steady state mode shape $Z_o(r)$ and integrate over the reactor volume V. Putting the resulting equation in terms of the total power P then gives

$$\Lambda \, dP/dt = P_o \rho + S \tag{14}$$

where P_o is the constant mean operating power, ρ the reactivity, and Λ the effective neutron lifetime. The term S has been added, an effective zero mean neutron noise source. Let $H_s(j\omega)$ be the transfer function from S to P in this equation, then

$$H(j\omega)/H_s(j\omega) = \Lambda / l \tag{15}$$

Using (13), (14), and (15), the white noise S is found to have a power spectral density

$$\langle S \rangle = E_f P_o D \Lambda^2 / l^2 \tag{16}$$

where l is the prompt neutron lifetime. Incidentally, this shows that fast reactors are much noisier than thermal reactors because the prompt neutron lifetime in fast reactors is shorter.

It is reasonable to assume that this noise is generated by the differences in the number of neutrons produced in individual fissions, rather than from cross-section probabilities, etc. Then this noise source S can be added only to the fast group equation. Furthermore, its power spectral density <S> from (16) is linear in the power P_o. Since the power is spatially distributed according to the fundamental $\Sigma_{fF}(r)\Phi_{fo}(r)$, then Equation (1) becomes

$$\frac{1}{v_F}\frac{\partial\phi_F}{\partial t} = [\nabla\cdot D_F\nabla - \Sigma_{aF} - \Sigma_R + (1-\beta)\nu\Sigma_{fF}]\,\phi_F$$

$$+ (1-\beta)\nu\Sigma_{fs}\phi_s + \lambda_c c + S\sqrt{E_f\Sigma_{fF}\Phi_{Fo}/P_o} \tag{17}$$

This is an approximation because the time spatial characteristics of the noise have not been considered in the derivation.

Because these fissions occur randomly in time, each occurrence is Poisson distributed. However, because so many of them occur, the central limit theorem assures their Gaussian distribution and whiteness.

The noise term S in Eq. (17) represents the process noise in the linearized state Eq. (12) only for the neutron noise associated with the neutron flux. From Eq. (13) it is seen that the noise associated with the power measurement is

$$P_m(t) = P(t) + m_p(t) \tag{18}$$

where $P_m(t)$ is the measured power, $P(t)$ is the true power, and $m_p(t)$ is a white Gaussian noise with power spectral density equal to E_fP_o/W.

There are sources of noise in the reactor model (10) in addition to the neutron flux noise. For example, the local coolant temperature T_{cm} can be measured. This measurement can be represented as the bulk fluid temperature T_c plus a zero mean random process m_T. If the outlet local temperature is measured, this is an output variable.

$$T_{cm}(t) = T_c(t) + m_{Tc}(t) \tag{19}$$

where m_{Tc} is a noise whose spectral density depends upon the local flow and temperature variations in the fluid and upon the temperature measurement device. If the inlet local temperature T_{om} is measured, then the inlet bulk temperature $T(o,t)$ in Eq. (3) must be represented as

$$T(o,t) = T_{om}(t) + m_{To}(t) \tag{20}$$

in which $T_{om}(t)$ is a known (measured) time function and $m_{To}(t)$ is a process noise entering into the state Eq. (12) in the same manner as S enters into (17). The power spectrum of m_{Tc} and m_{To} can be measured or estimated using measurements from similar reactors.

Assembling all the noise processes, the state Eq. (12) with noise terms can be represented as

$$\mathbf{U}(r)\frac{\partial}{\partial t}z(r,t) = \mathbf{L}(r)z(r,t) + \mathbf{G}(r)u(t) + w_p(r,t) \tag{21}$$

in which $w_p(r,t)$ is a sample function from a random field such that

$$E\{W_p(r,t)\} = 0 \tag{22}$$

and

$$E\{W_p(r,t)\, W_p^T(p,\sigma)\} = S(r,\rho)\delta(t-\tau) \tag{23}$$

where E is the expectation operation, superscript T denotes transpose of a vector, and $\delta(t)$ is the dirac Delta function. This formal white noise $w_p(t,r)$ is considered to be the "derivative" of a Weiner process heuristically generalized to have spatial dependence. The noise w_p is a formal derivative because the time derivative of any Weiner process does not exist with probability one at any time. However, using this formal derivative, which is a white noise, gives correct answers when dealing with linear systems. The results of this white noise analysis will be used here for simplicity and clarity to obtain formulae whose rigorous derivation is very complex and lengthy.

To have a complete input-output state space description, an equation for the output must be found in addition to the state equation (21). Since the power measurement from (18) and the temperature measurement from (20) are not functions of r, the output equation that assembles all of the equations of the form of (18) and (20) and relates them to the state z can be written as

$$y(t) = \int_V C(r)z(r,t)dr + W_m(t) \tag{24}$$

where C(r) is an m x n matrix with elements that are functions or symbolic functions of r. Also W_m is a spatially independent Gaussian white noise with zero mean, uncorrelated with W_p or any initial conditions, and variance

$$E\{W_m(t)W_m^T(\tau)\} = N\delta(t-\tau) \tag{25}$$

Equations (21) and (24) together describe the linearized space and time dependence of the input and output of a nuclear reactor in the presence of noise. However, this is not quite the description desired because the power control must be separate from the spatial control.

The control system studied here is a spatial control system, the purpose of which is to return all spatial deviations to a steady state. This spatial control system is somewhat at cross-purposes with the power control system, the purpose of which is to follow the changes of the desired value of the steady state. The spatial control system is a regulator, whereas the power control system is a servomechanism. If all other modes are controlled by the spatial system, the power control system need only detect and influence the fundamental mode $Z_o(r)$. This separation into spatial and power control systems permits minor variations in power due to the contributions of the higher modes, but these contributions quickly decay and so will be ignored. Therefore, the reactor can be controlled by one feedback system detecting and controlling solely the fundamental mode and another feedback system detecting and controlling all the other modes.

The spatial control system can be designed without regard to the exact nature of the power control system if the fundamental interacts only weakly, or does not interact at all, with the harmonics. This interaction will be examined in detail later, but now it will be assumed weak. In general, the fundamental will be denoted $Z_0(r)$, and $x(r,t)$ will be defined as the state variable to be controlled by the spatial, or harmonic, control system. Thus x is written as

$$x(r,t) = z(r,t) - a_0(t)Z_o(r) \tag{26}$$

Therefore, $x(r,t)$ is controlled by the harmonic control system, and $a_0(t)$, a scalar, is controlled by the power control system.

Since the harmonic control system may at most weakly influence the fundamental mode, the action of one harmonic system control rod must negate the effect of all the other harmonic system rods on the fundamental. The action of this one control rod may be found by substituting (26) in (12) and taking the inner product with respect to the adjoint of the fundamental:

$$\langle Z_o^{\dagger}(r), \mathbf{U}(r)\frac{\partial x}{\partial t}\rangle + \dot{a}_o(t)\ \langle Z_o^{\dagger}(r), \mathbf{U}(r)Z_o(r)\rangle$$

$$= \langle Z_o^{\dagger}(r), \mathbf{L}(r)x\rangle + a_o(t)\ \langle Z_o^{\dagger}(r), \mathbf{L}(r)Z_o(r)\rangle + \langle Z_o^{\dagger}(r), \mathbf{G}(r)\rangle\ u(t) \tag{27}$$

The first terms on either side are small with respect to the coefficients of a_o and $\dot{a}_o$ because of the weak interaction requirement, and so the effect of the control rods on the fundamental can be approximately negated if the last term is zero. Equating the last term of (27) to zero gives

$$0 = \langle Z_o^{\dagger}(r), \mathbf{G}(r)\rangle\ u(t) \tag{28}$$

If $\mathbf{G}(r)$ comes from Eq. (9), then

$$0 = \sum_{n=1}^{N} \int_V Z_o^{\dagger(2)}(r)\Phi_{s0}(r)\delta(r - r_n)d^3r\ u_n(t) \tag{28}$$

$$= \sum_{n=1}^{N} Z_o^{\dagger(2)}(r_n)\Phi_{s0}(r_n)\ u_n(t) \tag{29}$$

From this, the motion of one control rod may be expressed in terms of the motion of all the rest:

$$u_N(t) = \frac{-\sum_{N=1}^{N-1} Z_o^{\dagger(2)}(r_n)\Phi_{s0}(r_n)u_n(t)}{\Phi_{so}(r_N)\ Z_o^{\dagger(2)}(r_N)} \tag{30}$$

This condition is imposed on the spatial control system so that its performance will have little disturbing effect on the power control system. Now u and $\mathbf{G}$ in (12) may be redefined to incorporate (30) and will still be denoted u and $\mathbf{G}$. However, the state variable will now be denoted $x(r,t)$ as in (26). Therefore the system to be studied is written concisely from (21) and (24) as

$$\mathbf{U}(r)\,\frac{\partial}{\partial t}\,x(r,t) = \mathbf{L}(r)\,x\,(r,t) + \mathbf{G}(r)\,u(t) + w_p(r,t) \tag{31}$$

with measurement y(t) obtained from

$$y(t) = \int_V C(r)\,x\,(r,t)\,dr + w_m(t) \tag{32}$$

Before proceeding with the design of the spatial control system u based on observations of the measurements y, a discussion of controllability (and by duality, observability) is in order.

The spatial neutron flux can often be adequately approximated by a finite number of states. For example, these states may be the time-varying coefficients of a truncated expansion in spatial mode shapes, or the states may be flux values at a finite number of points in the reactor. The relationship between the state vector x(t), composed of the n states as elements, and the control vector u(t), composed of the r controls as elements, may be expressed as a lumped approximation to (31).

$$\frac{dx(t)}{dt} = Ax(t) + Bu(t) + f(t) \tag{33}$$

Here A and B are, respectively, nxn and nxr constant matrices. The input f(t) does not enter into the controllability of the system and is henceforth set equal to zero without loss of generality.

The problem is to determine the conditions under which u(t) can control x(t). Following Zadeh and Desoer (1963), the definition of controllability can be made more precise. The system, Eq. (33), is said to be controllable if, and only if, there exists some $u(t;x_o)$ that will force x(t) to the zero vector in a finite time for all initial conditions x_o. In this definition, the control is written as $u(t;x_o)$ because the particular control that is picked to force x(t) to the zero vector is allowed to depend on the initial condition x_o. Zadeh and Desoer (1963) show that this definition is equivalent to requiring that any state x_1 be reachable from any initial state x_o.

The general criterion for controllability of the system, Eq. (33), was first given by Pontryagin (1959), is discussed in detail by Zadeh and Desoer (1963), and was applied to a nuclear reactor by Wiberg (1967a). However, here the special case where A has distinct eigenvalues and u(t) is a scalar will serve as an illustration. Because A has distinct eigenvalues, there exists a similarity transformation T that reduces A to a diagonal matrix D. Changing variables in Eq. (33) from x to y, where

$$x = Ty \tag{34}$$

then substitution of Eq. (34) into Eq. (33) gives

$$\dot{y} = Dy + cu, \tag{35}$$

where

$$\mathbf{c} = T^{-1}B. \tag{36}$$

Solving Eq. (35) for each element of y results in

$$y_i(t_1) = y_i(0)\exp(d_{ii}t_1) + c_i \int_0^{t_1} \exp\left[d_{ii}(t_1-\tau)\right] u(\tau)d\tau. \tag{37}$$

If $y_i(t_1)$ is required to be the zero state to show controllability, then u(t) is chosen such that

$$-y_i(0) = c_i \int_0^{t_1} \exp(-d_{ii}\tau)u(\tau)d\tau. \tag{38}$$

This can always be done unless c_i is zero. Therefore, the condition for controllability for a plant with distinct eigenvalues is that there be no zero element of the c vector as defined by Eq. (36)

EXAMPLE 1

Consider a reactor in which the flux $\phi(r,t)$ obeys the equation

$$\frac{\partial\phi(r,t)}{\partial t} = \left(10\pi^2 + \frac{\partial^2}{\partial r2}\right) \phi(r,t) + \delta\left(r - \frac{1}{2}\right) u(t) \tag{39}$$

with the boundary conditions

$$\phi(0,t) = \phi(1,t) = 0 \tag{40}$$

Clearly this reactor is supercritical, and such a situation would never be allowed to happen in a practical case. However, this hypothetical example gives a simpler set of equations to use than a spatial xenon poisoning problem, which also exhibits the behavior of the example even when the reactor is not supercritical.

The solution to Eq. (39) can be expressed

$$\phi(r,t) = \sum_{n=1}^{\infty} a_n(t) \sin n\pi r. \tag{41}$$

Truncating this expansion to the first three modes and setting up the state equation as in Eq. (33) results in

$$\begin{Bmatrix} \dot{a}_1 \\ \dot{a}_2 \\ \dot{a}_3 \end{Bmatrix} = \begin{bmatrix} 9\pi^2 & 0 & 0 \\ 0 & 6\pi^2 & 0 \\ 0 & 0 & \pi^2 \end{bmatrix} \begin{Bmatrix} a_1 \\ a_2 \\ a_3 \end{Bmatrix} + \begin{Bmatrix} -2 \\ 0 \\ -2 \end{Bmatrix} u(t) \tag{42}$$

This happens to be in the diagonal form of Eq. (35) from which it can be seen that there is no way for u(t) to influence the second mode. Therefore, the system of Eq. (39) is uncontrollable.

EXAMPLE II

Suppose the delta function in Eq. (39) is moved from r = 1/2 to r = 1/4. Then Eq. (42) becomes

$$\begin{Bmatrix} \dot{a}_1 \\ \dot{a}_2 \\ \dot{a}_3 \end{Bmatrix} = \begin{Bmatrix} 9\pi^2 & 0 & 0 \\ 0 & 6\pi^2 & 0 \\ 0 & 0 & \pi^2 \end{Bmatrix} \begin{Bmatrix} a_1 \\ a_2 \\ a_3 \end{Bmatrix} + \begin{Bmatrix} \sqrt{2} \\ 2 \\ \sqrt{2} \end{Bmatrix} u(t) \tag{43}$$

This is controllable. If u(t) is required to be a function of the overall reactor power, then u(t) depends only on a_1 and a_3 because the overall reactor power is proportional to the volume integral of the flux for this reactor. Therefore, the input to the second mode is independent of a_2, and so the second mode remains unstable. The requirement that u(t) control the overall reactor power is a side condition, additionally imposed on any spatial control system. Imposition of this side condition results in the loss of feedback control of an originally controllable system.

EXAMPLE III

Suppose there are two controls, one at r = 1/4 and one at r = 1/2. Also, the side constraint of a separately controllable power level is imposed. The the state equation is

$$\begin{Bmatrix} \dot{a}_1 \\ \dot{a}_2 \\ \dot{a}_3 \end{Bmatrix} = \begin{Bmatrix} 9\pi^2 & 0 & 0 \\ 0 & 6\pi^2 & 0 \\ 0 & 0 & \pi^2 \end{Bmatrix} \begin{Bmatrix} a_1 \\ a_2 \\ a_3 \end{Bmatrix} + \begin{Bmatrix} \sqrt{2} \\ 2 \\ \sqrt{2} \end{Bmatrix} u_1 + \begin{Bmatrix} 2 \\ 0 \\ -2 \end{Bmatrix} u_2 \tag{44}$$

The power is approximately

$$p(t) = \int_0^1 \phi(r,t)dr \approx \frac{2}{\pi} \left\{ a_1 + \frac{a_3}{3} \right\} \tag{45}$$

Controlling the power level independently is equivalent to controlling a_1 independently of all the other modes, because a_1 can then be chosen as

$$a_1 = \frac{\pi}{2} p - \frac{a_3}{3} \tag{46}$$

To control the power level, define the power level control u_p as

$$u_p(t) = \sqrt{2}\, u_1(t) + 2u_2(t) \tag{47}$$

Then Eq. (44) becomes

$$\begin{Bmatrix} \dot{a}_1 \\ \dot{a}_2 \\ \dot{a}_3 \end{Bmatrix} = \begin{Bmatrix} 9\pi^2 & 0 & 0 \\ 0 & 6\pi^2 & 0 \\ 0 & 0 & \pi^2 \end{Bmatrix} \begin{Bmatrix} a_1 \\ a_2 \\ a_3 \end{Bmatrix} + \begin{Bmatrix} 0 \\ 2 \\ 2\sqrt{2} \end{Bmatrix} u_1 + \begin{Bmatrix} 1 \\ 0 \\ -1 \end{Bmatrix} u_p \tag{48}$$

Written in this form, it is apparent that the power control system u_p is separated from the spatial control system u_1. The spatial control system does not influence the side constraint of controllable power level because a_1 is controllable only by

u_p and not u_1. Since u_p is independent of u_1, the power level control is merely an independent input into the higher harmonics. Therefore, the control policy of u_1 can be chosen to control only the higher harmonics.

Note in this example that u_1 must be able to control all the higher harmonics. If the delta function of $r = 1/2$ were moved to $r = 3/4$, the third element in the vector multiplied by u_1 would be zero. This would mean there would be no way to control a_3. There are certain geometrical constraints, and these are explored by Wiberg (1967a).

EXAMPLE IV

Suppose (for simplicity) that in Eq. (44) the second control u_2, in the center of the reactor, was required to be zero by the demands of u_p. How can the single control u_1 control the two unstable modes a_2 and a_3?

Given the initial conditions a_{20} and a_{30}, u_1 can be chosen from Eq. (38).

$$\begin{aligned} -a_{20} &= 2\int_0^{t_1} \exp(-6\pi\tau)u_1(\tau)d\tau \\ -a_{30} &= \sqrt{2}\int_0^{t_1} \exp(-\pi\tau)u_1(\tau)d\tau \end{aligned} \tag{49}$$

The solution for $u_1(t)$ is not unique. For example, $u_1(t)$ can be chosen as a sum of Dirac delta functions

$$u_1(t) = a\delta(t) + \beta\delta(t - t_1) \tag{50}$$

where

$$\begin{Bmatrix} a \\ \beta \end{Bmatrix} = \frac{(1/2)}{\exp(-6\pi t_1) - \exp(-\pi t_2)} \begin{Bmatrix} \exp(-\pi t_1) & -\exp(-\pi t_2) \\ -1 & 1 \end{Bmatrix} \begin{Bmatrix} a_{20} \\ \sqrt{2}\, a_{30} \end{Bmatrix} \tag{51}$$

When t_1 is not zero, when the eigenvalues are distinct, and when the **c** vector from Eq. (36) contains no zero elements, then $u(t)$ can be chosen in this manner for a second-order system. Physically, this means that the two modes are given the first impulse to line them up in such a manner that the second impulse can make them both zero simultaneously. There should be no conceptual difficulty in this problem because, historically, systems with more than one unstable open loop pole have been controlled by feeding back a single output variable.

General conclusions may be drawn from these examples. All of the examples deal with controls located at isolated points in the reactor, and the controls are functions of time only, not of space. The first example illustrates how a control located on a node of a mode cannot control that mode. The second example illustrates that the number of side constraints can be separated from the remainder of the overall system. The fourth example illustrates how a number of unstable modes can be controlled by a smaller number of controls.

The above results can be extended to more complicated cases. The control need not be located at a point, so that the scalar control function of time can be multiplied by a general function of space, not merely a delta function. The examples can be extended to systems of partial differential equations, such as the multigroup model, and to any linear system with discrete eigenvalues. The extension to systems with continuous eigenvalue spectra is not known. Finally, the examples can also be extended to two and three spatial dimensions. In this case the controllability condition is that the vector point controls should not lie on a node line or plane of any possible combination of modes (Wiberg, 1967a).

The question of controllability of an infinite number of states is studied elsewhere in this volume. However, in reactor control it is usually sufficient to consider only a finite number of modes because the higher harmonics are very stable. Therefore, the investigation of controllability can generally be limited to the finite number of modes of interest.

A control system designer may, however, desire more controls than are strictly necessary for controllability. Hardware considerations such as safety, ease of fabrication, etc., often dictate the construction of more controls. Also, performance specifications or noise may require more controls.

The effect of more controls on performance can be seen as follows. Example IV illustrates that the control of two unstable modes by a single control requires two impulses separated by a finite time. Two separate controls can force both modes to zero immediately. Therefore, quicker response can be obtained from a system that has at least as many controls as unstable modes than from a system that does not.

Noise can also influence the spatial control. If a point control is very close to being on a node of an unstable mode, then the introduction of noise will cloud the location of the node and make control difficult. However, the subject of noise belongs to the areas of identification and observability, which are beyond the scope of this discussion. In general, it is advisable to locate a point control as far away as possible from a node.

Further applications of controllability theory are discussed in Wiberg (1967b).

NON-INTERACTING MODAL EXPANSIONS

The homogeneous part of the state equation (31) must first be considered before attempting solution with non-zero input and noise.

$$\mathbf{U}(r)\frac{\partial}{\partial t}x(r,t) = \mathbf{L}(r)\,x(r,t) \tag{52}$$

This can be solved by modal expansion, which will be either interacting or non-interacting.

As an example, consider states $x_1(r,t)$ and $x_2(r,t)$ obeying

$$\frac{\partial}{\partial t}\begin{Bmatrix} x_1 \\ x_2 \end{Bmatrix} = \begin{Bmatrix} \nabla^2 & \alpha \\ r\nabla & \nabla^2 \end{Bmatrix}\begin{Bmatrix} x_1 \\ x_2 \end{Bmatrix} \tag{53}$$

with boundary and initial conditions

$$x_1(o,t) = x_2(o,t) = x_1(1,t) = x_2(1,t) = 0 \tag{54}$$

$$x_1(r,o) = x_{10}(r),\; x_2(r,o) = x_{20}(r) \tag{55}$$

for $r\varepsilon(0,1)$, $t\varepsilon(o,\infty)$.

Considering the operator $\nabla^2 = \partial^2/\partial r^2$ in this case and the boundary conditions, it is reasonable to attempt the expansions

$$x_1(r,t) = \sum_{n=1}^{\infty} a_n(t) \sin n \pi r \qquad x_2(r,t) = \sum b_n(t) \sin n \pi r \tag{56}$$

Substituting these into (53), multiplying by sin m π r, and integrating from r = 0 to r = 1 gives

$$\frac{da_n}{dt} = -n^2\pi^2 a_m + \alpha b_n \qquad \frac{db_n}{dt} = -n^2\pi^2 b_n + \sum_{n=1}^{\infty} \gamma_{mn} a_n \tag{57}$$

where

$$\gamma_{mn} = n \pi \int_0^1 r \cos n \pi r \sin m \pi r \, dr \qquad 58)$$

Because each of the modal coefficients affect all the others in this expansion through the coupling coefficients γ_{mn}, this is an interacting modal expansion.

Now, consider modes $\phi_n(r) = (\phi_{1n}(r), \phi_{2n}(r))$ obeying

$$\lambda_n \phi_n(r) = \begin{Bmatrix} \nabla^2 & \alpha \\ r\nabla & \nabla^2 \end{Bmatrix} \phi_n(r) \tag{59}$$

with boundary conditions

$$\phi_n(0) = 0, \ \phi_n(1) = 0 \tag{60}$$

These modes can be found by many numerical techniques such as discretization and solution of the associated matrix eigenvalue problem, or modal expansion of $\phi_{1n}(r)$ and $\phi_{2n}(r)$ in terms of sin n π r, or, for this case, the following technique called "shooting".

From the top half of (59),

$$\alpha \ \phi_{2n}(r) = (\lambda_n - \nabla^2)\phi_{1n}(r) \tag{61}$$

Substituting this into the bottom half of (59) multiplied by α gives

$$\lambda_n(\lambda_n - \nabla^2)\phi_{1n}(r) = \alpha r \nabla \phi_{1n}(r) + \nabla^2(\lambda_n - \nabla^2)\phi_{1n}(r) \tag{62}$$

Starting from initial conditions

$$\phi_{1n}(0) = 0 \ \frac{d\phi_{1n}}{dr}(0) = 1 \tag{63}$$

different (possibly complex) values of λ_n can be tried to attain the "target" $\phi_{1n}(1) = 0$. Then $\phi_{2n}(r)$ can be found from (61) when $\phi_{1n}(r)$ is known.

Furthermore, solution of the adjoint equation

$$\lambda_n \phi_n^\dagger(r) = \begin{Bmatrix} \nabla^2 & -\nabla r \\ \alpha & \nabla^2 \end{Bmatrix} \phi_n^\dagger(r) \tag{64}$$

gives the biorthogonal set $\{\phi_n, \phi_n^\dagger\}$ corresponding to the discrete eigenspectrum $\{\lambda_n\}$.

The eigenspectrum is usually discrete and simple because of two physical restrictions on the class of mathematical operators. The first is that the operator domain (reactor extrapolated volume) is finite. The second is that the physical concept of criticality (same number of neutrons born as lost) occurs at a fundamental mode.

Even if the eigenfunctions ϕ_n are not complete, which would be an unusual situation in this case, they could form an adequate basis to represent the solution to the homogeneous state equation.

To solve the homogeneous state equation (53) in terms of the non-interacting modes ϕ, assume

$$x_1(r,t) = \sum_{n=1}^{\infty} c_n(t)\phi_{1n}(r) \qquad x_2(r,t) = \sum_{n=1}^{\infty} c_n(t)\phi_{2n}(r) \tag{65}$$

Substitution of this expansion into (53), multiplication by $(\phi_{1m}^{\dagger}(r)\cdot\phi_{2m}^{\dagger}(r))$ and integration over r from 0 to 1 gives

$$\frac{dc_m}{dt} = \lambda_m c_m \tag{66}$$

where from the initial conditions

$$c_m(o) = \int_o^1 (\phi_{1m}^{\dagger}(r)\phi_{2m}^{\dagger}(r))\begin{Bmatrix} x_{10}(r) \\ x_{20}(r) \end{Bmatrix} dr \tag{67}$$

Solution of (66) with initial condition (67) then gives the solution of the homogeneous state equation as

$$\begin{Bmatrix} x_1(r,t) \\ x_2(r,t) \end{Bmatrix} = \sum_{n=1}^{\infty} c_n(0)e^{\lambda_n t} \begin{Bmatrix} \phi_{1n}(r) \\ \phi_{2n}(r) \end{Bmatrix} \tag{68}$$

These modes do not interact with one another, as was the case for the expansion (56), and consequently the $\{\phi_n\}$ are called non-interacting modes.

To solve the state equation with non-zero control and/or noise, it must be assumed that the eigenfunction series is weakly convergent in r for all times greater than the initial time, for all inputs u, and for almost all noise sample functions w_p. For clarity, the state equation (31) to be solved is repeated here.

$$\mathbf{U}(r)\frac{\partial}{\partial t}x(r,t) = \mathbf{L}(r)x(r,t) + \mathbf{G}(r)u(t) + w_p(r,t) \tag{31}$$

The eigenproblem associated with this equation is

$$\lambda_n \mathbf{U}(r)\phi_n(r) = \mathbf{L}(r)\phi_n(r) \tag{69}$$

with $\phi_n(r) = 0$ on the boundaries of the reactor volume V. Using the same procedure as in the homogeneous case, i.e., solution of the adjoint eigenproblem, multiplication by the adjoint eigenvectors and integration over the reactor volume gives

$$x(r,t) = \sum_{n=1}^{\infty} c_n(t)\phi_n(r) \tag{70}$$

where

$$c_n(t) = e^{\lambda_n t}c_n(0) + \int_o^t e^{\lambda_n(t-\tau)}\nu_n(\tau)d\tau \tag{71}$$

in which

$$\nu_n(t) = \int_V \phi_n(r) \{\mathbf{G}(r)u(t) + w_p(r,t)\} \, dr \tag{72}$$

he expansion (70) is in terms of non-interacting modes even for the inhomogeneous equation (31).

The operators $\mathbf{U}(r)$, $\mathbf{L}(r)$, and $\mathbf{G}(r)$ can contain symbolic functions such as Dirac delta functions and their derivatives. This greatly increases their utility because boundary condition inputs, states on differing dimensional domains in space, etc., can be considered. An example for boundary condition inputs will be given in the optimal control and filtering problem considered later. The problem of states on differing dimensional domains will be illustrated here.

Consider states $x_1(r_1,r_2,t)$ and $x_2(r_2,t)$ on the unit square obeying

$$\frac{\partial}{\partial t} x_1(r_1,r_2,t) = (\partial^2/\partial r_1^2 + \partial^2/\partial r_2^2) \; x_1(r_1,r_2,t) \tag{73}$$

$$\frac{\partial}{\partial t} x_2(r_2,t) = \partial^2/\partial r_2^2 \; x_2(r_2,t) \tag{74}$$

with boundary conditions $x_1(r_1,0,t) = x_1(r_1,1,t) = x_1(1,r_2,t) = x_2(0,t) = x_2(1,t) = 0$, but

$$x_1(0,r_2,t) = x_2(r_2,t) \tag{75}$$

By following a technique explained more fully in Friedman (1956), the boundary condition (75) can be incorporated into (73) as

$$\partial x_1/\partial t = \nabla^2 x_1 + \delta'(r_1)x_2 \tag{76}$$

with all zero boundary conditions and where $\delta'(r_1)$ is the derivative of the Dirac $\delta(r_1)$. The relationship (75) can be obtained heuristically by integration of (76) twice with respect to r_1 from $r_1 = 0$ to $r_1 = \varepsilon$ and letting $\varepsilon \to 0$.

$$\lambda_n \begin{Bmatrix} \phi_1(r_1,r_2) \\ \phi_2(r_2) \end{Bmatrix} = \begin{Bmatrix} \partial^2/\partial r_1^2 + \partial^2/\partial r_2^2 & \delta'(r_1) \\ 0 & \partial^2/\partial r_2^2 \end{Bmatrix} \begin{Bmatrix} \phi_1(r_1,r_2) \\ \phi_2(r_2) \end{Bmatrix} \tag{77}$$

Symbolic functions are allowed to appear in this manner in the non-interacting modal equations (69). Furthermore, the operator $\mathbf{U}$ (r) in (69) must be kept in the general form because symbolic and other non-invertible functions are allowed to appear in it.

THE MODAL APPROACH TO FILTERING AND CONTROL

It is the purpose of this section to extend the separation principle in such a way that the feedback control can be found in terms of closed loop modes for the linear stochastic distributed state equation (31). The derivation is necessarily heuristic because of the lack of the extended Ito calculus and also because of the desire to treat a wide class of systems. Although the derivation is oriented towards boundary control, the results can easily be extended (Wiberg, 1967c) to the less practical case of distributed control. Here Sholar and Wiberg (1972) is the primary source of reference.

The state n-vector $x(r,t,\omega)$ obeys the formal stochastic partial differential equation (31). A more general boundary condition $N(s)\ x(.,t,\omega)=0$ is considered, where $N(s)$ is an $n \times n$ matrix with elements that are functions and are linear partial differential operators in r. Here s is the value of r on ∂V, the boundary of V, which must be sufficiently smooth to allow application of Green's theorem.

The initial condition $x_\tau(.,\omega) = x(.,\tau,\omega)$ is also a formal Gaussian n-vector random variable generalized to have spatial dependence such that the solution to the state equation (31) is well behaved. The notation f(.) denotes some function f, to make a distinction from the number f(r) which denotes the function f(.) evaluated at the particular point r. Furthermore, x_τ is characterized by $E\{x_\tau(r,\omega)\} = 0$,

$$E\{x_\tau(r,\omega)w_p^T(\rho,t,\omega)\} = 0$$

and

$$E\{x_\tau(r,\omega)x_\tau^T(\rho,\omega)\}= X(r,\rho) \tag{78}$$

where X(r,p) is a positive definite symmetric $n \times n$ matrix.

The boundary control l-vector u(y,t) is to be found as a function of time t and a functional of the measurement m-vector y defined in Eq. (32).

By choice of the functional u(y,t) the performance functional J is to be minimized

$$J(x,u) = \frac{1}{2t_e} E \left\{ \int_\tau^{t_f} \left[\int_V x^TQxdr + u^TRu \right] dt \middle| y_{(\tau,t)} \right\} \tag{79}$$

where $t_e = t_f - \tau$ keeps J finite even when $t_e \to \infty$ and where $E\{f(t,\omega)|y_{(\tau,t)}\}$ denotes the conditional expectation of the random process $f(t,\omega)$ given values $y(\sigma,\omega)$ for $\sigma\varepsilon(\tau,t)$. Q(r) is a symmetric positive definite $n \times n$ matrix with elements that are functions and symbolic functions of r, and R is a symmetric positive definite $l \times l$ matrix. These conditions on Q and R guarantee a solution in the limit $t_f \to \infty$, but by analogy with the finite dimensional case, Q(r) can probably be chosen non-negative definite. By choosing u(y,t) from a class U of nonlinear functionals, the optimization is more meaningful (Wonham, 1968) but many difficult mathematical problems arise, such as finding the restrictions on $\mathbf{M}$, $\mathbf{L}$, C, B, G, Q, and U such that a well-posed solution exists for the state x. The operator $\mathbf{U}$ in (31) is henceforth denoted $\mathbf{M}$.

The performance functional J given by (79) can be rewritten as

$$J(x,u) = \frac{1}{2t_e} E\left\{\int_\tau^{t_f}\left[\int_V \hat{x}^TQ\hat{x}dr - u^TRu\right]dt\, |y_{(\tau,t)}\right\} + \frac{1}{2t_e} E\left\{\int_\tau^{t_f}\int_V \tilde{x}^TQ\tilde{x}drdt\, |y_{(\tau,t)}\right\} \tag{80}$$

where $\hat{x} = E\{x|y_{(\tau,t)}\}$ and $\tilde{x} = x - \hat{x}$. This follows because $E\{\hat{x}^TA\tilde{x}|y_{(\tau,t)}\} = 0$. Furthermore, it can be shown (Bucy and Joseph, 1968, and Wonham, 1968) that the last term on the right in Eq. (80) is independent of u.

Given an optimal control law $u^o(y,t)$ where the superscript o denotes optimal, the optimal trajectory is the random process $x^o(.,t,\omega) = \phi(.,t,\omega;\ x_\tau,\tau)$. Consequently $\hat{x}^o$ depends implicitly upon $\hat{x}_\tau,\tau$, and t_f and not u. Therefore, consider the scalar

$\nu(\mathbf{M}\hat{x}_\tau,\tau;\ t_f)$ which is a functional of $\hat{x}_\tau$ and τ parametrized by t_f such that

$$E\left\{\int_V \nu(\mathbf{M}\hat{x}_\tau,\tau;\ t_f)dr\right\} = \frac{1}{2t_e} E\left\{\int_\tau^{t_f}\left[\int_V \hat{x}^{oT}Q\hat{x}^o dr + u^{oT}Ru^o\right]dt \middle| y_{(\tau,t)}\right\} \tag{81}$$

Denote $z(.,\tau,\omega) = \mathbf{M}(r)\hat{x}_\tau(.,\tau,\omega)$ to simplify the following notation. A one-to-one correspondence is assumed between z and $\hat{x}$ because z is the necessary initial data to start the time solution of the filter equation (89) which determines $\hat{x}$. Then the assumption of a well posed solution requires that $\hat{x}$ tend to $\hat{x}_\tau$ as $t \to \tau$. Consequently, z and x become interchangeable state variables of the filter.

Suppose $\nu(z,\tau;\ t_f)$ is non-negative, continuously differentiable in τ and twice continuously Frechet differentiable in z, has the final condition

$$\nu(z,t_f;\ t_f) = 0 \tag{82}$$

and satisfies the Hamilton-Jacobi equation

$$-2\frac{\partial\nu}{\partial\tau} = t_e^{-1}\hat{x}^TQ\hat{x} + \mathrm{tr}F\frac{\partial^2\nu}{\partial z^2} + 2\frac{\partial\nu}{\partial z}^T \mathbf{L}x - t_e\left\{\int_V B'^T\frac{\partial\nu}{\partial z'}dr'\right\}^T R^{-1}B^T\frac{\partial\nu}{\partial z} \tag{83}$$

where F(r,t) is an n x n matrix independent of $\hat{x}$ and u defined by

$$F(r,t) = \mathbf{M}(r)K(r,t)E\{q(t,\omega)q^T(t,\omega)\}(\mathbf{M}(r)K(r,t))^T \tag{84}$$

Here K(r,t) is the Kalman filter n x p gain matrix defined further in Eq. (100) and $q(t,\omega)$ is the white noise innovation process defined further in Eq. (90). It can now be heuristically demonstrated that, given $\nu(z,\tau,t_f) \geq 0$ satisfying (82) and (83), then the performance functional (79) is minimized by the choice of feedback control

$$u^o(\hat{x},t) = -t_eR^{-1}\int_V B^T(r)\frac{\partial\nu}{\partial z}(z(r,t,\omega),t;\ t_f)dr \tag{85}$$

For any $u(y,\tau)\epsilon U$ dependent on $y(\sigma)$ for $\sigma \leq \tau$ and for any τ,

$$t_e^{-1}\left\{u + t_eR^{-1}\int_V B^T\frac{\partial\nu}{\partial z}dr\right\}^T R\left\{u + t_eR^{-1}\int_V B^T\frac{\partial\nu}{\partial z}dr\right\} \geq 0 \tag{86}$$

because R is a positive definite matrix. Rearranging this and adding a term to both sides gives

$$t_e^{-1}u^TRu + t_e^{-1}\int_V \hat{x}^TQ\hat{x}dr \geq -\int_V\left\{2u^TB^T\frac{\partial\nu}{\partial z} + t_e\left[\int_V B'^T\frac{\partial\nu}{\partial z}dr'\right]R^{-1}B^T\frac{\partial\nu}{\partial z} - t_e^{-1}\hat{x}^TQ\hat{x}\right\}dr \tag{87}$$

Substituting for the last two terms from (83) gives for all τ,

$$t_e^{-1}u^TRu + t_e^{-1}\int_V \hat{x}^TQ\hat{x}dr \geq -2\int_V \left\{\frac{\partial\nu}{\partial\tau} + \frac{\partial\nu}{\partial z}^T \left[\mathbf{L}\hat{x} + Bu\right] + \frac{1}{2}\,\mathrm{tr}F\,\frac{\partial^2\nu}{\partial z^2}\right\}dr \quad (88)$$

But a result of Sholar, (1970) is the extension of the Kalman filter

$$\frac{\partial z}{\partial\tau} = \mathbf{M}\frac{\partial\hat{x}}{\partial\tau} = \mathbf{L}\hat{x} + Bu + \mathbf{M}K\left\{y - \int_V C'\hat{x}'dr'\right\} \quad (89)$$

However,

$$q(r,\omega) = y(t,\omega) - \int_V C(r')\hat{x}(r',t,\omega)dr' \quad (90)$$

is defined as the innovation process (Kailath, 1968) and is white. Consequently $F(r,t)$ is the covariance of the white noise input and, therefore, the term of (88) in brackets is the Itô time derivative (Skorokhod, 1965) of $\nu(z,\tau,t_f)$ extended as if valid in the functional case. Therefore (88) can be integrated forward in time τ to t with an expectation on the initial value $x(.,\tau,\omega)$ and conditioned on the knowledge of $y_{(\tau,t)}$. Use of (82) then gives

$$J(x,u) \geq E\int_V \nu(z,\tau;\,t_f)dr + \frac{1}{2t_e}E\left\{\int_\tau^{t_f}\int_V \tilde{x}^TQ\tilde{x}drdt\ y_{(\tau,t)}\right\} \quad (91)$$

The equality attained in (82) by the choice of u^o as given in (85) demonstrates the desired sufficient condition for optimal control.

To find a solution to the Hamilton-Jacobi equation (83), consider

$$2\nu(z,t;\,t_f) = \kappa(t,\omega) + z^T\int_V \mathbf{W}(r,r',t)z'dr' \quad (92)$$

where $\mathbf{W}(r,r',t)$ is a positive definite symmetric n x n matrix and $\kappa(t,\omega)$ is a scalar random process with $\kappa(t_f,\omega) = 0$.

Substitution of (92) into the Hamilton-Jacobi equation (83) and retention of only quadratic terms in $z = \mathbf{M}\hat{x}$ gives

$$-(\mathbf{M}\hat{x})^T\int_V \frac{d\mathbf{W}}{dt}\mathbf{M}'\hat{x}'dr' = t_e^{-1}\hat{x}^TQ\hat{x} + 2(\mathbf{L}\hat{x})^T\int_{V'}\mathbf{W}\mathbf{M}\hat{x}'dr' - t_e\left\{\int_{V'}B'^T\int_{V''}\mathbf{W}(r',r'',t)\mathbf{M}''\hat{x}''dr''\right\}^T dr'R^{-1}B^T\int_{V'''}\mathbf{W}\mathbf{M}'''\hat{x}'''dr''' \quad (93)$$

This quadratic operator equation holds for all $\hat{x}(.,t,\omega)$ except for a set of measure zero, and so holds for any dummy variable $\nu(.,t)$. Now define $\nu(.,t)$ by

$$\mathbf{M}(r)\frac{\partial\nu}{\partial t} = \mathbf{L}(r)\nu - B(r)R^{-1}\int_{V'} B^T(r')p'dr' \quad (94)$$

with some initial condition $\nu(.,\tau)$ and with the same boundary condition as on x. Since the object is to find $\mathbf{W}(r,r',t)$, define the adjoint canonical variable $p(r,t)$ as

$$p(r,t) = t_e\int_{V'}\mathbf{W}(r,r',t)\mathbf{M}'\nu'dr' \quad (95)$$

Using this, (85) and (92) give the optimal control as

$$u^o(\nu,t) = -R^{-1}\int_V B^T(r)\int_{V'}\mathbf{W}(r,r',t)\mathbf{M}(r')\nu(r',t)dr'drt_e \quad (96)$$

Taking the time derivative of p(.,t), multiplying by $(\mathbf{M}\nu)^T$, integrating over V, and using (93) and (94) then gives the adjoint canonical equation

$$\mathbf{M}^*(r)\,\frac{\partial p}{\partial t} = -\,\mathbf{L}^*(r)p - Q(r)\nu \tag{97}$$

Where the notation f^* denotes the adjoint (7,8,12) of the operator f. From (82) and (92) it is seen that $\mathbf{W}(r,r',t_f) = 0$ and, therefore, from (95) it is evident that

$$p(r,t_f) = 0 \tag{98}$$

Also, since x, $\hat{x}$ and ν have the same boundary conditions, then p(.,t) has the boundary conditions

$$\mathbf{N}^*(s)p(.,t) = 0 \tag{99}$$

Canonical equations for the filter have been obtained by Sholar (12). A summary of his results is given here. The distributed Kalman filter for the system (31) is given by Eq. (89). The estimate $\hat{x}(.,t,\omega)$ has the same boundary conditions as $x(.,t,\omega)$. The initial condition from (78) is that $\hat{x}(.,\tau,\omega) = 0$, and the gain matrix K(r,t) is found by

$$K(r,t) = \int_{V'} P(r,r't)C^T(r')dr'N^{-1} \tag{100}$$

where

$$P(r,r',t) = E\{\tilde{x}(r,t,\omega)\tilde{x}^T(r',t,\omega)\} \tag{101}$$

Using (78) gives the initial condition on P as

$$P(r,r',\tau) = X(r,r') \tag{102}$$

Furthermore, P(r,r',t) obeys the Riccati type matrix partial differential equation

$$\mathbf{M}\left(\mathbf{M}'\frac{\partial P}{\partial t}\right)^T = \mathbf{L}(\mathbf{M}'P)^T + \mathbf{M}(\mathbf{L}'P)^T + 1/2G(\mathbf{M}'G'S)^T + 1/2\,\mathbf{M}GSG'^T - \mathbf{M}KN(\mathbf{M}'K')^T \tag{103}$$

with boundary conditions

$$\mathbf{N}(s)\left|\mathbf{N}'(s')P(s,s',t)\right|^T = 0 \tag{104}$$

In a development similar to that of Eq. (93-99), dummy variables $\nu(.,t)$ and p(.,t) and the operator $\mathbf{P}(r)$ are defined by

$$p(r,t) = \mathbf{P}(r)\nu = \int_{V'} P(r,r',t)\mathbf{M}^*(r')\nu(r',t)dr' \tag{105}$$

Additionally, from (103)

$$P(r,r',t) = P(r',r,t) \tag{106}$$

Let $\nu(.,t)$ have arbitrary initial conditions $\nu(.,\tau) = \nu_\tau(.)$ so that p(.,t) has the initial condition

$$p(.,\tau) = \int_{V'} X(.,r')\mathbf{M}^*(r')\nu_\tau(r')dr' \tag{107}$$

Then the filter canonical equations corresponding to the Riccati equation (103) are

$$\begin{Bmatrix} \mathbf{M}^* & 0 \\ 0 & \mathbf{M} \end{Bmatrix} \frac{\partial}{\partial t} \begin{Bmatrix} \nu \\ p \end{Bmatrix}' = \begin{Bmatrix} \cdot\ -\mathbf{L}^* & C^T N^{-1} \int_{V'} C' - dr' \\ \frac{G}{2} \int_{V'} SG'^T \mathbf{M}^{*'} - dr' + \frac{\mathbf{M} G}{2} \int_{V'} SG'^T - dr' & \mathbf{L} \end{Bmatrix} \begin{Bmatrix} \nu \\ p \end{Bmatrix} \quad (108)$$

These can be found by multiplying the Riccati equation (103) on the left by $\nu^T(r,t)$ and on the right by $\nu(r',t)$ and integrating over V and V', and then using the fact that $\nu_\tau(.)$ is arbitrary. This procedure yields the boundary conditions on ν and p as

$$\begin{Bmatrix} \mathbf{N}^*(s) & 0 \\ 0 & \mathbf{N}(s) \end{Bmatrix} \begin{Bmatrix} \nu(.,t) \\ p(.,t) \end{Bmatrix} = 0 \quad (109)$$

on the boundary ∂V.

Both the control canonical equations (94 and 97) and the filter canonical equations (108) can be written in the form

$$\mathbf{F}(r) \frac{\partial}{\partial t} \begin{Bmatrix} \nu(.,t) \\ p(.,t) \end{Bmatrix} = \mathbf{H}(r) \begin{Bmatrix} \nu(.,t) \\ p(.,t) \end{Bmatrix} \quad (110)$$

where $\mathbf{F}(r)$ and $\mathbf{H}(r)$ are 2n x 2n matrix operators and are independent of t and t_f. Furthermore, the boundary conditions are given by Eq. (109) in the filter case and by the adjoint to (109) for the control case. The initial conditions on $\nu(.,\tau) = \nu_\tau(.)$ are arbitrary and the initial conditions on $p(.,t_f)$ equal zero from (87) for the control case.

Consider the eigenproblem associated with (110) and the corresponding boundary conditions.

$$\lambda_i \mathbf{F}(r) \begin{Bmatrix} \phi_i(.) \\ \theta_i(.) \end{Bmatrix} = \mathbf{H}(r) \begin{Bmatrix} \phi_i(.) \\ \theta_i(.) \end{Bmatrix} \quad (111)$$

Assume that the 2n-vector eigenfunctions $\{\phi_i(.),\theta_i(.)\}$ are complete, that the eigenvalues λ_i are discrete and simple, and that the expansion of $x(.,t,\omega)$, $\nu(.,t)$, and $p(.,t)$ in eigenfunction series is weakly convergent in r for all $t \geq t_o$ and almost all ω. Also choose $\{\phi_i(.),\theta_i(.)\}$ to satisfy the same boundary conditions as $\{\nu(.,t),p(.,t)\}$. Then

$$\begin{Bmatrix} \nu(.,t) \\ p(.,t) \end{Bmatrix} = \lim_{M \to \infty} \sum_{i=1}^{M} \gamma_i(t) \begin{Bmatrix} \phi_i(.) \\ \theta_i(.) \end{Bmatrix} \quad (112)$$

Since λ_i is an eigenvalue of $\{\mathbf{F},\mathbf{H}\}$ then λ_i is also an eigenvalue of $\{\mathbf{F}^*,\mathbf{H}^*\}$ so that

$$\lambda_i \mathbf{F}^*(r)\begin{Bmatrix}\phi_i^*(.)\\ \\ \theta_i^*(.)\end{Bmatrix} = \mathbf{H}^*(r)\begin{Bmatrix}\phi_i^*(.)\\ \\ \theta_i^*(.)\end{Bmatrix} \tag{113}$$

Then the 2n-vector eigenfunctions $\{\phi_i^*(.),\ \theta_i^*(.)\}$ form a biorthogonal set (7) with $\{\phi_i(.),\ \theta_i(.)\}$ when appropriately normalized, i.e.

$$\int_V (\phi_i^{*T}(r)\,\theta_i^{*T}(r))\mathbf{F}(r)\begin{Bmatrix}\phi_j(r)\\ \theta_j(r)\end{Bmatrix} dr = \delta_{ij} \tag{114}$$

To solve the canonical equations (110), substitute the expansion (112) into (110), use (113) and multiply on the right by $(\phi_j^{*T}, \theta_J^{*T})$. The integrate over V and use the biorthonormality property (114) to obtain

$$\frac{d\gamma_j}{dt} = \lambda_j \gamma_j \tag{115}$$

so that the open loop solution of (110) in terms of $\gamma_i(t_1)$ is

$$\begin{Bmatrix}\nu(.,t)\\ \\ p(.,t)\end{Bmatrix} = \lim_{M\to\infty} \sum_{i=1}^{M} \gamma_i(t_1)e^{\lambda_i(t-t_1)}\begin{Bmatrix}\phi_i(.)\\ \\ \theta_i(.)\end{Bmatrix} \tag{116}$$

Similar to the conditions found by Potter, (1966) in the finite dimensional case, the Hamiltonion nature of $\mathbf{F}$ and $\mathbf{H}$ and the corresponding boundary conditions insure that the eigenvalues of $\{\mathbf{F},\mathbf{H}\}$ are symmetric with respect to the imaginary axis, and it will be assumed that none are purely imaginary.

Define the superscripts - and + to pertain to quantities associated with the left and right half planes respectively. Order the left half plane eigenvalues as $Re\lambda_i \leq Re\lambda_{i+1}$. Then the expansion (113) can be written

$$\begin{Bmatrix}\nu(.,t)\\ \\ p(.,t)\end{Bmatrix} = \lim_{M\to\infty}\left\{\sum_{i=1}^{M}\gamma_i^-(t)\begin{Bmatrix}\phi_i^-(.)\\ \\ \theta_i^-(.)\end{Bmatrix} + \sum_{j=1}^{M}\gamma_j^+(t)\begin{Bmatrix}\phi_j^+(.)\\ \\ \theta_j^+(.)\end{Bmatrix}\right\} \tag{117}$$

From (115), it appears that either $\gamma_i^+(r)$ or $\gamma_i^-(t)$ becomes unbounded as $\lambda_i \to \infty$ for any $t \neq t_1$. However, this will be shown not to be the case in Eq. (123) because of the nature of $\gamma_i(t_1)$.

For clarity, this development will be divided into control and filter cases. Let the subscript c pertain to control and f pertain to filtering. Consider the control case first. Let the reciprocal basis to the vectors $\{\phi_{ci}^-(.)\}$ be denoted $\{\alpha_{ci}^-(.)\}$, i.e. any $\alpha_{ci}(.)$ has the property

$$\int_V \alpha_{cj}^T(r)\mathbf{M}(r)\phi_{ci}^-(r)dr = \delta_{ij}$$

The set $\{\phi_{ci}^-(.)\}$ is assumed complete, that is minimal, therefore the $\{\alpha_{ci}^-(.)\}$ exist (Singer, 1970) and are unique (Marti, 1970). The Gram-Schmidt process can be used to construct the $\{\alpha_{ci}(.)\}$ if the space is finite dimensional (Schmiedler, 1965), but not usually in the infinite dimensional case. However, note that only the existence of the $\{\alpha_{ci}(.)\}$ is needed in the development. Then the Dirac delta function can be expressed as

$$I_n\delta(r-r') = \lim_{M\to\infty} \sum_{i=1}^{M} \phi_{\bar{ci}}(r)\left\{\mathbf{M}^*(r')\alpha_{ci}(r')\right\}^T \tag{118}$$

where I_n is the n x n unit matrix. Consequently, $\nu(.,t)$ can be expanded in terms of $\{\phi_{ci}(.)\}$ from (117)

$$\nu(.,t) = \lim_{M\to\infty} \sum_{i=1}^{M} \left\{\gamma_{\bar{ci}}(t) + \sum_{j=1}^{M} \gamma^+_{ci}(t)\int_{V'}\{\mathbf{M}^*(r')\alpha_{ci}(r')\}^T\phi^+_{cj}(r')dr'\right\}\phi^-_{ci}(.) \tag{119}$$

From (98) the final condition $p(.,t_f) = 0$ can be imposed on (117) to get

$$0 = \lim_{M\to\infty}\left\{\sum_{i=1}^{M}\gamma_{\bar{ci}}(t_f)\theta_{\bar{i}}(.) + \sum_{j=1}^{M}\gamma^+_{cj}(t_f)\theta^+_j(.)\right\} \tag{120}$$

Let the set$\{\beta_{cj}(.)\}$ be the reciprocal basis to the vectors$\{\theta^+_{cj}(.)\}$, i.e.

$$\int_V \beta^T_{cj}(r)\mathbf{M}^*(r)\theta^+_{ci}(r)dr = \delta_{ij} \tag{121}$$

Similar to the$\{\alpha_{\bar{ci}}(.)\}$, the set$\{\beta_{cj}(.)\}$ is assumed to exist and be unique. Then use of the$\{\beta_{cj}\}$ gives

$$\gamma^+_{cj}(t_f) = -\lim_{M\to\infty}\sum_{i=1}^{M}\int_V[\mathbf{M}(r)\beta_{cj}(r)]^T\theta_{\bar{ci}}(r)dr\gamma_{\bar{ci}}(t_f) \tag{122}$$

Use of (115) and $\lambda_{\bar{i}} = -\lambda^+_j$ gives

$$\gamma^+_{cj}(t) = \lim_{M\to\infty}\sum_{i=1}^{M} \rho_{cji}(t-t_f)\gamma^-_{ci}(t) \tag{123}$$

where ρ_{cji} is defined as

$$\rho_{cji}(t-t_f) = -e^{-(\lambda_i^-+\lambda_j^-)(t-t_f)}\int_V [\mathbf{M}(r)\beta_{cj}(r)]^T\theta_{\bar{ci}}(r)dr \tag{124}$$

Substitution of (123) into (119), multiplication by α^T_{ck}, integration over V and use of the biorthogonality of the$\{\alpha_{ck}\}$ with respect to the$\{\phi_{\bar{ci}}\}$ gives

$$\int_V \alpha^T_{ck}(r)\mathbf{M}(r)\nu(r,t)dr = \gamma_{\bar{ck}}(t) + \lim_{M\to\infty}\sum_{j=1}^{M}\sum_{i=1}^{M}\rho_{cji}(t-t_f)\int_V\alpha^T_{ck}(r)$$

$$\mathbf{M}(r)\phi^+_{cj}(r)dr\gamma_{\bar{ci}}(t) \tag{125}$$

This can be written in matrix notation as

$$\left\{\int_V \alpha^T_{ck}\mathbf{M}\nu dr\right\} = \lim_{M\to\infty}\left\{I_M + V_{cM}\right\}\left\{\gamma^-_{ci}(t)\right\} \tag{126}$$

where the brackets denote infinite dimensional matrices and where the k, i-th element of V_{cM} is defined as

$$\{V_{cM}\}_{ki} = \sum_{j=1}^{M}\rho_{cji}(t-t_f)\int_V \alpha^T_{ck}(r)\mathbf{M}(r)\phi^+_{cj}(r)dr \tag{127}$$

This equation is a contraction mapping for $t_f - t$ large enough and can often be partitioned into a non-singular finite dimensional matrix and a contraction mapping for any $t < t_f$. Therefore the inverse of $\{I_M + V_{cM}\}$ usually exists for all M. It is conjectured that the existence of this inverse is related to the stabilizability of the system, as in the finite dimensional case. Also note

$$\lim_{t_f \to \infty} \{V_{cM}\} = \{0\} \tag{128}$$

Assuming the existence of the inverse of $\{I_M + V_{cM}\}$ in the limit as $M \to \infty$ permits solution of (126)

$$\gamma_{\bar{ci}}(t) = \lim_{M \to \infty} \sum_{k=1}^{M} \{I_M + V_{cM}\}^{-1}_{ik} \int_V \alpha^T_{ck}(r)\,\mathbf{M}(r)\nu(r,t)dr \tag{129}$$

Using (123), this equation can be substituted into the bottom half of (117) to obtain the desired feedback form of solution to the canonical equations.

$$p(.,t) = \lim_{M \to \infty} \sum_{i=1}^{M} \left|\theta_{ci}(.) + \sum_{j=1}^{M} \rho_{cji}(t-t_f)\theta^+_{cf}(.)\right| \sum_{k=1}^{M} \{I_M + V_{cM}\}^{-1}_{ik} \int_V \alpha^T_{ck}\,\mathbf{M}\nu dr \tag{130}$$

This is a representation for the operator $\mathbf{P}$ as given by (105). There is no need to compute $W(r,r',t)$ in the control case because combining (130) with (95) and (96) gives $u^o(\hat{x},t)$ in the correct operator form such that ν can be replaced directly by $\hat{x}$ as

$$u^o(\hat{x},t) = -\lim_{M \to \infty} R^{-1}\int_V B^T \sum_{i=1}^{M} \left|\theta_{\bar{ci}} + \sum_{j=1}^{M} \rho_{cji}\theta^+_{cj}\right| dr \sum_{k=1}^{M} \{I_M + V_M\}^{-1} \int_V \alpha^T_{ck}\,\mathbf{M}\,\hat{x}dr \tag{131}$$

There is, however, a need to compute the corresponding gain matrix $P(r,r,t)$ for the filter. Because of the adjoint nature of the filter canonical equations, it is desirable to give a separate development rather than simply to appeal to duality. Starting from Eq. (119), the set $\{\alpha_{fi}(.)\}$ is biorthogonal as

$$\int_V \alpha^T_{fi}(r)\,\mathbf{M}^*(r)\phi_{\bar{ff}}(r)dr = \delta_{ij} \tag{132}$$

In contrast to the control case, the adjoint of $\mathbf{M}$ and the right half plane ϕ's are of interest. Using an expansion similar to (119), Eq. (120) becomes, for the filter case,

$$\nu(.,t) = \lim_{M \to \infty} \sum_{j=1}^{M} \left|\gamma^+_{fj}(t) + \sum_{i=1}^{M} \gamma_{\bar{fi}}(t) \int_V [\mathbf{M}(r')\,\alpha_{fj}(r')]^T \phi_{\bar{fi}}(r')dr'\right| \phi^+_{fj}(.) \tag{133}$$

Now the initial condition (102) is used on (117), in the limit as $M \to \infty$.

$$\sum_{i=1}^{M} \gamma_{\bar{fi}}(\tau)\theta_{\bar{fi}}(.) + \sum_{j=1}^{M} \gamma^+_{fj}(\tau)\theta^+_{fj}(.) = \int_V X(.,r)\,\mathbf{M}^*(r)\left| \sum_{i=1}^{M} \gamma_{\bar{fi}}(\tau)\phi_{\bar{fi}}(r) + \sum_{j=1}^{M} \gamma^+_{fj}(\tau)\phi^+_{fj}(r)\right| dr \tag{134}$$

Then the $\{\beta_{fi}\}$ are biorthogonal as

$$\int \beta^T_{fi}(r)\,\mathbf{M}(r) \quad \{\theta_{\bar{fj}}(r) - \int_{V'} X(r,r')\,\mathbf{M}^*(r')\phi_{\bar{fj}}(r')dr'\} \quad dr = \delta_{ij} \tag{135}$$

Similar to (123), then $\gamma_{\bar{fi}}(t)$ can be obtained as

$$\gamma_{\bar{f}i}(t) = \lim_{M\to\infty} \sum_{j=1}^{M} \rho_{fij}(t-\tau)\gamma^{+}_{fj}(t) \tag{136}$$

where

$$\rho_{fij}(t-\tau) = -e^{-(\bar{\lambda}_i+\bar{\lambda}_j)(t-\tau)}\int_V \beta_{fi}(r)\,\mathbf{M}(r)\ \{\theta^{+}_{fi}(r)-\int_{V'} X(r,r')\,\mathbf{M}^{*}(r')\phi^{+}_{fj}(r')dr'\}\ dr \tag{137}$$

The development then proceeds similarly to Eqs. (125) through (130), resulting in

$$p(.,t) = \lim_{M\to\infty} \sum_{k=1}^{M}\{\theta^{+}_{fk}(.)+\sum_{i=1}^{M}\rho_{fik}(t-\tau)\theta^{-}_{fi}(.)\}\sum_{j=1}^{M}\{I_M + V_{fM}\}^{-1}_{kj}\int_V \alpha^{T}_{fj}\ \mathbf{M}^{*}\,\nu dr \tag{138}$$

where

$$\{V_{fM}\}_{kj} = \int_V \alpha^{T}_{fk}(r)\,\mathbf{M}^{*}(r)\sum_{i=1}^{M}\phi^{-}_{fi}(r)dr\rho_{fij}(t-\tau) \tag{139}$$

so that in the limit as $(t - \tau) \to \infty$ then $V_{fM} = 0$.

Because the expression (138) does not correspond to an operation on the measurement y, K(r,t) must be computed from P(r,r',t). Expand P(r,r',t) as

$$P(r,r',t) = \lim_{M\to\infty} \sum_{i=1}^{M}\sum_{j=1}^{M} \pi_{ij}(t)\theta^{+}_{fi}(r)\alpha^{T}_{fj}(r') \tag{140}$$

Substitution of this, (138) and (133) into (105) leads to

$$0 = \lim_{M\to\infty}\left|\sum_{i=1}^{M}\left(\theta^{+}_{fi} + \sum_{j=1}^{M}\rho_{fji}\theta^{-}_{fj}\right)\gamma^{+}_{fi} - \sum_{i=1}^{M}\sum_{j=1}^{M}\pi_{ij}\theta^{+}_{fi}\left(\gamma^{+}_{fj} + \sum_{k=1}^{M}\{V_{fM}\}_{ik}\gamma^{+}_{fk}\right)\right| \tag{141}$$

Multiplication by ξ^{T}_{i} orthonormal to θ^{+}_{fi} and integration over V then gives, in matrix notation,

$$\{\pi_{ij}(t-\tau)\} = \{I_M + A_M(t-\tau)\}\{I_M + V_{fM}(t-\tau)\}^{-1} \tag{142}$$

since (134) is true for all $\gamma^{+}_{i}(t)$. Here

$$\{A_M\}_{ij} = \rho_{fji}(t-t_f)\int_V \xi_i(r)\theta^{+}_{fj}(r)dr \tag{143}$$

Therefore, from (124)

$$\lim_{t\to\infty}\pi_{ij} = \delta_{ij} \tag{144}$$

By analogy with the lumped parameter case, the distributed closed loop system becomes time invariant as $t_f \to \infty$ and $\tau \to -\infty$. Then from (100, 140, and 144)

$$K(r,t) = \lim_{M\to\infty} \sum_{i=1}^{M}\theta^{+}_{fi}(r)\int_V \alpha^{T}_{fi}(r')C^{T}(r')dr'N^{-1} \tag{145}$$

and from (124, 128, and 131)

$$u^{o}(\hat{x},t) = -\lim_{M\to\infty} R^{-1}\sum_{i=1}^{M}\int_V B^{T}(r)\theta^{-}_{ci}(r)dr\ \int_{V'}\alpha^{T}_{ci}(r')\,\mathbf{M}(r')\hat{x}(r',t)dr' \tag{146}$$

This expression permits solution for the stochastic optimal control without inverting an infinite dimensional matrix, but its very simplicity invites the unwary user to forget the drastic assumptions made in its derivation.

If these are indeed the expressions for the optimum gain, then the filter error $\tilde{x} = x - \hat{x}$ can be found by subtracting (89) from (31) and using (32)

$$\mathbf{M}\frac{\partial \tilde{x}}{\partial t} = \mathbf{L}\tilde{x} - \mathbf{M}K \left\{ \int_{V'} C'\tilde{x}'dr' - w_m \right\} \tag{147}$$

However, from the filter canonical eigenproblem (108)

$$\left\{ \mathbf{L}^* - C^T N^{-1} \int_V CF - dr \right\} \phi_{fi}^+ = -\lambda_{fi}^+ \mathbf{M}^* \phi_{fi}^+ \tag{148}$$

The adjoint equation to this is

$$\left\{ \mathbf{L} - \mathbf{M}K \int_{V'} C' - dr' \right\} \phi_{fi}^{+*} = \lambda_{\bar{fi}} \mathbf{M} \phi_{fi}^{+*} \tag{149}$$

Since the ϕ_{fi}^{+*} can be normalized to form a biorthogonal set with ϕ_{fi}^+ and since the eigenvalues $\lambda_{\bar{fi}}$ are assumed simple, there is a correspondence between $\phi_{fi}^{+*}(.)$ and $\alpha_{fi}(.)$. However, it is not known how to find the $\phi_{fi}^{+*}(.)$ without first finding all the $\alpha_{fi}(.)$. Comparing (147) with (149) then allows the unforced ($w_m = 0$) filter error $\tilde{x}$ to be expressed as

$$\tilde{x}(.,t,\omega) = \lim_{M\to\infty} \sum_{i=1}^{M} \phi_{fi}^{+*}(.)e^{\lambda_{\bar{fi}}(t-\tau)} \int_V \phi_{fi}^{+T}(r)\,\mathbf{M}(r)x_\tau(r,\omega)dr \tag{150}$$

Also, the closed loop system (31) with control (146) becomes

$$\mathbf{M}\frac{\partial x}{\partial t} = \mathbf{L}x - BR^{-1}\int_{V'} B'^T \lim_{M\to\infty} \sum_{i=1}^{M} \theta_{\bar{ci}}'\, dr' \int_{V''} \alpha''_{ci}\, \mathbf{M}''(x''-\tilde{x}'')dr'' + Gw_p \tag{151}$$

With the forcing terms x and w_p set to zero, then

$$x(.,t,\omega) = \lim_{M\to\infty} \sum_{i=1}^{M} \phi_{\bar{ci}}(.)e^{\lambda_{\bar{ci}}(t-\tau)} \int_V \alpha_{ci}(r)\,\mathbf{M}(r)x_\tau(r,\omega)dr \tag{152}$$

From (150) and (152) the $\lambda_{\bar{ci}}$ and $\lambda_{\bar{fi}}$ are the closed loop system and filter poles, and the $\phi_{\bar{ci}}(.)$ and $\phi_{fi}^{+*}(.)$ are "closed loop modes".

As a first example, consider the time-invariant linear stochastic optimal control of the diffusion equation in one space dimension r. There is a boundary control at $r = 1$, a noise input at $r = 0$, and a sensor at $r = r_1$ where $0 \le r_1 \le 1$. The work of Sholar (1970) is summarized for this problem. The system is

$$\frac{\partial x(.,t,\omega)}{\partial t} = \frac{\partial^2 x(.,t,\omega)}{\partial r^2} \tag{153}$$

With $x(o,t,.) = w_p(t,.)$ and $x(1,t,.) = u(y,t)$. The output (32) is

$$y(t,\omega) = x(r_1,t,\omega) + w_m(t,\omega) \tag{154}$$

and it is desired to minimize the performance functional (79)

$$J\{x,u\} = \lim_{t_f\to\infty} \frac{1}{2t_f} E\left\{ \int_0^{t_f} \left(\int_0^1 x^2 dr + u^2 \right) dt \,\middle|\, y_{(o,t)} \right\} \tag{155}$$

Incorporation of the inhomogeneous boundary condition inputs is accomplished by the use of symbolic functions (Friedman, 1956, Wiberg, 1967c, Sholar, 1970) to put the system (153) into the form of Eq. (31)

$$\frac{\partial x(.,t,\omega)}{\partial t} = \frac{\partial^2 x(.,t,\omega)}{\partial r^2} + \frac{\partial\delta}{\partial r}(r-1)u(y,t) - \frac{\partial\delta}{\partial r}(r)w_p(t,\omega) \tag{156}$$

with the homogeneous boundary conditions $x(0,t,.) = 0 = x(1,t,.)$. From (146) the optimal control is

$$u^o(\hat{x},t) = \sum_{i=1}^{\infty} \frac{d\theta_{\overline{ci}}(1)}{dr} \int_{V'} \alpha_{ci}(r')\hat{x}(r',t,\omega)dr' \quad (157)$$

with the estimate $\hat{x}(.,t)$ generated by the Kalman filter (89) with gain given by (145)

$$\frac{\partial \hat{x}}{\partial t} = \frac{\partial^2 \hat{x}}{\partial r^2} + \frac{\partial \delta}{\partial r}(r-1)u + \frac{1}{N}\sum_{i=1}^{\infty} \theta_{\overline{fi}}(.)\alpha_{ti}(r_1)\{y(t,\omega) - \hat{x}(r_1,t,\omega)\} \quad (158)$$

The control canonical equations (94) and (97) for the eigenproblem (111) are

$$\begin{Bmatrix} \dfrac{d^2}{dr^2} & -\dfrac{\partial\delta}{\partial r}(r-1)\displaystyle\int_0^{1+}\frac{\partial\delta}{\partial r}(r'-1)\;-dr' \\ -1 & -\dfrac{d^2}{dr^2} \end{Bmatrix} \begin{Bmatrix} \phi_c \\ \theta_c \end{Bmatrix} = \lambda \begin{Bmatrix} \phi_c \\ \theta_c \end{Bmatrix} \quad (159)$$

with boundary conditions $\phi_{ci}(0) = \phi_{ci}(1) = \theta_{ci}(0) = \theta_{ci}(1) = 0$. This can be converted to be the equivalent form

$$d^4\theta_c/dr^4 = \lambda^2\theta_c \quad (160)$$

with boundary conditions

$$\theta_c(0) = d^2\theta_c/dr^2\,(0) = \theta_c(1) = 0,\; d^2\theta_c(1)/dr^2 + d\theta_c(1)/dr = 0 \quad (161)$$

Define $\eta^2 = \lambda$, where η is positive real, the eigenvalues λ can then be found from

$$\cot\eta = \coth\eta + 2\eta \quad (162)$$

The solutions are $\eta_i = i\pi + \varepsilon_i$, where $\varepsilon_i \sim e^{-i\pi}$ for i = 1, 2, ...

The eigenfunctions are

$$\phi_{\overline{ci}}(r) = 2\eta_i \sinh\eta_i \sin\eta_i r,\; \theta_{\overline{ci}}(r) = \sinh\eta_i \sin\eta_i r - \sin\eta_i \sinh\eta_i r \quad (163)$$

The completeness of these modes can be shown by comparison with the sequence $\{\sin i\pi r\}$ which is complete over $0 \le r \le 1$. The filter equation (108) for the eigenproblem (111) is

$$\begin{Bmatrix} -\dfrac{d^2}{dr^2} & \dfrac{1}{N}\delta(r-r_1)\displaystyle\int_0^t \delta(r'-r_1)-dr' \\ \dfrac{\partial\delta}{\partial r}(r)S\displaystyle\int_0^t \frac{\partial\delta}{\partial r}(r')-dr' & \dfrac{d^2}{dr^2} \end{Bmatrix} \begin{Bmatrix} \phi_f \\ \theta_f \end{Bmatrix} = \lambda \begin{Bmatrix} \phi_f \\ \theta_f \end{Bmatrix} \quad (164)$$

with zero boundary conditions. This can be solved in a manner similar to (159-163) giving

$$\phi_{\overline{f}}(r) = \frac{1}{N\tan\zeta} \begin{cases} \sin\zeta r(\sin\zeta r_1 - \tan\zeta\cos\zeta r_1) & \text{for } r_1 \ge r \\ \sin\zeta r_1(\sin\zeta r - \tan\zeta\cos\zeta r) & \text{for } r_1 \le r \end{cases} \quad (165)$$

$$\theta_{\overline{f}}(r) = S(\sin\zeta r_1 - \tan\zeta\cos\zeta r_1)(\cosh\zeta r - \coth\zeta\sinh\zeta r)$$

where $\zeta = \sqrt{\lambda}$, the left half λ plane solutions to

$$(N/S)\tan\zeta = (\sin\zeta r_1 - \tan\zeta\cos\zeta r_1)(\cosh\zeta r_1 - \coth\zeta\sinh\zeta r_1) \quad (166)$$

This relationship gives the closed loop pole locations as a function of N/S and the sensor location r_1. Since N/S and r_1 do not appear in the criterion (79), there is a correspondence between pole location and minimum criterion. The exact nature of this correspondence is complicated and will not be discussed here. However, it can be seen that the sensor location r_1 can be optimized as a function of N/S, the measurement to plant noise ratio.

As a second example, the finite dimensional case is considered. The finite dimensional control case is covered in Wiberg, 1967c and 1971. Only the solution to the finite dimensional Kalman filter via closed loop modes is given here. Corresponding to (117) is the matrix equation

$$\begin{Bmatrix} \nu(t) \\ p(t) \end{Bmatrix} = \begin{Bmatrix} F\gamma^-(t) + H\gamma^+(t) \\ G\gamma^-(t) + J\gamma^+(t) \end{Bmatrix} \quad (167)$$

where F, G, H, and J are n x n constant matrices whose columns are eigenvectors, and ν, p, γ^- and γ^+ are time-dependent n-vectors. At the initial time τ, because $p(\tau) = X\gamma(\tau)$

$$G\gamma^-(\tau) + J\gamma^+(\tau) = XF\gamma^-(\tau) + H\gamma^+(\tau) \quad (168)$$

Also,

$$\gamma^+(t) = e^{\bar{\Lambda}(t-\tau)}\gamma^+(\tau) = e^{-\Lambda^-(t-\tau)}\gamma^+(\tau) \quad (169)$$

and

$$\gamma^-(\tau) = e^{\Lambda^-(t-\tau)}\gamma^-(\tau) \quad (170)$$

where $\Lambda = \text{diag}(\lambda_1, \lambda_2, \ldots, \lambda_n)$. Then $\bar{\gamma}(t)$ can be eliminated using the above three equations, so that (167) becomes

$$\begin{Bmatrix} \nu(t) \\ p(t) \end{Bmatrix} = \begin{Bmatrix} \{H - Fe^{\Lambda^-(t-\tau)}(G - XF)^{-1}(J - XH)e^{\Lambda^-(t-\tau)}\}\gamma^+(t) \\ \{J - Ge^{\Lambda^-(t-\tau)}(G - XF)^{-1}(J - XH)e^{\Lambda^-(t-\tau)}\}\gamma^+(t) \end{Bmatrix} \quad (171)$$

Then $\gamma^+(t)$ is eliminated, giving the feedback control. In the case of the $\lim(t-\tau) \to \infty$, the feedback control simply becomes

$$p(t) = JH^{-1}\nu(t) \quad (172)$$

For the specific system

$$y = (0\ 1)x + w_m \qquad \frac{dx}{dt} = \begin{Bmatrix} 0 & 0 \\ 1 & -\sqrt{5} \end{Bmatrix} x + \begin{Bmatrix} 2 \\ 0 \end{Bmatrix} w_p \quad (173)$$

in which $E\{W_p(t)W_p(\tau)\} = \delta(t-\tau)$ and $E\{W_m(t)W_m(\tau)\} = \delta(t-\tau)$, the resulting Kalman filter is

$$\frac{d\hat{x}}{dt} = \begin{Bmatrix} 0 & 0 \\ 1 & -\sqrt{5} \end{Bmatrix} \hat{x} + k(y-(0\quad 1)\hat{x}) \quad (174)$$

Also, $k = PC^TN^{-1}$ as in (100) where P is found from $p = P\nu$ as in (105). To find ν and p, equation (110) is

$$\frac{d}{dt}\begin{Bmatrix} \nu \\ p \end{Bmatrix} = \begin{Bmatrix} 0 & -1 & 0 & 0 \\ 0 & \sqrt{5} & 0 & 1 \\ 4 & 0 & 0 & 0 \\ 0 & 0 & 1 & -\sqrt{5} \end{Bmatrix} \begin{Bmatrix} \nu \\ p \end{Bmatrix} \tag{175}$$

For this H matrix, $\lambda_1^+ = +1$, $\lambda_2^+ = +2$, $\lambda_1^- = -1$, and $\lambda_2^- = -2$ with

$$\begin{Bmatrix} \phi_1^+ \\ \theta_1^+ \end{Bmatrix} = \begin{Bmatrix} 1 \\ -1 \\ 4 \\ \sqrt{5}-1 \end{Bmatrix}, \begin{Bmatrix} \phi_2^+ \\ \theta_2^+ \end{Bmatrix} = \begin{Bmatrix} 1 \\ -2 \\ 2 \\ 2\sqrt{5}-4 \end{Bmatrix}, \begin{Bmatrix} \phi_1^- \\ \theta_1^- \end{Bmatrix} = \begin{Bmatrix} 1 \\ 1 \\ -4 \\ -\sqrt{5}-1 \end{Bmatrix} \begin{Bmatrix} \phi_2^- \\ \theta_2^- \end{Bmatrix} = \begin{Bmatrix} 1 \\ 2 \\ -2 \\ -2\sqrt{5}-4 \end{Bmatrix} \tag{176}$$

Then

$$k = JH^{-1}C^TN^{-1} = \begin{Bmatrix} 4 & 2 \\ \sqrt{5}-1 & 2\sqrt{5}-4 \end{Bmatrix} \begin{Bmatrix} 2 & 1 \\ -1 & -1 \end{Bmatrix} \begin{Bmatrix} 0 \\ 1 \end{Bmatrix} (1) = \begin{Bmatrix} 2 \\ 3-\sqrt{5} \end{Bmatrix} \tag{177}$$

Thus the error for the Kalman filter obeys

$$\frac{d\tilde{x}}{dt} = \left\{ \begin{pmatrix} 0 & 0 \\ 1 & -\sqrt{5} \end{pmatrix} - \begin{pmatrix} 2 \\ 3-\sqrt{5} \end{pmatrix} (0 \quad 1) \right\} \tilde{x} \tag{178}$$

which has poles at $-1 = \lambda_1^-$ and $-2 = \lambda_2^-$ and eigenvectors

$$\begin{Bmatrix} 2 & \vdots & -1 \\ 1 & \vdots & -1 \end{Bmatrix} = H^* = (H^{-1})^T \tag{179}$$

This method of "closed loop modes" for solution of the infinite dimensional Riccati equation has been extended to the stochastic case. Both stochastic optimal control and filtering along have been treated. The closed loop modes are the non-interacting modes applied to the state and co-state.

The advantages of the closed loop modes can now be enumerated. They provide an analytical solution in simple cases. Standard computer routines can be used when an analytical solution is not available.

To compare accuracy with other method, denote the number of terms in each expansion by Ξ. As Ξ becomes large, any consistent convergent method tends to the distributed solution. Any significant disparity in accuracy arises when Ξ is small. The closed loop modes retained are those associated with eigenvalues λ_i^- such that $|Re\lambda_i^-| \leq |Re\lambda_{i+1}^-|$ for i - 1, 2,...,Ξ. Therefore any error is known to decay with a time constant $\tau \leq |Re\lambda_{\Xi}|^{-1}$. When any other approximation is used, there is an interaction between the resulting ordinary differential equations. Then control of Ξ terms excites the neglected modes, leading to a larger error in most cases. In this case, the use of closed loop eigenfunctions is more accurate.

As for computational complexity, it may appear at first glance that closed loop modes are less efficient since a system of 2n equations must be solved. This is often, however, the most efficient method of solving the Riccati equations associated with the optimal control. Solving the Hamiltonian system (110) to obtain the closed loop modes is a step that would have to be done in any case, and actually saves steps computationally. As for conceptual complexity, the advantage of closed loop eigenfunctions is apparent. Compare any Ξ-th order polynomial characteristic function of the corresponding open loop eigenfunction approximation with the simple form of Eq. (159) or (166). A root locus for the parameter N/S can be plotted from Eq. (166). Also, a Galerkin technique could have been used to find a very close approximation to (166) in order to avoid computation of the eigenfunctions in all but the final design configuration. Because these are the closed loop system eigenvectors and eigenfunctions, more insight is given into the final system design. Finally, it is conceptually somewhat more appealing to use the closed loop eigenfunctions because these are the modes whose shapes do not change when singly excited, so that a person watching the closed loop system can actually see each mode decay. Therefore, in very many respects the closed loops eigenfunction method is to be preferred over any other method for solution of the stationary quadratic optimal control or filter problem.

The actual computation of the exact infinite dimensional feedback is impossible because of the necessity of finding the reciprocal bases $\{\alpha_i\}$, $\{\beta_i\}$ and $\{\xi_i\}$. However, since the actual mechanization must perforce be finite dimensional, these bases must be truncated. Then a number of methods are available for their computation, such as matrix inversion of a point mesh or the method described in Schmiedler, (1965).

The hardware of the optimal estimator or feedback controller is comparatively easy to realize. Since no realization can handle the infinite number of distributed parameters, the number Ξ of terms to be retained is chosen by consideration of the desired decay of the truncation error time constant

$$\tau \le |Re\lambda_{\Xi}|^{-1}$$

The corresponding modes are precalculated to obtain P(r,r',t) from (142) and thus K(r,t) from (100). Then K(r,t) is stored as a time-varying gain. In most cases, the stationary value obtained as $\tau \to \infty$ from (146) would be stored, rather than time varying gains. The noisy measurements from the plant (31) then pass through the gains and form an input to the Kalman filter (89). Note the interesting hardware possibility that the

$$\mathbf{M}\frac{\partial \hat{x}}{\partial \tau} = \mathbf{L}\,\hat{x} + Bu$$

part of this equation can be simulated by an analog realization to retain all the distributed parameters. Then the desired states $\hat{x}$ are picked off from the Kalman filter and are passed through the precalculated time-varying gain (131) or time-invariant gains (145) if $t_f \to \infty$. In summary, the optimal control consists of m measurements $y(t,\omega)$ fed into a box that transforms them into l controls $u(y,t)$.

FILTERING OF A LIQUID METAL FAST BREEDER REACTOR

Most present day power-producing reactors are thermal reactors in which fission produces high-speed neutrons that collide with other atoms until they are slowed to the velocities associated with thermal motion. The thermal reactor is designed to slow down most of the neutrons because only at thermal velocities is the fission cross-section large. In the thermal reactor, the fission process uses up the nuclear fuel which is often the rare isotope uranium 235.

A breeder reactor, on the other hand, uses the common isotope uranium 238. Absorption of a neutron transmutes uranium 238 to plutonium 239, which is then fissionable upon absorption of another neutron. A breeder reactor thus breeds fissionable fuel from material unuseable in a thermal reactor. Because the neutrons need not slow to thermal velocities, these reactors are called fast breeder reactors.

In a thermal reactor, the hydrogen in the water used to transfer power has the additional advantage of quickly slowing down the fast neutrons. In fast reactors, this advantage becomes a disadvantage and substances other than water must be used for cooling. A liquid metal, such as sodium, can transfer great amounts of heat without slowing the fast neutrons too much. This section is concerned with liquid metal fast breeder reactors (LMFBR's).

Here the filtering theory of the previous section is applied to a particular problem in an LMFBR. This application is NOT successful because the theory is incorrectly applied. The example is nonetheless a very good illustration of the problems involved in the spatial modeling and filtering of a nuclear reactor, even though the application is NOT successful for theoretical reasons that will be discussed thoroughly after the example is presented.

The particular problem studied is the propagation of coolant boiling. In an LMFBR it is possible (not probable) that coolant boiling can propagate as follows. Initially the liquid metal coolant is chemically pure. In reactor operation the sodium picks up contaminants from a number of sources. These contaminants might be substances escaping from improperly canned and failed fuel pins, substances leeched from stainless steel tubing and joints, reaction products from the surfaces of elements, pieces of structure, machines, and instruments that work loose over a period of time, etc. Some of these contaminants could possibly lodge or stick in such a way as to block coolant flow over a significant part of a particular fuel pin. Then the heat produced by the fuel pin could raise the liquid metal coolant temperature to boiling, or past a temperature that will damage the cladding on the fuel pin. When the liquid metal boils, in contrast to water, it cannot remove as much heat as in the liquid phase and consequently the fuel pin may melt. If it melts, or if the cladding is damaged, there could result a fuel-liquid metal interaction with a number of possible consequences.

The worst imaginable consequence is that this interaction would eventually expel a large quantity of coolant from a region of the core in which the reactivity coefficient of the coolant is positive. This might cause immediate core damage.

A much more probable consequence of the fuel-coolant interaction is that it would add more contaminant to the coolant. This would make other flow blockages more likely in the future. The action would then be similar to a slowly growing cancer, cutting off more and more flow over a time period that could be months in duration. The slowly growing coolant flow blockage would distort the core structure, and necessitate a long and costly reactor shut down.

This example will analyze the SNR-300 (Hafele and Kessler, 1972), a type of LMFBR. The SNR-300 has 151 subassemblies arranged in an approximately cylindrical manner in the reactor core. The liquid metal coolant is sodium. The sodium is pumped upward between and through the subassemblies to a plenum at the top of the reactor. From the plenum it is passed through heat exhangers and returned to bottom of the core.

A typical subassembly is hexagonal in cross-section, in which 165 fuel pins are arranged to form the hexagon. Each fuel pin is six millimeters in diameter and 0.95 meters in length. The fuel pins are parallel to one another in the subassembly and are spaced slightly apart from one another so that liquid sodium can flow lengthwise along the pins and remove the heat generated by them.

In normal speration the fuel pin center temperature is about 2300°C and the temperature near the edge of the fuel is about 700°C. The cladding temperature is about 380°C, and in normal operation the sodium exiting from the subassemblies has a temperature that ranges from 576°C at the core center to 530°C at the core edge. The coolant is pressurized to 2.5 atmospheres, where it would boil at about 1000°C.

Sodium boiling over one single fuel pin is extremely difficult to detect because there are 165 pins times 151 subassemblies equals 24,915 pins in the core. It is practically impossible to put at least one thermocouple on each pin. Instead, each subassembly is instrumented with four thermocouples, all reading the bulk sodium temperature at the exit. Logical comparison using a two-out-of-three rule, with the fourth thermocouple as a spare, then gives one subassembly bulk sodium exit temperature signal. Therefore 151 temperature signals emanate from the reactor. It is unlikely that fewer will suffice because each subassembly is somewhat thermally isolated from its neighbors by sodium coolant passing between the subassemblies.

The effect of sodium boiling over a few fuel pins in an otherwise unchanged subassembly is to increase the bulk sodium exit temperature only a few degrees. A small temperature change could also be caused by a blockage distributed over a number of coolant channels within the subassembly, in which case the sodium is not boiling. To prevent this situation from giving a false alarm, a small rise in temperature within a subassembly should be compared with other indications of sodium boiling, such as the noise spectra of the coolant, neutrons, reactivity, and acoustic noise.

Until sodium boiling has been proven innocous to structural integrity, methods of detection must be employed. Because no method of detection has been proven completely reliable, reactor shut down must be dictated by an evaluation of the complete reactor state. Use of the reactor equations together with the measurable quantities emanating from the reactor to form a Kalman filter will help evaluate the complete reactor state.

To obtain an estimation scheme that can be put into practice, a mathematical model for the LMFBR must be found that is a good compromise between simplicity and accuracy. In this example a rather gross approximation is made to obtain a very simple model, which should be improved upon in a complete study. All the flux equations (1) are lumped into one equation for the reactor powerP, then linearized about the mean value of the operating power which is assumed constant.

$$\Lambda \frac{dP}{dt} = P_o \sum_{i,j=1} \alpha_{ij} T_f^{ij} + S \tag{180}$$

where $P(t)$ = reactor power deviation from P_o, in mega watts

P_o = constant mean operating power, 723 MW

Λ = effective neutron lifetime, lumping fast and all delayed neutron lifetimes = 6.7 sec.

α_{ij} = temperature coefficient of reactivity $^oC^{-1}$

$S(t)$ = zero mean neutron noise source, MW

N = number of channels (groups of subassemblies)

M = number of axial zones

$T_f^{ij}(t)$ = fuel temperature deviation in the $i^{\underline{th}}$ channel and $j^{\underline{th}}$ axial zone, ^oC

The temperature coefficient of reactivity depends on the Doppler node fraction W_D^{ij}. Assuming that the overall temperature coefficient of reactivity is -0.005, then

$$\alpha_{ij} = -.005\, W_D^{ij} \qquad \sum W_D^{ij} = 1 \tag{181}$$

In the $i,j^{\underline{th}}$ node, for i = 1, N and j = 1, M, the average fuel temperature deviation T_f^{ij} obeys

$$C_{pf}\, N_P^{ij}\, \pi\, r_o^2\, H\rho_f \frac{dT_f^{ij}}{dt} = W_f^{ij} P - 2\pi r_o H N_P^{ij} h_T (T_f^{ij} - T_c^{ij}) \quad (182)$$

where

N_P^{ij} = total number of fuel pins per channel = $m_{ij}(\pi r_o^2 H\rho_f)^{-1}$

m^{ij} = fuel mass in the i, $j^{\underline{th}}$ node in grams

r_o = radius of fuel pin = .6 cm (note a more realistic number is .3 cm)

H = height of axial zone, = 95/M cm

ρ_f = density of fuel pin = 10 g/cm^3

C_{pf} = specific heat of fuel = .3 watt-sec/g ^oC

W_f^{ij} = power fraction in node $\sum W_f^{ij} = 1$

h_T = heat transfer coefficient = 1.0 watt/cm^2 ^oC

$T_c^{ij}(t)$ = sodium coolant temperature deviation in the i, $j^{\underline{th}}$ node, ^oC.

Assuming complete thermal isolation of each subassembly in the i,$j^{\underline{th}}$ node i = 1, N and j = 1, M, the coolant temperature deviation T_c^{ij} obeys

$$fN_p \pi r_o^2\, H\rho_f C_{P_c} \frac{dT_c^{ij}}{dt} = 2\pi r_o^2\, HN_p h_T (T_f^{ij} - T_c^{ij}) - fN_p \pi r_o^2 \rho_f C_{p_c} V (T_c^{ij} - T_c^{ij+1}) \quad (183)$$

where

f = coolant/fuel ratio = .127

V = coolant flow = 500 cm/sec.

$T_C^{iM+1}(t)$ = inlet sodium temperature, ^oC

Using the numerical values given for LMFBR typified by the SNR-300 gives the state equation

$$\frac{dP}{dt} = -.54 \sum_{i,j=1}^{N,M} W_D^{ij} T_f^{ij} + .15\, S, \quad \frac{dT_f^{ij}}{dt} = 3.3 (W_f/m)^{ij}\, P - 2(T_f^{ij} - T_c^{ij})$$

$$\frac{dT_c^{ij}}{dt} = 4(T_f^{ij} - T_c^{ij}) - 0.5\, M(T_c^{ij} - T_c^{ij+1}) \quad (184)$$

For N = 3 and M = 3 the reactor nodes are as pictured in Fig. 1 and the values of W_D^{ij}, W_f^{ij}, and m^{ij} are given in Table 1.

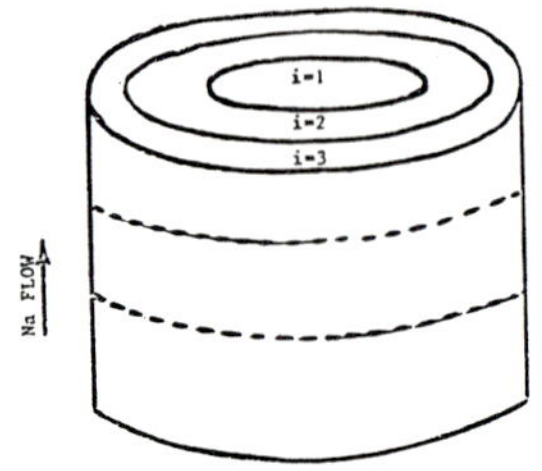

Figure 1. LMFBR core with 9 nodes

TABLE 1. Coefficients of a 9 Node LMFBR Model

i	j	Doppler Coefficient W_D^{ij}	Power Fraction W_f^{ij}	Fuel Mass x 10^6g m_{ij}
1	1	0.065	.032	.417
1	2	.167	.105	.380
1	3	.031	.025	.417
2	1	.135	.063	.925
2	2	.301	.195	.840
2	3	.040	.043	.925
3	1	.105	.119	9.145
3	2	.145	.346	8.292
3	3	.011	.072	9.145

The given numerical data are sufficient to calculate the dynamic behavior of a reactor model with N and M equal to three or less, for this example. A complete study would make use of much greater values of N and M. Because the 151 subassemblies are arranged in concentric annuli, the cylindrical symmetry necessitates the computation of, at most, N = 10 channels. This simplification makes the problem computationally feasible in a complete study.

However, rather than dealing with a large dimensional state vector, choose N or M unity. This lumps the reactor into one node and sacrifices accuracy for clarity. Then the equations (184) can be put into the vector matrix (state space) form

$$\frac{d}{dt}\begin{Bmatrix} P \\ T_f \\ T_c \end{Bmatrix} = \begin{Bmatrix} 0 & -.54 & 0 \\ .11 & -2 & 2 \\ 0 & 4 & -4.5 \end{Bmatrix} \begin{Bmatrix} P \\ T_f \\ T_c \end{Bmatrix} + \begin{Bmatrix} .15 & 0 \\ 0 & 0 \\ 0 & .5 \end{Bmatrix} \begin{Bmatrix} S \\ T_o \end{Bmatrix} \qquad (185)$$

where $T_o = T_c^{12}$ = inlet sodium temperature. This is the assumed state space equation for the operating LMFBR dynamics.

The only external measurements considered here are the 151 temperature signals indicating the bulk exit temperature of each subassembly and the signals from the neutron flux meters. Because spatial flux effects are negligible in the SNR-300, these flux meter signals can be logically combined to give one signal proportional to total power. In the following paragraphs the noise properties of these temperature and power signals are investigated.

The bulk exit sodium temperature of a subassembly of a prototype reactor was measured at zero reactor power. At zero power most of the measurement error is due to the turbulence in the sodium flow because the thermocouple measures local rather than bulk exit temperature. From the graph of the temperature measurement at zero power there appears a sine wave of 0.01°C amplitude at 0.1 Hz plus a hash of 0.005°C amplitude at frequencies greater than 5 Hz. For simplicity of modeling this behavior was taken to be a white noise with autocorrelation .003 $\delta(t-\tau)$ $°C^2$ sec. This number represents the lowest temperature noise the thermocouples measure in operation. As the reactor comes to full power, it is probable that more turbulence, cable pick-up, vibration, etc., will corrupt the measurements. If the industrial thermocouple standard of $\pm$ 1.5°C is taken as the worst case, then the temperature measurement white noise autocorrelation is at most 1.0 $\delta(t-\tau)$ $°C^2$ sec. Using numerical values in Eq. (13) gives

$$< P_m^2 > = 2.3 \times 10^{-5} + 3.9 \, |H_s(j\omega)|^2 \qquad (186)$$

Therefore the white noise S has autocorrelation 3.9 $\delta(t-\tau)$ $(MW)^2$ $sec.^{-1}$ and the neutron flux measurement noise has autocorrelation 2.3 x 10^{-5} $\delta(t-\tau)(MW)^2$ sec.

Substituting the measured value T_{om} of the inlet sodium temperature T_o into the state equations (185) gives the stochastic description of the LMFBR as

$$\frac{d}{dt}\begin{Bmatrix} P \\ T_f \\ T_c \end{Bmatrix} = \begin{Bmatrix} 0 & -.54 & 0 \\ .11 & -2 & 2 \\ 0 & -4 & -4.5 \end{Bmatrix} \begin{Bmatrix} P \\ T_f \\ T_c \end{Bmatrix} + \begin{Bmatrix} 0 \\ 0 \\ .5 \end{Bmatrix} T_{om} + \begin{Bmatrix} .15 & 0 \\ 0 & 0 \\ 0 & .5 \end{Bmatrix} \begin{Bmatrix} S \\ m_{T_o} \end{Bmatrix}$$

$$\begin{Bmatrix} P_m \\ T_{cm} \end{Bmatrix} = \begin{Bmatrix} 1 & 0 & 0 \\ 0 & 0 & 1 \end{Bmatrix} \begin{Bmatrix} P \\ T_c \end{Bmatrix} + \begin{Bmatrix} m_P \\ m_{T_c} \end{Bmatrix} \qquad (187)$$

where S, m_{T_o}, and m_{T_c} are independent zero mean white noises with

$$< S^2 > = 3.9 \quad (MW)^2 \, sec^{-1}. \qquad < m_{T_o}^2 > = .003 \text{ to } 1.0 \quad (°C)^2 \, sec.$$

$$< m_P^2 > = 2.3 \times 10^{-5} \quad (MW)^2 \, sec. \qquad < m_{T_c}^2 > = .003 \text{ to } 1.0 \quad (°C)^2 \, sec.$$

The solution of the steady state Riccati equation (172) was found for the three-dimensional state model (187) using the eigenvector approach of the previous section. The results are given in Table 2. Table 2 is arranged into four parts, labelled good or bad thermocouples and flux meter or no flux meter. Good thermocouples correspond to the research prototype white noise autocorrelation of .003 $\delta(t-\tau)$ $°C^2$ sec, and bad thermocouples to 1.0 $\delta(t-\tau)$ $°C^2$ sec. Flux meter refers to whether or not a flux meter was used in obtaining the state estimates. The standard deviations of the power, fuel temperature, and coolant temperature were found as the square root of the diagonal elements of P_{ss}, the steady state solution of the Riccati equation.

The computer program used to obtain the results in Table 2 was tested with both a two and an eight dimensional test subrouting and was found to be accurate to 0.5%. Unfortunately this accuracy is problem dependent, and the routine did not work for LMFBR models of more than three state variables. Because an accurate model must have more than 30 state variables, the theoretical procedures were at fault. The reasons for this theoretical difficulty are given in the next section, after the results of the three state variable analysis are presented.

Table 2 Results for the Three State Model

(P in MW, T_f in ^oC, T_c in ^oC)

Good Thermocouples and Flux Meter

TC read $\pm$ .05 C * Flux Meter Efficiency W 10^{-9}

Standard deviations:

P = 0.037 MW
T_f = 0.013 ^oC
T_c = 0.019 ^oC

$$P_{ss} = 10^{-4} \times \begin{Bmatrix} 14.1 & 0.5 & -0.4 \\ .05 & 1.7 & 1.7 \\ -.04 & 1.7 & 3.6 \end{Bmatrix}$$

Bad Thermocouples and Flux Meter

TC read $\pm$ 1 ^oC * W = 10^{-9}

Standard deviations:

P = 0.037 MW
T_f = 0.16 ^oC
T_c = 0.32 ^oC

$$P_{ss} = 10^{-4} \times \begin{Bmatrix} 14.4 & -7.3 & -9.0 \\ -7.3 & 261 & 355 \\ -9.0 & 335 & 1017 \end{Bmatrix}$$

Bad Thermocouples and No Flux Meter

TC read $\pm$ 1 ^oC W = 0

Standard deviations:

P = 0.76 MW
T_f = 0.27 ^oC
T_c = 0.35 ^oC

$$P_{ss} = 10^{-4} \times \begin{Bmatrix} 5287 & 363 & -72 \\ 363 & 732 & 656 \\ -72 & 656 & 1253 \end{Bmatrix}$$

Good Thermocouples and No Flux Meter

TC read $\pm$ .05 ^oC * W = 0

Standard deviations

P = 0.26
T_f = 0.063 ^oC
T_c = 0.044 ^oC

$$P_{ss} = 10^{-4} \times \begin{Bmatrix} 676 & 134 & 68 \\ 134 & 40 & 25 \\ 68 & 25 & 19 \end{Bmatrix}$$

* Really $< m_T^2 >$ = .003 and 1.0 respectively.

The Kalman filter corresponding to the three dimensional state model (187) with good thermocouples and a flux meter is pictured in Fig 2.

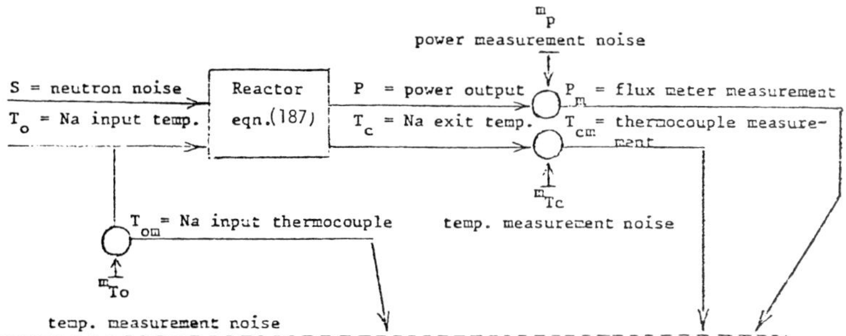

$$\frac{d}{dt}\begin{pmatrix}\hat{P}\\ \hat{T}_f\\ \hat{T}_c\end{pmatrix} = \begin{pmatrix}0 & -.54 & 0\\ .11 & -2 & 2\\ 0 & 4 & -4.5\end{pmatrix}\begin{pmatrix}\hat{P}\\ \hat{T}_f\\ \hat{T}_c\end{pmatrix} + \begin{pmatrix}0\\ 0\\ 1\end{pmatrix} T_{om} + \begin{pmatrix}61.3 & .0013\\ .22 & .056\\ -.17 & .119\end{pmatrix}\begin{pmatrix}P_m - \hat{P}\\ T_{cm} - \hat{T}_c\end{pmatrix}$$

COMPUTER

Figure 2. Mechanization of the Kalman Filter

The equations pictured withing the dotted lines are the Kalman filter, whose solutions P, T_f, and T_c are available to the reactor operators and the scram system in general.

Table 2 shows that the Kalman filter can estimate the unmeasurable components of the state vector with approximately the same variance as the measured components. The Kalman filter also significantly decreases the variance of the measured components beyond the accuracy of the measuring instruments. This can be done for all ranges of instrument accuracy, even for commercial as opposed to laboratory instruments. The Kalman filter gives promising results for this simplified model, but further research is needed to ascertain whether incipient sodium boiling can be detected in an operating reactor.

Because no results are available yet for higher dimensional LMFBR models, physical reasoning must be used to predict what will happen as the model dimensional increases. Adding more channels to the model, i.e. N = 15, makes the per channel contribution of the flux meter negligible. Therefore results for the temperature accuracies should tend toward those in the bottom half of Table 2. Adding more axial zones increases the sensitivity to changes in parameters of the state equation and decreases the variance from the physical models. With increased dimension, the spatial resolution will be much finer and the reactor operators will be able to tell where boiling occurred in the event of a scram. Thus the difficulty can be located quickly and the system brought back on line faster.

The mechanization in Figure 2 shows how the reactor state can be displayed to the reactor operators. Given some unforseen occurrence, this picture can help the operator assess the danger and determine the measures that must be taken to combat it. Thus a Kalman filter is useful for occurrences other than sodium coolant boiling. Indeed its ability to estimate the current reactor state and compare it with a normal operating state may be the most compelling reason for the installation of a Kalman filter on an operating LMFBR. The accident at Three-Mile Island would probably have been prevented if a Kalman filter had been installed on that thermal reactor.

The numerical values shown within the dotted lines of Figure 2 are the parametric values of the reactor equations. Because of fuel burn up, because of changes in the heat transfer coefficient with scale build-up, etc., these parameters are not constant as assumed. They vary slowly, over a period of days. Perhaps once a day the Kalman filter needs to be "tuned" to obtain new parameter values. A parameter evaluation routine using the Kalman filter and maximum likelihood methods (Goodwin and Payne, 1977) can obtain these parameter values and their variance. This can be done off-line on a large digital computer to obtain the most accuracy possible. Having estimates for these parameters on a daily basis will enable reactor operators to evaluate fuel burn-up, crud build-up, etc. Thus further efficiencies can be achieved by adjusting the loading schedules, etc., according to fuel burn up. Slow acting dangers such as crud build-up can also be assessed without inspecting reactor core components.

Thus Table 2 and Figure 2 indicate the following reasons a Kalman filter should be installed on an LMFBR:

1. A Kalman filter indicates incipient sodium boiling.
2. A Kalman filter tells where sodium boiling has occurred in the event of a scram.
3. "Tuning" of a Kalman filter indicates crud build-up and fuel burn-up distribution.
4. A Kalman filter gives the operators a picture of overall reactor dynamic operation to possibly forestall accidents such as at Three-Mile Island.

It must be remembered that the results of Table 2 were obtained for a very simple mathematical model. Thus P_{ss} represents the error variance only between the mathematical model and the Kalman filter. There is an additional error between the physical system and the mathematical model. Therefore the variances shown in Table 2 must be smaller than for a Kalman filter applied to the physical LMFBR system. Evaluation of the error between physical system and mathematical model requires further experiments using hardware as close as possible to that encountered in practice.

The preliminary results described in this section have shown the theoretical advantages of a Kalman filter applied to a nuclear reactor. Further investigation appears justified for both more detailed modeling and more hardware experiments.

PREMATURE DISCRETIZATION OF THE DELAY EQUATION

In the LMFBR filtering problem of the previous section, the computer routine for eigenvector solution of the steady state Riccati equation did not give a symmetric, positive definite answer for models of dimension greater than three. This difficulty stems from premature discretization of a delay-like equation (183). Before discretization and after cancelling some constants, this equation can be represented in the partial differential form:

$$fH\rho_f C_{p_c} \frac{\partial Tc}{\partial t} = 2Hh_T(T_f - T_c) - f\rho_f C_{p_c} V \frac{\partial Tc}{\partial x} \tag{188}$$

Rather than considering the filtering of this equation, the simpler first order delay equation

$$\frac{\partial x(r,t)}{\partial t} = \frac{\partial x(r,t)}{\partial r} \tag{189}$$

will be considered under control. The correspondence from control to filtering follows from duality. The boundary conditions are

$$x(0,t) = u(t) \tag{190}$$

$$x(1,t) = y(t) \tag{191}$$

Spatial discretization can be performed as in the previous section. Here the unit length will be broken into n discrete pieces. The approximation of the spatial partial differential is by backward difference

$$\frac{dx_1(t)}{dt} = n\left\{u(t) - x_1(t)\right\} \;\ldots\; \frac{dx_n(t)}{dt} = n\left\{x_{n-1}(t) - x_n(t)\right\} \tag{192}$$

$$y(t) = x_n(t) \tag{193}$$

where $x_k(t) = x(k,t)$. In state space form this is

$$\frac{dx}{dt} = Ax + bu \qquad y = cx \tag{194}$$

where

$$x = \begin{Bmatrix} x_1 \\ x_2 \\ \vdots \\ x_n \end{Bmatrix}, \; A = n\begin{Bmatrix} -1 & 0 & \cdots & 0 \\ 1 & -1 & \cdots & 0 \\ \vdots & & & \vdots \\ 0 & \cdots & 1 & -1 \end{Bmatrix}, \; b = n\begin{Bmatrix} 1 \\ 0 \\ \vdots \\ 0 \end{Bmatrix}, \; c = \begin{Bmatrix} 0 \\ \vdots \\ 0 \\ 1 \end{Bmatrix} \tag{195}$$

The controllability condition is

$$\text{rank}\left\{ b \;\vdots\; Ab \;\vdots\; \cdots \;\vdots\; A^{n-1}b \right\} = n \tag{196}$$

Substituting the values (195) gives

$$\text{rank}\begin{Bmatrix} n & -n^2 & n^3 & \cdots \\ 0 & n^2 & -2n^3 & \\ \vdots & 0 & n^3 & \\ \vdots & \vdots & \vdots & \\ 0 & \cdot & \cdots & \end{Bmatrix} = n \tag{197}$$

Clearly this is always satisfied for any n. Furthermore, the observability condition is similarly satisfied for any discretization n. This implies that the system is output controllable, and any change in this input is <u>immediately</u> noticed in the output. However, the solution to the distributed parameter representation (189) is

$$y(t) = u(t-1) \tag{198}$$

so that any change in the input u(t) is noticed only after one time unit delay. But any time-invariant finite dimensional realization is either contorllable or it is not i.e., any change in the input is immediately noticed in the output or it is not noticed at all. Therefore, it is not possible to represent such behavior by any order finite dimensional time-invariant representation.

The behavior of the original LMFBR Eq. (183) is similar to that of the simple delay

Eq. (189). The original LMFBR equation (183) becomes "closer and closer to unobservable" as the number of nodes and modes increases. This increase in model dimension causes the failure to compute a positive definite symmetric solution to the Riccati equation and is the reason theoretical procedures were at fault. Only a three dimensional representation of the LMFBR could be filtered.

This behavior gives rise to several important theoretical questions. Will the modal expansions of Section 5 permit solution of the partial differential state equation whose nodal approximations were used in Section 6 to represent the LMFBR? In other words, will this problem be avoided by keeping the partial differential formulation to the last and then making a discrete approximation? The modal expansions are the best possible finite dimensional representation of the system, as explained in the conclusions to Section 5. However, they are eventually a finite-dimensional approximation, whose solution must tend to that of a very high number of nodes. Therefore it is doubtful that any finite dimensional time-invariant representation will be free of the problem. Also, the reactor itself is not time-varying, so that no finite dimensional representation can suffice.

Will any infinite dimensional representation suffice, i.e., is the representation (189) observable? Clearly it is, after a delay of one time unit. A Kalman filter can be built for delay equations of the form of (198). Now the problem is clearly defined. What kind of distributed parameter Kalman filter can be implemented for the distributed parameter reactor equations with a delay type of equation in them? This problem is at present unsolved and poses an interesting and practical research problem with great theoretical importance.

REFERENCES

Borgwaldt, H. (1968). Neutron noise in a reactor with an external control loop. Nukleonik 11, 2, 76-84.

Bucy, R. S. and P. D. Joseph (1968). Filtering for Stochastic Processes with Applications to Guidance. Interscience, New York, NY.

Friedman, B. (1956). Principles and Techniques of Applied Mathematics. Wiley, New York, NY.

Glasstone, S. and M. C. Edlund (1952). The Elements of Nuclear Reactor Theory. Van Nostrand, Princeton, NJ.

Goodwin, G. C. and R. L. Payne (1977). Dynamic System Identification: Experiment Design and Data Analysis. Academic Press, New York, NY.

Hafele, W. and G. Kessler (1972). SNR: the German-Benelux fast breeder. Nuclear News. March.

Kailath, T. (1968). Innovations approach to least-squares estimation - 1. IEEE Trans. Autom. Control AC-13, 646-655.

Marti, J. T. (1970). Introduction to the Theory of Bases. Springer-Verlag, Berlin.

Pontryagin, L. S. (1959). Unspekh. Mat. Nauk 14,3 (in Russian)

Potter, J. E. (1966). Matrix quadratic solutions. SIAM J. Appl. Math. 14, 496-510.

Schmiedler, W. (1965). Linear Operators in Hilbert Space. Academic Press, New York, NY.

Sholar, M. S. (1970). Quadratic Optimal Control of Distributed Parameter Systems with Stochastic Inputs. Report No. 70-46, UCLA School of Engineering and Applied Science.

Sholar, M. S. and D. M. Wiberg (1972). Canonical equations for boundary feedback control of stochastic distributed parameter systems. Automatica, 8, 287-298.

Singer, I. (1970). Bases in Banach Space. Springer-Verlag, Berlin

Skorokhod, A. V. (1965). Studies in the Theory of Randon Processes. Addison-Wesley, Reading, MA.

Wiberg, D. M. (1967a). In C. T. Leondes (Ed.) Advances in Control Systems, Vol. 5. Academic Press, New York, NY.

Wiberg, D. M. (1967b). Controllability of the spatial flux shape. Nucl. Sci. & Eng. ASME Trans. Series D, 89. 379-384.

Wiberg, D. M. (1971). State Space and Linear Systems. McGraw Hill, New York, NY.
Wonham, W. M. (1968). In A. T. Bharucha - Reid (Ed.), Probabalistic Methods in Applied Mathematics, Academic Press, New York, NY.
Zadeh, L. and C. A. Desoer (1963). Linear System Theory. McGraw Hill, New York, NY.

SUBJECT INDEX